THE CAMBRIDGE HISTORY OF SCIENCE

VOLUME 5

The Modern Physical and Mathematical Sciences

Volume 5 is a narrative and interpretive history of the physical and mathematical sciences from the early nineteenth century to the close of the twentieth century. The contributing authors are world leaders in their respective specialties. Drawing upon the most recent methods and results in historical studies of science, they employ strategies from intellectual history, social history, and cultural studies to provide unusually wide-ranging and comprehensive insights into developments in the public culture, disciplinary organization, and cognitive content of the physical and mathematical sciences. The sciences under study in this volume include physics, astronomy, chemistry, and mathematics, as well as their extensions into geosciences and environmental sciences, computer science, and biomedical science. The authors examine scientific traditions and scientific developments; analyze the roles of instruments, languages, and images in everyday practice; scrutinize the theme of scientific "revolution"; and examine the interactions of the sciences with literature, religion, and ideology.

Mary Jo Nye is Horning Professor of the Humanities and Professor of History at Oregon State University in Corvallis, Oregon. A past president of the History of Science Society, she received the 1999 American Chemical Society Dexter Award for Outstanding Achievement in the History of Chemistry. She is the author or editor of seven books, most recently *From Chemical Philosophy to Theoretical Chemistry: Dynamics of Matter and Dynamics of Disciplines, 1800–1950* (1993) and *Before Big Science: The Pursuit of Modern Chemistry and Physics, 1800–1940* (1996).

THE CAMBRIDGE HISTORY OF SCIENCE

General editors
David C. Lindberg and Ronald L. Numbers

VOLUME 1: *Ancient Science*
Edited by Alexander Jones

VOLUME 2: *Medieval Science*
Edited by David C. Lindberg and Michael H. Shank

VOLUME 3: *Early Modern Science*
Edited by Lorraine J. Daston and Katharine Park

VOLUME 4: *Eighteenth-Century Science*
Edited by Roy Porter

VOLUME 5: *The Modern Physical and Mathematical Sciences*
Edited by Mary Jo Nye

VOLUME 6: *The Modern Biological and Earth Sciences*
Edited by Peter Bowler and John Pickstone

VOLUME 7: *The Modern Social Sciences*
Edited by Theodore M. Porter and Dorothy Ross

VOLUME 8: *Modern Science in National and International Context*
Edited by David N. Livingstone and Ronald L. Numbers

David C. Lindberg is Hilldale Professor Emeritus of the History of Science at the University of Wisconsin–Madison. He has written or edited a dozen books on topics in the history of medieval and early modern science, including *The Beginnings of Western Science* (1992). He and Ronald L. Numbers have previously coedited *God and Nature: Historical Essays on the Encounter between Christianity and Science* (1986) and *Science and the Christian Tradition: Twelve Case Histories* (2003). A Fellow of the American Academy of Arts and Sciences, he has been a recipient of the Sarton Medal of the History of Science Society, of which he is also past-president (1994–5).

Ronald L. Numbers is Hilldale and William Coleman Professor of the History of Science and Medicine at the University of Wisconsin–Madison, where he has taught since 1974. A specialist in the history of science and medicine in America, he has written or edited more than two dozen books, including *The Creationists* (1992) and *Darwinism Comes to America* (1998). A Fellow of the American Academy of Arts and Sciences and a former editor of *Isis*, the flagship journal of the history of science, he has served as the president of both the American Society of Church History (1999–2000) and the History of Science Society (2000–1).

THE CAMBRIDGE HISTORY OF SCIENCE

VOLUME 5

The Modern Physical and Mathematical Sciences

Edited by
MARY JO NYE

PUBLISHED BY THE PRESS SYNDICATE OF THE UNIVERSITY OF CAMBRIDGE
The Pitt Building, Trumpington Street, Cambridge, United Kingdom

CAMBRIDGE UNIVERSITY PRESS
The Edinburgh Building, Cambridge CB2 2RU, UK
40 West 20th Street, New York, NY 10011-4211, USA
477 Williamstown Road, Port Melbourne, VIC 3207, Australia
Ruiz de Alarcón 13, 28014 Madrid, Spain
Dock House, The Waterfront, Cape Town 8001, South Africa

http://www.cambridge.org

© Cambridge University Press 2003

This book is in copyright. Subject to statutory exception
and to the provisions of relevant collective licensing agreements,
no reproduction of any part may take place without
the written permission of Cambridge University Press.

First published 2003

Printed in the United States of America

Typeface Adobe Garamond 10.75/12.5 pt.   *System* LaTeX 2_ε [TB]

A catalog record for this book is available from the British Library.

Library of Congress Cataloging in Publication Data
(Revised for volume 5)
The Cambridge history of science
p. cm.
Includes bibliographical references and indexes.
Contents: – v. 4. Eighteenth-century science / edited by Roy Porter
v. 5. The modern physical and mathematical sciences / edited by Mary Jo Nye
ISBN 0-521-57243-6 (v. 4)
ISBN 0-521-57199-5 (v. 5)
1. Science – History.   I. Lindberg, David C.   II. Numbers, Ronald L.
Q125 C32   2001
509 – dc21
2001025311

ISBN 0 521 57199 5 hardback

CONTENTS

Illustrations	*page* xvii
Notes on Contributors	xix
General Editors' Preface	xxv
Acknowledgments	xxix

Introduction: The Modern Physical and Mathematical Sciences   1
   MARY JO NYE

PART I. THE PUBLIC CULTURES OF THE PHYSICAL SCIENCES AFTER 1800

1 **Theories of Scientific Method: Models for the Physico-Mathematical Sciences**   21
   NANCY CARTWRIGHT, STATHIS PSILLOS, AND HASOK CHANG
   Mathematics, Science, and Nature   22
   Realism, Unity, and Completeness   25
   Positivism   28
   From Evidence to Theory   29
   Experimental Traditions   32

2 **Intersections of Physical Science and Western Religion in the Nineteenth and Twentieth Centuries**   36
   FREDERICK GREGORY
   The Plurality of Worlds   37
   The End of the World   39
   The Implications of Materialism   43
   From Confrontation to Peaceful Coexistence to Reengagement   46
   Contemporary Concerns   49

3 A Twisted Tale: Women in the Physical Sciences in the Nineteenth and Twentieth Centuries 54
MARGARET W. ROSSITER

Precedents 54
Great Exceptions 55
Less-Well-Known Women 58
Rank and File – Fighting for Access 59
Women's Colleges – A World of Their Own 61
Graduate Work, (Male) Mentors, and Laboratory Access 62
"Men's" and "Women's" Work in Peace and War 63
Scientific Marriages and Families 65
Underrecognition 66
Post–World War II and "Women's Liberation" 67
Rise of Gender Stereotypes and Sex-Typed Curricula 70

4 Scientists and Their Publics: Popularization of Science in the Nineteenth Century 72
DAVID M. KNIGHT

Making Science Loved 74
The March of Mind 75
Read All About It 76
Crystal Palaces 77
The Church Scientific 78
Deep Space and Time 80
Beyond the Fringe 83
A Second Culture? 85
Talking Down 87
Signs and Wonders 88

5 Literature and the Modern Physical Sciences 91
PAMELA GOSSIN

Two Cultures: Bridges, Trenches, and Beyond 93
The Historical Interrelations of Literature and Newtonian Science 95
Literature and the Physical Sciences after 1800: Forms and Contents 98
Literature and Chemistry 99
Literature and Astronomy, Cosmology, and Physics 100
Interdisciplinary Perspectives and Scholarship 103
Literature and the Modern Physical Sciences in the History of Science 106
Literature and the Modern Physical Sciences: New Forms and Directions 108

PART II. DISCIPLINE BUILDING IN THE SCIENCES: PLACES, INSTRUMENTS, COMMUNICATION

6	**Mathematical Schools, Communities, and Networks**	113
	DAVID E. ROWE	
	Texts and Contexts	114
	Shifting Modes of Production and Communication	117
	Mathematical Research Schools in Germany	120
	Other National Traditions	123
	Göttingen's Modern Mathematical Community	127
	Pure and Applied Mathematics in the Cold War Era and Beyond	129
7	**The Industry, Research, and Education Nexus**	133
	TERRY SHINN	
	Germany as a Paradigm of Heterogeneity	134
	France as a Paradigm of Homogeneity	138
	England as a Case of Underdetermination	143
	The United States as a Case of Polymorphism	147
	The Stone of Sisyphus	152
8	**Remaking Astronomy: Instruments and Practice in the Nineteenth and Twentieth Centuries**	154
	ROBERT W. SMITH	
	The Astronomy of Position	154
	Different Goals	160
	Opening Up the Electromagnetic Spectrum	165
	Into Space	167
	Very Big Science	170
9	**Languages in Chemistry**	174
	BERNADETTE BENSAUDE-VINCENT	
	1787: A "Mirror of Nature" to Plan the Future	176
	1860: Conventions to Pacify the Chemical Community	181
	1930: Pragmatic Rules to Order Chaos	186
	Toward a Pragmatic Wisdom	189
10	**Imagery and Representation in Twentieth-Century Physics**	191
	ARTHUR I. MILLER	
	The Twentieth Century	193
	Albert Einstein: Thought Experiments	194
	Types of Visual Images	195
	Atomic Physics during 1913–1925: Visualization Lost	197
	Atomic Physics during 1925–1926: Visualization versus Visualizability	200

	Atomic Physics in 1927: Visualizability Redefined	203
	Nuclear Physics: A Clue to the New Visualizability	205
	Physicists Rerepresent	208
	The Deep Structure of Data	209
	Visual Imagery and the History of Scientific Thought	212

PART III. CHEMISTRY AND PHYSICS: PROBLEMS THROUGH THE EARLY 1900s

11	**The Physical Sciences in the Life Sciences**	219
	FREDERIC L. HOLMES	
	Applications of the Physical Sciences to Biology in the Seventeenth and Eighteenth Centuries	221
	Chemistry and Digestion in the Eighteenth Century	224
	Nineteenth-Century Investigations of Digestion and Circulation	226
	Transformations in Investigations of Respiration	230
	Physiology and Animal Electricity	233
12	**Chemical Atomism and Chemical Classification**	237
	HANS-WERNER SCHÜTT	
	Chemical versus Physical Atoms	238
	Atoms and Gases	239
	Calculating Atomic Weights	241
	Early Attempts at Classification	243
	Types and Structures	245
	Isomers and Stereochemistry	248
	Formulas and Models	250
	The Periodic System and Standardization in Chemistry	251
	Two Types of Bonds	254
13	**The Theory of Chemical Structure and Its Applications**	255
	ALAN J. ROCKE	
	Early Structuralist Notions	255
	Electrochemical Dualism and Organic Radicals	257
	Theories of Chemical Types	259
	The Emergence of Valence and Structure	262
	Further Development of Structural Ideas	265
	Applications of the Structure Theory	269
14	**Theories and Experiments on Radiation from Thomas Young to X Rays**	272
	SUNGOOK HONG	
	The Rise of the Wave Theory of Light	272
	New Kinds of Radiation and the Idea of the Continuous Spectrum	277
	The Development of Spectroscopy and Spectrum Analysis	280

	The Electromagnetic Theory of Light and the Discovery of X Rays	284
	Theory, Experiment, Instruments in Optics	287
15	**Force, Energy, and Thermodynamics**	289
	CROSBIE SMITH	
	The Mechanical Value of Heat	290
	A Science of Energy	296
	The Energy of the Electromagnetic Field	304
	Recasting Energy Physics	308
16	**Electrical Theory and Practice in the Nineteenth Century**	311
	BRUCE J. HUNT	
	Early Currents	311
	The Age of Faraday and Weber	312
	Telegraphs and Cables	314
	Maxwell	317
	Cables, Dynamos, and Light Bulbs	319
	The Maxwellians	321
	Electrons, Ether, and Relativity	324

PART IV. ATOMIC AND MOLECULAR SCIENCES IN THE TWENTIETH CENTURY

17	**Quantum Theory and Atomic Structure, 1900–1927**	331
	OLIVIER DARRIGOL	
	The Quantum of Action	332
	Quantum Discontinuity	334
	From Early Atomic Models to the Bohr Atom	336
	Einstein and Sommerfeld on Bohr's Theory	339
	Bohr's Correspondence Principle versus Munich Models	340
	A Crisis, and Quantum Mechanics	341
	Quantum Gas, Radiation, and Wave Mechanics	344
	The Final Synthesis	346
18	**Radioactivity and Nuclear Physics**	350
	JEFF HUGHES	
	Radioactivity and the "Political Economy" of Radium	352
	Institutionalization, Concentration, and Specialization: The Emergence of a Discipline, 1905–1914	355
	"An Obscure Oddity"? Radioactivity Reconstituted, 1919–1925	360
	Instruments, Techniques, and Disciplines: Controversy, 1924–1932	362
	From "Radioactivity" to "Nuclear Physics": A Discipline Transformed, 1932–1940	368
	Nuclear Physics and Particle Physics: Postwar Differentiation, 1945–1960	370

19 Quantum Field Theory: From QED to the Standard Model — 375
SILVAN S. SCHWEBER
Quantum Field Theory in the 1930s — 377
From Pions to the Standard Model: Conceptual Developments in Particle Physics — 382
Quarks — 388
Gauge Theories and the Standard Model — 391

20 Chemical Physics and Quantum Chemistry in the Twentieth Century — 394
ANA SIMÕES
Periods and Concepts in the History of Quantum Chemistry — 395
The Emergence of Quantum Chemistry and the Problem of Reductionism — 400
The Emergence of Quantum Chemistry in National Context — 404
Quantum Chemistry as a Discipline — 407
The Uses of Quantum Chemistry for the History and Philosophy of the Sciences — 411

21 Plasmas and Solid-State Science — 413
MICHAEL ECKERT
Prehistory: Contextual versus Conceptual — 414
World War II: A Critical Change — 417
Formative Years, 1945–1960 — 420
Consolidation and Ramifications — 425
Models of Scientific Growth — 427

22 Macromolecules: Their Structures and Functions — 429
YASU FURUKAWA
From Organic Chemistry to Macromolecules — 430
Physicalizing Macromolecules — 435
Exploring Biological Macromolecules — 437
The Structure of Proteins: The Mark Connection — 440
The Path to the Double Helix: The Signer Connection — 443

PART V. MATHEMATICS, ASTRONOMY, AND COSMOLOGY SINCE THE EIGHTEENTH CENTURY

23 The Geometrical Tradition: Mathematics, Space, and Reason in the Nineteenth Century — 449
JOAN L. RICHARDS
The Eighteenth-Century Background — 450
Geometry and the French Revolution — 454
Geometry and the German University — 458

	Geometry and English Liberal Education	460
	Euclidean and Non-Euclidean Geometry	462
	Geometry in Transition: 1850–1900	464

24 Between Rigor and Applications: Developments in the Concept of Function in Mathematical Analysis — 468
JESPER LÜTZEN

	Euler's Concept of Function	469
	New Function Concepts Dictated by Physics	470
	Dirichlet's Concept of Function	471
	Exit the Generality of Algebra – Enter Rigor	474
	The Dreadful Generality of Functions	477
	The Delta "Function"	479
	Generalized Solutions to Differential Equations	481
	Distributions: Functional Analysis Enters	484

25 Statistics and Physical Theories — 488
THEODORE M. PORTER

	Statistical Thinking	489
	Laws of Error and Variation	491
	Mechanical Law and Human Freedom	494
	Regularity, Average, and Ensemble	498
	Reversibility, Recurrence, and the Direction of Time	500
	Chance at the *Fin de Siècle*	503

26 Solar Science and Astrophysics — 505
JOANN EISBERG

	Solar Physics: Early Phenomenology	508
	Astronomical Spectroscopy	510
	Theoretical Approaches to Solar Modeling: Thermodynamics and the Nebular Hypothesis	512
	Stellar Spectroscopy	514
	From the Old Astronomy to the New	516
	Twentieth-Century Stellar Models	518

27 Cosmologies and Cosmogonies of Space and Time — 522
HELGE KRAGH

	The Nineteenth-Century Heritage	522
	Galaxies and Nebulae until 1925	523
	Cosmology Transformed: General Relativity	525
	An Expanding Universe	526
	Nonrelativistic Cosmologies	529
	Gamow's Big Bang	530
	The Steady State Challenge	531
	Radio Astronomy and Other Observations	532
	A New Cosmological Paradigm	533
	Developments since 1970	534

28 The Physics and Chemistry of the Earth — 538
NAOMI ORESKES AND RONALD E. DOEL

Traditions and Conflict in the Study of the Earth — 539
Geology, Geophysics, and Continental Drift — 542
The Depersonalization of Geology — 545
The Emergence of Modern Earth Science — 549
Epistemic and Institutional Reinforcement — 552

PART VI. PROBLEMS AND PROMISES AT THE END OF THE TWENTIETH CENTURY

29 Science, Technology, and War — 561
ALEX ROLAND

Patronage — 562
Institutions — 566
Qualitative Improvements — 568
Large-Scale, Dependable, Standardized Production — 569
Education and Training — 570
Secrecy — 571
Political Coalitions — 573
Opportunity Costs — 574
Dual-Use Technologies — 575
Morality — 577

30 Science, Ideology, and the State: Physics in the Twentieth Century — 579
PAUL JOSEPHSON

Soviet Marxism and the New Physics — 580
Aryan Physics and Nazi Ideology — 586
Science and Pluralist Ideology: The American Case — 589
The Ideological Significance of Big Science and Technology — 592
The National Laboratory as Locus of Ideology and Knowledge — 594

31 Computer Science and the Computer Revolution — 598
WILLIAM ASPRAY

Computing before 1945 — 598
Designing Computing Systems for the Cold War — 601
Business Strategies and Computer Markets — 604
Computing as a Science and a Profession — 607
Other Aspects of the Computer Revolution — 611

32 The Physical Sciences and the Physician's Eye: Dissolving Disciplinary Boundaries — 615
BETTYANN HOLTZMANN KEVLES

Origins of CT in Academic and Medical Disciplines — 617
Origins of CT in Private Industry — 621

	From Nuclear Magnetic Resonance to Magnetic Resonance Imaging	625
	MRI and the Marketplace	629
	The Future of Medical Imaging	631
33	**Global Environmental Change and the History of Science**	634
	JAMES RODGER FLEMING	
	Enlightenment	636
	Literary and Scientific Transformation: The American Case	638
	Scientific Theories of Climatic Change	641
	Global Warming: Early Scientific Work and Public Concern	645
	Global Cooling, Global Warming	648
	Index	651

ILLUSTRATIONS

8.1	The Dorpat Refractor, a masterpiece by Fraunhofer	*page* 157
8.2	The Leviathan of Parsonstown	161
8.3	The Hubble Space Telescope in the payload bay of the Space Shuttle *Enterprise*	171
10.1	An Aristotelian representation of a cannonball's trajectory	192
10.2	Galileo's 1608 drawing of the parabolic fall of an object	192
10.3	Representations of the atom according to Niels Bohr's 1913 atomic theory	198
10.4	The difference between visualization and visualizability	206
10.5	Representations of the Coulomb force	208
10.6	Representations of the atom and its interactions with light	210
10.7	Bubble chamber and "deep structure"	211
10.8	Images of data and their "deep structure"	213
10.9	Representations of the atom	214

NOTES ON CONTRIBUTORS

WILLIAM ASPRAY is Executive Director of the Computing Research Association in Washington, D.C. His studies of the history of mathematics and the history of computing include *John von Neumann and the Origins of Modern Computing* (1990) and *Computer: A History of the Information Machine* (1996), the latter book coauthored with Martin Campbell-Kelly.

BERNADETTE BENSAUDE-VINCENT is Professor of History and Philosophy of Science at the University of Paris X. She is the author of a number of articles on the history of chemistry. Among her recent books are *Lavoisier, mémoires d'une révolution* (1993); *A History of Chemistry* (English-language edition, 1996), coauthored with Isabelle Stengers; and *Eloge du mixte, matériaux nouveaux et philosophie ancienne* (1998).

NANCY CARTWRIGHT is Professor in the Department of Philosophy, Logic and Scientific Method at the London School of Economics and in the Department of Philosophy at the University of California, San Diego. She also directs the Centre for Philosophy of Natural and Social Science at the London School of Economics. She has written books about the role of theory in physics, about causality, about the politics and philosophy of Otto Neurath, and on the limits of scientific description.

HASOK CHANG is Lecturer in Philosophy of Science in the Department of Science and Technology Studies (formerly History and Philosophy of Science) at University College London. He received his PhD in philosophy at Stanford University in 1993. He has published various papers in the history and philosophy of modern physics. His current research interests are in historical and philosophical studies of the physical sciences, particularly in the eighteenth and nineteenth centuries.

OLIVIER DARRIGOL is a researcher at the Centre National de la Recherche Scientifique in Paris. In addition to several articles on the history of electrodynamics, quantum theory, and quantum field theory, he is the author of two books: *From c-Number to q-Numbers: The Classical Analogy in the History of Quantum Theory* (1992) and *Electrodynamics from Ampère to Einstein* (2000).

RONALD E. DOEL is Assistant Professor in the Departments of History and Geosciences at Oregon State University. He specializes in the history of the earth and environmental sciences in the twentieth century, as well as the international relations of science in the Cold War era. He is author of *Solar System Astronomy in America: Communities, Patronage and Interdisciplinary Research, 1920–1960* (1996).

MICHAEL ECKERT, formerly collaborator on the International Project in the History of Solid State Physics, is now working on the emergence of theoretical physics in Germany in the beginning of the twentieth century. He is editor of the scientific correspondence of Arnold Sommerfeld at the Institut für Geschichte der Naturwissenschaften of the University of Munich.

JOANN EISBERG teaches astronomy at Citrus College in Glendora, California. She earned her PhD at Harvard University with a dissertation on Arthur Stanley Eddington and early-twentieth-century models of stars. She is currently writing a biography of Beatrice Tinsley, who worked on cosmology and the evolution of galaxies.

JAMES RODGER FLEMING is Associate Professor and Director of the Science, Technology and Society Program at Colby College in Maine. His research interests include the history of the geophysical and environmental sciences, especially meteorology and climatology. His books include *Historical Perspectives on Climate Change* (1998) and *Meteorology in America, 1800–1870* (1990; paperback, 2000).

YASU FURUKAWA earned his PhD in history of science at the University of Oklahoma in 1983. He is Professor of the History of Science at Tokyo Denki University and the editor of *Kagakushi (Journal of the Japanese Society for the History of Chemistry)*. He has authored *Kagaku no shakai-shi* [*A Social History of Science*] (1989) and *Inventing Polymer Science: Staudinger, Carothers, and the Emergence of Macromolecular Chemistry* (1998).

PAMELA GOSSIN, Associate Professor of Arts and Humanities at the University of Texas–Dallas, teaches interdisciplinary courses in literature and the history of science. She is currently directing UTD's Undergraduate Medical and Scientific Humanities Program – a curriculum she developed. She is the editor of *An Encyclopedia of Literature and Science* (forthcoming) and the author of *Thomas Hardy's Novel Universe: Astronomy and the Cosmic Heroines of His Major and Minor Fiction* (forthcoming). She holds a dual PhD in English and the History of Science from the University of Wisconsin–Madison.

FREDERICK GREGORY is Professor of History of Science at the University of Florida and a past president of the History of Science Society. He is author of *Nature Lost? Natural Science and the German Theological Traditions of the Nineteenth Century* (1992). His research deals primarily with eighteenth- and nineteenth-century German science and with the history of science and religion.

FREDERIC L. HOLMES is chair of the Section of History of Medicine in the Yale University School of Medicine. He is a former president of the History of Science Society. He has written about Antoine Lavoisier and eighteenth-century chemistry, Claude Bernard and nineteenth-century physiology, Hans Krebs and intermediary metabolism, and the role of the Meselson-Stahl experiment in the formation of molecular biology.

SUNGOOK HONG teaches the history of physics at the Institute for the History and Philosophy of Science and Technology, University of Toronto. He is currently working on the history of the spectrum, the history of nineteenth-century electromagnetic theories, and the history of electrical engineering. He is the author of *Wireless: From Marconi's Black-Box to the Audion* (2001).

JEFF HUGHES is Lecturer in the History of Science and Technology in the Centre for History of Science, Technology and Medicine at the University of Manchester. His research has focused on the social and cultural history of radioactivity and nuclear physics during the period 1890–1940, and particularly on cultures of experiment and theory. He is currently completing books on the emergence of isotopy and on the rise of nuclear physics, and he is planning a volume on nuclear historiography.

BRUCE J. HUNT teaches in the History Department at the University of Texas at Austin. He is the author of *The Maxwellians* (1991). His current work concerns the relationship between telegraphy and electrical science in Victorian Britain.

PAUL JOSEPHSON writes about big science and technology. His last book was *Red Atom* (1999). He is now writing about technology and resource management practices in twentieth-century Russia, Norway, Brazil, and the United States. He teaches at Colby College, Waterville, Maine.

BETTYANN HOLTZMANN KEVLES's reviews of books on science and medicine have appeared often in the *Los Angeles Times* and on National Public Radio's *Science Friday*. Among her publications are *Females of the Species: Sex and Survival in the Animal Kingdom* (1986) and *Naked to the Bone: Medical Imaging in the Twentieth Century* (1997). She is currently working on a history of female astronauts.

DAVID M. KNIGHT completed a degree in chemistry and wrote a DPhil thesis at Oxford on the problem of the chemical elements in nineteenth-century Britain. He was appointed in 1964 as Lecturer in History of Science at the University of Durham and promoted to Professor in 1991. From 1982 to 1988 he edited the *British Journal for the History of Science*, and from 1994 to 1996 he was President of the British Society for the History of Science.

HELGE KRAGH is a member of the History of Science Department, Aarhus University, Denmark. He has contributed to the history of modern physics, chemistry, cosmology, and technology, and he also has an interest in historiography and philosophy of science. His most recent books are *Cosmology and Controversy: The Historical Development of Two Theories of the Universe* (1996) and *Quantum Generations: A History of Physics in the Twentieth Century* (1999).

JESPER LÜTZEN is Associate Professor in the Department of Mathematics at Copenhagen University's Institute for Mathematical Sciences. He is the author of *Joseph Liouville, 1809–1882: Master of Pure and Applied Mathematics* (1990) and of numerous articles in the history of mathematics, including studies on Heinrich Hertz, geometry, and physics.

ARTHUR I. MILLER is Professor of History and Philosophy of Science at University College London. Currently he is exploring creative thinking in art and science. He has written extensively on the history of the special theory of relativity. His most recent book is *Einstein, Picasso: Space, Time and the Beauty That Causes Havoc* (2001).

MARY JO NYE is Horning Professor of the Humanities and Professor of History at Oregon State University and a past president of the History of Science Society. Her research interests are in the history of chemistry and physics in the modern period, with attention to political and institutional contexts as well as the history of ideas. Her most recent book is *Before Big Science: The Pursuit of Modern Chemistry and Physics, 1800–1940* (1996), published in paperback edition in 1999.

NAOMI ORESKES is Associate Professor of History at the University of California, San Diego, and the author of *The Rejection of Continental Drift: Theory and Method in American Earth Science* (1999). She is currently working on a history of oceanography during the Cold War entitled *The Military Roots of Basic Science: American Oceanography in the Cold War and Beyond.*

THEODORE M. PORTER is Professor of History of Science in the Department of History at the University of California, Los Angeles. His books include *The Rise of Statistical Thinking, 1820–1900* (1986) and *Trust in Numbers: The Pursuit of Objectivity in Science and Public Life* (1995). He is currently working on a book on Karl Pearson and the sensibility of science and coediting (with Dorothy Ross) Volume 7 in *The Cambridge History of Science* on the social sciences.

STATHIS PSILLOS is Lecturer in the Department of Philosophy and History of Science, University of Athens, Greece. He was awarded his PhD in the Philosophy of Science from King's College, University of London. He was a British Academy Postdoctoral Fellow at the London School of Economics until July 1998. His book *Scientific Realism: How Science Tracks Truth* was published in 1999. He has published a number of articles in the philosophy of science, and he is currently working on an introductory book on *Causation and Explanation*.

JOAN L. RICHARDS is the author of *Mathematical Visions* (1988), a study of non-Euclidean geometry in late-nineteenth-century England, and the autobiographical *Angles of Reflection: Logic and a Mother's Love* (2000). She is currently working on a biography of Augustus and Sophia De Morgan and on a study of logic in England from 1826 to 1864. She is Associate Professor in the History Department at Brown University.

ALAN J. ROCKE, Henry Eldridge Bourne Professor of History at Case Western Reserve University, specializes in the history of European chemistry in the nineteenth century. His most recent books are *The Quiet Revolution: Hermann Kolbe and the Science of Organic Chemistry* (1993) and *Nationalizing Science: Adolphe Wurtz and the Battle for French Chemistry* (2001).

ALEX ROLAND is Professor of History at Duke University, where he teaches military history and the history of technology. His most recent research addresses military support of computer development in the United States in the 1980s and 1990s.

MARGARET W. ROSSITER is the Marie Underhill Noll Professor of the History of Science at Cornell University and the editor of *Isis*. She has published widely on the history of women in American science, including successive volumes on *Women Scientists in America* (1982 and 1995).

DAVID E. ROWE teaches in the Mathematics Department at the Johannes Gutenberg Universität Mainz after taking doctoral degrees in mathematics at the University of Oklahoma and in history at the City University of New York. His research and publications center on the mathematical work and cultural milieus of Felix Klein, David Hilbert, and, in his newest historical research, Albert Einstein.

HANS-WERNER SCHÜTT studied chemistry and earned his PhD in physical chemistry at the Christian Albrechts University, Kiel. He worked in a research department of Unilever and since 1979 has been Professor for the history of exact sciences and technology at the Technical University, Berlin. His main fields of interest are history of early-nineteenth-century chemistry, science and religion, and alchemy. His *Eilhard Mitscherlich: Prince of Prussian Chemistry* appeared in English translation in 1997. His most recent book is *Auf der Suche nach dem Stein der Weisen: Die Geschichte der Alchemie* (2000).

SILVAN S. SCHWEBER is Professor of Physics and Richard Koret Professor of the History of Ideas at Brandeis University and an Associate in the Department of History at Harvard University. His current historical research includes work on the introduction of probabilistic concepts into the biological and physical sciences and a scientific biography of Hans A. Bethe. His most recent books are *QED and the Men Who Made It: Dyson, Feynman, Schwinger, and Tomonaga* (1994) and *In the Shadow of the Bomb: Oppenheimer, Bethe, and the Moral Responsibility of the Scientist* (2000).

TERRY SHINN is Research Director at the Centre National de la Recherche Scientifique in Paris. He has been an editor and contributor for the *Sociology of Sciences Yearbook*. He has written extensively on nineteenth- and twentieth-century French technical education and engineering. His newest book, *Building French Research-Technology: The Bellevue Giant Electromagnet, 1900–1975*, will be published soon.

ANA SIMÕES received her PhD in history with a specialization in history and philosophy of science from the University of Maryland at College Park in 1993 for work on the genesis and development of quantum chemistry in the United States during the 1930s. She is currently Assistant Professor in the Departamento de Fisica at the Universidade de Lisboa, where she teaches history of science. Her contributions include articles on the history of quantum chemistry, as well as the history of science in Portugal. She was the leader of the Portuguese team for the European Community's Project Prometheus, a study of the reception of the ideas of the Scientific Revolution in countries at the periphery of Europe.

CROSBIE SMITH is Professor of History of Science and Director of the Centre for History & Cultural Studies of Science at the University of Kent at Canterbury. He coauthored (with M. Norton Wise) *Energy and Empire: A Biographical Study of Lord Kelvin* (1989) and coedited (with John Agar) *Making Space for Science* (1998). He is the author of *The Science of Energy: A Cultural History of Energy Physics in Victorian Britain* (1998), which won the History of Science Society's Pfizer Prize for 2000. His current research interests lie with the cultural history of late-nineteenth- and twentieth-century energy themes (especially Henry Adams).

ROBERT W. SMITH is Chair of the Department of History and Classics at the University of Alberta. He won the History of Science Society's Watson Davis Prize in 1990. He was Walter Hines Page Fellow at the National Humanities Center during 1992–3 and a Dibner Visiting Historian of Science in 1997. Among his recent works is *Reconsidering Sputnik: Forty Years after the Soviet Satellite* (2000), which he coedited with Roger Launius and John Logsdon.

GENERAL EDITORS' PREFACE

In 1993, Alex Holzman, former editor for the history of science at Cambridge University Press, invited us to submit a proposal for a history of science that would join the distinguished series of Cambridge histories launched nearly a century ago with the publication of Lord Acton's fourteen-volume *Cambridge Modern History* (1902–12). Convinced of the need for a comprehensive history of science and believing that the time was auspicious, we accepted the invitation.

Although reflections on the development of what we call "science" date back to antiquity, the history of science did not emerge as a distinctive field of scholarship until well into the twentieth century. In 1912 the Belgian scientist-historian George Sarton (1884–1956), who contributed more than any other single person to the institutionalization of the history of science, began publishing *Isis,* an international review devoted to the history of science and its cultural influences. Twelve years later he helped to create the History of Science Society, which by the end of the century had attracted some 4,000 individual and institutional members. In 1941 the University of Wisconsin established a department of the history of science, the first of dozens of such programs to appear worldwide.

Since the days of Sarton historians of science have produced a small library of monographs and essays, but they have generally shied away from writing and editing broad surveys. Sarton himself, inspired in part by the Cambridge histories, planned to produce an eight-volume *History of Science,* but he completed only the first two installments (1952, 1959), which ended with the birth of Christianity. His mammoth three-volume *Introduction to the History of Science* (1927–48), a reference work more than a narrative history, never got beyond the Middle Ages. The closest predecessor to *The Cambridge History of Science* is the three-volume (four-book) *Histoire Générale des Sciences* (1957–64), edited by René Taton, which appeared in an English translation under the title *General History of the Sciences* (1963–4). Edited just before the late-twentieth-century boom in the history of science, the Taton set quickly became dated. During the 1990s Roy Porter began editing the very useful Fontana History of Science (published in the United States as the Norton History of Science), with volumes devoted to a single discipline and written by a single author.

The Cambridge History of Science comprises eight volumes, the first four arranged chronologically from antiquity through the eighteenth century, the latter four organized thematically and covering the nineteenth and twentieth centuries. Eminent scholars from Europe and North America, who together form the editorial board for the series, edit the respective volumes:

Volume 1: *Ancient Science,* edited by Alexander Jones, University of Toronto
Volume 2: *Medieval Science,* edited by David C. Lindberg and Michael H. Shank, University of Wisconsin–Madison
Volume 3: *Early Modern Science,* edited by Lorraine J. Daston, Max Planck Institute for the History of Science, Berlin, and Katharine Park, Harvard University
Volume 4: *Eighteenth-Century Science,* edited by Roy Porter, Wellcome Trust Centre for the History of Medicine at University College London
Volume 5: *The Modern Physical and Mathematical Sciences,* edited by Mary Jo Nye, Oregon State University
Volume 6: *The Modern Biological and Earth Sciences,* edited by Peter Bowler, Queen's University of Belfast, and John Pickstone, University of Manchester
Volume 7: *The Modern Social Sciences,* edited by Theodore M. Porter, University of California, Los Angeles, and Dorothy Ross, Johns Hopkins University
Volume 8: *Modern Science in National and International Context,* edited by David N. Livingstone, Queen's University of Belfast, and Ronald L. Numbers, University of Wisconsin–Madison

Our collective goal is to provide an authoritative, up-to-date account of science – from the earliest literate societies in Mesopotamia and Egypt to the beginning of the twentieth century – that even nonspecialist readers will find engaging. Written by leading experts from every inhabited continent, the essays in *The Cambridge History of Science* explore the systematic investigation of nature, whatever it was called. (The term "science" did not acquire its present meaning until early in the nineteenth century.) Reflecting the ever-expanding range of approaches and topics in the history of science, the contributing authors explore non-Western as well as Western science, applied as well as pure science, popular as well as elite science, scientific practice as well as scientific theory, cultural context as well as intellectual content, and the dissemination and reception as well as the production of scientific knowledge. George Sarton would scarcely recognize this collaborative effort as the history of science, but we hope we have realized his vision.

David C. Lindberg
Ronald L. Numbers

ACKNOWLEDGMENTS

In writing this volume, both the contributing authors and I are indebted to criticism and comments from members of Volume 5's Board of Advisory Readers, each of whom read a subset of early versions of chapters for the volume. I express gratitude to these readers: William H. Brock (Eastbourne, U.K.), Geoffrey Cantor (University of Leeds), Elisabeth Crawford (Centre National de la Recherche Scientifique), Joseph W. Dauben (City University of New York), Lillian Hoddeson (University of Illinois), and Karl Hufbauer (Seattle, Washington). In addition, I thank four consulting readers, each of whom assisted with advice on a chapter in his area of expertise: Ronald E. Doel (Oregon State University), Dominique Pestre (Centre Alexandre Koyré, Paris), Alan J. Rocke (Case Western Reserve University), and David E. Rowe (Johannes Gutenberg-Universität Mainz).

I thank David C. Lindberg and Ronald L. Numbers for the invitation to edit Volume 5 on *The Modern Physical and Mathematical Sciences* in *The Cambridge History of Science,* and I am grateful to David C. Lindberg for his careful reading and comments on chapter drafts and on what has become the final text. A referee for Cambridge University Press was invaluable in suggesting revisions and improvements on an earlier version of the manuscript. Alex Holzman and Mary Child have been enthusiastic and expert as successive editors for the *Cambridge History of Science* project at Cambridge University Press. Mike C. Green, Helen Wheeler, and Phyllis L. Berk have provided Volume 5 with a high level of skillful editorial oversight.

J. Christopher Jolly and Kevin Stoller gave valuable assistance. I am grateful for continued research support from the Thomas Hart and Mary Jones Horning Endowment in the Humanities at Oregon State University. Some of the final work on the volume was done during the 2000–1 academic year, when I was a Senior Fellow at the Dibner Institute for the History of Science and Technology at the Massachusetts Institute of Technology. As always, Robert A. Nye has given moral and intellectual support and advice. Finally, I thank the contributing authors for their hard work, patience, and good humor in bringing this volume to fruition.

Mary Jo Nye

INTRODUCTION
The Modern Physical and Mathematical Sciences

Mary Jo Nye

The modern historical period from the Enlightenment to the mid-twentieth century has often been called an age of science, an age of progress or, using Auguste Comte's term, an age of positivism.[1]

Volume 5 in *The Cambridge History of Science* is largely a history of the nineteenth- and twentieth-century period in which mathematicians and scientists optimistically aimed to establish conceptual foundations and empirical knowledge for a rational, rigorous scientific understanding that is accurate, dependable, and universal. These scientists criticized, enlarged, and transformed what they already knew, and they expected their successors to do the same. Most mathematicians and scientists still adhere to these traditional aims and expectations and to the optimism identified with modern science.[2]

By way of contrast, some writers and critics in the late twentieth century characterized the waning years of the twentieth century as a postmodern and postpositivist age. By this they meant, in part, that there is no acceptable master narrative for history as a story of progress and improvement grounded on scientific methods and values. They also meant, in part, that subjectivity and relativism are to be taken seriously both cognitively and culturally, thereby undermining claims for scientific knowledge as dependable and privileged knowledge.[3]

---

[1] See, e.g., David M. Knight, *The Age of Science: The Scientific World View in the Nineteenth Century* (New York: Basil Blackwell, 1986). Comte's six-volume *Cours de philosophie positive* was published during 1830–42; for an abridged version, Auguste Comte, *The Positive Philosophy of Auguste Comte,* trans. Harriet Martineau (London: G. Bell & Sons, 1896).

[2] For the optimistic vision of unification and completeness, see Steven Weinberg, *Dreams of a Final Theory* (New York: Pantheon, 1992), and Roger Penrose, *The Emperor's New Mind* (New York: Oxford University Press, 1994). Against the possibility of completeness, see Nancy Cartwright, *The Dappled World: Essays on the Perimeter of Science* (Cambridge: Cambridge University Press, 1999).

[3] For a general discussion, Stephen Toulmin, *Cosmopolis: The Hidden Agenda of Modernity* (New York: Free Press, 1990). On "postmodernity" the classic text is Jean François Lyotard, *The Post-Modern Condition,* trans. Geoff Bennington and Brian Massumi (Minneapolis: University of Minnesota Press, 1984).

Historians of science have addressed these late-twentieth-century issues by greatly expanding their tools of study in terms of subjects, methods, themes, and interpretations. Most historians of science have come to believe that there can be no unified history of science predicated upon the assumption of a "logic" or "method" of science. Some historians have concluded that there is no longer any place for a grand narrative of science ("the history of science") or even of a single scientific discipline ("the history of chemistry"). As a result, much recent work in the history of science has focused on histories of scientific practices, scientific controversies, and scientific disciplines in very local times and spaces.[4]

Still, larger narratives persist, as demonstrated, for example, in the very successful series of single-authored Norton histories of science published in the 1990s, including *The Norton History of Chemistry* and *The Norton History of Environmental Sciences*.[5] Other examples of comprehensive histories include studies of twentieth-century physics, such as Helge Kragh's history of physics in the twentieth century and Joseph S. Fruton's history of biochemistry and molecular biology as the interplay of chemistry and biology.[6]

The chapters in Volume 5 of *The Cambridge History of Science* represent a variety of investigative and interpretive strategies, which together demonstrate the fertile complementarity in history of science and science studies of insights and explanations from intellectual history, social history, and cultural studies.

It should be noted that the biographical genre of history is explicitly excluded as a focus for any one chapter in the volume, although individual figures, not surprisingly, often loom large. Among these are William Whewell, Hermann von Helmholtz, William Thomson (Lord Kelvin), and Albert Einstein. In addition, none of the chapters has a specifically national focus, since Volume 8 in the *Cambridge History of Science* series concentrates precisely on the modern sciences in national and international contexts.[7]

---

[4] For an overview of assumptions and methodologies in the history of science and science studies, see Jan Golinski, *Making Natural Knowledge: Constructivism and the History of Science* (Cambridge: Cambridge University Press, 1998).

[5] William H. Brock, *The Norton History of Chemistry* (New York: W. W. Norton, 1992); Peter J. Bowler, *The Norton History of Environmental Sciences* (New York: W. W. Norton, 1993); Donald Cardwell, *The Norton History of Technology* (New York: W. W. Norton, 1995); John North, *The Norton History of Astronomy and Cosmology* (New York: W. W. Norton, 1995); Ivor Grattan-Guinness, *The Norton History of the Mathematical Sciences* (New York: W. W. Norton, 1998); Roy Porter, *The Greatest Benefit to Mankind: Medical History of Humanity* (New York: W. W. Norton, 1998); and Lewis Pyenson and Susan Sheets-Pyenson, *Servants of Nature: A History of Scientific Institutions, Enterprises, and Sensibilities* (New York: W. W. Norton, 1999).

[6] Helge Kragh, *Quantum Generations: A History of Physics in the Twentieth Century* (Princeton, N.J.: Princeton University Press, 1999), and Joseph S. Fruton, *Proteins, Enzymes, Genes: The Interplay of Chemistry and Biology* (New Haven, Conn.: Yale University Press, 1999).

[7] Ronald L. Numbers and David Livingstone, eds., *Modern Science in National and International Contexts*, vol. 8, *The Cambridge History of Science* (Cambridge: Cambridge University Press, forthcoming).

Most authors in this volume have provided a largely Western narrative of their subjects, suggesting to the reader that historians of science in the twenty-first century still have much to write about modern scientists and scientific work in non-Western cultures.[8]

Some common themes and interpretive frameworks run through the volume, as detailed in the following discussion. Perhaps most striking among leitmotifs is historians' continuing preoccupation with Thomas S. Kuhn's characterizations of everyday science and scientific revolutions. Historians' decisions to explain scientific traditions and scientific change in terms of gradual evolution or abrupt revolution remain at the core of interpretive frameworks in the history of science.[9]

PART I. THE PUBLIC CULTURE OF THE PHYSICAL SCIENCES AFTER 1800

The first section of the volume focuses on the public culture of the modern physical and mathematical sciences, with emphasis on the Western European and North American countries in which these physical sciences were largely institutionalized until the early twentieth century.

Nancy Cartwright, Stathis Psillos, and Hasok Chang lay out various expectations of modern philosophical writers and scientific practitioners about what they hoped to achieve by defining and employing "scientific method," whether inductive or deductive, empiricist or rationalist, realist or conventionalist, theory laden or measurement dependent in normative and operative outlines. Like Frederick Gregory in his discussion of the intersections of religion and science, the coauthors note the importance for many scientists (for example, Albert Einstein around 1900 or Steven Weinberg around 2000) of a Pythagorean-like belief in the mathematical structure of the world, or what Weinberg has called the kinds of law that correspond "to something as real as anything else we know."[10]

Gregory, like David M. Knight in his essay on scientists and their publics, describes a nineteenth-century European world in which religion and science

---

[8] However, see, e.g., Lewis Pyenson, *Civilizing Missions: Exact Sciences and French Overseas Expansion, 1830–1940* (Baltimore: Johns Hopkins University Press, 1993), and Zaheer Baber, *The Science of Empire: Scientific Knowledge, Civilization, and Colonial Rule in India* (Albany: State University of New York Press, 1996).

[9] Thomas S. Kuhn, *The Structure of Scientific Revolutions* (Chicago: University of Chicago Press, 1962). Among the many sources on Kuhn's work, see Nancy J. Nersessian, ed., *Thomas S. Kuhn*, special issue of *Configurations*, 6, no. 1 (Winter 1998). On "revolution," I. Bernard Cohen, *Revolution in Science* (Cambridge, Mass.: Harvard University Press, 1985). On the argument for ruptures and mutations (and against continuities and transitions), see Michel Foucault, *The Archaeology of Knowledge*, trans. A. M. Sheridan Smith (New York: Pantheon, 1972; 1st French ed., 1969).

[10] Quoted in Ian Hacking, p. 88, *The Social Construction of What?* (Cambridge, Mass.: Harvard University Press, 1999), from Steven Weinberg, "Sokal's Hoax," *New York Review of Books*, 8 August 1996, 11–15, at p. 14.

were held to be compatible in the face of increasing secularization. William Whewell stood almost alone among scientific intellectuals in opposing on religious grounds the hypothesis of the plurality of worlds. James Clerk Maxwell, the brothers William and James Thomson, Louis Pasteur, and Max Planck all found science and religion mutually supportive, once extreme statements of scientific materialism were eliminated. Gregory notes the paradox that scientists and theologians shared a belief in the existence of foundational principles for natural phenomena, while not always agreeing on how properly to characterize these first principles.

Gregory also notes a link between religion and science in a shared gender bias toward membership in the community of scientists, a theme taken up by Margaret W. Rossiter in her history of the exclusion of women from scientific education and scientific organizations. Although there have been relatively few women in the physical sciences in comparison to men, Marie Curie nonetheless is one of the best known of *all* scientists. Female physicists currently are found in much higher proportions in countries outside Japan, the United States, the United Kingdom, and Germany. Yet, this fact may not necessarily indicate greater opportunities for women so much as a gendered proletarianization of university educators in some countries.

Some of Rossiter's female scientists figure, as well, in Knight's discussion of the popularization of science, not because women were lecturing in public places like the Friday evening lectures of the Royal Institution, but because they were writing widely read and commercially successful books, such as Jane Marcet's *Conversations on Chemistry* (1807) and Mary Somerville's *Connexion of the Physical Sciences* (1834).

Knight notes, as does Pamela Gossin, the extraordinary popularity of the science of chemistry for the early-nineteenth-century imagination, a popularity that was eclipsed in the next decades by geology. Early in the nineteenth century, light, heat, electricity, magnetism, and the discovery of new elements – all parts of chemistry – excited attention. By century's end it was "auras" and table rapping that were the rage, along with x rays that could be used to see through human flesh.

We became familiar in the twentieth century with the idea of a polarization between the "two cultures" of the sciences and the humanities. Knight and Gossin remind us of the many scientists who have themselves written literature and poetry (among them Davy, Maxwell, C. P. Snow, Primo Levi, Carl Sagan, and Roald Hoffmann), as well as the novelists and poets who have studied the sciences and incorporated scientific elements into their work (Mary Shelley, Nathaniel Hawthorne, Edgar Allan Poe, Aleksandr S. Pushkin, Honoré de Balzac, Emile Zola, James Joyce, Virginia Woolf, Vladimir Nabokov). The science-educated novelist H. G. Wells appears and reappears in chapters of this volume. From Jonathan Swift and William Blake to Bertolt Brecht and Friedrich Dürrenmatt, scientists and their work have figured in the literary and artistic products of public culture.

Introduction 5

PART II. DISCIPLINE BUILDING IN THE SCIENCES: PLACES, INSTRUMENTS, COMMUNICATION

If natural philosophy, natural theology, chemical philosophy, and natural history were the fields of inquiry for the generalist savant who flourished in the eighteenth and early nineteenth centuries, scientific specialisms were to proliferate during the nineteenth century into disciplinary boundaries that enrolled professional "scientists" (the English term invented by William Whewell in 1833) in the classroooms, societies, and bureaucracies. The intricacies of discipline building have elicited considerable attention from historians of science in the last few decades, as has the construction of research schools and research traditions.

Among scientific disciplines, mathematics has been regarded as the foundational science since at least the time of Comte. Many mathematicians and historians of mathematics, as David E. Rowe points out, have never doubted the cumulative nature of mathematical knowledge and its reflection of a Platonic realm of permanent truths. Yet mathematics, too, is an intellectual and social activity that produces knowledge, sometimes by apparent revolutionary breakthroughs, as in the case of Georg Cantor's set theory, but also in the ongoing work of the normal production of university lecture notes, paradigmatic textbooks, and research journals. The result has been, as Rowe puts it, "vast quantities of obsolete materials," as well as revolutions, rediscoveries, and transformations of methods and insights long discarded.

Rowe insists particularly on the importance in the history of modern mathematics of the research seminars and of oral knowledge transmissions that took root in small German university towns in the early nineteenth century. These resulted in informal groups with intellectual orientation and loyalty to a particular mentor. National differences existed, for example, in the distinctive tradition of mixed mathematics in England.

National differences are at the heart of Terry Shinn's investigation of the relationships among science and engineering education, research capacity, and industrial performance in Germany, France, England, and the United States. Shinn takes the not-uncontroversial position that there *has been* a difference in economic achievement among these nations and that it *might be correlated* with the aims and structures of scientific education. Whereas Rowe emphasizes that neohumanist scholarship developed in Germany specifically in opposition to what post-Napoleonic Germans called the "school learning" of the French, Shinn emphasizes the successful linking of German scientific education and research with the needs of German industry, particularly in mechanics, chemistry, and electricity by the end of the nineteenth century.

At the heart of discipline building are not only the sites and spaces for the disciplines but also the array of instruments and the means of communication that define and mark off one intellectual field from another. Robert W. Smith's analysis of astronomical instrumentation notes striking changes in kind and scale that marked the history of astronomy from Giovanni Piazzi's 1801 discovery of an asteroid, using an altazimuth circle, to the 1990 launching of the Hubble Space Telescope. As Smith makes clear, the improvement of telescopes, both optical and radio, often was a goal in itself, rather than a means of addressing theoretical questions. Astronomy contributed its fair share in the nineteenth century to what historians have characterized as obsession with precision measurement.

As in other scientific disciplines in the twentieth century, the expense and the patronage of astronomy became ever greater after the Second World War. Like nuclear physicists, astronomers found themselves working in new kinds of organization, for example, the international university consortium, in which they collaborated with engineers, machinists, physicists, and chemists. In such large enterprises, as in smaller venues, communication patterns of scientists became crucial to disciplinary identities and distinctions, as well as to the accomplishment of original work.

Bernadette Bensaude-Vincent treats communication patterns and the construction of scientific languages in modern chemistry, while Arthur I. Miller focuses on changes in imagery and representation in modern physics, showing how language and image are instruments or tools for expressing theories and making predictions and discoveries, as well as for establishing group identity.

While some languages and images changed dramatically in intent and content over time, others remained remarkably stable. A small group of French chemists in 1787 famously created an artificial and theory-laden language for a new, antiphlogistonist chemistry, in which, as Bensaude puts it, the binomial name was to be a mirror image of the operations of chemical decomposition. This formalist and operationalist project succeeded quickly, despite objections to the French language from foreign chemists and opposition to theoretical names from pharmacists and artisans, who commonsensically preferred historical and descriptive names. Later projects for chemical nomenclature proved more conventional and pragmatic in design, perhaps because they were truly international and more consensual.

Miller's history of visual imagery in physics is similarly one of controversy and compromise among scientists. In this history, Miller distinguishes between visual images rooted in intuition (*Anschauung*) and visual images seated in perception (*Anschaulichkeit*). Hinting at parallels with the artistic forms developed by Pablo Picasso, Georges Braque, and, later, Mark Rothko, Miller details the increasingly abstract visualization adopted by Einstein, Werner Heisenberg and, later, Richard Feynmann. Yet, Miller argues, there is ontological realist content to Feynmann's diagrams. "All modern scientists," says Miller, "are scientific realists."

PART III. CHEMISTRY AND PHYSICS: PROBLEMS THROUGH THE EARLY 1900s

In turning to specific disciplinary areas of scientific study in the nineteenth and twentieth centuries, Parts III, IV, and V of this volume loosely employ the overlapping categories of chemistry and physics, atomic and molecular sciences, mathematics, astronomy, and cosmology, noting that these categories sometimes can be identified with professional disciplines and experts (chemistry, chemist) and sometimes not. Very different historical approaches are taken by the authors: intellectual history or social history, national traditions or local practices, gradual transitions or radical breaks.

Frederic L. Holmes disputes the long-standing claim, originated by scientists themselves, that nineteenth-century experimentalists, such as Helmholtz and Emile Dubois-Reymond, broke in the 1840s with vitalist presuppositions, providing a "turning point" for the reductionist application of the laws of physics and chemistry to living processes. On the contrary, Holmes argues that nineteenth-century scientists simply had more powerful concepts and methods available than had their predecessors for the exploration and characterization of digestion, respiration, nervous sensation, and other "vital" processes. Earlier investigators pursued similar aims, but with less satisfactory means at their disposal.

While historians and scientists often speak of a chemical revolution associated with the atomism of John Dalton, Hans-Werner Schütt notes the ongoing and unresolved discussions throughout the nineteenth century about the relationship between what chemists called "chemical atoms" (corresponding to chemical elements) and what natural philosophers and physicists treated as "physical atoms" (corresponding to indivisible corpuscles). Calculating relative atomic weights, defining the standard of comparison for atomic weights, classifying simple and complex substances and their behaviors by means of chemical symbols and systematic tables: All of these tasks were continuing challenges for chemists throughout the century.

What constituted a chemical fact or conclusive evidence for a formula, a classification, or a theory? Schütt relates Justus Liebig's conviction that "theories are expressions of contemporary views... only the facts are true." Alan J. Rocke notes August Kekulé's remark that it is an "actual fact," not a "convention," that sulfur and oxygen are each equivalent to two atoms of hydrogen. J. J. Berzelius distinguished between "empirical" and "rational" formulas for chemical molecules, one based in laboratory analysis and the second based in theory. These chemists were savvy about scientific epistemology. Yet they were not quick to adopt a new theory. Rocke has found that nearly all active organic chemists who were more than forty years old in 1858 ignored Kekulé's structure theory, while the younger generation took it on.[11] However, by the 1870s the structure theory provided a framework not only for academic chemistry but also for an expanding German chemical industry.

The reciprocal relationship between scientific innovation and industrial development is more fully developed in Crosbie Smith's study of energy and Bruce J. Hunt's analysis of electrical science. Sungook Hong also discusses the interplay among theoretical concept, laboratory effect, and technological artifact.

Hong challenges the usual history of nineteenth-century theories of light and radiation as a story of revolution. Many accounts of the wave versus particle theories of light attribute Fresnel's winning of the 1819 Academy of Sciences prize to his memoir's good fit with experimental data, in combination with the declining political and social fortunes of Laplacian physicists. Drawing upon an analysis by Jed Z. Buchwald, Hong concedes that Fresnel's mathematics fit the data, but adds that the prize-awarding jury at the time saw no significant physical hypothesis in Fresnel's work that would inhibit them from continuing to employ a ray (emission) analysis for studying light. In this case, as in the history of theories and experiments on the spectra of heat, light, and chemical (ultraviolet) radiations, Hong sees a process of "prolonged confusion" and gradual consensus, without crucial experiments, in the service of precise measurement.

Crosbie Smith addresses the question of simultaneous discovery, disputing Kuhn's presumption that energy was something in nature to be discovered. At the same time, Smith shows some of Kuhn's preoccupation with the means by which a paradigm is constituted. For Smith, it was North British (Scottish) cultures of engineering and Presbyterianism that made James Thomson and William Thomson determined to study the problem of the waste of useful work and to effect a reform of physical science, as they replaced the language and assumptions of action-at-a-distance and mechanical reversibility with a natural philosophy of energy and its transformations. In this aim, in Smith's analysis, the Thomson brothers were joined by Maxwell, most notably in his *Treatise on Electricity and Magnetism* (1873).

Hunt is less concerned with Presbyterianism than with technology, narrating, consistently with Crosbie Smith's account, the triumph of William Thomson's scientific approach to electrical engineering in the completion of Cyrus Field's venture for laying trans-Atlantic telegraphic cables during 1865–6. Hunt explains the influential reformulation of Maxwell's electromagnetic theory by Oliver Heaviside and by Heinrich Hertz in the 1880s, noting the gap between the continental action-at-a-distance approach to electromagnetism and Maxwell's field concept. An important linkage between the two was made in H. A. Lorentz's theory of tiny charges that are able

---

[11] See Max Planck's comment about generations in *Scientific Autobiography and Other Papers*, trans. F. Gaynor (New York: Philosophical Library, 1949), p. 33.

to move freely in conductors but are bound in place in material dielectrics. Thus, Hunt argues, Albert Michelson's anomalous failure to detect ether effects could also be seen as a confirmation of the electronic constitution of matter. In 1905 Einstein independently arrived at a new foundation for the electrodynamics of moving bodies.

Crosbie Smith's approach provides a good example of the contextualist and constructionist method of analyzing the history of science by way of focusing on scientific practitioners who construct knowledge concepts within local contexts for specific audiences, while drawing upon, or establishing, reputations for credibility and trustworthiness. Smith's approach fits squarely within the cultural studies of science. The approach is supported in striking manner by the excerpt in Smith's chapter from Joseph Larmor's obituary notice of Lord Kelvin, in which Larmor wrote that energy has "furnished a standard of industrial values... [of] power... measured with scientific precision as a commercial asset... [and] created the doctrine of inorganic evolution and changed our conceptions of the material universe."

PART IV. ATOMIC AND MOLECULAR SCIENCES IN THE TWENTIETH CENTURY

Relativity theory, quantum theory, and nuclear theory all departed radically in the early 1900s from textbook theories of matter and radiation. Although historians never deny the revolutionary contribution of Einstein to relativity theory, historical accounts of the early quantum theory differ in assessing the role of Max Planck's 1900 paper in breaking with classical physics. Kuhn provided a detailed historical argument that it was Einstein, not Planck, who realized the physical implications of Planck's first incomplete attempt at unifying the physics of radiation and of thermodynamics.

In Olivier Darrigol's analysis of the history of early quantum physics, it was Niels Bohr who was most radical of all. He quickly adopted Einstein's application of the light quantum to the emission and absorption of radiation by orbital electrons in the atom. In the early 1920s, Bohr was willing to embrace a statistical interpretation of energy conservation. In 1927 he abandoned visualizable electron orbits and waving radiation fields in favor of the complementarity (or, incompatibility) of the particle and wave pictures as two ways of describing the same thing.

Acausality, uncertainty, and indeterminism were said by Bohr to be in the nature of things. In contrast, Heisenberg attributed indeterminism to the operations of instruments. Least radically, Einstein was convinced that indeterminism results from the inadequate state of current knowledge. On the question of whether quantum mechanics in the late 1920s was a constructed response to antirational and antimaterialistic ideology in the Weimar Republic, as Paul Forman has argued, Darrigol sides with historians who see arguments internal to electron and radiation theory as sufficient to justify the radical move from determinism to indeterminism, and from a visualizable world to a phenomenological world. Some of the strongest proponents of the new physics came from outside Weimar political culture, including Bohr and Paul Dirac.[12]

If the history of quantum physics has been revised in the last decades, so too has the history of radioactivity and nuclear physics. What Jeff Hughes calls "bomb historiography" continues to have an important place in the history of modern physics. It now is supplemented by detailed studies of early centers of investigation of radioactivity in different locales (Paris, Berlin, Montreal, Vienna, Wolfenbüttel).[13] The production of radium for laboratory and medical markets, the training of personnel in radioactivity laboratory techniques and protocols, the negotiation of measurement standards and units, the improvement of instruments for counting radioactive and nuclear events, and the establishment of journals and conferences, by way of establishing a disciplinary field, constitute a recent historiographical approach.

If "bomb historiography" has been an understandable focus for nuclear physics, so has what Silvan S. Schweber calls the "inward bound" cognitive historiography of the search for smaller and smaller nuclear entities at higher and higher energies. Declining a Kuhnian approach (renormalization theory or broken-symmetry theory as "revolutions") or a Galison-like approach (studying the subcultures of experiment, theory, instruments, and their interfaces), Schweber adopts a narrative of the history of ideas that stresses the cumulative and the continuous, yet novel, developments in the history of particle physics.

Paradoxically, as in the history of the "atom," this is a history in which the "particles" have become increasingly phenomenological in character, described by field equations or S-matrix theory in the standard model, and including "quarks" with fractional charges that never have been observed. Schweber concludes that the standard model "is one of the great achievements of the human intellect," but that it is not a final theory.

In distinguishing physics and chemistry, it often is said that modern chemists concern themselves with molecules and atoms. In defining the relationship between quantum chemistry and chemical physics, Ana Simões explains that "understanding why and how atoms combine to form molecules is an intrinsically chemical problem, but it is also a many body problem."

Quantum chemistry as a discipline has both social and cognitive roots, with the conceptual origins strongly identified with Walter Heitler and Fritz London's application of Heisenberg's resonance theory to the hydrogen

---

[12] Paul Forman, "Weimar Culture Causality and Quantum Theory, 1918–1927: Adaptation by German Physicists and Mathematicians to a Hostile Intellectual Environment," *Historical Studies in the Physical Sciences*, 3 (1971), 1–116.

[13] In particular, on Berlin and the Kaiser Wilhelm Institute, see Ruth L. Sime, *Lise Meitner: A Life in Physics* (Berkeley: University of California Press, 1996).

molecule in 1927, and with Linus Pauling's and John Slater's independent (1931) characterizations of the carbon atom's bonds by "hybridized" electron orbitals.

Simões detects national styles at work in the application of quantum mechanics to chemistry, in this case, American pragmatism and German foundationalism. Yet she also finds that the competing methodologies of valence-bond theory and molecular-orbital theory crossed national lines, so that national styles are hardly the whole story. The early successes of valence-bond theory demonstrate the importance of model building, visualization, and approximative methods for chemists, as well as the power of charismatic personality (Pauling). Later successes of molecular-orbital theory are rooted similarly in personality and rhetorical skills (Charles Coulson), but equally in new instrumentation for fast computing and for molecular spectroscopy.

Michael Eckert argues similarly that plasma physics and solid-state physics acquired disciplinary identities less by differentiation from other fields than by integration. Elements of solid-state science can be found in the 1930s in Heisenberg's institute at Leipzig, Slater's department at MIT, or Nevill Mott's institute at Bristol. The study of "plasma" also had origins in industrial research laboratories, such as Irving Langmuir's at General Electric.

The Second World War created well-funded problems and communities for studying thermonuclear fusion and semiconductor electronics. After the war, fusion seminars were led by George Thomson at Harwell in Great Britain, Edward Teller at Los Alamos, and Andrei Sakharov and Igor Tamm at Arzamas 16 in the Soviet Union. Industry, along with governments and universities, encouraged these fields after the war. A course at Bell Laboratories in 1951 on transistor physics and technology was attended by 121 military personnel, 41 university scientists, and 139 industrial researchers. If Los Alamos and the nuclear atom were symbols of the Cold War, Silicon Valley and the silicon chip became symbols of the last *fin de siècle*.

Among the early leaders in this "solid-state science" were physical chemists and organic chemists who laid out the fundamentals of macromolecular and polymer science in the 1920s and 1930s, with Hermann Mark and Kurt Meyer at I. G. Farben prominent among them. The very idea that molecules might be very, very large was resisted by many organic chemists, but in the 1930s Wallace Carothers at Dupont Chemicals synthesized fibers with molecular weights in the tens of thousands, and Mark suggested that huge molecules might have coiled, spiraled, and flexible shapes that account for diverse physical properties in the solid state.

Yasu Furukawa's chapter notes the lack of communication between polymer scientists and contemporary protein researchers, a lacuna similar to the gap in communication between bacteriologists and geneticists in the history of molecular biology. It was Staudinger's student Rudolf Singer who prepared the polymeric substance DNA, estimating its molecular weight between 500,000 and 1,000,000, personally delivering a sample to Maurice

Wilkins at King's College in 1950. This was the sample from which Rosalind Franklin prepared the so-called B form for her revolution-making DNA x-ray diffraction patterns.[14]

PART V. MATHEMATICS, ASTRONOMY, AND COSMOLOGY SINCE THE EIGHTEENTH CENTURY

At the core of many developments in physical theories are mathematical methods of representation and investigation. Yet mathematics is not a mere handmaiden to physical theories but a science in its own right. Joan L. Richards and Jesper Lützen each emphasize continuities, transitions, and diversifications within mathematics. They also employ the terminology of discontinuity and revolution for late–nineteenth-century developments in which both geometry and analysis became emancipated, as Lützen puts it, from long-standing intuitive preconceptions of objective space.

Richards emphasizes the increasing freedom of geometry from concern with practical applications, a development that occurred in the German research universities in the 1830s. In counterpoint, Lützen stresses the constant interplay in mathematical analysis between demands for rigor and for application (in acoustics, hydrodynamics, electricity, and, later, quantum mechanics). While "rigor" is characteristic of the axioms of Euclidean geometry and of the foundations of A. L. Cauchy's mathematical analysis, analysis and its use of functions was pushed toward more and more faithful representations of the real world.

A striking example of application lies in the appropriation by Maxwell of the methods of statistics from the study of human populations to the study of molecular populations. Theodore M. Porter describes how the use of statistics, both in the study of death rates and the study of molecular motions, was used to demonstrate the existence of order in events that appear to be random. Yet, paradoxically, within a decade after its first development, statistical physics was on its way to undercutting confidence in the orderly determinism and necessity of the mechanical laws of the universe. For Maxwell and Boltzmann, who were not mathematicians but natural philosophers, these statistical models and mechanical models became useful, although they did not compel assent as perfect reflections of the natural world.

In the study of molecules and atoms, spectroscopy became an increasingly important tool. Joann Eisberg places spectroscopy at the focus of a diversification in astronomy from positional to descriptive astronomy in the course of the nineteenth century. Through spectroscopy, the stars became objects

---

[14] See Robert Olby, *The Path to the Double Helix: The Discovery of DNA* (London: Macmillan, 1974), and Maclyn McCarty, *The Transforming Principle: Discovering That Genes Are Made of DNA* (New York: W. W. Norton, 1985).

of laboratory science, as were atoms and molecules. Comte, who wrote early in the century that the physical and chemical nature and the temperatures of the stars would be forever unknown, was as wrong about distant stars as about invisible atoms.

The development of astronomy, as discussed in Part II by Robert W. Smith, was largely driven by improvements in telescopes and the invention of photography. Attempts to explain the origin of solar and stellar energy, as well as their past and future evolution, were rooted in physicists' gravitational and thermodynamic theories, but also in systems of classification characteristic of natural history. With photographic spectroscopy and bigger telescopes providing much more rapid means of accumulating information about larger numbers of stars ever more quickly, a factory system of division of labor began to develop within observatories, notably at Harvard Observatory from the 1880s to the 1920s, where a workforce of female plate readers, or computers, was employed by Edward Pickering. The gendering of labor, as discussed also by Rossiter in Part I, led to some unexpected results in the cases of Annie J. Cannon, Antonia Maury, Cecilia Payne, and (as mentioned also by Helge Kragh) Henrietta Leavitt, all of whom began drawing theoretical conclusions from the stars and the spectral lines that they were classifying.

The hypothesis that the sun and stars have detectable life sequences was common from the time of William Herschel, and it fit in well with later nineteenth-century notions of biological and thermodynamic evolution. Kragh argues that the notion that the universe is not static but is expanding was a novel idea, quickly embraced by astronomers in the 1930s. An even more truly novel theory was Einstein's general theory of relativity, which created a new science. Even though Einstein himself adopted a cosmology of linear time and "spherical" space, that is, a static universe, the expanding universe was easily adopted in the 1930s because it rested safely on Einstein's field equations.

In speaking of the "discovery" of an expanding universe, we run into difficulties, as is often the case in defining "discovery." Like Planck, Edwin Hubble was a reluctant revolutionary, if revolutionary he was, in emphasizing in 1929 the empirical nature of the galactic redshift, rather than immediately arguing for an expanding universe. In 1922, A. A. Friedmann had developed a general mathematical cosmology, which included static, cyclical, and expanding universes as special cases. Georges Lemaître specifically argued in 1927 that the physical universe is expanding. Thus, Kragh suggests, it is reasonable to credit Lemaître, not Hubble, with the "discovery."

If discoveries are hard to pin down, as Crosbie Smith and Darrigol also emphasize, so too are definitive solutions to problems ("how experiments end," in Galison's usage).[15] In the 1930s and 1940s, with calculations based

---

[15] See Peter Galison, *How Experiments End* (Chicago: University of Chicago Press, 1987), and *Image and Logic: A Material Culture of Microphysics* (Chicago: University of Chicago Press, 1997).

on the hypothesis of a "primeval atom," astrophysicists inferred that the age of the universe is less than the age of the stars. This problem, resolved some decades later to the satisfaction of the astronomical community, reappeared yet again after the processing of data from the Hubble Telescope in 1994.

Cosmology, writes Kragh, lacks disciplinary unity, and it has been insufficiently studied in its social, institutional, and technological makeup. In their chapter on the chemistry and physics of the earth, Naomi Oreskes and Ronald E. Doel take up these points of reference for the science of the earth, noting the competition in earth science between a physics tradition and a natural history tradition. By the mid-twentieth century the geophysics tradition was becoming ascendant over the natural-history geological tradition despite that fact that geologists had most often been right in disputes with physicists: The earth is much older than Kelvin had allowed, and the earth has experienced continental drift even though physicists had denied the existence of a plausible mechanism.

Oreskes and Doel root these changes not only in the epistemological prescriptions of influential scientists such as Charles van Hise for reducing geology to the principles of physics and chemistry, but also in shifting patterns of patronage. The Rockefeller Foundation funded geophysics, not geology; petroleum companies interested themselves in chemical analyses, not just the appearance of rocks and strata; the new military technologies of airplanes, missiles, submarines, and radar required geophysical, meteorological, and oceanographic knowledge for performance and protection.

PART VI. PROBLEMS AND PROMISES AT THE END OF THE TWENTIETH CENTURY

As mentioned by Oreskes and Doel, and by Alex Roland, one of the principal spurs to the science and technology of seismology came not from the need to study earthquakes or to understand the earth's interior, but from military and political requirements for detection of underground nuclear explosions. A minor theme in the chapter of Oreskes and Doel is the major theme for Roland's chapter, namely, the relationships among science, technology, and war. Roland's chapter, like others in Part VI, clearly addresses scientific and technological problems that are matters of state and business strategies, with direct implications for public welfare.

The Second World War was a turning point. The victors emerged with a completely different arsenal of weapons than they possessed when war began. More significantly, Roland argues, whereas a traditional conservatism in military forces had worked against the adoption of new technologies for many generations, the Second World War reversed this behavior. Governments in the United States, the Soviet Union, France, Great Britain, the People's Republic of China, India, and elsewhere imposed upon themselves the need for permanent military preparedness, requiring large outlays of monies for research and development for military purposes, as well as permanent protocols of secrecy for national security. Secrecy not only applied to nuclear energy and nuclear weapons but affected, as well, optics, computers, microelectronics, composite materials, superconductivity, and biotechnology. Universities, as well as private industry, became regular procurers of military contracts. In fiscal year 1995, Roland reports, MIT and the Johns Hopkins University were among the top fifty defense contractors in the United States in dollar volume. An assessment should be made, he suggests, of the cost of these developments to basic research and to socially needed programs, such as urban renewal and the reversal of environmental degradation.

If wartime needs have had significant effects on the conduct of scientific research, so too have national values and ideologies. It was not uncommon among historians after the Second World War to focus on the effects of "totalitarianism" on science and scientists in Stalin's Soviet Union or Hitler's national socialist Germany.[16] Recent historical work, including Paul Josephson's, enlarges the focus of ideology to include the democratic and pluralist United States during the Cold War and McCarthy period, as well as other countries.

Claims of ideology are difficult to sort out. Forman has argued that acausal quantum mechanics was welcomed in Germany by anti-Weimar intellectuals, yet promulgators of "Aryan" science, Philip Lenard and Johannes Stark best known among them, rejected quantum mechanics and relativity theory on the grounds that the new physics was insufficiently grounded in the real world. The "Mechanist" faction in the Soviet Union similarly spurned the new physics as "idealist" rather than "materialist," despite efforts by Boris Hessen and other members of the "Deborinite" faction to reconcile the new physics with dialectical materialism. Hessen disappeared in 1937 during the Great Terror in which some 10 million people died.

Josephson, like Loren R. Graham, Jessica Wang, and some other historians, concludes that most scientists try to avoid political commitments and to pursue their work the best they can, no matter what the political regime.[17] Nor can it be assumed that scientists are necessarily inclined toward democratic and inclusive political views. Indeed, Josephson argues that most German scientists distrusted the Weimar regime and welcomed the Nazis to power. There is considerable historical evidence that few non-Jewish German scientists protested the expulsion of Jewish colleagues.[18]

---

[16] David A. Hollinger, "Science as a Weapon in *Kulturkämpfe* in the United States during and after World War II," pp. 155–74, in *Science, Jews, and Secular Culture: Studies in Mid-Twentieth-Century American Intellectual History* (Princeton, N.J.: Princeton University Press, 1996).

[17] Loren R. Graham, *What Have We Learned about Science and Technology from the Russian Experience?* (Stanford, Calif.: Stanford University Press, 1998), and Jessica Wang, *American Science in an Age of Anxiety: Scientists, Anticommunism and the Cold War* (Durham: University of North Carolina Press, 1999).

[18] Ute Deichmann, "The Expulsion of Jewish Chemists and Biochemists from Academia in Nazi Germany," *Perspectives on Science*, 7 (1999), 1–86.

Of all the intellectual and social transformations wrought by the sciences and technology during the Second World War, perhaps the most astonishing is what William Aspray calls the "Computer Revolution," with no demurrer about using the term "revolution." Histories of computer science and computer culture have shifted attention from machine precursors, like Charles Babbage's Analytical Engine, which functioned as a stored-program computer, to the study of military, business, and scientific strategies for improving, programming, marketing, and using computer machines. The creation of academic "computer science" and "information science" programs in universities resulted at the end of the twentieth century from the integration of programs in engineering, mathematics, and cognitive science and artificial intelligence, with ever-increasing prestige for engineering.

Fast and precise computers were not only applied to modeling and to calculating previously intractable problems in theoretical chemistry and plasma physics, or in missile guidance and satellite orbits, but also for medical imaging and for global climate modeling, as described by Bettyann Holtzmann Kevles and by James Rodger Fleming in the concluding chapters of this volume. Kevles's account of the encounter in the 1970s of computers and medical instrumentation is another example of the integration of disparate disciplinary trajectories (solar astronomy, neurology, engineering, biochemistry, nuclear physics, solid-state physics) as individuals' interests converged on a single focus, in this case, medical applications. The 1979 Nobel Prize in Physiology or Medicine was shared by a nuclear physicist and an electrical engineer, the latter having funded his work from the British Department of Health and Social Security, in combination with the Electrical and Musical Industries' (EMI's) profits from the Beatles' records.

Small-scale research still could result, then, in unforeseen breakthroughs, as in the case of computerized tomography. In contrast, research on changes in the earth's climate, which began as small-scale record keeping in the seventeenth and eighteenth centuries, gave way by the 1970s to large-scale computerized projects like the RAND Corporation's program of climate dynamics for environmental security, relying on information from earth-orbiting satellites that were used for the dual purpose of monitoring nuclear weapons tests and global weather systems.

As Fleming shows, scientific and public interest in climate change goes back a long way, as does the conviction that the earth is getting warmer. Thomas Jefferson ascribed a warmer climate to the cutting down of trees and to increased agricultural cultivation. In the 1950s, C. S. Callendar's research concluded that atmospheric carbon dioxide from fossil fuels had increased the earth's temperature by 0.25 degrees in the previous fifty years. However, by the early 1970s, following the failure of Soviet grain harvests, public anxiety focused on the question of whether the earth is getting cooler.

Of particular historical interest is the cultural meaning of these concerns. For Jefferson, agriculture could be extended and improved in a warmer climate. For Svante Arrhenius in Sweden around 1900, glaciers were an unwelcome reminder of the earth's cold and uncivilized history. A modest increase in temperature of the earth's surface would be a good thing. However, by 1939 Callendar was concerned that humans were an unwelcome "agent of global changes" in the profligate production of carbon dioxide from fossil fuels. "Public-interest science" began to be defined by groups of citizens and scientists who advocated the promotion of science in the human interest, and even more broadly in the interest of the earth's diverse biological species.

Perhaps more than any other program of investigation within the physical and mathematical sciences, the presumptions, questions, methods, patronage, and applications of the science of global environment demonstrate the scale and complexity of materials, objects, and resources characteristic of the pursuit of knowledge at the end of the twentieth century. Simultaneously, critiques of modernity and of modern science often are integrated into social and ethical movements oriented toward global environmental and humanistic concerns.[19]

The histories in this volume demonstrate a wide and deep array of aims and strategies for studying the history of the physical and mathematical sciences in the modern period. The practice of history, like the practice of science, is a process that depends on conceptual reorientations and reinterpretations, as well as the invention of new research tools and the unearthing of new facts. This volume should orient the reader to much of what is known about the history of the modern physical and mathematical sciences, as well as to what is yet to be done.

---

[19] See Toulmin, *Cosmopolis* (cited note 3), p. 186.

Part I

THE PUBLIC CULTURES OF
THE PHYSICAL SCIENCES AFTER 1800

I

THEORIES OF SCIENTIFIC METHOD
Models for the Physico-Mathematical Sciences

Nancy Cartwright, Stathis Psillos, and Hasok Chang

Scientific methods divide into two broad categories: inductive and deductive. Inductive methods arrive at theories by generalizing from what is known to happen in particular cases; deductive methods, by derivation from first principles. Behind this primitive categorization lie deep philosophical oppositions. The first principles central to deductivist accounts are generally taken to be, as Aristotle described, "first known to nature" but not "first known to us." Do the first principles have a more basic ontological status than the regularities achieved by inductive generalization – are they in some sense "more true" or "more real"? Or are they, in stark opposition, not truths at all, at least for a human science, because always beyond the reach of human knowledge?

Deductivists are inclined to take the first view. Some do so because they think that first principles are exact and eternal truths that represent hidden structures lying behind the veil of shifting appearances; others, because they see first principles as general claims that unify large numbers of disparate phenomena into one scheme, and they take unifying power to be a sign of fundamental truth.[1] Empiricists, who take experience as the measure of what science should maintain about the world, are suspicious of first principles, especially when they are very abstract and far removed from immediate experience. They generally insist on induction as the gatekeeper for what can be taken for true in science.

Deductivists reply that the kinds of claims we can arrive at by generalizing in this way rarely, if ever, have the kind of precision and exceptionlessness that we require of exact science; nor are the concepts that can be directly tested in experience clear and unambiguous. For that we need knowledge that is expressed explicitly in a formal theory using mathematical representations and theoretical concepts not taken from experience. Those who maintain the centrality of implicit knowledge, who argue that experiment and model

---

[1] For defense of the importance of unification, cf. P. Kitcher, "Explanatory Unification," *Philosophy of Science*, 48 (1981), 507–31.

building have a life of their own only loosely related to formal theory, or who aim for the pragmatic virtues of success in the mastery of nature in contrast to an exact and unambiguous representation of it, look more favorably on induction as the guide to scientific truth.

The banners of inductivism and deductivism also mark the divide between the great traditional doctrines about the source of scientific knowledge: empiricism and rationalism. From an inductivist point of view, the trouble with first principles is in the kind of representations they generally involve. The first principles of our contemporary physico-mathematical sciences are generally expressed in very abstract mathematical structures using newly introduced concepts that are characterized primarily by their mathematical features and by their relationships to other theoretical concepts. If these were representations taken from experience, inductivists would have little hesitation in accepting a set of first principles from which a variety of known phenomena can be deduced. For induction and deduction in this case are just inverse processes. When the representations are beyond the reach of experience, though, how shall we come to accept them? Empiricists will say that we should not. But rationalists maintain that our capacity for thought and reason provide independent reasons. Our clear and distinct ideas are, as René Descartes maintained, the sure guide to truth; or, as Albert Einstein and a number of late-twentieth-century mathematical physicists urge, the particular kind of simplicity, elegance, and symmetry that certain mathematical theories display gives them a purchase on truth.

These deeper questions, which drive a wedge between deductivism and inductivism, remain at the core of investigation about the nature of the physico-mathematical sciences. They will be grouped under five headings below: I. Mathematics, Science, and Nature; II. Realism, Unity, and Completeness; III. Positivism; IV. From Evidence to Theory; V. Experimental Traditions. It is usual in philosophy to find that the principal arguments that matter to current debates have a long tradition, and this is no less true in theorizing about science than about other topics. Thus, an account of contemporary thought about scientific method for the physico-mathematical sciences necessarily involves discussion of a number of far older doctrines.

MATHEMATICS, SCIENCE, AND NATURE

How do the claims of mathematics relate to the physico-mathematical sciences? There are three different kinds of answers:

Aristotelianism[2]

Quantities and other features studied by mathematics occur in the objects of perception. The truths of mathematics are true of these perceptible

---

[2] Cf. Aristotle, *Metaphysics* μ–3.

quantities and features, which are further constrained by the principles of physics. Thus, Aristotle can explain how demonstrations from one science apply to another: The theorems of the first science are literally about the things studied in the second science. The triangle of optics, for instance, is a perceptible object and as such has properties like color and motion. In geometry, however, we take away from consideration what is perceptible (by a process of *aphairesis* or abstraction) and consider the triangle merely "*qua* triangle." The triangle thus considered is still the perceptible object before us (and need not be in the mind), but it is an object of thought.

This doctrine allows Aristotelians to be inductivists. The properties described in the first principles of the mathematical sciences literally occur in the perceptible world. Yet it dramatically limits the scope of these principles. How many real triangles are there in the universe, and how does our mathematics apply where there may be none at all, for example, in the study of rainbows? The same problem arises for the principles of the sciences themselves. Theories in physics are often about objects that do not exist in perceptible reality, such as point masses and point charges. Yet these are the very theories that we use to study the orbits of the planets and electric circuits. The easy answer is that the perceptible objects are "near enough" to being true point masses or true triangles for it not to matter. But what counts as near enough, and how are corrections to be made and justified? These are the central issues in the current debate among methodologists over "idealization" and "de-idealization."[3]

Pythagoreanism

Many modern physicists and philosophers (Albert Einstein being a notable example) maintain, with the early Pythagoreans, that nature is "essentially" mathematical. Behind the phenomena are hidden structures and quantities. These are governed by the principles of mathematics, plus, in current-day versions, further empirical principles of the kind we develop in the physico-mathematical sciences. Some think that these hidden structures are "more real" than what appears to human perception. This is not only because they are supposed to be responsible for what we see around us but, more importantly, because the principles bespeak a kind of necessity and order that many feel reality must possess. Certain kinds of highly abstract principles in modern physics are thought to share with those of mathematics this special necessity of thought.

Pythagoreanism is a natural companion to rationalism. In the first place, if a principle has certain kinds of special mathematical features – for example, if the principle is covariant or it exhibits certain abstract symmetries – that is supposed to give us reason to believe in it beyond any empirical evidence. In the second, many principles do not concern quantities that are measurable

---

[3] Cf. the series *Idealization* I–VIII in *Poznan Studies in the Philosophy of the Sciences and the Humanities*, ed. J. Brzezinski and L. Nowak, etc. (Amsterdam: Rodopi, 1990–7).

in any reasonable sense. For instance, much of modern physics studies quantities whose values are not defined at real space-time points but instead in hyperspaces. Pythagoreans are inclined to take these spaces as real. It is also typical of Pythagoreans to discuss properties that are defined relative to mathematical objects as if they were true of reality, even when it is difficult to identify a measurable correlate of that feature in the thing represented by the mathematical object. (For example, what feature must an observable have when the operator that represents it in quantum mechanics is invertible?) Current work in the formal theory of measurement develops precise characterizations of relationships between mathematical representations on the one hand and measurable quantities and their physical features on the other, thus providing a rigorous framework within which these intuitive issues can be formulated and debated.[4]

Instrumentalism and Conventionalism

The French philosopher, historian, and physicist Pierre Duhem (1861–1916) was opposed to Pythagoreanism. Nature, Duhem thought, is purely qualitative. What we confront in the laboratory, just as much as in everyday life, is a more or less warm gas, Duhem taught.[5] Quantity terms, such as "temperature" (which are generally applied through the use of instruments), serve as merely *symbolic* representations for collections of qualitative facts about the gas and its interactions. This approach makes Duhem an instrumentalist both about the role of mathematics in describing the world and about the role of the theoretical principles of the physico-mathematical sciences: These serve not as literal descriptions but, rather, as efficient instruments for systematization and prediction. The methods for coming to an acceptance or use of the theoretical principles of physics, then, will clearly not be inductive. Duhem advocated instead the widely endorsed hypothetico-deductive method. He noted, however, that the method is, by itself, of no help in confirming hypotheses, a fact which lends fuel to instrumentalist doctrines (see the section "From Evidence to Theory"). Duhem's arguments still stand at the center of debate about the role of mathematics in science.

Alternative to the pure instrumentalism of Duhem is the conventionalism of his contemporary, Henri Poincaré (1854–1912), whose work on the foundations of geometry raised the question "Is physical space Euclidean?" Poincaré took this question to be meaningless: One can make physical space possess *any* geometry one likes, provided that one makes suitable adjustments to one's physical theories. To show this, Poincaré described a possible world in which the underlying geometry is indeed Euclidean, but due to the existence of a strange physics, its inhabitants conclude that the geometry of their world is non-Euclidean. There are then two empirically equivalent theories

---

[4] See, for instance, D. H. Krantz, R. D. Luce, P. Suppes, and A. Tversky, *Foundations of Measurement* (New York: Academic Press, 1971).
[5] P. Duhem, *Aim and Structure of Physical Theory* (New York: Atheneum, 1962).

to describe this world: Euclidean geometry plus strange physics versus non-Euclidean geometry plus usual physics. Whatever geometry the inhabitants of the world choose, it is not dictated by their empirical findings. Consequently, Poincaré called the axioms of Euclidean geometry "conventions."

Poincaré's conventionalism included the principles of mechanics as well.[6] They cannot be demonstrated independently of experience, and they are not, he argued, generalizations of experimental facts. For the idealized systems to which they apply are not to be found in nature. Nor can they be submitted to rigorous testing, since they can always be saved from refutation by some sort of corrective move, as in the case of Euclidean geometry.

So, Poincaréan conventions are held true, but their truth can be established neither a priori nor a posteriori. Are they then held true merely by definition? Poincaré repeatedly stressed that it is experience that "suggests," or "serves as the basis for," or "gives birth to" the principles of mechanics, although experience can never establish them conclusively. Nevertheless, like Duhem and unlike either the Aristotelians or the Pythagoreans, for Poincaré and other conventionalists the principles of geometry and the principles of physics serve as symbolic representations of nature, rather than literally true descriptions (see the next section).

REALISM, UNITY, AND COMPLETENESS

These are among the most keenly debated topics of our day. One impetus for the current debates comes from the recent efforts in the history of science and in the sociology of scientific knowledge to situate the sciences in their material and political setting. This work reminds us that science is a social enterprise and thus will draw on the same kinds of resources and be subject to the same kinds of influences as other human endeavors. Issues about the social nature of knowledge production, though, do not in general make special challenges for the physico-mathematical sciences beyond those that face any knowledge-seeking enterprise and, hence, will not be focused on here.

For many, knowledge claims in the physico-mathematical sciences do face special challenges on other grounds: (1) The entities described are generally unobservable. (2) The relevant features are possibly unmeasurable. (3) The mathematical descriptions are abstract; they often lack visual and tangible correlates, and thus, many argue with Lord Kelvin and James Clerk Maxwell, we cannot have confidence in our understanding of them.[7] (4) The theories

---

[6] Cf. H. Poincaré, *La Science et L'Hypothèse* (Paris: Flammarion, 1902).
[7] See C. Smith and M. N. Wise, *Energy and Empire: A Biographical Study of Lord Kelvin* (Cambridge: Cambridge University Press, 1989), and J. C. Maxwell, "Address to the Mathematical and Physical Section of the British Association," in *The Scientific Papers of James Clerk Maxwell*, ed. W. D. Niven, 2: 215–29; *Treatise on Electricity and Magnetism*, vol. 2, chap. 5.

often seem appropriate as descriptions only of a world of mathematical objects and not of the concrete things around us. These challenges lie at the core of the "realism debate."

On a *realist* account, a theory purports to tell a literally true story as to how the world is. As such, it describes a world populated by a host of unobservable entities and quantities. Instrumentalist accounts do not take the story literally. They aim to show that all observable phenomena can be embedded in the theory, which is then usually understood as an uninterpreted abstract logico-mathematical framework. Currently another view has been gaining ground.[8] One may, with the realist, take the story told by the theory literally: The theory describes how the world might be. Yet, one can at the same time suspend one's judgment as to the truth of the story. The main argument for this position is that belief in the truth of the theoretical story is not required for the successful use of the theory. One can simply believe that the theory is empirically adequate, that is, that it saves all observable phenomena. It should be noted that "empirically adequate" here is to be taken in a strong sense; if we are to act on the theory, it seems we must expect it to be correct not only in its descriptions of what has happened but also about what will happen under various policies we may institute.

Realists argue that the best explanation of the predictive successes of a theory is that the theory is true. According to the *inference to the best explanation,* when confronted with a set of phenomena, one should weigh potential explanatory hypotheses and accept the best among them as correct, where "bestness" is gauged by some favored set of virtues. The virtues usually cited range from very general ones, such as simplicity, generality, and fruitfulness, to very subject-specific ones, such as gauge invariance (thought to be important for contemporary field theory), or the satisfaction of Mach's principle (for theories of space and time), or the exhibition of certain symmetries (now taken to be a sine qua non in fundamental particle theories).

Opponents of realism urge that the history of physics is replete with theories that were once accepted but turned out to be false and have been abandoned.[9] Think, for instance, of the nineteenth-century ether theories, both in electromagnetism and in optics, of the caloric theory of heat, of the circular inertia theories, and of the crystalline spheres astronomy. If the history of science is the wasteland of aborted best explanations, then current best theories themselves may well take the route to this wasteland in due course.

Realists offer two lines of defense, which work in tandem. On the one hand, the list of past theories that were abandoned might not after all be very big, or very representative. If, for instance, we take a more stringent account of empirical success – for example, we insist that theories yield novel predictions – then it is no longer clear that so many past abandoned

---

[8] See especially B. C. van Fraassen, *The Scientific Image* (Oxford: Clarendon Press, 1980).
[9] Cf. L. Laudan, "A Confutation of Convergent Realism," *Philosophy of Science,* 48 (1981), 19–49.

theories were genuinely successful. In this case, the history of science would not after all give so much reason to expect that those of our contemporary theories that meet these stringent standards will in their turn be abandoned.

On the other hand, realists can point to what in theories is not abandoned. For instance, despite the radical changes in interpretation, successor theories often retain much of the mathematical structure of their predecessors. This gives rise to a realist position much in sympathy with the Pythagoreanism discussed in the first section. According to "structural realism," theories can successfully represent the *mathematical structure* of the world, although they tend to be wrong in their claims about the entities and properties that populate it.[10] The challenge currently facing structural realism is to defend the distinction between *how an entity is structured* and *what this entity is*. In general, realists nowadays are at work to find ways to identify those theoretical constituents of abandoned scientific theories that contributed essentially to their successes, separate these from others that were "idle," and demonstrate that the components that made essential contributions to the theory's success were those that were retained in subsequent theories of the same domain. The aim is to find exactly what it is most reasonable to be a scientific realist about.

Closely connected with, but distinct from, realism are questions about the *unity* – or unifiability – of the sciences and about the *completeness* of physics. It is often thought that if the theories of physics are true, they must fix the behavior of all other features of the material universe. Thus, unity of the sciences is secured via the reducibility of all the rest to physics. Opposition views maintain that basic theories in physics may be true, or approximately so, yet not *complete:* They tell accurate stories about the quantities and structures in their domains, but they do not determine the behavior of features studied in other disciplines, including other branches of physics.[11] Whether reductions of one kind or another are possible "in principle," there has over the last decade been a strong movement that stresses the need for *pluralism* and interdisciplinary cooperation in practice.[12]

---

[10] Cf. J. Worrall, "Structural Realism: The Best of Both Worlds?" *Dialectica,* 43 (1989), 99–124; P. Kitcher, *The Advancement of Science* (Oxford: Oxford University Press, 1993); S. Psillos, "Scientific Realism and the 'Pessimistic Induction,'" *Philosophy of Science,* 63 (1996), 306–14.

[11] For classic loci of these opposing views, see P. Oppenheim and H. Putnam, "Unity of Science as a Working Hypothesis," in *Concepts, Theories and the Mind-Body Problem,* ed. H. Feigl, M. Scriven, and G. Maxwell (Minneapolis: University of Minnesota Press, 1958), pp. 3–36; and J. Fodor, "Special Sciences, or the Disunity of Science as a Working Hypothesis," *Synthese,* 28 (1974), 77–115. For contemporary opposition to doctrines of unity, see J. Dupré, *The Disorder of Things: Metaphysical Foundations of the Disunity of Science* (Cambridge, Mass.: Harvard University Press, 1993); for arguments against completeness, see N. Cartwright, *The Dappled World* (Cambridge: Cambridge University Press, 1999).

[12] Cf. S. D. Mitchell, L. Daston, G. Gigerenzer, N. Sesardic, and P. Sloep, "The Why's and How's of Interdisciplinarity," in *Human by Nature: Between Biology and the Social Sciences,* ed. P. Weingart et al. (Mahwah, N.J.: Erlbaum Press, 1997), pp. 103–50, and S. D. Mitchell, "Integrative Pluralism," *Biology and Philosophy,* forthcoming.

POSITIVISM

All varieties of positivism insist that *positive knowledge* should be the determinant of what practices and claims are accepted in science. Differences arise over two issues: (a) What is positive knowledge? and (b) What are the principles of determination? We shall focus on the Vienna Circle here since most of the positivist legacy in current Anglo-American thinking about science has been inherited through it.[13] The Vienna Circle offered special forms of the two dominant kinds of answers to both questions.

Members of the Circle met in Vienna from 1925 until the group was broken up by Nazi oppression in 1935. Their ideas were influenced by the new physics, particularly Einstein's theory of relativity. A number of Circle members, especially Otto Neurath (1882–1945) and Edgar Zilsel (1891–1944) and to a lesser extent Rudolf Carnap (1891–1970), were politically active and held strong socialist views. In general, they saw their belief in socialism and their advocacy of a scientific style of philosophy as closely allied. (Neurath, for instance, embraced a scientifically interpreted version of Marxist materialism.)

What is positive knowledge? It is knowledge of what can be really known, where "what can be really known" is what happens in the real world. But how shall we characterize the kinds of things that happen in the real world? This problem arises as much for the physicalism and philosophical naturalism of the 1990s as it did for earlier positivists. Physicalism maintains that all true descriptions of the world are fixed by the physical descriptions true of it – the main target of concern being mental states and emotions and the features and norms of social groups. Its companion, philosophical naturalism, urges that philosophy has no special subject matter other than what is already studied in science. But what constitutes a physical description, or the proper subject matter of science?

The positivism of the Vienna Circle took a double stand: a materialist "metaphysics" and a "verificationist" epistemology. Their materialism dictated either that all there is is what physics studies ("physics-ism"), or that what there is is what occurs in space and time ("physicalism"). Their verificationism dictated that what is really true is what can be verified in experience. By taking these stands, they aimed to rule out from the realm of positive knowledge both religion and Hegelian idealism. Religion was attacked for its mystical characters and moral injunctions; Hegelian idealism, for its philosophical obscurities, its realm of pure ideas, and its teleological account of the history of humanity; and both, for their contempt for the physico-mathematical sciences.

Both of these stands were motivated by the positivists' aim to answer the question of what can be really known. The central epistemic problem

---

[13] For a general discussion of the logical positivists, see T. Uebel, ed., *Rediscovering the Forgotten Vienna Circle: Austrian Studies on Otto Neurath and the Vienna Circle* (Dordrecht: Kluwer, 1991).

is whether knowledge is conceived of as private or as public. Traditional empiricism assumes that all one can really be sure of are facts about one's own experience. Thus, Ernst Mach's (1838–1916) defense of a positivist reading of physics is titled *The Analysis of Sensations*. Following John Locke, George Berkeley, and David Hume, it also assumes that the only concepts that can be meaningfully spoken of should be built out of sensory experience. Notoriously, Hume (1711–1776) used this restriction to undermine the concept of causality, the concept of one thing's making another happen in contrast to that of mere regular association. Many modern positivists continue this attack. They insist that physics has no place for causality. This is not just because causality is not part of our observable experience but also because of the "theory-dominated" assumption that physics knowledge equals physics equations (an assumption that excludes knowledge of how things work) and that physics equations record mere association. Concerns about causality in physics have become prominent recently, both because of the possibility of nonlocal causal influences in quantum mechanics raised by J. S. Bell's work on the Einstein-Podolsky-Rosen experiment and because of a renewed interest in how physics is put to work to intervene in the world.[14]

On the side of the private view of knowledge is the claim that our individual experiences are the only plausible candidates for nonanalytic knowledge of which we can be certain; and if we do not found our scientific claims in something of which we can be reasonably certain, we have no genuine claim to knowledge at all. The entire edifice of modern knowledge, even in physics and other exact sciences, may be a chimera. Opposed to this is the view that knowledge is necessarily a public, cooperative enterprise to which a great number of persons must contribute and of which a single person can possess only a minuscule part. This claim, which is clearly closer to science as we see it practiced, is one of the central tenets of studies in the sociology of knowledge of the 1980s and 1990s. The public view of knowledge can also count on its side the private-language argument, in establishing that the idea of private knowledge does not make sense.[15]

FROM EVIDENCE TO THEORY

What are the principles that allow us to deduce higher-level knowledge from lower? Rudolf Carnap first proposed an *Aufbau* – a way to construct new knowledge from some given positive base methodically, whether the base is

---

[14] J. S. Bell, *Speakable and Unspeakable in Quantum Mechanics* (Cambridge: Cambridge University Press, 1987); M. S. Morgan and M. C. Morrison, eds., *Models as Mediators* (Cambridge: Cambridge University Press, 1999).

[15] L. Wittgenstein, *Philosophical Investigations*, 3d ed. (New York: Macmillan, 1958); see also S. A. Kripke, *Wittgenstein on Rules and Private Language* (Oxford: Blackwell, 1982).

private or public.[16] But many believe that scientific knowledge clearly goes far beyond a mere reassemblage of what is given in the positive base. Carnap himself later offered a theory of confirmation to show how and to what degree evidence can make further scientific hypotheses probable, and the hunt for a viable theory of confirmation is still on.[17] The problem is to find something that can fix the probability. Carnap took the probabilistic relation between evidence and hypotheses to be a logical one; hence, "inductive logic." One of the troubles with inductive logics, from Carnap till now, is that they require that the evidence and hypotheses be expressed in a formal language. Some view the requirement of formality as an advantage, since knowledge claims must be both exact and explicit to count as genuinely scientific. Others claim, however, that it places undue constraints on the expressive power of science; in addition, the probability assignments that emerge tend to be highly sensitive to the choice of language.

One major approach to confirmation is the hypothetico-deductive method. Scientific claims are put forward as hypotheses from which are deduced empirical consequences that can be compared with experimental results. Clearly this requires that both the hypotheses and the evidence be described formally enough for deduction to be possible. The most telling objection to the hypothetico-deductive method is the so-called Duhem-Quine problem: Scientific theories never imply testable empirical consequences on their own but only when conjoined with a (usually elaborate) network of auxiliary assumptions. If the empirical consequences are not borne out, one of the premises must be rejected, but nothing in the logic of the matter decides whether it is the theory or an auxiliary that should go.

But even if the empirical consequences of a theory T are borne out, does this provide support for T? To infer T from E and "T implies E" is to commit the fallacy of affirming the consequent. This problem is known as the "problem of underdetermination of theory by evidence": that T determines E does not imply that only T does so; any number of hypotheses contradictory to T may do so as well. This bears on the realist claim that it is rational to infer to the best explanation. If all we require to say that T explains E is that T imply E, then the ability of a theory to explain the evidence does not logically provide any reason to believe in that theory over any of the indefinite number of other theories (most unknown and unarticulated) that do so as well. The problem of underdetermination was the reason that Karl Popper (1902–1994) insisted that theories can never be confirmed, but can only be shown to be false.[18] But the Duhem-Quine problem remains, for it obviously affects attempts to falsify single hypotheses as much as attempts to confirm them.

[16] R. Carnap, *Der Logische Aufbau der Welt* (Berlin: Weltkreis, 1928), translated as *The Logical Structure of the World* (Berkeley: University of California Press, 1967).
[17] R. Carnap, *Logical Foundations of Probability* (Chicago: University of Chicago Press, 1950).
[18] Karl R. Popper, *The Logic of Scientific Discovery* (London: Hutchinson, 1959).

The basic assumption of the hypothetico-deductive method – that theories should be judged by their testable consequences – no longer seems sacrosanct in contemporary physics. Many of the new developments in high theory are justified more by the mathematical niceties they exhibit than by the positive consequences they imply. String theory is the central example of the 1990s, with some physicists and philosophers suggesting that mathematics is the new laboratory site for physics.[19] This is, however, still a slogan and not a developed methodological or epistemological position. Other equally notable philosophers and physicists oppose this dramatic departure from even the weakest requirements of empiricism. Does the existence of a flourishing physics community pursuing this mathematics-based style of theory development provide on-the-ground evidence against the epistemological and ontological arguments that support empiricism? Or do the positivist arguments show that these new theories will have to make a real contribution to positive knowledge before they can be adopted? Debate at this time is at a standoff.

There are two further main contemporary theories of confirmation. The first is bootstrapping; the second, Bayesian conditionalization. Bootstrapping is the one that on the face of it looks closest to what happens in contemporary physics.[20] In a bootstrap, the role of antecedently accepted old knowledge looms large in confirmation. The inference to a new hypothesis is reconstructed as a deduction from the evidence plus the background information. Thus, the question "Why do the data cited count as evidence for the hypothesis?" has a trivial answer – because, given what we know, the data logically imply the hypothesis. The method is dependent on our willingness to take the requisite background information as known, and on our justification for doing so. How well justified are the kinds of premises generally used in bootstrap confirmations? A cautious inductivist who wishes to stay as close to the facts as possible may be wary, since the premises almost always include assumptions far stronger and far more general than the hypothesis to be confirmed. For example, in order to infer the charge of "the" electron in an experiment designed to provide new levels of precision, we will assume that all electrons have the same charge.

On the Bayesian account of confirmation, the probabilistic relation between evidence for a theoretical hypothesis and the hypothesis itself is not seen as a logical relation, as with Carnap, but rather as a subjective estimate. Nevertheless, the axioms of probability place severe constraints on the estimates. The probability of a hypothesis H, in the light of some evidence e, is given by Bayes's theorem:

$$\text{prob}(H/e) = \frac{\text{prob}(e/H)\,\text{prob}(H)}{\text{prob}(e)}$$

---

[19] Cf. P. Galison's discussion "Mirror Symmetry: Persons, Objects, Values," in *Growing Explanations: Historical Reflections on the Sciences of Complexity*, ed. N. Wise, in preparation.
[20] C. Glymour, *Theory and Evidence* (Princeton, N.J.: Princeton University Press, 1980).

Bayesians take the degree of belief in a hypothesis H to be the subjective estimate of its probability ($prob(H)$). But they insist that it should be revised in accord with Bayes's formula as evidence accumulates. In recent years, the Bayesian approach has been extended to cover a large number of issues, including the Duhem-Quine problem, the problem of underdetermination, and questions of why and when experiments should be repeated.[21]

Although Bayesianism is gaining currency, not only among philosophers but also among statisticians, both specific Bayesian recommendations and the general approach are highly controversial.[22] The most general criticism is that too much is left to subjectivity: New probability assessments of hypotheses depend on original subjective assessments, both on the prior degree of belief in a hypothesis ($prob\ (H)$) and on the likelihood of the evidence given the hypothesis ($prob(e/H)$). Realists in particular would prefer to find some way to maintain that the degree to which a piece of evidence confirms a hypothesis is an objective matter.

EXPERIMENTAL TRADITIONS

Nowadays it is common to complain about the "theory-dominated" approach in the history and philosophy of science. This domination by theory springs from the long-standing assumption, advocated at various periods in the history of the physico-mathematical sciences and widespread since World War II, that the ultimate aim of science is to produce satisfactory theories. One corollary of this assumption is that the primary purpose of observation and experimentation is to validate or test theories. Then the central issue becomes how well observations can ground theories. The doctrine that all observation is "theory-laden," developed during the 1960s and 1970s, gave observation an even weaker role by suggesting that observations could not be made at all unless they were framed by theories and not accepted unless they were validated by theories.[23]

Against this perspective, more recent work maintains that "experimentation has a life of its own," to borrow a now-famous slogan.[24] (In this chapter, we focus on experimentation, rather than observation in general, since a number of interesting issues come out more clearly when we consider explicitly experimental situations, involving conscious planning and contrivance

---

[21] C. Howson and P. Urbach, *Scientific Reasoning: The Bayesian Approach* (La Salle, Ill.: Open Court, 1989).
[22] Cf. C. Glymour, "Why I Am Not a Bayesian," in *Theory and Evidence* (Princeton, N.J.: Princeton University Press, 1980), pp. 63–93, and D. Mayo, *Error and the Growth of Experimental Knowledge* (Chicago: University of Chicago Press, 1996).
[23] Cf. N. R. Hanson, *Patterns of Discovery* (Cambridge: Cambridge University Press, 1958); T. S. Kuhn, *The Structure of Scientific Revolutions* (Chicago: University of Chicago Press, 1962; 2d ed. 1970); P. K. Feyerabend, *Against Method* (London: New Left Books, 1975).
[24] I. Hacking, *Representing and Intervening* (Cambridge: Cambridge University Press, 1983), p. 150.

on the part of the observers.) First of all, many argue that the purpose of experimentation is not confined to theory testing. Experiment may be an end in itself or, more likely, serve some other purposes than those of theoretical science, ranging from public entertainment to technological control; the contexts giving rise to these aims could be as grand as imperial world domination or as immediate as brewing.[25]

Whatever one thinks about the aim of experimentation, the question about validity must be addressed. How do we ensure that our observations are valid? Or, at least, how do we judge how valid our observations are? The relevant notion of validity will certainly depend on the aims of those who are making and using the observations, but the least common denominator is probably some weak sense of truth or correctness. This kind of notion of validity is contrary to radical relativism, but it does not involve any commitment to realism concerning theories.

Conscientious practitioners have long been clear about the extraordinary difficulty of achieving high-quality observations. In the context of a quantitative science, observation means measurement. Whenever an instrument is used, the question arises about the correctness of its design and functioning – something painfully clear to those who have tried to improve measurement techniques.

Strategies for achieving validity in observations can be classified into two broad groups: theory dominated and theory independent. Theory-dominated strategies attempt to give theoretical justifications of measurement methods. For instance, in a physiology laboratory, we trust that a nerve impulse is being recorded correctly because we trust the principles of physics underlying the design of the electrical equipment. This, however, only pushes the problem out of sight, as Duhem recognized clearly.[26] Any conscientious investigator must ask how the theoretical principles justifying the measurement method are themselves justified. By other measurements? And what shows that *those* measurements are valid?

These worries have fueled attempts to formulate theory-independent strategies for achieving validity in observations. Many positivistic philosophers made a retreat to sense-data, but even sense-data came to be seen as less than assuredly certain. Currently it does not seem plausible that theory ladenness in its most fundamental sense can be escaped, because any concepts used in the description of observations carry theoretical implications and expectations (and are therefore open to revision). More recently, many methodologists have sought to base validity on independent confirmation: It would be a highly unlikely coincidence for different methods to give the same results, unless the results were accurate reflections of reality. Although

---

[25] For discussions of the various purposes and uses of experimentation, see D. Gooding, T. Pinch, and S. Schaffer, eds., *The Uses of Experiment* (Cambridge: Cambridge University Press, 1989), and M. N. Wise, ed., *The Values of Precision* (Princeton, N.J.: Princeton University Press, 1995).
[26] P. Duhem, *Aim and Structure of Physical Theory* (New York: Atheneum, 1962), part II, chap. 6.

intuitively persuasive and reflected widely in experimental practice, this line of argument fails to go beyond the pragmatic, as exhibited nicely in the inconclusive results of recent debates regarding the reality of invisible structures observed to be the same through different microscopes.[27]

In the remainder of this section we examine two of the more plausible attempts to eliminate theory dependence in measurements from the history of physics, one by Victor Regnault (1810–1878) and another by Percy Bridgman (1882–1961). Although virtually forgotten today, perhaps because he did not make significant theoretical contributions, Regnault was easily considered the best experimental physicist in all of Europe during his professional prime in the 1840s. His fame and authority were built on the extreme precision that he was able to achieve in many fields of physics, particularly in the study of thermal phenomena. In his vast output, we find very little explicit philosophizing, but some important aspects of his method can be gleaned from his practice.

For Regnault, the search for truth came down to "replacing the axioms of the theoreticians with precise data."[28] For instance, others before him had made thermometers on the basis of the assumption that one knew the pattern of thermal expansion (usually assumed to be uniform) of some material or other. This was justified by an appeal to various theories, such as basic calorimetry (Brook Taylor, Joseph Black, Jean-André De Luc, Adair Crawford) or various versions of the caloric theory (John Dalton, Pierre-Simon Laplace). Regnault rejected this practice, arguing that it was impossible to verify theories about the thermal behavior of matter unless one already had a trusted thermometer.

How, then, did Regnault manage to design thermometers without assuming any prior knowledge of the thermal behavior of matter? He employed the criterion of "comparability," which required that all instruments of the same type give the same value in a given situation, if that type of instrument is to be trusted as correct. Regnault recognized comparability as a necessary, but not sufficient, condition for correctness. This recognition made Regnault ultimately pessimistic about guaranteeing the correctness of measurement methods, in contrast to the recent advocates of independent confirmation. However, a more pragmatic and positive reading of Regnault is possible. Although comparability did not guarantee correctness, it did give stability to experimental results. Regnault had little faith in the stability of anything founded on theory, having done much work himself to show that the simple and universal laws believed to govern the behavior of gases were mere approximations.[29]

[27] I. Hacking, "Do We See Through a Microscope?" in *Images of Science*, ed. P. M. Churchland and C. A. Hooker (Chicago: University of Chicago Press, 1985), pp. 132–52, and B. C. van Fraassen's reply to Hacking in the same volume, pp. 297–300.

[28] J. B. Dumas, *Discours et éloges académiques* (Paris: Gauthier-Villars, 1885), 2: 194.

[29] V. Regnault, "Relations des expériences...pour déterminer les principales lois et les données numériques qui entrent dans le calcul des machines à vapeur," *Mémoires de l'Académie Royale des*

Regnault's inclination to eliminate theory from the foundations of measurement was shared by Percy Bridgman, American scientist-turned-philosopher and pioneer in experimental high-pressure physics. In one crucial way, Bridgman was more radical than Regnault. What came to be known as Bridgman's "operationalism" eliminated the thorny question of validity altogether, by defining concepts through measurement operations: "In general, we mean by any concept nothing more than a set of operations; *the concept is synonymous with the corresponding set of operations.*"[30] Then, at least in principle, any assertion that a measurement method is correct becomes tautologically true.

Bridgman's thought was stimulated by two major influences. One was his methodological interpretation of Albert Einstein's special theory of relativity, which to him taught the lesson that we will get into errors and meaningless talk unless we specify our concepts by reference to concrete measurement operations. When Einstein gave a precise definition of distant simultaneity by specifying precise operations for its determination, it became clear that observers in relative motion with respect to each other would disagree about which events were simultaneous with which. Bridgman argued that physicists would not have gotten into such errors if they had adopted the operational attitude from the start.

The other formative influence on Bridgman's philosophy was his own Nobel Prize–winning work in high-pressure physics, which emphasized to him how much at sea the scientist was in realms of new phenomena. His experience of creating and experimenting with pressures up to an estimated 400,000 atmospheres, where all previously known methods of measurement and many previously known regularities ceased to be applicable, supported his general assertion that "concepts...are undefined and meaningless in regions as yet untouched by experiment."[31]

Appraisals of Bridgman's thought on measurement have differed widely, but it would be fair to say that there has been a general acceptance of his insistence on specifying the concrete operations involved in measurement as much as possible. On the other hand, attempts to eliminate nonoperational concepts altogether from science (such as extreme behaviorism in psychology) are generally considered to have failed, as it is easily agreed that theoretical concepts are both useful and meaningful.[32] But the rejection of operationalism as a theory of meaning also implies the rejection of Bridgman's radical solution to the problem of the validity of measurement methods, which remains a subject of open debate.

Sciences de l'Institut de France, 21 (1847), 1–748; see p. 165 for a statement of the comparability requirement.
[30] P. W. Bridgman, *The Logic of Modern Physics* (New York: Macmillan, 1927), p. 5; emphasis original.
[31] Ibid., p. 7.
[32] C. G. Hempel, *Philosophy of Natural Science* (Englewood Cliffs, N.J.: Prentice Hall, 1966), chap. 7.

2

INTERSECTIONS OF PHYSICAL SCIENCE AND WESTERN RELIGION IN THE NINETEENTH AND TWENTIETH CENTURIES

Frederick Gregory

When we consider issues in science and religion in the nineteenth century and even in subsequent years, we naturally think first of the evolutionary controversies that have commanded public attention. However, there are important ways in which developments in physical science continued to intersect with the interests of people of all religious beliefs. Indeed, the closer one approached the end of the twentieth century, the more the interaction between science and religion was dominated by topics involving the physical sciences, and the more they became as important to non-Christian religions as to various forms of Christianity. For the nineteenth century, most issues were new versions of debates that had been introduced long before. Because these reconsiderations were frequently prompted by new developments in physical science, forcing people of religious faith into a reactive mode, the impression grew that religion was increasingly being placed on the defensive. For a variety of reasons, this form of the relationship between the two fields changed greatly over the course of the twentieth century until, at the dawn of the third millennium of the common era, the intersection between science and religion is currently being informed both by new theological perspectives and by new developments in physical science.

Religion intersects with the physical sciences primarily in questions having to do with the origin, development, destiny, and meaning of matter and the material world. At the beginning of the period under review, the origin of matter itself was not regarded as a scientific question. The development of the cosmos, however, or how it had acquired its present contours and inhabitants, was a subject that had been informed by new telescopic observations and even more by the impressive achievements of Newtonian physical scientists of the eighteenth century. The Enlightenment had also produced fresh philosophical examinations of old religious questions and even of religious reasoning itself. As a result, the dawn of the nineteenth century brought new answers to questions about humankind's uniqueness in the universe and about the ultimate fate of physical nature, topics that are discussed in the first two sections of this chapter. Not surprisingly, old questions about the sufficiency of explanations regarding matter and its properties resurfaced, appearing to force a choice between science and religion. Aspects of this confrontation are treated here in a separate section on the implications of materialism. It would take another century before developments within science and within religion would produce reengagement in the present. An abbreviated chronicle of these intersections forms the final two segments of this contribution.

●

THE PLURALITY OF WORLDS

By the dawn of the nineteenth century, the notion that planets other than Earth were inhabited by intelligent beings had become a dogma taught in scientific books and preached from pulpits. Long before, a related theological question had been raised and dealt with: How did the possibility of the existence of beings other than humans affect an understanding of the doctrines of divine incarnation and redemption? The answer that emerged was that although extraterrestrial creatures could not have sinned as Adam did since they did not come from Adam, Christ's death was effective for their redemption without his having to go to another world to die again.[1] By the second half of the eighteenth century, theologians were in the main agreed that the existence of life elsewhere added to nature's testimony to the greatness of God, while prominent secular thinkers, such as the philosopher Immanuel Kant (1724–1804), the astronomer William Herschel (1738–1822), and the physicist/astronomer Pierre-Simon Laplace (1749–1827), had their own reasons for joining the many others who asserted their belief in the existence of life on other worlds.[2]

The happy accommodation of science and religion that had been achieved with respect to the plurality of worlds came crashing down with the publication of Thomas Paine's (1737–1809) *Age of Reason* in 1796. A few years before this date, Paine had worked on the book while in France during the radical phase of the French Revolution. The chief source of his radical critique of Christianity was, in fact, his inability to accept that one could simultaneously hold to pluralism, the belief that there are many inhabited worlds, and Christianity.[3] In fact, Paine accepted that there were other inhabited worlds. What he could not abide was the "conceit" that the redemptive scheme on earth was somehow paradigmatic for all of creation. To Paine, acceptance of

---

[1] This solution to the question was first enunciated by the French theologian William Vorilong, who died in 1463. Cf. Michael J. Crowe, *The Extraterrestrial Life Debate, 1750–1900: The Idea of a Plurality of Worlds from Kant to Lowell* (Cambridge: Cambridge University Press, 1986), pp. 8–9.
[2] Ibid., p. 161.
[3] Marjorie Nicolson, "Thomas Paine, Edward Nares, and Mrs. Piozzi's Marginalia," *Huntington Library Bulletin*, 10 (1936), 107.

life elsewhere in the universe rendered Christianity's claim to be the exclusive means of redemption absurd.

In his magisterial study of the history of the extraterrestrial debates, Michael Crowe chronicles their intensification in the wake of Paine's salvo against Christianity. By far the majority of responses rejected Paine's conclusion in favor of renewed arguments that extraterrestrials served as evidence of God's greatness, a circumstance that confirms historian John Brooke's observation about the resilience of natural theology in the face of challenging new developments in science and thought in the nineteenth century.[4] A turning point occurred at midcentury with the anonymous publication of William Whewell's (1794–1866) *Of the Plurality of Worlds*. In this 1853 book Whewell, a mineralogist, philosopher, and Anglican cleric at Cambridge and the university's most prominent figure, reversed his earlier acceptance of pluralism because he came to believe that it could not, in fact, be reconciled with Christianity. Whewell's identity as author of the book did not remain a secret for long. His reviewer in the *London Daily News* expressed astonishment that anyone, let alone the Master of Trinity College, would attempt to restore "the exploded myth of man's supremacy over all other creatures in the universe."[5]

While others saw the rejection of life elsewhere in the universe as myopic egoism, Whewell took seriously the dichotomy Paine had presented more than fifty years earlier. His conclusion was to opt for the alternative Paine had thought absurd; namely, Whewell simply rejected "the assertions of astronomers when they tell us that [the earth] is only one among millions of similar habitations."[6] In order to counter the pluralism that had become solidly ensconced within the English tradition of physico-theology, Whewell chose to cast his argumentation primarily in a scientific and philosophical mode. But its motivation derived from religion. Out of eternal wisdom and grace, God had suffered and died so that human beings could be saved; there could be no more than one great drama of God's mercy; there could be but one savior. To imagine something analogous existing on other worlds was repugnant to Whewell.

By accepting Paine's dichotomy of choices, Whewell was opposing the tack taken by natural theologians in their treatment of the celebrated deist. The Scottish clergyman Thomas Chalmers and others had responded to Paine by denying that they had to choose between pluralism and Christianity because the two could be shown to be compatible. By forcing the issue as he did, Whewell, in fact, did not persuade the majority to go with him. When the dust had settled on the heated series of debates that Whewell's book generated in the 1850s, pluralism remained the consensus view among scientists and theologians.[7]

---

[4] John Brooke, *Science and Religion: Some Historical Perspectives* (Cambridge: Cambridge University Press, 1991), chap. 6.
[5] Crowe, *Extraterrestrial Life Debate*, pp. 267, 282.
[6] Quoted by Crowe, *Extraterrestrial Life Debate*, p. 285, from *Of the Plurality of Worlds*.
[7] Ibid., pp. 351–2.

As the century wound down, a growing number of individual celestial bodies were eliminated as fit sites of possible life, and a limited pluralism replaced the more enthusiastic versions of earlier decades. In 1877 the Italian astronomer Giovanni Schiaparelli, in the course of testing a new telescope's capacity to observe a planetary surface, characterized dark lines he was able to detect on the surface of Mars as channels (*canali*). Thus opened a debate over Martian "canals," which lasted into the second decade of the twentieth century and captured the attention of the international public. Before it was over, one observer, who was reported in the 2 June 1895 *San Francisco Chronicle* as an agnostic and therefore unbiased by religion, claimed to have detected in a map of a canal-studded Mars the Hebrew letters making up the word for the Almighty.[8]

The intertwining of the religious and scientific has been and remains a characteristic feature of considerations of the question of life in the universe. As the citation from the *San Francisco Chronicle* illustrates, the pluralist controversy resembled, especially from the beginning of the nineteenth century, "a night fight in which the participants could not distinguish friend from foe until close combat commenced."[9] This entanglement of science and religion, while true of the engagement between professional scientists and theologians, is particularly evident whenever the issue spills over into the popular imagination. At the beginning of the twenty-first century, the conclusion that pluralism has become a modern myth or an alternative religion has been asserted with respect to the claims of science, but it has also received shocking confirmation in the willingness of ordinary citizens even to surrender their lives in the expectation that extraterrestrial life would provide the means of securing final religious fulfillment.[10]

THE END OF THE WORLD

In addition to concern about the ultimate destiny of humankind, people of faith have also frequently inquired about the fate of the universe itself. Convictions about how the world would end had also undergone considerable change by 1800. Since at least the mid-seventeenth century, natural philosophers had begun to counter the commonly held assumption that the end times were at hand and that as a consequence nature was deteriorating as the Psalmist had foreseen it would.[11] In its place appeared the idea that

---

[8] Cited in William Sheehan, *The Planet Mars: A History of Observation and Discovery* (Tucson: University of Arizona Press, 1996), pp. 88–90. Ronald Doel observes that the Martian canal controversy became problematic for American astronomers in the early twentieth century because it threatened to split them over the issue of extraterrestrial life. See *Solar System Astronomy in America* (Cambridge: Cambridge University Press, 1996), pp. 13–14.
[9] Crowe, *Extraterrestrial Life Debate*, p. 558.
[10] Ibid, p. 645, n. 22.
[11] Psalm 102:26: "The heavens shall wax old as doth a garment." The function of this interpretation was to oppose the heathenish doctrine of Aristotle, in which the world was regarded as eternal. For

nature was a law-bound system. One continued to assume that the cosmos was subject, as Isaac Newton (1642–1727) had observed, to occasional correction by its divine superintendent, but in the main it could be regarded as a stable machine. Some bold minds were even prompted to speculate on ways in which the solar system might have come about by means of God's secondary or indirect supervision, as opposed to a direct divine intervention. This tendency blossomed in the eighteenth century into a willingness to consider a natural cosmogony, a creation of the cosmos by natural law.[12] It would be left to the nineteenth century to deal with the implications of all this for God's relationship to nature. How, for example, could one resolve the internal tensions between a naturalistic account of creation and development, which involved apparently irreversible processes, and a scientific representation of nature as a mechanically reversible machine?

What was perhaps unexpected as the nineteenth century began was the role that would be played by physicists specializing in the new science of thermodynamics. The increasing acceptance of the notion of nature bound by natural law implied that in the minds of scientists, the future was not threatened by a final physical denouement such as that which was predicted in the Bible to accompany the Battle of Armageddon.[13] But if scientists and theologians were coming to regard the world as a perfect machine that would operate forever in accordance with law (law which still for most had been imposed on it by God), how could such a notion square with descriptions of the end times in which "the heavens shall pass away with a great noise and the elements shall melt with fervent heat, the earth also and the works that are therein shall be burned up"?[14]

The Laplacian notion of a stable and eternal cosmos therefore ran counter to traditional religious teaching. It also appeared to contradict a scientific conviction of natural philosophers from the seventeenth century onward. Because natural philosophers since Simon Stevin (1548–1620) and Galileo Galilei (1564–1642) had developed numerous arguments against the possibility of a perpetual motion machine, it was inevitable that sooner or later they would have to reconcile this conviction with the alleged eternal stability of the heavens.[15] Recognition of the need for reconciliation was delayed until the middle of the nineteenth century for at least two reasons. First, although the Laplacian cosmos was a system in which observed, irreversible physical

---

an account of the Renaissance notion of the running down of the physical world, see "The Decay of Nature," chap. 2 in Richard Foster Jones, *Ancients and Moderns: A Study of the Rise of the Scientific Movement in Seventeenth Century England* (Berkeley: University of California Press, 1965).

[12] See Ronald L. Numbers, *Creation by Natural Law: Laplace's Nebular Hypothesis in American Thought* (Seattle: University of Washington Press, 1977).

[13] Revelation 16:18, 20. Old Testament references to the demise of the original creation are paralleled in the final book of the New Testament. Compare Isaiah 65:17 and Revelation 21:1.

[14] 2 Peter 3:10.

[15] Cf. Arthur W. J. G. Ord-Hume, *Perpetual Motion: The History of an Obsession* (New York: St. Martin's Press, 1977), pp. 32ff.

processes were exposed as merely apparent and not permanent, the impression given by the French scientist's idea of creation by natural law was one of *development*. Indeed, Laplace's hypothesis went a long way toward preparing the ground for later evolutionary claims in biology. Discussions of perpetual motion had been traditionally carried out with respect to a purely mechanical context, not one in which growth or development was involved.[16] Second, Laplace did not eliminate God completely from a supervisory role over nature. He located God's concern with the world not at the level of individual planets, but with the more general laws that governed all the possible specific arrangements planets could assume. Although Laplace himself did not assume that God necessarily intended the solar system to last forever, the impression left by his *System of the World* was that the planets constituted a stable arrangement.[17] The notion that God's direct involvement with nature was to be found in the design of the most general laws, in other words, had implications that could work in opposite directions. On the one hand, it could reassure scientists that the cosmos was in fact divinely superintended, but on the other, it could postpone the question of why the eternal motion of the heavens did not force a concession that perpetual motion was in fact possible.

As a result of investigations into various transformations of one kind of "force" into another (for example, chemical force into electrical force, electrical force into heat force), numerous figures in the nineteenth century began to consider whether the general capacity to do work was conserved in the universe. In the course of making fundamental contributions to thermodynamics in the 1820s, Sadi Carnot (1796–1832) had assumed that heat "force" was conserved when it was used to produce mechanical effects; that is, no heat force was transformed *into* mechanical motion. By the 1840s some physicists were conjecturing that although there was no net loss of nature's total quantity of force, heat was in fact not conserved when mechanical motion was produced; that is, heat force *became* mechanical force – there was a mechanical equivalent of heat. Separate from this question, however, was another, one particularly relevant to the eternal working of the heavens: Were there physical contexts in which "force" might have to be created?

During the 1840s, when what later came to be known as the conservation of energy was being formulated, at least one of the contributors to the discovery,

---

[16] This is not to suggest that mechanical explanations of living things were absent at the beginning of the century, nor that they would not become central to the eventual resolution of the problem raised by an eternally stable cosmos. Cf. my " 'Nature is an Organized Whole': J. F. Fries's Reformulation of Kant's Philosophy of Organism," in *Romanticism in Science,* ed. S. Poggi and M. Bossi (Amsterdam: Kluwer, 1994), pp. 91–101. For the relevance of the understanding of the solar system as an organism to the debate over perpetual motion, see Kenneth Caneva, *Robert Mayer and the Conservation of Energy* (Princeton, N.J.: Princeton University Press, 1993), p. 146.

[17] "Could not the supreme intelligence, which Newton makes to interfere, make [the arrangement of the planets] depend on a more general phenomenon? . . . Can one even affirm that the preservation of the planetary system entered into the views of the Author of Nature?" Quoted from Laplace's *System of the World,* by Numbers, *Creation by Natural Law,* p. 126.

Robert Mayer (1814–1878), initially thought that while the destruction of force was impossible, the eternal motion of the heavens indicated that force was in fact being created by God. After consensus had emerged that force could be *neither* created nor destroyed, a property which the physicist William Thomson (1824–1907) associated with God's immutability, there emerged the recognition that what Thomson began calling "energy" was nevertheless subject to what he called "dissipation." Energy that had been dissipated continued to exist but was no longer available to do work. Through the work of Rudolf Clausius (1822–1888) and others, physicists realized that since such dissipation unavoidably accompanied the transformation of heat into other forms of energy, the amount of dissipated energy in the universe was gradually increasing. Logic dictated what seemed a tragic conclusion, one enunciated most powerfully by Hermann von Helmholtz (1821–1894) in a public lecture in Königsberg early in 1854: If there was a fixed total of energy in the universe and if portions of that total were increasingly becoming unavailable to do work, then the day would come when all of the energy would be unavailable and no more work could be done.[18] An argument could be made from physics that there *was* a final denouement coming, even if it was far in the future and even if it would be a whimper rather than the bang implied by biblical prophecy.

The theological implications of discoveries being made in thermodynamics ran in the opposite direction from the conclusions that had been drawn by some geologists of the time. From the 1830s on, the noted scientist Charles Lyell (1797–1875) had been teaching that a careful reading of the evidence from geological strata in Europe supported the conclusion not only that the earth was enormously old but that geological processes occurred in the context of steady state rather than of development. In other words, were one to be transported far back in time, one would be able to recognize the geological terrain because it was subject to local and temporary but not universal and permanent change. Lyell's conclusions were later used by Charles Darwin's (1809–1882) supporters to justify the vast time scale that evolution by natural selection required. The geological evidence, while irrelevant to theological issues of eschatology, was enlisted in support of a conception of evolutionary development that challenged traditional religious explanations of origin.

Physicists such as Thomson resented the claim that geological change was ultimately nondirectional, because Lyell's view persisted in spite of the theoretical work in thermodynamics that marked the decades around midcentury. Thomson challenged Lyell's view in public, even to the point of opposing the theory of evolution by natural selection. On the basis of thermodynamical calculations of the rate at which the earth had cooled from an uninhabitable

---

[18] Cf. "On the Interaction of Natural Forces," in H. von Helmholtz, *Popular Scientific Lectures* (New York: Dover Publications, Inc., 1962), pp. 59–90, at pp. 73–4.

molten mass to the solid crust on which life was thriving, Thomson, who became Lord Kelvin in 1892, concluded that the time that had passed *since* the earth was cool enough for the earliest life to have survived was insufficient to have permitted evolution by natural selection. From his first estimate of 100 million years, Kelvin kept revising his calculations downward until in his last public pronouncement on the subject in 1897, he was willing to grant but a scant 24 million years to Darwin and the evolutionists. While his Scottish Protestantism did not require that he reject evolution, he could not accept the dependence on chance required by natural selection. God was in control of Thomson's universe, including the fact that it was running down. Thomson scholar Crosbie Smith has noted that Thomson's understanding of matter and energy "kept constantly in mind the relationship of these concepts to a wider theological dimension throughout the long and difficult construction of this system."[19]

THE IMPLICATIONS OF MATERIALISM

Eschatology, however, was not the only theological area affected by the newly established laws of thermodynamics. Most controversial, perhaps, was the attempt to relate these laws to the question of whether an explanation based on mechanical interactions of matter was adequate to exhaust all of nature's secrets, including those accompanying organic and psychical processes. Were life and mind subject to the laws of conservation of matter and energy that had become fundamental truths of physics? In an 1861 address to the Royal Institution, Helmholtz left little doubt about his view that they were, a sentiment echoed and brought to a wider audience in 1874 in a famous presidential address to the British Association by the physicist John Tyndall (1820–1893). Tyndall's materialistic campaign even exposed prayer to public ridicule by making it the object of a scientific test.[20]

Others betrayed a more ambiguous position about the relationship between religion and science. Building on the perspective of the theologian Ludwig Feuerbach (1804–1872), who explained the origin of historical Christian doctrine as a projection born of human needs, the popular scientific materialist Ludwig Büchner (1824–1899) urged his readers to face courageously the negative consequences of science for traditional religious belief. Yet Büchner and other scientific materialists retained their conviction

---

[19] Crosbie Smith, "Natural Philosophy and Thermodynamics: William Thomson and the 'Dynamical Theory of Heat,'" *British Journal for the History of Science*, 9 (1976), 315. Cf. also Joe D. Burchfield, *Lord Kelvin and the Age of the Earth* (Chicago: University of Chicago Press, 1990), pp. 72–3.

[20] Stephen Brush, "The Prayer Test," *American Scientist*, 62 (1974), 561–3. Helmholtz's address is "On the Application of the Law of Conservation of Force to Organic Nature," *Proceedings of the Royal Institution*, 3 (1858–62), 347–57. Tyndall's so-called Belfast address is found in *British Association for the Advancement of Science Report*, 44 (1874), lxvii–xcvii, and was also published separately as *Advancement of Science* (New York: A. K. Butts, 1874).

that the cosmos reflected an ultimate purpose that incorporated human goals but was not limited to them. In the aftermath of Darwin's book on evolution by natural selection, Helmholtz's fellow countryman Ernst Haeckel (1834–1919) appealed to the unifying capacity of energy conservation to support a monistic religion in which outdated doctrines such as freedom of the will, immortality of the soul, and existence of a personal deity were abandoned. In their place Haeckel put belief in the "law of substance," a law he felt incorporated into one precept the individual conservation principles of matter and energy and which articulated for him the religious meaning inherent in nature.

Not everyone, of course, agreed with the wholesale surrender of traditional doctrine to the dictates of the new laws of thermodynamics. A number of prominent figures in Britain, including Thomson, James Clerk Maxwell (1831–1879), Thomson's brother James (1822–1892), and others, discussed whether a mind with free will could direct the energies of nature, possibly even to the point of reversing the effects of dissipation.[21] Some Catholic theologians rejected the claim that physiological and especially psychophysical systems had been shown to be subject to energy conservation. They argued that the human soul could in fact act on matter, not by any mechanical interaction but in a manner that could only be grasped by synthesizing scientific and religious interpretations. Body and soul were coprinciples, with neither outside the other. One could neither permit the soul to be reduced to matter or energy nor deny that the soul could affect the body.[22]

The science of chemistry produced its own heroic defender of traditional religious belief against materialism in the person of Louis Pasteur. Historically, investigations into interactions of matter had intersected with religious concerns in discussions about alchemy and in debates about atomism. In the nineteenth century, a more publicly visible interaction took place over the issue of spontaneous generation, a subject that included discussions both of the origin of life on earth from lifeless matter (abiogenesis) and of the spontaneous production of microorganisms from organic matter (heterogenesis).

For religiously orthodox people, the beginning of life on earth was unquestionably due to God's direct creative act as described in the Genesis creation account. More religiously liberal minds and many scientists thought the matter involved a much more complex decision. While there were few, if any, who asserted that the origin of life occurred apart from God's intent and

---

[21] See the excellent treatment of the extended development of these issues, including Maxwell's introduction of what Thomson named a demon, in Crosbie Smith and M. Norton Wise, *Energy and Empire: A Biographical Study of Lord Kelvin* (Cambridge: Cambridge University Press, 1989), pp. 612–33.

[22] Erwin Hiebert, "The Uses and Abuses of Thermodynamics in Religion," *Daedalus*, 95 (1966), 1063ff. A different approach to the question of spirit was taken by physicist Oliver Lodge and others involved in the scientific investigation of psychical phenomena. Cf. John D. Root, "Science, Religion, and Psychical Research: The Monistic Thought of Oliver Lodge," *Harvard Theological Review*, 71 (1978), 245–63.

control, there were those who included it under the Laplacian notion of creation by natural law. To suggest that the origin of life itself was part of a larger developmental process like that described by the nebular hypothesis was, for example, appealing early in the century to J.-B. Lamarck (1744–1829) in France and G. H. Schubert (1780–1860) in Germany, while in the 1840s it was accepted by Robert Chambers (1802–1871) in England. However, none of these men were regarded during their lifetimes as representative of a scientific mainstream in their respective countries; consequently, they contributed little to an acceptance of abiogenesis.[23]

After midcentury the focus of the debate lay with the alleged production of microorganisms from organic matter. Here the situation became further confused. The antireligious German scientific materialists of the 1850s, for example, did not enjoy at all that they were on the same side of the issue of abiogenesis as the discredited *Naturphilosoph* Schubert. Regarding heterogenesis, they disagreed among themselves, from Karl Vogt's (1817–1895) doubt of its possibility to Ludwig Büchner's confidence that it would be proven true.[24] In France Félix Pouchet argued that heterogenesis could be demonstrated by experiment and that it could be reconciled with traditional Christian views. As far as the conservative French public under Louis Napoleon was concerned, however, spontaneous generation, evolution, and pantheistic materialism were all German evils that had to be resisted in the Second Republic, just as they had been a generation earlier during the Restoration reign of Charles X. There the hero had been Georges Cuvier in his debate with Etienne Geoffroy Saint-Hilaire. In the 1860s the Académie des Sciences appointed two commissions to examine spontaneous generation, each one concluding that the highly regarded chemist Louis Pasteur (1822–1895) had shown conclusively that Pouchet was wrong about his claim to have produced heterogenesis experimentally. Pasteur, who deliberately cast the issue of spontaneous generation as a confrontation with materialism, successfully demonstrated that experimental science could be convincingly enlisted in defense of religion.[25]

---

[23] For acceptance of abiogenetic spontaneous generation in the speculative evolution of Lamarck, cf. the 1809 *Zoological Philosophy*, trans. Hugh Elliot (New York: Hafner, 1963), pp. 236–7; in the *Naturphilosophie* of Schubert, cf. the 1808 *Ansichten von der Nachtseite der Naturwissenschaft*, 4th ed. (Dresden: Arnoldische Buchhandlung, 1840), p. 115; and in the evolutionary musings of Chambers, cf. the 1844 *Vestiges of the Natural History of Creation* (New York: Humanities Press, 1969), p. 58.

[24] Cf. Frederick Gregory, *Scientific Materialism in Nineteenth Century Germany* (Dordrecht: Reidel, 1977), pp. 169–75.

[25] Cf. Gerald L. Geison, *The Private Science of Louis Pasteur* (Princeton, N.J.: Princeton University Press, 1995), chap. 5. Geoffrey Cantor's impressive study of Michael Faraday provides a different example of how science mediated the private and public life of a highly respected experimentalist who was also religiously conservative. Cantor's analysis of Faraday's simultaneous devotion to natural science and to the strictly biblical views of the Sandemanian sect helps to clarify the role of metascientific principles in dealing with issues of science and religion. Cf. Geoffrey Cantor, *Michael Faraday: Sandemanian and Scientist: A Study of Science and Religion in the Nineteenth Century* (New York: St. Martin's Press, 1991).

FROM CONFRONTATION TO PEACEFUL COEXISTENCE TO REENGAGEMENT

Pasteur's public critique of materialism was but one indication of the increasing tendency over the course of the nineteenth century for scientists to usurp the social role enjoyed by clergy in earlier times to coordinate the meaning residing in nature with the meaning of human existence. It seemed, however, that for every Pasteur or Kelvin who came down on the side of a traditional religious perspective, there were twice as many Tyndalls and Büchners who proclaimed the need to abandon old views. If the scientist was now the recognized authority on nature, it appeared that once-popular theological arguments, such as those profitably utilized in natural theology, had lost their persuasive power. The new authority of science was a contributing factor to the larger process of secularization that was affecting traditional beliefs of all religious persuasions.[26] By the second half of the nineteenth century, the old easy association of religious and scientific enterprises had given way to a complicated series of attitudes about the relationship between science and religion.

Two different approaches characterized the various positions taken. Those utilizing the first approach assumed that science and religion shared common territory and that the way in which disagreements were to be handled was clear. Within this approach there were, to be sure, several different ways of resolving disagreements between scientific and religious claims when they occurred. Hard-line representatives of orthodoxy, for example, continued to insist that scientific explanations simply had to give way to religious doctrine when there was a conflict. More liberal minds believed that compromise was necessary on both sides and that an accommodation would be possible when both the scientific and theological implications were better known. Finally, more extreme scientific naturalists resolved differences by insisting that theological doctrine defer to the results of science when there was a contradiction between the two. All three groups agreed, however, that there was but one truth to be found. At issue was who had correctly identified the way to get at it.[27]

Others preferred a second approach stemming from the thought of Immanuel Kant at the end of the eighteenth century and revived in the second half of the nineteenth by German theologians. In this approach, the quest for nature's one truth was abandoned as a goal of metaphysics because it was deemed impossible to achieve. Natural science was recharacterized as a strictly utilitarian enterprise the task of which was to master the world for use

---

[26] This tendency was particularly evident in France under the Third Republic, where widespread anticlericalism caused Catholic Church leaders to encourage work by Catholic scientists who had retained their faith. Cf. Harry Paul, *The Edge of Contingency: French Catholic Reaction to Scientific Change from Darwin to Duhem* (Gainesville: University Presses of Florida, 1979), pp. 181ff.

[27] Cf. Frederick Gregory, *Nature Lost? Natural Science and the German Theological Traditions of the Nineteenth Century* (Cambridge, Mass.: Harvard University Press, 1992), chaps. 3–5.

by humans. While freedom was given to science to explain nature however it wished, such explanations provided no metaphysical understanding at all, since their intent lay elsewhere. But if science must be purged of metaphysical claims, so too must theology. Neither could get at nature's truth. The understanding of religion also had to be recharacterized; religion must be restricted to the realm of the moral. In this approach, which would be shared by the burgeoning community of existentialist thinkers in the new century, science and religion were assumed not to intersect on common ground. All familiar references to an intimate relationship between God and nature disappeared.[28]

The growing confidence among laypeople and some scientists in the second half of the nineteenth century that knowledge of nature's fundamental physical laws was nearing completion ran counter to the neo-Kantian interpretation of science and religion just described.[29] It supported the traditional view, the so-called Platonic ideal in which "all genuine questions must have one true answer and one only."[30] But theologians such as Karl Barth and Rudolf Bultmann, who embraced the neo-Kantian depiction of the relationship between science and religion as the foundation for their own existential systems, were not the only ones to question the Platonic ideal in the new century. Developments within physics at the end of the nineteenth century led to the formulation of relativity theory and quantum mechanics in the twentieth, both of which led scientists to acknowledge that the theoretical representation of reality was a far more complex enterprise than the one inherited from their predecessors. Gone was the deterministic mechanical view of the world that had reigned since Laplace. In its place appeared an uncertain world in which paradox accompanied all attempts to inquire about nature's most basic entities. What resulted was a new willingness, at least among many physical scientists and theologians, to pursue separate goals in a peaceful juxtaposition of endeavors.[31] This mutual distancing of scientists and theologians continued to characterize their relationship until well after the new century's midpoint.[32]

---

[28] Ibid., chaps. 6–7. While the French physicist Pierre Duhem also emphasized that scientific propositions do not refer to objective existence and therefore cannot intersect with metaphysical doctrines, his embrace of Catholicism differentiated him from the German neo-Kantians. On Duhem see Harry Paul, *Edge of Contingency,* chap. 5.

[29] Herrmann was critical of the theologian who was waiting for natural scientists to finish their work before undertaking a new confession of faith. Cf. *Nature Lost,* p. 244. For a discussion of a related sentiment among some scientists, cf. Lawrence Badash, "The Completeness of Nineteenth-Century Science," *Isis,* 63 (1972), 48–58.

[30] Isaiah Berlin, *The Crooked Timber of Humanity: Chapters in the History of Ideas,* ed. Henry Hardy (New York: Knopf, 1991), p. 5.

[31] Cf. Ueli Hasler, *Beherrschte Natur: Die Anpassung der Theologie an die bürgerliche Naturauffassung im 19. Jahrhundert* (Bern: Peter Lang, 1982), p. 295. Cf. also Keith Yandell, "Protestant Theology and Natural Science in the Twentieth Century," in *God and Nature,* ed. David Lindberg and Ronald Numbers (Berkeley: University of California Press, 1986), pp. 448–71.

[32] The lack of formal engagement by practitioners of the two fields may be one of the reasons that some statistical measures of the personal religious belief of scientists, at least in the United States,

If the twentieth century brought intellectual developments in physical science and theology that eroded the older confidence of practitioners from both disciplines, so too did events outside the scholarly community. The occurrence of two world wars and the immediate onset of a global nuclear threat contributed in their own ways to a new sense of uncertainty, bringing in its wake an openness to the questioning of the foundations of modernity itself. From new work on the history of science (largely physical science) by Thomas Kuhn came the call to place the *context* of historical developments in science on at least an equal footing with the cognition of their contents. Kuhn dissociated himself from those who came to focus in their historical treatments almost exclusively on the social or cultural context; nevertheless, among the ramifications of Kuhn's achievement that made their way into public debates was the claim that historians and scientists have to modify the conviction, historically common to both disciplines, that theirs is a business of finding truth. In the words of one analyst of Kuhn's impact, humankind has had to learn to bear the tension between not knowing truth and having to aim at it anyway.[33]

The postmodern view that has blossomed since Kuhn's seminal work has especially affected discussions about science and religion, since postmodern thinkers typically are critical of even aiming at truth. Richard Rorty attacks what he regards as the assumption of the last three centuries that through philosophical exploration one can, at least in theory, "touch bottom." Rorty's critique of the attempt to ground truth claims from various fields of discourse in an overarching metatheory of universal relevance has been dubbed "antifoundationalism."[34] In Rorty's view, one simply should not ask questions about the nature of truth any longer, because humans do not have the ability to move beyond their beliefs to something that serves as a legitimating ground. In this perniciously relativistic perspective, an inquiry about the rights of science and religion loses all meaning in the face of an "anything goes" mentality where the only matter of interest is power. Historically, scientists and theologians have shared the belief in the existence of a foundation, although they have disagreed on how properly to characterize it. In their attempts either to integrate or to respond to postmodern critiques, however, representatives of science and religion are discovering that their shared determination to pursue truth has the potential to make them more allies than enemies. The result has been a greater willingness to engage each other.

---

show no appreciable change between 1916 and 1996. Cf. Edward J. Larson and Larry Whitman, "Scientists Are Still Keeping the Faith," *Nature,* 386 (1997), 435–6. In popular and public culture, however, several issues between science and religion were forced by the onset of the atomic age. Cf. James Gilbert, *Redeeming Culture: American Religion in an Age of Science* (Chicago: University of Chicago Press, 1997).

[33] Cf. David A. Hollinger, *In the American Province: Studies in the History and Historiography of Ideas* (Bloomington: Indiana University Press, 1985), p. 128.

[34] Richard Rorty, *Philosophy and the Mirror of Nature* (Princeton, N.J.: Princeton University Press, 1979), pp. 5–6. For the characterization of Rorty's view as "antifoundationalism," cf. Karen L. Carr, *The Banalization of Nihilism: Twentieth Century Responses to Meaninglessness* (Albany: State University of New York Press, 1992), p. 88.

CONTEMPORARY CONCERNS

A glance at developments in Roman Catholic thought in the twentieth century reveals one example of the new engagement between science and religion. Pope Pius XII's concession in the 1950 encyclical *Humani generis* that the human body may have resulted from evolutionary development opened a half century of reconsideration within Catholic thought. Under Pope Paul VI, the Church affirmed in 1965 "the legitimate autonomy of human culture and especially of the sciences," and Pope John Paul II continued moving in the new direction through his own involvement with the subject of evolution and with his thirteen-year study of the Church's condemnation of Galileo. The pope's declaration in 1992 that the Church had erred in condemning Galileo for disobeying its orders is but one of the initiatives that he and other Catholic thinkers have undertaken to reassess the Church's position on the relationship of religion and science.[35]

Meanwhile, professional scientists and Protestant theologians enjoyed a peaceful coexistence during the first half of the twentieth century, enabled both by the development of the new physics and by the dominance among theologians of the Barthian view that God was not to be sought in nature. Reengagement has occurred as especially the latter view has been challenged. In 1961 the theologian Langdon Gilkey argued that there was an internal contradiction at the heart of Barthian neoorthodoxy. Barth had insisted that God is "wholly other." While in this context orthodox language was appropriate, Barth implicitly assumed a classical view of nature as a closed, causal continuum. What resulted was a contradiction between orthodox language and liberal cosmology.[36]

Since this time, there has been renewed interest in resuscitating the relationship between God and nature that had been cut off and even mishandled in neoorthodoxy. In these recent attempts is an evident willingness to abandon the classical mechanical worldview in favor of dynamic alternatives in which old metaphors are deemed simply no longer adequate. Characteristic of many of the newer approaches is a depiction of divine

---

[35] The relevant section of the papal encyclical *Humani generis* is 36. The encyclical is reprinted in *The Papal Encyclicals* (Ann Arbor, Mich.: The Pierian Press, 1990), 4: 175ff. For the relevant section of Paul VI's promulgation of the pastoral constitution on the Church in the modern world, see *Gaudium et spes* (Washington, D.C.: U.S. Catholic Conference, 1965), par. 59. John Paul II's address to a 1985 conference in Rome on "Evolution and Christian Thought," along with the contributions of Catholic participants in the conference, is given in *Evolutionismus und Christentum*, ed. Robert Spaemann, Reinhard Löw, and Peter Koslowski (Weinheim: Acta humaniora, VCH, 1986). For thoughts on the reassessments of John Paul II, see *John Paul II on Science and Religion: Reflections on the New View from Rome*, ed. Robert John Russell, William R. Stoeger, and George V. Coyne (Notre Dame, Ind.: University of Notre Dame Press, 1990).

[36] Cf. Robert John Russell, "Introduction," *Quantum Cosmology and the Laws of Nature: Scientific Perspectives on Divine Action*, ed. R. J. Russell, Nancey Murphy, and C. J. Isham (Notre Dame, Ind.: University of Notre Dame Press, 1993), p. 7. Gilkey's article was "Cosmology, Ontology, and the Travail of Biblical Language," *Journal of Religion*, 41 (1961), pp. 194–205.

action in metaphors of personal agency. In some systems, God is represented as external to nature; in others, new biological and feminine analogies stress a more intimate connection to the world. In virtually all, however, the challenges raised by the rise of quantum theory in physics lie at the heart of the reformulation of the theological conclusions. Not surprisingly, one area of particular focus has involved work in theoretical physics bearing on cosmology.

Physical scientists themselves have produced two restatements of one of the classic contentions in science and religion, the argument from design. Restrictions of space do not permit treatment of recent contentions about the irreducibility of biochemical complexity; consequently, only the so-called anthropic principle will be discussed here.[37] As its name implies, this principle appeals to evidence from the physical world purportedly suggesting that the presence of humans had been anticipated when the cosmos was formed. Such reasoning links modern forms of the argument to an important thread of the well-established tradition of natural theology. From at least the seventeenth century on, natural theologians have made claims of this kind.[38]

Early in the twentieth century, some physicists had noted the repeated presence of certain large numbers in nature that resulted from dimensionless ratios involving atomic and cosmological constants. In the wake of separate contributions to this subject by Arthur Eddington, Paul Dirac, Robert Dicke, and others, conclusions specifically involving the gravitational constant and the age of the universe have emerged that attempt to draw out implications for the way the universe has developed.[39] Had the value of the gravitational constant, for example, been a greater or smaller number than it is, then either the universe would have ceased expanding before elements other than hydrogen had been able to form or it would have expanded as a gas without creating galaxies. In either case, there would have been no observers produced to ask why the gravitational constant has the very convenient value (for them) that it does. Dicke concluded in 1961 that the universe appeared to be "somewhat limited by the biological requirements to be met during the epoch of man."[40] More recent investigations have produced greater than a dozen coincidental physical and cosmological quantities the values of which

---

[37] Biochemist Michael Behe, while not an orthodox creationist, maintains with an impressive argument that the irreducible complexity of the biochemical mechanisms operating in vital functions could not have been produced by evolutionary processes as we know them. Cf. *Darwin's Black Box: The Biochemical Challenge to Evolution* (New York: Free Press, 1996).

[38] Throughout John Ray's work on natural theology, for example, there appear repeated notations of the way in which the physical cosmos has been arranged to serve human ends. Cf. *The Wisdom of God Manifested in the Works of the Creation* (New York: Arno Press, 1977), p. 66. This is a facsimile reprint of the seventh edition, which appeared in 1717. The first edition was 1691.

[39] A discussion of this work can be found in the definitive book on the subject by John D. Barrow and Frank J. Tipler, *The Anthropic Cosmological Principle* (Oxford: Clarendon Press, 1986), pp. 224–55.

[40] Robert Dicke, "Dirac's Cosmology and Mach's Principle," *Nature*, 192 (1961), 440. Dirac's original letter is entitled "The Cosmological Constants" and is found in *Nature*, 139 (1937), 323.

seem to be circumscribed by the requirements for life. Theoretical physicist John Wheeler has summarized the anthropic principle to say that "a life-giving factor lies at the center of the whole machinery and design of the world."[41]

It should be noted that just because one invokes the final causation embedded in the anthropic principle, one does not thereby necessarily commit oneself to belief in the existence of a transcendent God who designed the universe. According to one critic, however, an appeal to the anthropic principle is merely a secularized version of the old design argument. Physicist Heinz Pagels maintains that because they are loath to resort to religious explanations, some atheist scientists find that the anthropic principle is as close as they can get to God. In spite of what defenders of the argument might say, they are, according to Pagels, motivated by religious reasons. They should be willing openly to take the leap of faith that other more honest proponents of the anthropic principle take and say that "the reason why the universe seems tailor-made for our existence is that it was tailor-made."[42]

Yet the same critics who view the value of the gravitational constant as purely accidental and of no "explanatory" value whatever are frequently uncomfortable with one possible implication of their position; namely, if there is no reason the constant has the value that it does, then presumably there are other universes where it has a different value and where life as we know it has not developed. When such critics reject out of hand any talk about other universes, a subject that also crops up in the so-called many worlds interpretation of quantum mechanics, they can appear to be insisting on a closed set of beliefs about science that are defined in as dogmatic a manner as any other narrowly conceived religious interpretation.[43]

In bringing this survey to a close, mention should be made of an evaluation of modern physics based on religious considerations that has been directed more to a popular audience than to professional scientists and theologians. Using as a point of departure the historical and current relative absence of women in physics, especially theoretical physics, some have attempted to explain this circumstance by establishing a common link in the missions of

---

[41] John Wheeler, "Foreword," in Barrow and Tipler, *Anthropic Cosmological Principle*, p. vii. In 1979 Freeman Dyson said: "The more I examine the universe and the details of its architecture, the more evidence I find that the universe in some sense must have known we were coming." Quoted from Dyson's *Disturbing the Universe* by John Polkinghorne, *The Faith of a Physicist: Reflections of a Bottom-Up Thinker* (Princeton, N.J.: Princeton University Press, 1994), p. 76.

[42] Heinz Pagels, as quoted by Martin Gardner, "WAP, SAP, PAP, & FAP," *New York Review of Books* (3 May 1986), p. 22. For their part, Barrow and Tipler seem content to reject traditional theism, in which God is regarded as wholly separate from the physical universe, in favor of pantheism, the doctrine that holds that the physical universe is *in* God, but that God is more than the universe. (Cf. *The Anthropic Cosmological Principle*, p. 107) Of the many systems they discuss they appear to draw most from the thought of the French Jesuit theologian Pierre Teilhard de Chardin. (Cf. pp. 195–205, 675–7). For Barrow and Tipler's rejection of the possibility of extraterrestrial intelligent life, cf. chap. 9.

[43] Cf. B. S. DeWitt and N. Graham, eds., *The Many-Worlds Interpretation of Quantum Mechanics* (Princeton, N.J.: Princeton University Press, 1973).

Western religion and science.[44] Although the argument depends on sweeping historical generalizations that have been objected to, there is no denying the resonance of this gender-based analysis of science and religion with the values of postmodern Western culture.

Central to the approach are two claims on which the general thesis is based. First, it is asserted that there is nothing essential to Christianity about the dominant role men have acquired. A male celibate clergy successfully rose to dominance only in the second millennium of the Church's history, as a patriarchal ideal finally defeated the androgynous ideal with which it had been in competition. Second, proponents assert that one by-product of the rise of the mechanical worldview in the Scientific Revolution was the availability of a means by which the established clerical order could resist forces that threatened to reform it. One aspect of the general outbreak of heresy in the Renaissance and Reformation periods, they argue, was the rise of a religiously based magical tradition which, although it shared with Aristotelian science an organic conception of nature, sought to know the Divine intellect through means unacceptable to Church practice. By opposing the organic conception of nature with a mechanical view, the men of the Scientific Revolution, despite giving the appearance of challenging the existing Church powers, functioned to consolidate a new male priestly order. The view of nature as a self-developing autonomous organism was discredited and replaced with a nature controlled and ruled by God the giver of fixed mechanical law.[45]

These two claims, that male ecclesiastical power was a late addition to Christianity and that nature as mechanism functioned as a creative defense of established order, are the foundation of a more general thesis. The argument is that, in putting the lid on post-Reformation disorder with the help of the new mechanistic science of laws, the same male-dominant structure that had earlier characterized the religious establishment became part of the new science. Further, in spite of impressions to the contrary, science continued to retain the trappings of a religious mission and, as had been the case since the tenth century whenever humans have presumed to engage the holy, it continued to retain a privileged position for men.

The clearest expression of this modern "religious" mission can be recognized wherever one encounters the ancient Pythagorean search for nature's mathematical symmetry and harmony. This Pythagorean religion was transformed by early mechanists into a search for the mind of the Christian God. That quest has been tempered since the seventeenth century by a concern

---

[44] A growing interest in non-Western religion and science has been in evidence at the turn of the twenty-first century. A challenge for scholars is the completion of a work parallel to *The History of Science and Religion in the Western Tradition: An Encyclopedia*, ed. Gary B. Ferngren, Edward J. Larson, and Darrell W. Amundsen (New York: Garland, 2000).

[45] Cf. Margaret Wertheim, *Pythagoras's Trousers: God, Physics, and the Gender Wars* (New York: Times Books, 1995), chap. 4; David F. Noble, *A World Without Women: The Christian Clerical Culture of Western Science* (New York: Knopf, 1993), chap. 9.

to find more practical mathematical relationships in nature, but it has not disappeared. In fact, wherever the religious mission has been retained in its pure form, as, for example, in the quest for a Theory of Everything in theoretical physics, fewer women scientists will be found. Since the nature of science "is determined by what a society *wants* from its science, what a society decides it *needs* science to explain, and finally what society decides to *accept* as a valid form of explanation," the meaning of science would be more socially responsible if we rid it of the outdated religious virus that too long has infected it from within.[46]

Throughout the last two centuries in virtually all cases of interaction between physical science and religion, the diversity of opinion displayed has stemmed from the variety of assumptions that have been *brought to* the issues by the participants. Always, however, there has been a basic question, the answer to which has been decisive in the past and will continue to be so for future explorations of issues in physical science and religion: "Is the Person or is matter in motion the ultimate metaphysical category? There really is no third."[47]

---

[46] Wertheim, *Pythagoras's Trousers*, p. 33. Although she is obviously sympathetic to a cultural analysis of science, Wertheim does not subscribe to the radical relativism of some postmodernists where science is concerned. Cf. p. 198.

[47] Erazim Kohak, *The Embers and the Stars: A Philosophical Inquiry into the Moral Sense of Nature* (Chicago: University of Chicago Press, 1984), p. 126.

3

A TWISTED TALE
Women in the Physical Sciences in the Nineteenth and Twentieth Centuries

Margaret W. Rossiter

Dismissed as inconsequential before the 1970s, the history of the contributions of women to the physical sciences has become a topic of considerable research in the last two decades. Best known of the women physical scientists are the three "great exceptions" from central Europe – Sonya Kovalevsky, Marie Sklodowska Curie, and Lise Meitner – but in recent years, other women and other countries and areas have been receiving attention, and more is to be expected in the future. The overall pattern for most women in these fields, the nonexceptions, has been one of ghettoization and subsequent attempts to overcome barriers.

PRECEDENTS

Before 1800 there were several self-taught and privately-tutored "learned ladies" in the physical sciences. Included were the English self-styled "natural philosopher" Margaret Cavendish (1623–1673), who wrote books and in the 1660s visited the Royal Society of London, which had not elected her to membership; the German astronomer Maria Winkelmann Kirch (1670–1720), who worked for the then-new Berlin Academy of Sciences in the early 1700s; the Frenchwoman Emilie du Chatelet (1706–1749), who translated Newton's *Principia* into French before her premature death in childbirth in 1749; the Italians Laura Bassi (1711–1778), famed professor of physics at the University of Bologna, and Maria Agnesi (1718–1799), a mathematician in Bologna; Ekaterina Romanovna Dashkova (1743–1810), the director of the Imperial Academy of Sciences in Russia; and Marie Anne Lavoisier (1758–1836), who helped her husband Antoine with his work in the Chemical Revolution.[1]

---

[1] Lisa T. Sarasohn, "A Science Turned Upside Down: Feminism and the Natural Philosophy of Margaret Cavendish," *Huntington Library Quarterly,* 47 (1984), 289–307; Londa Schiebinger, "Maria

Women's scattered contributions to the physical sciences became more numerous and less aristocratic around 1800 in Britain when Jane Marcet (1769–1858) started her series of famous popular textbooks, as *Conversations on Chemistry,* and Caroline Herschel (1750–1848) helped her brother William with his astronomy and, on her own, located eight comets.[2] In France, Sophie Germain (1776–1831) read physics books in her father's library, used the pseudonym "Henri LeBlanc" on bluebooks submitted surreptitiously to the men-only Ecole Polytechnique, and corresponded with Karl Friedrich Gauss. In 1831 Scotswoman Mary Somerville (1780–1872) translated Laplace's *Mécanique céleste* into English, and in the 1840s Nantucket astronomer Maria Mitchell (1818–1889) discovered a comet.[3]

Later in the nineteenth century, when higher education opened to women, many more began to study the physical sciences. But inasmuch as higher education placed certain restrictions on their entrance and participation, full careers in the physical sciences opened to only a few. They generally had a higher threshold of entry than the more accessible field of natural history. By the late nineteenth century, a career in the physical sciences required such credentials as higher degrees, often obtainable only at foreign universities, and scientific publications, usually requiring long stays in distant laboratories. In fact the rise of the laboratory, generally acclaimed in the history of the physical sciences, can be seen as a new level of exclusion, creating new male retreats or preserves to which women gained entry only by special permission.

GREAT EXCEPTIONS

The history of women in the physical sciences in the nineteenth and twentieth centuries is dominated by the careers and legends of the three great exceptions

---

Winkelman at the Berlin Academy, A Turning Point for Women in Science," *Isis,* 78 (1987), 174–200; Mary Terrall, "Emilie du Chatelet and the Gendering of Science," *History of Science,* 33 (1995), 283–310; Paula Findlen, "Science as a Career in Enlightenment Italy, The Strategies of Laura Bassi," *Isis,* 84 (1993), 441–69; Paula Findlen, "Translating the New Science: Women and the Circulation of Knowledge in Enlightenment Italy," *Configurations,* 2 (1995), 167–206; A. Woronzoff-Dashkoff, "Princess E. R. Dashkova: First Woman Member of the American Philosophical Society," *Proceedings of the American Philosophical Society,* 140 (1996), 406–17. On the others, see Marilyn Bailey Ogilvie, *Women in Science: Antiquity Through the Nineteenth Century: A Biographical Dictionary with Annotated Bibliography* (Cambridge, Mass.: MIT Press, 1986; 1990). Her *Women and Science: An Annotated Bibliography* (New York: Garland, 1996) is also indispensable.

[2] Susan Lindee, "The American Career of Jane Marcet's *Conversations on Chemistry,* 1806–1853," *Isis,* 82 (1991), 8–23; Marilyn Bailey Ogilvie, "Caroline Herschel's Contributions to Astronomy," *Annals of Science,* 32 (1975), 149–61.

[3] Louis L. Bucciarelli and Nancy Dworsky, *Sophie Germain: An Essay in the History of the Theory of Elasticity* (Dordrecht: Reidel, 1980); Elizabeth C. Patterson, *Mary Somerville and the Cultivation of Science, 1815–1840* (The Hague: Nijhoff, 1983); Sally Gregory Kohlstedt, "Maria Mitchell and the Advancement of Women in Science," in *Uneasy Careers and Intimate Lives: Women in Science, 1789–1979,* ed. Pnina G. Abir-Am and Dorinda Outram (New Brunswick, N.J.: Rutgers University Press, 1987), pp. 129–46.

who played prominent roles in mainstream European mathematics and science: Sonya Kovalevsky (1850–1891), the Russian mathematician who was the first woman to earn a PhD (at the University of Göttingen in absentia in 1874) and the first woman in Europe to become a professor (at the University of Stockholm in 1889); Marie Sklodowska Curie (1867–1934), the Polish-French physicist-chemist who discovered radium and won two Nobel Prizes; and Lise Meitner (1878–1968), the Austrian physicist who participated in the discovery of nuclear fission together with Otto Hahn and Fritz Strassmann, but who did not share in Hahn's 1944 Nobel Prize in chemistry and spent her later years in exile in Sweden.[4]

Biographies written on these three figures highlight their subjects' uniqueness and specialness. Each woman seemed, for inexplicable reasons, to rise and achieve at a time when few other women did. Few if any had ties to one another or to any women's movement, or so we are told in these works about them, but they did benefit from openings made by other women and probably others have benefited from their "firsts." Generally they worked to make themselves so outstanding as to be worthy of a personal favor or exemption or exception, rather than to build ties and alliances that would effect permanent institutional change. They squeezed through but left the pattern intact.

Perhaps it is unfair to expect a biographer of one woman in one or several countries and fields to link her subject to other women in other fields in other countries. But this leads to contradictions. Sonya Kovalevsky, we are told, was known throughout Europe in the 1880s, but then there is no evidence in works about Marie Curie that while growing up in Russian-dominated Poland in the 1880s, she ever heard of Kovalevsky, let alone modeled her own career on hers, as she might well have done.[5]

Most of what has been written about these exceptional women has been in a heroic mode or revolves around a central message, such as a love story. Studies of Curie still are based on limited primary materials and are heavily influenced by Eve Curie's sentimental best-selling biography of her mother in the late 1930s, later made into a wartime movie.[6] But other scholars, notably

---

[4] There are several biographies of Kovalevsky; the most recent is by Ann Hibner Koblitz, *A Convergence of Lives: Sofia Kovalevskaia: Scientist, Writer, Revolutionary* (New Brunswick, N.J.: Rutgers University Press, 1993; rev. ed.). The latest biography on Curie is by Susan Quinn, *Marie Curie* (New York: Simon & Schuster, 1994), reviewed by Lawrence Badash in *Isis* in 1997. See also Ruth Sime, *Lise Meitner: A Life in Physics* (Berkeley: University of California Press, 1996); Elvira Scheich, "Science, Politics, and Morality: The Relationship of Lise Meitner and Elisabeth Schiemann," *Osiris*, 12 (1997), 143–68. For more details on the scientific work of the women physicists mentioned here and of others, see Marilyn Ogilvie and Joy Harvey, eds., *The Biographical Dictionary of Women in Science, Pioneering Lives from Ancient Times to the mid-20th century,* 2 vols. (New York: Routledge, 2000), and the website maintained by Nina Byers, "Contributions of Women to Physics" at <http://www.physics.ucla.edu/~cwp>.

[5] Quinn, *Marie Curie*.

[6] Eve Curie, *Madame Curie,* trans. Vincent Sheean (Garden City, N.Y.: Doubleday, Doran, 1938); and the movie *Madame Curie,* starring Greer Garson and Walter Pidgeon (1943).

Helena Pycior and J. L. Davis, are now studying aspects of Curie's scientific work and research school.[7]

Most satisfactory to date is the biography of Lise Meitner by Ruth Sime, who shows in some detail how much preparation and intelligence (in the espionage sense) it took to be in the right place at the right time.[8] While there are such things as coincidences, a series of them often indicates careful planning. And a successful career in the sciences for a woman required not only luck but a lot of strategic planning to know where to make one's own opportunities and how to avoid dead ends, hopeless battles, and insuperable obstacles.

These women were able to obtain correct information about their best opportunities, and they contrived to come up with the resources (wealthy parents, earnings as a governess, or a "fictitious" marriage to a fellow student) to get there at a time when it was rare even for more mobile male students to do so. As daughters, these women might also have been expected to stay at home and take care of aging parents. Yet the "exceptions" managed to disentangle themselves from this filial obligation and to have innovative family arrangements.

The main reason to leave home and family and to migrate was to find world-class mentors, whom they chose wisely, and who, being insiders, helped them to jump barriers, work on interesting problems, and become exceptions to the many petty rules and exclusions that would have daunted them otherwise. Kovalevsky left Russia with her fictitious husband Vladimir to study mathematics in Germany with Karl Weierstrass, who was devoted to her and assisted her later career, as also did Gösta Mittag-Leffler in Stockholm. Marie Sklodowska traveled to Paris to study physics at a time when various German universities, which did physics better, were still largely closed to women. In Paris she wisely sought out Pierre Curie, married him, and worked with him on her radium research. Lise Meitner studied with Ludwig Boltzmann in Vienna in the first years when women were allowed in Austrian universities and then, encouraged by none other than Max Planck, was allowed by Emil Fischer to work with Otto Hahn at the Kaiser Wilhelm Institute for chemistry outside Berlin – if she used the side door and kept out of sight. Later she became head of the physics section within it. These women all showed extraordinary, even legendary, levels of perseverance and determination.

Though foreign women were often granted educational opportunities denied to local women (who might then expect a job in the same country), their situation could and did become difficult if they stayed on and held

---

[7] Helena M. Pycior, "Reaping the Benefits of Collaboration While Avoiding Its Pitfalls: Marie Curie's Rise to Scientific Prominence," *Social Studies of Science*, 23 (1993), 301–23; Helena M. Pycior, "Pierre Curie and 'His Eminent Collaborator Mme. Curie,'" in *Creative Couples in the Sciences*, ed. Helena Pycior, Nancy Slack, and Pnina Abir-Am (New Brunswick, N.J.: Rutgers University Press, 1996), pp. 39–56; and J. L. Davis, "The Research School of Marie Curie in the Paris Faculty, 1907–1914," *Annals of Science*, 52 (1995), 321–55.

[8] Sime, *Lise Meitner: A Life in Physics*.

a job in that country. Then sexual indiscretions might be reported in the press, as happened to Marie Curie in Paris in 1911. Worse, if the economy soured and/or right-wing movements arose, as occurred in Germany, Austria, Spain, and elsewhere in the 1930s, those who were Jewish, were particularly vulnerable and could become targets of the press or political regime and even forced to flee at a moment's notice, as many did.

Though they defied all stereotypes and rose to become unique and memorable figures, these "exceptions" did not change the stereotypes and the norms (to which we turn in a moment) that have worked to keep most women out of sight in their own time and throughout history.[9]

LESS-WELL-KNOWN WOMEN

Beyond the exceptions was a host of other female physical scientists of possibly similar caliber who are not as well known. These include the French chemist Irène Joliot-Curie (1897–1956), daughter of Marie and Pierre Curie, who shared the Nobel Prize in chemistry with her husband Frédéric (1900–1958) in 1935 for work on artificial radioactivity; the German-American physicist Maria Goeppert-Mayer (1906–1972), who shared the 1963 Nobel Prize in physics with two others for her work on magic numbers in spin ratios in atoms; and Dorothy Crowfoot Hodgkin (1910–1994), an English crystallographer and biochemist who won the Nobel Prize alone in 1964 for determining the structure of a series of complex biological molecules.[10] Still others who should have won it include Rosalind Franklin (1920–1958), the English crystallographer of nucleic acids; crystallographer Kathleen Lonsdale (1903–1971), who discovered that the benzene ring was flat; and C. S. Wu (1912–1997), the Chinese-American physicist who showed in 1957 that parity was not conserved.[11] Also notable were the astronomers Annie Jump Cannon (1863–1941), Henrietta Leavitt (1868–1921), and the British-born Cecilia Payne-Gaposchkin (1900–1979), all of the Harvard College Observatory.[12] Beyond these would be Agnes Pockels (1862–1935),

---

[9] Margaret Rossiter, "The Matthew Matilda Effect in Science," *Social Studies of Science*, 23 (1993), 325–41.

[10] Margaret Rossiter, "'But She's an Avowed Communist!' *L'Affaire Curie* at the American Chemical Society, 1953–55," *Bulletin for the History of Chemistry*, no. 20 (1997), 33–41; Bernadette Bensaude-Vincent, "Star Scientists in a Nobelist Family: Irène and Frédéric Joliot-Curie," in *Creative Couples*, ed. Helena Pycior, Nancy Slack, and Pnina Abir-Am, chap. 2. See also Karen E. Johnson, "Maria Goeppert Mayer: Atoms, Molecules and Nuclear Shells," *Physics Today*, 39, no. 9 (September 1986), 44–9; Joan Dash, *A Life of One's Own* (New York: Harper and Row, 1973), and Peter Farago, "Interview with Dorothy Crowfoot Hodgkin," *Journal of Chemical Education*, 54 (1977), 214–16.

[11] Anne Sayre, *Rosalind Franklin & DNA* (New York: W. W. Norton, 1975); Maureen M. Julian, "Dame Kathleen Lonsdale," *Physics Teacher*, 19 (1981), 159–65; N. Benczer-Koller, "Personal Memories of Chien-Shiung Wu," *Physics and Society*, 26, no. 3 (July 1997), 1–3.

[12] John Lankford, *American Astronomy, Community, Careers, and Power, 1859–1940* (Chicago: University of Chicago Press, 1997), p. 53; *Cecilia Payne-Gaposchkin: An Autobiography* (Cambridge: Cambridge University Press, 1984).

the German housewife whose letter to Lord Kelvin about soap bubbles helped to launch the study of thin films; Julia Lermontova (1846–1919), the first Russian woman to earn a doctorate in chemistry; physicists German Ida Noddack (1896–1978) and Canadian Harriet Brooks (1876–1933); and Swiss chemists Gertrud Woker (1878–1968) and Erika Cremer (b. 1900).[13]

These less-well-known women merit study because their careers should show us more about everyday science and the opportunities open and closed to most women. In addition, their presence, usually controversial, so strained the levels of tolerance of the time that by the 1920s, when faculty positions had opened to more than a trickle of women, the increase in numbers provoked strong opposition and produced a reaction or backlash, which was especially pronounced in Germany but also of note in Spain and Austria. There, fascist groups, fueled by widespread fears and resentments of many kinds, rose up, seized power, and drove out many of these women, often Jewish, who were just getting a foothold in university faculties in the physical sciences. Mathematicians Emmy Noether and Hilda Geiringer von Mises fled into exile, and French historian of chemistry Hélène Metzger disappeared forever on the way to Auschwitz. The Nazis were relentless and, unlike others, made no exceptions, especially not for these otherwise nearly exceptional women.[14]

RANK AND FILE – FIGHTING FOR ACCESS

The history of women in science, particularly in the physical sciences, is unbalanced in that it centers largely on a few famous women who were pretty much exceptions to the prevailing norms in their society at the time. (This is also true of the history of men in science, which emphasizes the work of the Nobelists, even though it is logically and pedagogically incorrect to discuss the exceptions to a rule before stating what that rule or norm is.) This focus or emphasis on the exceptions and near exceptions is particularly unfortunate in the history of women in science, for it overlooks and so minimizes or dismisses the far more common patterns of exclusion, marginalization,

---

[13] M. Elizabeth Derrick, "Agnes Pockels, 1862–1935," *Journal of Chemical Education,* 59 (1982), 1030–1; Charlene Steinberg, "Yulya Vsevolodovna Lermontova (1846–1919)," *Journal of Chemical Education,* 60 (1983), 757–8; Fathi Habashi, "Ida Noddack (1896–1978)," *C[anadian] I[nstitute] of M[etals] Bulletin* 78, no. 877 (May 1985), 90–3; Ralph E. Oesper, "Gertrud Woker," *Journal of Chemical Education,* 30 (1953), 435–7; Marelene F. Rayner-Canham and Geoffrey W. Rayner-Canham, *Harriet Brooks: Pioneer Nuclear Scientist* (Montreal: McGill-Queen's University Press, 1992); Jane A. Miller, "Erika Cremer (1900–)," in *Women in Chemistry and Physics: A Biobibliographic Sourcebook,* ed. Louise S. Grinstein, Rose K. Rose, and Miriam H. Rafailovich (Westport, Conn.: Greenwood Press, 1993), pp. 128–35. This biobibliography is one of a new genre of useful reference works.

[14] Noether and Joan L. Richards, "Hilda Geiringer," in *Notable American Women: The Modern Period, A Biographical Dictionary,* ed. Barbara Sicherman and Carol Hurd Green (Cambridge, Mass.: Harvard University Press, 1980), pp. 267–8; Suzanne Delorme, "Metzger, Hélène," in *Dictionary of Scientific Biography,* IX, 340.

underemployment and unemployment, underrecognition, demoralization, and suicide. But it is hard to correct this imbalance, for little is known about these generally obscure women. Thus, in a further twist – that might please the whimsical British mathematician Lewis Carroll, who wrote about Alice in Wonderland – the exceptions have in a sense become the norm, since we seldom hear of the rank and file, who have been largely obliterated from history.[15] This distortion has led to an imbalance in current knowledge about women's place in the physical sciences.

The focus on the exceptions, who experienced few problems, particularly omits the long struggle for higher degrees faced by women aspiring to be scientists or even just wanting to study science. Universities were founded beginning in the mid-twelfth century in Europe, but women were not admitted to any institutions for higher education until 1865 when Vassar College opened in the United States. Thus, women were not allowed to study at the university level for nearly seven centuries, despite Laura Bassi's presence on the Bologna faculty in the mid-eighteenth century.

It was only with the opening of higher education to women – first at mid-nineteenth century in the United States, but in the 1880s in Britain, in France in the 1890s, and finally in Austria in 1897 and Germany in 1908 – that there were to be more than a few women in science. For several decades, there was such an uneven level of educational and occupational opportunity in Western countries that women in search of greater opportunities often had to leave home and travel abroad. Some stayed only a few years; others spent their entire careers abroad. Much progress had been made by the 1930s, so much, in fact, that the women's more visible presence provoked the backlash mentioned earlier, especially against Jewish women. Some were expelled, but, unable to return home, they were then forced to seek refuge in another foreign country. Others faced worse. Much more progress was made after World War II, when many ex-colonial and newly socialist nations, such as China and those in Eastern Europe, made female literacy and education a priority.

A lot of what is written about women "in science" is really about gaining access to its institutions, because while individuals might have a variety of attitudes toward women in science, most institutions were exclusionary, either deliberately – in written policies or in unwritten traditions – or inadvertently, as when there was simply no precedent, for no women had applied before or been present at its creation. This institutional barrier was a big hurdle for the first women who later sought entrance; in some cases, this was a very long struggle that dissipated energies that in a more egalitarian society could have been spent on other ventures. England and Germany, where so much of the world's science was done and taught in the nineteenth and twentieth

---

[15] In addition to exclusionary barriers, women scientists were also held to a higher level of expectations. (See Margaret W. Rossiter, *Women Scientists in America: Struggles and Strategies to 1940* [Baltimore: Johns Hopkins University Press, 1982], p. 64.)

centuries, were (and still are) particularly restrictive about admitting women to educational and scientific institutions.

Women's entrance into the older British universities was glacially slow and proceeded incrementally, with admission to examinations (including the natural sciences Tripos at Cambridge), the creation of separate women's colleges, the awarding of certificates and then actual degrees, and finally admission to the traditional colleges.[16] In the United States, the movement started in the 1830s with the establishment of many women's seminaries, some of which later became colleges.

WOMEN'S COLLEGES – A WORLD OF THEIR OWN

Separate, independent colleges for women, as well as coordinate colleges for women affiliated with men's universities, have played a large role in the training and especially the employment of female physical scientists, primarily in the United States and England. Astronomer Maria Mitchell, for example, became the first woman science professor in the United States when she was hired at Vassar College in the 1860s. Among her students were chemist Ellen Richards (1842–1911), one of the founders of the field of home economics; Mary Whitney (1847–1921), her successor in astronomy at Vassar; and Christine Ladd-Franklin (1847–1930), a physicist-turned-psychologist of note. Several of these colleges had science departments that were (and still are) quite strong in chemistry, such as Mount Holyoke, which remains into the new millennium the largest producer of female PhDs in chemistry in the United States. Sophie Newcomb College in New Orleans was also strong in chemistry, while Bryn Mawr College, the only separate women's college with a graduate school that awarded doctorates in the physical sciences, also trained a string of notable women geologists. Wellesley College was important in several fields, including astronomy, mathematics, and physics. Notable among the faculty with long careers at American colleges for women were physicists Frances Wick at Vassar; Sarah Whiting (1847–1927) and Hedwig Kohn (1887–1965) at Wellesley; Rose Mooney at Newcomb and Hertha Sponer-Franck (1895–1968) at Duke University's women's college; and chemists Emma Perry Carr (1880–1972), Mary Sherrill (1888–1968), Lucy Pickett (b. 1904), and most recently Anna Jane Harrison (1912–1998) at Mt. Holyoke College.[17]

---

[16] Roy MacLeod and Russell Moseley, "Fathers and Daughters: Reflections of Women, Science, and Victorian Cambridge," *History of Education*, 8 (1979), 321–33; Carol Dyhouse, *No Distinction of Sex? Women in British Universities 1870–1939* (London: UCL Press, 1995).

[17] Marie-Ann Maushart, *"Um mich nicht zu vergessen:" Hertha Sponer – Ein Frauenleben für die Physik im 20. Jahrhundert* (Bassum: Verlag für Geschichte der Naturwissenschaften und der Technik, 1997); Carol Shmurak, "Emma Perry Carr: The Spectrum of a Life," *Ambix*, 41 (1994), 75–86; Carol Shmurak, " 'Castle of Science': Mount Holyoke College and the Preparation of Women in Chemistry, 1837–1941," *History of Education Quarterly*, 32 (1992), 315–42.

There were also a few important colleges for women in England. Dorothy Hodgkin spent her long career in crystallography at Somerville College, Oxford, where one of her chemistry students was Margaret Thatcher, whose subsequent career took a different turn. Rosalind Franklin was a graduate of Newnham College, Cambridge, in chemistry.

Elsewhere, American missionaries established colleges for women in Istanbul, Beirut, and India, but such colleges never caught on in Germany, where separate institutions for women were considered inferior. Nevertheless, in France Marie Curie taught for a time at the normal school for female teachers at Sèvres.[18]

To a certain extent these colleges trained women for burgeoning areas of "women's work" (as we shall see), but their alumnae include a relatively large proportion of the pioneers and subsequent, even current, participants in most of the physical sciences, often as many as from the far larger "coeducational" universities that in reality had very few women majors in the physical sciences. Agnes Scott College in Georgia, for example, had by 1980 graduated fifteen women who later earned PhDs in chemistry – the same number as the far larger Massachusetts Institute of Technology, where relatively few women completed majors in chemistry.[19]

The role of the women's colleges in the United States has diminished in recent decades, because around 1970 the trustees at some colleges voted to admit men. At about the same time, their counterparts at many previously all-male institutions (Caltech, Princeton, Amherst, the Jesuit institutions, the military and naval academies, and others) admitted women for the first time. Yet single-sex education is hardly dead, as currently there is in the United States a resurgence in all-girl schools at the primary and secondary school level, and it is widely known that they prepare women better in nontraditional areas, including the physical sciences.

GRADUATE WORK, (MALE) MENTORS, AND LABORATORY ACCESS

Switzerland was unusually important for women in science and medicine because its educational institutions, especially the University of Zurich, were staffed largely by liberal faculty members ousted from Germany after the 1848 revolution. They admitted large numbers of female students starting in the

---

[18] James C. Albisetti, "American Women's Colleges Through European Eyes, 1865–1914," *History of Education Quarterly*, 32 (Winter 1992), 439–58; Jo Burr Margadant, *Madame le Professeur: Women Educators in the Third Republic* (Princeton, N.J.: Princeton University Press, 1990). Nuclear physicist Salwa Nassar (Berkeley PhD, 1944) chaired the physics department at the American University of Beirut and in 1966 became head of the Beirut College for Women ("We See by the Papers," *Smith College Alumnae Quarterly*, 57 [1965–6], 163).

[19] Alfred E. Hall, "Baccalaureate Origins of Doctorate Recipients in Chemistry: 1920–1980," *Journal of Chemical Education*, 62 (1985), 406–8.

1860s when no other European universities would do so. Hardly any of these early students were Swiss; most were from Russia, France, Germany, England, and the United States.[20] Also in Zurich around 1900 was the Serbian Mileva Marić (1875–1948), who has since gained fame as Albert Einstein's fellow student at the Eidgenössische Technische Hochschule (ETH) and as his first wife.[21]

Starting in the late nineteenth century, work at certain laboratories in physical sciences became important, though at first these were male spaces. Yet some professors heading these world-famous laboratories accepted women, and a trickle of female students and researchers began to work with them. Starting in the 1880s, for example, a series of female physicists worked at the famous Cavendish Laboratory at Cambridge University. Among these were Rose Paget, who later married its director J. J. Thomson; the Canadian Harriet Brooks, whom Ernest Rutherford invited to follow him when he became the laboratory's director; the American Katharine Blodgett (1898–1979), the first woman to earn a doctorate at Cambridge University and later the collaborator of Irving Langmuir at General Electric, in the 1920s; and Joan Freeman of Australia in the late 1940s.[22]

Some mentors welcomed female students, worked with them, and supported their subsequent careers. Madame Curie welcomed students from Eastern Europe at her Radium Institute, and physiological chemist Lafayette B. Mendel (1872–1937) trained forty-eight women PhDs at Yale University in the 1920s and 1930s.[23]

"MEN'S" AND "WOMEN'S" WORK IN PEACE AND WAR

Women are generally quite rare in what can be considered "men's work" – mainstream university departments and large industrial laboratories, often supported by defense budgets and infused with a military ethos – and very

---

[20] Ann Hibner Koblitz, "Science, Women, and the Russian Intelligentsia: The Generation of the 1860s," *Isis*, 79 (1988), 208–26. See also Thomas N. Bonner, *To the Ends of the Earth: Women's Search for Education in Medicine* (Cambridge, Mass.: Harvard University Press, 1993).

[21] Gerald Holton, "Of Love, Physics and Other Passions: The Letters of Albert [Einstein] and Mileva [Marić]," *Physics Today*, 47 (August 1994), 23–9, and (September 1994), 37–43; *Albert Einstein/Mileva Marić: The Love Letters*, ed. J. Renn and R. Schulman (Princeton, N.J.: Princeton University Press, 1992).

[22] Paula Gould, "Women and the Culture of University Physics in Late Nineteenth-Century Cambridge," *British Journal for the History of Science*, 30 (1997), 127–49; Marelene F. Rayner-Canham and Geoffrey W. Rayner-Canham, *Harriet Brooks*; Kathleen A. Davis, "Katharine Blodgett and Thin Films," *Journal of Chemical Education*, 61 (1984), 437–9; Joan Freeman, *A Passion for Physics: The Story of a Woman Physicist* (Bristol, England: Adam Hilger, 1991).

[23] Marelene F. Rayner-Canham and Geoffrey W. Rayer-Canham, sr. authors and eds., *A Devotion to Their Science: Pioneer Women of Radioactivity* (Philadelphia: Chemical Heritage Foundation; and Montreal: McGill-Queen's University Press, 1997); Margaret Rossiter, "Mendel the Mentor: Yale Women Doctorates in Biochemistry, 1898–1937," *Journal of Chemical Education*, 71 (1994), 215–19.

predominant in the two kinds of "women's work."[24] Jobs deemed suitable for women have often been low-level, subordinate, dead-end, invisible, and monotonous staff and service positions, such as technical assistants of various sorts, chemical librarians, chemical secretaries, calculators or computers, computer programmers, and astronomical counters. Among the more famous women in these positions were Annie Jump Cannon of the Harvard College Observatory and Jocelyn Bell Burnell (b. 1943) of the United Kingdom, who participated in the discovery of pulsars that won Anthony Hewish and Martin Ryle the Nobel Prize for physics in 1974.[25]

The somewhat different jobs deemed suitable for women are often situated away from the men, usually in a slightly removed location or discipline, such as teaching a science at a women's college, serving as a dean of women, or working in the field of "home economics," a branch of nutrition and domestic science developed for female chemists in the United States in the late nineteenth century.[26] Unlike the assistants mentioned previously, some women have held high rank in these womanly jobs. This pattern of sex-typing has spread to some other countries as well, and female physical scientists, such as Rachel Makinson of Australia, have been employed in the area of "textile physics."[27]

Some female physical scientists have held government jobs, as with the Commonwealth Scientific and Industrial Research Organization (CSIRO) in Australia; various agencies of the American government, such as the U.S. Geological Survey and the National Bureau of Standards; and the Geological Survey and the Dominion Observatory in Canada.[28] Historically, these organizations have paid lower salaries to women than to men, refused to hire married women, and offered little advancement, but there have been some reforms in recent decades. In the early 1970s Anglo-American astronomer E. Margaret Burbidge (b. 1919) even served briefly as Astronomer Royal of the Royal Greenwich Observatory in the United Kingdom.

---

[24] Ellen Gleditsch (1879–1968) became in 1929 the first female professor at the University of Oslo. See Anne-Marie Weidler Kubanek, "Ellen Gleditsch (1879–1968), Nuclear Chemist," in *Notable Women in the Physical Sciences*, ed. Benjamin F. Shearer and Barbara S. Shearer (Westport, Conn.: Greenwood Press, 1997), pp. 127–31. This very useful biobibliographical work has information on 96 women. For data on the proportion of women employed in particular subfields of the physical sciences in the United States in 1956–8, see Margaret Rossiter, "Which Science? Which Women?" *Osiris*, 12 (1998), 169–85.

[25] Margaret Rossiter, "Women's Work in Science, 1880–1910," *Isis*, 71 (1980), 381–98. See also Margaret Rossiter, "Chemical Librarianship: A Kind of 'Women's Work' in America," *Ambix*, 43 (March 1996), 46–58. On Jocelyn Bell, see Sharon Bertsch McGrayne, *Nobel Prize Women in Science: Their Lives, Struggles, and Momentous Discoveries* (Secaucus, N.J.: Carol Publishing, 1993), which includes several other near-Nobelists.

[26] See Sarah Stage and Virginia Vincenti, eds., *Rethinking Women and Home Economics in the Twentieth Century* (Ithaca, N.Y.: Cornell University Press, 1997).

[27] Nessy Allen, "Textile Physics and the Wool Industry: An Australian Woman Scientist's Contribution," *Agricultural History*, 67 (1993), 67–77.

[28] See, for example, Nessy Allen, "Achievement in Science: The Careers of Two Australian Women Chemists," *Historical Records of Australian Science*, 10 (December 1994), 129–41.

It was the pressing manpower needs of World War I that opened jobs for women in chemistry and engineering in Canada, Australia, England, Germany, and elsewhere. Marie Curie, Lise Meitner, and other physical scientists made themselves useful as x-ray technicians – a new job at the time – during the war. At the other extreme, German chemist Clara Immerwahr (1870–1915), Fritz Haber's wife at the time, committed suicide, perhaps in protest of his development of poison gases.[29]

In World War II, several immigrant female physicists (such as Maria Goeppert Mayer and Leona Woods Marshall Libby (1919–1986) worked on the atomic bomb project in the United States, while others filled in for male professors at the universities and otherwise "kept the seat warm" for the men's eventual return. Lise Meitner, one of the discoverers of nuclear fission, was one of the very few physicists who refused an invitation to Los Alamos to work on the atomic bomb. Other scientists with antiwar political views were the English crystallographers Dorothy Crowfoot Hodgkin and Kathleen Lonsdale. The latter, a Quaker, developed a reputation as a pacifist and protester of nuclear testing in the 1950s and 1960s. By contrast, French-woman Irène Joliot-Curie was pro-Communist in the 1940s and 1950s and helped to train some of the Chinese physicists who would later build China's hydrogen bomb. As such, she was unwelcome in the United States and not even acceptable as a member of the American Chemical Society despite her Nobel Prize in chemistry.[30]

SCIENTIFIC MARRIAGES AND FAMILIES

Because female scientists have often married male scientists, there is a phenomenon of "endogamy," or marrying within the tribe. Most famous are the two Curie couples – Marie and Pierre and then Irène and Frédéric Joliot. Others of note were the American chemists Ellen and Robert Richards, Irish and English astronomers Margaret (1848–1915) and William Huggins, British mathematicians Grace Chisholm (1868–1944) and Will Young, Czech-American biochemists Gerty (1896–1957) and Carl Cori, German-American physicist Maria and American chemist Joseph Mayer, and Chinese-American physicists C. S. Wu and Yuan (Luke) Wu, to name just a few.[31]

---

[29] Gerit von Leitner, *Der Fall Clara Immerwahr: Leben für eine humane Wissenschaft* (Munich: Beck, 1993); Haber's second wife Charlotte published an autobiography, *My Life with Fritz Haber* (1970).
[30] Gill Hudson, "Unfathering the Thinkable: Gender, Science and Pacificism in the 1930s," in *Science and Sensibility: Gender and Scientific Enquiry, 1780–1945*, ed. Marina Benjamin (Oxford: Blackwell, 1991). See n. 10.
[31] Several are in Helena Pycior et al., *Creative Couples*. There are lists of American couples in Margaret W. Rossiter, *Women Scientists in America*, p. 143, and Margaret W. Rossiter, *Women Scientists in America: Before Affirmative Action, 1940–1972* (Baltimore: Johns Hopkins University Press, 1995), pp. 115–20. All the couples listed were heterosexual.

Beyond the mother–daughter relationship of Marie and Irène Curie have been father–daughter combinations, as the chemists Edward and Virginia Bartow; mother–son sets, as among astronomers Maria Winkelmann Kirch (1670–1720) and Christoph Kirch; and brother–sister combinations, as astronomers William and Caroline Herschel and chemists Chaim and Anna Weizmann (?–1963) in England and Israel; and sister–sister dyads, such as the Anglo-Irish popularizers of astronomy Ellen (1840–1906) and Agnes Clerke (1842–1907), the Americans astronomer Antonia (1866–1952) and paleontologist Carlotta Maury (1874–1938), and the American-French neuroanatomist Augusta Déjerine-Klumpke (1859–1927) and astronomer Dorothea Klumpke Roberts (1861–1942).[32]

UNDERRECOGNITION

Many scientific societies, starting with the very first, the Royal Society of London in 1662, long refused to admit women as members. The Royal Society relented in the late 1940s after decades of struggle and admitted three outstanding women, including crystallographer Kathleen Lonsdale.[33] Practices at other younger and more specialized societies varied. Ellen Richards and a few others were present at the founding of the American Chemical Society in 1876; Charlotte Angas Scott (1858–1931) was elected a member of the council at the first meeting of the American Mathematical Society in 1894; and Sarah Whiting (1847–1927) was a charter member of the American Physical Society in 1899. But even when women became members, it was often a long time – a century with the chemists and longer with the mathematicians – before any woman became president. In this there were wide national differences. In Britain, the Chemical Society of London was among the laggards.[34] The American and French national academies were also very slow. The first female physical scientists elected to the U.S. National Academy of Sciences, which was established in 1863, were physicists Maria Goeppert Mayer in 1956 and C. S. Wu in 1958. The Académie des Sciences did not elect its first woman until physicist Yvonne Choquet-Bruhat in 1979.[35]

Female physical scientists have probably been more active over the years in local and regional groups than in national or international ones, but

---

[32] Meyer W. Weisgal, "Prof. Anna Weizmann," *Nature*, 198 (1963), 737; for the others, see Ogilvie and Harvey, eds., *The Biographical Dictionary of Women in Science*.

[33] Joan Mason, "The Admission of the First Women to the Royal Society of London," *Notes and Records of the Royal Society of London*, 46 (1992), 279–300. On Lonsdale, see n. 11; on Stephenson, see Robert E. Kohler, "Innovation in Normal Science: Bacterial Physiology," *Isis*, 76 (1985), 162–81.

[34] Joan Mason, "A Forty Years' War," *Chemistry in Britain*, 27 (1991), 233–8, is on women's admission to the Chemical Society of London.

[35] Jim Ritter, "French Academy Elects First Woman to Full Membership," *Nature*, 282 (January 1980), 238.

the former groups are less often studied.³⁶ In the eighteenth century, social settings like salons or coffee houses were conducive to women's participation, but more recently, even local organizations, such as campus clubs, were for a long time staunchly male-only. This had adverse consequences for female students or professionals, for a lot of "informal communication" took place at rathskellers, men's clubs, other smoke-filled rooms, and sacrosanct places, such as the bar at the Chemists' Club in New York City.³⁷

Two American organizations have responded to the general underrecognition of women by scientific societies by establishing separate women's prizes, for example, the Annie Jump Cannon Prize of the American Astronomical Society (AAS) and the Garvan Medal of the American Chemical Society (ACS). The Cannon Prize was started in the early 1930s when Annie Jump Cannon received an award from the Association to Aid Women in Science shortly before it went out of existence. Not agreeing with the association's leaders that women's problems in science had then been solved, Cannon donated the $1,000 to the AAS that set up a woman's prize. It was offered at three-to-five-year intervals until the early 1970s when Anglo-American astronomer E. Margaret Burbidge caused a bit of a stir by refusing to accept it on the grounds that a separate prize for women was discriminatory. A committee was set up to investigate this problem, and it recommended using the funds for a fellowship for a young female astronomer, to be administered by the American Association of University Women.³⁸

Similarly, the Garvan Medal was started in the late 1930s when foundation official Francis P. Garvan was overheard in an elevator saying that there had never been any female chemists. When corrected by an indignant woman, he agreed to underwrite a special ACS prize for a distinguished female chemist. It has since been supported by the W. R. Grace Company and is awarded annually by the ACS.³⁹

POST–WORLD WAR II AND "WOMEN'S LIBERATION"

After World War II, two developments affected opportunities for female scientists. In many countries, including India, Vietnam, and Israel, as they became independent nations, the literacy rate and educational level of women

---

³⁶ Icie Macy Hoobler was in 1930 the first woman to head a section of the American Chemical Society. See Icie Gertrude Macy Hoobler, *Boundless Horizons: Portrait of a Pioneer Woman Scientist* (Smithtown, N.Y.: Exposition Press, 1982).
³⁷ See Margaret W. Rossiter, *Women Scientists in America... to 1940*, chaps. 4, 10, and 11, and *Women Scientists in America, ... 1940–1972*, chap. 14.
³⁸ Margaret W. Rossiter, *Women Scientists in America... to 1940*, pp. 307–8; Rossiter, *Women Scientists in America, ... 1940–1972*, pp. 352–3; E. Margaret Burbidge, "Watcher of the Skies," *Annual Reviews of Astronomy and Astrophysics*, 32 (1994), 1–36.
³⁹ Rossiter, *Women Scientists in America... to 1940*, p. 308; Rossiter, *Women Scientists in America, ... 1940–1972*, pp. 342–5; Molly Gleiser, "The Garvan Women," *Journal of Chemical Education*, 62 (1985), 1065–8.

rose dramatically. Other countries, especially in Eastern Europe, were taken over by Communist governments, which accorded women more education and higher status than had often been true earlier. Other governments have also made literacy and numeracy for women a high priority. Little has yet been written about any of this, but it should have been a golden age for the higher education of women.[40]

Nevertheless, female physical scientists, such as physicists Joan Freeman and Yuasa Toshiko (1909–1980) and astronomer Beatrice Tinsley (1941–1981), have felt it necessary to leave their home countries, Australia, Japan, and New Zealand, for greater educational and employment opportunities in the United Kingdom, the United States, and France, respectively. Because the only job Yuasa, trained in France by the Joliot-Curies, could get in her homeland in the late 1940s was in a women's college, and because the American occupation forces prohibited nuclear research in Japan at the time, she returned to France and spent her whole career at the Centre National de la Recherche Scientifique (CNRS).[41]

As funding for the physical sciences skyrocketed in the post–World War II era, largely as a result of the Cold War between the United States and "the Communist bloc," women in many countries found new opportunities in different kinds of scientific employment.[42]

In the United States between 1969 and 1972, a branch of the "women's liberation" movement was devoted to science. Vera Kistiakowsky (b. 1928) led the move to start a women's committee within the American Physical Society, and Mary Gray (b. 1939) was one of the founders of the independent Association for Women in Mathematics, both of which still exist. In the 1980s various well-publicized Women in Science and Engineering (WISE) and "new blood" schemes made news in England, and in Australia and Germany in the 1990s. Since 1992, the European Union has awarded fellowships named for Marie Sklodowska Curie (who left Poland for France) to scientists who will go to other European countries.[43]

---

[40] John Turkevich, *Soviet Men [sic] of Science, Academicians and Corresponding Members of the Academy of Sciences of the USSR* (Princeton, N.J.: D. Van Nostrand, 1963), includes meteorologist Ekaterina Blinova, chemists Rakhil Freidlina and Aleksandra Novoselova, and hydrodynamicist (and biographer of Sonya Kovalevsky) Pelageya Kochina. On Soviet women astronomers, see A. G. Masevich and A. K. Terentieva, "Zhenshchiny-astronomy," *Istoriko-Astronomischeskie Issledovaniia*, 23 (1991), 90–111.

[41] Joan Freeman, *A Passion for Physics;* Edward Hill, *My Daughter Beatrice: A Personal Memoir of Dr. Beatrice Tinsley, Astronomer* (New York: American Physical Society, 1986); and Eri Yagi, Hisako Matsuda, and Kyomi Narita, "Toshiko Yuasa (1909–1980), and the Nature of her Archives at Ochanomizu University in Tokyo," *Historia Scientarum*, 7 (1997), 153–63.

[42] On the United States, see Margaret W. Rossiter, *Women Scientists in America, . . . 1940–1972*.

[43] David Dickson, "France Seeking More Female Scientists with Offer of $4,500 Scholarships," *Chronicle of Higher Education*, 25 September, 1985; Allison Abbott, "Europe's Poorer Regions Woo Researchers," *Nature*, 388 (1997), 701. The Marie Curie Fellowship Association of current and former fellows has a website: www.mariecurie.org.

Although as stated at the outset, most of what is written about women in physical sciences centers on the United States and Western Europe (as does most history of science), some data published in 1991 is already helping to broaden scholarly concern to female physical scientists in other places. In 1991 physicist W. John Megaw of York University, Canada, presented data on the worldwide distribution of female physicists in 1988, which have been widely cited since then.[44] His study shows dramatically that women account for the highest proportion of physics faculties in Hungary (47%), followed by Portugal (34%), the Philippines (31%), the USSR (30%), Thailand (24%), Italy (23%), Turkey (23%), France (23%), China (21%), Brazil (18%), Poland (17%), and Spain (16%). East Germany at 8% outranked Japan (6%), the United Kingdom and West Germany (4%), and the United States (3%). Megaw's data may attract more scholarly interest to the history in these countries of female physical scientists about whom little is known, but who are faring and succeeding better institutionally than their counterparts in presumably enlightened Western Europe and the United States.[45] Among the reasons for these wide national differences are historical issues, such as the modernization of Kemal Ataturk in Turkey in the 1930s, the amount of scientific training required of both sexes in secondary schools (as in Italy and Turkey), and the status and monetary compensation of the scientific profession in general.[46] For example, in Latin America and the Philippines, private corporations hire and pay men so well that the universities must hire women.[47]

International comparisons may help to further gender analysis of the physical sciences, for once it is shown that many countries do it all differently, it will be easy to supersede Western-based essentialist arguments of what is "manly" and what women do "differently." Getting beyond the "great exceptions" and into the many other responses to patriarchy provided by

---

[44] W. John Megaw, "Gender Distribution in the World's Physics Departments," in National Research Council, *Women in Science and Engineering: Increasing Their Numbers in the 1990s* (Washington, D.C., 1991), p. 31; a special issue of *Science,* 263 (11 March, 1994); Mary Fehrs and Roman Czujko, "Women in Physics: Reversing the Exclusion," *Physics Today,* 45 (1992), 33–40; "Global Gaps and Trends," *World Science Report, 1996* (Paris: UNESCO Publications, 1996), p. 312.

[45] For starters, see Carmen Magallon, "Mujeres en Las Ciencias Fisico-Quimicas en Espana: El Instituto Nacional de Ciencias y el Instituto Nacional de Fisica y Quimica (1910–1936)," *Llull,* 20 (1997), 529–74; Monique Couture-Cherki, "Women in [French] Physics," in Hilary Rose and Steven Rose, *The Radicalisation of Science: Ideology of the Natural Sciences* (New York: Holmes & Meier, 1976), chap. 3. On East Germany, see H. Tscherisch, E. Malz, and K. Gaede, "Sag mir, wo die Frauen sind!" *Urania,* 28, no. 3 (March 1965), 178–89; on Australia, Ann Moyal, "Invisible Participants: Women Scientists in Australia, 1830–1950," *Prometheus,* 11, no. 2 (December 1993), 175–87.

[46] Chiara Nappi, "On Mathematics and Science Education in the U.S. and Europe," *Physics Today,* 43, no. 5 (1990), 77–8; Albert Menard and Ali Uzun, "Educating Women for Success in Physics: Lessons from Turkey," *American Journal of Physics,* 61, no. 7 (July 1993), 611–15.

[47] Marites D. Vitug, "The Philippines: Fighting the Patriarchy in Growing Numbers," *Science,* 263 (1994), 1492.

international comparisons promises to open up fascinating and long-overdue new insights into the worldwide history of women in the physical sciences.

RISE OF GENDER STEREOTYPES AND SEX-TYPED CURRICULA

In the seventeenth and eighteenth centuries, mathematics and physics had not been typed by sex – Bernard de Fontenelle's classic *Conversations on the Plurality of Worlds* (1686) has as its leading figure the Marquise, a bright, witty, and attractive lady, and Francesco Algarotti's *Newtonianism for the Ladies* (1737) was aimed at a similar audience. There was also the curiously titled magazine *The Ladies' Diary* that lasted throughout most of the eighteenth century in England, though only about 10 percent of its contributors were women. All offered entertainment as well as popular education in elementary science and mathematics.[48]

But by the 1820s, sex-typing of the physical sciences was common, and arithmetic, physics, chemistry, and to a lesser extent astronomy were considered masculine.[49] Recent work has shown that nineteenth-century American academies taught mathematics and science to boys and girls, but around 1900, when girls began to outnumber boys in the American public schools, efficiency experts armed with IQ and interest tests were introduced in order to limit the student's training to his or her appropriate future. Since women were deemed unlikely to make much use of advanced high school mathematics, it was dropped from the curricula offered them. Social practices arose (such as asking "What is a nice girl like you doing in physics class?") that deterred many bright women from high school physics and steered them toward Latin, biology, or home economics. Similarly with the college curriculum, women were induced to think that they would be happier or more successful in the humanities or social or biological sciences than in the physical sciences.[50]

Since then, whole areas of educational research have been devoted to why students pick the majors they do or why in the course of their four years at college so many drop their initial intentions to major in physical sciences. Even when the American government was offering fellowships in

---

[48] Bernard de Fontenelle, *Conversations on the Plurality of Worlds,* introduction by Nina Gelbart (Berkeley: University of California Press, 1990); Teri Perl, "The Ladies' Diary or Woman's Almanack, 1704–1841," *Historia Mathematica,* 6 (1979), 36–53; Ruth and Peter Wallis, "Female Philomaths," *Historia Mathematica,* 7 (1980), 57–64.

[49] Patricia Cline Cohen, *A Calculating People: The Spread of Numeracy in Early America* (Chicago: University of Chicago Press, 1983).

[50] Kim Tolley, "Science for Ladies, Classics for Gentlemen: A Comparative Analysis of Scientific Subjects in the Curricula of Boys' and Girls' Secondary Schools in the United States, 1794–1850," *History of Education Quarterly,* 36 (1996), 129–53.

these very areas because the nation was having scientific manpower shortages, relatively few fellowships went to women. More than stereotyping was at work here; there was active disrecruitment in almost every physical science classroom.

Yet feminist philosophers have had little success in analyzing the gender components in the physical sciences. A few have tried or are trying. Meanwhile, anthropologist Sharon Traweek has published an ethnography of the Stanford Linear Accelerator in California and Ko-Enerugie butsurigaku Kenkyusho (KEK) in Japan in the 1980s, which describes a great deal of gender bias in the workplace and more importantly in the minds of the workers in both countries, though it manifests itself in different ways.[51]

In many ways, women's experience in the physical sciences has been the obverse of the usual history of physical sciences: There have been relatively few female physical scientists (unlike the many in the biological and social sciences), but a few of them, such as Marie Curie, are the best known of *all* scientists. Back in the seventeenth and eighteenth centuries when the sciences, including especially the physical sciences, were struggling to identify themselves, their methods, and their terrain, women were deliberately excluded from participation. They seemed to represent all that "science," whatever it was, was claiming not to be: Science portrayed itself as rational, unemotional, and logical. By the nineteenth century when many institutions had been created to embody these earlier masculine attitudes, women found that they had to fight to participate – in nearly every country and at every university. Even the victors were marginalized or ghettoized in segregated employment. Only the three Great Exceptions reached the highest levels and made important scientific and mathematical discoveries that have withstood subsequent attempts to drop even them from the historical record.

The fight for access was long but successful enough for a new cohort of younger women both to participate in World War I and then afterward to incur the attention, wrath, and brutality of the Nazis in the 1930s and 1940s. Since then, with women's liberation movements in many countries, women have been making progress in the physical sciences. Recently they have been doing best numerically and proportionally in socialist and Latin countries, but there, too, they have encountered a so-called glass ceiling or limitation on their advancement. Their failure during the last twenty-five years to make as much quantitative progress in the United States as have women in the biological, geological, and other sciences is also a cause for concern.[52]

---

[51] Sharon Traweek, *Beamtimes and Lifetimes: The World of High Energy Physicists* (Cambridge, Mass.: Harvard University Press, 1988). See also Robyn Arianrhod, "Physics and Mathematics, Reality and Language: Dilemmas for Feminists," in *The Knowledge Explosion: Generations of Feminist Scholarship*, ed. Cheris Kramarae and Dale Spender (New York: Teachers College Press, 1992), chap. 2.

[52] Mary Fehrs and Roman Czujko, "Women in Physics: Reversing the Exclusion."

4

SCIENTISTS AND THEIR PUBLICS
Popularization of Science in the Nineteenth Century

David M. Knight

In 1799 the Royal Institution was founded in London, in the wake of various provincial literary and philosophical societies; in 1851, under Prince Albert of Saxe-Coburg's aegis, the Great Exhibition attracted vast crowds to London, yielding profits to buy land in South Kensington for colleges and museums; and in 1900 the Paris Exposition heralded a new century of scientific and technical progress. There were prominent critics, but the wonders of science proved throughout the nineteenth century to be attractive to audiences of the aristocracy and gentry, of working men, and of everybody in between – which was fortunate, because in this world of competing beliefs and interests, of markets and industrial capitalism, those engaged in science needed to arouse the enthusiasm of people who would support them. Popularization started in Europe but was taken up in the United States, in Canada and Australasia, in India and other colonies, and in Japan.[1]

We shall focus upon Britain because of its place as the first industrial nation, where cheap books and publications emerged early, and scientific lectures were a feature of intellectual life. Specialization came relatively late to British education, so that until the end of the nineteenth century, university graduates shared to a great extent a common culture. Great Britain contained two nations, the English and the Scots, whose educational histories were very different; and Ireland was another story. Scotland had been, ever since its Calvinist Reformation, a country where education was valued and could be had cheaply in parochial schools and at the universities: It was throughout the eighteenth and nineteenth centuries an exporter of talent, to England, the Continent, and North America. Anglican England saw education as a privilege, and the English were also concerned that too much education would produce overqualified and unemployable people. In the face

---

[1] D. Kumar, "The Culture of Science and Colonial Culture," *British Journal for the History of Science* (hereafter *BJHS*), 29 (1996), 195–209.

of economic expansion in the Industrial Revolution, this perception gradually changed, and by 1850 the churches were giving an elementary education to most children. But there was a strong tradition of minimum government, of laissez-faire.

It was not until 1870 that compulsory state education was introduced, about a century after most other Western European states. At about the same time, provincial universities began to take off, the great stimulus being the Franco-Prussian War of 1870, where the better-educated Prussians defeated a France that had seemed the more formidable military power. Down to that date, the medieval universities of Oxford and Cambridge, with religious tests to exclude non-Anglicans and an ideal of "liberal education," had been dominant, despite a slowly growing challenge from the secular University of London formally chartered in the 1830s. Only at the very end of the century did universities in Great Britain get state funds.

Ireland, not strictly a colony, was throughout the nineteenth century part of the United Kingdom with England and Scotland; but its many problems meant that, like India, it was used as a laboratory for social experiments, notably the "Queen's University," with constituent colleges in different cities and (because of Catholic–Protestant tensions) secular syllabuses. Popular science was featured in Ireland, most notably in lively Dublin, but elementary education, especially in the impoverished countryside, was weak. In years of oppression and famine, countless Irish emigrated to Great Britain, North America, and Australia – usually to humbler jobs than those of the better-educated Scots. Overall, the British experience was unique, but not untypical.

The word "science" in English in 1800 covered all organized knowledge, whereas "arts" included manufacturing and engineering. The word "scientist" was coined by the Cambridge polymath William Whewell (1794–1866) in the 1830s, but it did not come into general use for half a century or so. It came to mean a specialist, a kind of professional, and by the early twentieth century, popularizing was rather despised, bringing no credit within a scientific community oriented toward research and perhaps formal teaching.[2] Popular writings were (and are) rated even below textbooks by scientific mandarins, and were often written by specialist writers, rather than by eminent scientists. It was different in the nineteenth century, when a scientific reputation – such as that of Humphry Davy (1778–1829), Michael Faraday (1791–1867), T. H. Huxley (1825–1895), Justus von Liebig (1803–1873), or Hermann von Helmholtz (1821–1894) – was enhanced by a capacity to get ideas across in public lectures or in essays.[3]

---

[2] D. M. Knight and H. Kragh, eds., *The Making of the Chemist* (Cambridge: Cambridge University Press, 1998); on textbooks and popular books, see A. Lundgren and B. Bensaude-Vincent, eds., *Communicating Chemistry* (Canton, Mass.: Science History Publications, 2000).

[3] D. M. Knight, "Getting Science Across," *BJHS,* 29 (1996), 129–38.

MAKING SCIENCE LOVED

The French Revolution of 1789 was identified not only with youth but also with science, liberating everyone from kingcraft and priestcraft. Terror, the execution of A. L. Lavoisier (1743–1794), war, and the rise of Napoleon led to revulsion from the left-wing dream, especially in Britain; and Joseph Priestley (1733–1804), the great advocate of science and reform, found himself driven into unhappy exile in the United States in the reaction of the 1790s.[4] France around 1800 led the world in science, but Britain led the world in technology: And just as the French needed men of science to help with the war effort – for instance, to supervise the recasting of church bells as cannon – so in Britain in these hungry years, agriculture as well as industry seemed ripe for scientific improvement.

In 1799 the American Tory Benjamin Thompson (1753–1814), created Count Rumford for his services to Bavaria, succeeded in getting Sir Joseph Banks (1743–1820) and other grandees to back his proposals for a Royal Institution promoting science in London.[5] In fashionable Albemarle Street, it would have lectures to interest and enthuse the opulent; a laboratory; and also classes associated with exhibitions of machinery to educate artisans. In January 1802, Davy made himself famous with a polished introductory lecture to his course on chemistry. Rumford departed for France with the short-lived Peace of Amiens in that year. Without him, the Royal Institution lost interest in the artisans (and the manufacturers who wanted to keep their machinery and processes secret anyway) and became a center for popular lectures of high caliber delivered to prominent men and women, whose membership fees supported a research laboratory. Throughout the century, here as elsewhere, the performers were male, the audience mixed.

Davy's sometimes colorful rhetoric was suited to his audience: Science depended upon the unequal distribution of property, but its application would bring great benefits to all of Britain's inhabitants.[6] These were not delusive dreams, like those of alchemical visionaries, his hearers could look forward to a bright day, of which they already beheld the dawn, as men of science (filled with reverence and with awe) penetrated to the bosom of the earth and searched the bottom of the ocean to allay the restlessness of their desires. Davy excited his hearers with reports of his research on tanning, fertilizers, geology, electrochemistry, and acidity; when he lectured in Dublin, there was a black market in tickets. The pattern changed over the years. Faraday started the Christmas Lectures for children, and eminent scientists were also invited to lecture accessibly about their own work, with

---

[4] B. Bensaude-Vincent and F. Abbri, eds., *Lavoisier in European Context* (Canton, Mass.: Science History Publications, 1995).
[5] M. Berman, *Social Change and Scientific Organization* (London: Heinemann, 1978), pp. 1–32.
[6] D. M. Knight, *Humphry Davy* (Cambridge: Cambridge University Press, 2d ed. 1998), pp. 42–56.

well-contrived demonstration experiments, on Friday evenings during the London season (the winter and spring). But the form of the Institution remained what Davy had made it, and it helped to ensure that science was seen as a part of high culture.

Chemistry, in the wake of Priestley and Lavoisier, promised both intellectual excitement and usefulness. These were picked up by Jane Marcet (1769–1858) in her *Conversations on Chemistry* (1807), written for girls who wanted to know more detail than they could acquire from lectures such as Davy's, and by Samuel Parkes (1761–1825) in his *Chemical Catechism* of the same date, written with boys in mind who would, like the author, work in a chemical trade. Parkes has extensive annotations, some of which amount to encomia upon the wisdom and goodness of the Creator; he was an enthusiastic Unitarian, and especially in Britain and the United States, natural theology was an important part of popular science.[7] Both these books sold well in successive editions throughout two decades.

THE MARCH OF MIND

In 1807, books were still a luxury item. They were hand printed and expensive, and they came in paper wrappers or in thin boards ready to be taken to a bookbinder (like the young Faraday). Illustrations made from copperplate engravings added to the price. But at this time, wood engraving on the hard end grain of boxwood, as in the popular natural histories of Thomas Bewick (1753–1828), made pictures much cheaper, and they could also (unlike copperplates) be set into the text. Wood engravings were durable, but for long runs, casts, called clichés, were made from them. For bigger pictures, lithographs drawn with wax crayons on stone, which was then wetted and inked with greasy ink, were much cheaper than engravings. From the 1820s, steam presses, stereotyping, wood-pulp paper (chemically bleached), and case bindings of decorated cloth made books much cheaper, better illustrated, and accessible to a mass market.

Although – especially in backward England – many people were still illiterate, there was growing demand for reading matter, and popular science appealed to those with an elementary education. Mechanics' institutes, for artisans like those dropped by the Royal Institution and its imitators, grew up in industrial towns and cities, offering lectures and libraries. Young men, like Faraday when an apprentice or Benjamin Brodie (1783–1862) – a future president of the Royal Society – as a well-connected medical student in London, joined less-formal self- improvement societies where science was prominent.[8]

---

[7] F. Kurzer, "Samuel Parkes: Chemist, Author, Reformer," *Annals of Science*, 54 (1997), 431–62.
[8] F. A. J. L. James, ed., *The Correspondence of Michael Faraday* (London: Institution of Electrical Engineers, 1991–), letters 3–29.

Among the elite, the Cambridge Philosophical Society brought together those in that churchy and conservative university who were interested in advancing mathematics and science. There was no place for women or fashion there, but everywhere, mind or intellect was seen to be on the march: Especially in Parisian and then German and London medical circles, interest in science went with contempt for the "Establishment" and a vision of a meritocratic future.[9] Parliamentary reform, achieved in part in Britain in 1832 and attempted all over Europe in 1848, went with this program and was associated with the increasing gathering and use of statistics. Inventions, such as Davy's safety lamp for miners, promoted a vision of science as something carried on by men of genius in the metropolis.[10] But while a conservative image of science was available, especially in connection with natural theology – and it would be difficult to overemphasize the importance of religion (especially in the Anglo-Saxon world) in the nineteenth century – there was again by the 1820s and 1830s a radical alternative, modernizing, view.

READ ALL ABOUT IT

Lorenz Crell (1745–1816) helped form the chemical community in eighteenth-century Germany with his journal *Chemische Annalen,* and Lavoisier disseminated his innovations through his *Annales de Chimie.*[11] These publications were aimed at scientific practitioners, but in Britain the *Philosophical Magazine* and *Nicholson's Journal* competed for a wider market of those interested in science and perhaps practicing it. Their chatty tone, with reviews and translations, octavo format, and cheaper crowded paper, contrasted with the august volumes published by the Royal Society; and they were commercial propositions, like most popular science. Other journals were published in Edinburgh and in Glasgow, mostly absorbed in the end by the *Philosophical Magazine,* which also formed a model for the *American Journal of Science* of Benjamin Silliman (1779–1864). In natural history, as well as the splendid *Transactions of the Linnean Society,* there were such popular publications as the *Magazine of Natural History,* whose editors encouraged controversy and published articles without the formality of peer review or refereeing. In some cases, as with *The Chemist* of 1824, a journal explicitly appealed to readers excluded from the genteel world; it mocked the pretensions of Davy in its first editorial, recommended cheap forms of apparatus, and paid contributors so that the editors could decide what topics should be covered. Not surprisingly, a journal freighted with such utopian hopes speedily sank.

[9] A. Desmond, *The Politics of Evolution* (Chicago: University of Chicago Press, 1989); T. L. Alborn, "The Business of Induction," *History of Science,* 34 (1996), 91–121.
[10] J. Golinski, *Science as Public Culture* (Cambridge: Cambridge University Press, 1992), pp. 188–235.
[11] M. P. Crosland, *In the Shadow of Lavoisier* (London: British Society for the History of Science, Monograph 9, 1994).

Books were also crucial. The Society for the Diffusion of Useful Knowledge began publishing during the 1820s, galvanizing the older rival Society for the Promotion of Christian Knowledge. In the 1830s Dionysius Lardner (1793–1859) edited a series of little books called The Cabinet Cyclopedia. These included a noteworthy *Preliminary Discourse* by John Herschel (1792–1871) – a discussion of scientific method by a great generalist and natural philosopher, who also contributed a *Treatise on Astronomy* – as well as other workmanlike volumes on the various sciences; and a curious set on biology by William Swainson (1809–1883), an advocate for the Quinarian System of classifying organisms in patterns of circles. Among the most successful publishers of information books were the Chambers brothers, William and Robert, in Edinburgh. Robert (1802–1871), in 1844, published anonymously his *Vestiges of the Natural History of Creation,* which became notorious for its evolutionary perspective (from galaxies to humans) and was very widely attacked, and read.[12]

CRYSTAL PALACES

The British Museum, founded in the eighteenth century, contained both beautiful historic artifacts and specimens of natural history, but it did not much welcome the general public and had no formal educational program. In contrast, in revolutionary Paris, the Museum of Natural History became a great center for research and lectures. Exhibitions and museums were a feature of the early nineteenth century, but the former were often of freaks and wonders, and the latter might be professional, like that at the Royal College of Surgeons in London. Learned societies held "conversaziones," open to members and their guests (including ladies), where objects, experiments, or devices of interest would be on display; but these were again a part of high culture.

As Europe emerged from the hungry forties, the threat of revolution lifting with economic boom, a Great Exhibition of the Works of all Nations in London was planned for 1851. Its most dramatic feature was its building: the Crystal Palace, an enormous glass house (enclosing large trees) put up in Hyde Park. Designed in nine days by Joseph Paxton (1801–1865), a former gardener's boy, when previous plans had been rejected and with only nine months to go before opening day, it was ready on time – an amazing feat of the railway age, assembled from accurately standardized components brought to the site from distant factories and coordinated there. The hugely successful exhibition drew orderly crowds from all over Britain and overseas to see the latest industrial and aesthetic creations: The only such exhibition so far to make a profit, it made palpable a vision of technical progress.

[12] R. Chambers, *Vestiges,* ed. J. Secord (Chicago: University of Chicago Press, 1995).

While Britain was clearly the leading industrial nation, perceptive commentators (including Henry Cole [1808–1882] and Lyon Playfair [1818–1898], the main organizers) saw the "American System" of mass production and interchangeable parts, and French industrial design, as signs that this pre-eminence was soon to end, and they urged better scientific and technical education. South Kensington, and comparable districts in other great cities like Berlin, developed into centers for both formal education and rational amusement – popular science in museums.[13]

Provincial cities, too, established museums of science, arts, and natural history, sometimes associated with collections of pictures and statuary and often founded in conjunction with a visit by a peripatetic Association for the Advancement of Science. Festivals and exhibitions depend upon ballyhoo and excitement, but museums have permanent collections, and their directors faced the difficult task of balancing the wants of casual visitors and of children with the needs of those undertaking research.

Natural history has always involved important collections of specimens. For museums of physical sciences, the problem became more acute as their exhibits of apparatus or machinery turned with the passage of time into collections of historic importance – hard to display excitingly and unavailable for hands-on play.[14] Visiting museums, which even in Sabbatarian countries like England opened on Sundays, was an important and improving leisure activity in the earnest nineteenth century. Architecturally they came to resemble classical temples dedicated to the Muses, or gothic cathedrals, thus representing classical order or spiritual aspirations. The scientists of the nineteenth century were the heirs, after all, of both the Enlightenment and the Romantic Movement, and a kind of pantheism or nature worship came easily to them. Museums might be associated with libraries and with botanic and zoological gardens dedicated to classifying plants and animals and "acclimatizing" them: transferring merino sheep to Australia, rubber trees to Malaysia, quinine to India, and so on.[15] These benefits of science, at which we sometimes now look askance, were lauded as great improvements to the world.

THE CHURCH SCIENTIFIC

The word "scientist" was coined in a discussion at the Cambridge meeting of the British Association for the Advancement of Science in 1833. Reflecting on his confrontation with Samuel Wilberforce (1805–1873) at

---

[13] S. Forgan and G. Gooday, "Constructing South Kensington," *BJHS*, 29 (1996), 435–68; John R. Davis, *The Great Exhibition* (Stroud, England: Sutton Publishing, 1999).

[14] N. Jardine, J. Secord, and E. C. Spary, *Cultures of Natural History* (Cambridge: Cambridge University Press, 1996), pt. III; A. Wheeler, "Zoological Collections in the Early British Museum," *Archives of Natural History*, 24 (1997), 89–126.

[15] H. Ritvo, *The Platypus and the Mermaid* (Cambridge, Mass.: Harvard University Press, 1997), pp. 1–50.

the Oxford meeting of 1860, Huxley declared that had a Council of the Church Scientific been called then, it would probably have condemned the Darwinian heresy. He had in mind a meeting of academicians and professors, like the bishops and abbots who attended the Vatican Council of 1869–70; his metaphor is striking, because science did develop rather like a religion, with a clerisy addressing laymen at evangelistic meetings like those of the BAAS.[16] Their presidents and councils came to join those of academies as exponents of the scientific point of view, with access to government and the media.

The British Association did not meet mainly in famous old university cities but all around the British Isles and even in Canada, South Africa, and Australia.[17] It was not the first peripatetic body; its model was from Germany, a constellation of large and small states until the empire was formed in 1870, and even to some extent after that. In the 1820s, Lorenz Oken (1779–1851) organized annual meetings of *Naturforscher,* each year in a different state; after all, there was then no national capital like Paris or London. After some initial unease, the various governments came to welcome the men of science and to compete culturally – in their universities, opera houses, and hosting of such meetings – thus popularizing science for their citizens.

Foreigners were also welcome, and some who went from Britain were much impressed, seeing the opportunity to wrest science from the effete grasp of Londoners and place it in the strong hands of provincials and Dissenters. That was not quite what happened, although sometimes a provincial amateur, such as James Joule (1818–1889), succeeded in getting the eminent to listen to his work on thermodynamics. But the meetings, which began at York in 1831, proved very popular and attracted large crowds of men and women. Cities competed to attract them, offering both to host civic receptions and to build a museum or other scientific institution; and local societies for astronomy, natural history, or other sciences were duly promoted. People could see and hear Faraday and Huxley in the flesh, rather than just read about them; and sometimes there were angry debates – good to watch – which proved that science was not just a dispassionate exercise of reasoning upon facts, as Baconian apologists would have it. The Association, in its turn, became a model for those in the United States, France, and Australasia.[18]

The sublime science of astronomy had a large amateur following, though a telescope was a large investment; and for the working class, natural history had the advantages of cheapness, sociability, and fresh air. Field trips, and sessions perhaps in a room above the public bar, went with the identifying of species, at

---

[16] A. Desmond, *Huxley: Evolution's High Priest* (London: Michael Joseph, 1997); P. White, *Huxley* (Cambridge: Cambridge University Press, forthcoming).

[17] J. Morrell and A. Thackray, *Gentlemen of Science* (Oxford: Clarendon Press, 1981).

[18] S. G. Kohlstedt, *The Formation of the American Scientific Community* (Urbana: University of Illinois Press, 1976); R. MacLeod, ed., *The Commonwealth of Science* (Oxford: Oxford University Press, 1988); R. W. Home, ed., *Australian Science in the Making* (Cambridge: Cambridge University Press, 1988).

which members sometimes became very expert.[19] In both these fields, the gap between the advancement of knowledge and popular science became blurred: Great observatories were restricted by their long-term research programs, and any careful observer with a telescope might see some new planet swim into his ken or, anyway, a comet. In 1820 the Royal Astronomical Society was formed, one of the earliest to be concerned with a physical science; and it flourished, bringing together people with a wide range of interests.

DEEP SPACE AND TIME

By the later eighteenth century, there were no significant believers in the Aristotelian or Ptolemaic world, with the Earth at its center; the vast spaces that had frightened Pascal had come to be accepted. Great reflecting telescopes, like that of William Herschel (1738–1822) at Windsor, the six-foot mirror of Lord Rosse (1800–1867) in Ireland, and then the giant telescopes in the United States, enabled the heavens to be gauged, revealed spiral nebulae, and made our planet feel even smaller. We can see this in popular books: Herschel's *Astronomy* in Lardner's series was a solid but unmathematical read, unrelieved by pictures or invocations of sublimity. J. P. Nichol (1804–1859), on the other hand, published in 1850 a magnificent volume, *The Architecture of the Heavens,* with dark-ground plates of Rosse's discoveries and allegorical illustrations by the Scottish painter David Scott. Robert Ball (1830–1919) of Dublin published in 1886 his *Story of the Heavens,* which was strikingly illustrated; and the writings of Richard Proctor (1837–1888), especially his *Half-hours with a Telescope,* 1868, were beautifully clear and sold extremely well. Proctor left Britain, settling in America, and his output was popular on both sides of the Atlantic. He, and earlier Thomas Dick (1774–1857) in his *Sidereal Heavens* of 1840, argued for a plurality of inhabited worlds.

The idea that God would have put inhabitants only on the Earth, given a vast universe, seemed absurd in the midcentury. Only Whewell emerged as a prominent opponent of the idea, as earlier he had been critical of the "deductive" arrogance of P. S. Laplace (1749–1827) who had no need of God. Whewell feared that (as in *Vestiges*) those who supported plurality would have to deny the special status of mankind so crucial to Christianity, and accept some kind of evolutionary picture in which life emerged from inorganic matter whenever and wherever the time was ripe.

In 1874 the BAAS met in Belfast, and John Tyndall (1820–1893), who was president, took the opportunity not simply to dilate upon science and its possibilities but to present a worldview based upon atomic theory, luminiferous ether, and Darwinism, which among them would account for everything. This caused an immense scandal: His program of wresting the whole of

---

[19] A. Secord, "Artisan Botany," in Jardine, *Cultures,* pp. 378–93.

cosmology from the clergy was denounced from the pulpits of Belfast and elsewhere. The Belfast Address, an eloquent appeal, it seemed, for a materialistic worldview, was very widely read and commented upon – and disliked by mandarin physicists, such as Lord Kelvin (1824–1907) and Maxwell, who disdained Tyndall's windy popularizing rhetoric.

Astronomical observations were crucial for determining longitude and latitude as the wide-open spaces on Earth were being formally and scientifically explored. Accounts of the voyages and travels of James Cook (1728–1779), Galaup de la Pérouse (1741–1788), P. S. Pallas (1741–1811), Matthew Flinders (1774–1814), Meriwether Lewis (1774–1809) and William Clark (1770–1838), and many others aroused great enthusiasm and sold well; and the objects they brought back swelled collections of natural history and ethnography. Scientific academies in France, Britain, Russia, the United States, and other countries promoted expeditions, so that areas hitherto blank on the map were gradually filled in, coastlines and estuaries charted, and magnetic data collected and mapped – maps and atlases seen as both high-level and popular science. Alexander von Humboldt (1769–1859) introduced thematic maps, which could, for example, represent isotherms, in writing up his Latin American journeys.

Humboldt's books were very popular with armchair travelers, who relished his enthusiastic prose and scientific accuracy, and were an exemplar for the young Charles Darwin (1809–1882).[20] His voyage on HMS *Beagle* was one in a great international series of projects, the scientific results of which could be accessibly presented to a public hungry for such things. Such reports often led to missionary activity, which saw a tremendous boom in the nineteenth century, as well as to colonization, by design or sometimes almost by accident, as naval or army officers of European nations assumed powers to pacify and govern those they deemed incapable of governing themselves.[21] The inhabitants and raw materials of these colonies then interested their new masters, governments, and peoples in Europe, who might also from time to time become excited and angry about injustices committed in their name in distant lands. Colonies were always controversial.

Deep time was also controversial.[22] When the eminent surgeon James Parkinson (1755–1824) began publishing his three-volume *Organic Remains of a Former World* in 1804, his frontispiece with Noah's Ark, a rainbow, and some fossil creatures (which had missed the boat and become extinct) was already out of date. His later volumes took into account the researches of Georges Cuvier (1769–1832), who had found a series of faunas beneath the hill at Montmartre, demonstrating that a single flood could not account for extinction. A longer time scale than the seventeenth-century Irish Archbishop Ussher's, in which the world began in 4004 B.C., was required; this, and the

[20] J. Browne, *Charles Darwin: Voyaging* (London: Cape, 1995), pp. 236–43.
[21] M. T. Bravo, "Ethnological Encounters," in Jardine, *Cultures*, pp. 338–57.
[22] M. J. S. Rudwick, *Scenes From Deep Time* (Chicago: University of Chicago Press, 1992).

reconstruction of extraordinary fossil creatures, was a source of enormous excitement.

Numerous authors took literalist, liberal, or what we could call agnostic lines, and indeed, one of the functions of the BAAS had been to recognize geologists as scientists and to protect them from supposedly ignorant attacks. The Geological Society of London was famous for its debates, whereas other societies did their best to stifle controversy or keep it behind closed doors. Geology also depended on visual language: Buckland's *Bridgewater Treatise* (1836), demonstrating the goodness and wisdom of God, had pictures of dinosaur tracks and also a handsome colored fold-out plate illustrating the Earth's history through the geological epochs and their characteristic species.[23] Illustrations of extinct animals and plants began to exert their uncanny fascination, as "dragons of the prime that tare each other in their slime" moved the imagination of Alfred Tennyson (1809–1892).

Deep time thus became familiar, but actually thinking in terms of millions of years, like millions of miles, was and is not easy. And the ancestry of man was an explosive topic: Were we just animals? Were some peoples more akin to apes than others? Our dignity and morality were threatened; hairy, stooping, grunting ancestors who had made their way in the struggle for existence did not worry Huxley, whose book *Man's Place in Nature* (1862) was a great feat of popularization – but many were uneasy.[24]

Huxley found himself locked in controversy with Kelvin over deep time. Darwinians assumed that they could extrapolate from changes in river deltas and exposed coastlines over hundreds of years to the raising and erosion of rock formations over hundreds of millions of years. Kelvin reminded them of the laws of thermodynamics. He computed the age of the Sun, assuming that it was composed of the best-quality coal, and was also getting energy from meteor collisions and gravitational collapse. Making the most favorable assumptions, this led to an age of around a hundred million years. Then he computed the age of the Earth, assuming that it was slowly cooling and applying the mathematics he had picked up from J. B. J. Fourier (1768–1830) on heat flow. This led to a comparable figure; and physicists are always delighted to find two lines of reasoning concordant. Kelvin took some pleasure in reminding brash colleagues like Huxley that physicists could quantify; and his addresses, originally delivered in the late 1860s, were republished in 1894 in his *Popular Lectures and Addresses*. Darwinians could only reply that natural selection must work faster than they had thought, or that something was perhaps wrong with the calculations. When the latter were found to be right, with the discovery of radioactivity, geophysics was set back for a generation.

---

[23] N. A. Rupke, " 'The End of History' in the Early Pictures of Geological Time," *History of Science*, 36 (1998), 61–90.

[24] A. P. Barr, ed., *Thomas Henry Huxley's Place in Science and Letters* (Athens: University of Georgia Press, 1997).

BEYOND THE FRINGE

The reconstruction of the fossil record was respectable science, and Darwinian evolution became so despite resistance. But right through the nineteenth century, as before and since, there were would-be sciences that often attracted enormous public attention but never achieved the magical status of *scientia*. Indeed, popular science always includes such features, despite the efforts of professionals to purge them away and get the public interested exclusively in those questions with which professors are concerned. In the late eighteenth century, Anton Mesmer (1734–1815) had sent people into trances by passing magnets over them, and animal magnetism, or mesmerism, became a matter of furious controversy and enormous interest first in Vienna and then in Paris. A committee of the French Academy of Sciences, including Benjamin Franklin (1706–1790), established that magnetism was not involved and dismissed the whole phenomenon, but mesmerists continued unabashed throughout the nineteenth century.

Electricity and magnetism were also popular features of the alternative medicine of the early nineteenth century. Established therapies were never very effective (opium, quinine, and alcohol were said to comprise the doctor's armory), and whatever orthodox practitioners might say, desperate diseases demanded desperate remedies.[25] Many people were thus attracted (like Darwin) to water cures and to homeopathy, with its principle that minute doses of what caused a disease would cure it. And in the first decades of the nineteenth century, another new science appeared: craniology, or phrenology, the study of the bumps on the head. Starting again in Germany with F. J. Gall (1758–1828), it spread to France, and his disciple J. G. Spurzheim (1776–1832) brought it to Britain. The crucial idea was that the baby's skull was soft and took up the form of the brain beneath. The faculties were located in different regions of the brain, and correlating bulges and concavities in the cranium with strengths and weaknesses in mind would make it possible to read character. Especially in Edinburgh, with its great medical school and educational tradition, the science caught on, and a society and journal were founded.[26]

The founders hoped that phrenology would speedily be incorporated into the medical curriculum, but a murderer was found to have a big bump of benevolence, and for most, the science became a parlor game. For the widely read educationalist George Combe (1788–1858), however, it was essential for teachers assessing the capabilities of pupils; it was also taught to artists and was popular in mechanics' institutes – the language of "bumps" entered the language, though the science never entered the pantheon.

---

[25] E. Shorter, "Primary Care," in *The Cambridge Illustrated History of Medicine*, ed. R. Porter (Cambridge: Cambridge University Press, 1996), pp. 118–53.
[26] R. Cooter, *The Cultural Meaning of Popular Science* (Cambridge: Cambridge University Press, 1984); L. J. Harris, "A Young Man's Critique of an 'Outré' Science: Charles Tennyson's 'Phrenology,'" *Journal for the History of Medicine & Allied Sciences*, 52 (1997), 485–97.

More mysterious were the auras that Karl, Baron von Reichenbach (1788–1869) detected around "sensitive" persons – usually women, and especially pregnant women. These auras also could be seen around magnets and crystals. He was a chemist by training and practice and an expert on meteorites, and his book was translated into English by William Gregory (1803–1858), professor of chemistry in Edinburgh and the translator also of works by Liebig, which we would consider mainstream science. A curious substance called "odyle" was responsible for the manifestations and played a very important role in the economy of the universe.

Nineteenth-century credulity was mocked, for example, by Charles Mackay (1814–1889) in *Extraordinary Popular Delusions* (1841, with a new edition in 1852); but by the 1850s, a new craze had reached Europe from America – spiritualism. In semidarkness, tables rocked, ouija boards spelled out messages, and mediums might levitate or emit ectoplasm taking the form of somebody deceased. Mediums were usually female, and séances provided opportunities (generally unavailable in Victorian England or New England) for holding not merely hands but also arms and legs. These phenomena engaged the interest (intellectual and emotional) of various men of science, especially in Britain and usually after a bereavement. William Crookes (1832–1919) concluded from various experiments that new forces of nature had been revealed, but his paper submitted to the Royal Society's journal was rejected after a row, and he had to publish it in a more popular periodical.[27]

In 1882 the Society for Psychical Research was founded under the aegis of Henry Sidgwick (1838–1900), the amazingly well-connected Cambridge philosopher.[28] Sidgwick was a notable but reluctant agnostic, who had resigned his post because he could no longer subscribe to orthodox Christianity and hoped that if survival after death could be proved, then religion would be put onto a firmer basis. The Society included two future presidents of the Royal Society, Crookes and J. J. Thomson (1856–1940), and two prime ministers, J. H. Gladstone (1809–1898) and Arthur Balfour (1848–1930), as well as William James (1842–1910) and various bishops and professors. We would have to say that in the years around 1900, psychical research counted as respectable science; and certainly phantasms, hauntings, and mysterious happenings were soberly investigated by empirically minded men and women.

It seemed that more often than could be easily put down to chance, people saw a phantasm of someone they loved who was at that moment in mortal danger, and telepathy sometimes really seemed to happen. After all, radio waves, cathode rays, and x rays were just being investigated; the world was more perplexing than Tyndall had dreamed of in Belfast. Psychical research was a field in which there was nothing deep and recondite, where the common sense of ordinary people might be more appropriate than the learned

---

[27] H. Gay, "Invisible Resource: William Crookes and his Circle of Support," *BJHS*, 29 (1996), 311–36.
[28] J. Oppenheim, *The Other World* (Cambridge: Cambridge University Press, 1985); D. M. Knight, *Science in the Romantic Era* (Aldershot, England: Ashgate Variorum, 1998), pp. 317–24.

ignorance of trained scientists; and thus it was popular. The accounts of phantasms and ghosts give extraordinary glimpses into the lives of our ancestors, their dangerous travels, and their sudden deaths, as well as the assumptions they made about whose testimony was trustworthy and whose was not. Just as extraordinary stories about visitors from outer space, reincarnations, and miracle cures arouse more excitement in our day than orthodox and intellectually demanding science and medicine, so in the nineteenth century the various fringe sciences claimed a giant's share of attention.

A SECOND CULTURE?

Davy, and later Faraday, Huxley, and Tyndall at the Royal Institution, presented science as a part of high culture, where the imagination of the man of genius was kept under control by experiment, rather as the poet's was by the exigencies of meter and rhyme. It was not too arcane; science was trained and organized common sense, as Huxley famously put it – both adjectives being important. Davy wrote poetry, admired in his day, as Erasmus Darwin had done at the end of the eighteenth century, Davy's verse being effusive and romantic rather than didactic. In the early nineteenth century, there was no professional science, and thus no "culture," no scientific community with its shared education and values to set against the literary culture, as C. P. Snow (1905–1980) did in his controversial lecture on "the two cultures" amid educational debates in the 1950s.

For Matthew Arnold (1822–1888), Victorian aristocrats were "barbarians," hunting and fighting, while those involved in industry and commerce were smug "philistines," uninterested in cultural activity unless it was safely domesticated. As industrial revolutions opened new avenues of social mobility, those who lacked the familiarity with literature, music, painting, and sculpture that went with inherited wealth sought in science – especially astronomy and natural history – something beyond mere business. Snow found that in the mid-twentieth century, scientists found solace in music rather than literature or the visual arts. If that was true then, or is true today, it was not so in the nineteenth century. Helmholtz wrote a famous work, *Sensations of Tone*, about the physics of music, which was accessible to musicians and remains a classic; but he also studied and wrote popularly on color and our perception of it.[29] Chemists like Davy worked on pigments ancient and modern, while physicists like John Herschel and Maxwell wrote poetry.

Science was prominent in some nineteenth-century poetry, most notably Tennyson's *In Memoriam*, which gave us the haunting phrase "nature red in tooth and claw" and memorable stanzas about geological time. Tennyson had picked up his knowledge from reading Lyell and *Vestiges* – his readers would have become aware of current scientific thinking, partly as a threat, in

[29] D. Cahan, ed., *Hermann von Helmholtz* (Berkeley: University of California Press, 1993).

reading the poem.³⁰ Huxley considered *In Memoriam* an example of scientific method and admired Tennyson's other writings also.

Science is similarly to be found in women's writing: in *Frankenstein* by Mary Shelley (1797–1851) and in *Middlemarch* by George Eliot (Marianne Evans, 1819–1880), who had previously translated, from the German, rationalistic works by David Strauss and Ludwig Feuerbach. Mary (Mrs. Humphry) Ward's (1851–1920) best-selling novel about religious doubt, *Robert Elsmere*, 1888, given away to promote soap in what must have been a very literate America, contains surprisingly little science. The hero's faith is chiefly undermined by historic doubts, rather than concern about miracles, but science is in the background, and the book created an enormous furor following upon a review of it by Gladstone.³¹

Reviews were prominent in the intellectual life of the nineteenth century.³² Indeed, they were the main humanistic journals until historical, literary, and philosophical publications on the lines of scientific periodicals appeared late in the century. In Continental Europe, eighteenth-century reviews made thought in one language accessible in another. In Britain, the *Monthly Review* consisted of book reviews that were essentially paraphrases or lengthy quotations – the object was to convey the writer's style and conclusions, and critical appraisal was generally secondary. The *Edinburgh Review* changed all that: Its articles, written from a Whig viewpoint, were trenchant commentaries of twenty or thirty closely printed pages on books of all kinds, including scientific works, monographs, textbooks, and even issues of journals. They are what we would call essay-reviews, written for the well-informed but unspecialized reader; and sometimes the essayist would go off on a tangent, so that the book reviewed became a point of departure, as with Henry Holland (1788–1873) discussing "Modern Chemistry" in the rival *Quarterly Review* in 1847. The *Quarterly* was Tory; the *Westminster*, radical; and the *North British* represented the Free Presbyterian Church of Scotland.

Whatever their political or religious stance (and the two generally went together in Britain), these quarterlies would normally have at least one essay in every issue concerned with science or technology. Contributions were anonymous, and so editors could amend them (though they did this at their peril if they blue-penciled an eminent author), and reviewers could speak their minds in the small intellectual world of the day – when authorship (as with Samuel Wilberforce's essay on Darwin) was in fact often an open secret. They were an expression of high culture, often outspoken in criticism when dealing with literature (attacks on William Wordsworth and John Keats are

---

³⁰ A. J. Meadows, "Tennyson and Nineteenth-Century Science," *Notes and Records of the Royal Society*, 46 (1993), 111–18.
³¹ J. Sutherland, *Mrs Humphry Ward* (Oxford: Oxford University Press, 1990).
³² J. Shattuck and M. Wolff, eds., *The Victorian Periodical Press: Samplings and Soundings* (Leicester, England: Leicester University Press, 1982).

notorious) or religion, but usually respectful about science, seeing a duty in getting the latest ideas across without jargon or excessive detail.

The question was whether this was enough by the 1870s. In his monthly *Nineteenth Century* from 1877, James Knowles (1831–1908) provoked lively debates with signed articles; among his coups was bringing Huxley and Gladstone into public conflict about science.[33] But in 1864, Crookes played an important part in launching the *Quarterly Journal of Science*. This was to be a kind of review, devoted to science and appearing at a time when specialization meant that those active in one science did not necessarily understand what those in other fields were up to. They thus needed up-to-date popular writing, just as much as those outside the scientific community. But this journal, which went monthly in 1879, was superseded by weeklies, such as Crookes's *Chemical News* and *Nature*, edited by Norman Lockyer (1836–1920), which brought prestige, but not money, to Macmillan, its London publisher. By the end of the century, one can speak of a scientific "culture."

TALKING DOWN

Textbooks and works of popular science were written by notable researchers, such as Huxley, Tyndall, and Kelvin, but increasingly such writing came to be seen as a distinct activity with its own particular skills. Huxley hoped in his popular lectures to convey "scientific method," and with it, in his case, an agnostic attitude toward anything dogmatic or metaphysical.[34] In his wake, scientism – the idea that only empirical scientific explanations are genuine – gained ground, especially among popular writers, to the distaste of fastidious prominent scientists, who often then (as since) retained or found religious belief and metaphysical interests. Thus, Balfour Stewart (1828–1887) and P. G. Tait (1831–1901) popularized thermodynamics in their *Unseen Universe*, which was also a work of religious apologetics, while Balfour's philosophical writings were designed to establish that science, like everything else, rested upon belief.

Darwin's cousin Francis Galton (1822–1911) studied the careers and relationships of scientists (and other eminent men) as a contribution to the long-running "nature or nurture" debate. He, as an adherent of "scientific naturalism," also investigated the efficacy of prayer, comparing the life spans of the royal family, often prayed for in church, with those of aristocrats; there was no difference. Popular writers, such as Jules Verne (1828–1905) in France, revived the genre of science fiction to present a picture of high adventure amid technical progress.

---

[33] P. Metcalf, *James Knowles* (Oxford: Clarendon Press, 1980), pp. 274–351.
[34] B. Lightman, *The Origins of Agnosticism* (Baltimore: Johns Hopkins University Press, 1987), pp. 7–15.

Faith in science was on the increase as death rates fell, with scientific medicine at last making a real impact and religion seeming to be fuddy-duddy and old-fashioned. The public turned to journals, including the *Popular Science Monthly*, the *Scientific American*, the *English Mechanic*, and *Science Gossip*.[35] Self-improvement and an interest in nature were now also expressed in magazines with a technological bent, accompanied by advertisements. Optimism was everywhere.

Thermodynamics, however, was delivering another message: that the Sun could not burn forever, and that the Earth was steadily cooling down. In a few tens of millions of years, according to Kelvin's calculations, life here will have become impossible, and all the achievements of mankind will have turned to dust.[36] This idea was taken up by H. G. Wells (1866–1946) in his novel *The Time Machine*, in which the time traveler going forward finds that the human race has evolved into two species (one from effete aristocrats, the other from ferocious proletarians), and then further on that all intelligent life has disappeared from the cooling Earth. A deep pessimism about science and technology similarly permeates the novels of Thomas Hardy (1840–1928).

A fascination with degeneration and degradation was thus allied with the sciences in the popular mind, leading to widespread anxiety about whether disorder in society, as in the physical world, was inevitably increasing. Darwinian development, too, was not necessarily progressive, and for Cesare Lombroso (1836–1909) and his many popular echoes, criminals and the unintelligent represented throwbacks to primitive ancestors. All the gains of civilization might be lost in atavism. Galton was a pioneer of eugenics, hoping to promote good breeding by ensuring that the more intelligent had larger families than the foolish and improvident.[37] Such ideas, commonplace in the opening years of the twentieth century, were acted upon by governments, democratic as well as dictatorial, who sterilized the unfit: Popular science could issue in policy.

SIGNS AND WONDERS

By 1800 newspapers had been around for a long time, but the coming of cheap paper and steam presses, and the lifting of "the tax on knowledge" to which they had been subject, meant that Britain was early in the field of mass-circulation papers. The building of the railway system, and the electric telegraph that developed hand in hand with it, meant that national newspapers carrying up-to-date international material became ever more

---

[35] R. Barton, "The Purposes of Science and the Purposes of Popularization," *Annals of Science*, 55 (1998), 1–33.
[36] C. Smith and N. Wise, *Energy and Empire* (Cambridge: Cambridge University Press, 1989), pp. 524–645.
[37] J. Pickstone, "Medicine, Society and the State," in Porter, *Medicine*, pp. 304–41.

important. What newspapers have always wanted were stories, though they were prepared to carry rather dull information as well. Sometimes the sciences provided excitement, although the most famous case was a hoax: John Herschel had gone to South Africa in 1833–8 to observe the southern stars, and a New York newspaper reported that he had seen inhabitants on the moon.

This duly boosted sales, but usually newspapers had to rely upon events such as the meetings of the British Association or major exhibitions to get something newsworthy. Even so, the debate between Huxley and Bishop Samuel Wilberforce at Oxford in 1860 was not properly reported because it happened on a Saturday afternoon when the main BAAS meeting was over and the reporters had gone home. Accounts of lectures, the opening of new buildings, real or imagined medical advances, and obituaries of men of science occupied an important place in newspapers. Huxley's review of the *Origin of Species* appeared in the London *Times* and was important in making the book known, and Faraday's letter to the *Times* exposing table turning was another celebrated landmark. The more popular newspapers usually carried less science. Armaments and innovations therein, the ironclad warship, the breech-loading gun, gun-cotton, and other explosives duly got into the news, as also did pollution from sewage and chemical works and accounts of vivisections. Popular stories about science were not all positive.

As well as newspapers there were magazines. *Punch*, with its lighthearted editorial matter and its cartoons, did get across aspects of science, especially Darwinism and our relationship with monkeys. The caricatures of the eminent (including leaders of science) in *Vanity Fair* were and are much prized; they were kinder than the caricatures of Priestley, Banks, Davy, and others around 1800. Wood engraving, lithography, and photography (often combined) meant that pictures became increasingly prominent; the slabs of text characteristic of newspapers and magazines in the early years of the century gave way to a livelier look. And science got in because of its importance, and sometimes its aesthetic quality.[38]

Science in the 1790s was harmless, perhaps useful, its image somewhat tarnished by memories of projectors and by association with revolution. By 1900 it was formidable, playing a major part in education and in economic life, for the equation of technology with applied science was accepted by readers of popular science. At the Paris Exposition of 1900, electricity, now providing the energy that was recently proved to underlie matter, was the great novelty.[39] Crowds flocked again to innovations, hoping that science would usher in a new century of peace and progress. The wonders of science were

---

[38] L. P. Williams, *Album of Science: the Nineteenth Century* (New York: Scribner's, 1978).
[39] R. Brain, *Going to the Fair* (Cambridge: Whipple Museum, 1993); R. Fox, "Thomas Edison's Parisian Campaign," *Annals of Science*, 53 (1996), 157–93.

there as before (there were even tribesmen in exotic villages), but brought up to date as the world hustled down the ringing grooves of change. It was a splendid spectacle; the nineteenth century had been an age of science, and the twentieth would be even more so. As we know, first the *Titanic* disaster revealed the dangers of hubris, and then between 1914 and 1918, in World War I ("the chemists' war"), developments in aircraft and poisonous gas proved both the alarming power of science and society's need for it.

5

LITERATURE AND THE MODERN PHYSICAL SCIENCES

Pamela Gossin

Richard Feynman (1918–1988) loved to tell the story of his close encounter with poetry while a graduate student in physics at Princeton. Sitting in on a colloquium in which "somebody" analyzed the structural and emotional elements of a poem, Feynman was set up as an impromptu respondent by the graduate dean, who was confident that the situation would elicit a strong reaction. To the literary scholar's inquiry, "Isn't it the same in mathematics . . . ?" Feynman was asked to relate the problem to theoretical physics. He tells us about his reply:

> "Yes, it's very closely related. In theoretical physics, the analog of the word is the mathematical formula, the analog of the structure of the poem is the interrelationship of the theoretical bling-bling with the so-and-so" – and I went through the whole thing, making a perfect analogy. The speaker's eyes were *beaming* with happiness.
>
> Then I said, "It seems to me that no matter *what* you say about poetry, I could find a way of making up an analog with *any* subject, just as I did for theoretical physics. I don't consider such analogs meaningful."[1]

Like other anecdotes in Feynman's memoirs, this story – in both its enactment and retelling – is framed upon a frequently recurrent motif of clever one-upmanship that displays several constituent characteristics of his psychology and personality. The special notice he takes of the smile he is about to wipe off the speaker's face participates in the kind of intellectual sadism that

---

[1] Richard P. Feynman, *"Surely You're Joking, Mr. Feynman!"* (New York: Bantam, 1989), p. 53. The standard reference tool for literature and science studies is *The Relations of Literature and Science: An Annotated Bibliography of Scholarship, 1880–1980*, edited by Walter Schatzberg et al. (New York: Modern Language Association, 1987). Since 1993, annual bibliographies appear in *Configurations: A Journal of Literature, Science, and Technology*. The *Encyclopedia of Literature and Science* (edited by Pamela Gossin, forthcoming, Greenwood Press) contains seven hundred entries on the interrelations of literature and science, approximately one-fifth of which will treat the interrelations of literature and the physical sciences. For support of my work, I express gratitude for a Research Fellowship in the History of Science funded by the George and Eliza Gardner Howard Foundation (Brown University).

Feynman later enjoyed as the perpetrator of elaborate practical jokes, some with near life-and-death consequences. His killing deflation of the unnamed poetry scholar (had he been a physicist, would his name have been more memorable?) may indicate the depth of Feynman's uneasiness around the "fancy" artistic pursuits that he admittedly perceived as less masculine and, therefore, less admirable and worthwhile than mechanical abilities and blue-collar occupations. Although he detects the speaker's eagerness to locate common ground between humanistic and scientific endeavors, Feynman's comment resists that objective. His expression of general disdain for the subjectivity and apparent arbitrariness of literary knowledge and poetic interpretation responds more directly to the unexpressed attitudes and expectations of the audience he hopes most to impress, namely, the other scientists present.

Still and all, Feynman had curiosity enough about the humanities to attend the literary talk. In graduate school, and in later life, he made conscious efforts to seek out opportunities to explore unfamiliar scientific disciplines, as well as philosophy, music, and art. Many of his stories express concern about his negative attitudes toward the arts, offer possible explanations for why he developed them, and recount the ways he went about testing their validity and reforming them. Whatever his youthful reactions toward the humanities as intellectual disciplines, literary and artistic expression *were* central to his own creative endeavors, including his eccentric extended investigations into human behavior and his search for means alternative to mathematics for encapsulating and communicating his understanding of nature. Ironically, five pages after his declaration that he finds abstract analogies between literature and science devoid of meaning, Feynman employs a practical comparison between himself and Madame Bovary's husband in order to convey the significance of an instance in which his enthusiastic, but amateur, approach to scientific research failed and what he learned from the failure. Indeed, throughout his memoirs, Feynman self-consciously describes analogy building as essential to his analytical approach to physics itself. He recognizes also that analogies are essential and powerful components of his much-heralded lectures and famed teaching.

In many ways, Feynman's individual experience is emblematic of the larger complex of uneasy cultural relations between literary scholars and scientists – public and professional tension broken intermittently by direct antagonism; private recognition and eclectic exploration of commonalities of intellectual processes, practice, and expression. Feynman's self-education models an important means by which members of one "culture" can overcome personal, social, and professional prejudices and develop an appreciation for the "other." As a master storyteller (in several senses of the word), Feynman recognized that the creative arts, music, literature, and science all participate in the common endeavor of telling stories about the universe. Whatever their discipline, practitioners engaged in the process of investigating, recording, and disseminating their observations and discoveries about natural phenomena share a vital need to experiment with their language of choice – whether artistic, poetic, or mathematical. As the following account suggests, Feynman has been far from alone in his attempts to explore the interrelations of literature and the modern physical sciences.

TWO CULTURES: BRIDGES, TRENCHES, AND BEYOND

Virtually any discussion today of the interrelations of literature and science still necessarily reflects the wave of influence generated by the notion of "two cultures." Perhaps having originated in attitudes recorded in texts as early as Plato's *Republic,* philosophical arguments regarding the relative virtues and values of literature and science as ways of knowing and as modes of expression oscillate across Western intellectual tradition, often in tandem with equally powerful conceptions of their essential unity. In the early modern period, Renaissance men and women of letters and sciences nonetheless distinguished between the fictive and factual elements of their intellectual pursuits. Isaac Newton personally eschewed poetry, yet Newtonian science demanded the muse. British Romantic poets proposed toasts against science while studying the astronomy, chemistry, and physiology of the natural investigators (not yet "scientists") they befriended. In their famous exchange late in the nineteenth century, Matthew Arnold and T. H. Huxley debated the historical worth and contemporary educational benefits of classical literary and cultural bodies of knowledge versus the modern, scientific, mathematical, and mechanical. Their heated debate flared again in the postatomic era in lectures and essays by Jacob Bronowski, C. P. Snow, F. R. Leavis, Michael Yudkin, Aldous Huxley, and others.

The construct of two cultures has been dramatically played out between creative literature and the physical sciences in many settings. Educators deem the skills and talents necessary for success in the two fields to be so incommensurable that they have developed segregated courses to teach students who are proficient in one area something of the other ("Physics for Poets," "Poetry for Physicists"). Indeed, poets and physicists are depicted as occupying such remote positions on the literature–science continuum that even an atomic blast could effect only their temporary fusion. In the chilling heat of that moment, a verse from the Bhagavad-Gita flashed across the mind of J. Robert Oppenheimer. In the aftermath, physics remained directly implicated in the two cultures debates. Snow and Leavis argued whether the second law of thermodynamics was as important a contribution to knowledge and culture as the works of Shakespeare.[2] Bronowski urged that the

---

[2] C. P. Snow, "The Two Cultures and the Scientific Revolution," *Encounter,* 12, no. 6 (1959), 17–24, and 13, no. 1 (1959), 22–7 (repr. Cambridge: Cambridge University Press, 1959). For historical and cultural perspectives, as well as reprints of Snow's essays, see also *The Two Cultures,* Canto edition

human values essential to the practice of both the arts and sciences must rise with renewed unity of purpose from the ashes of Nagasaki.[3] As crusaders for interdisciplinary understanding between the two cultures, Bronowski and Snow fought with equal bravery on both sides. They were not only accomplished scientists, essayists, and popularizers of science but also able writers of novels, poetry, drama, and literary criticism. The facility with which they were able, at midcentury, to move between cultures and to combine them provided living proof that mutual appreciation and participation across the humanities and sciences were possible. The apparent ease with which they did so, however, may have led them to underestimate the difficulties others would encounter in trying to follow their lead.

During the last quarter of the twentieth century, literary scholars and theorists, historians, philosophers and sociologists of science, and scientists embarked upon ambitious enterprises to "bridge the gap" between the sciences and humanities. As a result, the body of interdisciplinary scholarship exploring points of connection between literature and science increased exponentially. Numerous interdisciplinary curricula and programs were established at major research universities, at technological institutions, and on liberal arts campuses. One of the principal motivations for the founding of SLS, the Society for Literature and Science (1985), was the perceived need to develop a grand unified theory of literature and science. Efforts to unite literature and science via theory, both within and outside SLS, have generally entailed the analysis of one in terms of the other. These efforts have included experiments with the development of "scientific" literary criticism; the application of literary theory and criticism to scientific practice, methods, and methodologies; consideration of the "literary" output of scientific communities, with special attention to the rhetoric and narrative structure of scientific texts and their audiences; and expansion of the construct of "literature" to include science as writing or linguistic production.[4]

Despite the long historical interrelations of literature and science and significant steps toward developing a postdisciplinary concept of "one culture," deep, seemingly unresolvable differences between literary and scientific communities reverberated, newly amplified, in the science–culture "wars" of the 1990s. In the face of cultural critiques of science, some scientists vocally

---

with introduction by Stefan Collini (Cambridge: Cambridge University Press, 1993); F. R. Leavis, *Two Cultures: The Significance of C. P. Snow* (New York: Pantheon, 1963).

[3] J. Bronowski, *Science and Human Values* (New York: Harper and Row, 1956).

[4] For example: Roger Seamon, "Poetics Against Itself: On the Self-Destruction of Modern Scientific Criticism," *PMLA*, 104 (1989), 294–305; Stuart Peterfreund, ed., *Literature and Science: Theory and Practice* (Boston: Northeastern University Press, 1990); George Levine, ed., *One Culture: Essays in Science and Literature* (Madison: University of Wisconsin Press, 1987); Joseph W. Slade and Judith Yaross Lee, eds., *Beyond the Two Cultures: Essays on Science, Technology, and Literature* (Ames: Iowa State University Press, 1990); Frederick Amrine, ed., *Literature and Science as Modes of Expression* (Dordrecht: Kluwer, 1989); Charles Bazerman, *Shaping Written Knowledge: The Genre and Activity of the Experimental Article in Science* (Madison: University of Wisconsin Press, 1988); David Locke, *Science as Writing* (New Haven, Conn.: Yale University Press, 1992).

expressed dismay with the inaccurate use and misleading appropriation of scientific concepts by literary theorists, writers, and artists, suggesting that those within scientific communities were, or should be, the best cultural interpreters and critics of their own work.[5] Despite the positive, mutually educational effects of "peace" conferences on the local scholarly communities who participated in them, controversial and very public exchanges between Alan Sokal and Andrew Ross and between Sharon Traweek and Sidney Perkowitz raised further serious questions about whether "interdisciplinarity" should be declared a failed experiment.[6]

For many interdisciplinary travelers, the classically fabled gates of ivory and horn still mark the horizon for the integrated study of literature and science.[7] For others, however, now as always, the navigational signposts of binary oppositions and disciplinary boundaries appear as but curious relics of distant relevance. Deeply engaged in personal synthesis of the humanities and sciences or fruitfully involved in cross-disciplinary collaborations, they fix their sights sharply on the open waters beyond.

THE HISTORICAL INTERRELATIONS OF LITERATURE AND NEWTONIAN SCIENCE

Most historians of science are aware of the extent to which Isaac Newton and Newtonian science inspired contemporary literary responses, both positive and negative, sometimes within the works of the same literary writer. Alexander Pope (1688–1744) translated Newton to heaven in a tour-de-force couplet, but a few years later implicated him in the ultimate social and moral decay of the world (*The Dunciad*). For every admiring ode by a James Thomson or physico-theological poet, there is satiric critique from Jonathan Swift or another wit. The common perception of strong polarity among early-eighteenth-century literary attitudes toward Newtonianism may well have contributed to the later development of the two cultures mentality, in general, and to the particularly antithetical relationship of literature and physics. Indeed, the extent of Newton's personal influence in dismissing the poetic arts (however offhanded his comments may originally have been) should not be underestimated. As the next generation of natural philosophers sought

---

[5] Paul Gross and Norman Leavitt, *Higher Superstition: The Academic Left and Its Quarrels with Science* (Baltimore: Johns Hopkins University Press, 1994); John Brockman, *Third Culture: Beyond the Scientific Revolution* (New York: Simon and Schuster, 1995).
[6] "Science Wars," special issue of *Social Text*, 46–7 (1996), 1–252; keynote address of Sharon Traweek and personal exchange with Sidney Perkowitz following, SLS Conference, Atlanta, Georgia, fall 1996. A "peace" conference was held at Southampton, July 26–8, 1997; personal communication with Jay Labinger, Beckman Institute, California Institute of Technology. See also *Physics World* (Sept. 1997), 9.
[7] In classical mythology, the two gates of the unconscious through which dreams of illusion and fantasy, or those of real predictive value, respectively, arrived (see Homer's *Odyssey*, Book XIX).

to emulate his mathematical description of nature, they made ontological cuts in their cultural universes in the places they thought he had as well. Of course, they could not imitate the private Newton they never knew. Recent studies of the "other" Newton by such scholars as Margaret Jacob, Betty Jo Teeter Dobbs, Kenneth Knoespel, and Robert Markley have begun to demonstrate how textual exegesis and historical, literary, metaphorical, and natural philosophical ways of knowing were deeply integrated in his mind and work.

Contemporary literary representations of Newton and Newtonian science also do not support a hypothesis of incipient "two culturism." During the last half of the eighteenth century, Newton's ideas influenced both his scientific and literary descendants' views of the natural and supernatural. Samuel Johnson and William Blake offer a telling study in contrasts for the immediate post-Newtonian period. Neither participated in the unrelenting scorn or high-flown deification that have so long been thought to characterize literary reactions to Newtonian science. Their complex personal syntheses of literature and scientific knowledge and practice provided, respectively, models of measured moral response and powerful poetic alternatives for literary and cultural consideration of the physical sciences in the nineteenth and twentieth centuries.

As argued by Richard B. Schwartz in *Samuel Johnson and the New Science* (1971), Johnson (1709–1784) did not share the antiscience views of earlier wits and satirists. Carefully compiling evidence for Johnson's substantial personal interest and reading in both "ancient" and "modern" natural philosophy, including Newtonian conceptualizations of matter, vacuity, and the plenum, Schwartz demonstrated that Johnson consistently encouraged natural investigation, albeit within a larger moral frame of human behavior. Although Johnson's popular magazine essays in *The Rambler, Adventurer,* and *Idler* depict virtuosi, collectors, and projectors (typical targets for satirists' ridicule), he judges their activities according to their immediate or potential utility to humankind, the degree to which the actors fully engage their God-given time and talents, and the extent to which their actions lead to salvation. The spendthrift collector of natural trifles, the achievements and promise of the Royal Society, the medical practice and chemical experiments of Boerhaave, the electrical investigations of Stephen Gray, the Newton-Bentley correspondence, and the observational astronomy and telescopic improvements of William Herschel all represent, for Johnson, opportunities to reconcile moral and natural philosophy and direct his readers to lead upright lives of active intellectual inquiry.

Far from reacting against Newtonianism, Johnson embraced the essence of its methods in his literary style and moral views. His essays reflect his use of skepticism in moderation, careful observation, and empirical testing of his theories about human behavior. His gentlemanly literary style, under the influence of scientific essays by Francis Bacon, Thomas Sprat, and others, emphasizes the recoverable histories and verifiable information about his subjects and avoids repeating legend, rumors, and speculation.

We do not know to what extent William Blake (1757–1827) had direct knowledge of Newtonian science, but scholars are confident that he was conversant in the details of contemporary observational astronomy, including new eighteenth-century asterisms, as well as the local technologies of craftsmen like himself and the wider effects – both technical and social – of the Industrial Revolution.[8] In symbol and metaphor, Blake held the Newtonian worldview of materialism, mechanism, and rationalism responsible for the worst consequences of the spread of industry – humans were becoming the machines they made ("dark Satanic mills"). For Blake, the "successes" of Newtonian law, order, and mathematical description imprison the world within one man's vision of it, reducing its infinite complexity to Newton's limited powers of observation and reason. In long visionary poems, *Vala, Jerusalem, Book of Urizen, Milton, The Marriage of Heaven and Hell,* among others, Blake creates a poetic cosmos in which the material and spiritual realms and the representatives of reason and imagination are in existential opposition. For Blake, however, opposition is "true friendship," and it is out of the tensions of "contraries" that progress and energy are generated. In this sense, he sees himself as Newton's necessary contrary, challenging the Newtonian system and its definitions of space, time, change, motion, the material, and perception with his own.

In the short poem "Mock on, Mock on, Voltaire, Rousseau," Blake encapsulates his philosophy against the "action-reaction" world of rational materialism. In Blake's view, particulate matter appears in the form of physical entities because observers limit themselves to physical seeing. Illuminated by divine imagination, "every sand becomes a Gem" and "The Atoms of Democritus/ and Newton's Particles of Light/Are sands upon the Red seashore/Where Israel's tents do shine so bright," that is, symbolic of redemptive promise and revealing of the spiritual reality of nature that he believes it is his prophetic duty to recover. In both long and short verse, Blake creates his own symbolic system and poetic forms of expression, blending art and technology, and thus participates in the system building of which he accuses Newton. Significantly, however, Blake's vortical cosmos is constructed free of the geometrical constraints that in his eyes damn the Newtonian universe.

Johnson and Blake are but two of many creative writers in the immediate post-Newtonian period who experienced new ideas and discoveries in the physical sciences as integral parts of their culture – not separate entities in opposition to it. While there has been strong precedent within literary history for doing so, labeling literary writers, such as Johnson or Blake, "pro-science" or "antiscience" says very little about how they responded to the

---

[8] Donald Ault, *Visionary Physics: Blake's Response to Newton* (Chicago: University of Chicago Press, 1974); Jacob Bronowski, *William Blake and the Age of Revolution* (New York: Harper and Row, 1965).

scientific enterprise of their day. Additionally, within internalist history of science and history of ideas traditions, scholars have tended to analyze literary texts with a "science in literature" approach that has reinforced the perception of the two as separate cultural phenomena. Such studies have often focused on identifying the "accuracy" or "inaccuracy" of literary representations of science and assessing the degree of direct correspondence they bear to their original scientific contexts and meanings. Although such evaluations can be extremely useful in tracing the popular dissemination of science through culture, as we see further illustrated by the discussion of literature and the modern physical sciences in the sections that follow, the ongoing fascination lies with the details of how and why literary writers understand, interpret, and represent scientific concepts and discoveries in the various ways they do.

LITERATURE AND THE PHYSICAL SCIENCES AFTER 1800: FORMS AND CONTENTS

For historians of science and students of the history of science who are exploring the interrelations of literature and science for the first time, the complex array of relevant primary texts can appear daunting. It should be somewhat reassuring to recall, however, that literature and the physical sciences have shared much of the same history in many of the same texts. For at least two thousand years, in fact, poetry was the genre of choice for writing about physical nature, especially astronomy, astrology, meteorology, and cosmology. For historians of ancient, medieval, and early modern science, philosophical verse and lyric and epic poetry have long been essential – if not definitive – texts. To the present day, concepts and discoveries in chemistry, mathematics, astronomy, and physics have continuously been disseminated at the popular level through various forms of creative writing.

Poetry, drama, novels, short stories, prose essays in popular science or nature writing, scientific biography and autobiography, professional scientific articles and textbooks, journals, and diaries are all forms of literature (fairly traditionally defined) that can serve as rich resources for investigating the humanistic and cultural relations of the modern physical sciences. Less traditional "texts" include film, television, museum display, instruments, experiments, laboratory journals, oral tradition, popular music, graphic novels (comic books), computer programs and games, websites, art exhibits, dance, and other forms of performance art. While the two cultures rhetoric of academic exchanges between scientists and creative writers may have sounded increasingly convincing from the beginning of the nineteenth into the late twentieth centuries, many forms of literature and popular culture exhibit creative consideration of the meaning of science, the range and depth of which belie the notion.

Despite some science/culture warriors' rhetoric to the contrary, most literary allusions to modern science are not irresponsibly casual, and most twentieth-century literary scholars and creative writers *are* aware that Einstein's theories of relativity cannot be aptly summarized by the catchall phrase "everything is relative." Most literary craftspeople who incorporate physical sciences into their writing actively seek and achieve at least a respectable popular level of understanding of the concepts of the astronomy, physics, chemistry, mathematics, or chaos sciences they use. Many have extensive education and training in the sciences; others are working scientists. The final products of the creative writing process, however, also reflect aesthetic, philosophical, social, spiritual, and emotional requirements and choices. From the most minute particular to the most general abstract law, from quarks to string theory, scientific allusions, metaphors, analogies, and symbols permeate the literature of their age, often extending beyond the constraints of their strict scientific definitions, connotations, and chronologies. Challenged and inspired by the difficulty of satisfying the demands of literary form and expression in tandem with the technical aspects of science, writers resolve the tensions between "beauty and truth" in a wide variety of ways. Different sets of generic conventions and audience expectations operate in different types of texts, and so the space science in Gene Roddenberry's science fiction sagas, for instance, cannot fairly be judged by the same exacting standards that might apply to the military technological content of one of Tom Clancy's historical novels.

To begin to construct a useful understanding of the mutual relations of science and literary forms, themes, imagery, diction, and tropes as they have developed over time, firsthand experience with primary materials is indispensable. Assuming that many of this volume's readers will not have extensive previous knowledge of nineteenth- and twentieth-century creative literature, the following two sections offer brief overviews of literary texts that especially engage chemistry, astronomy, cosmology, and physics.

LITERATURE AND CHEMISTRY

Surprisingly (especially to the authors of monographs on literature and Darwinism or evolutionary theory), J. A. V. Chapple identifies chemistry as the "most exciting" science to the nineteenth-century British popular imagination.[9] Discoveries about the nature of light, heat, electricity, magnetism, and the identification of new elements, as well as the theoretical and experimental work of Antoine-Laurent Lavoisier, John Dalton, Alessandro Volta, Luigi Galvani, Humphry Davy, Michael Faraday, and

---

[9] J. A. V. Chapple, *Science and Literature in the Nineteenth Century* (London: Macmillan, 1986), esp. pp. 20–45.

William Thomson (Lord Kelvin), all inspired interest in natural forces and the ability of human beings to describe them mathematically, to understand and control them. Poets and novelists incorporated a wide variety of chemical concepts into their works' subject matter, plot structure, and philosophical and social themes. Among the most prominent topics are the notion of a cosmic web of correspondences between natural phenomena, chemical transformation and catalysts, and the concepts of affinity, attraction–repulsion, energy, force, and activity. Literary writers built images and metaphors from their knowledge of observed phenomena, such as electrical storms, cloud formations, the wind, and rainbows, and from such equally vivid theoretical concepts as heat-death, miasma theory, the transference and conservation of energy, and radiation.

The relation of living to nonliving through organic and inorganic chemistry was especially fascinating to nineteenth-century writers. Mary Shelley drew upon her knowledge of the details and implications of experimental chemistry, Galvanism, and vitalism in *Frankenstein*. Electricity, magnetism, chemical interactions and compounds, as well as astronomical discoveries and theories inform P. B. Shelley's verse. Samuel Taylor Coleridge attempted to create a poetical and philosophical synthesis of chemistry, physics, astronomy, and cosmology. Davy and James Clerk Maxwell included versifying among their experimental endeavors. Although their poetry has not proven quite as immortal as their science, many of their concepts, discoveries, and philosophical ideas took on lives of their own in literary metaphor. Maxwell's work, and his Demon, in particular, appear in the work of such diverse authors as Paul Valéry, Stéphane Mallarmé, and Thomas Pynchon. Perhaps the most sophisticated use of the organic "web" motif occurs in the fiction of George Eliot, who stretches its implications across psychological, social, and national frames. In American literature, Nathaniel Hawthorne, Herman Melville, James Fenimore Cooper, and Edgar Allan Poe variously draw upon alchemy, chemistry, metallurgy, Cartesian, Newtonian, and Laplacian cosmology, magnetism, electrical experiments, and the related concepts of vitalism and mesmerism. Literary investigation of chemical science is also strongly evident in the writings of W. B. Yeats, Goethe, Novalis, and E.T.A. Hoffmann.

LITERATURE AND ASTRONOMY, COSMOLOGY, AND PHYSICS

The mathematical confirmation of Newtonian astronomy and physics fascinated nineteenth-century writers as much as the new telescopic discoveries and theories of the Herschels. Poetry and novels include allusions and metaphors drawn from a wide range of astronomical phenomena and concepts, including the stability of the solar system, comets, nebulae and the nebular hypothesis, variable stars and multiple star systems, the voids of deep space, stellar distances and proper motion, the relation of distance and time in telescopic observation, the "new" night sky in the Southern Hemisphere, the plurality of worlds, extraterrestrials, entropy, and the life cycle of the sun. Writers call upon their interest in astronomy to consider such themes as the argument from design, the role of the supernatural in the establishment and maintenance of natural law, the relation of humanity to nature, and astronomers' roles as interpreters of universal history and creation, especially in relation to cosmology, evolution, and geology.

William Wordsworth, Walt Whitman, and Emily Dickinson each recorded poetic responses to observational astronomy ("Star-Gazers," "When I Heard the Learn'd Astronomer," "Arcturus"). Coleridge, P. B. Shelley, and Ralph Waldo Emerson responded more broadly to Newtonian astronomy and cosmology. Alfred Tennyson, a student of Whewell with a strong amateur interest in astronomy, synthesized his understanding of cosmology and evolutionary theory in *In Memoriam*. Thomas Hardy wrote several poems commemorating his firsthand observations, and the scene-setting, timekeeping, and foreshadowing devices in most of his major novels depend heavily on astronomical phenomena. His *Two on a Tower* is the most "astronomical" novel of the age, featuring an astronomer as the main character and a comet, lunar eclipse, the Milky Way, and variable stars as plot devices, themes, and analogies.[10] In his aesthetically and historically perverse response to nineteenth-century astronomy, Algernon Swinburne drew upon Greek atomism and Lucretius to fuse sound and sense, poetry and cosmology ("Hertha" and "Anactoria"). In "Meditation Under Stars," George Meredith explored the common chemical origins of human life with the inorganic stars. The poet Francis Thompson created a complex analogy between the forces of faith and grace and planetary dynamics ("A Dead Astronomer"). In their poetry and fiction, such diverse writers as José Martí, Alexander Pushkin, Honoré de Balzac, Stendhal, Charles Baudelaire, Arthur Rimbaud, and Emile Zola explored the meaning of cosmological theories and the operation of physical laws (thermodynamics, chance, complexity) within humanistic contexts and the social realm.

As a newly redeveloping genre, nineteenth-century science fiction became increasingly sophisticated in its blending of contemporary science with social commentary (Mark Twain's *A Connecticut Yankee in King Arthur's Court,* Jules Verne's *From Earth to the Moon,* H. G. Wells's *First Men in the Moon*). Wells's work especially draws upon technical detail both past and present (Keplerian elements in *In the Days of the Comet;* heat-death and evolutionary theory in *The Time Machine*). Edwin A. Abbott's *Flatland* is a rare fictional treatment of geometry and mathematics. At the *fin de siècle*, optimistic visions of space

---

[10] Pamela Gossin, *Thomas Hardy's Novel Universe: Astronomy, and the Cosmic Heroines of His Minor and Major Fiction* (Aldershot, England: Ashgate Publishing, forthcoming 2002).

and time travel are countered by bleak treatments of the laws of physics, particularly the theme of entropy (Joseph Conrad's *Heart of Darkness*). Indeed, literary exploration of the utopian/dystopian possibilities of the physical sciences would remain a central concern for science fiction writers for at least the next one hundred years (Evegeny Zamiatin, Arkady and Boris Strugatsky, Isaac Asimov, Arthur Clarke, and Ursula Le Guin).

Early in the twentieth century, novelists and dramatists developed experimental literary forms that modeled Einsteinian concepts of space and time, relations of subject and observer, uncertainty, indeterminacy, and complexity (James Joyce's *Ulysses, Finnegans Wake;* Virginia Woolf's *To the Lighthouse, The Waves;* Vladimir Nabokov's *Bend Sinister, Ada;* virtually any of Samuel Beckett's works).[11] Jorge Luis Borges, Julio Cortazár, Umberto Eco, Italo Calvino, and Pynchon again created remarkable innovations in structural and narrative uses of entropy, non-Euclidean geometry, relativity theory, quantum mechanics, and information theory, as have Robert Coover, Penelope Fitzgerald, Don DeLillo, and Alan Lightman. Similarly, twentieth-century poets have drawn inspiration in both form and content from astronomy and space sciences, entropy, relativity, postatomic and quantum physics (Mary Barnard, *Time and the White Tigress;* Diane Ackerman, *The Planets: A Cosmic Pastoral;* T. S. Eliot, *The Waste Land;* William Carlos Williams, "St. Francis Einstein of the Daffodils," "Paterson"; John Updike, "Cosmic Gall"; Robinson Jeffers's "Star-Swirls"; Ernesto Cardenal, *Cosmic Canticle*).

Key figures from the history of the physical sciences play important roles in twentieth-century drama, as in Bertolt Brecht's *Life of Galileo* and Friedrich Dürrenmatt's *The Physicists,* as well as in historical novels, such as those by John Banville and Arthur Koestler. In *Mason and Dixon* (1997), a unique combination of historical novel and magic realism, Pynchon offers significant insights into the invention of narrative for the history of astronomy, exploring the possibilities and limitations of authorial perception and voice, historical characterization, the relation of plot to space-time, as well as the nature and use of chronologies and other technologies of measurement. Other frequently recurrent themes in twentieth-century literature of the physical sciences include radiation, radioactivity, and x-ray technology (H.G. Wells, Karel Capek, Russell Hoban, and Thomas Mann); gender relations in the postatomic world (Margaret Atwood, Ursula Le Guin); mathematics, game theory, cybernetics, artificial intelligence, virtual reality, and information technology (DeLillo, Gary Finke, Richard Powers, Marge Piercy, William Gibson, Neal Stephenson).

Anthologies of scientific poetry and collections of literary writing about science can be useful for the initial identification of relevant texts for

---

[11] Alan J. Friedman and Carol C. Donley, *Einstein As Myth and Muse* (Cambridge: Cambridge University Press, 1985), pp. 67–109.

interdisciplinary research or classroom use. Some care should be exercised in using such materials, however, as they often give a misleading impression that all or most literary uses of science are literal, overt references to and descriptions of "real" or at least realistic scientific concepts or practice, science for science's sake.[12] Creative writers invent sophisticated scientific allegories and symbolic systems of meaning within their texts. They employ deep structural scientific metaphors and extended conceits, often creating vast fictional or poetic worlds in which they test and explore science's power and meaning. While identification and analysis of science in literature will always prove valuable for constructing an understanding of the interrelations of literature and the physical sciences, students of literature and science can quickly discover that there is a lot more to the story by engaging interdisciplinary critical and interpretative studies.

INTERDISCIPLINARY PERSPECTIVES AND SCHOLARSHIP

Interdisciplinary scholarship that explores the interrelations of creative literature and astronomy, physics, mathematics, and chemistry consists primarily of particulate, local investigations that do not expressly contribute to the construction, reinforcement, reformation, or replacement of a generally acknowledged master narrative (either positively or negatively construed). Many working within literature and science, in fact, celebrate the orderly disorder of this scattershot, chaotic, scholarly productivity as indicative of the new creative energy inherent in any emergent intellectual enterprise. Indeed, they have tried purposefully, creatively, and actively to avoid both the process and products of traditional historical generalizations, believing them to be, at best, inauthentic and prescriptive; at worst, falsifying and restrictive. In their attempts to resist constructing (and being constructed by) the content of totalizing histories, such scholars have turned away from traditional forms of historiography and criticism as well, preferring to generate nonnarrative, nonlinear literary artifacts to represent their fields (e.g., encyclopedias, dictionaries, compendia, panel discussions, and volumes of individually authored essays, rather than monographs).

As a result, broad chronological surveys that treat the interrelations of literature and the modern physical sciences as a whole are rare.[13] Many interdisciplinary studies, however, do provide historical perspectives of the literary relations of a single aspect or branch of the physical sciences,

---

[12] Walter Gratzer, ed., *A Literary Companion to Science* (New York: W. W. Norton, 1990); Bonnie Bilyeu Gordon, ed., *Songs from Unsung Worlds: Science in Poetry* (Boston: Birkhäuser, 1985); John Heath-Stubbs and Phillips Salman, eds., *Poems of Science* (New York: Penguin, 1984).
[13] The most notable exception: Noojin Walker and Martha Gulton's *The Twain Meet: The Physical Sciences and Poetry* (American University Studies, Series XIX: General Literature, 23) (New York: Lang, 1989), which offers a wide chronological survey.

or science as represented within a particular genre of literature. By focusing on the figure of the scientific practitioner in *From Faust to Strangelove* (1994), Roslynn Haynes is able to trace literary and cultural representations and their changing forms and significance over several centuries. Martha A. Turner examines concepts of mechanism as they appear in two hundred years of novel writing, from Jane Austen to Doris Lessing.[14] A. J. Meadows's *The High Firmament* (1969) surveys the presence of astronomy, with special attention to the use of astronomical imagery, in literature from the fifteenth into the early twentieth century. The author of this chapter works on the interdisciplinary cross-influences of literary writers and astronomers from the Scientific Revolution to the present, with particular attention to their perceptions of "revolutionary" astronomical developments, aesthetic sensibility, and representations of women in their philosophies and cosmologies.[15] In *Cosmic Engineers: A Study of Hard Science Fiction* (1996), Gary Westfahl traces the development of a subgenre of science fiction and the significant roles of science "faction" within it.

Other scholars offer specialized studies of literature and science in a single national context or within a carefully defined time period, such as Soviet science and fiction after Stalin, the reception of quantum theory in German literature and philosophy, and French literature in relation to the science of Newton and Einstein.[16] Robert Scholnick's edition of scholarly essays offers historical and literary analyses of the engagement of science by American writers over three and a half centuries, from Edward Taylor's Paracelsian medical poetry, through the unique responses to the positive and negative potential of science and technology by Mark Twain, Hart Crane, and John Dos Passos, to examinations of cybernetics and turbulence in contemporary American fiction. The collective effect of such volumes suggests the dynamic interrelations of letters and sciences in America from just after the Scientific Revolution to the present day.[17]

---

[14] Martha A. Turner, *Mechanism and the Novel: Science in the Narrative Process* (Cambridge: Cambridge University Press, 1993).

[15] Pamela Gossin, "'All Danaë to the Stars': Nineteenth-Century Representations of Women in the Cosmos," *Victorian Studies*, 40, no. 1 (Autumn 1996), 65–96; "Living Poetics, Enacting the Cosmos: Diane Ackerman's Popularization of Astronomy in *The Planets: A Cosmic Pastoral*," *Women's Studies*, 26 (1997), 605–38; "Poetic Resolutions of Scientific Revolutions: Astronomy and the Literary Imaginations of Donne, Swift and Hardy," PhD diss., University of Wisconsin–Madison, 1989. See also "Literature and Astronomy," pp. 307–14 in *History of Astronomy: An Encyclopedia*, ed. John Lankford (New York: Garland, 1996), and "Literature and the Scientific Revolution," in *The Scientific Revolution: An Encyclopedia*, ed. Wilbur Applebaum (New York: Garland, 2000).

[16] Rosalind Marsh, *Soviet Fiction Since Stalin: Science, Politics and Literature* (London: Croom Helm, 1986); Elisabeth Emter, *Literature and Quantum Theory: The Reception of Modern Physics in Literary and Philosophical Works in the German Language, 1925–70* (Berlin: de Gruyter, 1995); Ruth T. Murdoch, "Newton and the French Muse," *Journal of the History of Ideas*, 29, no. 3 (June 1958), 323–34; Kenneth S. White, *Einstein and Modern French Drama: An Analogy* (Washington, D.C.: University Press of America, 1983).

[17] Robert J. Scholnick, ed., *American Literature and Science* (Lexington: University of Kentucky Press, 1992); Joseph Tabbi, *Postmodern Sublime: Technology and American Writing from Mailer to Cyberpunk*

Perhaps predictably, the literature and science of nineteenth-century Britain has generated more secondary studies than those of any other time and place to date.[18] Tess Cosslett, through case studies of Tennyson, George Eliot, Meredith, and Hardy, identifies prominent characteristics of the era's "scientific movement" and demonstrates how both science and literature participated in the creation of Victorian notions of scientific truth, law, and organic kinship.[19] J. A. V. Chapple surveys British literature in relation to the major thematic developments of virtually every science extant in the nineteenth century, including astronomy, physics, chemistry, meteorology, various branches of natural history and the life sciences, psychology, anthropology, ethnology, philology, and mythology.[20] Peter Allan Dale investigates scientific positivism and literary realism as responses to Romanticism in the philosophy, aesthetics, literature, and culture of the era.[21] Jonathan Smith analyzes the influence of Baconian inductivism upon nineteenth-century Romantic poetry and chemistry, narratives of uniformitarianism, geometry, and the methods of literary "scientific" detection.[22] Full-length case studies in literature and science are available on innumerable nineteenth-century figures, including both Shelleys, William Wordsworth, Goethe, Thoreau, Emerson, Herman Melville, George Eliot, Tennyson, Verne, Whitman, and Twain, to name a few. Trevor Levere's study of Coleridge and Davy, *Poetry Realized in Nature* (1981) is a masterful example of the extent to which historical contextualizations and close reading of primary essays, notebooks, and poetry work together to illuminate the interrelations of literature and science on personal, social, philosophical, and international levels.

In twentieth-century texts, "literature" and "science" became so multivalent that most interdisciplinary scholars found it necessary to carefully define and delimit their subject matter in working with them. Some did so by offering close interpretative analyses of literature and science as defined by, and within, the works of individual writers (such as Theodore Dreiser, G. M. Hopkins, James Joyce, and Samuel Beckett). Others concentrated

---

(Ithaca, N.Y.: Cornell University Press, 1995); John Limon, *The Place of Fiction in the Time of Science: A Disciplinary History of American Writing* (Cambridge: Cambridge University Press, 1990); Lisa Steinman, *Made in America: Science, Technology and American Modernist Poets* (New Haven, Conn.: Yale University Press, 1987); Ronald E. Martin, *American Literature and the Universe of Force* (Durham, N.C.: Duke University Press, 1981).

[18] James Paradis and Thomas Postlewait, eds., *Victorian Science and Victorian Values: Literary Perspectives* (New Brunswick, N.J.: Rutgers University Press, 1985); Patrick Brantlinger, ed., *Energy and Entropy: Science and Culture in Victorian Britain* (Bloomington: Indiana University Press, 1989); Gillian Beer, *Open Fields: Science in Cultural Encounter* (Oxford: Clarendon Press, 1996), chaps. 10–14.

[19] Tess Cosslett, *The "Scientific Movement" and Victorian Literature* (New York: St. Martin's Press, 1982).

[20] Chapple, *Science and Literature in the Nineteenth Century*.

[21] Peter Allan Dale, *In Pursuit of a Scientific Culture: Science, Art and Society in the Victorian Age* (Madison: University of Wisconsin Press, 1989).

[22] Jonathan Smith, *Fact and Feeling: Baconian Science and the Nineteenth-Century Literary Imagination* (Madison: University of Wisconsin Press, 1994).

on the various constructions of particular developments, such as quantum physics or "quantum poetics."[23] Interdisciplinary criticism was also fruitfully directed toward the rhetorical structures and strategies of antinuclear fiction; the literature of "modern" alchemy, hermeticism, and occultism; literary interrelations with information technology; Einstein's theories of relativity in literature and culture; literature and scientific field models; and the interactions of chaos sciences with contemporary fiction, poetry, and literary theory.[24]

LITERATURE AND THE MODERN PHYSICAL SCIENCES IN THE HISTORY OF SCIENCE

The necessarily limited scope of the foregoing discussion should not be allowed to reinforce the all-too-common perception that literature and science studies are exclusively produced by scholars trained in literary theory and criticism. Although the Society for Literature and Science and its journal, *Configurations,* have certainly given disciplinary form and structure to "literature and science" (and the clear majority of SLS members *do* teach and publish within literature and language studies), history of science is – in and of itself – a major mode of, and central contributor to, studies of the interrelations of literature and the modern physical sciences. Historians of science have long expressed pride in the inherent interdisciplinarity of their enterprise, which requires deep engagement with the methods, methodologies and content of at least two professional fields. Although there may be compelling personal and professional reasons for not marketing their scholarship as such, by engaging multiple layers of meaning, by attending to the rhetorical style, audiences, and linguistic construction of the primary texts they interpret and analyze, many historians of science have always already been doing "literature and science" (some all of their careers, without knowing it).

Indeed, studies of "literature and science" and "cultural influences" upon and within science have been officially incorporated into the history of science

---

[23] Susan Strehle, *Fiction in the Quantum Universe* (Chapel Hill: University of North Carolina Press, 1992); Robert Nadeau, *Readings from the New Book on Nature: Physics and Metaphysics in the Modern Novel* (Amherst: University of Massachusetts Press, 1981); Daniel Albright, *Quantum Poetics: Yeats, Pound, Eliot, and the Science of Modernism* (Cambridge: Cambridge University Press, 1997).

[24] Patrick Mannix, *The Rhetoric of Antinuclear Fiction: Persuasive Strategies in Novels and Films* (Lewisburg, Pa.: Bucknell University Press, 1992); Timothy Materer, *Modernist Alchemy: Poetry and the Occult* (Ithaca, N.Y.: Cornell University Press, 1995); William R. Paulson, *The Noise of Culture: Literary Texts in a World of Information* (Ithaca, N.Y.: Cornell University Press, 1988); Alan J. Friedman and Carol C. Donley, *Einstein As Myth and Muse* (Cambridge: Cambridge University Press, 1985); N. Katherine Hayles, *The Cosmic Web: Scientific Field Models and Literary Strategies in the Twentieth Century* (Ithaca, N.Y.: Cornell University Press, 1984); Hayles, *Chaos Bound: Orderly Disorder in Contemporary Literature and Science* (Ithaca, N.Y.: Cornell University Press, 1990); Hayles, ed., *Chaos and Order: Complex Dynamics in Literature and Science* (Chicago: University of Chicago Press, 1991); Alexander Argyros, *A Blessed Rage for Order: Deconstruction, Evolution and Chaos* (Ann Arbor: University of Michigan Press, 1991).

as a professional discipline since its first establishment in the early years of the twentieth century. Thanks in large part to the astute sensibilities of John Neu, longtime editor of the *Isis Cumulative Bibliography,* historians of science have annually been made mindful of the "humanistic relations" of their fields. Historians of the physical sciences, in particular, have maintained steady interest in tracing literary, artistic, and broader cultural references to physics and chemistry, sharing their findings regularly in professional publications, such as the *Journal for the History of Astronomy,* the popular *Sky and Telescope* and *Star-Date,* and most recently, on HASTRO, an electronic listserv for topics related to the history of astronomy. Full-length studies in the history of science by such well-known scholars as Thomas Kuhn, Marie Boas, and Gerald Holton have been informed by critical biographies and interpretative analyses of individual literary writers who achieved a high degree of scientific literacy and employed sophisticated scientific images and themes in their work. Owen Gingerich, best known as an historian of physical science, occasionally publishes his "transdisciplinary" research, exploring literary works with astronomical content.[25] The subfield within the history of the modern physical sciences that has most traditionally and most consistently utilized works of creative literature as central source materials has been the popularization of science. Recently, historians of science investigating the rhetorical and social construction of chemistry and physics have successfully applied methodologies primarily developed within "literature and science."

As theoretical trends within literature and science studies move beyond poststructuralist views, interdisciplinary scholars are engaging historical considerations of the interrelations of science and culture with renewed interest and understanding, recognizing the presence of theory within historical methods and practice. Historically informed criticism directs attention toward the ways in which cultural influences, including literary products and practices, shape the development of science through the influence of language and metaphor, or by actively participating in its popularization and cultural construction (see recent studies by James J. Bono, David Locke, and N. Katherine Hayles, for example). Such studies tend to analyze literature and science in relation to a third concern, such as an interest in the formation of discursive communities and their linguistic practices, rhetorical strategies, issues of gender, race and class, as well as social and political power. The extent to which historical contexts and methodologies play increasingly vital roles within these formulations may serve as an early indication that interdisciplinary scholars are turning toward "history" as a promising mediating term between literature and science, and between the

---

[25] Owen Gingerich, "Transdisciplinary Intersections: Astronomy and Three Early English Poets," in *New Directions for Teaching and Learning: Interdisciplinary Teaching,* no. 8, ed. A. White (San Francisco: Jossey-Bass, Dec. 1981), pp. 67–75, and "The Satellites of Mars: Prediction and Discovery," *Journal of the History of Astronomy,* 1 (1970), 109–15 (in relation to *Gulliver's Travels*).

two cultures, more generally. With education and training in theories of historiography, textual analysis, and knowledge of scientific developments and concepts within intellectual, historical, philosophical, and social contexts, historians of science are well situated to participate in, and shape, such discussions.

LITERATURE AND THE MODERN PHYSICAL SCIENCES: NEW FORMS AND DIRECTIONS

As we venture into the twenty-first century, conventional forms of print literature (including professional academic writing) are likely to represent a smaller and smaller percentage of the media through which the interrelations of literature and science will be expressed. Poets such as Elizabeth Socolow, Siv Cedering, Richard Kenney, and Rafael Catalá will continue to invent new verse forms and structures to contain and represent their understanding of science. Physicists and chemists (such as Fay Ajzenberg-Selove, Roald Hoffmann, Nicanor Parra, Carl Djerassi) will follow the examples of their colleagues – past and present – to publish innovative autobiographies, memoirs, poetry, and novels about their own scientific work and insights. Scholars working in literature and science studies will also find increasing interest in, and need for, inventive forms of analysis and interpretation. As we are further called upon to adapt to the needs of the ever-changing classroom, to new opportunities for public education, and to a shrinking market for scholarly books, we may find ourselves encouraged to experiment with new literary and artistic forms of representation and expression, such as innovative science textbooks that present the concepts of the physical sciences in historical and cultural contexts, "popular" histories of science, imaginative biographies of science, historical novels of science, television documentaries, screenplays and films, educational CDs and DVDs, scientific visualizations, virtual reality simulations, and interactive websites. Errol Morris's ingenious cinematographic use of "time's arrow" in his film version of "A Brief History of Time," Robert Kanigel's adaptation of infinite series to the formal structure of his telling of Ramanujan's work and life, and Dava Sobel's hybrid historical novel/memoir of science in *Galileo's Daughter* each model exciting new ways to combine the two cultures that teach us something about both.

As this chapter's opening discussion of Richard Feynman suggests, some of the most compelling "texts" for future studies of literature and the modern physical sciences may indeed be the human individuals who personally embody interdisciplinary and cross-disciplinary learning. No doubt some of his peers dismissed Feynman's interest in literature, art, and music as just as embarrassingly irrelevant to physics as his frequenting of stripclubs. To others, such investigations represent external manifestations of his mind at work, providing insight into the ways in which his exercise of impassioned open-mindedness, uninhibited inventiveness, and playful pattern seeking may also have enabled the development of his famous diagrams and unique problem-solving capabilities in physics. Prominent figures like Feynman, Snow, Bronowski, Hoffmann, Pynchon, et al. may serve as the subjects of rich case studies for examining the integration of the two cultures; yet Nobel laurels are not required for the fostering of such interests. By studying the quiet lives of interdisciplinarity led by laboratory researchers, studio artists, creative writers, scholars, and classroom teachers, we may find that an analysis of their personal experiments and mutual collaborations will yield unexpected insights into the ways in which individuals teach themselves – and their colleagues across the cultures – to integrate art, literature, and science.

We may still be some years away from the time when cognitive scientists or neuroscientists will be able to tell us with some confidence the extent to which we are born with a genetic gift of interdisciplinarity and/or an ability to promote the growth of our own "bicultural" brain structures through an eclectic engagement with life and the world around us. We have already arrived at a moment, however, when we can reconfigure our narrative accounts of such minds, no longer regarding them as exceptional aberrations to arbitrarily constructed monocultural norms, but instead appreciating the integrative thought processes they display for their own sake. Through continued studies of cognitive, personal, and interpersonal engagements of the creative arts, literature, humanities, and sciences, we may discover new ways for them to perform important cultural work together.

Part II

DISCIPLINE BUILDING IN THE SCIENCES
Places, Instruments, Communication

6

MATHEMATICAL SCHOOLS, COMMUNITIES, AND NETWORKS

David E. Rowe

Mathematical knowledge has long been regarded as essentially stable and, hence, rooted in a world of ideas only superficially affected by historical forces. This general viewpoint has profoundly influenced the historiography of mathematics, which until recently has focused primarily on internal developments and related epistemological issues. Standard historical accounts have concentrated heavily on the end products of mathematical research: theorems, solutions to problems, and the technical difficulties that had to be mastered before a well-posed question could be answered. This kind of approach inevitably suggests a cumulative picture of mathematical knowledge that tells us little about how such knowledge was gained, refined, codified, or transmitted. Moreover, the purported permanence and stability of mathematical knowledge begs some obvious questions with regard to accessibility – known to whom and by what means? Issues of this kind have seldom been addressed in historical studies of mathematics, which often treat priority disputes among mathematicians as merely a matter of "who got there first." By implication, such studies suggest that mathematical truths reside in a Platonic realm independent of human activity, and that mathematical findings, once discovered and set down in print, can later be retrieved at will.

If this fairly pervasive view of the epistemological status of mathematical assertions were substantially correct, then presumably mathematical knowledge and the activities that lead to its acquisition ought to be sharply distinguished from their counterparts in the natural sciences. Recent research, however, has begun to undercut this once-unquestioned canon of scholarship in the history of mathematics. At the same time, mathematicians and philosophers alike have come increasingly to appreciate that, far from being immune to the vicissitudes of historical change, mathematical knowledge depends on numerous contextual factors that have dramatically affected the meanings and significance attached to it. Reaching such a contextualized understanding of mathematical knowledge, however, implies taking into account the variety of *activities* that produce it, an approach that necessarily deflects attention from the finished products as such – including the "great works of the masters" – in order to make sense of the broader realms of "mathematical experience." As Joan Richards has observed, historians of mathematics have resisted many of the trends and ignored most of the issues that have preoccupied historians of science in recent decades.[1] A wide gulf continues to separate traditional "internalist" historians of mathematics from those who, like Richards, favor studies directed at how and why mathematicians in a particular culture attach meaning to their work. On the other hand, an actor-oriented, realistic approach that takes mathematical ideas and their concrete contexts seriously offers a way to bridge the gulf that divides these two camps. Such an approach can take many forms and guises, but all share the premise that the type of knowledge mathematicians have produced has depended heavily on cultural, political, and institutional factors that shaped the various environments in which they have worked.

TEXTS AND CONTEXTS

In his influential *Proofs and Refutations,* the philosopher Imre Lakatos offered an alternative to the standard notion that the inventory of mathematical knowledge merely accumulates through a collective process of discovery.[2] While historians have not been tempted to adopt this Lakatosian model whole cloth, its dialectical flavor has proven attractive even if its program of rational reconstruction has not. For similar reasons, the possibility of adapting T. S. Kuhn's ideas to account for significant shifts in research trends has been debated among historians and philosophers of mathematics, although no clear consensus has emerged from these discussions. Advocates of such an approach have tried to argue that contrary to the standard cumulative picture, revolutionary changes and major paradigm shifts *do* take place in the history of mathematics. Thus, Joseph Dauben pointed to the case of Cantorian set theory – which overturned the foundations of real analysis and laid the groundwork for modern algebra, topology, and stochastics – as the most recent major mathematical revolution. Judith Grabiner made a similar argument by drawing on the case of Cauchy's reformulation of the conceptual foundations of the calculus, whereas Ivor Grattan-Guinness has described the tumultuous mathematical activity in postrevolutionary France in terms of convolutions rather than revolutions, arguing that his term

---

[1] Joan L. Richards, "The History of Mathematics and 'L'esprit humain': A Critical Reappraisal," in *Constructing Knowledge in the History of Science,* ed. Arnold Thackray, Osiris, 10 (1995), 122–35. An important exception is Herbert Mehrtens, *Moderne-Sprache-Mathematik* (Frankfurt: Suhrkamp, 1990), a provocative global study of mathematical modernity that focuses especially on fundamental tensions within the German mathematical community.

[2] Imre Lakatos, *Proofs and Refutations,* ed. John Worrall and Elie Zahar (Cambridge: Cambridge University Press, 1976).

better captures the complex interplay of social, intellectual, and institutional forces.[3]

Most of the discussions pertaining to revolutions in mathematics have approached the topic from the rather narrow standpoint of intellectual history. Advocates of this approach might well argue that if scientific ideas can be viewed à la Kuhn in the context of competing paradigms, then why not treat volatile situations like the advent of non-Euclidean geometry in the nineteenth century similarly? Nevertheless, it cannot be overlooked that comparatively little attention has been paid to other components of a Kuhnian-style analysis. Research trends, in particular, need to be carefully scrutinized before the roles of the historical actors – the mathematicians, their allies and critics – can be clearly understood. Contextualizing their work and ideas means, among other things, identifying those mainstream areas of research that captivated contemporary interests: the types of problems they hoped to solve, the techniques available to tackle those problems, the prestige that mathematicians attached to various fields of research, and the status of mathematical research in the local environments and larger scientific communities in which higher mathematics was pursued. In short, a host of issues pertaining to "normal mathematics" as seen in the actual research practices typical of a given period need to be thoroughly investigated.[4] Perhaps then the time will be ripe to look more carefully at the issue of "revolutions" in mathematics.

This is not meant to imply that the conditions that shape mathematical activity deserve higher priority than the knowledge that ensued from it. On the contrary, the concrete forms in which mathematical work has been conveyed pose an ongoing challenge to historians. Enduring intellectual traditions have centered traditionally on paradigmatic texts, such as Euclid's *Elements* and Newton's *Principia*. After Newton, works of a comprehensive character continued to be produced, but with the exception of P. S. Laplace's (1749–1827) *Mécanique céleste*, such synthetic treatments were necessarily more limited in their scope. Thus, C. F. Gauss's (1777–1855) *Disquisitiones arithmeticae* (1801) gave the first broad presentation of number theory, whereas Camille Jordan's (1838–1922) *Traité des substitutions et les équations algébriques* (1870) did the same for group theory.

No field of mathematical research is likely to endure for long without the presence of a recognized paradigmatic text that distills the fundamental results

---

[3] Joseph W. Dauben, "Conceptual Revolutions and the History of Mathematics: Two Studies in the Growth of Knowledge," in *Revolutions in Mathematics,* ed. Donald Gillies (Oxford: Clarendon Press, 1992), pp. 49–71; Judith V. Grabiner, "Is Mathematical Truth Time-Dependent?" in *New Directions in the Philosophy of Mathematics,* ed. Thomas Tymoczko (Boston: Birkhäuser, 1985), pp. 201–14; Ivor Grattan-Guinness, *Convolutions in French Mathematics, 1800–1840* (Science Networks, vols. 2–4) (Basel: Birkhäuser, 1990).

[4] For a model study exemplifying how this can be done for the case of topology, see Moritz Epple, *Die Entstehung der Knotentheorie: Kontexte und Konstruktionen einer modernen mathematischen Theorie* (Braunschweig: Vieweg, 1999).

and techniques vital to the subject. Euler's *Introductio in analysin infinitorum* (1748) fulfilled this function for those who wished to learn what became the standard version of the calculus in the eighteenth century. Throughout the nineteenth century, new literary genres emerged in conjunction with the vastly expanded educational aims of the period. French textbooks set the standard throughout the century, most of them developed from lecture courses offered at the Ecole Polytechnique and other institutions that cultivated higher mathematics. A. L. Cauchy's (1789–1859) *Cours d'analyse* (1821), the first in a long series of French textbooks on the calculus that bore the same title, gave the first modern presentation based on the limit concept. In the United States, S. F. Lacroix's (1765–1843) *Traité du calcul différentiel et du calcul intégral* was introduced at West Point, displacing more elementary British texts. When E. E. Kummer (1810–1893) and K. Weierstrass (1815–1897) founded the Berlin Seminar in 1860, the first books they acquired for its library were, with the exception of Euler's Latin calculus texts, all written in French.[5]

Most students in Germany, however, spent relatively little time studying published texts since the lecture courses they attended reflected the whims of the professors who taught them. By the nineteenth century, the old-fashioned *Vor-Lesungen,* where a gray-bearded scholar stood at his lectern and read from a text while his auditors struggled to stay awake, had largely disappeared. The *Vorlesungen* of the new era, though generally based on a written text, were delivered in a style that granted considerable latitude to spontaneous thought and verbal expression. Some *Dozenten* committed the contents of their written texts to memory, while others improvised their presentations along the way. But however varied their individual approaches may have been, the modern *Vorlesung* represented a new didactic form that strongly underscored the importance of oral communication in mathematics. It also led to a new genre of written text in mathematics: the (usually) authorized lecture notes based on the courses offered by (often) distinguished university mathematicians, a tradition that began with C. G. J. Jacobi (1804–1851). Thus, to learn Weierstrassian analysis, one could either go to Berlin and take notes in a crowded lecture hall or else try to get one's hands on someone else's *Ausarbeitung* of the master's presentations. Printed versions, like the textbook of Adolf Hurwitz and Richard Courant, only appeared much later. Monographic studies of a more systematic nature continued to play an important role, but the growing importance of specialized research journals, coupled with institutional innovations that fostered close ties between teaching and scholarship, served to undermine the once-dominant position of standard monographs. This trend reached its apex in Göttingen, where from 1895 to 1914 the lecture courses of Felix Klein (1849–1925) and David Hilbert (1862–1943) attracted talented students from around the world.

---

[5] Kurt-R. Biermann, *Die Mathematik und ihre Dozenten an der Berliner Universität, 1810–1933* (Berlin: Akademie Verlag, 1988), p. 106.

Hilbert's intense personality left a deep imprint on the atmosphere in Göttingen, where mathematicians mingled with astronomers and physicists in an era when all three disciplines interacted as never before. Yet Hilbert exerted a similarly strong influence through his literary production. As the author of two landmark texts, he helped inaugurate a modern style of mathematics that eventually came to dominate many aspects of twentieth-century research and education. Hilbert's *Zahlbericht*, which appeared in 1897, assimilated and extended many of the principal results from the German tradition in number theory that had begun with Gauss. Just two years later, he published *Grundlagen der Geometrie*, a booklet that eventually passed through twelve editions. By refashioning the axiomatic basis of Euclidean geometry, Hilbert established a new paradigm not only for geometrical research but also for foundations of mathematics in general. Three decades later, inspired by the work of Emmy Noether (1882–1935) and Emil Artin (1898–1962), B. L. van der Waerden's (1903–1993) *Moderne Algebra* (1931) gave the first holistic presentation of algebra based on the notion of algebraic structures. As Leo Corry has shown, van der Waerden's text served as a model for one of the century's most ambitious enterprises: the attempt by Nicolas Bourbaki, the pseudonym of a (primarily French) mathematical collective, to develop a theory of mathematical structures rich enough to provide a synthetic framework for the main body of modern mathematical knowledge.[6]

Yet even this monumental effort, which left a deep mark on mathematics in Europe and the United States in the period from roughly 1950 to 1980, eventually lost much of its former allure. Since then, mathematicians have made an unprecedented effort to communicate the gist of their work to larger audiences. Relying increasingly on expository articles and informal oral presentations to present their findings, many have shown no reluctance to convey new theorems and results accompanied by only the vaguest of hints formally justifying their claims. This quite recent trend reflects a growing desire among mathematicians for new venues and styles of discourse that make it easier for them to spread their ideas without having to suffer from the strictures imposed by traditional print culture as defined by the style of Bourbaki. Since the 1980s, some have even begun to question openly whether the ethos of rigor and formalized presentation so characteristic of the modern style makes any sense in the era of computer graphics. The historical roots of this dilemma, however, lie far deeper.

SHIFTING MODES OF PRODUCTION AND COMMUNICATION

Looking from the outside in, no careful observer could fail to notice the striking changes that have affected the ways mathematicians have practiced their

---

[6] Leo Corry, *Modern Algebra and the Rise of Mathematical Structures* (Science Networks, 16) (Boston: Birkhäuser, 1996).

craft over the course of the last two centuries. Long before the advent of the electronic age and the information superhighway, a profound transformation took place in the dominant modes of communication used by mathematicians, and this shift, in turn, has had strong repercussions not only for the conduct of mathematical research but also for the character of the enterprise as a whole. Stated in a nutshell, this change has meant the loss of hegemony of the written word and the emergence of a new style of research in which mathematical ideas and norms are primarily conveyed orally. As a key concomitant to this process, mathematical practices have increasingly come to be understood as group endeavors rather than activities pursued by a handful of geniuses working in splendid isolation. When seen against the backdrop of the early modern period, this striking shift – preceding the more recent and familiar electronic revolution – from written to oral modes of communication in mathematics appears, at least in part, as a natural outgrowth of broader transformations that affected scientific institutions, networks, and discourse.

Working in the earlier era of scientific academies dominated by royal patrons, the leading practitioners of the age – from Newton, Leibniz, and Euler, to Lagrange, and even afterward Gauss – understood the activity of doing and communicating mathematics almost exclusively in terms of the symbols they put on paper. Epistolary exchanges – often mediated by correspondents such as H. Oldenburg and M. Mersenne – served as the main vehicle for conveying unpublished results. To the extent that the leading figures did any teaching at all, their courses revealed little about recent research-level mathematics, nor were they expected to do so. The savant mathematician wrote for his peers, a tiny elite. As fellow members of academies and scientific societies during the seventeenth and eighteenth centuries, European mathematicians and natural philosophers interacted closely. By the end of this period, however, polymaths like J. H. Lambert had become rare birds, as the technical demands required to master the works of Euler and Lagrange were imposing. Still, higher mathematics had never been accessible to more than a handful of experts, and to learn more than the basics one generally had to seek out a master: Just as Leibniz sought out Huygens in Paris, so Euler turned to Johann Bernoulli in Basel. Mathematical tutors remained the heart and soul of Cambridge mathematical education throughout the nineteenth century, a throwback to an earlier, more personalized approach.

After 1800, mathematical affairs on the Continent underwent rapid transformation in the wake of the French Revolution, which sparked a series of profound political and social changes that reconfigured European science as well as its institutions of higher education. Enlightenment ideals of social progress based on the harnessing of scientific and technological knowledge animated the educational reforms of the period, which unleashed an unprecedented explosion of scientific activity in Paris during the Napoleonic period. Such mathematicians as Lazare Carnot, M. J. A. Condorcet, Gaspard Monge

(1746–1818), Joseph Fourier (1768–1830), and even the aged J. L. Lagrange (1736–1813) played a prominent part throughout. If leading figures like Carnot and Monge rallied to the Revolution's cause during the years of peril, most later directed their energies to scientific rather than political causes. Monge fell from power along with his beloved emperor, but Laplace managed to find favor with each passing regime. Only the staunch Bourbon sympathizer Cauchy, the most prolific writer of the century, found the new regime of Louis Philippe so distasteful that he felt compelled to leave France. Teaching and research remained largely distinct activities, but the once-isolated academicians were thrust into a new role: to train the nation's technocratic elite, a task that set a premium on their ability to convey mathematical ideas clearly.

In the German states, particularly in Prussia, a strong impulse arose to counter the rationalism and utilitarianism associated with the French Enlightenment tradition. To a large extent, modern research institutions emerged as an unintended by-product of this Prussian attempt to meet the challenge posed by France. Drawing on a Protestant work ethic and sense of duty so central to Prussian military and civilian life, scholarship *(Wissenschaft)* gained a deeper, quasi-religious meaning as a calling. Somewhat ironically, this reaction was coupled with a neohumanist approach to scholarship that proved highly conducive to the formation of modern research schools – in contrast to the "school learning" that continued to dominate at many European universities throughout the eighteenth century. Against the background of Romanticism, neohumanist values based on a revival of classical Greek and Latin authors permeated German learning from the founding of Berlin University in 1810 up until the emergence of the Second Empire in 1871.[7]

German scientists seldom faced the problem of having to justify their work to theological and political authorities. In their own tiny spheres of activity, scholars reigned supreme, while in exchange for this token status of freedom, they were expected to offer unconditional and enthusiastic allegiance to the state. Such were the terms of the implicit contract that bound the German professoriate to honor king and *Kaiser*. In return, they enjoyed the privileges of limited academic freedom and disciplinary autonomy, along with a social status that enabled them to hobnob with military officers and aristocrats. If French savant mathematicians bore responsibility for training a new profession of technocrats, German professors mainly taught future Gymnasien teachers, a position that carried considerable social prestige itself. Indeed, a number of Germany's leading mathematicians, including Kummer and Weierstrass, began their careers as Gymnasien teachers, but scores of others who never dreamt of university careers published respectable work in leading academic journals.

---

[7] See Lewis Pyenson, *Neohumanism and the Persistence of Pure Mathematics in Wilhelmian Germany* (Philadelphia: American Philosophical Society, 1983).

Mathematicians have often found ways to communicate and even to collaborate without being in close physical proximity. Nevertheless, *intense* cooperative efforts have normally necessitated an environment where direct, unmediated communication could take place, and precisely this kind of atmosphere arose quite naturally in the isolated settings of small German university towns. Indeed, Germany's decentralized university system, coupled with the ethos of *Wissenschaft* that pervaded the Prussian educational reforms, created the preconditions for a new "research imperative" that provided the animus for modern research schools.[8] Throughout most of the nineteenth century, these schools typically operated in local environments, but with time, small-scale research groups began to interact within more complex organizational networks, thereby stimulating and altering activity within the localized contexts. Collaborative efforts and coauthored papers, still comparatively rare throughout the nineteenth century, became increasingly popular after 1900. By the mid-twentieth century, papers with multiple authorship, and often acknowledging the assistance of numerous other individuals, were at least as numerous as those composed by a single individual. Such collaborative research presupposes suitable working conditions and, in particular, a critical mass of researchers with similar backgrounds and shared interests. A work group may be composed of peers, but often one of the individuals assumes a leadership role, most typically as the academic mentor to the junior members of the group. This type of arrangement – the modern mathematical research school – has persisted in various forms throughout the nineteenth and twentieth centuries.

MATHEMATICAL RESEARCH SCHOOLS IN GERMANY

The emergence of distinct mathematical research schools and traditions in the nineteenth century and their rapid proliferation in the twentieth accompanied a general trend toward specialization in scientific research. Recently, historians of the physical sciences have focused considerable attention on the structure and function of research schools, undertaking a number of detailed case studies aimed at exploring their finer textures.[9] At the same time, they have tried to understand how the kind of locally gained knowledge produced by research schools becomes "universal," a process that involves analyzing all the various mechanisms that produce consensus and support within broader scientific networks and circles. Similar studies of mathematical schools, on

---

[8] For the preconditions, see Steven R. Turner, "The Prussian Universities and the Concept of Research," *Internationales Archiv für Sozialgeschichte der deutschen Literatur,* 5 (1980), 68–93. For an overview of historiographic trends, see John Servos, "Research Schools and their Histories," *Research Schools: Historical Reappraisals,* ed. Gerald L. Geison and Frederic L. Holmes, *Osiris,* 8 (1993), 3–15.

[9] See, for example, the essays in *Research Schools: Historical Reappraisals,* ed. Gerald L. Geison and Frederic L. Holmes, *Osiris,* 8 (1993), 227–38.

the other hand, have been lacking, a circumstance no doubt partly due to the prevalent belief that "true" mathematical knowledge is from its very inception universal and, hence, stands in no urgent need to win converts. As suggested earlier, this standard picture of mathematics as an essentially value-free discipline hampers any serious attempt to understand mathematical practices historically. As a relatively stable element within the complex fluctuating picture of mathematical activity over the last two centuries, research schools offer historians a convenient category for better understanding *how* mathematicians produce their work, rather than focusing exclusively on the end products of these efforts, mathematical texts. Nevertheless, a word of caution must be added with respect to "research programs" in mathematics, since these have often transcended the local environments of schools, a tendency that can easily blur important conceptual distinctions about schools and knowledge that ought to be maintained.

Unlike seminars or mathematical societies, which were organizations governed by statutes or written regulations, mathematical research schools emerged as purely spontaneous arrangements with no such formal structures. Thus, determining membership in a school or even the very existence of such a setting can be quite problematic, owing to the voluntary character of the enterprise. Clearly, the leader of a school had to be not only an acknowledged authority in the field but also someone capable of imparting expertise to pupils. Leadership carried with it, among other demands, an obligation to supervise doctoral dissertations and postdoctoral research. This supervisory function, however, could take almost any form, depending on the working style adopted by the school's leader. In the final instance, this kind of arrangement depended on an implicit reciprocal agreement between the professor and his students, as well as among the students themselves, to form a symbiotic learning and working environment based on the research interests of the professor. Unlike that of laboratory research schools, however, the principal aim of a typical mathematical school was neither to promote a specific research program nor to engage in a concerted effort to solve a problem or widen a theory. These were merely potential means subordinate to the real end, which was to produce talented new researchers. For the strength of a mathematical school depended mainly on the quality of its younger members as gauged by their later achievements as mature, creative mathematicians.

The appellation "school" has traditionally been used by mathematicians to describe various groups of individuals who share a general research interest or, perhaps, a particular orientation to their subject. This usage places the accent on intellectual affinities and implies nothing more than a loosely shared intellectual context. It has been commonplace, for example, to speak of practitioners of Riemannian or Weierstrassian function theory as members of two competing "schools," despite the fact that B. Riemann (1826–1866) never drew more than a handful of students, whereas Weierstrass was the acknowledged leader of the dominant school of his day (even France's leading analyst, Charles Hermite [1822–1901], was supposed to have said "Weierstrass est notre maître à tous"). Clearly, the spheres of activity and influence of Riemann and Weierstrass were radically dissimilar, suggesting that here, as elsewhere, it would be more apt to distinguish between two competing mathematical *traditions* rather than schools, particularly in view of the complex methodological issues involved.

Since the personality and working style of its leader played such an important part in shaping the character of a school, general patterns are difficult to discern. Still, one typical, though by no means universal, feature of mathematical research schools in nineteenth-century Germany was the strong sense of loyalty their members displayed. If obsequious behavior and subservient attitudes were commonplace in Prussian society, the more extreme forms of discipleship practiced in some mathematical schools clearly constituted a special kind of spiritual attachment, as illustrated by the following three examples. Jacobi's Königsberg school could never have retained its lasting fame had not Friedrich Richelot, his *Lieblingsschüler*, assistant, and later successor, assumed the role of keeper of the flame. Robert Fricke, pupil, protégé, and later nephew of Felix Klein, spent his most productive years writing the four massive volumes (Klein-Fricke, *Theorie der elliptischen Modulfunktionen*, and Fricke-Klein, *Theorie der automorphen Funktionen*) that extended and refined his teacher's earlier work. Friedrich Engel went to Norway as a kind of mathematical Boswell to Klein's friend, Sophus Lie (1842–1899). Engel spent the next ten years writing Lie's three-volume *Theorie der Transformationsgruppen* and, after Lie's death, another twenty years preparing his collected works in seven volumes. These three disciples devoted most of their careers to glorifying the reputations of the schools they represented. But while the names of their teachers would be familiar to every educated mathematician today, these ultradevout pupils have been all but forgotten.

In nineteenth-century Germany, the prominently situated schools associated with Jacobi, A. Clebsch (1833–1872), Weierstrass, and Klein all interacted with one another, spawning larger networks and spheres of influence. A common trait among less successful schools, on the other hand, was the pursuit of research in a fairly narrow field, as such work often failed to attract interest outside of a small corps of experts. This situation could easily spell the death of a research school once its mentor passed from the scene. Institutional power generally implied a long-term affiliation with a single base of operations, and with few exceptions, mathematical research schools have never thrived without stable leadership. Many of the more successful, however, also maintained strong ties with other centers, thereby gaining employment opportunities for their students as well as a support network within the research community. The dangers of a diaspora effect were largely mitigated by the sense of loyalty toward their former mentors that was felt by those who went on to other institutions. Research schools thus eventually became integrated into more complex networks, some operating within national communities, others involving international institutions or contacts abroad.

OTHER NATIONAL TRADITIONS

Throughout the first half of the nineteenth century, the networks linking mathematicians overlapped with those that had formed between astronomers and physicists. The strength of these ties was especially pronounced in the work of older figures, such as Laplace and Gauss, but some younger contemporaries, Jacobi and Cauchy among them, maintained similarly strong interests in mathematical physics.[10] Mathematicians had long enjoyed an exalted position within the French scientific community, but it was during the Napoleonic era that they first assumed an important role as educators. The curriculum instituted at the newly founded Ecole Polytechnique bore the strong imprint of Monge's vision. Its elite corps of engineering students imbibed huge quantities of mathematical knowledge, with a special emphasis on analysis and that Mongian specialty, descriptive geometry. Monge, whose teaching talents equaled his abilities as a researcher, inspired a generation of researchers – J. Hachette, V. Poncelet, M. Chasles, et al. – who, along with J. D. Gergonne (1771–1859), went on to lay the groundwork for the nineteenth-century renaissance in geometry. The physicalist side of this Mongian legacy was upheld by E. Malus, C. Dupin, and P. O. Bonnet, who made vital contributions to geometrical optics and differential geometry. Meanwhile, the French tradition in analysis, stretching from Lagrange, Adrien-Marie Legendre (1752–1833), and Laplace to S. D. Poisson (1781–1840), Fourier, and Cauchy, was even more dominant. Little wonder that until the 1830s, Parisian mathematics remained practially a world unto itself. Unfortunately, this exclusivity sometimes led to a callous neglect of budding talent, the two most dramatic cases being Evariste Galois (1811–1832) and the Norwegian Niels Henrik Abel (1802–1829).

A more open attitude toward the work of "outsiders" emerged, however, with the two figures who assumed Cauchy's mantle, Joseph Liouville and C. Hermite.[11] The latter pair, along with the aged Legendre, gave enthusiastic support to the new theory of elliptic functions and higher transcendentals cofounded by Jacobi and Abel. Owing to its rich connections with number theory and algebra, this theory quickly assumed a central place not only within analysis but in nearly all parts of pure mathematics as well. The famous Jacobi inversion problem posed one of the era's major challenges,

---

[10] On Laplace's physical research program, see Robert Fox, "The Rise and Fall of Laplacian Physics," *Historical Studies in the Physical Sciences*, 4 (1974), 89–136.

[11] See Jesper Lützen, *Joseph Liouville, 1809–1882* (Studies in the History of Mathematics and Physical Sciences, 15) (New York: Springer-Verlag, 1990).

prompting contributions by Weierstrass and Riemann that garnered nearly instantaneous fame for both. Thus, by midcentury, research trends in the pacesetting mathematical communities of France and Germany had begun to shift toward fields located on the "pure" end of the mathematical spectrum. By 1900, the Ecole Normale Supérieure had replaced the Ecole Polytechnique as the principal training ground for France's new generation of elite mathematicians: Gaston Darboux (1842–1917), Henri Poincaré (1854–1912), Emile Picard (1856–1941), and others. But the French pedagogical system was still dominated by drill and technical proficiency, apparently reinforced by the assumption that mathematical creativity constituted an innate talent that could neither be taught nor nurtured. Even in Paris, where Poincaré and Picard regularly offered courses on advanced topics, students could find nothing comparable to the German-style seminars that served as a bridge to the world of research mathematics.

Much as the defeat of Napoleon's forces led to a flowering of intellectual pursuits in Germany, the liberation of Italy from Austria and the ensuing *Risorgimento* led to renewed activity that found ample expression in mathematical spheres.[12] A signal event for this revival occurred in 1858 when Francesco Brioschi (1824–1897) founded the *Annali di matematica pura et applicata*. By century's end, Italian mathematicians came to be recognized as the world's leading authorities in most areas of geometry. As director of the engineering school in Rome after 1873, Luigi Cremona (1830–1903) stood at the heart of geometrical research in Italy during its period of ascendancy. Cremona transformations became a major tool in the new birational geometry suggested by Riemann's work, a field that eventually overshadowed projective geometry in the work of the Italian tradition. In Turin, Corrado Segre (1863–1924) founded a school in algebraic geometry that built on the earlier work of Julius Plücker and Klein but also extended the results of the Clebsch school.

In differential geometry, Luigi Bianchi's (1856–1928) three-volume *Lezioni di geometria differenziale* (1902–1909) provided a worthy sequel to Darboux's monumental *Leçons sur la théorie générale des surfaces* (4 vols., 1887–96). Building on another Riemannian legacy, the theory of quadratic differential forms as first elaborated by the Germans E. B. Christoffel and R. Lipschitz, as well as E. Beltrami's theory of differential parameters, in 1884 the Paduan Gregorio Ricci-Curbastro (1853–1925) developed a so-called absolute differential calculus. At first shunned by leading differential geometers as an abstract symbolism that failed to produce concrete geometrical results, the absolute differential calculus was elaborated with applications to elasticity theory and hydrodynamics by Ricci and his pupil Tullio Levi-Civita (1873–1941) into what became modern tensor calculus. Still, interest in this subject remained

---

[12] See Simonetta Di Sieno et al., eds., *La Matematica Italiana dopo L'Unita* (Milan: Marcos y Marcos, 1998).

confined to a few experts before 1916, the year in which Einstein gave a lengthy presentation of the tensor calculus as a prelude to his first extensive exposition of the general theory of relativity.[13]

Whereas large-scale institutional reforms in France, Germany, and Italy created favorable preconditions for the formation of three vibrant national research communities, Britain remained out of step with these developments throughout the entire century. In 1812 the Cambridge Analytical Society, led by George Peacock, Charles Babbage, and John Herschel, attempted to reform calculus instruction by shunting aside Newtonian fluxions in favor of the Leibnizian notation that had long since won sway on the Continent. This movement met with some modest success, but hardly brought sweeping changes even at Cambridge, the only English university that offered serious mathematical instruction. Its old-fashioned Tripos system, where future Wranglers honed their skills in the rooms of their tutors, looked quaint indeed outside the world of Victorian England.[14] Throughout the century, England's amateur tradition continued to pervade much scientific research, as mathematics remained the handmaiden of natural philosophy. Meetings of the London Mathematical Society, founded in 1865, resembled the casual gatherings of a typical gentleman's club. Cambridge remained a mathematical backwater until well after the turn of the century when G. H. Hardy (1877–1947) joined forces with J. E. Littlewood (1885–1977) and the geometer H. F. Baker.

Nevertheless, several remarkably creative mathematicians emerged from this antiquated system, including Arthur Cayley (1821–1895) and J. J. Sylvester (1814–1897), both of whom made important contributions to algebra and geometry. Algebraic invariant theory gained much of its impetus from projective geometry, a subject that received its first thorough analytical treatment in the textbooks written by the Irish geometer George Salmon (1819–1904). The welcome Salmon's work received in Germany can be seen from the numerous editions of the Salmon-Fiedler texts that appeared during the last four decades of the century. With Cayley's avid assistance, Wilhelm Fiedler greatly amplified the later editions of these monographs with material drawn from more recently published research on algebraic curves and surfaces. Thus, it was primarily by means of such mediated literary transmission, rather than through direct oral discourse, that Cayley, Sylvester, and Salmon influenced subsequent developments. Deeper and more enduring was the impact exerted on mathematics by contemporary British natural philosophers, especially the Irish astronomer-mathematician William Rowan Hamilton (1805–1865) and the Scottish physicists William Thomson (1824–1907), Peter Guthrie Tait (1831–1901), and James Clerk Maxwell (1831–1879). Among the

[13] Karin Reich, *Die Entwicklung des Tensorkalküls: Vom absoluten Differentialkalkül zur Relativitätstheorie* (Science Networks, vol. 11) (Basel: Birkhäuser, 1992).
[14] Joan L. Richards, *Mathematical Visions: The Pursuit of Geometry in Victorian England* (Boston: Academic Press, 1988).

distinguished array of Wranglers and physicists who embodied this British national style were such leading figures as J. W. Strutt (Lord Rayeligh), Arthur Schuster, Robert S. Ball, John Perry, J. H. Poynting, Horace Lamb, and Arthur Eddington.[15] Like Cayley and Sylvester, however, none of these champions of the dominant applied style became the head of a research school.

Throughout the course of the nineteenth century, Britain managed to develop a distinctly mixed mathematical tradition quite its own. Every modern calculus text contains some version of the theorems of George Green (1793–1841) and George Gabriel Stokes (1819–1903), results of fundamental importance for theoretical physics. Yet before 1900, relatively few students would have had more than a passing familiarity with these theorems, which only entered the core of the mathematical curriculum when vector analysis came into ascendancy. Up until the outbreak of World War I, the merits of so-called direct methods were hotly debated by traditionalists, who opposed them in favor of good old-fashioned Cartesian coordinates, and those who advocated various special systems. Tait fervently championed W. R. Hamilton's quaternions, a system challenged in turn by a small but vocal band of German mathematicians who preferred H. G. Grassmann's (1809–1877) approach. While the mathematicians were busy squabbling, two physicists decided that a system even simpler than quaternions would suit them just fine, and in the 1890s J. W. Gibbs (1839–1903) and Oliver Heaviside (1850–1925) found what they were looking for by fashioning modern vector analysis.[16] The notion of a vector field that emerged soon afterward was rooted in the kinds of physical speculations pursued by Thomson, Tait, and Maxwell in their thermo- and electrodynamical investigations. After H. Hertz's direct experimental verification of Maxwell's theory and the elegant presentation of Maxwell's equations in vector form by Heaviside, field physics and vector analysis, with their evident advantages over coordinate methods, quickly gained ground on the Continent.

During the late Wilhelmian era, the German universities reached the pinnacle of their international influence, exerting an especially strong impact on the younger generation of mathematicians in the United States. Their preferred mentor Felix Klein managed to attract more talented American youth during the late 1880s and early 1890s than all other mathematicians in Europe combined.[17] His pupils spearheaded a successful effort to build viable graduate programs at the three universities that would dominate American mathematics for years to come: Chicago, Harvard, and Princeton.

---

[15] See the essays in P. M. Harman, ed., *Wranglers and Physicists* (Manchester: Manchester University Press, 1985).

[16] Michael J. Crowe, *A History of Vector Analysis* (Notre Dame, Ind.: University of Notre Dame Press, 1967).

[17] Karen H. Parshall and David E. Rowe, *The Emergence of the American Mathematical Research Community, 1876–1900: J. J. Sylvester, Felix Klein, and E. H. Moore* (Providence, R.I.: American Mathematical Society, 1994), pp. 175–228.

At the University of Chicago, which opened in 1892, Eliakim Hastings Moore (1862–1932) was joined by two of Klein's former German students, Oskar Bolza (1877–1942) and Heinrich Maschke (1853–1908). This triumvirate quickly established itself as the nation's dominant school by the turn of the century. Even though Chicago's mathematicians failed to score any dramatic research breakthroughs, their work was situated close to some of the era's most active developments. Moore, in particular, had a sharp eye for new trends, and Chicago's better students soon found themselves working at the fast-moving frontiers of modern mathematics. A few of those students went on to help change not only the way mathematics looked but also the manner in which it was done at their respective institutions. Indeed, five star graduates of the Chicago school emerged as dominant figures during the 1920s and 1930s: George D. Birkhoff (1884–1944) solidified Harvard's position as the leading center for research in analysis and mathematical physics; Oswald Veblen (1880–1960) provided leadership for the flourishing of geometry and topology at Princeton; Leonard Dickson (1874–1954) and Gilbert Ames Bliss (1876–1951) carried on their mentors' legacies in Chicago; and Robert Lee Moore (1881–1974) founded a research school in point-set topology at the University of Texas that spawned several generations of academic progeny.

GÖTTINGEN'S MODERN MATHEMATICAL COMMUNITY

By 1900, independent mathematical research communities with their own organizations had formed in Britain, France, Germany, Italy, and the United States; soon thereafter, research schools and mathematical societies would emerge in Russia, Poland, Sweden, and other countries. Through visits and by attending conferences and international congresses, members of these communities and local centers began to intensify their contacts, building new networks of power and influence. These developments gradually led to a transformation in conventional modes of production and communication among mathematicians, a shift shaped by a complex variety of factors – technical, social, political, educational – that affected nearly all forms of scientific endeavor in various ways. At the same time, the social status and function of mathematicians, as producers and purveyors of mathematical knowledge, underwent significant changes. Reforms in higher education went hand in hand with new institutions that placed a premium on various forms of mathematical knowledge taught by professional pedagogues. Traditional links with subjects like astronomy, geodesy, and mechanics survived, but after 1900 these were recast in accord with rapidly diverging professional research interests.

Between 1900 and the outbreak of World War I, these new forces found their boldest expression in Göttingen, where Klein and Hilbert headed a new kind of center for the mathematical sciences that significantly altered established norms for the conduct of research.[18] Although a multifaceted enterprise, the Göttingen experiment was particularly influential owing to the atmosphere that surrounded Hilbert's dynamic research group, which burst the mold of the traditional mathematical school. Hilbert began his career as an algebraist with strong interests in algebraic number theory, but after 1900 his research interests underwent a dramatic shift. Thereafter, he and his army of students began concentrating on various topics in analysis (integral equations, calculus of variations, and so forth), work strongly linked with Hilbert's interests in mathematical physics. At the same time, Göttingen emerged as a hub of activity within the widening networks of international contacts. Klein had a burning ambition to turn Göttingen into a microcosm of the mathematical world, a center offering mathematicians a platform for gaining access to the major trends in research. His principal literary vehicle in bringing this about after 1895 was the *Encyklopädie der mathematischen Wissenschaften*, a mammoth project that enlisted the services of leading scholars from Italy, Great Britain, France, and the United States. Its main thrust and none-too-hidden agenda involved articulating the role of mathematics in those disciplines most heavily dependent on sophisticated mathematical techniques. Taking aim at theoretical physics, Klein gained the support of such luminaries as Arnold Sommerfeld, Paul Ehrenfest, H. A. Lorentz, and Wolfgang Pauli. Even better known are the many trails of influence in pure mathematics that can be traced back to Hilbert's pupils and disciples. When seen in these terms, the Göttingen mathematical community of Klein and Hilbert emerges as nothing less than a watershed phenomenon. Its participants experienced a new kind of working environment characterized by intense social interaction, collaboration, and cutthroat competition.

The events of World War I shattered the always fragile relations between French and German mathematicians. Symptomatic of the prevailing embitterment was the decision to hold the postponed fifth international congress in Strasbourg in 1920 and to exclude participation by Germans. By the time of the 1928 Bologna congress, few favored prolonging the boycott, but the Germans themselves were divided about whether to participate. Ludwig Bieberbach (1886–1982), with backing from the Dutch topologist and intuitionist L. E. J. Brouwer (1881–1966), tried to mount a counterboycott, an effort that fell flat when Hilbert decided to lead a German delegation to the congress.[19] This encounter presaged Bieberbach's activities as the leading spokesman for "Aryan mathematics" during the Nazi era. For a half century or more, the German universities had attracted mathematical talent from around the world, but once Hitler seized power, his regime prompted a brain drain of staggering proportions. Some found temporary refuge in the Soviet

---

[18] David E. Rowe, "Klein, Hilbert, and the Göttingen Mathematical Tradition," in *Science in Germany: The Intersection of Institutional and Intellectual Issues*, ed. Kathryn M. Olesko, *Osiris*, 5 (1989), 189–213.
[19] On Brouwer, see Dirk van Dalen, *Mystic, Geometer, and Intuitionist: The Life of L. E. J. Brouwer*, vol. 1 (Oxford: Clarendon Press, 1999).

Union, others in England, but the bulk of those who were lucky enough to escape the terror emigrated to the United States.

PURE AND APPLIED MATHEMATICS IN THE COLD WAR ERA AND BEYOND

The exodus of European mathematicians to the United States during the 1930s involved a concomitant transformation of research interests that affected the émigrés and native Americans alike.[20] Before the outbreak of World War II, pure mathematics totally dominated the North American scene, led by the centers at Chicago, Harvard, and Princeton. Applied mathematics gained considerable momentum, however, from wartime research. Two leading outposts were founded by former Göttingen figures: Richard Courant (1888–1972) built a mathematical institute at New York University practically from the bottom up, while Theodor von Kármán (1881–1963) created an institute for aerodynamical research at California Institute of Technology. Under the leadership of R. G. Richardson, Brown University emerged as another major center for applied research. Mathematicians conducted ballistics tests at the Aberdeen Proving Grounds; some worked on developing radar systems at MIT's Radiation Laboratory; others, such as Stanislaw Ulam, joined J. Robert Oppenheimer's research team at Los Alamos. MIT's Vannevar Bush played an instrumental part in launching the U.S. Navy's Office of Scientific Research and Development, as well as the Applied Mathematics Panel that was established under the directorship of Warren Weaver in 1942. Many of the members of this transformed American mathematical community thus became deeply engaged in war-related research, just as had been the case during World War I. But unlike their predecessors, practically all of whom returned to pursue pure mathematics after 1919, a considerable number of Americans remained actively engaged in applied research after World War II, much of which was government funded.[21]

With the demise of viable working conditions in Europe came the emergence of a second mathematical "super power" in the Soviet Union.[22] Russia's first active research school had been founded in Moscow around 1900 by Dmitri Egorov (1869–1931) and expanded by his student N. N. Lusin (1883–1950), who specialized in the theory of real functions. A. Ya. Khinchin and M. Ya. Suslin followed his lead, contributing to the renovation of real analysis and Fourier series that took place during this

---

[20] See Reinhard Siegmund-Schultze, *Mathematiker auf der Flucht vor Hitler* (Dokumente zur Geschichte der Mathematik, Band 10) (Braunschweig: Vieweg 1998).
[21] See Amy Dahan Dalmedico, "Mathematics in the Twentieth Century," in *Science in the Twentieth Century*, ed. John Krige and Dominique Pestre (Paris: Harwood Academic Publishers, 1997), pp. 651–67.
[22] See Loren Graham, *Science in Russia and the Soviet Union* (Cambridge: Cambridge University Press, 1993).

period. The Russian school of analysts closely followed the work of leading French figures, including Emile Borel (1871–1956) and Henri Lebesgue (1875–1941), who had created modern theories of measure and integration based on Cantorian set theory. Lusin played a role comparable to that of E. H. Moore in the United States, training a number of gifted students, two of whom far surpassed their teacher: A. N. Kolmogorov (1903–1987), a pioneering figure in probability theory and dynamical systems, and Paul S. Alexandroff (1896–1982), who made seminal contributions to algebraic topology. Kolmogorov's school pursued the two principal directions that guided its leader's research, stochastics and dynamical systems. Modern stochastics began with Kolmogorov's axiomatization of probabilistic systems in the 1930s; in the 1940s he worked on statistical methods for studying turbulence; and in the 1950s he launched the now-famous theory of perturbed Hamiltonian systems that has come to be called KAM (Kolmogorov-Arnold-Moser) theory, one of the cornerstones in the theory of dynamical systems. Among the many distinguished figures associated with the Kolmogorov school were Y. Manin, V. I. Arnold, and S. P. Novikov. Mathematical researchers enjoyed a privileged status in Soviet society, where like star athletes and chess masters, they were nurtured in a system that cultivated their talents from an early age. Beginning in 1936, mathematics olympiads were held every year, a form of competition that served to identify the likely members of the next generation's mathematical elite.

The Cold War and ensuing space race between the United States and the Soviet Union meant large military budgets and lavish support for scientific and technical programs. In the wake of these political events, new organizations, like the National Science Foundation, opened numerous opportunities for professional mathematicians. One of the few American leaders who took a critical view of this encroachment of government agencies on mathematical research was MIT's Norbert Wiener (1894–1964). Steve Heims contrasted Wiener's attitude with that of John von Neumann (1903–1957), the brilliant Hungarian émigré who worked closely with military leaders in the United States and later served as a member of the Atomic Energy Commission.[23] With von Neumann, American mathematics passed into the era of the electronic computer.

Still, the 1960s and 1970s witnessed a major resurgence of interest in pure mathematics, as the "new math" movement swept through American education and as graduate schools began granting nearly a thousand doctoral degrees per year. This was the era that coined the watchword of every tenure-track assistant professor – "publish or perish" – the name Michael Spivak chose for his low-budget mathematical publishing house. A younger generation pushed for a new purist style; inspired by Bourbaki and abstract structures, they produced a mountain of new results, many dealing with highly

---

[23] Steve J. Heims, *John von Neumann and Norbert Wiener* (Cambridge, Mass.: MIT Press, 1980).

esoteric problems intelligible only to the initiated specialist. It is ironic that the founders of the Bourbaki movement were among the few who recognized the debt their ideas owed to the largely forgotten accomplishments of the past century.

As political tensions gradually subsided during the 1980s and 1990s, new communities interacted with old ones in an atmosphere in which national boundaries no longer constrained discourse as they had throughout most of the Cold War era. International congresses became truly international events, drawing mathematicians from Eastern Asia, South America, and Africa. A new wave of Russian émigrés enriched the North American community, whose membership increasingly came to resemble the diverse ethnic mix characteristic of late-twentieth-century culture. Like many other segments of contemporary life, mathematical research has been profoundly affected by the electronic revolution, leading to vast new networks of communication and collaboration among mathematicians around the world. In the wake of this upheaval, the significance of traditional mathematical schools, around which so much teaching and research activity had centered in the past, would now appear doubtful for the future.

If more recent events defy capsule summary, they at least reveal that research trends and fashions in mathematics, like those in other disciplines, can and do undergo abrupt changes. In the 1960s and 1970s, category theory, point-set topology, and catastrophe theory were all the rage; by the 1990s they had all but disappeared from the scene, as fractals and computer graphics captured mathematicians' fancies. Indeed, specialization and the urgency to publish new findings have quickly generated an immense wealth of information in recent decades. But the oft-repeated claim that the overwhelming preponderance of known mathematical results has been attained only rather recently – after 1950 or so – merely reinforces the illusion that this boom constitutes the latest phase in a steadily rising growth curve. In terms of their broader significance for mathematical culture, one might more plausibly argue that this explosion of new results merely parallels another well-known phenomenon of modern-day life: instantaneous (unplanned) obsolescence.

Seen from this vantage point, the issues of accessibility and retrievability appear in a very different light. As in any other human endeavor, the mythic stockpile of once-found mathematical results contains vast quantities of obsolete materials of no conceivable use to or interest for present-day working mathematicians or anyone else. For modern research mathematicians, reference works like *Mathematical Reviews* and *Zentralblatt* have become indispensable tools for gaining access to the latest findings in the published literature. But this hardly means that these types of resources make the bulk of present-day mathematical knowledge potentially retrievable to any trained mathematician who bothers to flip through enough pages. For all practical purposes, the collective culture of professional mathematicians in the postmodern era has only a rather limited access to the work and ideas of their predecessors. If the history of mathematics demonstrates anything, it is that mathematical results can just as easily be forgotten as found. Sometimes old results are discovered anew, but when this happens, the ideas involved rarely ever reemerge exactly as before. Such processes of rediscovery and transmission are nearly always accompanied by more or less subtle transformations that may significantly alter the meanings that a later generation or a different culture attaches to its findings.

7

THE INDUSTRY, RESEARCH, AND EDUCATION NEXUS

Terry Shinn

This chapter explores the impact of science and technology research capacity and educational change on industrial performance in the century and a half since 1850. Analysis covers four countries remarkable for their industrial achievement, England, France, Germany, and the United States. It is important to note that for each of these countries, economic growth has often been organized around contrasting systems of education and research.

Today, most scholars agree that education, as a general phenomenon, does not constitute a linear, direct determinant of industrial growth. For example, Fritz Ringer has shown that although German and French education had numerous parallels in the nineteenth and early twentieth centuries, such as per capita size of cohorts, the economic development of the two nations was extremely different.[1] Peter Lundgreen, who has compared the size of France's and Germany's engineering communities and the character of training, has come to much the same conclusion.[2] Robert Fox and Anna Guagnini, in a comparative study of education and industry in six European countries and the United States for the pre–World War I decades, demonstrate that although nations had contrasting rates of industrial growth, their educational policies and practices nevertheless frequently converged.[3]

The existence of a direct and linear connection between research and industry is also viewed as doubtful today. For example, during the decades immediately preceding and following World War I, very few French firms possessed any research capacity, and with scant exception, neither was applied research present inside the educational system. Still, France's industry advanced at a

---

[1] Fritz K. Ringer, *Education and Society in Modern Europe* (Bloomington: Indiana University Press, 1979), pp. 230–1 and 237.
[2] Peter Lundgreen, "The Organization of Science and Technology in France: A German Perspective," in *The Organization of Science and Technology in France, 1808–1914*, ed. Robert Fox and George Weisz (Cambridge: Cambridge University Press, 1980), pp. 327–30.
[3] Robert Fox and Anna Guagnini, *Education, Technology and Industrial Performance, 1850–1939* (Cambridge: Cambridge University Press, 1993), p. 5.

steady albeit slow pace, thanks largely to alternative innovation-acquisition practices, such as patent procurement, licensing, and concentration on low-technology sectors.[4] In large measure, France's industrial capacity was derivative, often depending on the importation of technology from abroad.[5]

I will argue that while industrial performance is rarely coupled directly either to research or to education, it is nevertheless the case that economic development is strongly associated with a bimodal factor of research/education. Only when interacting in a particular fashion does their potential to promote industrial innovation emerge. I will furthermore suggest that in order to be effective, research must be vested with specific structural attributes that enable industry to benefit, and that the same holds for science and technical education. A range of historical mechanisms, some positive and others inhibiting, will be set forth.

GERMANY AS A PARADIGM OF HETEROGENEITY

Scholars agree that the final third of the nineteenth century saw a sharp change in the relations of capitalistic industrial production, in effect, the birth of the "capitalization of knowledge."[6] Systematic and formalized learning emerged as a crucial component of industrial processes, alongside the existing key elements of capital, equipment, labor, and investment. Before midcentury, technical training had largely taken the form of apprenticeship. The elaboration of industrial novelty had been left to chance and frequently originated in sources exogenous to industry. With the capitalization of knowledge, however, scientific and technical capacity acquired the guise of formal learning, which assumed a central role within firms; and appropriately differentiated education arose that offered the required concepts, technical information, and skills. Similarly, industrial innovation was no longer left to isolated, private inventors. Applied research was increasingly promoted inside firms, and government and academia also sponsored applied science and engineering-related investigations. By all accounts, Germany was the first nation to move toward the capitalization of knowledge, and accordingly, it developed a range of well-adapted educational sites and research establishments.

In the half century before World War I, German industrial performance was truly staggering on numerous counts. It suddenly moved ahead of England and France at midcentury. Germany spearheaded the second industrial revolution, and in doing so, it set historical record after record for

---

[4] Terry Shinn, "The Genesis of French Industrial Research – 1880–1940," in *Social Science Information*, 19, no. 3 (1981), 607–40.

[5] Robert Fox, "France in Perspective: Education, Innovation, and Performance in the French Electrical Industry, 1880–1914," in Fox and Guagnini, *Education, Technology,* pp. 201–26, particularly pp. 212–14.

[6] H. Braverman, *Labor and Monopoly Capital: The Degradation of Work in the Twentieth Century* (New York: Monthly Review Press, 1974).

economic growth. But precisely to what degree was this impressive achievement dependent on education- and research-associated elements? The renowned Technische Hochschulen are often portrayed as the linchpin of German educational service to industry in the late nineteenth and early twentieth centuries, and beyond this as an exemplar of what education-industry relations can achieve.[7] Between 1870 and 1910, three new schools were added (Aachen, Danzig, and Breslau) to eight previously established institutions in Prussia and the other *Länder* (Berlin, Karlsruhe, Munich, Dresden, Stuttgart, Hanover, Braunschweig, and Darmstadt). They provided technical education in science, engineering, and applied research to tens of thousands of industry-minded men. By around 1900, instruction at the Technische Hochschulen had become four-pronged: (1) deduction of technical rules from industrial activities; (2) deduction of technical rules from natural laws; (3) adaptation of sometimes abstruse calculating techniques for industrial needs; (4) systematic research into materials and processes applicable to industry.

Between 1900 and 1914 alone, the Technische Hochschulen graduated more than 10,000 exceptionally qualified students who flooded an already saturated labor market. Alumni became engineers in manufacturing firms in areas associated with chemistry, electricity (and later also electronics), optics, and mechanics. Many rose to positions of top management, and some became directors of firms. Technische Hochschulen offered five to seven years of instruction, after 1899 optionally leading to a doctorate degree. The right to grant this diploma was hard won and achieved only after a bitter twenty-year struggle against the nation's well-entrenched universities. Until the end of the century, the German university had enjoyed an uncontested monopoly over doctoral education. The victory of the Technische Hochschulen was singularly important, for it was emblematic of the newly acquired high status of engineering and technical learning and represented tacit admission of the crucial position of industry in the rapidly modernizing German social order.

Historians have noted that the late-nineteenth-century emergence of Germany's highly acclaimed Technische Hochschulen, whose reputation was entwined with industrial success, was part of a broader educational and cultural transformation. Until midcentury, classical humanistic education, *Bildung*, had comprised the foremost and almost uncontested form of education in Germany. Classical learning was the hallmark of the educated, traditional bourgeoisie, and such learning was acquired in the very exclusive Gymnasien and universities. Humanistic training alone had conferred social legitimacy. After 1850, however, a measure of "modern" learning began to penetrate Germany's educational system. The Realgymnasien, which stressed pragmatic, utilitarian curricula, such as science, technology, and modern languages, began to rival the humanistic Gymnasien, and it was from these

---

[7] Lundgreen, "The Organization of Science and Technology in France"; Ringer, *Education and Society*, pp. 21–54.

schools that the Technische Hochschulen recruited their students. During the latter decades of the century, the students enrolled in modern secondary schools far outnumbered those in the classical Gymnasien, and the employment opportunities linked to the modern technological and industrial stream were growing rapidly both in number and prestige. In the latter third of the nineteenth century, then, science- and technology-related learning had come to occupy a place near the summit of the educational hierarchy alongside erstwhile humanistic learning. Industrial technology had become a mechanism for achieving considerable social and political legitimacy.[8]

However, recent historiography has cast doubt on the causal character of the Technische Hochschulen in late-nineteenth-century German industrial performance. Wolfgang König claims that before 1900, it was not highly advanced technical learning that spearheaded industry, but instead intermediate technical skills. Thus, the Technische Hochschulen played a less central role in German economic growth than is generally considered to be the case. Their primary objective was competition with the traditional universities, as they sought to climb in the educational hierarchy. To achieve the wanted end, it had been necessary to demonstrate competence in relatively academic, in contrast to more utilitarian, industrial fields of teaching and research. It was only after 1900, when the Technische Hochschulen had successfully challenged the universities, that they turned their full attention to concrete industrial development, and with remarkable success.[9]

König insists that before 1890, it was not the Technische Hochschulen but rather a range of mixed, somewhat lower-level institutions of technical education that drove the expansion of Germany's economy, namely, the Technische Mittelschulen. This constellation of schools prospered particularly in the 1870s and 1880s. The constellation was composed mainly of innumerable local, small training institutes that had flourished in the many *Länder* during the entirety of the century. Unlike the Technische Hochschulen, during this critical period the Technische Mittelschulen catered specifically and exclusively to industry, and König claims that their graduates (and often not those of the Technische Hochschulen) temporarily comprised the key source of technical innovation in the traditional domain of mechanics, as well as the science/technology-intensive domains of chemistry and electricity. They offered full-time instruction in eminently practical topics. The duration of courses was generally twelve to eighteen months, after which graduates immediately entered industrial employment. They were acknowledged as high-quality technicians, and many became in-house engineers. Their worth lay in the rare capacity to combine skill and utilitarian knowledge. Significantly, König's conclusions complement the argument of Ringer, who sees in the

---

[8] Ringer, *Education and Society*, pp. 73–6.
[9] Wolfgang König, "Technical Education and Industrial Performance in Germany: A Triumph of Heterogeneity," in Fox and Guanini, *Education, Technology*, pp. 65–87.

Oberrealschulen and their like (higher primary education) the bulwark of Germany's modernization process.[10]

For the end of the century, however, there is agreement that it had become the Technische Hochschulen that supplied much of the scientific and technological knowledge entailed in the continuing growth of industry; and the Technische Hochschulen continued to perform this role until late into the interwar era. The topography of higher German technical learning has changed relatively little. Today, the Technische Hochschulen still furnish firms in advanced and traditional technology with armies of highly trained engineers. To this cluster of schools must be added a new group – the technical universities – which arose in the 1960s. The latter perform the same cognitive and professional functions as the Technische Hochschulen, and they constitute the German university's strategic reaction to a situation in which it was losing a growing number of talented students. Another cluster of technical institutions arose in the 1960s, the Fachhochschulen.[11] These schools have taken the place of the former Technische Mittelschulen. They offer a moderately long cycle of instruction, four years versus the six or seven years in Technische Hochschulen. The German technical education system continues to be characterized, however, not only by its remarkable heterogeneity but also by the existence of relatively supple boundaries between institutions. It is, hence, quite possible for students in the lower-level Fachhochschulen to transfer without penalty either to the higher status Technische Hochschulen or to a university. In sum, pliable transverse structures underpin heterogeneity, while redefinable hierarchic structures guarantee its perpetuation. The result is that German industry has, since the middle of the nineteenth century, had an immense diversity of institutions of technical education from which to draw. Such diversity has allowed high industrial performance, as firms can recruit new employees in response to changing technology and shifting economic opportunities.[12]

But the might of nineteenth- and twentieth-century German industry has not been based solely on scientific and technical training. The capitalization of knowledge in the modern economic order also requires innovation through research. In the person of Justus von Liebig (1803–1873), Germany possessed a progenitor of modern university/industry research and knowledge relations. Even in the first half of the nineteenth century, the Fatherland could boast exceptional industrial performance in agricultural chemistry and pharmacy, thanks to linkage between academic and entrepreneurial research. Numerous historians have convincingly shown that during the last 150 years, German

---

[10] Ringer, *Education and Society*, pp. 21–64.
[11] B. B. Burn, "Degrees: Duration, Structures, Credit, and Transfer," in *The Encyclopedia of Higher Education*, ed. Burton R. Clark and Guy Neave (Oxford: Pergamon Press, 1992), 3: 1579–87.
[12] Max Planck Institute, *Between Elite and Mass Education: Education in the Federal Republic of Germany*, trans. Raymond Meyer and Adriane Heinrichs-Goodwin (Albany: State University of New York Press, 1992), vol. 1, chap. 1.

chemistry has owed much of its incontestable successes to a combination of endogenous and exogenous applied science.[13] As early as 1890, Bayer possessed a full-time staff of industrial research chemists and a well-equipped laboratory that was fully integrated into the giant firm's complex bureaucratic structure.[14] From midcentury onward, the Zeiss Jena optics works thrived on the basis of massive in-house research, and on research imported from Germany's universities and Technische Hochschulen. The same was true for the nation's expanding electrical and electromechanical sector.

The empire's industrial performance also benefited from indirect research contributions. The Physikalisch-Technische Reichsanstalt's second section, specializing in technology, assisted enterprise in two strategic fashions.[15] Research carried out there paved the way for German-based technological standards that sometimes prevailed in world competition. Equally important, the second section undertook research in the field of instrumentation.[16]

FRANCE AS A PARADIGM OF HOMOGENEITY

In comparison with that of Germany, French industry developed more slowly and less cyclically. Over the span of the nineteenth century, the economy grew at about 1% annually, while that of its neighbor rose by an additional 50%, sometimes attaining a growth rate of over 6%. Germany has continued to outstrip France during most of the present century as well.[17] France's more gradual expansion has been ascribed to a number of factors, such as banking policy, savings patterns, and problems in raw materials, as well as to certain mental, ideological, and cultural inclinations. These considerations are clearly relevant to France, but much of the country's sluggish development is also associated with educational institutions of a particular configuration, and with particular structures connected with applied research. France's system of higher scientific and technical education is doubtless the most segmented, stratified, and hierarchic of all economically advanced nations. Structural rigidities in education, and also in firms, long generated awkward and often impenetrable boundaries. Until recently, public research agencies had turned their back on enterprise. This state of affairs is historically underpinned by a cleavage between, on the one hand, a form of social and political legitimacy

---

[13] Ludwig F. Haber, *The Chemical Industry: 1900–1930: International Growth and Technological Change* (Oxford: Clarendon Press, 1971).

[14] Georg Meyer-Thurow, "The Industrialization of Invention: A Case Study from the German Chemical Industry," *Isis*, 73 (1982), 363–81.

[15] David Cahan, *An Institute for an Empire: The Physikalisch-Technische Reichsanstalt, 1871–1918* (Cambridge: Cambridge University Press, 1989).

[16] Terry Shinn, "The Research-Technology Matrix: German Origins, 1860–1900," in *Instrumentation between Science, State and Industry,* ed. Bernward Joerges and Terry Shinn (Dordrecht: Kluwer, 2001), pp. 29–48.

[17] Rondo E. Cameron, "Economic Growth and Stagnation in France, 1815–1914," *Journal of Modern History*, 30 (1958), 1–13.

bound to antiutilitarian, high-minded science and esoteric high mathematics and, on the other hand, much lower status, gritty, empirical and manual skills linked to economic matters.

France's system of higher scientific and technical education contains four acutely differentiated strata: the traditional grandes écoles, the lower grandes écoles, the national engineering institutes that have historically been connected to the science faculties, and the new grandes écoles. While each segment possesses educational virtues and certain potential for industry, the specific clusters stand isolated. For technical students, transverse and vertical movement is precluded. Moreover, historically, no form of educational or institutional hybridization occurred.

The traditional grandes écoles were established during the course of the eighteenth century – the Ecole des Mines, the Ecole des Ponts-et-Chaussées, the Ecole d'Artillerie, the Ecole de Génie Militaire – and lastly the Ecole Polytechnique, set up in the midst of the 1789 Revolution. The explicit function of this constellation of schools was to secure and protect the prerogatives and powers of the French state. Although the schools trained *ingénieurs*, these were not engineers in either the German or Anglo-American sense of the term. Alumni were guardians of the state's interests, becoming either topranking military officers or high civil servants. Civil servants were planners and supervisors in areas related to infrastructure development, exploitation of mineral resources, and the like. Traditional grandes écoles graduates thus became "social engineers," rather than industrial personnel, and direct actors in the process of economic growth.[18] This was fully consistent with their training in mathematical analysis, a narrowly deductive epistemology, and Greek and Latin. Indeed, it was not until well into the twentieth century that the modern scientific subjects of mechanics, electricity, and the like penetrated the Ecole Polytechnique, and that research became a priority.

France nevertheless required technical personnel to staff its nascent industries. Pragmatic technical education emerged in the early nineteenth century with the foundation of the Ecoles des Arts et Métiers, which were a key component in France's system of lower grandes écoles.[19] Established by Napoleon for the orphans and sons of soldiers, these schools provided short-term training in fields such as woodcraft, metalworking, plumbing, mechanics, and so on. Quickly, however, the number of institutions in the constellation grew, courses became more advanced, and students were drawn from the petite bourgeoisie and lower middle classes. Instruction developed into a two-year program that included elementary mathematics and elementary science. The thrust of learning was consistently practical. By the end of the nineteenth century, recruitment was being regulated through a national *concours*. With

---

[18] Terry Shinn, *Savoir scientifique et pouvoir social; L'école polytechnique, 1789–1914* (Paris: Presse de la fondation nationale des sciences politiques, 1980).
[19] Charles R. Day, *Les Écoles d'Arts et Métiers L'enseignement technique en France, XIXe–XXe siècle* (Paris: Belin, 1991).

few exceptions, graduates went into industry where they became technicians, production foremen, and engineers. Some rose to administrative positions in firms, but this was relatively infrequent. Throughout much of the nineteenth and twentieth centuries, graduates of the Ecoles des Arts et Métiers thus comprised the middle-level technical cadre of French enterprise.

While the services rendered by the lower grandes écoles and their graduates have proven crucial to France's industrial performance, their contributions have been limited. The schools were created in an age of mechanics, and the institutions proved very slow to move into new technical sectors such as chemistry, electricity, and electronics. Moreover, the Ecoles des Arts et Métiers failed to incorporate research into their programs, or to consider engineering as a science. The approach of the schools and their alumni has been pragmatic, yet not exploratory. Innovation has never become a component of practice or thought. Indicative of the fragile position and status of these schools, it was not until the eve of World War I that they were permitted to award the title of *ingénieur industriel*. This constituted an important victory, for it marked the point at which an educational cluster managed to embrace officially the same nomenclature as the nonindustrial (anti-industry?) traditional grandes écoles. While industrial education and science continued to lack the immense legitimating advantages conferred by the Ecole Polytechnique and by esoteric mathematics and high science, a measure of social status and influence was nevertheless slowly accruing to technology. Despite this, the achievement pales when compared with the 1899 victory of the Technische Hochschulen, which simultaneously raised the academic status of industrial knowledge and prepared the way for much more effective relations with enterprise.

A second stream of relatively low-level technical learning arose during the period 1875 to 1900 – the dawning of republican science. When in 1871 the Second Empire succumbed to Prussia and the Third Republic was established, intellectuals and university professors figured among the ranks of the victorious republicans. A succession of governments revitalized the science faculties, providing them with new buildings, comfortable laboratories, large staffs, and the recruitment of an unprecedented number of students. Research thrived. For the first time, industry was authorized to invest in the science faculties, and the latter were permitted to become involved in local industrial activities. It was in this context that strong university/industry ties came about.[20]

In fewer than twenty-five years, the regalvanized faculties set up almost three dozen institutes of applied science. Their function was twofold: (1) to assist regional industry to solve pressing technical problems, frequently accompanied by academic research in applied science; (2) to offer training at

---

[20] Mary Jo Nye, *Science in the Provinces: Scientific Communities and Provincial Leadership in France, 1870–1930* (Berkeley: University of California Press, 1986); Terry Shinn, "The French Science Faculty System, 1808–1914: Institutional Change and Research Potential in Mathematics and the Physical Sciences," *Historical Studies in the Physical Sciences*, 10 (1979), 271–332.

the technician level – courses running between one and two years – for the offspring of the provincial lower middle classes interested in taking employment with expanding local firms. The institutes covered technical areas as diverse as brewing, wine making, food, paints and lacquers, photography and photometry, electricity and electromechanics, organic and inorganic chemistry, and so on. Significantly, the birth and rise of the science faculty–related technical schools took place against the backdrop of a profound economic recession. Between roughly 1875 and 1902, the usually stable French economy experienced difficulties. If viewed from a purely economic perspective, this period was one of low demand by industry for technical manpower and for new products and processes. Despite this fact, industry participated actively in the rise of the new technical institutes. Why? It may have been motivated more by political factors than economic ones. In this case, structural integration between learning and industry and research may have been more apparent and ephemeral than real and consequential.

Between 1875 and the outbreak of war in August 1914, these institutes educated many thousands of technicians. In industry they supplied the low-level cadre required for manufacture. In some important respects, alumni formed the backbone of France's second industrial revolution. But three serious problems quickly impaired the operation of these institutions: (1) On the eve of the 1914 war, a growing number of enterprises distanced themselves from the faculties and their applied science institutes. Industry investment in them declined. (2) On the morrow of the war, with the exception of Strasbourg, France's faculties crumbled, and the institutes were the first bodies to be disbanded or cut back. In the 1920s and 1930s, they constituted little more than a shadow of their former selves – few graduates, no more research, disinterest on the part of business.[21] (3) From about 1900 to the 1930s, a spate of very small, private engineering schools appeared in France, as well as innumerable correspondence courses for engineering. By the late 1920s, the engineering market had become glutted by a mass of people possessing a variety of training (much of it poor), and this rapidly provoked an acute crisis in the engineering occupation.[22] Who exactly was an engineer, and what institutions had the right to confer the title?

Schools and graduates battled with each other. In 1934 a state commission convened to regulate the profession. As in the case of the rapid educational expansion of the 1870s and 1880s, the flurry of activity did not coincide with a phase of industrial growth and demand for technical expertise. Once again, the important questions of technical education and certification were not synchronous with economic growth. The French technical community turned in on itself, rather than facing outward in the direction of enterprise. By contrast,

---

[21] Dominique Pestre, *Physique et physiciens en France 1918–1940* (Paris: Editions des Archives Contemporaines, 1984). See chaps. 1 and 2.
[22] André Grelon, *Les ingénieurs de la crise: Titre et profession entre les deux guerres* (Paris: Editions de l'Ecole des Hautes Études en Sciences Sociales, 1986).

in Germany, engineers identified themselves with the Verein Deutscher Ingenieure, which negotiated with educational institutions on one side and with firms on the other, in order to strengthen the technical profession and to form a cohesive national technical/industrial system. In France, however, professional engineering associations were numerous, often small, fragmented, and weak. The identity of engineers lay principally with the schools that formed them. Their logic was often in the first place, that of their alma mater, in the second place, that of their technical occupation, and, only last, the logic of enterprise.

Finally, the new grandes écoles, established in the three decades preceding World War I (the Ecole Supérieure de Physique et de Chimie, the Ecoles Supérieure d'Électricité, and the Ecole Supérieure d'Aéronautique), became France's equivalent to Germany's Technische Hochschulen. Each of the establishments was fathered by eminent scientists and engineers, whose intellectual and professional trajectories included both academic endeavors and industrial involvement. In the decades immediately following their foundation, the new grandes écoles provided instruction in elementary mathematics, applied science, and engineering. Soon, however, the curriculum became more advanced and complex. Higher applied mathematics, pure science and applied science, and engineering were taught. This greater mathematization of learning drew students from ever higher social classes, and it also raised the position of the schools in the formal national educational hierarchy.[23]

From the outset, the new grandes écoles engaged in research, and after 1945, research increasingly became a focal point of the teaching program. In the case of the Ecole de Physique et de Chimie, the Curies did all of their pioneering work in radioactivity at the school, which became associated with industrial uses of radiation. The three schools that form this constellation have figured centrally in the research and advanced engineering of most post-1945 industrial achievements in electricity and electronics, aeronautics, synthetic chemistry, and technical sectors linked to classical macroscopic physics, such as fluid mechanics. It is impossible to exaggerate the contribution in engineering and research of these institutions.

Until the 1960s and 1970s, openness to research within industry was rare, and even fewer were the firms that possessed a research capacity. French industry was singular for its indifference, or even hostility, to science. In a survey of more than a score of France's technically leading firms in the 1920s, only a quarter had a significant research capacity. The other companies depended on the purchase of patents and licensing for innovation.[24] In the 1890s, several companies had temporarily opened small research laboratories, but they were quickly abandoned. It was not until the post–World War II era that an authentic groundswell in favor of industrial research developed,

---

[23] Terry Shinn, "Des sciences industrielles aux sciences fondamentales: La mutation de l'Ecole supérieure de physique et de chimie," *Revue française de Sociologie*, 22, no. 2 (1981), 167–82.

[24] Shinn, "The Genesis of French Industrial Research."

and it was precisely at this juncture that the new grandes écoles intensified their mixture of high engineering and experimental research orientation.

To palliate deficiencies in applied research, government intervened. France was pressured by the events of World War I to coordinate extant research, and to finance fresh projects for national defense. But for all practical purposes, this project did not survive the war. As indicated, after 1918 France's science faculties had largely collapsed, and with them much of the country's research potential. Government belatedly recognized this problem and grew concerned in the late 1920s. Throughout the 1930s, the need to reenforce the nation's war-making technical potential, as well as pure science, led the government to found a series of research agencies. These were poorly funded, yet they did offer scholarships to promising young scientists and provided some funding for laboratories. The discourse underpinning the agencies emphasized a mix of industrial knowledge and basic research. The Centre National de la Recherche Scientifique Appliquée was founded in 1938, with the express aim of assisting French enterprise and helping prepare for eventual war with Germany. In 1939 it was superseded by the Centre National de la Recherche Scientifique, which today remains France's premiere research institute. After World War II, other national research institutes were revitalized or established – the Commissariat à l'Energie Atomique, the Institut National de la Santé et de la Recherche Médicale, and so on. The state's goal was always technological and pure knowledge. Despite this, technology, applied science, and the engineering sciences have generally been marginal, with very little integration with technical education, and with little involvement in enterprise. Although research in fundamental science prospered, up until the 1970s France failed in numerous economic sectors to formulate a systematic multicomponent innovation program capable of enhancing industrial capacity.[25]

ENGLAND AS A CASE OF UNDERDETERMINATION

Of the four countries dealt with in this chapter, the operation of English research and technical education is doubtless the historically most indefinite case. The ambiguity and inconclusiveness is tied to three considerations: (1) The remarkable industrial performance of England in areas of mechanics-related production in much of the eighteenth and early nineteenth centuries (textiles, pumping of mines, railways, etc.) suggests to some analysts that the country possessed an adapted program of technical education in the field, and perhaps some research capacity. (2) For the late nineteenth and twentieth centuries, England exhibited a considerable number and variety of initiatives in technical training and investigation, which are sometimes regarded as evidence of achievement. (3) Since England and the United States are associated culturally and industrially, it is sometimes

[25] Fox, "France in Perspective," p. 212.

inferred that because England developed certain initiatives derived from those of America, the English counterparts functioned as effectively as the U.S. programs.

Fritz Ringer states that England only acquired a fully integrated universal primary and secondary school system in the early twentieth century. The Education Act of 1902 established effective compulsory education for all social classes. A range of curricula was offered, extending from the classics to modern science and technology and to more immediately practical training. For the first time, the country could boast a quality system beyond the "ancient nine" very outstanding "public schools" that had traditionally prepared the social and political elites, and that had for some opened the way to Oxbridge.[26] Indeed, until the establishment in 1836 of University College London, which offered instruction in science and modern topics, Cambridge and Oxford constituted the sole universities in England. While comprehensive schooling is perhaps not entirely a prerequisite to an efficacious research and technology training program, it is nevertheless an immense benefit. The fact that both Germany and France introduced strong and differentiated public education systems roughly fifty years before England almost certainly gave the two nations an edge, at least in general literacy, and by dint of this, also in technical literacy.

Ringer indicates that it was not until 1963, with the Technical Education Act, that England organized a coherent system of higher technical education. The Technical Education Act linked secondary schooling to higher education, permitted some movement of students within various constellations of higher training, and established important areas of differentiation inside higher formal learning, with a measure of legitimacy for technological and industrial education.[27] Indeed, it was not until after World War II that England's university capacity began to expand commensurately with that of other nations. In the 1880s and 1890s, a few new universities were created, among them Birmingham, Leeds, and Bristol. During the entire interwar era, only one new university was opened, Reading in 1926. By contrast, after 1945, English higher education expanded rapidly. Five former university colleges were transformed into universities: Nottingham, Southampton, Hull, Exeter, and Leicester. Seven entirely new universities were set up: Sussex, York, East Anglia, Lancaster, Essex, Kent, and Warwick. The year 1963 may be regarded as the emblematic date for the systematization and integration of English industry-related education, and for the full social recognition and legitimation of technical learning – as was the date 1899 for the German Technische Hochschulen and the date 1934 for the French engineering community. Again, English achievement came late when compared to other industrially advanced nations.

---

[26] Ringer, *Education and Society*, pp. 208–10.
[27] Ibid., p. 220.

From the 1820s onward, England, more than any other country, boasted a host of mechanics institutes located in a large number of provincial industrial sites. The schools at Manchester are the best known and most fully studied.[28] Thousands of English technicians passed through such schools in the course of the nineteenth century. But in substantive terms, what were these mechanical institutes? First, according to all accounts, they recruited their students from the lower social classes – classes whose level of primary education was very modest. The kind of instruction offered was often haphazard – a little arithmetic, design, work with motors and mechanisms, and so on. While the level of training varied considerably from institute to institute, it was by and large rather low. Perhaps most important of all, the vast majority of those who entered mechanics institutes did not remain for the full program.[29] Some students attended courses for a few months or a year. Many others attended only night courses, and then disappeared from the school registry. This unstructured and intermittent mode of training contrasts with France and Germany in the domain of mechanics. France's Ecoles des Arts et Métiers comprised a two-year coherent program of full-time instruction. The German Technische Mittelschulen drew students who already had a sound higher primary education, and then gave them an additional twelve to eighteen months of full-time training. Until the eve of World War I, in England on-the-job experience and apprenticeship prevailed over formal learning in mechanics. But after 1900 industrial technology and formal technical learning began to gain in status.

The field of industrial chemistry (that is, autonomous, academic, industrially relevant science) also emerged rather late, and in specific arenas, even after World War II. This happened much later than in Germany, and France, too, had already developed considerable expertise by this time. But the situation in England proved extremely complex, characterized by multiple tentative projects and by confused and sometimes contradictory currents.

The Royal College of Chemistry opened in London in 1845, but its mandate remained ambiguous – chemical analysis versus descriptive data. According to R. Bud and G. K. Roberts, the battle between pure chemistry and pragmatic chemistry was fought between the 1850s and 1880s, and the conflict was settled in favor of the former.[30] During this period, English science colleges represented abstract knowledge, and the polytechnics represented utilitarian chemistry. The battle was resolved in 1882 with the opening of the Kensington Normal School, where applied chemistry was taught, but with a status lower than that of pure chemistry. While historians agree about

[28] Colin Divall, "Fundamental Science versus Design: Employers and Engineering Studies in British Universities, 1935–1976," *Minerva*, 29 (1991), 167–94.
[29] Anna Guagnini, "Worlds Apart: Academic Instruction and Professional Qualifications in the Training of Mechanical Engineers in England, 1850–1914," in Fox and Guagnini, *Education, Technology*, pp. 16–41.
[30] R. Bud and G. K. Roberts, *Science versus Practice: Chemistry in Victorian Britain* (Manchester: Manchester University Press, 1984).

the lower status of applied chemistry, disagreement persists over its position in academia, and over academia/industry relations.

Some historians point to the multifaceted aspects of English applied chemistry and to the contradictions of chemistry teaching. Although pure chemistry reigned inside the university, attitudes of staff toward applied studies and research and toward industry were often heterogeneous, and thus difficult to define. Universities, like the University of Leeds, provided instruction in fundamental chemistry, and some staff clearly stated that applied chemistry was important to graduates who would become teachers at normal schools and polytechnics, and whose task it was to prepare industrial personnel. It is implicit that although the university did not legitimate applied learning, it was nevertheless open to teaching it – graduates could thereby take up careers that demanded pragmatic knowledge. Academia's distance from application was protected by the fact that in England, it was not a university diploma in chemistry that legitimated an employee in the eyes of an employer, but rather the certificate accorded by the Institute of Chemistry, a professional body. Finally, as late as 1911, it was neither the university system nor the professional Institute of Chemistry that sought to establish standards of industrial chemistry. Instead, the Association of Chemical Technologists, an industry body, struggled to impose its will. Here, the landscape of actors, interests, and institutions was varied, often dispersed and tangled – a jigsaw landscape! There existed no system, and no integration.[31] It is as if initiatives were consistently underdetermined, lacking extension and provisions that would enable them to interlock with other projects.[32]

The problematic uncertainty and industry/research/education mismatch seen here in applied chemistry persisted into the 1930s and beyond. In 1939 the whole of England could claim only 400 students training in chemical engineering.[33] For the same year in the United States, there were more than that number enrolled in the discipline at MIT alone. But the fundamental difference between the two countries was not that of scale, but rather the organizational structures of industry, research, and learning. The developing British chemical engineering community struggled to persuade both business and academia of the importance of fostering their speciality. It had to make industry grasp that chemical engineering procedures had far more profit potential than traditional applied chemistry. It had to convince academia to replace, at least in part, instruction in traditional industrial chemistry. While both before and after World War II there was some scattered backing for chemical engineering in academia and enterprise, support remained

---

[31] J. F. Donnelly, "Representations of Applied Science: Academics and Chemical Industry in Late Nineteenth-Century England," *Social Studies of Science*, 16 (1986), 195–234.

[32] Michael Sanderson, *The Universities and British Industry, 1850–1970* (London: Routledge and Kegan Paul, 1972).

[33] Colin Divall, "Education for Design and Production: Professional Organization, Employers, and the Study of Chemical Engineering in British Universities, 1922–1976," *Technology and Culture*, 35 (1994), 258–88, especially 265–6.

desultory. There was no driving force capable of consolidating interest or of bringing groups together. Enterprise and education were each isolated, and sometimes reciprocally alienating. In the case at hand, the initiatives of a professional applied chemistry body very gradually brought the two forces, business and universities, into alignment. It is not as if there were no initiatives in behalf of chemical engineering. They were abundant. The difficulty lay in the fact that efforts were hit-and-miss, often of short duration, and rarely coordinated.

In spite of myriad initiatives in the domains of research and of technical, engineering, and scientific education, little was achieved. Each program, albeit in itself rich, often failed to embrace a comprehensive vision. When such a vision did arise, it was practice that proved too fragile and fragmented. The fundamental problem of English underdetermination, though, was the failure to arrive at "extension," that is, the capacity for one subscheme to move beyond its narrow base, to transcend and to intermesh with other schemes.

THE UNITED STATES AS A CASE OF POLYMORPHISM

While Germany was the first nation to organize fully the capitalization of knowledge, the United States quickly followed. Before World War I, the performance of numerous U.S. industries depended, on the one hand, on the rational organization of innovation in the form of endogenous and exogenous research and, on the other hand, on a strong and finely structured convergence between industry's growing requirement for technical and scientific learning and a "suitable orientation" of America's universities. In effect, the conscious and careful orchestration of extant and fresh knowledge had become a crucial component of the U.S. capitalistic economic and social order. Technical knowledge, as labor before it, had become an entity for investment, surplus value, profit, and exploitation.

The historiography of U.S. industry, research, and education relations in the late nineteenth and early twentieth centuries falls into three families: (1) Some historians argue that U.S. corporate capitalism has long possessed both the power and organizational capacity to shape the cognitive focus and norms of the technical professions, and has had the foresight and influence to determine the intellectual and vocational policies and practices of universities. In this view, corporate requirements, logic, and structures have successfully dictated university and professional activities. (2) The American university landscape has long been extremely varied, particularly with respect to the balance among engineering, applied science, and fundamental science. While engineering and applied science sometimes prevail, they do not enjoy hegemony. Government, philanthropy, and autonomous currents committed to fundamental science frequently resist the logic and influence of applied learning and research. (3) The technical professions in the form of engineering and science societies, jealous of their autonomy and potential for a key position in American society, have negotiated effectively both with the university, which trains and certifies them, and with industry, which cannot function without their specialist skills. According to this interpretation, professional demands, even more than industry, have shaped university and business operations. Although on many levels and at first blush these three historiographies certainly appear divergent and even contradictory, Nathan Rosenberg and Richard Nelson propose a synthesis that offers at least a measure of reconciliation.

Between 1890 and 1920, many of America's biggest chemical and electrical companies became large and complex corporations. Internal organization was increasingly bureaucratic and rationalized. This trait extended to labor, equipment, the acquisition of raw materials, investment, management, manufacture, and markets. The organization of scientific and technical knowledge also soon succumbed to this logic, and by necessity the organization of innovation was rationalized. No longer was invention to be left to circumstance; it was to be subjected to the control and laws of enterprise.[34] According to David Noble, this bureaucratization of learning and ever-growing ability to institutionalize and integrate research inside firms constituted outcomes of the American corporation's hegemony over higher scientific and technical education, and over the conduct of the technical professions in the early twentieth century.[35] Symptomatic of this trend, companies like General Electric (1900), Westinghouse (1903), American Telephone and Telegraph (1913), Bell Telephone (1913), Dupont (1911), Eastman-Kodak (1912), Goodyear (1908), General Motors (1911), U.S. Steel (1920), Union Carbide (1921), and so on set up big, well-organized research laboratories.[36] The purpose was twofold: first, to compete effectively with other firms through development of novel products or more efficient manufacturing methods; second, to patent new products or methods, but without putting them on the market, thereby blocking competitors from gaining in turn. By the 1920s, each of the laboratories had staffs in the hundreds. The phenomenon of corporate research continued to expand during much of the interwar era, and according to a business poll taken in 1937, more than 1,600 firms had a research unit. But what was the source of the science and engineering personnel required by these laboratories, and the source of the technicians responsible for the ever-more-specialized tasks of manufacture?

Antebellum American colleges had perceived their principal role to be the teaching of philosophy, moral rectitude, and civic responsibility. They

---

[34] Alfred D. Chandler, *The Visible Hand: The Managerial Revolution in American Business* (Cambridge, Mass.: Belknap Press, 1977).

[35] David Noble, *America by Design: Science, Technology, and the Rise of Corporate Capitalism* (New York: Knopf, 1977), pp. vi and xxii–xxxi.

[36] Ibid., pp. 110–16; Leonard S. Reich, *The Making of American Industrial Research: Science and Business at GE and Bell, 1876–1926* (Cambridge: Cambridge University Press, 1985).

prepared the nation's social and political elites. To the extent that natural philosophy figured in the curriculum, it was taught in the spirit of a "liberal arts education," and not as technology or experimentation.[37] However, indifference to experimental science, engineering, and technology was not everywhere the rule in early-nineteenth-century America. In the first half of the century, two Hudson Valley institutions, West Point Academy and Rensselaer Polytechnic Institute, specialized in engineering and technology. The Massachusetts Institute of Technology (MIT), founded in 1862, soon followed, and on its heels Yale University set up an engineering department.[38] The 1862 Land Grant Act directly involved state government in sponsorship of the applied sciences and teaching at the newly created state universities – first in agriculture and then quickly thereafter in mechanics, chemistry, and electrical technology. By the end of the century, America had some eighty-two engineering schools. Yet even this proved insufficient to sate corporations' need for scientists and engineers.

Business consequently initiated two strategies. In the 1890s, and to a diminishing degree in the next decade, firms introduced company schools. They thereby sought to train their own technical personnel. Big corporations like General Electric and Bell provided scientific and engineering instruction for new employees and offered some advanced courses to older staff. There was also to be a second payoff. Through a company school, it was hoped that the firm could inculcate its own special corporate culture, thereby moving toward the solution of certain managerial problems as well as technical ones. Yet this scheme was short-lived. Companies could not span the breadth of required courses. Business soon admitted that industrial training was best carried out inside America's colleges and universities.[39] To push university educators in the appropriate direction, in 1893 business, some colleges, and a few engineering groups founded the Society for the Promotion of Engineering Education. The society's goal was threefold: (1) to promote a liberal arts college education; (2) to lobby in behalf of science courses that were adapted to engineering rather than pure knowledge; and (3) to ensure that engineering instruction genuinely addressed current industrial issues. The goal here was not just to transform American universities into docile institutions of applied learning sensitive to the changing demands of enterprise, however. The university was also intended to become an annex of the industrial research laboratory. Firms quickly grasped that not all research should, or could, be done inside the company. Universities possessed special expertise and equipment that could also be harnessed to entrepreneurial innovation. Again, according to

---

[37] Arthur Donovan, "Education, Industry, and the American University," in Fox and Guagnini, *Education, Technology*, pp. 255–76; Paul Lucier, "Commercial Interests and Scientific Disinterestedness: Consulting Geologists in Antebellum America," *Isis*, 86 (1995), 245–67.
[38] Henry Etzkowitz, "Enterprises from Science: The Origins of Science-based Regional Economic Development," *Minerva*, 31 (1993), 326–60.
[39] Noble, *America by Design*, pp. 212–19; Donovan, "Education, Industry, and the American University."

this view, although the Society for the Promotion of Engineering Education included some professional engineering groups, these were little more than a passive intermediary body intended to pressure educators further. At bottom, the Society was most definitely a corporate pressure group whose aim was to bend U.S. higher scientific and technical learning to business's particular ends. Here then, in the twentieth century the American university became by and large a research university, and to a considerable degree a university of applied research.[40]

Developments at MIT are frequently invoked to demonstrate how technology and applied science have become all-pervasive. On the eve of World War I, a young but highly talented chemist, A. A. Noyes, became professor of chemistry at MIT. He soon emerged as head of the department, whose experimental and theoretical research results achieved prominence in the United States and beyond. Four years later, a second young chemist, W. Walker, joined the MIT chemistry department staff – his specialty was in applied chemistry. The field of Noyes was basic research, while that of Walker was exclusively industrial science. Owing to corporate thirst for chemical engineers, Walker rapidly acquired a considerable following, both in business and inside MIT. The technical demand sparked by the 1914–18 war further reinforced his influence.[41] On the morrow of the war, conflict between the two men and their respective paradigms flared. When Walker demanded a new, separate applied chemistry facility, Noyes threatened to resign. He insisted that any university in which applied science fully eclipsed basic research and learning was starkly incomplete and did not deserve the title "university." MIT accepted Noyes's resignation. The latter moved to Caltech, where he set up that institution's chemistry department.

This victory of corporate technology over fundamental science at MIT set the context for a second important development. In the 1930s, the economy of the Boston region slumped not simply because of the depression but also because of a more general, structural flight of capital. Firms closed and unemployment rose. However, immediately after World War II, local financiers and industrialists, working closely with administrators and scientists at MIT, strove to reverse this threatening current through the establishment of a new knowledge-enterprise category. A form of partnership was proposed between the scientific expertise of the university and the entrepreneurial expertise of local businessmen. In this spirit, in 1946 the MIT-based American Research and Development Corporation was founded. Its purpose was to solicit technically and economically viable projects from regional groups (businessmen or scientists) to help organize the venture, and to provide limited seed money. The MIT American Research and Development Corporation served entrepreneurial interests by intermeshing knowledge and

---

[40] Roger L. Geiger, *To Advance Knowledge: The Growth of American Research Universities, 1900–1940* (New York: Oxford University Press, 1986). By the same author, *Research a Relevant Knowledge: The American Research Universities since World War II* (Oxford: Oxford University Press, 1993).
[41] Etzkowitz, "Enterprises from Science."

capitalistic projects. It was the progenitor of the modern venture capital system.

Alternatively, although a sizable portion of U.S. university research and teaching is inarguably linked to enterprise, John Servos insists that any claim suggesting that corporate interests totally drive university activities is wrongminded, for it disregards key features of the American knowledge system. Servos's study of the emergence of chemical engineering at MIT in the early decades of the twentieth century provides nuances in the interpretation that corporate imperatives proved unconditionally victorious. Indeed, Walker and applied chemistry took over much of MIT chemistry, and Noyes was forced to leave. However, this did not spell the closure of fundamental scientific research and instruction at the university. In order to balance the influence of corporations, university administrators looked to nonbusiness sources of funding. In particular, philanthropic organizations, such as the Rockefeller Foundation, and government agencies, such as the National Research Council, were contacted and invited to contribute grants expressly for basic science. (This was crucial not least of all because during the Great Depression, corporate contributions to MIT had fallen sharply.) Here then, an institution admittedly committed to industrial applications nevertheless decided on a multipronged strategy that enabled it to succeed both with industry and in areas of fundamental knowledge.[42]

On a complementary register, the workings of professional bodies in engineering and science are seen by some scholars as constituting an additional key factor in the triangle of industry/research/education relations in America, and sometimes as comprising a check on corporate hegemony. Various American engineering societies have in the course of the last century pursued independent lines of action that have not always coincided with corporate objectives – "the revolt of the engineers."[43] The American Physical Society constitutes another instance. In this century, the Society expanded from just a few hundred members to over 10,000. Some practitioners have been employed in industry, but many others have held academic positions. On certain occasions, a cleavage has arisen between entrepreneurial and university demands and, more often than not, the American Physical Society has jealously protected what it regarded as its specific professional prerogatives and the ideals of independent academic research.[44] There thus exist a number of decisive historical cases in which the logic of professional autonomy has countered enterprise, rather than functioning either as an agency for the execution of business policy or as a relay mechanism between corporations and education.[45]

---

[42] John W. Servos, "The Industrial Relations of Science: Chemical Engineering at MIT, 1900–1939," *Isis*, 71 (1980), 531–49.
[43] Edwin Layton, Jr., *The Revolt of the Engineers: Social Responsibility and the American Engineering Profession* (Baltimore: Johns Hopkins University Press, 1986).
[44] Daniel Kevles, *The Physicists: The History of a Scientific Community in Modern America* (New York: Knopf, 1978).
[45] Donovan, "Education, Industry, and the American University."

Finally, as mentioned earlier, in a highly thoughtful article Nathan Rosenberg and Richard Nelson have presented an argument that helps align what are often divergent analyses of the dynamics among American industrial performance, research capability, and the evolution of technical education. Rosenberg and Nelson accept the claim that since late in the nineteenth century, American science and academic life have been colored by a concern with utility. But the authors are equally quick to point out that a broad cultural propensity toward utility does not necessarily signify that education and research are all applied and organized to serve enterprise. Indeed, they suggest that in American culture, the dichotomy is not between applied learning and antiapplied learning. A consensus exists in favor of utility. The relevant cleavage lies between short-term research and long-term research. Short-term research is carried out either in the corporate setting or inside academia, but in connection with business. Long-term research, say Rosenberg and Nelson, is not the purview of enterprise. It is conducted within academia. Its practitioners are not opposed to the eventual application of their findings – quite the contrary. However, academic practitioners of long-term science require a special intellectual and social climate, and they possess a set of expectations and a value system (and sometimes also need special kinds of resources) not available outside academia. Rosenberg and Nelson thereby plead for a division of intellectual labor; however, the distinction is not one of utility versus nonutility, but instead a long-term time scale and strategy versus short-term response to immediate entrepreneurial demand.[46]

THE STONE OF SISYPHUS

In a short chapter it is not possible to introduce even highly telling nuances; only the salient features of the industry/research/education triangle could be raised. While some of the key literature figures here, this too sometimes has had to be curtailed. Four analytic parameters are presented in this text: underdetermination, heterogeneity, homogeneity, and polymorphism. There are sound historical and sociological grounds for arguing that these analytic parameters constitute more effective devices for assessing economic transformations than some erstwhile devices, such as the size of an economy and its education/research system, degrees of centralization, or the relative weight of planning.[47]

---

[46] Nathan Rosenberg and Richard Nelson, "Universities and Technical Advance in Industry," *Research Policy*, 23 (1994), 323–47.

[47] R. R. Nelson, *National Innovation Systems: A Comparative Analysis* (New York: Oxford University Press, 1993); Henry Etzkowitz and Loet Leÿdesdorf, eds., *Universities and the Global Knowledge Economy; a Triple Helix of University-Industry-Government Relations* (London: Pinter, 1997); M. Gibbons, C. Limoges, H. Novotny, S. Schwartzmann, P. Scott, and M. Trow, *The New Production of Knowledge: The Dynamics of Science and Research in Contemporary Societies* (London: Sage, 1994).

Perhaps the most important aspect of any industry/research/education system is its instability in the face of shifting internal and external priorities, needs, and social-political options. The performance of economies, the capacity of educational systems, and the level of innovation and the adaptiveness of research also act in conformity with the law of Brownian motion. Achievement remains precarious. The stone of Sisyphus has always had to be shoved upward, against the force of gravity. There are no ready-made formulas for making this easier, although certain structures may, in specific contexts, prove more helpful than others. Pluralism, inventiveness, and adaptability appear in many of the historical cases presented here as positively correlated with economic development. But this pluralism is not synonymous with liberalism. To the contrary. Structural pluralism requires social and political framing in order to survive and flourish. This framing is the necessary precondition of a truly pluralistic order, and this order is in turn the precondition for modern economic capacity.

8

REMAKING ASTRONOMY
Instruments and Practice in the Nineteenth and Twentieth Centuries

Robert W. Smith

In the years between 1800 and the end of the twentieth century, astronomy was fundamentally transformed. That which had been at heart a science of position, in which astronomers strove to say where an object is, not what it is, became in many ways a vastly more wide-ranging and large-scale enterprise in terms of the questions asked of nature, the number of astronomers it engaged, the level of public and private support it enjoyed, and the size and sophistication of the instruments employed, as well as the remarkable extension of observations beyond the narrow window of optical wavelengths.

The focus of this chapter is on those changes in observational astronomy that comprised central elements in this remaking of astronomy between 1800 and 2000: the reform of positional astronomy in the early nineteenth century, the rise of astrophysics (although little attention is devoted to the study of the Sun, as that is discussed elsewhere in this volume), and the ways in which shifting forms of patronage provided new opportunities for state-of-the-art instruments. In all of these areas, the history of the telescope, the key instrument in observational astronomy in the last four centuries, will be key. We shall also see that the improvement of telescopes was often not tied to answering particular theoretical questions, but was rather seen as a worthy goal in itself that would, in turn, lead to novel results. It should be noted, too, that we are concerned with astronomy in the Western world at the cutting edge of research. I shall therefore not address very interesting questions about the development of instruments for use primarily in demonstrations (e.g., planetaria) or employed chiefly for recreational uses.

THE ASTRONOMY OF POSITION

On the evening of 1 January 1801, Giuseppe Piazzi (1749–1826) readied his instruments in the Santa Ninfa tower at the royal palace of Palermo to examine a region of the sky in the constellation of Taurus. His aim was to locate a faint star listed in a catalog by the French astronomer Abbé Nicholas-Louis de Lacaille (1713–1762). For those interested in the motions of the members of the solar system, the stars presented the backdrop against which such motions could be traced and then interpreted in terms of Newtonian gravitation. For Piazzi, Lacaille's star would be one more addition to an accurate reference system. Piazzi and other astronomers determined the celestial coordinates of objects by observing their passage across the meridian. Astronomers, both professionals and amateurs, exploited a variety of meridian instruments, but increasingly in the early nineteenth century they employed German-built transit circles (or "meridian circles" as they were also called). Measurements were made in the celestial coordinates of declination – with the aid of the instrument's divided circle – and right ascension, determined by an accurate clock. Astronomers, then, were preoccupied with the astronomy of position.

On that night of 1 January 1801, however, Piazzi chanced upon an object that did not stay fixed with respect to its neighboring stars. What could it be? Was it a comet? It did not display a comet's fuzzy appearance. Piazzi therefore termed it a "starlike comet." In fact, it would turn out that Piazzi had made a major discovery by stumbling upon the first of what astronomers would call an "asteroid" or "minor planet," a very important addition to the sorts of bodies to be found in the solar system and one that sparked enormous interest.

Piazzi was able to employ in his observations of the minor planet one of the most important instruments of his time, an altazimuth circle by the great London instrument maker Jesse Ramsden (1735–1800). After two abortive efforts, this circle had been finally completed in 1789. Braced with ten conical spokes in a manner similar to Ramsden's "Great Theodolite," the five-foot-diameter vertical circle was divided to 6 minutes and to 6 minutes, too, for the three-foot-diameter azimuth circle. The azimuth circle was read by a micrometer microscope to one second of arc; two micrometer microscopes were used to read the vertical circle.[1] Much the finest complete circle made to that date, this instrument was the first of its kind to be used for serious research, and one of the outstanding achievements of eighteenth-century technology.[2] It also marked the beginning of a succession of large observatory circles that were fully divided and possessed telescopic sights.[3]

The increasing use of circles to fix the position of objects was one important element in the changes in astronomical practice in the early nineteenth

---

[1] On circles in the late eighteenth and early nineteenth centuries, see William Pearson, *An Introduction to Practical Astronomy* (London: Longman and Green, 1829); J. A. Bennett, *The Divided Circle: A History of Instruments for Astronomy, Navigation and Surveying* (Oxford: Phaidon-Christie's, 1987); and Allan Chapman, *Dividing the Circle: The Development of Critical Angular Measurement in Astronomy 1500–1850* (New York: Horwood, 1990).

[2] Allan Chapman, "The Astronomical Revolution," in *Möbius and his Band: Mathematics and Astronomy in Nineteenth-Century Germany*, ed. John Fauvel, Raymond Flood, and Robin Wilson (Oxford: Oxford University Press, 1993), pp. 34–77, at p. 46.

[3] Bennet, *The Divided Circle*, p. 128.

century. It was, moreover, through a variety of technological developments allied with greater attention to, and new methods of dealing with, observational errors that positional astronomy was reformed in the first half of the nineteenth century. The leading exponent of this new sort of positional astronomy was F.W. Bessel (1784–1846). In 1806, he became an assistant at J. H. Schröter's (1745–1816) well-equipped private observatory in Lilienthal, but it was the thirty-six years he spent as director and professor at the Königsberg Observatory that really set his stamp on nineteenth-century astronomy. Bessel arrived at Königsberg in 1810. The observatory was completed three years later and was, as he put it, a "new temple of science." He and his family lived on the top floor. The first floor was devoted to astronomy and possessed a cruciform design. The main instruments were situated in the apse, over which a dome was later placed, further underlining its resemblance to a church.[4]

Although Bessel had rejected a business career after serving a commercial apprenticeship, he "retained a business mentality," Kathryn Olesko has emphasized, "by accounting for every value in a transaction, even in his measurements of the heavens."[5] In fact, through a relentless desire for precision – in observations, computations, and scientific instruments – Bessel and the other members of the German school of practical astronomy set the science onto a new footing. Bessel and his colleagues decided that observers had to be calibrated, not just instruments, and differences among individual observers taken into account (when we turn shortly to the first determination of stellar parallax, we shall see where these methods paid dividends).[6] The new methodology was soon entrenched at other important German observatories established or revamped early in the century.[7]

Germany became increasingly important for the production of instruments, too. In the late eighteenth century, the superiority of English instrument manufacturers, such as Ramsden, had been unquestioned. But in the first quarter of the nineteenth century, leadership in the production of optical instruments passed from London opticians to the German school, the most notable representative of which was the Munich optical shop of Joseph von Utzschneider (1763–1840) and Joseph Fraunhofer (1787–1826).

A long-standing problem in glassmaking was how to make large blanks of homogeneous glass free of striae. In the 1780s, a Swiss artisan, Pierre-Louis Guinand (1748–1824), had begun experimenting with the manufacture of flint glass. While he had taken big strides by the end of the 1790s, Guinand's

---

[4] Kathryn M. Olesko, *Physics as a Calling: Discipline and Practice in the Königsberg Seminar for Physics* (Ithaca, N.Y.: Cornell University Press, 1991), pp. 26–31.
[5] Olesko, *Physics as a Calling*, p. 26.
[6] Simon Schaffer, "Astronomers Mark Time: Discipline and the Personal Equation," *Science in Context*, 2 (1998), 115–45.
[7] Chapman, "The Astronomical Revolution," provides a useful overview of astronomy in Germany in the early nineteenth century. Still helpful is Robert Grant, *A History of Physical Astronomy*... (London: Henry G. Bohn, 1852).

Figure 8.1. The Dorpat Refractor, a masterpiece by Fraunhofer. (Courtesy of the Library of the Institute of Astronomy.)

major step came in 1805 when he used fireclay rods, in place of the usual wooden ones, to stir the liquid glass. This improved the quality of the blanks he produced, as well as enabling him to fashion larger ones. In the same year, Guinand moved from Switzerland to Munich and carried with him closely guarded secrets of glassmaking, but he eventually returned to Switzerland in 1814. Fraunhofer worked with Guinand for five years, so that to Fraunhofer's formidable expertise in optical techniques and theory were added the skills and knowledge in practical glassmaking. He also applied himself to securing better determinations of optical constants for different sorts of glass, knowledge of which was key for the design of various astronomical instruments. These more accurate constants, for example, were very important for Fraunhofer's improvements in achromatic telescopes.[8]

Perhaps Fraunhofer's most important telescope was the 9.6-inch refractor completed in 1824 for the Dorpat Observatory in the Russian province of Livonia (see Figure 8.1). For some years it was the largest refractor in the world and possessed a very fine "aplanatic" object-glass and an "equatorial mount," the latter feature a characteristic of Fraunhofer's telescopes.[9] The

---

[8] For Fraunhofer's own publications, see Eugen C. J. Lommel, ed., *Joseph Fraunhofer's gesammelte Schriften* (Munich: Verlag der K. Akademie, 1888).
[9] Henry C. King, *The History of the Telescope* (London: Charles Griffin, 1955), pp. 180–4.

type of mounting employed for the Dorpat refractor meant that when the telescope was in use, its weight was counterpoised on the equatorial axis so that the telescope tube could be readily moved into position. The polar axis then rotated at the rate of the sidereal day. Fraunhofer also exploited a falling weight, controlled by a clock mechanism, to drive the polar axis (a feature that would become standard for large refractors). With the aid of this instrument, F. G. W. Struve (1793–1864), among other things, surveyed the entire sky from the celestial pole to a declination of $-15°$, and as a result some 120,000 stars were examined.

In 1830, Struve was appointed director of the newly established observatory at Pulkovo, a small village to the south of St. Petersburg. Under Struve, both its instrumentation and its approach to research were German, and after the official opening in 1839, it was often seen as the most complete and significant astronomical observatory in the world, a position it held for much of the rest of the century. The observatory has also been the subject of an excellent analysis by M. E. W. Williams that links the architectural design with its astronomical functions. Equipped with living facilities and an extensive library, as well as rooms that housed both permanent and portable instruments and spaces for the astronomers to perform calculations, Pulkovo, Williams emphasizes, set new standards for careful design.[10]

Struve was far from alone in his preference for German-built instruments. German manufacturers, for example, were widely accepted as the best makers of heliometers. Hence, when in 1842 the Oxford University Observatory ordered one of these kinds of telescopes, it placed its order with Repsold of Hamburg. The first types of heliometers were built in the eighteenth century to measure the diameter of the Sun's disk by bringing images formed by two objectives to coincidence. It was later revised so as to become a powerful tool for determining the angular separation of two stars, and Bessel exploited a heliometer by Fraunhofer to solve one of the long-standing and most prized problems in astronomy: the determination of stellar parallax. It is worth emphasizing, however, that while Bessel's heliometer was important for his success, what proved decisive in measuring the distance to the star 61 Cygni was Bessel's mathematical skill in reducing his observations, including, for example, accounting accurately for atmospheric refraction.[11] It was, John Herschel (1792–1871) contended when he awarded Bessel the Gold Medal

---

[10] Mari E. W. Williams, "Astronomical Observatories as Practical Space: The Cast of Pulkowa," in *The Development of the Laboratory: Essays on the Place of Experiment in Industrial Civilization*, ed. Frank A. J. L. James (Basingstoke, England: Macmillan Press Scientific and Medical, 1989), pp. 118–36. See also A. H. Batten, *Resolute and Undertaking Characters: The Lives of Wilhelm and Otto Struve* (Dordrecht: Reidel, 1988), and A. N. Dadaev, *Pulkovo Observatory*, trans. Kevin Krisciunas, NASA Technical Memorandum –75083 (Washington, D.C., 1978) from *Pulkova Kaya Observtoricheskaia* (Leningrad: NAUKA, 1972).

[11] Mari E. W. Williams, "Attempts to Measure Annual Stellar Parallax: Hooke to Bessel," unpublished PhD thesis, University of London, 1981.

of the British Royal Astronomical Society, "the greatest and most glorious triumph which practical astronomy has ever witnessed."[12]

In the early nineteenth century, then, there was great enthusiasm for, and enormous prestige attached to, precision astronomy. But not all of its practitioners had a clear sense of the goals that lay behind their labors. There were ready justifications in terms of utility and practical navigation, surveying, geography, and the construction of an ever more exact reference system for astronomy. However, J. A. Bennett has challenged the limits of these utilitarian rationales. He argues:

> The functions of official observatories in particular – those founded by national or local governments or by universities – were to a large degree symbolic. An impressive observatory building indicated an enlightened interest in the highest form of science. Inside were placed the latest models of precision instruments, and the staff settled down to extended programmes of meridian observations. There was generally no clear theoretical aim to the exercise, and the observations themselves often remained unpublished, or if published remained unused.[13]

Positional astronomy, in Bennett's interpretation, was driven not only by utilitarian goals but also by the notion of the official observatory as "a token of stability, integrity, order, permanence," as well as by the moral benefits to be gained from generating accurate star-catalogs.[14]

Not all observatories were run in a rigorous manner, however. The Paris Observatory, for example, was notorious for the laxity of its operations, and unreduced observations piled up for years. This situation only changed with the appointment in 1854 of the autocratic and hugely demanding U. J. J. LeVerrier as director.[15] In contrast to Paris before LeVerrier, the Royal Observatory at Greenwich, as run by George Biddell Airy (1801–1892), exemplified mid-nineteenth-century British views of efficiency. There, Airy not only incorporated German methods but also extended them so that the observatory ran with what some historians have judged to be factory-like precision, including a rigid hierarchy of staff and a tight division of labor. Observation was mechanized, and observers themselves were now subjected to as much scrutiny as the stars.[16]

---

[12] John Herschel, *Monthly Notices of the Royal Astronomical Society,* 5 (1841), 89.
[13] Bennett, *The Divided Circle,* p. 165. See also J. A. Bennett, "The English Quadrant in Europe: Instruments and the Growth of Consensus in Practical Astronomy," *Journal for the History of Astronomy,* 23 (1992), 1–14.
[14] Bennett, "English Quadrant," p. 2.
[15] P. Levert, F. Lamotte, and M. Lantier, *Urbain Le Verrier, savant universel, gloire nationale, personnalité contentine* (Coutances, France: OCEP, 1977).
[16] For a guide to the literature on Airy's Greenwich, see Robert W. Smith, "National Observatory Transformed: Greenwich in the Nineteenth Century," *Journal for the History of Astronomy,* 22 (1991), 5–20. Among the most important works are A. J. Meadows, *Greenwich Observatory.* Vol. 2: *Recent History (1836–1975)* (London: Taylor and Francis, 1975), and Schaffer, "Astronomers Mark Time."

DIFFERENT GOALS

To at least one astronomer, the ambitions of astronomers in the late eighteenth and nineteenth centuries were too limited. William Herschel (1738–1822) devoted his energies to elaborating a bold alternative that centered on what he termed the "natural history of the heavens," in which he saw himself as a sort of celestial botanist in pursuit of "specimens." In his quest for ever more light, Herschel built a series of reflecting telescopes, his most famous being, ironically, what is generally agreed to be one of his least successful, the giant 40-foot reflector completed in 1789 at a cost of over £4,000.

Despite Herschel's efforts and the admiration with which he was viewed, at the time of his death in 1822 astronomy was usually synonymous with positional astronomy, whether pursued by professionals or amateurs; when a serious astronomer turned a telescope to the skies it was generally a refracting telescope as these were judged to be much more dependable than reflectors, as well as far better suited to the needs of positional astronomy. Herschel had nevertheless advertised the potential of big reflectors, and others tried to follow his lead. One such was an Irish nobleman, the third Earl of Rosse (1800–1867), who began to experiment with reflectors in the 1820s. Herschel had left some details of his procedures in telescope making, but these were far from comprehensive, and of his optical techniques he was silent (Herschel had sold many telescopes and he regarded some of his techniques as trade secrets). In a number of areas, therefore, Rosse had to think out matters anew.

After completing a 36-inch reflector in 1840, Rosse turned to what would be his boldest project, a reflecting telescope with a mirror an unprecedented 72 inches in diameter and some four tons in weight.[17] His giant telescope – it came to be known as the "Leviathan of Parsonstown" – was completed in 1845 (see Figure 8.2). With its aid, and despite the awkwardness of its operation and its location under the often cloudy skies of central Ireland, Rosse and his observing colleagues soon made a very important find: Some nebulae possess a spiral structure. Little else of astronomical importance emerged, however.

The Liverpool brewer and telescope maker William Lassell (1799–1880) built smaller instruments than did Rosse, but in some ways his efforts were clearer markers of future developments, and his telescopes were more successful as astronomical tools. He constructed two sizable reflectors, a 24 inch and a 48 inch. Both were equatorially mounted, and this was to become standard for large reflectors for more than a century.[18] For Rosse and Lassell,

---

[17] On the Rosse reflectors, see J. A. Bennett, *Church, State, and Astronomy in Ireland: 200 Years of Armagh Observatory* (Armagh: Queen's University of Belfast, 1990), and King, *The History of the Telescope*, pp. 206–17.

[18] On Lassell, see Robert W. Smith and Richard Baum, "William Lassell and the Ring of Neptune: A Case Study in Instrumental Failure," *Journal for the History of Astronomy*, (1984), 1–17, and Allan Chapman, "William Lassell (1799–1880): Practitioner, Patron, and 'Grand Amateur' of Victorian Astronomy," *Vistas in Astronomy*, 32 (1988), 341–70. On large reflectors and their British and Irish context, see J. A. Bennett, "The Giant Reflector, 1770–1870," in *Human Implications of Scientific Advance*, ed. Eric Forbes (Edinburgh: Edinburgh University Press, 1978), pp. 553–8, at p. 557.

Figure 8.2. The Leviathan of Parsonstown. The telescope tube can be seen slung between two masonry piers. (Courtesy of the Director and Trustees of the Science Museum, London.)

devising big reflectors essentially involved schemes of "cut and try," and generally driving their construction was the logic of developing a technology and seeing what could be done with it, rather than fashioning an instrument in response to the urge to answer specific theoretical questions.[19]

Lassell, like Rosse and Herschel, also employed speculum metal mirrors for his reflectors, but the era of this technology was nearing its end. Its last hurrah was represented by the 48-inch Melbourne telescope completed by Thomas (1800–1878) and Howard Grubb (1841–1931) at the Grubb works in Dublin in 1868. A committee established by the Royal Society also played a significant role in its design, but the completed reflector never lived up to expectations.[20]

One issue that committee members debated was whether or not to advocate a silver-on-glass primary mirror. They rejected this choice as too risky and opted for the supposedly safer choice of speculum. The first glass astronomical mirrors to receive a silver coating were, in fact, produced in 1856 by K. A. von Steinheil (1801–1870) and the French physicist Léon Foucault (1819–1868). A major advantage of such mirrors was that they were lighter

---

[19] Robert W. Smith, "Raw Power: Nineteenth Century Speculum Metal Reflecting Telescopes," in *Cosmology: Historical, Literary, Philosophical, Religious, and Scientific Perspectives*, ed. Norris Hetherington (New York: Garland, 1993), pp. 289–99.

[20] Bennett, *Church, State, and Astronomy*, pp. 130–4, and *Correspondence Concerning the Great Melbourne Reflector*... (London: Royal Society, 1871).

than equivalent speculum metal mirrors and so posed lesser problems for the support systems. Also, glass disks were easier to figure and polish than metal mirrors. Although the silver coatings tarnished quickly in a damp atmosphere, replacing such coatings was simpler than refiguring and repolishing a speculum metal mirror.[21] The future of big reflectors, in fact, would lie with silver-coated glass mirrors, but until the silvering process could be applied to large mirrors, and a compelling use found for such inevitably costly instruments, refractors continued to be the telescopes of choice for professionals who prized their stability, rigidity, and dependability over the potentially powerful, but often awkward, puzzling, and idiosyncratic large reflectors.[22]

The compelling use of big reflectors came later in the century with the rise of astrophysics and the development, most crucially, of astronomical spectroscopes to examine the spectra of heavenly bodies (allied with theories to interpret the observations of spectra) and sensitive photographic plates that would record objects and features of objects invisible to the naked eye.[23] But astrophysicists (as they would later be called) needed more equipment than just a spectroscope attached to a telescope. A pioneer later recalled astrophysics in the 1860s:

> Then it was that an astronomical observatory began, for the first time, to take on the appearance of a laboratory. Primary batteries, giving forth noxious gases, were arranged outside one of the windows; a large induction coil stood mounted on a stand on wheels so as to follow the positions of the eye-end of the telescope, together with a battery of Leyden jars; shelves with Bunsen burners, vacuum tubes, and bottles of chemicals especially of specimens of pure metals, lined its walls.[24]

This sort of observatory was quite unlike the traditional observatory directed toward positional astronomy, and many professional astronomers were ambivalent, others hostile, toward the enterprise of "astrophysics" (a term generally attributed to Johann Carl Friedrich Zöllner [1834–1882] in 1865). A number of nations, nevertheless, soon established astrophysical observatories, as well as incorporating astrophysics into the activities of their existing astronomical institutions. The first observatory specifically established by a state for the pursuit of astrophysics was the one founded in 1874 by

---

[21] King, *History of the Telescope*, p. 262.
[22] Albert Van Helden, "Telescope Building, 1850–1900," in *Astrophysics and Twentieth-Century Astronomy to 1950*, Part A. Vol. 4 of *The General History of Astronomy*, ed. Owen Gingerich (Cambridge: Cambridge University Press, 1984), pp. 40–58.
[23] On the rise of astrophysics, see, among others, D. B. Herrmann, *Geschichte der Astronomie von Herschel bis Hertzsprung* (Berlin: VEB Deutscher Verlag der Wissenschaften, 1975) and the works cited in the chapter by J. Eisberg.
[24] William Huggins, "The New Astronomy: A Personal Retrospect," *The Nineteenth Century*, 91 (1897), 907–29, at p. 913.

Kaiser Wilhelm I at Potsdam, Germany.[25] Other state-supported astrophysical observatories soon followed at Meudon in France and South Kensington in London.

Many of the practitioners of astrophysics of necessity also became skilled photographers, as the adoption of photographic techniques transformed both astrophysical investigation and astronomy in the late nineteenth and early twentieth centuries. Instead of having to rely on visual observations, astronomers were able to record permanently the light from sources, inspecting the photographic plates at their leisure. The first astronomical photographs were taken by use of the daguerreotype process, and during the 1840s, photographs were secured of the Sun, Moon, and solar spectrum. In 1850, the first successful star photograph was obtained at Harvard. With the discovery of the wet collodion process in 1851, more sensitive plates were made available, although they were limited to an effective exposure time of about ten minutes. During the late 1870s and 1880s, the wet plate was in turn supplanted by the dry plate, ushering in the era of astronomical photography, inasmuch as the exposure times of the dry plates could be extended almost indefinitely.[26]

Photography also made possible very extensive observing programs on stellar spectra. Just as positional astronomers during the nineteenth century had collected massive amounts of data on star positions, so some astrophysicists exploited photographic methods to amass huge quantities of data on the spectra at the end of the century and in the early decades of the twentieth. We have noted the importance of factory-like methods to the running of the Royal Observatory at Greenwich. The Lick Observatory in California adopted similar methods for the collection of the radial velocities of stars.[27] But it was E. C. Pickering (1846–1919) at the Harvard Observatory in Cambridge, Massachusetts, who took Airy's approach to new levels in terms of a division of labor – a division that was often gender specific – and a rigid hierarchy in pursuit of the collection and analysis of the light from hundreds of thousands of stars.[28]

By about 1910, astrophysics, then, had progressed to the stage at which it possessed clearly formulated research methods, a number of journals,

---

[25] D. B. Herrmann, "Zur Vorgeschichte des Astrophysikalischen Observatorium Potsdam (1865 bis 1874)," *Astronomischen Nachrichten*, 296 (1975), 245–59. Much of early astrophysics was devoted to the study of the Sun. By far the best overview and analysis of solar studies in the nineteenth and twentieth centuries is Karl Hufbauer, *Exploring the Sun: Solar Science since Galileo* (Baltimore: Johns Hopkins University Press, 1991).

[26] John Lankford, "The Impact of Photography on Astronomy," in *Astrophysics and Twentieth-Century Astronomy to 1950*, Part A. Vol. 4 of *The General History of Astronomy*, ed. Owen Gingerich (Cambridge: Cambridge University Press, 1984), and references cited therein.

[27] Donald E. Osterbrock, John R. Gustafson, and W. J. Shiloh Unruh, *Eye on the Sky: Lick Observatory's First Century* (Berkeley: University of California Press, 1988).

[28] Howard Plotkin, "Edward C. Pickering," *Journal for the History of Astronomy*, 2 (1990), 47–58, and B. Z. Jones and L. G. Boyd, *The Harvard College Observatory: The First Four Directorships, 1839–1919* (Cambridge, Mass.: Belknap Press of Harvard University Press, 1971). A very important work on the overall development of astronomy in the United States is John Lankford, *American Astronomy: Community, Careers, and Power 1859–1940* (Chicago: University of Chicago Press, 1997).

defined programs of research, and solid institutional support. It was still a relatively small field when compared with the more traditional astronomy, but the institutionalization of astrophysics had taken its biggest strides in the United States where positional astronomy was less strongly entrenched than in Europe. Also, by the early twentieth century, the rise of the United States as a major economic power was mirrored by its growing importance in the manufacture and use of very large instruments. These instruments, too, were the most visible signs of the success American astrophysicists had in securing research funds from the newly rich.[29]

It is now well established that George Ellery Hale (1868–1938) was the most effective advocate and "salesman" for large astrophysical enterprises. His ability to deal successfully with philanthropic foundations and wealthy individuals led to a string of giant telescopes, and he was the pivotal figure in the shifts of the ways in which U.S. astronomers ran observatories and planned, built, and operated big instruments.[30] After having brought into being the 40-inch refractor of the Yerkes Observatory of the University of Chicago, completed in 1897, Hale grew frustrated with the lack of support he reckoned he was receiving in Chicago. In 1904, he left for California and set about establishing the Mount Wilson Solar Observatory (in 1920 it was to become the Mount Wilson Observatory) for the newly founded and flush-with-dollars Carnegie Institution of Washington, one of the foundations that would transform the prospects of American science. Under Hale's directorship, Mount Wilson became in many ways a concrete manifestation of the cooperative research (what would later be termed "interdisciplinary" research) in which he, and many American scientists of his generation, believed. He brought physicists into the observatory by establishing a physical laboratory, as well as machine shops, in nearby Pasadena. As Albert Van Helden points out in an excellent overview of telescope building in the first half of the twentieth century, Hale had done much to solve the problems that had traditionally plagued astronomers in terms of obtaining funds for new instruments, maintaining existing ones adequately, and staffing an observatory. Mount Wilson became the leading astrophysical observatory in the world, and by the second decade of the twentieth century, the United States had become the leading power in observational astrophysics.[31]

By the late 1910s, moreover, the large reflecting telescope had been transformed. It was no longer recognizable as the instrument of the

---

[29] Howard S. Miller, *Dollars For Research: Science and Its Patrons in Nineteenth Century America* (Seattle: University of Washington Press, 1970).

[30] Helen Wright, *Explorer of the Universe: A Biography of George Ellery Hale* (New York: Dutton, 1966), and Donald E. Osterbrock, *Pauper and Prince: Ritchey, Hale, and Big American Telescopes* (Tucson: University of Arizona Press, 1993). It should nevertheless be noted that in his thinking he was often well in advance of other American observatory directors.

[31] Albert Van Helden, "Building Telescopes, 1900–1950," in *Astrophysics and Twentieth-Century Astronomy to 1950*, ed. Owen Gingerich, pp. 134–52, at p. 138.

mid-nineteenth century that, while capable of gathering plenty of light, was often very clumsy to use, as well as prone to optical defects. The earlier heavy speculum metal mirrors had, for example, been replaced by lighter glass mirrors coated with silver (later they would be aluminized, which improved their performance still further). Reflectors also held definite advantages over refractors for photographic investigations in that they did not suffer from chromatic aberration.

By 1917 and the completion of the 100-inch Hooker reflector at Mount Wilson, at the time much the biggest in the world, large reflectors had become very powerful and reliable research tools. It was chiefly with the 100-inch reflector, for example, that Edwin Hubble (1889–1953), aided by Milton Humason (1891–1972), conducted his extremely important investigations of galaxies in the 1920s and 1930s.[32] By the late 1920s, in fact, refractors had by and large been relegated to a few specialized activities – such as the measurement of double stars and stellar parallax – and were no longer the workhorses of the pacesetting observatories. Refractors, it was widely agreed, had reached their practical limits with the giant 40-inch refractor of the Yerkes Observatory. To go beyond this size of objective posed enormous practical problems, particularly in terms of the support of, as well as the great loss of light passing through, such a larger objective.

Big reflectors, nevertheless, were far from problem free. The glass used for the 100 inch, for instance, limited the accuracy of its polishing as well as its use. The employment of new sorts of glass that expanded or contracted by only minute amounts with temperature changes was crucial to the further improvement of reflectors. The giant 200-inch reflector for which Hale won funding from the Rockefeller Foundation in 1928 (although it was not completed until 1948) had a pyrex mirror. The fashioning of this mirror also proved important in solving various technical problems that had prevented the proliferation of very large telescopes, and so the research performed to produce it had repercussions far beyond Mount Palomar where the 200 inch was sited.[33]

OPENING UP THE ELECTROMAGNETIC SPECTRUM

The big reflector came of age in the United States in the first decades of the twentieth century and did so with the aid of private support. The next major development in observational astronomy came with the remarkable advance of astronomical observations into regions of the electromagnetic spectrum beyond the visible, a shift fueled largely by government monies. In the early

---

[32] Robert W. Smith, *The Expanding Universe: Astronomy's 'Great Debate' 1900–1931* (Cambridge: Cambridge University Press, 1982).
[33] Van Helden, "Building Telescopes, 1900–1950," pp. 147–8.

1930s, astronomical measurements were restricted to wavelengths between 3×10^{-7} and 10×10^{-7} m, but by the early 1960s such measurements extended from 10^{-12} to beyond 10 m, an enormous and utterly unexpected increase in the size of the "window" on the universe. The rapidity of this change owed much to developments set in motion or hastened by World War II, and the conflict had a profound effect on astronomy in three main areas: (1) greatly increased funding, with government monies available for the physical sciences as never before; (2) new technologies; and (3) new kinds of astronomers.

One branch of astronomy to benefit from all these factors was practically unknown before the war, radio astronomy. Before the early 1930s, astronomers secured almost all of their observational information from visible wavelengths in the electromagnetic spectrum, but the detection of cosmic radio waves extended this narrow window on the universe. Although pioneers had tried to detect radio emissions from the Sun around 1900, not till 1932 did Karl Jansky (1905–1950), a radio physicist at the Bell Telephone Labs in New Jersey, accidentally discover radio waves from the Milky Way. This find caused little stir among what would later be termed optical astronomers (at the time, the term did not exist; only with the later development of radio astronomy did it come into use). But during the mid-1930s, radar techniques were very widely studied, principally because of their potential applications for war making. The further development of these techniques was spurred enormously by the war itself and led to great advances in electronics. The war also produced many people skilled in electronics, some of whom later became astronomers, thereby extending the astronomical community beyond the previously narrow range of optical astronomers.[34] Some astronomers also strove to develop new sorts of light detectors in order to take advantage of these novel technological possibilities.[35]

Researchers from outside mainstream astronomy were critical in the establishment of radio astronomy, and this new field sprang chiefly from a symbiosis of radio physicists and electrical engineers, with relatively little help from astronomers.[36] One of the leading practitioners of radio astronomy was the British scientist Bernard Lovell (b. 1913). Trained as a physicist before the war, he spent the war years developing radar techniques. This experience provided him not only with skills that he could apply initially to radar studies of meteors and then to radio astronomy, but also with expertise in "science politics" and awareness of what it takes to secure funding for large-scale scientific enterprises. Soon after war's end, Lovell conceived

---

[34] Woodruff T. Sullivan, ed., *The Early Years of Radio Astronomy: Reflections Fifty Years After Jansky's Discovery* (Cambridge: Cambridge University Press, 1984), and *Classics in Radio Astronomy* (Dordrecht: Reidel, 1982).

[35] David DeVorkin, "Electronics in Astronomy: Early Application of the Photoelectric Cell and Photomultiplier for Studies of Point-Source Celestial Phenomena," *Proceedings of the IEEE*, 73 (1985), 1205–20.

[36] Woodruff T. Sullivan, "Early Radio Astronomy," in *Astrophysics and Twentieth-Century Astronomy to 1950*, ed. Owen Gingerich, pp. 190–8, at p. 190.

the idea of building a very large steerable "dish" to collect radio waves from astronomical bodies. But the wavelengths of radio waves are many times larger than the wavelengths of visible light. For a radio telescope to have resolving power comparable to that of an optical telescope, it must be many times larger than its optical equivalent. Lovell's goal was a dish some 76 m in diameter, a size that posed enormous design demands. Also, Lovell expected its cost to be far higher than could be afforded by his university alone, the University of Manchester in England. Hence, what became known as the Mark I radio telescope required substantial government funding, as well as the involvement of teams of scientists and engineers.[37] Even a 76-m dish, with radio waves of, say, 1 m in wavelength, produces a resolving power of about one degree, one-twentieth as good as the naked eye, and a value far too large to be of much help in establishing which optical objects correspond to which radio sources.

To overcome this handicap, radio astronomy groups at Sydney, Australia, and the University of Cambridge in England began to develop "interferometers," that is, instruments in which the radiation from a radio source is examined by widely spaced antennae and then combined. The development of ever more powerful interferometers, in fact, became a central goal of radio astronomers and led, for example, to the completion in 1981 of the "Very Large Array," a linked network of twenty-seven antennae in a desert in New Mexico, made possible by federal support. Through the exploitation of electronic computers to integrate the data secured by a number of individual aerials, results could be obtained that were equivalent in some ways to that secured by a single, much bigger antenna, but without the same level of expense or the same sort of engineering challenge. The VLA, for example, has the resolution of an antenna twenty-two miles across. Even bigger baselines have been exploited as radio astronomers have linked radio telescopes that were very widely separated, sometimes even on different continents, by the use of atomic clocks to provide extremely accurate measurements.[38]

INTO SPACE

The development of astronomy from above the Earth's atmosphere by use of rockets and satellites opened up yet more regions of the electromagnetic spectrum, as well as offering the prospect to optical astronomers of

---

[37] Bernard Lovell, *The Story of Jodrell Bank* (New York: Oxford University Press, 1968), and *The Jodrell Bank Telescopes* (New York: Oxford University Press, 1985). See also Jon Agar, "Making a Meal of the Big Dish: The Construction of the Jodrell Bank Mark I Radio Telescope as a Stable Edifice," *British Journal for the History of Science,* 27 (1994), 3–21, and *Science and Spectacle: The Work of Jodrell Bank in Post-War British Culture* (Amsterdam: Harwood, 1998).

[38] It should also be added that while the early years of radio astronomy have attracted considerable attention, historians have yet to tackle in-depth these more recent developments. Perhaps the most innovative of these works is David Edge and Michael Mulkay, *Astronomy Transformed: The Emergence of Radio Astronomy in Britain* (New York: Wiley, 1976).

securing images and spectra of astronomical objects far sharper than those to be achieved with ground-based instruments. Like radio astronomy, space astronomy in many ways sprang from various technological, scientific, and social developments that owed much to World War II. Most significantly, the German-built V-2 rocket provided an emphatic demonstration that the enormous and complex engineering problems posed by the construction and guidance of rockets powerful enough to lift astronomical instruments above the atmosphere had, to a large degree, been solved.[39]

After the war, the United States and, to a lesser extent, the Soviet Union made use of German experts to advance their own rocket programs. With the onset of the Cold War and, most notably, the drive to build intercontinental ballistic missiles to carry nuclear warheads, research expanded on both missiles themselves and the medium through which such missiles would travel, the upper atmosphere. As David DeVorkin has demonstrated, scientists became closely involved in both of these aspects of missile research. In so doing, they often flew scientific instruments in the rockets and, for example, performed observations of the Sun in wavelength ranges of ultraviolet light that never reach the ground owing to the blocking effect of the atmosphere.[40]

With the leap of the Cold War into outer space with the launch of *Sputnik I* in 1957, space astronomy became one weapon in the battle for international prestige between the United States and the Soviet Union. This led to millions, and in time, billions of dollars being devoted to astronomy from above the atmosphere. Space astronomy, both literally and metaphorically, took off.

In the United States, it was NASA, the National Aeronautics and Space Administration, that managed the great majority of programs to study astronomical objects from space. These programs exploited a variety of technologies, ranging from balloons to spacecraft sent to other bodies in the solar system, although, to begin with at least, space projects were often viewed diffidently and, at times, with hostility by some ground-based astronomers who thought the monies directed to such enterprises could be better spent on earth-bound telescopes.[41]

One of the new fields was x-ray astronomy. X rays are absorbed by air, and those who wish to pursue x-ray astronomy must fly their instruments above the obscuring layers of the Earth's atmosphere. X-ray astronomy began in the early 1950s with a research group at the Naval Research Laboratory in Washington, D.C., led by Herbert Friedman (b. 1916). To begin with, this

---

[39] Michael J. Neufeld, *The Rocket and the Reich: Peenemünde and the Coming of the Ballistic Missile Era* (New York: Free Press, 1995).

[40] David DeVorkin, *Science with a Vengeance: How the U.S. Military Created the U.S. Space Sciences after World War II* (New York: Springer-Verlag, 1992). On solar research in the space age, with a strong emphasis on the instruments employed, see Hufbauer, *Exploring the Sun*, pp. 211–305.

[41] This is a theme of the early part of Homer E. Newell, *Beyond the Atmosphere: Early Years of Space Science* (Washington, D.C.: NASA, 1980). See also Robert W. Smith, *The Space Telescope: A Study of NASA, Science, Technology, and Politics* (Cambridge: Cambridge University Press, 1993), paperback edition, pp. 44–7.

group, for the most part, flew Geiger counters aboard rockets to examine the x rays emitted by the Sun in the few minutes of available observing time before the detectors arched back into the atmosphere under the pull of gravity. But during the 1960s, x-ray astronomy became an international enterprise with active research groups in Europe and the Soviet Union, as well as in the United States. In many ways, x-ray astronomy came of age with the launch in 1970 of *Uhuru*, the first satellite dedicated to x-ray astronomy. With the flight of a satellite, x-ray astronomers were no longer restricted to a few minutes of observing time made available by a rocket flight.

In 1963, Riccardo Giacconi (b. 1931), the Italian-born leader of a research group at a company known as American Science and Engineering based near Boston, had conceived of a plan for such an x-ray satellite. The group was already very experienced in launching satellites with x-ray instruments inside them, as its members had worked in 1961 and 1962 for the U.S. Department of Defense, measuring the bursts of nuclear weapons at high altitudes. Hence, as often in space astronomy, a scientific project, *Uhuru*, owed much to instruments and techniques developed for national security purposes. *Uhuru* was designed by the American Science and Engineering group to scan the skies and produce a catalog of x-ray sources, as well as to examine individual sources in some detail. The satellite detected 339 x-ray sources in a little over two years. Such attempts to map the sky at x-ray wavelengths were pursued by later satellites, too.[42]

In the 1960s, NASA missions also gave a major boost to studies of the solar system, most spectacularly with the flights of spacecraft to the Moon and planetary flybys.[43] Both the United States and Soviet Union also sought to land spacecraft on the planets. The first successful landing was that of the Soviet *Venera 7* on Venus in 1970. Despite the incredibly hot and hostile conditions of the Venusian atmosphere and surface, *Venera 7* radioed back twenty-three minutes worth of data. A later version of the same craft, *Venera 13*, soft-landed on Venus in 1982 and returned a color picture, as well as securing other data during its descent. In mid-1975, the United States launched two craft to Mars, *Viking 1* and *Viking 2*. Each was a kind of double spacecraft, as each carried both a "lander" and an "orbiter," the lander to touch down on the Martian surface, and the orbiter to orbit the planet and

---

[42] On the early years of x-ray astronomy, Richard F. Hirsh, *Glimpsing an Invisible Universe: The Emergence of X-ray Astronomy* (Cambridge: Cambridge University Press, 1983), and Wallace Tucker and Riccardo Giacconni, *The X-Ray Universe* (Cambridge, Mass.: Harvard University Press, 1985). Also important on the NRL group is "Basic Research with a Military Context: The Naval Research Laboratory and the Foundations of Extreme Ultraviolet and X-Ray Astronomy," unpublished PhD diss., Johns Hopkins University, 1987.

[43] The scholarly literature on the history of planetary astronomy and planetary science in the latter part of the century, particularly outside of the United States, is very sparse. See, however, William E. Burrows, *Exploring Space: Voyages in the Solar System* (New York: Random House, 1990); Ronald E. Doel, *Solar System Astronomy in America: Communities, Patronage, and Interdisciplinary Research, 1920–1960* (Cambridge: Cambridge University Press, 1996); and Joseph N. Tatarewicz, *Space Technology and Planetary Astronomy* (Bloomington: Indiana University Press, 1990).

return data, including images of the surface. The first lander descended to Mars's Chryse Plain on 20 July 1976. It was followed a little later by the second lander on the Utopian Plain, thousands of kilometers away from *Viking 1* and much nearer to the Martian North Pole. In addition to returning some 4,500 images from the planet's surface, each lander performed several experiments. One involved a retractable boom. Controlled from the Earth, the boom could be extended to scoop up and collect samples of Martian material. These samples were then carried by the boom to a number of experimental packages aboard the lander so that they could be analyzed by means of a variety of techniques.

One goal of the *Viking* lander experiments was to search for life. The results were somewhat ambiguous, but the conclusion of the *Viking* scientists was that they had not found evidence of life, at least not in the immediate vicinity of *Landers 1* and *2*.[44] The data provided by the *Viking* spacecraft had nevertheless transformed many of the ideas of planetary scientists about Mars, not only on the question of life but also on the planet's chemical and geological composition, orbital characteristics, atmosphere, and weather. The *Viking* missions also serve as examples of large-scale space science missions in which many hundreds, if not thousands, of scientists and engineers are involved in very complex processes of deciding on scientific goals, designing instruments, and securing and analyzing the scientific data that eventually result. For such "Big Science" missions, the scientists form into various teams, and so identifying one or a few astronomers or planetary scientists as *the* key figure or figures becomes very hard, if not impossible, despite the prize system in science, which is still geared to the individual investigator.

Viking was an American mission, and although the space sciences were dominated at first by the United States and Soviet Union, other nations played increasingly larger roles during the 1970s. By the following decade, the European Space Agency possessed a very significant space science program. Japan, too, was building and launching space missions by the 1980s. Hence, when in 1985 and 1986 Halley's Comet made one of its regular returns to the inner regions of the solar system, it was met by a small armada of craft from the European Space Agency, the Soviet Union, and Japan, but not the United States.[45]

VERY BIG SCIENCE

The enormous cost of ambitious space projects has also encouraged international partnerships. One such project is the Hubble Space Telescope, an enterprise of NASA with the European Space Agency as a junior partner.

---

[44] See, for example, E. C. Ezell and L. Ezell, *On Mars: Exploration of the Red Planet, 1958–1978* (Washington D.C.: Scientific and Technical Information Branch, NASA, 1984), and William E. Burrows, *Exploring Space: Voyages in the Solar System and Beyond* (New York: Random House, 1990).

[45] John M. Logsdon, "Missing Halley's Comet: The Politics of Big Science," *Isis,* 80 (1989), 254–80.

Figure 8.3. The Hubble Space Telescope in the payload bay of the Space Shuttle *Enterprise*. (Courtesy of the National Aeronautics and Space Administration.)

Essentially an orbiting observatory at the heart of which is a reflecting telescope with a primary mirror 2.4 meters in diameter, the Hubble Space Telescope was designed and built by many thousands of people at a cost of over $2 billion and was launched into space in 1990. It is, thus, the most expensive scientific instrument ever constructed, let alone the most expensive telescope, and both it and the program to build it were, as the author has argued at length elsewhere, the product of a great range of forces: scientific, technical, social, institutional, economic, and political.[46]

For an astronomical enterprise of this scale, the only possible source of funds was government money, and in order for the telescope to come into being, enormous efforts had to be devoted to making it politically feasible, not just technically and scientifically feasible. For example, in the years between 1974 and 1977, many hundreds of astronomers, led by two prominent Princeton astronomers, John Bahcall (b. 1934) and Lyman Spitzer, Jr. (1914–1997, widely regarded as the main champion of the telescope), worked through various lobbying campaigns with a range of allies to convince the U.S. Congress and White House of the telescope's worth (see Figure 8.3). Winning political approval for the Hubble Space Telescope meant the mobilization

---

[46] This is a central theme of Smith, *The Space Telescope*.

of a scientific community, not just the negotiations of a few power brokers with the wealthy and the generous in the manner of George Ellery Hale. With a very big nationally or internationally supported instrument costing many tens or perhaps hundreds of millions or billions of dollars, it is not simply a question of getting a "yes" or "no" decision to proceed from, say, the White House and Congress and then, if the answer is yes, building the chosen instrument. The job of winning this political approval can have profound effects for the technologies chosen, the engineering approach, the choice of builders, the operation of the instrument, the siting of its operational base, and the scientific problems to be tackled. Instruments of the biggest sort can reconstitute the institutional organization of the astronomical enterprise and be themselves reconstituted "by the necessary financial, political, and ideological ties to society at large."[47]

Operation of the telescope has been the charge of a "university consortium." These management organizations came into being in the wake of the enormous expansion of the role of the federal government in scientific research and development after World War II. University consortia became the means by which the concept of "national" laboratories and observatories – national in the sense of being built with federal monies and open to all qualified scientists – was transformed into working institutions. Unlike the situation before the war when the biggest telescopes were exploited by very few astronomers, federal monies have helped to "democratize" astronomy and open up the use of many of the most powerful instruments, such as the Hubble Space Telescope and the Very Large Array. Private patronage has nevertheless continued to play an extremely important part in the development of ground-based optical astronomy, particularly in the United States. The roughly two hundred million dollars provided by the Keck Foundation for the two giant telescopes atop Mauna Kea in Hawaii is the prime example, and certainly the advent of space astronomy has not slowed the construction of big, and often very innovative, telescopes on the ground.

The Hubble Space Telescope has been typical, too, of many new instruments in that even if a patron requires a detailed accounting of the scientific problems a new instrument might tackle, the builders are often motivated by a more general sense that constructing an instrument more powerful in some way or ways than competing devices will surely lead to discoveries. By the late twentieth century, instruments had become viewed as engines of discovery, and without a steady supply of ever more powerful instruments, the accepted wisdom has it, the research enterprise will wither.

The study of large-scale astronomy projects, be they space telescopes or observatories on the ground, also underlines the fact that with the growth of the astronomical enterprise in the twentieth century has come an

---

[47] James Capshew and Karen Rader, "Big Science: Price to the Present," *Osiris*, 2nd ser., 7 (1992), 3–25, at p. 16.

increasing division of labor. It is now a commonplace that the building and operation of the biggest sorts of astronomical instruments engage a wide range of specialists, including, for example, software and thermal engineers, computer scientists, data analysts, and, in the case of space astronomy and planetary science missions, spacecraft operators and sometimes astronauts. The increasing division of labor has, in considerable part, been driven by the increasingly large scale of projects (large here in terms of the physical size of instruments, number of people involved, and management complexity, as well as cost). This is a basic shift away from the way things were done by professional astronomers for much of the nineteenth century. Instruments were generally purchased from makers, even if astronomers had been closely involved in the design, and astronomers were then responsible for their use.

From the vantage point of the end of the twentieth century, it is clear that the scope of the astronomical endeavor has greatly expanded from that of the early nineteenth century, not only in terms of the questions astronomers ask about the universe but also the kinds of information they collect and employ in their arguments, in addition to the range of people engaged in observational astronomy. This transformation contains perhaps two main elements or drivers, for both of which expanded opportunities for patronage were crucial: (1) the institutionalization of astrophysics on a large scale in the first two or three decades of the twentieth century (which also saw the relative decline of positional astronomy); and (2) the rapid expansion of observational astronomy beyond the optical range of the electromagnetic spectrum to a series of new wavelength ranges. In the first, private monies were important and in the second, government support was dominant.

In looking at these changes, we have seen that the particular developments undergone by observational astronomy were not due simply to some inexorable working out of the internal demands of the science. Rather, the history of observational astronomy in the nineteenth and twentieth centuries has been shaped by the wider scientific enterprise of which it has been a part, as well as by those societies in which it has been pursued.

9

LANGUAGES IN CHEMISTRY

Bernadette Bensaude-Vincent

Language plays a key role in shaping the identity of a scientific discipline. If we take the term "discipline" in its common pedagogical meaning, a good command of the basic vocabulary is a precondition to graduation in a discipline. When disciplines are viewed as communities of practitioners, they are also characterized by the possession of a common language, including esoteric terms, patterns of argumentation, and metaphors.[1] The linguistic community is even stronger in research schools, as a number of studies emphasize.[2] Sharing a language is more than understanding a specific jargon. Beyond the codified meanings and references of scientific terms, a scientific community is characterized by a set of tacit rules that guarantee a mutual understanding when the official code of language is not respected.[3] Tacit knowing is involved not only in the understanding of terms and symbols but also in the uses of imagery, schemes, and various kinds of expository devices. A third important function of language in the construction of a scientific discipline is that it shapes and organizes a specific worldview, through naming and classifying objects belonging to its territory. This latter function is of special interest in chemistry.

According to Auguste Comte, the method of rational nomenclature is the contribution of chemistry to the construction of the positivistic or scientific method.[4] Although earlier attempts at a systematic nomenclature were made in botany, the decision by late-eighteenth-century chemists to build up an

---

[1] Mary Jo Nye, *From Chemical Philosophy to Theoretical Chemistry* (Berkeley: University of California Press, 1993), pp. 19–24.
[2] *Research Schools: Historical Reappraisals,* ed. Gerald L. Geison and Frederic L. Holmes, *Osiris,* 8 (1993). See especially R. Steven Turner, "Vision Studies: Helmholtz versus Hering": 80–103, on pp. 90–3.
[3] K. M. Olesko, "Tacit Knowledge and School Formation," in *Research Schools, Osiris,* 8 (1993), 16–49
[4] Auguste Comte, *Cours de philosophie positive,* vol. 2 (Paris, 1830–42; reedition, Hermann, 1975), vol. 1, pp. 456, and 584–5.

artificial language based on a method of nomenclature played a key role in the emergence of modern chemistry.

Adam gave names to all the animals in the biblical book of *Genesis*. Naming is the necessary activity of the intellect that is confronted with a variety of beings. Chemists ordinarily deal with the individual properties of many different substances. As the population of substances dramatically increased in the late eighteenth century, thanks to improved analytic methods, chemists more and more needed stable and systematic names for communicating and for teaching.

In the late nineteenth century, innumerable organic compounds were created by synthesis. This expanding population, which is both the fruit of the chemists' creativity and a terrible burden, required subject indexing in publications, such as Beilstein's *Handbuch der organischen Chemie* or the *Chemical Abstracts*. Today, chemists have to provide names for approximately 6 million substances. The main problem is that the need for names always anticipates the prescribing of rules for names. Chemists have had to invent strategies for facing this challenge.

Whereas working chemists are extremely concerned with their language and are fond of stories behind the names in use, few historians of chemistry have ventured into this domain.[5] Maurice Crosland's classic *Historical Studies in the Language of Chemistry*, first published in 1962, remains the major reference on the nomenclature that was set up at the time of the chemical revolution.[6] Strangely enough, later reforms of the chemical nomenclature have not attracted much scholarship and are known to us only thanks to chemists who were active participants in the reforms.[7] Their historical accounts most often emphasize the role of individualities and the difficulties of

---

[5] See, for instance, Roald Hoffmann and Vivian Torrence, *Chemistry Imagined: Reflections on Science* (Washington, D.C.: Smithsonian Institution Press, 1993), and Primo Levi, "The Chemists' Language," I and II, from *L'Altrui Mestiere*, Giulio Einaudi Editore, 1985, Eng. trans., in *Other People's Trades* (New York: Summit Books, 1989).

[6] Maurice P. Crosland, *Historical Studies in the Language of Chemistry*, 2d. ed. (New York: Heinemann, 1978). For a literary analysis of the chemical language within the framework of the French Enlightenment culture, see Wilda Anderson, *Between the Library and the Laboratory: The Language of Chemistry in 18th-Century France* (Baltimore: Johns Hopkins University Press, 1984).

[7] Pieter Eduard Verkade, *A History of the Nomenclature of Organic Chemistry* (Dordrecht: Reidel, 1985); R. S. Cahn and O. C. Dermer, *An Introduction to Chemical Nomenclature*, 5th ed. (London: Butterworth Scientific , 1979); Alex Nickon and Ernest Silversmith, *Organic Chemistry: The Name Game, Modern Coined Terms and their Origins* (New York: Pergamon Press, 1987); James G. Traynham, "Organic Nomenclature: The Geneva Conference 1892 and the Following Years," in *Organic Chemistry: Its Language and the State of the Art*, ed. M. Volkan Kisakürek (Basel: Verlag Helvetica Chimica Acta, 1993), pp. 1–8; Kurt L. Loening, "Organic Nomenclature: The Geneva Conference and the Second Fifty Years: Some Personal Observations," in *Organic Chemistry*, ed. M. Volkan Kisakürek, pp. 35–46; Vladimir Prelog, "My Nomenclature Years," in *Organic Chemistry*, ed. M. Volkan Kisakürek, pp. 47–54. However, a fruitful collaboration between two historians and two chemists should be mentioned, which unfortunately was not followed by later attempts: W. H. Brock, K. A. Jensen, C. K. Jorgensen, and G. B. Kauffman, "The Origin and Dissemination of the Term 'Ligand' in Chemistry," *Ambix*, 21 (1981), 171–83.

consensus. So omnipresent remain the difficulties of naming, that the past still belongs to the chemists' memory, rather than to formal history. It is also strange that, apart from one volume by François Dagognet thirty years ago, chemical language has been virtually unexplored by philosophers, despite the fashion for the philosophy of language over the past decades.[8]

Rather than trying to reconstruct the whole process of the evolution of chemical language over the past two centuries, this chapter will focus on three key episodes that can be seen as crucial moments when decisions were made that shaped the current language of chemistry. For the language of chemistry, the nineteenth century started a few decades before 1800. The first tableau takes place in Paris in 1787, where the so-called modern language of chemistry was submitted to the Royal Academy of Sciences. The second is located at Karlsruhe in September 1860, when for the first time an international meeting of chemists was organized to make decisions about terminology and formulas. The third tableau will bring us to Liège in 1930, when the Commission on Nomenclature of Organic Chemistry, appointed by the Union internationale de chimie pure et appliquée, met and voted on rules for naming organic compounds.

All three episodes depict the attempts made by groups of chemists to clarify the language and facilitate communication in chemistry. However, the three reforms took place in different settings and displayed various strategies that reflect the state of the discipline of chemistry at these times.

1787: A "MIRROR OF NATURE" TO PLAN THE FUTURE

In January 1787, Louis-Bernard Guyton de Morveau (1737–1816), a lawyer and well-known chemist from Dijon, arrived at the Paris Academy of Sciences in the midst of a controversy between phlogistonists and antiphlogistonists. Since Guyton was in charge of the chemistry dictionary for the *Encyclopédie méthodique,* and in correspondence with various foreign chemists, he was extremely attentive to invitations to build up a universal and systematic language for chemistry. Throughout the eighteenth century, chemists had been increasingly dissatisfied with their language. They wanted to rid themselves of the alchemical heritage of names with mythological references, and they complained that various names were being used for one single substance or, symmetrically, that one name referred to different substances. Exchanges between chemists throughout Europe, coupled with intense translating activity, made these defects particularly visible. Pierre-Joseph Macquer (1718–1784) and Torbern Bergman (1735–1784) made timid attempts at systematizing,

---

[8] François Dagognet, *Tableaux et langages de la chimie* (Paris: Vrin, 1969). See also a semiologic approach to chemical nomenclature: Renée Mestrelet-Guerre, *Communication linguistique et sémiologie: Etude sémiologique des systèmes de signes chimiques,* PhD diss., Universitat Autonoma de Barcelona, 1980.

especially in naming the substances recently identified, like gases, or classified, like salts and minerals.[9]

In 1782, Guyton had authored a bolder project for reforming the whole of chemical nomenclature.[10] His project, like the botanical nomenclature set up by Carl Linnaeus, was based on the assumption that denominations should reveal "the nature of things," although Guyton chose Greek rather than Latin etymologies (presumably due to his strong opposition to the language of the Jesuits). Guyton's general principle was: simple names for simple substances and compound names for chemical compounds in order to express composition. When the composition is uncertain, Guyton proposed, an arbitrary and meaningless term is to be preferred. In itself, the project of making an artificial language for chemistry, breaking with the traditional language forged by the users of chemical substances over centuries, was ambitious and revolutionary. However, Guyton's goals were extremely modest. In keeping with earlier attempts, his reform was clearly designed to reach a consensus among European chemists.

Six months later, however, the project was deeply changed. When Guyton came to submit his project at the Paris Royal Academy, in early 1787, he found the chemistry class divided by the controversy over phlogiston chemistry. He met with Antoine-Laurent Lavoisier (1743–1794), Claude-Louis Berthollet (1748–1822), and Antoine-François de Fourcroy (1755–1809), all three partisans of the antiphlogistonist theory. They "converted" him to the new doctrine and persuaded him to revise his project in accordance with it. In a few weeks, the four of them had transformed Guyton's earlier outline of a new language into a weapon against phlogistonists.[11] The word "phlogiston" was eradicated, while terms such as "hydrogen" (generator of water) and "oxygen" (generator of acids) reflected Lavoisier's alternative theory. Lavoisier also provided a philosophical legitimation for the new language by referring to

---

[9] Torbern Bergman, "Meditationes de systemate fossilium naturali," in *Nova Acta Regiae Societatis Scientarum Upsaliensis*, 4 (1784), 63–128; see M. Beretta, *The Enlightenment of Matter: The Definition of Chemistry from Agricola to Lavoisier* (Canton, Mass.: Science History Publications, 1993), pp. 147–9; see also W. A. Smeaton, "The Contributions of P. J. Macquer, T. O. Bergman and L. B. Guyton de Morveau to the Reform of Chemical Nomenclature," *Annals of Science*, 10 (1954), 97–106, and M. P. Crosland, *Historical Studies* (1962), 144–67.

[10] Louis-Bernard Guyton de Morveau, "Mémoire sur les dénominations chymiques, la nécessité d'en perfectionner le système et les règles pour y parvenir," *Observations sur la Physique, sur l'histoire naturelle et sur les arts* (19 Mai 1782), 370–82. Also published as a separate brochure in Dijon, 1782, see Georges Bouchard, *Guyton de Morveau, chimiste et conventionnel* (Paris: Librairie académique Perrin, 1938). In 1787, an alternative and more traditional nomenclature was built up by a Belgian chemist and never adopted. See B. Van Tiggelen, "La Méthode et les Belgiques: l'exemple de la nomenclature originale de Karel van Bochaute," in *Lavoisier in European Context: Negotiating a Language for Chemistry*, ed. B. Bensaude-Vincent and F. Abbri (Canton, Mass.: Science History Publications, 1995), pp. 43–78.

[11] L. B. Guyton de Morveau, A. L. Lavoisier, C. L. Berthollet, and A. F. de Fourcroy, *Méthode de nomenclature chimique* (1787; reprint, Philadelphia: American Chemical Society, 1987); all quotations will refer to the 1994 edition, with introduction and notes (Seuil, Paris), translated by the author unless otherwise noted.

Condillac's philosophy of language.[12] He assumed that words, facts, and ideas were like three various faces of one single reality and that a well-shaped language was a well-shaped science. Linguistic customs and chemical traditions carried only errors and prejudices. By contrast, a language proceeding from the simple to the complex would keep the chemists on the track of truth. The language of analysis that Lavoisier and his collaborators promoted was more a "method" than a "system," and it was said to reflect nature itself. Actually, nature was identified with the products of chemical manipulations performed in the laboratory. Every compound name was the mirror image of the operations of its decomposition.

Like other nomenclatures, this one rested on an implicit classification. Instead of the traditional naturalists' taxonomic categories of genus, species, and individual, the chemists' classification was structured like a language with an alphabet of thirty-three simple substances distributed into four sections: (1) "simple substances belonging to the three realms of nature" (including caloric, oxygen, light, hydrogen, and nitrogen); (2) "nonmetallic, oxidizable, acidifiable simple substances"; (3) "metallic, oxidizable, acidifiable simple substances"; and (4) "earthy, salifiable, simple substances." This classification was a compromise between the old notion of universal principles and the definition of element as a unit of combination. The simple substances only made the first column of a synoptic table summarizing the whole system.[13] Tables were a favorite means of representation, which Foucault depicted as the center of knowledge in the "classic era."[14] However, the table displayed at the Academy of Sciences in 1787 and the tables published in the second section of Lavoisier's *Elements of Chemistry* differed from the previous "tables of relations" used by the eighteenth-century chemists.[15] Affinity tables condensed a knowledge painstakingly acquired through thousands of experiments. Lavoisier's tables incorporated empirical knowledge, but were rather aimed at ordering the material world like a language, an analytical language modeled after Condillac's *Logic*. The grammar of this language was derived from a dualistic theory of combinations. It was implicitly assumed that chemical compounds were formed by two elements or two radicals acting as elements.

While Lavoisier pretended that the new language mirrored nature, many of his contemporaries objected that such terms as oxygen were theory laden,

---

[12] Lavoisier,"Mémoire sur la nécessité de réformer et de perfectionner la nomenclature de la chimie," in Lavoisier et al., *Méthode de nomenclature,* pp. 1–25.

[13] The second column included the combinations of simple bodies with caloric (i.e., put in gaseous state); the third column included the compounds of simple substances with oxygen; column 4, the compounds of simple substance with oxygen plus caloric; column 5 included oxygenated simple substances combined with bases (i.e., neutral salts); column 6 is a small division for "simple substances combined in their natural state" (see Fourcroy, *Méthode de nomenclature,* pp. 75–100).

[14] Michel Foucault, *Les mots et les choses* (Paris: Gallimard, 1968), pp. 86–91.

[15] Lissa Roberts, "Setting the Tables: The Disciplinary Development of Eighteenth-Century Chemistry as Related to the Changing Structure of Its Tables," in *The Literary Structure of Scientific Argument,* ed. Peter Dear (Philadelphia: University of Pennsylvania Press, 1991), pp. 99–132.

rather than mere expressions of well-established facts. From all over Europe, chemists discussed the reform and tried to improve a number of names. Alternative proposals were made for oxygen, because Lavoisier's theory of acid was not accepted, and for *azote* (a-zoon, meaning "not for animals") because many other gases are not fit for animal life. This is why English chemists adopted nitrogen instead of azote. The French chemists led an intensive campaign of persuasion by mobilizing Madame Lavoisier for translations and for dinners; they created their own journal, the *Annales de chimie* in 1789. Finally, thanks to many translations of the textbooks written by Fourcroy, Chaptal, Lavoisier, and Berthollet, the French nomenclature was widely adopted by 1800. Adoption implied various strategies of linguistic adaptations. A number of chemists resented the French hegemony in a domain that, in principle, should be universal. German chemists, like the Polish, chose to translate the French-Greek terms into German (for instance, *Sauerstoff* for oxygen and *Wasserstoff* for hydrogen), whereas English and Spanish chemists simply changed the spelling and the endings of the terms.

The ongoing triumph of the nomenclature contrasts with the later abandonment of the graphic symbols proposed to replace the old alchemical symbols still in use in the affinity tables. The system of "characters" designed by Pierre-Auguste Adet (1753–1834) and Jean-Henri Hassenfratz (1755–1827), two young disciples of Lavoisier, closely followed the analytical logic of the nomenclature and provided pictorial symbols for the composition of substances. Why was it ignored? Apparently, Lavoisier was more concerned with changing the language in accordance with his theoretical views than with promoting a convenient symbolism for filling the tables. As pointed out by François Dagognet, the "new chemistry," based on the algebraic analytical interpretation of chemical reactions, favored a "vocal-structural" system, instead of a geometrical pictorial representation of chemical reactions.[16]

Because the reform of the nomenclature played a key role in the chemical revolution, it has often been described as the outcome of the revolutionary process. It is important, however, to reconsider this reform in the broader perspective of the history of chemistry as a long-term project that mobilized the chemical community through the course of the eighteenth century and, more importantly for our present purpose, to appreciate its impact on the discipline of chemistry. The descriptive names indicating the proportion of the constituents in a compound aided memorization. Moreover, as Lavoisier pointed out in his *Elements of Chemistry*, the analytical language, by inviting the chemical student to proceed from the simple to the complex, facilitated the teaching of chemistry.

However, this language, forged by academic chemists, prompted a divorce between them and the dyers, the glassmakers, and the manufacturers who kept the traditional terms inherited from artisanal traditions. Certainly

---

[16] Dagognet, *Tableaux et langages de la chimie*, pp. 45–52.

compositional names (e.g., "iron sulfate" and "iron sulfite"), as well as the constitutional formulas that were later derived from them, provided significant information for chemists whose main program was to determine the nature and the proportion of the constituents of inorganic or organic compounds. Nevertheless, these names deprived the pharmacists of knowledge about the medical properties embedded in many traditional terms. Thus, the new nomenclature contributed to the subordination of pharmacy to chemistry and, more broadly, to the redefinition of chemical arts as applied chemistry.[17]

The chemical language built up by the four French chemists was an integral part of Lavoisier's attempts to promote and legitimize a new practice of chemistry. Analytical procedures controlled by the balance displaced and discredited experimental results based on qualitative data, whereas phenomenological features, such as odors, colors, taste, or aspect, were discarded from the nomenclature. For instance, the "white of lead" and the "Prussian blue" used by dyers, became, respectively, "lead oxide" and "iron prussiate"; "stinking air" was renamed "sulfuretted hydrogen gas." The new language not only ignored the chemists' senses but also deprived the chemical substances of their history by banishing all reference to their geographical origins or to their discoverers.

In fact, when considered over the next decades, the principles of the new language were never strictly applied. A first and decisive step aside was made in the early nineteenth century when, after isolating chlorine, Humphry Davy (1778–1829) established that some substances – hydrochloric acid, for instance – presented characteristic acidic properties even though oxygen did not enter into their composition. Oxygen should have been renamed, but custom took over the imperative of systematicity. Over time, many elements isolated with the help of electrical analysis brought back odors and colors into the nomenclature. For instance, chlorine, bromine, and iodine were coined after the Greek terms *chloros* meaning green, *bromos* meaning stink, and *iodes* denoting violet. However, it is interesting to note that not all sensible qualities regained acceptability. Though elements went on being named after colors (thallium), smells (bromine), and countries (gallium), no one used tastes anymore.

Mythological references flourished again, too, with the term "morphine" coined in 1828 by Péligot, after the god Morpheus. Geographical data resurfaced: The term "benzene," for instance, reminds us of the resin produced by the bark of a tree native to Sumatra and Java with the name *Styrax benzoin;* gutta-percha, a gum that played a crucial part in the development of the electric telegraph, was named after the Malay *getha percha* tree in 1845.

---

[17] A. C. Déré, "La réception de la nomenclature réformée par le corps médical français," in *Lavoisier in European Context,* ed. B. Bensaude-Vincent and F. Abbri, pp. 207–24, and J. Simon, "The Alchemy of Identity: Pharmacy and the Chemical Revolution (1777–1809)," PhD diss., Pittsburgh University, 1997.

Nineteenth-century nationalism pervaded chemical language: Gallium, discovered by a French chemist, and germanium for another element predicted by Mendeleyev, were followed by scandium and polonium. Even the banished Latin language came back with the alphabetic symbols, initials of Latin names, that were introduced by Berzelius (1779–1848) in 1813.[18] The objective of systematicity, which governed the creation of an artificial language for chemistry, thus remained an ideal contradicted by the daily practice of language.

1860: CONVENTIONS TO PACIFY THE CHEMICAL COMMUNITY

In early September 1860, 140 chemists from all over Europe (and one from Mexico) met in the capital of the Grand Duke of Baden for a three-day meeting. The initiative for this extraordinary congress, the first international meeting of chemists, came from three professors: the young Friedrich-August Kekulé (1829–1896), a professor at the University of Ghent; Charles Adolphe Wurtz (1817–1884) at the University of Paris; and Karl Weltzien at the Polytechnik School in Karlsruhe. Teaching and communicating chemistry was their main concern. As in 1787, the chemists complained of a great disorder in their language. The divergence, however, was more concerned with formulas than with names. The debate thus shifted from nomenclature to graphical representation.

In his *Lehrbuch der Organischen Chemie*, Kekulé reported nineteen different formulas for acetic acid. There were many different ways to write the formula of a common substance such as water. The formula HO, introduced by John Dalton (1766–1844), was still in use in 1860 with 8 for the atomic weight of oxygen based on $H = 1$. Some chemists, however, adopted Berzelius's notation. They determined atomic weights on the basis $o = 100$, which meant that $H = 6.24$, $C = 75$, and $N = 88$. Berzelius wrote water H̶O. The H̶ (barred H) meant a double atom of hydrogen, that is, two atoms of the same element that combined by pairing. Berzelius similarly wrote H̶Cl̶ for hydrochloric acid, because a double atom of hydrogen combined with a double atom of chlorine, and N̶H̶³ for ammonia because three double atoms of hydrogen combined with one double atom of nitrogen. In the 1840s, the notation HO, grounded on equivalent weights and recommended by Gmelin, Liebig, and Dumas, prevailed. In the 1850s, Charles Gerhardt (1816–1856), considering volumes and weights together, suggested doubling

---

[18] It was Berzelius who rejected the pictograph symbols used by Dalton and introduced the letters of the alphabet, index numbers, dots, and bars. Proportions were indicated by superscript figures or symbols. On the debates caused by the introduction of symbols, see T. L. Alborn, "Negotiating Notation: Chemical Symbols and British Society, 1831–35," *Annals of Science*, 46 (1989), 437–60, and Nye, *From Chemical Philosophy*, pp. 91–102.

the equivalent weights and proposed $O = 16$, $C = 12$. Water, he said, was made of two atoms of hydrogen plus one of oxygen and occupies two volumes, if one atom of hydrogen occupies one volume. Similarly, HCl, hydrochloric acid, made of one atom (or volume) of hydrogen and one atom of chlorine, occupies two volumes; and ammonia, NH^3, formed of one atom (or volume) of nitrogen and three atoms of hydrogen, occupies two volumes. Gerhardt had noticed that in many reactions between organic compounds, one never obtained one equivalent of water but two. The quantities formed were H^4O^2 and C^2O^4 for carbonic acid. Consequently, Wurtz wrote water H^4O^2 in a memoir published in 1853. Gerhardt, however, strongly recommended halving organic formulas because they doubled the real equivalents (Gerhardt still wrote equivalents instead of molecules). This proposal condemned the dualistic interpretation of many organic compounds in favor of a "unitary" view of composition, based on reactions of substitution where one atom of hydrogen was replaced by an atom of chlorine, for instance.

With the unitary view, a new style of graphism prevailed. First, Auguste Laurent (1808–1853) based his theory of substitution on the image of a molecular architecture with a rigid arrangement of atoms inside the molecule.[19] He favored pictorial representations of the spatial configuration of molecular structure, similar to the pictures used by crystallographers to represent the geometrical structure of crystals. Second, the type theory initiated by August W. Hofmann (1818–1892) and Alexander W. Williamson (1824–1904) and extended by Gerhardt provided the basis of a classification of organic compounds. Vertical formulas, modeled after the types hydrogen, water, and ammonia, flourished:

$$\left\{ \begin{array}{c} H \\ \\ H \end{array} \right. \quad O \left\{ \begin{array}{c} H \\ \\ H \end{array} \right. \quad N \left\{ \begin{array}{c} H \\ H \\ H \end{array} \right.$$

The formulas for acids, for instance, were derived from the water type by substituting radicals (groupings of atoms) for hydrogen.

Clearly, the confusion in formulas and notation, which prompted the Karlsruhe Congress, emerged from theoretical conflicts about the composition of organic compounds. The letter sent by the organizers to potential participants explicitly acknowledged the theoretical dimension of the debates over the language of chemistry: "Dear Distinguished Colleagues: The great development that has taken place in chemistry in recent years, and the differences in theoretical opinions that have emerged, make a Congress,

---

[19] Laurent's nucleus model led him to a complex nomenclature for organic compounds. However, his rules of nomenclature would provide a basis for the Geneva proposals in 1892. About Laurent's tentative nomenclature of organic compounds published posthumously in the *Méthode de chimie, Paris* (1854), see M. Blondel-Mégrelis, *Dire les choses, Auguste Laurent et la méthode chimique* (Paris: Vrin, 1996).

whose goal is the discussion of some important questions as seen from the standpoint of the future progress of the science, both timely and useful."[20]

Three questions were submitted to the deliberation of the assembly:

- definition of important chemical notions such as those expressed by the words: atom, molecule, equivalent, atomic, basic.
- examination of the issue of equivalents and of chemical formulas.
- establishment of a notation and of a uniform nomenclature.

Was the agreement on formulas and notation subordinated to a theoretical agreement on the basic notions of chemistry? If the organizers expected to decide by a vote on the atomic structure of matter, it is not surprising that the Congress failed to reach an agreement on these matters, although Stanislao Cannizzaro (1826–1910) convinced many participants to adopt Avogadro's distinction between atom and molecule and Gerhardt's formulas revised by Cannizzaro.

The Karlsruhe Congress is usually described as a crucial episode in the history of chemical atomism because it was the moment when the modern distinction between atoms and molecules based on Avogadro's law was adopted.[21] This dominant perspective, like the analogous historical treatment of the 1787 reform as an aspect of the chemical revolution, emphasizes the heavy dependence of language on theoretical assumptions.

It was not the existence of atoms that was at stake in Karlsruhe, however, as has been clearly established in recent scholarship.[22] Staunch advocates of atomism, such as Gerhardt, Kekulé, and even Wurtz, never claimed that matter was "really" organized into atoms and molecules. Gerhardt's type formulas were not meant as representations of the molecular architecture, but rather simply as "the relationship according to which certain elements or groups of elements substitute for or transport each other from one body to another in double decomposition."[23] Gerhardt thus refused to think of the radicals isolated by his type formulas as real separable entities. Kekulé, who is often considered the father of structural chemistry because he discovered the benzene ring, nevertheless refused to give any ontological meaning to structures:

> The question whether atoms exist or not has but little significance from a chemical point of view: its discussion rather belongs to metaphysics.... I

---

[20] *Compte-rendu des séances du congrès international de chimistes réuni à Karlsruhe le 3, 4, 5 septembre 1860* (1892; translated by J. Greenberg) in Mary Jo Nye, *The Question of the Atom, from the Karlsruhe Congress to the First Solvay Conference, 1860–1911* (Los Angeles: Tomash, 1984), p. 6. See also B. Bensaude-Vincent, "Karlsruhe, septembre 1860: l'atome en congrès," *Relations internationales* (Les Congrès scientifiques internationaux), 62 (1990), 149–69.

[21] Mary Jo Nye, *The Question of the Atom*, pp. xiii–xxxi.

[22] Alan J. Rocke, *From Dalton to Cannizzaro: Chemical Atomism in the Nineteenth Century* (Columbus: Ohio State University Press, 1984), pp. 287–311, and *Nationalizing Science: Adolphe Wurtz and the Battle for French Chemistry* (Cambridge, Mass.: MIT Press, 2001).

[23] Gerhardt, *Traité de chimie organique*, vol. 4 (Paris: Firmin-Didot, 1854–6), pp. 568–9.

> have no hesitation in saying that, from a philosophical point of view, I do not believe in the actual existence of atoms, taking the word in its literal signification of indivisible particles.... As a chemist, however, I regard the assumption of atoms, not only advisable, but as absolutely necessary in chemistry."[24]

Karlsruhe was not a battle between realistic atomists and positivistic or idealistic equivalentists. In fact, few equivalentists answered the invitation to take part in the discussions at Karlsruhe. Moreover, as convincingly argued by Alan Rocke, the battle was more or less already won in 1860.[25] Rather, Karlsruhe was an attempt to reach a consensus on formulas beyond divergent theoretical commitments. Even Cannizzaro, the advocate of atoms and molecules, did not expect any "conversion" from the meeting, as is clear from the conclusion of his speech:

> And if we are unable to reach a complete agreement upon which to accept the basis for the new system, let us at least avoid issuing a contrary opinion that would serve no purpose, you can be sure. In effect, we can only obstruct Gerhardt's system from gaining advocates every day. It is already accepted by the majority of young chemists today who take the most active part in advances in science. In this case, let us restrict ourselves to establishing some conventions for avoiding the confusion that results from using identical symbols that stand for different values. Generalising already established custom, it is thus that we can adopt barred letters to represent the double atomic weight.[26]

These words, aimed at divorcing theory from language, typically illustrate a conventionalist attitude in regard to language, in stark contrast to Lavoisier's reference to "nature." In this respect, Karlsruhe marks the acme of conventionalism in chemistry. It was a common attitude shared by most atomists and equivalentists. A kind of diplomatic compromise, based on the conservation of common words together with some degree of arbitrariness, but not even formulated as a rule, was adopted at the conclusion of the Congress: "The Congress consulted by the chairman expresses the wish that the use of barred symbols, representing atomic weights twice those that have been assumed in the past, be introduced into science."[27] However, Cannizzaro's speech prompted conversions to Gerhardt's system and played a crucial part as one origin of the periodic system in the evolution of chemistry. Dmitry Mendeleyev, like Julius Lothar Meyer, often declared that Cannizzaro's speech

---

[24] A. Kekulé, "On Some Points of Chemical Philosophy," *The Laboratory*, 1 (27 July 1867); reprint in R. Anschütz, *August Kekulé*, 2 vols. (Berlin: Verlag Chemie, 1929).

[25] Alan J. Rocke, "The Quiet Revolution of the 1850s: Social and Empirical Sources of Scientific Theory," in *Chemical Sciences in the Modern World*, ed. S. H. Mauskopf (Philadelphia: University of Pennsylvania Press, 1993), pp. 87–118, and *The Quiet Revolution: Hermann Kolbe and the Science of Organic Chemistry* (Berkeley: University of California Press, 1993).

[26] *Compte-rendu*, translated by J. Greenberg in Nye, *The Question of the Atom*, p. 28.

[27] Ibid.

and his *Sunto d'un corso di filosofia chimica* (1858) were the key factors that led to the discovery of the periodic law and consequently to the well-known system that ordered the "building blocks of the universe."[28]

It is also worth noting that even Laurent and Cannizarro, both believers in the physical reality of chemical atoms, never referred names and formulas to any real entities. Distance between words and things, between formulas and the reality referred to, is precisely one major feature of this chemical language.[29] In Kekulé's view, it was one distinctive feature of the identity of chemistry. Whereas the kinetic theory of gases envisaged molecules as real micro balls moving around in a flask, the chemist's molecule – defined as the smallest unit of a substance to enter into a combination – might never exist as an isolable entity. When chemists started representing the bonds that link atoms together in the molecule, as when J. H. van't Hoff (1852–1911) developed a three-dimensional image of the atom of carbon in the shape of the tetrahedron, the representations were by no means intended as images of molecular reality. Spatial formulas, as well as type formulas and structural formulas, were above all instruments for the chemists. They were, first, tools of classification of reactivity, helping the chemist to find analogies; they were also tools of prediction, guiding synthesis, especially for dyeing molecules.[30] The formulas both anticipated and assisted the making of real substances. Similarly, the balls-and-rods molecular models introduced by August Wilhelm Hofmann were built for didactic purposes. They were not naive representations of atoms as colored balls, but pragmatic macroscopic *analogons* or images helpful for dealing with a reality that was usually viewed as beyond reach.

The conventionalist attitude culminated in France, the last bastion of equivalentism. The French chemical establishment, epitomized by Marcellin Berthelot in particular, who refused the atomist notation until the end of his life, is often portrayed as made up of stubborn and conservative minds tied up by the dogmas of positivism.[31] In fact, as Mary Jo Nye pointed out, Berthelot considered that the dilemma was by no means vital for the progress of chemistry and that the choice was "a matter of taste."[32] When the issue of choice was raised at the Ecole Polytechnique in the 1880s, because one

---

[28] See W. van Spronsen, *The Periodic System of the Chemical Elements: A History of the First Hundred Years* (Amsterdam: Elsevier, 1969), and B. Bensaude-Vincent, "Mendeleev's Periodic System of Chemical Elements," *British Journal for the History of Science*, 19 (1986), 3–17.

[29] M. G. Kim, "The Layers of Chemical Language," *History of Science*, 30 (1992), 69–96, 397–437; "Constructing Symbolic Spaces: Chemical Molecules in the Académie des Sciences," *Ambix*, 43 (1996), 1–31; M. Blondel-Mégrelis, *Dire les choses*, pp. 266–73.

[30] August Wilhelm Hofmann, for instance, built up his research program on type formulas. See Christoph Meinel, "August Wilhelm Hofmann – 'Reigning Chemist-in-Chief,'" *Angewandte Chemie* (international edition), 31 (1992), 1265–1398.

[31] See for instance, Jean Jacques, *Marcellin Berthelot: Autopsie d'un mythe* (Paris: Belin, 1987), pp. 195–208. Berthelot and Jungfleisch finally adopted the atomic notation in the fourth edition of their *Traité de chimie*, Paris (1898–1904).

[32] Mary Jo Nye, "Berthelot's Anti-atomism: A Matter of Taste?" in *Annals of Science*, 38 (1981), 586–90.

professor used the equivalentist system while Edouard Grimaux taught the atomic notation, the members of the Council regarded the alternative as a "purely pedagogical issue," analogous to the choice of a system of coordinates in mathematics.[33]

1930: PRAGMATIC RULES TO ORDER CHAOS

In 1787, the reform of language was achieved in less than six months by a small group of four chemists clearly identified as French scientists. In 1860, a collection of individuals met together in Karlsruhe to make a transnational decision on the best language for communicating and teaching chemistry. In 1930, a permanent commission prescribed dozens of rules aimed at standardizing the nomenclature of organic compounds. The reform of nomenclature was no longer an extraordinary event. Rather, it had become a continuous process of revision and an integral part of what is called "normal science." The language of chemistry is no longer a national or a transnational issue in the hands of a few motivated individuals.[34] It is an international enterprise, fully integrated in the process of internationalization of science, which developed in the late nineteenth century. The commission for nomenclature, first coordinated by the Union of the Chemical Societies, became a permanent institution in the context of the Union internationale de chimie pure et appliquée (UICPA) created in 1919, with French as its official language and without Germany because of the decision of the allied nations after World War I to boycott German science. The international union was reestablished after World War II as the International Union of Pure and Applied Chemistry (IUPAC), with English as the official language. The Commissions on Nomenclature were much more than simple by-products of the internationalization of science. As emphasized in a number of studies, the Commissions acted as a driving force, though the concern for international coordination never completely abolished national rivalries.[35]

The first attempt at reform followed the first International Conference of Chemistry held in Paris in 1889. A special section was appointed under the leadership of Charles Friedel (1832–1899), who was in charge of preparing a

---

[33] Edouard Grimaux, *Théorie et notation chimiques*, Paris (1883), and Catherine Kounelis, "Heurs et malheurs de la chimie: La réforme des années 1880," in B. Belhoste, A. Dahan-Dalmedico, and A. Picon, eds., *La Formation polytechnicienne 1794–1994* (Paris: Dunod, 1994), pp. 245–64.

[34] Christoph Meinel, "Nationalismus und Internationalismus in der Chemie des 19 Jarhunderts," in P. Dilg, ed., *Perspektiven der Pharmaziegeschichte: Festschrift für Rudolf Schmitz* (Graz: Akademische Druck-u. Verlagsanstalt 1983), pp. 225–42.

[35] B. Schroeder-Gudehus, "Les congrès scientifiques et la politique de coopération internationale des académies des sciences," *Relations internationales*, 62 (1990), 135–48; E. Crawford, *Nationalism and Internationalism in Science, 1880–1939* (Cambridge: Cambridge University Press, 1992); A. Rasmussen, "L'internationale scientifique, 1890–1914," PhD diss., Ecole des Hautes Etudes en Sciences Sociales, Paris, 1995.

set of recommendations to be voted upon during an international conference on chemical nomenclature held in Geneva, April 1892. Why organize a special event devoted to language issues? Since Karlsruhe, a number of individual initiatives had attempted to systematize the nomenclature of organic compounds: For instance, Williamson introduced parentheses into formulas to enclose the invariant groups – for example, $Ca(CO_3)$ – and proposed the suffix *ic* for all salts.[36] Hofmann introduced the systematic names for hydrocarbons, using suffixes following the order of the vowels in order to indicate the degree of saturation: *ane, ene, ine, one, une.*[37] A great confusion once again reigned in the language of chemistry. Instructions were being given by the various scientific journals that had sprung up in the late nineteenth century.

The aim of the Geneva Nomenclature was mainly to standardize terminology and to make sure that a compound would appear under one single heading in catalogs and dictionaries. The Commission on Nomenclature felt legitimized enough to propose an official name for each organic compound. Official names were built upon the molecular structure and were to be as revealing of constitution as were chemical formulas. Names were based on the longest continuous chain of carbons in the molecule, with suffixes designating the functional groups and prefixes denoting substituent atoms. Sixty-two resolutions were adopted by the Geneva group, which considered only acyclic compounds. The official names were never applied practically, although they are still mentioned in modern textbooks because they provide governing principles. Yet Geneva still is present in chemists' minds as a founding event. Had I retained only well-remembered events to mark the evolution of chemical nomenclature, the Geneva Conference would no doubt have been preferable. The Liège Conference, though less well known, is nevertheless more characteristic of the new regime of nomenclature.

In many respects, the conference held in Liège in 1930 contrasted with the Geneva Conference.[38] It was the end result of a long process of elaboration. The first impulse from the International Association of Chemical Societies, founded in 1911, was disrupted by the war and resumed by the UICPA. Two permanent commissions were set up. The Commission for the Nomenclature of Inorganic Chemistry appointed the Dutch chemist W. P. Jorissen as chairman, and the Commission for the Nomenclature of Organic Chemistry also appointed a Dutch chemist, A. F. Holleman, as chairman. The choice of chairmen belonging to a minor linguistic area clearly indicated an attempt to construct a universal language that would not reflect the hegemony of any one nation.

[36] W. H. Brock, "A. Williamson," in *Dictionary of Scientific Biography*, XIV, 394–6.
[37] A. W. Hofmann, *Proceedings of the Royal Society*, 15 (1866): 57, quoted by James Traynham, in *Organic Chemistry*, ed. M. Volkan Kisakürek, p. 2.
[38] A detailed account of this conference is to be found in Verkade, *A History of the Nomenclature*, pp. 119–78.

In 1922, both commissions formed a Working Party, with representatives of various linguistic areas to prepare the rules. Not only professors but also journal editors were invited to join the Working Party. Following regular meetings in 1924, 1927, 1928, and 1929, the Working Party in charge of organic chemistry issued reports that were submitted for criticism and then amended before the final vote in Liège. The Working Party in charge of inorganic chemistry also met several times before issuing its final rules at the Tenth Conference of the UICPA at Rome in 1938. The new regime of naming was thus characterized by a long process of negotiations, which allowed both for the making of new terms familiar to chemists before their official adoption and for the reaching of consensus before the final vote.[39] Whereas the Geneva Conference, presided over by Friedel, was dominated by the "French spirit and the French logic," the Liège rules codified suggestions by American chemists, particularly A. M. Patterson, who was directly connected with *Chemical Abstracts*.[40] The Germans, though excluded because of the boycott, were consulted, however, and finally invited to Liège.[41]

The style of the Liège nomenclature is quite different from that of Geneva: No more official names. The committee report, unanimously adopted in Liège, conformed to the linguistic customs of *Beilstein* and *Chemical Abstracts*, with minor corrections. Rule 1 reads as follows: "The fewest possible changes will be introduced into the universally adopted terminology." Liège, however, broadened the scope of the Geneva nomenclature. Rules were set up for naming the "functionally complex compounds," that is, those bearing more than one type of function. The final vote allowed both the official Geneva names and the Liège nomenclature to be used. The ideal of systematicity thus gave way to a more pragmatic strategy. Flexibility and permissibility were considered to be the most efficient means for favoring a general adoption of the standard language in the daily practices of chemistry, whether in textbooks or journals, in the classrooms or the factories.

Since Liège, this pragmatic attitude has prevailed in all successive revisions, in Lucerne (1936), in Rome (1938), and after World War II, in Paris (1957). The current nomenclature is by no means as systematic as what was dreamed by the 1787 reformers. Trivial names – not referring to the structure of the compounds – coexist with systematic names, conforming to the rules. In fact, both in inorganic and organic chemistry, a majority of names are semitrivial, that is, a mixture of anecdote and of constitution.[42]

---

[39] Verkade, *A History of the Nomenclature*, p. 127.
[40] Ibid., p. 8.
[41] On the boycott of Germany, Hungary, and Austria by the allied nations after World War I, see B. Schroeder-Gudehus, *Les Scientifiques et la paix* (Montreal: Les presses de l'université de Montréal, 1978), pp. 131–60.
[42] IUPAC, *Nomenclature of Inorganic Chemistry*, 2d ed. (London: Butterworths Scientific Publications, 1970); IUPAC, *Nomenclature of Organic Chemistry* (the so-called Blue Book) (London: Pergamon Press, 1979); B. P. Black, W. H. Powell, and W. C. Fernelius, *Inorganic Chemical Nomenclature*,

TOWARD A PRAGMATIC WISDOM

Three major points can be emphasized in conclusion. "Chemistry is structured like a language." This assertion, paraphrasing what the French psychoanalyst Jacques Lacan stated about the unconscious, is the main feature of the successive reforms of language. Since 1787, it has been tacitly assumed that chemical compounds are formed like words and phrases out of an alphabet of elemental units, whose combinations allow the building up of an indefinite number of compound words, according to a complex syntax. Whatever the identity of the basic units – were they elements, radicals, functions, atoms, ions, molecules – the linguistic metaphor still inspires contemporary chemists. Pierre Laszlo, for instance, has collected his chemical views under the title *La parole des choses* (The speech of things).[43]

The three tableaux here described suggest that the establishment and standardization of chemical language actively contributed to the cementing together of a chemical community in various ways. Although the first systematic nomenclature was elaborated in a specific cultural context, in the midst of a controversy heavily laden with nationalistic interests, it rapidly reached a quasi-universal status. The construction of universality was first achieved through the solidarity of the antiphlogistonists, which helped overcome divergent views among the founders of the nomenclature, and then through an active campaign of translations and linguistic adaptations, which helped spread the local and artificial language around the world.

Whereas a local community created a universal language in prerevolutionary France, the Karlsruhe Congress, for the first time, convened an international community of chemists, who in a three-day meeting reached consensus about conventions for the formulas and symbols of their language. From universality to internationalism, the language of chemistry followed a globally changing attitude toward the project of a universal language. The construction of an artificial language had been abandoned in the late nineteenth century, while most efforts converged toward the construction of international languages based on existing natural languages, such as Esperanto.[44] Later on, such projects were, in turn, abandoned in favor of more pragmatic attitudes. In Liège, an international community was represented by the permanent members of the Working Party set up by the IUPAC. By the 1930s, so strong was the structure of the international chemistry community that errors

---

Principles and Practice (Washington D.C.: American Chemical Society, 1990); P. Fresenius and K. Görlitzer, *Organic Chemical Nomenclature* (Chichester, England: Hellis Harwood, 1989). On more recent developments of the method for indexing chemical formulas, see W. H. Brock, *The Fontana/Norton History of Chemistry* (New York: W. W. Norton, 1993), pp. 453–4.

[43] Pierre Laszlo, *La parole des choses* (Paris: Hermann, 1995).

[44] See Anne Rasmussen, "A la recherche d'une langue internationale de la science, 1880–1914," in *Sciences et langues en Europe*, ed. Roger Chartier and Pietro Corsi (Paris: Centre Alexandre Koyré, EHESS, 1996), pp. 139–55.

and defects in current terminology could be tolerated. Indeed, flexibility now reinforces a community spirit because it creates a kind of connivance among experts who know what it is about.

Finally, this rapid survey exemplifies two alternative strategies for controlling the language. One is the legislative attitude, illustrated by the 1787 founders who sought to build up a new artificial and systematic language on a tabula rasa, as in the attempt at prescribing official names in 1892. By contrast, the Karlsruhe Congress and the Liège Conference illustrate a conventionalist attitude, more skeptical, practical, and respectful of customs. This strategy has been the dominating one up to the end of the twentieth century and reveals a deep change of attitude toward the chemical heritage received from the past. Clarence Smith, a member of the Working Party for the Liège nomenclature, suggested in 1936: "Could we but wipe out all existing names and start afresh, it would not be a very difficult task to create a logical system of nomenclature. We have, however to suffer for the sins of our forefathers in chemistry."[45] This "chemical wisdom," deeply contrasting with the revolutionary attitude of 1787, results from the increasing difficulty of keeping up with systematicity when the compounds are extremely complex. How long can it prevail? The next century might well bring back the need for a radical change and a more systematic language in chemistry.

---

[45] Clarence Smith, *Journal of the Chemical Society* (1936), 1067, quoted by James G. Traynham, "Organic Nomenclature," in *Organic Chemistry*, ed. M. Volkan Kisakürek, p. 6.

10

IMAGERY AND REPRESENTATION IN TWENTIETH-CENTURY PHYSICS

Arthur I. Miller

Scientists have always expressed a strong urge to think in visual images, especially today with our new and exciting possibilities for the visual display of information. We can "see" elementary particles in bubble chamber photographs. But what is the deep structure of these images? A basic problem in modern science has always been how to represent nature, both visible and invisible, with mathematics, and how to understand what these representations mean. This line of inquiry throws fresh light on the connection between common sense intuition and scientific intuition, the nature of scientific creativity, and the role played by metaphors in scientific research.[1]

We understand, and represent, the world about us not merely through perception but with the complex interplay between perception and cognition. Representing phenomena means literally *re*-presenting them as either text or visual image, or a combination of the two. But what exactly are we *re*-presenting? What sort of visual imagery should we use to represent phenomena? Should we worry that visual imagery can be misleading?

Consider Figure 10.1, which shows the visual image offered by Aristotelian physics for a cannonball's trajectory. It is drawn with a commonsensical Aristotelian intuition in mind. On the other hand, Galileo Galilei (1564–1642) realized that specific motions should not be imposed on nature. Rather, they should emerge from the theory's mathematics – in this way should the book of nature be read. Figure 10.2 is Galileo's own drawing of the parabolic fall of an object pushed horizontally off a table. It contains the

---

[1] In recent years, exploring the use of visual imagery in science has turned into a growth area. Among others studying this subject in physics I mention Peter Galison, *Image and Logic: A Material Culture of Microphysics* (Chicago: University of Chicago Press, 1997); Gerald Holton, *Einstein, History and Other Passions* (Woodbury, N.Y.: AIP Press, 1995); David Kaiser, "Stick Figure Realism: Conventions, Reification, and the Persistence of Feynman Diagrams, 1948–1964," *Representations*, 70 (2000), 49–86; and Sylvan S. Schweber, *QED and the Men Who Made It: Feynman, Schwinger, and Tomonaga* (Princeton, N.J.: Princeton University Press, 1994). A selection of my own publications, as well as lengthy bibliographies on imagery studies, are in Arthur I. Miller, *Imagery in Scientific Thought: Creating 20th Century Physics* (Cambridge, Mass.: MIT Press, 1986) and *Insights of Genius: Imagery and Creativity in Science and Art* (Cambridge, Mass.: MIT Press, 2000).

191

Figure 10.1. An Aristotelian representation of a cannonball's trajectory. It illustrates the Aristotelian concept that the trajectory consists essentially of two separate motions, unnatural (away from the ground) and natural (toward the ground). In this figure the transition between unnatural and natural motions is a circular arc. (From G. Rivius, *Architechtur, Mathematischen, Kunst,* 1547).

noncommonsensical axiom of his new physics that all objects fall with the same acceleration, regardless of their weight, in a vacuum. Yet in Galileo's day, no one had yet produced a vacuum, a notion considered as absurd in Aristotle's physics. After all, every observed motion is continuous. If the object happened to encounter an evacuated portion of space, then its trajectory would become erratic. Since this had not been observed, ergo, no vacuums.

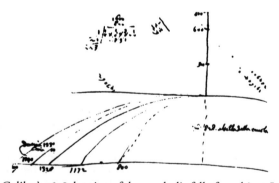

Figure 10.2. Galileo's 1608 drawing of the parabolic fall of an object. It can be interpreted as his experimentally confirming conservation of the horizontal component of velocity, and of the decomposition of the vertical and horizontal components to give the parabolic trajectory of a body projected, in this case, horizontally. Galileo was beginning to think along the lines of free fall through a vacuum. (Biblioteca Nazionale Centrale, Florence.)

Galileo's message is that understanding nature requires abstraction beyond the world of sense perceptions into other possible worlds, for example, a world in which there are vacuums. The breathtaking extent of Galileo's abstraction is clear from the stunning difference between Figures 10.1 and 10.2. Whereas Figure 10.1 is a tranquil landscape drawing, Galileo's displays a curve deduced from a mathematical formalism and drawn on a two-dimensional axis according to distance as a function of time.

In Galilean-Newtonian physics, our notion of what is commonsense intuition becomes transformed into a higher level. Yet despite the new way in which we "see" trajectories, what is being imaged are objects amenable to our sense perceptions. With this background, let us move to the twentieth century, in which what is counterintuitive would reach levels undreamed of, yet eventually become to scientists as commonsensical as Galileo's.

THE TWENTIETH CENTURY

The onset of the twentieth century was one of optimism in science. The *fin de siècle* malaise was exploded by three monumental discoveries: x rays (1895), radioactivity (1896), and the electron (1897). Scientists crashed into the new century full of enthusiasm toward exploring this new cache of riches. Although most scientists suspected that these new effects might be caused by entities invisible to direct viewing, their mode of representation remained grounded in phenomena actually observed. So, for example, electrons were depicted as billiard balls possessing charge. This mode of representation was extended into the subatomic world and turned out to be sufficient for the class of empirical data being explored at the time. Nothing succeeds like success, and so no extreme changes in representation were deemed necessary. In this sense, scientists lagged somewhat behind artists who were already experimenting with abstract representations of nature.

Representation versus abstraction was a topic of great interest to artists and scientists at the turn of the twentieth century. In the sciences, the Viennese scientist-philosopher Ernst Mach's (1838–1916) positivism prevailed. According to Mach, the serious scientist should focus on experimental data reducible to perceptions. So Mach considered atoms to be merely auxiliary hypotheses, helpful perhaps for calculational purposes. Although the electron's discovery led some scientists to question positivism, there were many prominent supporters. Consequently, the electron's mode of representation remained firmly attached to the world of sense perceptions.

This was becoming less the case in art where a countermovement to the figuration and perspective that had held sway ever since the Renaissance surfaced forcefully in the Postimpressionism of Paul Cézanne (1839–1906). The trend toward abstraction would continue in art during the first decade and a half of the twentieth century owing mainly to the explorations of space by Pablo Picasso (1881–1973). In science, at first, the move toward abstraction would be of a less visual sort. Rather, it was toward exploring phenomena that lay beyond sense perceptions, while somewhat ironically using visual imagery abstracted from phenomena we have actually witnessed. Albert Einstein (1879–1955) was the catalyst for this movement.

ALBERT EINSTEIN: THOUGHT EXPERIMENTS

Key ingredients to Einstein's creative thinking were thought experiments framed in vivid visual imagery. He realized a preference for this mode of thought while attending a preparatory school in Switzerland during 1895–6, which emphasized the power of visual thinking. In 1895 the precocious 16-year-old boy framed the key problem in nineteenth-century physics in a bold new way.

In thought experiments, scientists imagine physical phenomena in their "mind's eye" as they occur in an idealized manner, abstracted from prevailing physical conditions. Initially, all experiments are thought experiments. For example, Galileo imagined what it is like for objects to fall freely with no wind resistance from the mast of a moving ship. In this way, he could eventually transfer the situation to objects falling through a vacuum. But Galileo's thought experiments, as well as those of most later scientists, were used to present arguments for hypotheses that had already been proposed. For example, Galileo's thought experiments that we just mentioned "tested" his hypothesis that all objects fall through a vacuum with the same acceleration regardless of weight. Einstein's thought experiments were different: His were unplanned and they were insights that resulted in discoveries. I have in mind his two great thought experiments of 1895 and 1907.

On the basis of his readings in electromagnetic theory, the young Einstein conceived in his mind's eye what it would be like to catch up with a point on a light wave.[2] According to Newtonian mechanics and its accompanying intuition, this should be possible. In this case, the velocity of light measured by the thought experimenter ought to decrease as he catches up with the point on the light wave. But this conclusion violates the principle of relativity since, by measuring a variable velocity of light, the thought experimenter could detect whether he or she is in an inertial reference system.

Although, at first, Einstein did not know quite what to make of the problem situation, by 1905 he concluded that according to the thought experimenter's "intuition," the velocity of light ought to be independent of any relative motion between source and observer – because any violation of the principle of relativity would be counterintuitive.[3] Consequently, the principle of

---

[2] Albert Einstein, "Autobiographical Notes," in *Albert Einstein: Philosopher: Scientist*, ed. P. A. Schilpp, (Evanston, Ill.: Open Court, 1949), pp. 2–94.
[3] Einstein, "Autobiographical Notes," p. 53.

relativity should be raised to the exalted status of an axiom, which means, for example, that the velocity of light is independent of any relative motion between the light's source and the observer. So, no matter what velocity the thought experimenter's laboratory attains, the measured velocity of light will remain the same. This result is terribly counter to Galilean-Newtonian intuition and comes about because time is a relative quantity. Just as the consequences of Galilean-Newtonian physics became intuitive, so have the results of Einstein's special theory of relativity.

In summary, Einstein's move to raise the principle of relativity to an axiom was an audacious one because by 1905, he had realized that his 1895 thought experiment encapsulated all possible ether-drift experiments. The ones actually performed were magnificent failures because they measured no variation of the velocity of light. Physicists offered scores of hypotheses to explain the dramatic difference between what was expected and what is observed. Einstein's Gordian resolution asserted that these beautiful state-of-the-art experiments were, in fact, foredoomed to failure.[4]

Another key thought experiment occurred to Einstein in 1907, while working at the Swiss Federal Patent Office in Bern. This experiment led to a basic part of the general theory of relativity, the equivalence principle. In this situation, the thought experimenter jumps off the roof of a house and simultaneously drops a stone. He realizes that the stone falls at relative rest with respect to him, while they both fall under the influence of gravity. It seems, therefore, as if in his vicinity there is no gravity. Einstein's great realization is that the thought experimenter can consider himself and the stone to be at relative rest by replacing the Earth's gravitational field with an inertial force causing acceleration equal in magnitude but oppositely directed – this is the equivalence principle of 1907, a basis of the 1915 general theory of relativity.[5]

TYPES OF VISUAL IMAGES

Despite the startling changes in intuition and common sense, both the special and general theories of relativity are based on visual imagery abstracted from phenomena we have actually witnessed. Throughout the first decade of the twentieth century, scientists assumed that this would always be the case. But a cloud on this horizon had already appeared in 1905, when in another memorable paper of his annus mirabilis, Einstein proposed that for studying certain processes, it is useful to assume that light can also be a particle, or light quantum, as he called it.[6] Just about every other scientist considered this

---

[4] For details see Arthur I. Miller, *Albert Einstein's Special Theory of Relativity: Emergence (1905) and Early Interpretation (1905–1911)* (New York: Springer-Verlag, 1998).

[5] Arthur I. Miller, "Einstein's First Steps towards General Relativity: *Gedanken* Experiments and Axiomatics," *Physics in Perspective,* 1 (1999), 85–104.

[6] Albert Einstein, "Über einen die Erzeugung und Verwandlung des Lichtes betreffenden heuristischen Gesichtspunkt," *Annalen der Physik,* 17 (1905), 132–48.

totally bizarre. After all, was there not a perfectly viable wave theory of light that explained in a very satisfying way such phenomena as interference? By very satisfying, everyone meant a visual model of how light waves interfere with one another, abstracted from observed interference phenomena with water waves. Physicists lamented that no such visual model seemed possible for light quanta.

Another hint of startling developments just over the horizon appeared in Einstein's 1909 article on radiation, entitled "On the development of our intuition [*Anschauung*] of the existence and constitution of radiation," where he further explored the wave/particle duality of light, according to which light can be a wave and a particle at the same time.[7] Einstein emphasizes two points: First, according to relativity theory, light can exist independent of any medium. This is rather counterintuitive because if we speak of light waves, we have in mind something that "waves," just as there cannot be water waves without water. The ether of nineteenth-century physics had served this purpose for light waves. Second, light can have a particle mode of existence for which no visual imagery can be constructed to explain interference. All of this clashes with our intuition or *Anschauung*. Since German terminology played an extremely important role in developments in twentieth-century physics, let us discuss it.

Modern physics has linked intuition with visual imagery partly through the rich philosophical lexicon of the German language. This came about because relativity and atomic physics were formulated almost exclusively by scientists educated in the German scientific-cultural milieu. Philosophy was an integral part of learning in the German school system, particularly the ideas of Immanuel Kant (1724–1804). Kant spun an intricate philosophical system, the goal of which was to place Newtonian physics on a firm cognitive foundation.

Kant's philosophical system is set out in his monumental book of 1781, *The Critique of Pure Reason*, where he carefully separated intuition from sensation.[8] His ultimate goal was to differentiate higher cognition from the processing of mere sensory perceptions. In German, the word for intuition is *Anschauung*, which can be translated equally well as "visualization." To Kant, intuitions or visualizations can be abstractions from phenomena we have actually witnessed. So the visual images of relativity are visualizations, while relativity becomes the new scientific intuition that replaces the Newtonian one, which, in turn, had replaced Aristotle's.

Consider, for example, the light wave in Einstein's 1895 thought experiment. The visual imagery is one of visualization because no one has ever actually "seen" light waves. Rather, light waves are visual representations of light that are abstracted from phenomena concerning water waves.

---

[7] Albert Einstein, "Entwicklung unserer Anschauungen über das Wesen und die Konstitution der Strahlung," *Physikalische Zeitschrift*, 10 (1909), 817–25.

[8] Immanuel Kant, *Critique of Pure Reason*, trans. N. K. Smith (New York: St. Martin's Press, 1929).

This visualization is imposed on their mathematical representation, which emerges from solutions to equations of optics, not surprisingly called wave equations.

On the other hand, we can try to investigate physical phenomena in a more concrete way. For example, magnetic lines of force are demonstrated directly by the disposition of iron filings placed on a sheet of paper held over a bar magnet. The next step is to abstract the rough lines of force given by the patterns formed by iron filings to continuous lines that fill all of space and can be mathematically described by certain symbols in the equations of electromagnetic theory. The latter imagery is visualization or Anschauung. The former is what Kant calls "visualizability" or *Anschaulichkeit*. For example, at the turn of the century, the nature of the Anschauung of magnetic lines of force was hotly debated in the German physics and engineering communities.[9]

In Kantian terminology, we say that Anschaulichkeit is what is immediately given to the perceptions or what is readily graspable in the Anschauung: visualizability [*Anschaulichkeit*] is less abstract than visualization [*Anschauung*]. Strictly speaking then, visualizability is a property of the object itself, and visualization of an object results from the cognitive act of knowing the object. In Kant's philosophy, the visual imagery of visualizability (Anschaulichkeit) is inferior to the images of visualization (Anschauung).[10] Anschauung can also be translated as "intuition," by which is meant the intuition of phenomena that results from a combination of cognition and perception. Consistently with philosophic-scientific meanings of Anschauung and Anschaulichkeit, I will render the adjective *anschaulich* as "intuitive." Translating this formalism to the way in which scientists in the German-language milieu understood it is to say that the Anschauung of an object or phenomenon is obtained from a combination of cognition and mathematics.

In classical physics, visualization and visualizability are synonymous because there is no reason to believe that experimenting on a system in any way alters the system's properties. So far so good. But scientists assumed that this applied also to objects that, right from the start, were never visible, such as electrons. This was the case for Niels Bohr's (1885–1962) theory of the atom, to which we now turn.

ATOMIC PHYSICS DURING 1913–1925: VISUALIZATION LOST

Drawing mainly upon Ernest Rutherford's (1871–1937) experiments of 1909–11 in which he discovered the atom's nucleus, Bohr in 1913 proposed an

---

[9] Arthur I. Miller, "Unipolar Induction: A Case Study of the Interaction between Science and Technology," *Annals of Science,* 38 (1981), 155–89.
[10] Miller, *Insights of Genius,* chap. 2.

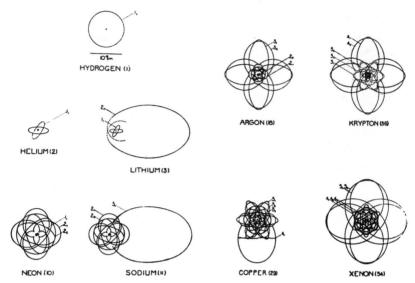

Figure 10.3. Representations of the atom according to Niels Bohr's 1913 atomic theory. (H. Kramers, *The Atom and the Bohr Theory of Its Structure* [London: Glyndendal, 1923].)

atomic theory based on the pleasing visualization of the atom as a minuscule solar system (see Figure 10.3). His was a bold theory, built, as it was, specifically on violations of such time-honored notions of classical physics as continuity and visualization of trajectories: Bohr's atom emits radiation in discontinuous bursts as the atomic electron makes an unvisualizable jump between allowed orbits. The atomic electron disappears and reappears like the Cheshire cat. But what remained essentially classical in Bohr's theory was its visual imagery, which was imposed upon it owing to its use of symbols from classical celestial mechanics, suitably altered. For example, use of such symbols as the radius of an orbit permitted imposition of the solar system visualization.

This technique of extrapolating concepts from the macroscopic to the microscopic was not new. A central aspect of scientific creativity is the scientist's ability to create something new by relating it to something already understood. This is the goal of metaphors, an extremely important facet of scientific research.[11] The interaction metaphor is a good approximation to the reasoning often used in scientific research. Basically, an interaction metaphor is of the form:

$$\{x\} \text{ acts as if } \textit{it were a } \{y\}$$

where the instrument of metaphor – *as if* – relates the primary subject x to the secondary subject y. The curled brackets around x and y signal a collection of

[11] Miller, *Insights of Genius*, chap. 7.

properties. Connections between the collection {*y*} and the primary subject are usually not obvious and may not even hold. This is where high creativity enters because in certain circumstances, scientists use metaphors to create similarity.

Although I am paraphrasing, it is crystal clear that Bohr was using the following visual metaphor of an interaction sort:

> *The atom behaves* as if *it were a minuscule solar system.*

The instrument of metaphor – *as if* – signals a mapping, or transference, from the secondary subject (classical celestial mechanics with its accompanying visual imagery, all of which is suitably altered with the axioms of Bohr's theory), for the purpose of exploring the not yet well-understood primary subject (atom).

To get further insight into why metaphors have become of interest to the study of scientific creativity and to the meaning of science itself, let us explore the "deep structure" of Bohr's metaphor a bit further. The primary subject is the key here. Scientists explore the essence of the term "atom." In order to get at this, they work in successive approximations. So, in 1913, the term atom stood for the Bohr atom of that era, which was studied by using appropriate forms of classical physics, suitably modified.

The deeper process here is one of using scientific theories to probe worlds beyond sense perception. Einstein did this with the special and general theories of relativity, which revealed such phenomena as the relativity of space and time, and the specific geometry of curved space-time. The consequences of special relativity are based on the taking into account of effects produced by the very high but finite velocity of light, instead of assuming the velocity to be infinite as in Newtonian physics, as seems to be the case perceptually. Bohr teased out effects due to the very small but nonzero value of Planck's constant, another universal constant of nature. Just as setting the velocity of light to infinity permits passage from special relativity to Newtonian physics, setting Planck's constant equal to zero permits transition from the quantum to classical realms, in which, for example, there is no wave/particle duality of light. These limiting statements are known as "correspondence principles," or "correspondence limit cases."

In summary, metaphor is the tool by which scientists can pass between possible worlds, sometimes using correspondence principles. Since we aim to understand the essence of the primary subject, which is the atom in the case in question, the primary subject remains fixed while we pass from theory to theory. In philosophy of science, this is known as scientific realism: Invisible entities postulated by theories exist *independently* of the theories themselves. The opposing view is scientific antirealism, in which invisible entities, or those not open to direct observation, do not exist. All modern scientists are scientific realists. In any case, as we discussed, what direct observation means is basically unclear because we never observe anything directly with our perceptions.

That goes for everyday observations, too, in which the equation – understanding = perception plus cognition – is the key even to our daily lives.

Change of metaphor rescued the Bohr theory during 1923–5. The reason is that by 1923, data had accrued that atoms do not respond to light *as if* they were minuscule solar systems. Visual imagery was abandoned and mathematics became the guide. The term "image" was shifted to the new mathematical framework of Bohr's theory, in which atomic electrons were described according to the following metaphor which has no visual imagery:

The atom behaves as if each of its electrons were replaced by a collection of "substitution" electrons attached to springs.

The physics of objects attached to springs is well known. There is no visual imagery here because Bohr's theory required each real atomic electron to be replaced by as many "substitution" electrons as there are possible atomic transitions, of which there are an infinite number. Through Bohr's correspondence principle, it was possible to link up the mathematical formalism of substitution electrons on springs with the fundamental axioms of Bohr's theory and produce certain results in agreement with data.

ATOMIC PHYSICS DURING 1925–1926: VISUALIZATION VERSUS VISUALIZABILITY

By early 1925, Bohr's theory had collapsed entirely and atomic physics lay in ruins. Most physicists do not thrive in situations such as this. Werner Heisenberg (1901–1976) did and, in June 1925, formulated the modern atomic physics called quantum mechanics. Heisenberg based quantum mechanics on unvisualizable electrons whose properties emerged from a nonstandard mathematics, in which quantities like momentum and position do not generally commute. The essential clue for Heisenberg's discovery is rooted in clever manipulation of the substitution electrons in Bohr's 1923 metaphor. Heisenberg claimed that his theory contained only measurable quantities, a programmatic intent that physicists in Bohr's circle had adopted since 1923. Consequently, a description in space and time was avoided. The atomic electron was "described" by the radiation it emitted during transitions, which is measured by its spectral lines.

As we might have expected, however, Heisenberg was dissatisfied with this state of affairs. In 1926 he wrote that the present theory labored "under the disadvantage that there can be no directly intuitive [*anschaulich*] geometrical interpretation," and that a key point is to explore "the manner in which symbolic quantum geometry goes over into intuitive classical geometry."[12]

---

[12] Max Born, Werner Heisenberg, and Pasqual Jordan, "Zur Quantenmechanik. II," *Zeitschrift für Physik*, 35 (1926), 557–615.

In general, praise for Heisenberg's new theory was tempered by its lack of any visual imagery. But how to regain visual imagery?

In 1926, problem after problem that had resisted solution in the old Bohr theory was solved. Yet what bothered physicists of the ilk of Bohr and Heisenberg was that not only were the intermediate steps in calculations not well understood but, even more fundamentally, the atomic entities themselves were of unfathomable counterintuitivity. In addition to the wave/particle duality of light proposed by Einstein in 1905 and 1909, the French physicist Louis de Broglie (1892–1987) suggested that electrons also have a dual nature.[13] So, like the peculiar situation of light behaving as particles, physicists had to imagine electrons as waves.

In 1926 Erwin Schrödinger (1887–1961) offered a way to restore visual imagery. He proposed a wave mechanics in which atomic entities are represented as charged waves whose properties emerged from the familiar mathematics of differential equations and which, he claimed, avoided the discontinuities inherent in Heisenberg's quantum mechanics, for example, quantum jumps between permitted energy states. Schrödinger made it abundantly clear why he decided to formulate a wave mechanics:

> My theory was inspired by L. de Broglie ... and by short but incomplete remarks by A. Einstein. ... No genetic relation whatever with Heisenberg is known to me. I knew of his theory, of course, but felt discouraged, not to say repelled, by the methods of the transcendental algebra, which appeared very difficult to me and by lack of visualisability [*Anschaulichkeit*].[14]

Consistently with his view of the credibility of extrapolating classical concepts into the atomic realm, Schrödinger equates Anschaulichkeit with Anschauung. He continues in this paper by expressing his disapproval of a physical theory based on a "theory of knowledge," in which we "suppress intuition [*Anschauung*]." Although objects that have no space-time description may exist, Schrödinger was adamant that "from the philosophical point of view," atomic processes are not in this class. His version of atomic physics offered the possibility of using the visual imagery of classical physics, that is, Anschauung, suitably reinterpreted. He went on to drive his proof of the equivalence between the wave and quantum mechanics to what he considered the logical conclusion: When speaking of atomic theories, one "could properly use the singular."

Physicists of the older generation, such as Einstein and H. A. Lorentz (1853–1928), praised Schrödinger's theory. On 27 May 1926, Lorentz wrote to

---

[13] Louis de Broglie, "Recherches sur la théorie des quanta," *Annles de Physique,* 3 (1925), 3–14.
[14] Erwin Schrödinger, "Über das Verhältnis der Heisenberg-Born-Jordanschen Quantenmechanik zu der meinen," *Annalen der Physik,* 70 (1926), 734–56. Translated by the author unless otherwise noted. See also, A. I. Miller, "Erotica, Aesthetics, and Schrödinger's Wave Equation," forthcoming in Graham Farmeloe, ed., '*It Must Be Beautiful': Great Equations of the Twentieth Century* (London: Granta, 2002).

Schrödinger, agreeing with the latter's wave mechanics: "If I had to choose between your wave mechanics and the [quantum] mechanics, I would give preference to the former, owing to its greater visualisability [*Anschaulichkeit*]."[15]

Heisenberg was privately furious over Schrödinger's work and the rave reviews it received from the scientific community. To his colleague Wolfgang Pauli (1900–1958), Heisenberg wrote on 8 June 1926: "The more I reflect on the physical portion of Schrödinger's theory the more disgusting I find it. What Schrödinger writes on the visualisability [*Anschaulichkeit*] of this theory... I consider trash."[16]

Clearly, the stakes were high in this dispute because the issue was nothing less than the intuitive understanding of physical reality itself, replete with visual imagery.

Heisenberg recalled the psychological situation at this time as extremely disturbing. In print, he objected to Schrödinger's *imposing* on quantum theory "intuitive [*anschaulich*] methods" of the sort that previously had led to confusion.[17] Heisenberg suggested limitations on any discussion of the "intuition problem [*Anschauungsfrage*]."[18]

In a paper of September 1926, Heisenberg began to focus on what he took to be the central issue:

> [T]he electron and the atom possess not any degree of physical reality as the objects of daily experience.... Investigation of the type of physical reality which is proper to electrons and atoms is precisely the subject of atomic physics and thus also of quantum mechanics."[19]

Thus, the basic problem facing atomic physics was the concept of physical reality itself. Compounding the situation was that physicists must use everyday language, with its perceptual baggage, to describe atomic phenomena, which are not only beyond perception but whose entities are terribly counterintuitive.

In summary, whereas by the beginning of 1925 atomic physics was in shambles, by mid-1926 there were two apparently dissimilar theories: Heisenberg's was based on nonvisualizable particles and couched in a difficult and unfamiliar mathematics; Schrödinger's claimed a visualization and was set on more familiar mathematics. And yet a gnawing problem emerged: No one really understood what either formalism meant. Although Schrödinger

---

[15] K. Prizbaum, ed., *Letters on Wave Mechanics: Schrödinger, Planck, Einstein, Lorentz*, trans. M. J. Klein (New York: Philosophical Library, 1967).
[16] Wolfgang Pauli, *Wissenschaftlicher Briefwechsel mit Bohr, Einstein, Heisenberg, u.a. I: 1919–1929*, ed. A. Hermann, K. von Meyenn, and V. F. Weisskopf (Berlin: Springer, 1979).
[17] Archive for History of Quantum Physics: Interview of Heisenberg by Thomas S. Kuhn, 22 February 1963; on deposit at the Niels Bohr Library located in the American Institute of Physics, College Park, Md.
[18] Werner Heisenberg, "Mehrkörperproblem und Resonanz in der Quantenmechanik," *Zeitschrift für Physik*, 38 (1926), 411–26.
[19] Werner Heisenberg, "Zur Quantenmechanik," *Die Naturwissenschaften*, 14 (1926), 889–994.

claimed to have proven the equivalence between the wave and quantum mechanics, Heisenberg and Bohr disagreed, owing to what they considered to be Schrödinger's erroneous claims for his theory's interpretation. The only thing on which Heisenberg and Schrödinger agreed was that basic issues in physics verged on the philosophical and centered on the concept of intuition and visual imagery.

ATOMIC PHYSICS IN 1927: VISUALIZABILITY REDEFINED

Heisenberg wrote to Pauli on 23 November 1926 of his intense discussions with Bohr to come to grips with these problems: "What the words 'wave' and 'corpuscle' mean we know not anymore."[20] Linguistic difficulties were not new to the quantum theory. They had surfaced along with the wave/particle duality of light, in which the wave and particle attributes are related by Planck's constant. But equating energy, which connotes localization, with frequency, which connotes nonlocalization, is like trying to equate apples with fish. How can something be continuous and discontinuous at the same time, like light and then electrons are supposed to be? Using thought experiments, Bohr and Heisenberg struggled with questions like this, and others such as how light quanta can produce interference.

In early 1927, Heisenberg produced a classic paper in the history of ideas in which he proposed a way out of this morass: "On the intuitive [*anschauliche*] content of the quantum-theoretical kinematics and mechanics."[21] The importance of the concept of intuitivity is clear from its use in the title. Immediately Heisenberg launched into a linguistic analysis: "The present paper sets up exact definitions of the words velocity, energy, etc. (of the electron)." In Heisenberg's view, from the peculiar mathematics of the quantum mechanics, in which momentum and position generally do not commute, already "we have good reason to be suspicious about uncritical application of the words 'position' and 'momentum.'" Heisenberg's resolution of the paradoxes involved in extrapolating language from the world of sense perceptions into the atomic domain is to let the mathematics of quantum mechanics be the guide, since it produces, among other results, the uncertainty relations. The mathematics of quantum mechanics defines how "we understand a theory intuitively [*anschaulich*]," which is separate from the visualization of atomic processes.

In the course of this paper, Heisenberg went on to demonstrate the incorrectness of certain of Schrödinger's physical interpretations of his theory that Schrödinger thought could bring back the old visualization

---

[20] Pauli, *Wissenschaftlicher Briefwechsel.*
[21] Werner Heisenberg, "Über den anschaulichen Inhalt der quantentheoretischen Kinematik und Mechanik," *Zeitschrift für Physik*, 43 (1927), 172–98.

or Anschauung, for example, that discontinuities in atomic transitions exist also in wave mechanics, and that it is incorrect to regard the waves in Schrödinger's theory as representing particles in the sense of classical physics. Rather, the waves in Schrödinger's theory are probabilities for the occurrence of certain phenomena.

Bohr disagreed vehemently with Heisenberg's paper for two principal reasons: Heisenberg focused on particles, to the exclusion of waves, thereby considering one-half of the quantum mechanical situation; and Heisenberg seemed to renounce visual imagery altogether.

Bohr offered another approach, which he called complementarity. It is a generalization of Heisenberg's considerations on visualizability.[22] Instead of choosing one mode of existence over another, Bohr took on both as acceptable. Bohr reasoned that the seemingly paradoxical situation of waves and particles arises only if we understand "particle" and "wave" to refer to objects and phenomena from the world of sense perceptions. Bohr found that the clue to an understanding resides in Planck's constant, which links particle and wave concepts. The extremely small but nonzero value of Planck's constant signals that we cannot rely on our sense perceptions to understand atomic phenomena.

According to complementarity, the wave and particle attributes of light and matter are complementary in the sense that both are necessary to characterize the atomic entity. But they are mutually exclusive because in any experiment, only one side will reveal itself. If the experiment is set up to measure particle properties, then the atomic entity will behave like a particle. What about the power of prediction, which is central to any viable physical theory and is linked in classical physics to a description in space and time, that is, to a visual imagery? Complementarity shifts prediction, and so causality as well, of fundamental processes to the conservation laws of energy and momentum. Bohr's message is that you can draw pictures, if you wish, but remember that they are naive representations. In this way, Bohr succeeded in finessing the problem of visual imagery. This did not satisfy Heisenberg.

In summary, Heisenberg proposed that the mathematics of quantum mechanics had decided the theory's "intuitive content," as well as the notion of visualizability in the atomic realm. This was an important step because in the atomic domain, visualization and visualizability are mutually exclusive. Visualization is an act that depends on cognition. So visualization is what Heisenberg referred to as the "ordinary intuition [*Anschauung*]" that could not be extended into the atomic domain. Visualizability concerns the intrinsic properties of elementary particles that may not be open to our perceptions, and so to which mathematics is the key. The uncertainty relations illustrate this well. Atomic physics reverses the original Kantian order

---

[22] Niels Bohr, "The Quantum Postulate and the Recent Development of Atomic Theory," *Nature* (Supplement), (14 April 1928), 580–90.

of Anschauung and Anschaulichkeit and, so too, transforms the concept of intuitive [*anschaulich*] once again.

From 1927 through 1932, however, Heisenberg resisted any imagery of atomic phenomena – that is, for Heisenberg and other quantum physicists, in the atomic domain visualizability did not yet possess a unique depictive or visual component. Through his work on nuclear physics in 1932, Heisenberg realized a way to generate the new visual imagery of visualizability in quantum physics. From 1932 on, Heisenberg used the term Anschaulichkeit for the visual imagery of quantum mechanics. For example, in 1938, he wrote of universal constants, such as the velocity of light and Planck's constant, as designating the "limits which are set in the application of intuitive [*anschaulich*] concepts," and so signaling, as well, transformations in the concept of intuition.[23] To study this sweeping change and its ramifications, we turn to Heisenberg's 1932 theory of the nuclear force.

NUCLEAR PHYSICS: A CLUE TO THE NEW VISUALIZABILITY

Consider a situation in which a concept can be neither introduced by a laboratory demonstration nor even discussed with existing terminology. In such cases, the function of catachresis can be played by a metaphor that sets a reference (or definition) for such a term, which philosophers of science refer to as a natural kind term because it is part of the fabric of nature.[24] The term *nuclear force* is a natural kind term. It was introduced in 1932 to denote the attractive force between a neutral neutron and a positively charged proton. But in classical physics there are only two sorts of attractive forces: gravitational and electromagnetic. The term nuclear force, therefore, poses an extremely nonclassical situation for which no language existed – by language I mean the language of theoretical physics. But even ordinary language is problematic here wherein opposites attract while likes repel. Another of Heisenberg's great scientific discoveries was the proper metaphor for the nuclear force.

As a clue to a theory of the nuclear force, Heisenberg recalled one of his dazzling discoveries in quantum mechanics. In order to explain certain properties of the helium atom, a system that resisted solution in the old Bohr theory, he postulated in 1926 a force between the atom's two electrons that depended on their being indistinguishable. Under this so-called exchange force, the indistinguishable electrons exchange places at a rapid rate. This situation is clearly unvisualizable.

---

[23] Werner Heisenberg, "Über die in der Theorie der Elementarteilchen auftretende universelle Länge," *Annalen der Physik*, 32 (1938), 20–33; translated in Arthur I. Miller, *Early Quantum Electrodynamics* (Cambridge: Cambridge University Press, 1994).

[24] Richard Boyd, "Metaphor and Theory Change: What Is 'Metaphor' a Metaphor For?" in *Metaphor and Thought*, ed. Andrew Ortony, 2d ed. (Cambridge: Cambridge University Press, 1993).

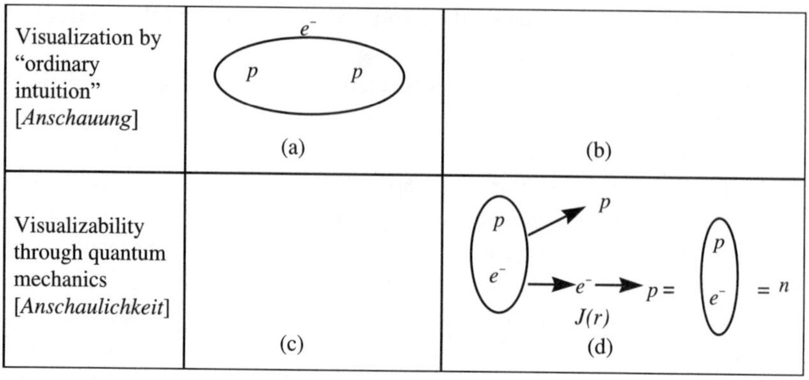

Visualization by "ordinary intuition" [Anschauung]	(a) [p — e⁻ — p in ellipse]	(b)
Visualizability through quantum mechanics [Anschaulichkeit]	(c)	(d) [diagram with $J(r)$ and $= n$]

Figure 10.4. The difference between visualization and visualizability. Frame (a) depicts the solar system H_2^+ ion, which is the visual imagery imposed on the mathematics of Bohr's atomic theory, where the ps denote protons about which the electron (e^-) revolves. But Bohr's theory could not produce proper stationary states for this entity. Frame (c) is empty because quantum mechanics gives no visual image of the exchange force. Frame (b) is empty because classical physics yields no visualization for the nuclear exchange force. Frame (d) is the depiction of Heisenberg's nuclear force, which is generated from the mathematics of his nuclear theory, where n is a neutron assumed to be a proton-electron bound state, and e is the electron carrying the nuclear force. (*Source:* Arthur I. Miller, *Insights of Genius: Imagery and Creativity in Science and Art* [Cambridge, Mass.: MIT Press, 2000], p. 241.)

The success of the exchange force for the helium atom led physicists to extend it to molecular physics. Of particular interest was another bane of the old Bohr theory, the H_2^+ ion, depicted in Figure 10.4(a). According to quantum mechanics, the exchange force for the H_2^+ ion operates through the electron being exchanged between the two protons at the rate of 10^{12} times per second. Clearly this process is unvisualizable, and so the box in Figure 10.4(c) is empty.

In 1932 Heisenberg decided to take the exchange force inside the nucleus by formal analogy, as he writes:

> If one brings a neutron and a proton to a spacing comparable to nuclear dimensions, then – in analogy to the H_2^+ – a migration of negative charge will occur.... The quantity $J(r)$ [in Figure 10.4(d)] corresponds to the exchange or more correctly migration [of an electron resulting from neutron decay]. The migration can be made more intuitive [*anschaulich*] by the picture [of spinless electrons].[25]

Had Heisenberg tried to visualize the nuclear exchange force generated by the mathematics of his nuclear theory, the image would have looked like the one in Figure 10.4(d). Heisenberg assumed that inside the nucleus the neutron is a

---

[25] Werner Heisenberg, "Über den Bau der Atomkerne. I," *Zeitschrift für Physik, 77* (1932), 1–11.

compound object consisting of an electron and a proton. He was untroubled about a spinless nuclear electron because at this point, Heisenberg and Bohr were willing to entertain the notion that quantum mechanics was invalid inside the nucleus.

Consequently, for Heisenberg, what began as a mere *analogy* with the H_2^+ ion became a more general visual *metaphor* in the nuclear case, which we may paraphrase from the quotation from his 1932 paper:

The nuclear force acts as if *a particle were exchanged.*

The secondary subject (particle exchanged) sets the reference for the primary subject (nuclear force). In Heisenberg's nuclear exchange force, the neutron and proton do not merely exchange places. The metaphor of motion is of the essence here because the attractive nuclear force is carried by the spinless nuclear electron. Although Heisenberg's nuclear theory did not agree with data on the binding energies of light nuclei, his play with analogies and metaphors generated by the mathematics of quantum mechanics was understood to be a key to extending the concept of intuition in the subatomic world.

By November of 1934, Heisenberg's concept of the nuclear force being carried by an improper electron had been discarded, owing to the work of the Japanese physicist Hideki Yukawa (1907–1981).[26] Yukawa returned to the mathematical formulation of Heisenberg's 1932 theory and replaced the functional form for $J(r)$ with one suitable for exchange of a proper particle, eventually called a meson. The amazing result was that the entity basic to the secondary subject – exchanged particle – turned out to apply also to the primary subject and the exchanged particle turned out to be physically real. Coincidentally, the proper terminology for the attractive force between neutral and charged particles as due to particles being exchanged was thus established.

The complex web of research initiated by Heisenberg's and Yukawa's nuclear physics led, in 1948, to a dramatic advance with the emergence of two apparently different theories of quantum electrodynamics, the theory of how electrons and light interact.[27] The version of Julian Schwinger (1918–1994) and Sin-itoro Tomonaga (1906–1979) was mathematically elegant and difficult to use, while Richard P. Feynman's (1918–1989) was based on a diagrammatic description that originated in certain mathematical rules whose origin was not rigorous.[28] When, in 1949, Freeman J. Dyson (b. 1923) proved the equivalence of the two formulations, just about every physicist switched to Feynman's visual methods. Such is the importance of visual representations to physicists.

---

[26] Hideki Yukawa, "On the Interaction of Elementary Particles. I," *Proceedings of the Phsyico-Mathematical Society of Japan*, 3 (1935), 48–57.
[27] Arthur I. Miller, *Early Quantum Electrodynamics: A Source Book* (Cambridge: Cambridge Universtiy Press, 1994).
[28] Schweber, *QED and the Men Who Made It*.

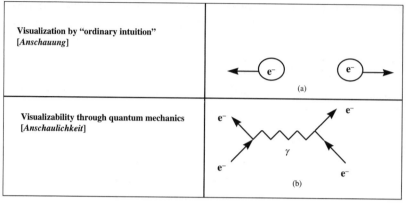

Figure 10.5. Representations of the Coulomb force. (a) The Coulomb force from elementary physics textbooks. (b) The Feynman diagram, which is the appropriate representation of the Coulomb force, in which two electrons interact by exchanging a light quantum (γ).(*Source:* Arthur I. Miller, *Insights of Genius: Imagery and Creativity in Science and Art* [Cambridge, Mass.: MIT Press, 2000], p. 248).

Feynman's formulation is based on the visual imagery of visualizability, and not visualization. The difference can be understood by comparing different representations of the Coulomb interaction between two electrons (Figure 10.5).

The visual imagery in Figure 10.5(a) is abstracted from phenomena that we have actually witnessed: Electrons are depicted as distinguishable billiard balls possessing electrical charge. This imagery was *imposed* on classical electromagnetic theory and turned out to be incorrect for use in the atomic domain, where electrons are simultaneously wave and particle.

Figure 10.5(b) is a Feynman diagram for the repulsive force between two electrons that is carried by a light quantum. Details are not required for an appreciation of the central point, which is that we would not have known how to draw Figure 10.5(b) without the mathematics of the quantum mechanics that *generates* it. This is visualizability. Thus, we can assume that the mathematics of quantum mechanics offers a glimpse of the subatomic world, where entities can be simultaneously continuous and discontinuous. Feynman diagrams *represent* interactions among elementary particles in a realistic manner – that is, there is ontological content to these diagrams. That we must draw them with the usual figure and ground distinctions is owing to limitations of our senses. By figure and ground I mean the simple distinction between a well-defined structure set against a background of secondary importance. Today physicists visualize in Feynman diagrams.

PHYSICISTS REREPRESENT

At this juncture we can tie together much of what we have said regarding intuition and visualization under the more general concept of *representation*.

Whereas the representation of the atom as a minuscule solar system could not be maintained [see Figure 10.6(a)], the more abstract representation in Figure 10.6(b) for energy levels still holds for atomic physics. In 1925, another representation of the material in Figure 10.6(b) appeared in what Heisenberg and Hendrik Kramers (1894–1954) referred to as a "term diagram" in Figure 10.6(c), which is generated from the mathematics of the last throes of Bohr's dying atomic theory.[29]

We discussed how Heisenberg's first move toward a theory of nuclear physics contained the seeds of a visual representation of atomic phenomena [see Figure 10.4(d)]. Paramount in analyzing this work was the concept of intuition coupled with visual imagery. This required physicists to distinguish between visualization and visualizability. Heisenberg's early results on nuclear physics, which assume that particles carry forces, culminated in the Feynman diagrams, which made their appearance in 1948 (see Figure 10.5(b)). With hindsight, Heisenberg wrote that the "term diagrams were like Feynman diagrams nowadays" because they were suggested by the mathematics of phenomena treated within the old Bohr theory.[30] Figure 10.6(d) is a Feynman diagram that replaces the term diagram in Figure 10.6(c).

Feynman diagrams offer a means of transforming the concept of naturalistic representation into one offering a glimpse of a world beyond the intuition of Galilean-Newtonian and relativity physics. They offer the proper visualizability of atomic physics which, alas, we can render only with the usual figure and ground distinction. These diagrams are presently the most abstract way of glimpsing an invisible world. In 1950, Heisenberg welcomed Feynman diagrams as the "intuitive [*anschaulich*] contents" of the new theory of quantum mechanics.[31] Once again, for Heisenberg, theory decided what is intuitive, or visualizable.

THE DEEP STRUCTURE OF DATA

Parallel to the way in which Galileo's theory of physics leads to the "deep structure" of projectile motion, Feynman diagrams provide the deep structure of a world beyond appearances, the world of elementary particles. They offer a representation of nature from available data, for example, bubble chamber photographs.

Consider the famous bubble chamber picture in Figure 10.7(a), which is the scattering of two elementary particles: a muon antineutrino from an electron. It is a major discovery that went far to substantiate the so-called

---

[29] Hendrik A. Kramers and Werner Heisenberg, "Über die Streuung von Strahlung durch Atome," *Zeitschrift für Physik*, 31 (1925), 223–57.
[30] Archive for History of Quantum Physics: Interview of Heisenberg by Thomas S. Kuhn, 13 February 1963.
[31] Werner Heisenberg, "Zur Quantentheorie der Elementarteilchen," *Zeitschrift für Naturforschung*, 5 (1950), 251–9.

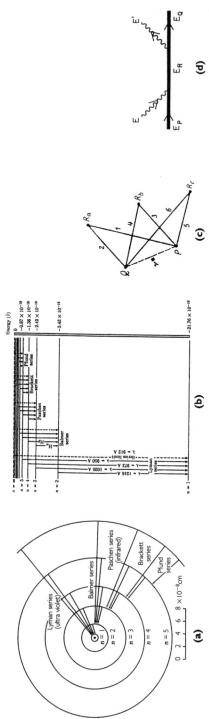

Figure 10.6. Representations of the atom and its interaction with light. (a) A more detailed version of the hydrogen atom in Bohr's theory as depicted in Figure 10.3. The number n is the principal quantum number and serves to tag the atomic electron's permitted orbits. Lyman, Balmer, etc., are names for the series of spectral lines the atom emits when its electron drops from higher to lower orbits. (b) Another way of representing the Bohr atom. The horizontal lines are energy levels corresponding to permitted orbits, but are more general. The representation in (b) survived the demise of Bohr's atomic theory and remained essential to atomic theory. (c) Another manner of visually representing some of the information in (b). It is taken from a 1925 paper of Hendrik Kramers and Heisenberg, published shortly before Heisenberg formulated the quantum mechanics. Kramers and Heisenberg referred to the diagram in (c) as a "term diagram," in which R_a, R_b, R_c, Q, and P are energy levels in an atom struck by light. The incident light causes the atom to make transitions from a state P to a state Q via intermediate states R. The energy difference between the states P and Q is hv^*, where the frequency of the incident light is much greater than v^* in order to promote the atom to its excited states. (d) A Feynman diagram for the processes in (a) to (c), for the case in which they were all caused by the interaction of atoms with light. In (d), $E(E')$ is the energy of the incident (scattered) light, E_P and E_Q are the energies of the atom's initial and final states, and E_R is the energy of possible intermediate states. The atom's trajectory in space-time is taken to be horizontal. (Arthur I. Miller, *Insights of Genius: Imagery and Creativity in Science and Art* [Cambridge, Mass.: MIT Press, 2000], p. 398.)

Imagery and Representation in Twentieth-Century Physics 211

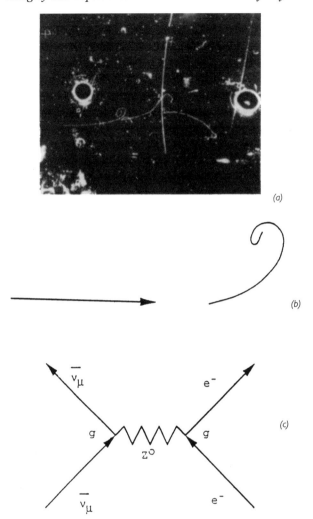

Figure 10.7. Bubble chamber and "deep structure." (a) The first bubble chamber photograph of the scattering of a muon antineutrino ($\bar{\nu}_\mu$) from an electron (e^-). (b) The "deep structure" in (a) according to the electroweak theory. Instead of two electrons interacting by exchanging a light quantum [Figure 10.5 (b)], according to the electroweak theory an antineutrino ($\bar{\nu}_\mu$) and an electron (e^-) interact by exchanging a Z^0 particle, where g is the charge (coupling constant) for the electroweak force. (Arthur I. Miller, *Insights of Genius: Imagery and Creativity in Science and Art* [Cambridge, Mass.: MIT Press, 2000], p. 407.)

electroweak theory, which unifies the weak and electromagnetic forces, and was formulated in 1968 by Steven Weinberg (b. 1932) and Abdus Salam (1926–1996).[32] A key to the theoretical basis of the electroweak theory was

---

[32] Arthur I. Miller and Frederik W. Bullock, "Neutral Currents and the History of Scientific Ideas," *Studies in History and Philosophy of Modern Physics*, 6 (1994), 895–931.

comparison with the Feynman diagram in Figure 10.5(b) for the way in which two electrons interact. Arguing metaphorically with this process as the secondary subject, Weinberg and Salam were able to construct the Feynman diagram in Figure 10.7(b) which, in turn, they used to predict the event subsequently discovered and illustrated in Figure 10.7(a).[33] This is good evidence to argue that the Feynman diagram in Figure 10.7(b) is a glimpse into the deep structure of the *real* world of particle physics. To which we add that the hypothesized intermediate Z^0 was subsequently discovered.

In summary, this is another instance where visual representations are crucial for scientific discovery and the understanding of physical reality, in addition to their usefulness for calculational purposes. It is of interest to juxtapose in Figure 10.8 "data" for the Coulomb repulsion between electrons and for the electroweak theory. Figure 10.8(a) is datum that we assume nature gives to us. Actually, it is a naive commonsensical representation of the Coulomb force. Figure 10.8(c) is actual data from a bubble chamber and is many layers removed from the "raw" primordial process. Figures 10.8(b) and 10.8(d) are the deep structure of those data.

VISUAL IMAGERY AND THE HISTORY OF SCIENTIFIC THOUGHT

We have explored the importance of visual thinking to Bohr, Einstein, Feynman, Heisenberg, Schrödinger, Salam, and Weinberg. These cases contain conclusions about visual imagery in scientific research and so in creative scientific thinking: (1) Visual imagery plays a causal role in scientific creativity (Einstein's thought experiments); (2) Visual imagery is usually essential for scientific advance (Bohr, Einstein, Feynman, Heisenberg, Schrödinger, Salam, and Weinberg); and (3) Visual imagery generated by scientific theories can carry truth value (Feynman diagrams). Conclusions (1) to (3) go far toward substantiating that visual images are not epiphenomena and are essential to scientific research. Consequently, they play a role in supporting results from cognitive science that indicate the importance of visual thinking.[34]

Let us explore this point a bit further on the basis of what we have already learned. The development of quantum physics is an especially interesting case because it displays the dramatic transformations in visual imagery resulting from advances in science. The reason, basically, is the transition from classical to nonclassical concepts. The solar system imagery for Bohr's original theory was imposed on its foundation in classical mechanics. This phase of development of Bohr's theory concerned the *content* of a visual representation, which is what is being represented.

---

[33] Miller, *Insights of Genius*, chap. 7.
[34] Ibid., chap. 8.

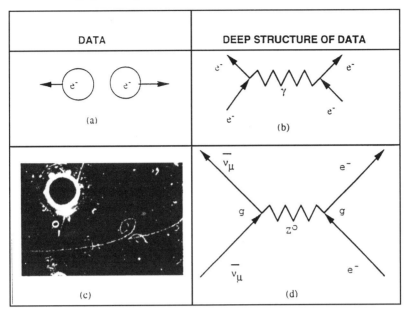

Figure 10.8. Images of data and their "deep structure." Data is exhibited in (a) and (c). (a) The situation where two electrons are depicted as two like-charged macroscopic spheres that move apart because like charges repel. The arrows indicate their receding from one another. (b) A glimpse into this process. (c) The deep structure in the bubble chamber photograph from (c) is given by the Feynman diagram in (d).

Beginning in 1923, the visual imagery of the solar system atom was discarded in favor of permitting the available mathematical framework itself to *represent* the atom. This phase focuses upon the *format* of a representation, or the representation's encoding. Mathematics was the guide and led to Heisenberg's breakthrough in 1925. Soon after, in 1927, the quest began for a new representation of the atomic world that culminated in Feynman diagrams. This transition in visual imagery is depicted in Figure 10.9.

Whereas imagery and meaning were imposed on physical theories prior to quantum physics, the reverse occurred after 1925. Quantum theory presented to scientists a new way of "seeing" nature. Heisenberg began to clarify the new mode of seeing with the uncertainty principle, while Bohr's complementarity principle approached the problem from a wider viewpoint that included an analysis of perceptions. The principal issue turned out to be the wave/particle duality, which rendered such terms as position, momentum, particle, and wave ambiguous. An upshot of the content-format-content shift in Figure 10.9(c) is the ontological status accorded to Feynman diagrams, which are the new visual imagery, the new *Anschaulichkeit*.

Visual representations have been transformed by discoveries in science and, in turn, have transformed scientific theories. They offer a glimpse of an invisible

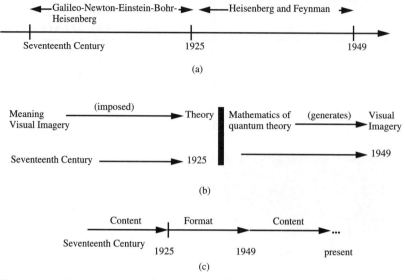

Figure 10.9. Representations of the atom. (a) The major figures in the conceptual transition in theorizing during the seventeenth century until 1925, and then from 1925 to 1949. (b) The major change from visual imagery and its meaning being imposed on physical theories (seventeenth century through 1925) to the mathematics of quantum physics, generating the relevant physical imagery with its meaning. (c) This is the transition from content to format to content. (*Source*: Arthur I. Miller, *Insights of Genius: Imagery and Creativity in Science and Art* [Cambridge, Mass.: MIT Press, 2000], p. 322.)

world in which entities are simultaneously wave and particle, and so cannot even be imagined. Entities in this domain are desubstantialized, as we have come to understand this concept.

Like scientists, artists also explore worlds that are visible and invisible. And so not coincidentally, a similar trend toward desubstantiation occurred in art almost coincidentally with science. In the early part of the twentieth century, artists were somewhat ahead of scientists in the trend toward increased abstraction and so away from classical representations.

The rise of Cubism presents an interesting case because it was programmatic in intent and achieved its goal by single-minded artists, such as Picasso and Georges Braque (1882–1963). Its aim, as set out by Picasso, was gradually to reduce form to geometry.[35] Yet although Cubism is abstract, one can still recognize body parts and other objects. Picasso never crossed the Rubicon into Abstract Expressionism.

The core issue is that at the beginning of the twentieth century, art and science moved toward increasing abstraction. Why this was the case and

---

[35] See Arthur I. Miller, *Einstein, Picasso: Space, Time and the Beauty That Causes Havoc* (New York: Basic Books, 2001).

what it had to do with the avant-garde culture is an issue I cannot go into here. It is relevant to what we have discussed, however, that it took until 1948 for transformation of representation in physics to the more abstract visualizability with its accompanying desubstantiation. On the other hand, the Russian artist Wassily Kandinsky and the Dutch artist Piet Mondrian worked along these lines since the second decade of the twentieth century, while developing offshoots of Cubism. With little understatement we can say that the visual representation in a Feynman diagram is an advance in visual imagery in science akin to a jump from the art of Giotto's predecessors to the modern Abstract Expressionism of a Mark Rothko, whose canvases display subtly vibrating large strips of colors, one flowing into the other, that is, complete desubstantiation. Thus have visual representations increased in abstraction since the late nineteenth and early twentieth centuries.

Part III

CHEMISTRY AND PHYSICS
Problems Through the Early 1900s

II

THE PHYSICAL SCIENCES IN THE LIFE SCIENCES

Frederic L. Holmes

The historical relations between the physical sciences and the life sciences have often been framed in terms of overarching conceptions about the nature of vital processes. Thus, in antiquity, the mechanistic viewpoint of the atomists, represented in physiological thought by the Alexandrian anatomist Erasistratus, is contrasted with the teleological foundations of Aristotle's biology, defended in late antiquity by Galen. For the early modern period, the Aristotelian framework within which William Harvey (1578–1657) discovered the circulation of the blood is contrasted with the "mechanical conception of life," introduced in the new "mechanical philosophy" of René Descartes (1596–1650), and a chemical conception of life, associated with the iconoclastic Renaissance physician Paracelsus (1493–1541).[1]

For the nineteenth century, the cleavage between the "vitalist" views of physiologists early in the century and the "reductionist" views of physiologists coming of age in the 1840s, who aimed to reduce physiology to physics and chemistry, has been treated as the most significant turning point in the relation between the physical and biological sciences. The views of these, mainly German, physiologists are often compared with those of the most prominent French physiologist, Claude Bernard (1813–1878), who also opposed vitalism but believed, nevertheless, that life is something more than the physical-chemical manifestations through which it must be investigated.

Without denying the broad philosophical and historical interest that these conceptions of life held, and still hold, I will shift emphasis away from them here, on the grounds that these views did not determine the pace or nature of the application to the life sciences of explanations and investigative methods

---

[1] The "life sciences" is a late-twentieth-century term, used to refer collectively to the many disciplines that treat aspects of living organisms. The phrase was not commonly used during the historical periods discussed in this chapter, but can be used here to avoid more pointed anachronisms. On its twentieth-century history, see, for example, Garland Allen, *Life Science in the Twentieth Century* (New York: Wiley, 1975).

based in the physical sciences. Whether maintaining an identity between vital processes and those of the inorganic realms of nature, or insisting on differences, those who have studied living nature have always recognized that some basic phenomena of life, such as movement and the transformation of matter, are shared with other natural events. The interpretations that researchers gave to these processes have always been dependent on the conceptions available from concurrent thought and investigation about the rest of the physical world. The teleological outlook of Aristotle did not differentiate life from the inanimate world, because he thought that the movements of the heavenly bodies were as ordered and purposeful as were those of living creatures. The same principles – form and matter, the four elements, and the rules for their transformations – that ordered terrestrial change in general, also explained for him such processes as the generation and nutrition of animals.[2]

The relation between the physical and life sciences changed fundamentally during the seventeenth century, because of the emergence of two new sciences – mechanics and chemistry – which provided new methods and concepts, derived primarily from the study of inanimate objects, that offered new sources of insight for understanding plants, animals, and humans in health and disease. According to a persistent historical tradition, these two resources were applied separately by two groups who held contrasting worldviews on the question. In a set of *Lectures on the History of Physiology*, first published in 1901, the physiologist Michael Foster (1836–1907) wrote that

> the school of physiology proper, the school of Vesalius and Harvey, was split up into the school of those who proposed to explain all the phenomena of the body and to cure all its ills on physical and mathematical principles, the iatro-mathematical or iatro-physical school, and into the school of those who proposed to explain all the same phenomena as mere chemical events, the iatro-chemical school.

Foster's division has echoed through more recent treatments of the period, and the ideological tone that he attributed to the two schools has stuck with them. Thus, Richard Westfall wrote in 1971 that "[i]atromechanism did not arise from the demands of biological study; it was far more the puppet regime set up by the mechanical philosophy's invasion.... One can only wonder in amazement that the mechanical explanations were considered adequate to the biological facts, and in fact iatromechanics made no significant discovery whatever."[3]

---

[2] Aristotle, *Parts of Animals,* trans. A. L. Peck (Cambridge, Mass.: Harvard University Press, 1955), p. 73.

[3] Michael Foster, *Lectures on the History of Physiology during the Sixteenth, Seventeenth, and Eighteenth Centuries* (Cambridge: Cambridge University Press, 1924), p. 55; Richard S. Westfall, *The Construction of Modern Science: Mechanisms and Mechanics* (New York: Wiley, 1971), p. 104. For a more subtle interpretation, see Mirko D. Grmek, *La première révolution biologique* (Paris: Payot, 1990), pp. 115–39.

APPLICATIONS OF THE PHYSICAL SCIENCES TO BIOLOGY IN THE SEVENTEENTH AND EIGHTEENTH CENTURIES

The most prominent of the iatromechanists was Giovanni Alfonso Borelli (1608–1679). Born in Italy in 1608, and an admirer of Galileo, Borelli made important contributions to celestial mechanics before turning, late in his career, to the study of motion in animals. His massive work on the subject, *De Motu Animalium,* was published in 1683, three years after his death. In his introduction, Borelli stated that no one before him had solved the difficult problems of the physiology of movement in animals "by using demonstrations based on Mechanics." This invocation, and the fact that Part I, on the "external motions of animals," was mainly an application of mechanical laws to analyze the motions of muscles and bones as systems of levers, appears to confirm his reputation as the preeminent "iatromechanist." The picture becomes more complex, however, when we read attentively Part II, "On the Internal Motions of Animals and their Immediate Causes." There, Borelli adduced anatomical evidence, including microscopical discoveries by his younger colleague, Marcello Malpighi (1628–1694); chemical analysis of the blood by Robert Boyle (1627–1691) and others; and the discoveries of the circulation by Harvey and of the lacteal ducts by Jean Pecquet (1622–1674), as well as mechanical arguments. He provided a comprehensive interpretation of circulation, respiration, and the traditional stages of digestion and nutrition, as well as the processes of secretion newly generalized from recent discoveries of the ducted glands.[4]

In the familiar style of seventeenth-century "mechanical philosophy," Borelli often depicted these internal processes in terms of the shapes and movements of particles composing the fluids of the body. But chemical phenomena, such as acid–alkali reactions, were also being reinterpreted at just this time in similar terms. The mechanism of muscular contraction that Borelli developed in *De Motu Animalium* illustrates well the interplay of physical, chemical, and mechanical reasoning in his physiology. The actual shortening of the muscle he attributed to the inflation of a series of tiny rhomboidal-shaped cavities postulated to make up the length of the individual fibers shown anatomically to constitute muscle. By mechanical analysis, he showed how such little chambers would shorten as they were inflated. For the cause of the inflation, however, he rejected theories, such as that of Descartes, requiring the movement of a substance through the nerves or blood into the muscle. None of these physical mechanisms could account for the instantaneous contraction of a muscle or its immediate relaxation afterward. "We should have thought it impossible" to understand these

---

[4] Thomas Settle, "Borelli, Giovanni Alfonso," *Dictionary of Scientific Biography,* II, 306–14; Giovanni Alfonso Borelli, *On the Movement of Animals,* trans. Paul Maquet (Berlin: Springer Verlag, 1989).

instantaneous actions swelling and deflating the muscles, Borelli wrote, "if chemical operations had not suggested that similar operations are carried out by Nature everywhere." Mixing acid solutions with alkaline salts causes rapid ebullition, which also rapidly subsides. The blood "is abundantly provided with alkaline salts." The mechanism Borelli proposed provided that alkaline salts derived from the blood mixed with a spirituous juice released from the ending of the nerve in the muscle when an impulse sent by the will reached it. This mixture "thus can provoke ebullition and effervescence in the fibers almost instantly."[5]

I have dwelt at length on this example because it is representative of the early application of the new physical sciences of the seventeenth century to physiological explanation. Borelli was not doctrinaire, nor did he attempt to explain all the phenomena according to physical and mathematical principles. He used all the empirical knowledge of the body and explanatory resources available to him. He judged astutely the realms appropriate to physical interpretations and the boundaries between physical and chemical events. That his mechanisms appear to twentieth-century readers as "speculative" and inadequate to the complexity of the "biological facts" is not due to facile reasoning or to an invasion of physiology by "iatromechanism," but to the differences between the state of the physical and chemical knowledge he could bring to the difficult physiological problems with which he grappled, and the knowledge available to those who investigated these problems in later centuries.

The most effective applications of the physical sciences to the study of vital processes during the seventeenth and early eighteenth century were those dealing with the mechanics of circulation. Following the discovery of that phenomenon by William Harvey, the visible movements of the heart, and of the blood through the arteries and veins, provided the one obvious opportunity to subject a physiological process to the kinematic and dynamic principles of the new science of mechanics. The first step in treating the circulation as such a system was taken in Paris in 1653 by Jean Pecquet, the discoverer of the lacteal vessels and the flow of chyle through them into the vena cava. Drawing on new concepts of the weight and pressure of the air derived from barometric experiments, Pecquet argued that blood is circulated by the impulsion of the systole of the heart and by contraction of the blood vessels under the pressure exerted on them by the air.[6]

In England, Richard Lower (1631–1691) published in 1669 an analysis of the movement of the heart based on more detailed observations of the ventricular muscles than had been known to Harvey, and offered a new calculation of the rapidity of the circulation, according to which the "whole mass of the blood is ejected from the heart not once or twice within an hour, but many

---

[5] Borelli, *Movement*, pp. 205–42. See Leonard G. Wilson, "William Croone's Theory of Muscular Contraction," *Notes and Records of the Royal Society of London*, 16 (1961), 158–78, for sources and background of Borelli's theory of muscular contraction.

[6] John Pecquet, *New Anatomical Experiments* (London: T. W., 1653), pp. 91–140.

times." Borelli also analyzed the movements of the heart muscles, and by comparison with mechanical models concluded that the heart propels blood by bringing the lateral walls of the ventricles closer together. By comparison with the weight that can be lifted by the masseter muscles of the jaw, Borelli estimated that the force exerted by the muscles of the heart "can be more than 3000 pounds." More realistically, he explained how the blood can move "continuously and uninterruptedly through the body of the animal," even though the compression of the heart is discontinuous. Because the arteries themselves are constricted by the contraction of their circular fibers, and by contractions of the other muscles of the body, the blood keeps flowing through the arteries even during the diastole of the heart.[7]

In 1717, James Keill (1673–1719) calculated the "force of the heart in driving the blood" on the basis of a proposition from Newton's *Principia*, relating the velocity of a fluid to the height from which it falls. The velocity of the blood he measured by the quantity that ran from the cut artery of a dog in ten seconds. His result, that the "force of the heart is equal to the weight of five ounces," led him to comment on "how vastly short this force falls of that determined by Borelli." Keill showed also, by calculating the increase in cross-sectional area of the arteries at each branching, that the velocity of the blood greatly decreases as it moves from the aorta to the capillaries.[8]

These analyses of circulation were successful, not in the sense that they were definitive, but in that they dealt with phenomena amenable to observation and experimentation, and to the forms of mathematical analysis of which the new mechanics was capable. They fit most easily with the conviction of those, such as Keill, that "[t]he animal body is now known to be a pure machine." The narrow limits of the approach are better illustrated by efforts to explain secretions by mechanical means. Seventeenth-century mechanists, such as Descartes and Borelli, likened the secretory glands to sieves, through which particles whose size and shape fit the pores in the gland were selectively separated from the blood. Keill saw the inadequacy of such models and proposed one in their place that relied on Newtonian conceptions of short-range attractive forces between particles in the blood. Neither type of explanation, however, could be brought into detailed relation with the observed anatomy or function of the secretory glands, and such speculations led nowhere except to the later eighteenth-century vitalist reaction against simplistic mechanical explanations.[9]

The most auspicious outcome of the efforts of Borelli and others to estimate the force of the blood in the heart and arteries was their provoking the

---

[7] Richard Lower, *De Corde,* trans. K. J. Franklin, in *Early Science in Oxford,* ed. R. T. Gunther, vol. 9 (London: Dawsons, 1932), chaps. 1–3.; Borelli, *Movement,* pp. 242–73.
[8] James Keill, *Essays on Several Parts of the Animal Oeconomy,* 2d ed. (London: George Strahan, 1717), pp. 64–94.
[9] Ibid., p. iii; René Descartes, *Treatise of Man,* trans. Thomas Steele Hall (Cambridge, Mass.: Harvard University Press, 1972), p. 17; Borelli, *Movement,* pp. 345–8, 356–7.; Keill, *Essays,* pp. 95–202.

Reverend Stephen Hales (1677–1761), an English country parson, to undertake one of the most productive experimental investigations of the eighteenth century. The efforts of these "ingenious persons," Hales wrote in 1728, "have differed as widely from one another as they have from the truth, for want of a sufficient number of data to argue from." Believing that the "animal fluids move by hydraulic and hydrostatical laws," Hales made "some enquiry into the nature of their motions by a suitable series of experiments." Eschewing the indirect methods of his predecessors, he determined the force of the blood in the arteries by the most immediate (and as he himself acknowledged, disagreeable) means possible. Tying down a horse, he inserted a long, vertical glass tube into its crural artery and observed the height to which the blood rose in the tube. Repeating this basic experiment on various animals under a range of conditions, Hales observed that the force "is very different, not only in animals of different species, but also in animals of the same kind[;] . . . the force is continually varying."[10]

His investigation of such variations, in different parts of the circulation as well as in different conditions, led Hales to a further development of Keill's analysis of the decrease in the velocity of the blood in the branches of the arteries; to a development of Borelli's view that the elasticity of the arteries converts the intermittent propulsion of the heart into an "almost even tenor of velocity" of the blood in the finer capillaries; to measurements of the resistance "which the blood meets with in passing in the capillary arteries" that explained "the great difference in the force of the blood in the arteries to that in the veins"; and to investigations of the effects of the viscosity of the blood on its motions. By adapting his experiments to "hydraulick and hydrostatic laws," Hales not only vindicated the efforts of half a century to apply mechanics to the "animal oeconomy," but provided, alongside his similar experiments on "vegetable statics," a model for the role of the physical sciences in the life sciences the impact of which lasted into the nineteenth century.[11]

CHEMISTRY AND DIGESTION IN THE EIGHTEENTH CENTURY

The science of chemistry provided no such enduring experimental achievements in the life sciences until near the end of the eighteenth century. The analysis of plant and animal matter occupied much of the efforts of chemists from the early seventeenth century on. These results, together with the emergence of a well-defined chemistry of acids, bases, and neutral salts, did enable physiologists to form chemical images of the processes of digestion, nutrition, secretion, and excretion. For example, Hermann Boerhaave (1668–1738) depicted these processes in his lectures in the early eighteenth century as a

---

[10] Stephen Hales, *Statical Essays: Containing Haemastaticks* (1733; reprint, New York: Hafner, 1964), pp. xlv–xlvi, 1–37.
[11] Ibid., pp. 22–3, 37–186.

gradual conversion of the "acidescent" plant matters serving as nutrients to "alkalescent" end products, a view that echoed frequently throughout the century. But these images could not be turned into the foundations of a progressive research program. Both the general potential and the specific limitations of chemical explanation are highlighted in the experiments on digestion published in 1752 by René-Antoine Réaumur (1683–1757).

Like Hales, Réaumur devised new experimental approaches to a problem first set forth by Borelli. According to Borelli, food was digested in a different manner by birds with muscular stomachs and animals with membranous stomachs. In the former, the internal walls of the stomach grind the food like millstones. Although interested as usual in the force exerted by such stomachs, he could not measure it directly, but "surmised" the force from that which the human jaw can exert in the similar function of breaking open hard foods. Animals with a membranous stomach, on the other hand, "digest meat and bones with some very powerful ferment as corrosive water [i.e., acid] corrodes and dissolves metals."[12]

Eighty years later, physicians and scientists were still divided over the question of whether digestion was caused by "trituration" (grinding) or the action of a solvent, or both. Réaumur settled this question by means of one of the most engaging experimental investigations in the early modern life sciences. That the stomachs of birds with gizzards crush hard food he proved by feeding them hollow metal tubes, which he retrieved from their excrement and found flattened or otherwise distorted. By flattening similar tubes with a pair of pliers, he was able to estimate the force the stomachs could exert. With a bird of prey known to regurgitate the indigestible materials it swallowed, Réaumur inserted pieces of meat and other foods into hollow tubes, the ends of which were enclosed by threads wound around them, permitting fluids to enter. When retrieved, the meat contained in the tubes had been reduced, partially or wholly, to a semifluid state. The experiments offered decisive evidence that birds with thin-walled stomachs digested their food by means of a solvent. "To which of the many solvents that chemistry furnishes us," he asked, "can it be compared?" Collecting gastric juice from the stomach by placing sponges in the hollow tubes, he could establish little more than that it tasted salty and reddened "blue paper" – that is, that it was acidic. In the end, he could offer no more specific a chemical description of digestion than to repeat the comparison Borelli had made between digestion and the action of an acid on a metal.[13]

Later in the century, the prolific Italian experimentalist Lazzaro Spallanzani (1729–1799) greatly extended Réaumur's experiments on digestion. He succeeded where his predecessor had failed, in digesting food outside the animal with gastric juice procured from its stomach. But Spallanzani got little further than Réaumur had with the chemical characterization of the process. He and

---

[12] Borelli, *Movement*, pp. 402–3.
[13] R.-A. de Réaumur, "Sur la digestion des oiseaux," *Mémoires de l'Académie Royale des Sciences* (1752, pub. 1756), 266–307, 461–95.

several other investigators who took up the problem during the 1770s and 1780s could not even agree on whether gastric juice was acidic or neutral.[14]

The inability of eighteenth-century experimentalists to define the chemical nature of digestion is particularly telling, because of all vital processes, digestion appeared most immediately accessible to chemical analysis. It took place within a container where its progress could be observed. As food passed through the stomach and intestines, and was absorbed into the lacteal vessels, it underwent visible changes in color and consistency. Already in antiquity, Galen had observed the movement of these contents by cutting open the stomach and intestines of living animals. Despite these advantages, chemical analysis in the eighteenth century could not pick out substances or changes distinctive enough to specify, beyond the simple analogy used by Réaumur, what the chemical process of digestion might be. That is not to say that chemistry was helpless or futile in its quest for further meaning. The comparative analyses of the gastric juice of several animals carried out in 1786 by the French chemist L. C. H. Macquart applied a systematic repertoire of extractions and reagents, from which he could identify and give the quantitative proportions of a "lymphatic substance" like that in blood, several salts, and phosphoric acid. If they had not yet been able to answer the question "by what mechanism can the stomach carry out this indispensable preparation" of foods for the sustenance and repair of the animal body, that was, Macquart affirmed, because "we are only beginning to fix our attention" on the problem. In a full century since Borelli had posed the problem, progress had been slow, but it accelerated rapidly during the next half century.[15]

NINETEENTH-CENTURY INVESTIGATIONS OF DIGESTION AND CIRCULATION

The implication of the foregoing treatment of the role of the physical sciences in the life sciences is that the era in which this role has commonly been thought to have been established – the nineteenth century – was not a departure from, but was built upon, earlier foundations. What marked the more auspicious successes of nineteenth-century applications of physics and chemistry to the study of life was not a new attitude of physiologists or physicians toward physical laws, but the emergence within the physical sciences of more powerful concepts and methods adaptable to the exploration of vital phenomena. This contention can be illustrated by following into the nineteenth century the investigation of the mechanics of the circulation and the chemistry of digestion.

---

[14] Lazzaro Spallanzani, *Dissertations Relative to the Natural History of Animals*, vol. 1 (London: J. Murray, 1784). For a contemporary review of these efforts, see M. Macquart, "Sur le suc gastric des animaux ruminans," *Mémoires de l'Académie Royale des Sciences* (1786, pub. 1790), 355–78.

[15] Galen, *On the Natural Faculties*, trans. Arthur John Brock (Cambridge, Mass.: Harvard University Press, 1952), pp. 241–3.; Macquart, "Suc gastric," pp. 361–78.

In 1823 the Académie des Sciences of Paris announced that the "prix de physique" for 1825 would be awarded for the determination, "through a series of chemical or physiological experiments," of the processes of digestion. In justification of this choice, the announcement declared:

> Up to now the imperfection of the procedures of chemical analysis has not permitted us to acquire exact notions of the phenomena that take place in the stomach and intestines during the work of digestion. The observations and experiments, even those made with the utmost care, have led only to superficial knowledge of this subject of such direct interest to us.
>
> Today, when the procedures for the analysis of animal and plant matters have acquired more precision, one can hope that with suitable care one can reach important ideas about digestion.[16]

It is notable that this statement referred not to novel procedures resulting from a new chemistry, but only to the greater precision of procedures that had earlier been "imperfect." The analytical methods in question were, in fact, not products of the mutations wrought by the recent "chemical revolution," but the outcome of a gradual development since the mid-eighteenth century of methods for the extraction, isolation, and characterization of plant and animal matters. During the decade preceding the announcement, the Swedish chemist Jöns Jacob Berzelius (1779–1848) had become the leading practitioner of such methods.

The most important submission for the prize (which was not awarded to anyone) came from Germany. At Heidelberg the anatomist and physiologist Friedrich Tiedemann (1781–1861) had already begun in 1820, together with the chemist Leopold Gmelin (1788–1853), an extended investigation of digestion and related processes. By the time they conformed with the specification for the prize that experiments be extended to all four classes of vertebrates, Tiedemann and Gmelin had devoted five years to a monumental research program. Before they could identify chemical changes associated with digestion, it was necessary for them to analyze each of the digestive fluids, saliva, gastric juice, pancreatic juice, and bile. To study the digestive changes, they fed animals "simple nutrients" – albumin, casein, fibrin, and starch. To identify substances that might appear or disappear along the digestive tract, they removed the contents found in the stomach and intestines of animals at given time intervals after feeding, and subjected them to a standardized sequence of extractions with solvents and treatment with reagents.

Definitive results were not easy to attain. The chemically very similar simple nutrients "are not marked by such distinct characteristics that they can be easily recognized in the different sections of the nutritive canal, mixed with digestive fluids, by means of the addition of chemical reagents." Their

---

[16] Quoted in Friedrich Tiedemann and Leopold Gmelin, *Die Verdauung nach Versuchen*, vol. 1 (Heidelberg: Karl Groos, 1826), pp. 1–2. Translated by the author unless otherwise noted.

most general result, that the foodstuffs are dissolved in the stomach, was a conclusion that, as Tiedemann and Gmelin acknowledged, many before them had already reached. Only in the case of starch were there available identification tests that enabled them to demonstrate its conversion to sugar during digestion – probably the first specific chemical reaction shown to take place within the animal organism. They viewed their investigation as a continuation of a long tradition, citing predecessors as far back as the seventeenth century. They credited Réaumur and Spallanzani, for example, with the proof that peristaltic motions of the stomach were not essential to digestion. Nevertheless, despite its general inconclusiveness, their massive work went so far beyond all previous experiments and analyses that it became, at the same time, a culmination and the starting point for a new phase in the history of digestion. Those who extended such experiments and analyses during the next decade rapidly produced more novel results. Tiedemann and Gmelin had ended the conflicts over whether gastric juice was neutral or acidic by showing that it was neutral in an empty, unstimulated stomach, but that gastric secretions produced by stimulating the stomach contained a free acid.[17]

Different investigators still disagreed on the specific acid secreted, but by the early 1830s, it was becoming clear that both an acid and an organic substance were essential to the action of gastric juice. In the anatomical museum directed by Johannes Müller (1801–1858) in Berlin, his assistant Theodor Schwann (1810–1882) was able to characterize the organic matter as distinct from all known animal matters by testing it with the standard reagents, even though he could not isolate it. Here, too, new concepts arising in general chemistry were brought quickly to bear on the life sciences. In the conversion of alcohol to ether, a reaction frequently studied by organic chemists, sulfuric acid activated the process without being consumed. Eilhard Mitscherlich (1794–1863) called the role of such agents, which did not enter the products of the reaction, "contact" actions. Drawing on this idea, Schwann asked whether the organic digestive principle acted by "contact." Although he could not establish that the principle was not consumed, he did find that it acted in such small quantities that it must be a contact process. Comparing it to the alcoholic "ferment" that similarly acted in minute quantities relative to the quantity of alcohol produced, Schwann defined a general class of "ferments." The digestive ferment he named "pepsin."[18]

From Schwann's discovery of pepsin, one can trace a continuous investigation of its digestive action throughout the nineteenth and into the twentieth century. Moreover, his redefinition of a ferment broadened into a growing

---

[17] Ibid., pp. 4, 295–6, 146–7.
[18] Theodor Schwann, "Ueber das Wesen des Verdauungsprocesses," *Archiv für Anatomie* (1836), 90–138. On Mitscherlich, see Hans Werner Schütt, *Eilhard Mitscherlich: Prince of Prussian Chemistry*, trans. William E. Russey ([Philadelphia]: American Chemical Society and Chemical Heritage Foundation, 1997), pp. 147–58.

class of ferment actions that were viewed by the mid-nineteenth century as fundamental to many life processes. By the twentieth century, when the demonstration of cell-free alcoholic fermentation by Eduard Buchner (1860–1917) had resolved a long debate over whether fermentation required a living organism, and ferments had been renamed enzymes, these studies further broadened into one of the main foundations of biochemistry.[19]

Just as the more precise chemical methods available in the early nineteenth century enabled physiologists to penetrate more deeply into the chemical events of digestion than could their predecessors of the eighteenth century, so too did more rigorous standards of physical measurement and the development of theoretical and experimental hydrodynamics enable them to improve on Stephen Hales's account of the mechanism of circulation. By measuring the resistance of fluids through tubes of very small diameter, the English natural philosopher Thomas Young (1773–1829) concluded in 1808 that the friction in vessels approaching the size of capillaries was much greater than that in those the size of the aorta. Young thus provided new evidence for a view that Hales had already maintained eighty years earlier. Young also relied on Hales's measurements of the forces and motions in the blood vessels themselves.[20]

In France, Jean Léonard Marie Poiseuille (1797–1869) designed a new instrument, which he named the hemadynamometer, to measure more accurately the pressures in various parts of the circulatory system. The U-shaped tube was filled with mercury, and the horizontal extension of the shorter of its two vertical arms was inserted directly into a vein or artery. Finding that the arterial pressure was equal in different arteries at different distances from the heart, Poiseuille attributed this unexpected result to the elasticity of the arterial walls. His instrument and his measurements opened a period of extensive quantitative experimentation on the dynamics of circulation.[21]

In 1827, the anatomist Ernst Heinrich Weber (1795–1878) applied what he had learned about wave motion through studies of the movements of water in glass-sided troughs with his brother, the physicist Wilhelm Weber (1804–1891), to a reexamination of the nature of the arterial pulse. Hitherto, physiologists had assumed that the impulse imparted to the blood by the contraction of the heart expanded the arteries simultaneously throughout their length, and that the movement of the blood through the arteries was inseparable from the arterial expansion. On the basis of the hydrodynamic

---

[19] For a survey of these developments, see Joseph S. Fruton, *Molecules and Life* (New York: Wiley-Interscience, 1972), pp. 22–85.
[20] Thomas Young, "Hydraulic Investigations, Subservient to an Intended Croonian Lecture on the Motion of the Blood," *Philosophical Transactions of the Royal Society* (1808), 164–86; Thomas Young, "The Croonian Lecture, on the Functions of the Heart and Arteries," ibid. (1809), 1–31.
[21] J.-L.-M. Poiseuille, "Recherches sur la force du coeur aortique," *Journal de Physiologie*, 8 (1828), 272–305; Poiseuille, "Recherches sur l'action des artères dans la circulation artérielle," *Journal de Physiologie*, 9 (1829), 44–52; Poiseuille, "Recherches sur les causes du mouvement du sang dans les veines," *Journal de Physiologie*, 10 (1830), 277–95.

experiments, however, Weber was able to distinguish the very rapid wave motion through the blood, which caused the pulse as it traveled along the arteries, from the much slower motion of the blood itself through the arteries. This new insight transformed the investigation of the mechanics of the circulation. In the 1840s, two German physiologists, Alfred Volkmann and Carl Vierordt, took up measurements of the movements and pressures of the blood in the heart, arteries, and veins. When in 1847 Carl Ludwig (1816–1895) invented an instrument that enabled him to record rapid changes in blood pressures on a revolving drum, the modern era of the investigation of the hydrodynamics of circulation was well under way, and it has continued ever since to build on the foundations thus established.[22]

TRANSFORMATIONS IN INVESTIGATIONS OF RESPIRATION

Links between the application of concepts and methods from the physical sciences to the study of physiological functions in the nineteenth century, and investigations of the same functions during the preceding two centuries, are most obvious in the two cases described, because the circulation of the blood and the digestive action of the stomach were the two functions most easily recognized from the seventeenth century onward as special manifestations, respectively, of more general mechanical and chemical phenomena. The connections are more subtle when we turn to other functions, such as respiration or animal electricity, the nature and significance of which were in large part revealed by transformations in chemistry and physics that themselves began only during the late eighteenth century.

Galen asked the question "What is the use of breathing?" in the second century A.D. In attempting to answer it, he likened respiration to a flame. During the seventeenth century, a group centered around Robert Boyle strengthened the analogy between respiration and combustion by showing that both an animal and a burning candle consumed a small portion of the air in an enclosed space over water. One of their number, John Mayow (ca. 1641–ca. 1679), proposed a comprehensive theory of respiration, according to which animals consumed "nitro-aerial" particles contained in the atmosphere. By the mid-eighteenth century, however, this theory had faded from discussion, along with the nitro-aerial particles.[23]

---

[22] Ernst Heinrich Weber and Wilhelm Weber, *Wellenlehre auf Experimente gegründet* (Leipzig: Fleischer, 1825); Ernst Heinrich Weber, *Friedrich Hildebrandt's Handbuch der Anatomie des Menschen*, vol. 3 (Braunschweig: Schulbuchhandlung, 1831), pp. 69–70; Ernst Heinrich Weber, "Ueber die Anwendung der Wellenlehre auf die Lehre vom Kreislauf des Blutes und insbesondere auf die Pulslehre," *Berichte über die Verhandlungen der Königlichen Sächsischen Gesellschaft der Wissenschaften zu Leipzig* (1850), 164–6; Carl Ludwig, "Beiträge zur Kenntniss des Einflusses der Respirationsbewegungen auf den Blutlauf im Aortensystems," *Archiv für Anatomie* (1847), 242–302.

[23] Galen, "On the Use of Breathing," in *Galen on Respiration and the Arteries*, ed. David J. Furley and J. S. Wilkie (Princeton, N.J.: Princeton University Press, 1984), pp. 81–133. For descriptions of these developments, see Leonard G. Wilson, "The Tranformation of Ancient Concepts of Respiration in

The advent of "pneumatic chemistry" during the 1760s allowed a fresh start toward understanding respiration. Joseph Black (1728–1799), the discoverer of "fixed air," showed that both respiration and combustion produce that substance. Joseph Priestley (1733–1804) asserted that respiration produced phlogiston. But all previous views on the subject were superseded by the theory of respiration that Antoine Lavoisier (1743–1794) developed in intimate connection with the theory of combustion that initiated the chemical revolution.[24]

In 1774, when Lavoisier had already shown that phosphorus and sulfur gain weight when they burn and that metals gain weight when they are calcined, he explained both processes by the "fixation" of either the air of the atmosphere or some portion of it. Not yet able to identify the components of the atmosphere, he defined them by means of respiration as the "respirable" and "irrespirable" portions. By 1777 he had identified the respirable portion as what he named one year later "oxygen," and he was then able to understand respiration as the combination of oxygen with carbon to form fixed air. Just as combustion produced heat, so did respiration release "animal heat."[25]

In collaboration with the mathematician Pierre Simon Laplace (1749–1827), Lavoisier devised, in 1782, an ice calorimeter with which they could measure the quantity of heat released during a physical or chemical change. With this apparatus, they showed in 1783 that a guinea pig melted approximately the same quantity of ice in a given time as the combustion of charcoal melted in producing the same quantity of fixed air. This result they took to be a confirmation that respiration is the slow combustion of carbon. Shortly afterward they found that charcoal contains inflammable air as well as fixed air, and discovered that water is composed of inflammable air and oxygen. In 1785 Lavoisier modified his theory of respiration to include the combustion of both carbon and inflammable air, the latter producing water.

In 1789 Lavoisier began a series of experiments on respiration in which he was assisted by a young follower named Armand Seguin (1767–1835). Finding that a guinea pig respired more rapidly when in digestion than when in abstinence, and that Seguin's respiration increased markedly when he performed physical work that could be measured as the lifting of a weight to a given

---

the Seventeenth Century," *Isis*, 51 (1959), 161–72; Robert G. Frank, *Harvey and the Oxford Physiologists* (Berkeley: University of California Press, 1980); and Diana Long Hall, *Why Do Animals Breathe?* (New York: Arno Press, 1981).

[24] Of the numerous historical discussions of Lavoisier's theory of respiration and its impact on later investigation, see especially Everett Mendelsohn, *Heat and Life: The Development of the Theory of Animal Heat* (Cambridge, Mass.: Harvard University Press, 1964), pp. 134–83; Charles A. Culotta, "Respiration and the Lavoisier Tradition: Theory and Modification, 1777–1850," *Transactions of the American Philosophical Society*, n.s. 62 (1972), 1–41; François Duchesneau, "Spallanzani et la physiologie de la respiration: Revision théorique," in *Lazzaro Spallanzani e la Biologia del Settecento*, ed. Walter Bernardi and Antonella La Vergata (Florence: Olschki, 1982), pp. 44–65; and Richard L. Kremer, *The Thermodynamics of Life and Experimental Physiology: 1770–1880* (New York: Garland, 1990).

[25] This and the following paragraphs summarize a detailed account of the steps in the development of Lavoisier's theory given in Frederic Lawrence Holmes, *Lavoisier and the Chemistry of Life* (Madison: University of Wisconsin Press, 1985).

height, Lavoisier not only confirmed but also expanded the scope of his theory of respiration. He viewed the respiratory combustion as the source both of animal heat and of work. Moreover, he now saw the respiratory combustion as integral to the overall exchange of matter between the organism and its surroundings. The carbon and hydrogen (the new name given to inflammable air in the reform of the chemical nomenclature that Lavoisier and his associates had in the meantime devised) consumed must be replaced through digestion if the animal is to remain in material equilibrium. "The animal machine," he wrote,

> is governed mainly by three types of regulators: respiration, which consumes hydrogen and carbon, and furnishes caloric; digestion, which replenishes, through the organs which secrete chyle, that which is lost in the lungs; and transpiration, which augments or diminishes according as it is necessary to carry off more or less caloric.

Lavoisier's mature theory of respiration still left many unanswered questions, most conspicuously about the site within the animal at which the hydrogen and carbon were burned, the nature of the substance or substances that contain the carbon and hydrogen burned, and the relationship between their combustion and the change of color when venous blood becomes arterial. These questions occupied experimentalists for several generations. They sought also to demonstrate more conclusively than had Lavoisier and Laplace that the heat an animal produces is equal to that which an equivalent quantity of carbon and hydrogen produce in combustion. Despite these ongoing uncertainties, Lavoisier's theory of respiration deeply and permanently transformed the relationship between the physical and the life sciences. For the first time, the material exchanges of the organism could be understood in a way that integrated the traditional physiological functions of digestion, nutrition, respiration, and the formation of animal heat within a framework of specific chemical and physical processes.

Lavoisier also initiated the elementary analysis of plant and animal substances. Half a century later, when his methods had been made capable of measuring with precision the quantities of carbon, hydrogen, oxygen, and nitrogen composing organic compounds, and when the three basic classes of compound – carbohydrates, fats, and what were later called proteins – composing foodstuffs and the animal body had been distinguished, it was possible to give a far more complex picture of the relations between the assimilation of foodstuffs, their breakdown to provide heat and mechanical work, the respiratory gaseous exchanges, and substances excreted. During the 1840s, the two most prominent organic chemists of their time, Jean-Baptiste Dumas (1800–1884) in Paris and Justus von Liebig (1803–1873) in Germany, provided images of these processes that were in part speculative, but which stimulated extensive further investigation. More lasting was the connection that Lavoisier's theory of respiration permitted during the 1840s between physiology and one of the most far-reaching physical laws to emerge in the nineteenth century, that of the conservation of energy.

Hermann von Helmholtz (1821–1894), who gave that law its first rigorous mathematical formulation in 1847, succinctly summarized its application to living organisms at the end of his famous treatise *Die Erhaltung der Kraft*. Animals, he wrote,

> take up oxygen and the complicated oxidizable compounds created by plants, give these out again mostly burned, as carbonic acid and water, in part reduced to simpler compounds, consuming, therefore a certain quantity of chemical potential force, and create in its place heat and mechanical force. As the latter represents a small amount of work relative to the heat, the question of the conservation of force reduces nearly to that of whether the combustion and transformations of the materials serving for nutrition create the same quantity of heat that the animals give off.[26]

On the basis of existing experiments, Helmholtz concluded that the "approximate" answer to this question was "yes." Nearly half a century of experimentation later, this affirmation could be made with precision. During the intervening years, the application of the law of the conservation of energy to the exchanges between plants and animals had already become one of the most powerful arguments for assimilating life to the general physical laws of nature.[27]

PHYSIOLOGY AND ANIMAL ELECTRICITY

Space permits only brief mention of the emergence of the phenomenon known in the nineteenth century as "animal electricity." Experimentation and theoretical explanation of the phenomena associated with electrical charge, discharge, attraction, repulsion, and conduction constituted one of the dominant activities of eighteenth-century physics. The discovery that several kinds of fish, including the torpedo and an eel, can deliver discharges similar to an electric shock, led some natural philosophers to speculate that these creatures were "animal phials," or living Leyden jars. Various effects of electrification on the growth or fructification of plants, as well as the obvious effects of electric discharges on humans, fueled speculation that electricity constituted the nervous fluid, or even the fundamental principle of life. To some observers, the power of electricity enlarged the possibilities for explanation of vital phenomena beyond the narrow bounds of mechanics and chemistry. When Luigi Galvani (1737–1798) discovered by accident in 1792 that frog

---

[26] H. Helmholtz, *Über die Erhaltung der Kraft: eine physikalische Abhandlung* (Berlin: Reimer, 1847), p. 70.
[27] Frederic L. Holmes, "Introduction," to Justus Liebig, *Animal Chemistry*, trans. William Gregory (New York: Johnson Reprint, 1964), pp. i–cxvi.

legs twitched under the influence of lightning discharges, and was able to reproduce the phenomenon by touching an isolated nerve and muscle with a combination of two metals, he interpreted these results by postulating that muscles contain electricity stored as in a Leyden jar, and that the discharge of this electricity causes the contractions.[28]

Older histories of science viewed Galvani's explanation of the effects he had observed as a mistake corrected by Alessandro Volta (1745–1827), who showed that the electric current was generated by a difference of potential between the two metals included in the circuit. On this basis, Volta devised a pile, consisting of repeated series of the two metals, separated by moist paper, which could generate the electrical current independently of the frog. More recently, historians have noted that neither Volta nor Galvani won this debate, because both were partly right. Some of Galvani's phenomena were due to electricity generated by the "voltaic" pile, but some observations, such as the muscular contractions caused by forming a loop in which the cut end of a nerve touched the muscle to which it was attached, were independent of metals. Nevertheless, an active experimental effort by a number of scientists to repeat and extend Galvani's observations faded after two decades, probably because of the failure to attain decisive new results. Meanwhile, the "Galvanic" currents generated by voltaic cells acquired an important role as a tool for investigating the nervous system.[29]

When François Magendie (1783–1855) discovered in 1822 that the posterior roots of the spinal nerves are sensory, and the anterior roots are motor, he first distinguished them by noting the loss of these functions when he severed the nerves at their point of exit from the vertebrae of the spinal cord. In his second paper on the subject, he added as a counterproof the reappearance of the functions when he stimulated the nerves after having separated them from the spinal cord. As earlier physiologists had done, he pinched, pulled, and pricked the nerves to irritate them. But he added "still another genre of proof to which to submit the spinal roots; that is galvanism." By touching the spinal nerves with electrodes connected to a voltaic battery, he passed an electric current through them and confirmed that the resulting contractions were much stronger for the anterior roots than for the posterior roots. Electric stimulation proved quickly to be so much more effective and

---

[28] The most comprehensive of the historical accounts of this activity is J. L. Heilbron, *Electricity in the 17th and 18th Centuries* (Berkeley: University of California Press, 1979). See also Philip C. Ritterbush, *Overtures to Biology: The Speculations of Eighteenth Century Naturalists* (New Haven, Conn.: Yale University Press, 1964), pp. 15–56.

[29] Ritterbush, *Overtures to Biology*, pp. 52–6, wholeheartedly echoed the older view. On Volta's discovery of the battery, see Giuliano Pancaldi, "Electricity and Life: Volta's Path to the Battery," *Historical Studies in the Physical and Biological Sciences*, 21 (1990), 123–60; Marcello Pera, *The Ambiguous Frog: The Galvani-Volta Controversy on Animal Electricity*, trans. Jonathan Mandelbaum (Princeton, N.J.: Princeton University Press, 1992); J. L. Heilbron, "The Contributions of Bologna to Galvanism," *Historical Studies in the Physical and Biological Sciences*, 22 (1991), 57–82; Maria Trumpler, "Questioning Nature: Experimental Investigations of Galvanism in Germany, 1791–1810," unpublished PhD diss., Yale University, 1992.

controllable than the older means that it played a major role in the numerous experiments, following Magendie's discovery, that he and other physiologists conducted to map out the sensory and motor nerves of the peripheral nervous system.[30]

The fact that an electrical current could stimulate the transmission of a nerve impulse revived speculation that the impulse was itself electrical. In his authoritative *Handbuch der Physiologie des Menschen,* Johannes Müller argued against such views by enumerating the differences between the properties of the electric currents used to stimulate nerve impulses and the nature of the conduction of the impulses along the nerves.[31]

Further advances in the physical sciences again impinged on this biological question. The discovery of electromagnetism provided a new means by which to detect very small electrical currents. Galvanometers were quickly introduced into physiological experimentation. It was by means of an extremely sensitive galvanometer that Emil Du Bois-Reymond (1818–1896) was able, during the 1840s, to detect, when a frog nerve was stimulated, a "negative swing" of the needle of the instrument, which he interpreted as evidence that the nerve impulse consisted of the propagation of a change in the electric charge along the nerve. It was also with the aid of a galvanometer that Helmholtz was able, in 1850, to determine the velocity of a nerve impulse, a process that had hitherto been regarded as either instantaneous, or at least too rapid to be measured.[32]

Helmholtz, Du Bois-Reymond, Ernst Brücke (1819–1892), and Carl Ludwig (1816–1895) met in Berlin in 1847, and are said to have agreed there on a program the aim of which was to reduce physiology to physics and chemistry. The next year, in the introduction to his *Untersuchungen über thierische Elektricität,* Du Bois-Reymond made a statement of his scientific creed, which has been taken by historians as the "manifesto" of the "1847 group." In it he defined the ultimate objective of physiology as the reduction of vital processes to the interactions of the elementary particles of matter under the influence of attractive and repulsive forces. He included in his discussion also a refutation of the idea of a vital force independent of physical or chemical forces.[33]

---

[30] François Magendie, "Expériences sur les fonctions des racines des nerfs rachidiens," *Journal de Physiologie expérimentale et pathologique,* 2 (1822), 276–9; Magendie, "Expériences sur les fonctions des nerfs qui naissent de la moelle épinière," *Journal de Physiologie,* 366–71.
[31] Johannes Müller, *Handbuch der Physiologie des Menschen,* 3d ed., vol. 1 (Koblenz: Hölscher, 1838), pp. 645–7.
[32] Emil Du Bois-Reymond, *Untersuchungen über thierische Elektricität,* vol. 1 (Berlin: Reimer, 1848); H. Helmholtz, "Messungen über den zeitlichen Verlauf der Zuckung animalischer Muskeln und die Fortpflanzungschwindigkeit der Reizung in den Nerven," *Archiv für Anatomie* (1850), 276–364; Kathryn M. Olesko and Frederic L. Holmes, "Experiment, Quantification, and Discovery: Helmholtz's Early Physiological Researches, 1843–1850," in *Hermann von Helmholtz and the Foundations of Nineteenth Century Science,* ed. David Cahan (Berkeley: University of California Press, 1993), pp. 50–108.
[33] Du Bois-Reymond, *Untersuchungen,* pp. xlix–L.

The association of Du Bois-Reymond's advocacy of the reduction of physiology to physics and chemistry with his attack on vital forces has contributed to a general historical impression that the modern successes of the physical sciences in the experimental investigation of life processes required the overthrow of a pervasive vitalism that had previously blocked progress. Thirty years ago, however, Paul Cranefield pointed out that the members of the 1847 group were never able to fulfill Du Bois-Reymond's criterion for the reduction of a physiological process to a molecular mechanism. What they did do very effectively was to apply physical and chemical methods to the investigation of, and physical and chemical laws to the interpretation of, phenomena that remained also biological. As this account suggests, they were, in this regard, only building on foundations gradually laid over two centuries. There is little evidence that any new opportunity to apply theories or investigative tools originating in the physical sciences had been effectively delayed by vitalistic opposition. On the contrary, these cases suggest not only that such theories and tools were exploited in the life sciences as soon as they became available, but also that in the case of combustion and electricity, the life sciences were deeply involved in the emergence of the physical and chemical advances themselves.[34]

---

[34] Paul F. Cranefield, "The Organic Physics of 1847 and the Biophysics of Today," *Journal of the History of Medicine*, 12 (1957), 407–23.

12

CHEMICAL ATOMISM AND CHEMICAL CLASSIFICATION

Hans-Werner Schütt

During the past decades, the historiography of nineteenth-century chemistry has become increasingly complex and at the same time more interesting. Rising on the sound "internalistic" foundations laid by Aaron J. Ihde's *The Development of Modern Chemistry* (1964) and, of course, by James R. Partington's *A History of Chemistry* (four volumes, 1961–1970), the edifice of more recent historiography depicts chemistry as an endeavor in close interaction with the cultural and intellectual currents of that time.[1] Departing from the question of the disciplinary identity of chemistry during the nineteenth century and thus joining "the separate analyses of schools, disciplines, and traditions into an integrated analytical matrix," Mary Jo Nye has with a sure hand sketched out the framework of this chemical edifice.[2] As reflected in her book, new problems have moved to the forefront, among them the question of the disciplinary development of chemistry and its subdisciplines, such as biochemistry, stereochemistry, and physical chemistry, as well as questions of the emergence of scientific schools and of science policy in general. Last but not least are the questions of the metaphysical background

---

[1] For more recent comprehensive "Histories of Chemistry" with very informative sections on the nineteenth century, see William H. Brock, *The Fontana History of Chemistry* (London: Fontana Press, 1992), pp. 128–664, and – considerably shorter – John Hudson, *The History of Chemistry* (London: Macmillan, 1992), pp. 77–243, and – even shorter and more like a collection of essays – David M. Knight, *Ideas in Chemistry: A History of the Science* (Cambridge: Athlone, 1992). See also Bernadette Bensaude-Vincent and Isabelle Stengers, *Histoire de la chimie* (Paris: Edition la Découverte, 1993). There is no comparable comprehensive work in recent German literature.
  Good reference books in respect to primary sources are Henry M. Leicester and Herbert M. Klickstein, eds., *A Source Book in Chemistry*, 4th ed. (Cambridge Mass.: Harvard University Press, 1968), and David M. Knight, ed., *Classical Scientific Papers: Chemistry*, 2 vols. (New York: American Elsevier, 1968, 1970). For a chemical bibliography of primary sources with short introductions, cf. Sieghard Neufeldt, *Chronologie der Chemie, 1800–1980* (Weinheim: Verlag Chemie, 1987). It is always useful to consult the – very internalistic – histories of organic chemistry by Graebe (from the end of the eighteenth century to about 1880) and Walden (from 1880 to the 1930s): Carl Graebe, *Geschichte der organischen Chemie* (Berlin: Springer, 1920; repr. 1971), and Paul Walden, *Geschichte der organischen Chemie seit 1880* (Berlin: Springer, 1941; repr. 1972).
[2] Mary Jo Nye, *From Chemical Philosophy to Theoretical Chemistry* (Berkeley: University of California Press, 1993), p. 19.

and of the internal discourses among chemists. In this context, two questions are of eminent importance: What is "chemical" in chemistry, that is, what distinguishes chemistry from its neighbor sciences? And how does chemistry arrange the objects of its scientific experiences?

CHEMICAL VERSUS PHYSICAL ATOMS

One may well say that it is the notion of units of matter showing certain measurable relations of their weight and possessing certain characteristics to explain the specificity of chemical reactions (that is, the notion of "chemical atomism") that distinguished chemistry from its close relative, physics. This may sound a little strange, as we are used to seeing atoms as the same entities in all sciences. From the time of Democritus (ca. 410 B.C.) up to the first decades of the nineteenth century, atoms were considered to be little indivisible lumps of matter, which, when forming compounds, are attached to one another either by their respective shapes or by "affinities." In the eighteenth century, the notion of elective affinities and a qualitative judgment of the respective strengths of those affinities was used by scientists like Etienne Geoffroy (1672–1731) to classify substances topologically.[3] But it was John Dalton (1766–1844), in his famous book *A New System of Chemical Philosophy* (three parts, two volumes, 1808–10, 1827), who bound together atomism and chemistry in a new way by defining chemical elements as matter consisting of atoms of the same relative weight in respect to the atomic weight of other elements. As a consequence of his theory, Dalton stated that if there are different relative weights for the same combination, then the ratio of those weights must be small integral numbers (the law of multiple proportions).[4]

On this basis, the regularities in the respective weights of chemical elements in compounds could be explained by chemical atomism, as defined by Alan Rocke in the first comprehensive monograph devoted entirely to this subject: "There exists for each element a unique 'atomic weight,' a chemically indivisible unit, that enters into combination with similar units of other elements in small integral multiples."[5]

That chemists throughout the nineteenth century had difficulties with atoms and their ontological meaning, with affinity, valence, and so forth, is well known to historians of chemistry, but it was Rocke who showed in

---

[3] Ursula Klein, *Verbindung und Affinität, Die Grundlegung der neuzeitlichen Chemie an der Wende von 17. zum 18. Jahrhundert* (Basel: Birkhäuser, 1994), pp. 250–86.

[4] The classic monograph on Dalton is Arnold Thackray, *Atoms and Powers: An Essay on Newtonian Matter-Theory and the Development of Chemistry* (Cambridge, Mass.: Harvard University Press, 1970); see also D. S. L. Cardwell, ed., *John Dalton and the Progress of Science* (Manchester: Manchester University Press, 1986).

[5] Alan J. Rocke, *Chemical Atomism in the Nineteenth Century* (Columbus: Ohio State University Press, 1984), p. 12; for primary sources of the late nineteenth century, cf. Mary Jo Nye, *The Question of the Atom: From the Karlsruhe Congress to the First Solvay Conference, 1860–1911* (Los Angeles: Tomash Publishers, 1984).

detail how the way in which chemists struggled with the notion of atoms set them apart from physicists. Some scientists, among them chemists, believed in the reality of atoms as indivisible particles that cannot be split up by any means. Other scientists, among them chemists, thought of atoms as a mere notion of convenience. Dalton tended to be a realist who thought that his assumptions about atomic weights and the indivisibility of atoms were highly probable. Yet, at the same time, Dalton may be called as much a "chemical atomist" as scientists like William Hyde Wollaston (1766–1828), who thought that the only empirical basis for calculating formulas was equivalent weights and that the translation of equivalent weights into atomic weights is an act of convention. As can be ascertained by chemical means, the difference between the physicists' atoms and the chemists' atoms lies in the fact that the chemical atom is "something plus something". There are discontinuous properties like elective affinities and multiple valencies that cannot be explained by the mass, motion, and gravitational forces of the physical atom.

During the entire nineteenth century there were chemists and physicists who refused to enter into debates on the ontology of atoms. But in the second half of that century, many antiatomists or, rather, antiontologists at least recognized the heuristic value of an atomic-molecular hypothesis. Nor were conventional viewpoints arbitrary. Such notions as affinity or chemical atoms – though unexplained throughout the whole century – were forced upon chemists by experimental facts, chemistry being an experimental, laboratory science par excellence; or as Frederic L. Holmes put it so aptly: "The ideas go into and come out of investigations."[6] The empirical data of stoichiometry needed an explanatory basis to be of any predictive value, and chemical atoms provided such a basis. Thus, the aim of the nineteenth-century chemists was not to explain matter and affinity per se but to forge theoretical tools in order to arrange the many empirical data coming out of the laboratory in such a way as to explain both the chemical behavior of known substances and predict new substances.

ATOMS AND GASES

After the 1840s, the kinetic theory of gases as proposed by Rudolf Clausius (1822–1888), James Clerk Maxwell (1831–1879), Ludwig Boltzmann (1844–1906), and others both confirmed certain assumptions of chemical atomism and drew the chemical atom and the physical atom more closely together. This theory treated heat not as the effect of an imponderable material called caloric but as the result of a collision of particles. Even though they were skeptical about atoms, scientists like Marcellin Berthelot (1827–1907), who

---

[6] Frederic L. Holmes, *Lavoisier and the Chemistry of Life: An Exploration of Scientific Creativity* (Madison: University of Wisconsin Press, 1985), p. xvi. There are other examples to prove the point; cf. Nye, *From Chemical Philosophy*, p. 52.

made synthesis instead of analysis the foundation of chemical research, could successfully tackle the problems of why many organic reactions take time and why there is an equilibrium in incomplete reactions between all partners of the reaction.[7] This research established a connection between the physical theory of particles in motion and the chemical behavior of those particles. The atomic-molecular hypothesis gained in plausibility when, at the end of the century, Jacobus H. van't Hoff (1852–1911), Svante Arrhenius (1859–1927), Johannes Diderik van der Waals (1837–1923), and others demonstrated that the gas laws also apply to solutions.

In 1860, at the famous first international congress of chemists in Karlsruhe, where the participants discussed the various definitions of "equivalent," "atom," and "molecule," there were sharp disagreements about the future of chemical atomism. Stanislao Cannizzaro (1826–1910) denied any meaningful distinction between a chemical atom and a physical atom, while August Kekulé (1829–1896) maintained that for the chemists, the notion of atom and molecule should be inferred solely from chemical laws. Many chemists shared the opinion that except for mass, other properties of the atom that might be inferred from physical hypotheses cannot explain chemical behavior. Therefore, in the eyes of physicists like Pierre Duhem (1861–1916), chemistry was a science closer to zoology and botany than to physics, or at least to Newtonian natural philosophy. In the late nineteenth century, French scientists such as Berthelot echoed this opinion, which had also been propagated by Jean-Baptiste Dumas (1800–1884) in the 1830s. It should be noted, however, that chemical natural history was not a mere counting of substances that chemists considered similar. It was based on theories that tried to explain "the activities of chemical molecules in the biological language of form and function rather than in the mechanical language of matter, motion and force."[8]

By the end of the century the picture had changed: Physics and chemistry had drawn together in thermodynamics, kinetics, and an advanced electrochemistry. Furthermore, the first attempts made by the physicist Joseph John Thomson (1856–1940) and others to deduce the periodic chemical properties as expressed in the periodic table from the inner structure of the atom proved fruitful. Such phenomena as ionization, whereby the ion behaved so differently from the atom; cathode rays, which demonstrated that the atoms emit material particles; spectroscopy, which suggested that chemical elements have a capacity for internal vibrations; and, last but not least, radioactive decay paved the way for the atomic theories of the twentieth century. A distinction between chemical atom and physical atom became obsolete as a new research

---

[7] Jutta Berger, *Affinität und Reaktion: Über die Entstehung der Reaktionskinetik in der Chemie des 19. Jahrhunderts* (Berlin: Verlag für Wissenschaffs-u. Regionalgeschichte, 2000) pp. 126–55; Mary Jo Nye, "Bertholet's Anti-Atomism: A 'Matter of Taste'?" *Annals of Science*, 38 (1981), 585–90.

[8] Mary Jo Nye, *Before Big Science: The Pursuit of Modern Chemistry and Physics, 1800–1940* (London: Twayne Publishers, Prentice Hall Int., 1996), p. 121.

program overcame the skepticism of the antiatomists. The studies of reaction mechanisms by Christopher Ingold (1893–1970) and others in the early twentieth century, bringing about a new classification according to types of reactions, may be seen as the final point in the unification of chemical and physical worldviews.[9]

Yet even at the end of the century, the idea of submicroscopic particles in motion did not stay uncontested, either on the side of the physicists or on the side of the chemists. From a positivistic point of view, physicists like Ernst Mach (1838–1916) and chemists like Wilhelm Ostwald (1853–1932) stressed that the existence of these particles could not be proved empirically and that the thermodynamics of chemical reaction should be treated solely in terms of the metabolism of energy. Thermodynamics, as Ostwald saw it, appeared superior to atomism in that its second law could explain irreversible processes, while the concept of atoms in motion could not explain the distinction between past and present.

CALCULATING ATOMIC WEIGHTS

So the question of atomism, whether chemical or not, was still open at the end of the nineteenth century. Nevertheless, all chemists agreed on its keystone – the principle that chemical elements have weights specific to the elements. From an epistemological standpoint, however, these weights could not be considered invariants.[10] The numerical value of both the equivalent weight and the atomic weight rested not only on experimental data but also on certain hypothetical assumptions and rules.

During the first half of the nineteenth century, the question in debate had been about how to determine atomic or equivalent weight relative to weights of other elements in combination. Some chemists like Wollaston considered this relative weight to be equivalent to a standard weight in the most simple combination. The problem: What is the most simple combination?

Dalton assumed that the simplest binary compound consists of just one atom of each of the two elements. Others tried to deduce the weight from the number of gaseous volumes of a certain vapor density reacting with a standard volume. Here the problem lay in the difficulty of proving that under the same external conditions, the same volume of different gases contains the same number of particles. The law of gaseous volumes that Joseph-Louis Gay-Lussac (1778–1850) had found in 1808 seemed to suggest just that.

[9] Kenneth T. Leffek, *Sir Christopher Ingold: A Major Prophet of Organic Chemistry* (Victoria, B.C.: Nova Lion Press, 1996).

[10] One big step in determining invariants – along with Theodor Svedberg – was carried out by Jean Perrin (1870–1942), who in 1908 found a method of estimating Avogadro's number from the Brownian motion of gamboge particles. Cf. Mary Jo Nye, *Molecular Reality: A Perspective on the Scientific Work of Jean Perrin* (London: Macdonald, 1972), pp. 97–142.

Gay-Lussac, who in respect to atomism felt close to his mentor Claude-Louis Berthollet (1748–1822), refrained from drawing any conclusions from his purely empirical findings. Berthollet not only rejected the law of simple proportions but also tried to steer clear of the quicksand of atomism, preferring to classify elements according to affinity toward reference substances like oxygen. Gay-Lussac shared Berthollet's opinion that Dalton's whole theory was based on an arbitrary rule – that of simplicity.

Nevertheless, in the 1820s most chemists, led by Jöns Jacob Berzelius (1778–1848), felt free to use the word atom, even though it was not at all clear that the chemical atoms were really indivisible, as the law of gaseous volumes suggested. Nor was it clear what the term atom or related terms like "molecule constituent" actually meant. To Auguste Laurent (1807–1853), the chemical atom meant the smallest quantity of a simple body that is necessary to operate in a combination.

Still, the behavior of gaseous volumes when undergoing chemical reactions posed puzzles. Even if one assumes that equal volumes contain equal numbers of particles – which Dalton denied – it is not at all clear whether Gay-Lussac's law relates only to elements or also to compounds, and whether the volumes of gases produced by the reaction also follow the law. The debate over what Gay-Lussac's law really means did not end till the Karlsruhe Congress, when Cannizzaro persuaded the delegates – several of them after the Congress when they read his pamphlet – that all problems could be solved if one accepted Amedeo Avogadro's (1776–1856) hypothesis of 1811 that in general, elementary gases consist of diatomic molecules.[11] But speculations on the divisibility of atoms and on compounds made of several atoms of the *same* element seemed to be so absurd that most chemists separated their efforts to systematize facts from their speculations about physical atoms, inasmuch as different methods, such as the determination of vapour densities and specific heats, employed to determine ultimate atoms like that of sulphur, yielded different results.

Even the great master of analytical chemistry Berzelius passed over Avogadro's hypothesis. For his determination of atomic weights, he relied instead on a combination of certain rules about the contents of oxygen in the acidic and basic parts of salts; on Gay-Lussac's law with respect to simple gases; on Eilhard Mitscherlich's (1794–1863) rule on the isomorphism of chemical compounds having the same rational formula (1818/19); and on the rule of Pierre Louis Dulong (1785–1838) and Alexis Thérèse Petit (1791–1820), which states that atomic heats – the products of gram atom and specific heat – of many heavy elements are inversely proportional to their atomic weights (1819).[12]

---

[11] In 1814 André-Marie Ampère (1775–1836) proposed a similar hypothesis based on crystallographical considerations.

[12] Hans-Werner Schütt, *Eilhard Mitscsherlich: Prince of Prussian Chemistry* (Washington, D.C.: American Chemical Society, Chemical Heritage Foundation, 1997), pp. 97–109.

It must be added that Berzelius did not adopt a seemingly very attractive assumption put forward in 1815/16 by William Prout (1785–1850). Prout postulated a kind of "proto hyle," a basic lump of matter in all elements whose atomic weights then should add to integral multiples of the weight of this lump. Prout tentatively identified the basic lump of matter with the hydrogen atom.[13] Alongside Berzelius, chemists like Jean Servais Stas (1813–1891) rejected Prout's hypothesis on analytical grounds, while chemists such as Thomas Thomson (1773–1852) and Dumas tended to support it.

EARLY ATTEMPTS AT CLASSIFICATION

Not only did Berzelius provide the most trustworthy analytical data for determining atomic weights, but he also was pivotal in classifying chemical substances. Attempts at classification were the focus of chemical discourse. The splendor and misery of this discourse, in respect to both chemical atomism and chemical classification, was that chemists had generalities and laws as intermediaries between facts and causes, but they had no method of unequivocally determining causes. As Mi Gyung Kim has shown with clear analytical insight, the discourse may be arranged in three different layers, namely *natural philosophy* dealing with theories of matter and attraction, and with calculation; *chemistry proper* dealing with substances, affinities, and compositions, and with experiments; and, as already mentioned, *natural history* dealing with relationships and with observation.[14]

The platform on which all attempts at chemical classification rested was stoichiometry, as developed by Berzelius and others after the term itself and the law of equivalents had been introduced in 1792/4 by Jeremias Benjamin Richter (1762–1807) in a publication full of "Pythagorean" speculations.[15] In 1813 Berzelius proposed a system of chemical notations as a shorthand for the language of stoichiometry that in itself was a classification system capable of expressing groups of similar phenomena and of displaying their interactions. By and large, the symbols of this system were the same as those we use today.[16]

The combination of letters that Berzelius proposed certainly does not represent any physical reality in nature, any more than the brackets and lines of later theories. But the letters surely said something about nature: They were an instrument of order. In 1832 Berzelius proposed two types of formulas. The

---

[13] William H. Brock, *From Protyle to Proton: William Prout and the Nature of Matter 1785–1985* (Bristol: Adam Hilger, 1985).
[14] Mi Gyung Kim, "The Layers of Chemical Language," I: "Constitution of Bodies v. Structure of Matter," II: "Stabilising Atoms and Molecules in the Practice of Organic Chemistry," *History of Science*, 30 (1992), 69–96, 397–437.
[15] Jeremias Benjamin Richter, *Anfangsgründe der Stöchyometrie oder Messkunst chymischer Elemente*, 2 vols. in 3 parts (Breslau and Hirschberg, 1792–4; repr. Hildesheim: Olms, 1968).
[16] Maurice P. Crosland, *Historical Studies in the Language of Chemistry* (London: Heinemann, 1962), pp. 265–81.

"empirical formulas" show only the quantitative result of an analysis, whereas the "rational formulas" show the "electrochemical division" of the molecules. In particular, the rational formulas reveal the close interdependence between chemical language and theory. If, for instance, one knows the correct rational formula of a crystallized compound and one has an isomorphous substance, that is, a chemically different substance which nevertheless has the same crystal form, containing an element of unknown atomic weight, one can prognosticate from the rational formula how many units of the element in question take part in one unit of the compound.

Berzelius deduced his rational formulas from his theory of electrochemical dualism, which served as a tool for all chemical classification. Even before the invention of the voltaic pile (1800), Johann Wilhelm Ritter (1776–1810) in 1798 had found that different metals display the same order with regard to their electrical effects and their affinities for oxygen. In 1807, Humphry Davy (1778–1829) produced elementary potassium and sodium by "electrochemical decomposition," and in 1818/19 Berzelius linked chemical dualism, as proposed by Antoine Lavoisier (1743–1794), to Davy's electrochemical ideas in a comprehensive theory of chemical combinations. The theory implied that all "acids" are dualistic compounds of nonmetals and oxygen, and it took some time until all chemists accepted the existence of acids without oxygen, which meant classification of acids as hydrogen compounds.[17]

Berzelius attempted to extend the principles of his taxonomy to organic compounds. In this context, one must add that it was not always clear what was to be considered "organic." During the first half of the nineteenth century, a demarcation criterion seemed to be the presence of "vis vitalis" in organic meterial. In this scheme, relatively simple substances, which do not belong to series of complex compounds and are just excretion products of organs, were not considered to be truly organic or possessing "vis vitalis." Thus, Berzelius characterized urea as having a composition at the borderline between the organic and the inorganic.[18] In historical introductions to chemistry courses, we still today find the opinion that the notion of vitalism was disproved by Friedrich Wöhler's (1800–1882) synthesis of urea in 1828. This legend, introduced by August Wilhelm Hofmann (1818–1892) in an obituary for Wöhler in 1882, was refuted by Douglas McKie in 1944.[19] The debate on

---

[17] Justus Liebig, "Über die Constitution der organischen Säuren," *Annalen der Pharmacie*, 26 (1838), 1–31.

[18] Jöns Jacob Berzelius, *Lehrbuch der Chemie*, trans. Friedrich Wöhler, 10 vols. (Dresden: Arnoldische Buchhandlung, vol. 1 1833, vol. 2 1833, vol. 3 1834, vol. 4 1835, vol. 5 1835, vol. 6 1837, vol. 7 1838, vol. 8 1839, vol. 9 1840, vol. 10 1841), 9: 434.

[19] Douglas McKie, "Wöhler's 'Synthetic' Urea and the Rejection of Vitalism: A Chemical Legend," *Nature*, 153 (1944), 608–10; cf. John H. Brooke, "Wöhler's Urea, and Its Vital Force? – A Verdict from the Chemist," *Ambix*, 15 (1968), 84–114; repr. in John H. Brooke, *Thinking about Matter: Studies in the History of Chemical Philosophy* (Great Yarmouth, Norfolk, England: Variorum, 1995), chap. 5; Hans-Werner Schütt, "Die Synthese des Harnstoffs und der Vitalismus," in Hans Poser and Hans-Werner Schütt, eds., *Ontologie und Wissenschaft: Philosophische und wissenschaftshistorische Untersuchungen zur Frage der Objektkonstitution* (Berlin: TUB Publikationen, 1984), pp. 199–214. Wöhler's

vitalism did not at all stop after 1828, but owing to new developments in chemistry – among them the recognition of the role of catalysis, and the laboratory synthesis of organic compounds from the elements – vitalism in the course of the century was slowly ousted from chemistry.

A genuinely chemical demarcation criterion rests in the fact that organic substances always contain hydrogen and carbon, sometimes in stoichiometrically large amounts. In their efforts to put organic compounds into groups and thus somehow to systematize the ever-growing field of known carbon-rich substances, chemists in the early years of the nineteenth century relied on chemical "standard behavior" or on the recurrence of certain uniform components in the compounds under investigation. Fats, for instance, as characterized by Carl Wilhelm Scheele (1742–1786) in the eighteenth century and by Michel-Eugène Chevreul (1786–1889) in the two decades after 1810, consisted of "a sweet principle of oils and fats," as Scheele called it – that is, glycerol – and of compounds that give a sour reaction and form salts – that is, fatty acids.

In order to put classification within organic chemistry on a sounder theoretical basis, and in order to harmonize the classification of inorganic and organic substances, Berzelius assumed that organic substances contain "radicals" of carbon compounds that behave just as elements behave in inorganic compounds. Gay-Lussac's discovery of cyanogen (1815), and especially Justus Liebig's (1803–1873) and Wöhler's discovery of the benzoyl radical, which proved to be a stable subgroup in many organic compounds (1832), motivated Berzelius to elaborate his theory.

TYPES AND STRUCTURES

The hypothesis of organic radicals opened a path through the "dark forest" of chemistry, as Wöhler put it, but it had no explanatory function in the sense that it could not show how and why certain elements come together to form element-like substances.

The long story of types and structures in nineteenth-century chemistry cannot be retold here in detail. So it must suffice to mention that after 1834, a new phenomenon that ran contrary to the hypothesis of electrochemical dualism brought about a revision of the whole system of classification. In this year, Dumas found that in chloroform, an electropositive atom of hydrogen can be replaced by an electronegative atom of chlorine or other halogens.

Chemical substitution allowed Dumas to propagate a natural classification with *chemical types,* which exhibit the same fundamental chemical properties and may be assembled in genera, and *molecular types,* which, like the chemical types, possess the same number of equivalents but do not display the same

---

synthesis was actually a rearrangement reaction. By 1824 Wöhler had already synthesized oxalic acid without causing the slightest sensation.

fundamental properties and may be classified in families. In this taxonomy, the guiding factors were the inner relations between constituents rather than their electrochemical nature.

In 1835/36, Dumas's former assistant Laurent put forward a theory, according to which chemically analogous substances like naphthalene and its halogen derivatives may be considered as one group, which consists of a hydrocarbon compound as "fundamental radical" and its substitution products as "derived radicals." In this context, the term "radical" was no longer used in a dualistic sense but in the sense of "unitary types." This offered a method of classification of organic substances in (hypothetically) isomorphous groups having the same carbon skeleton, which together possess their own modes of reactivity and have a certain resistance to fundamental chemical modifications. Inasmuch as the structure of the molecules of a given substance is reflected in its crystal shape, Laurent's approach linked chemical to crystallographic considerations.

To Laurent's concept, Charles Gerhardt (1816–1856) added a theory of "residues," stating that when two complex molecules combine, they eliminate a simple molecule like water and at the same time copulate together. Seen this way, the products of substitution reactions are unitary molecules and do not consist of two parts held together electrostatically.[20]

In 1853, to permit a general classification of all organic compounds, Gerhardt proposed four basic types of molecules. In 1846 Laurent had already suggested a "water type" of compounds analogous to water, and Alexander Williamson (1824–1904) had used this notion to explain the relationship between alcohols and both symmetric and asymmetric ethers. Around 1850, research by Adolphe Wurtz (1817–1884) and Hofmann led to the concept of an ammonia type. Gerhardt added the hydrogen and the hydrogen chloride type and introduced the concept of "homologous series" to account for the slight and serial alteration of properties in certain groups of substances. William Odling (1829–1921) added methane and its derivatives as a fifth type, and in 1853 he extended the water type to double and triple multiple types.

In Gerhardt's eyes, the types were heuristic classificatory devices and had no structural significance because the riddle of the ultimate nature of molecular arrangements would never (in his opinion) be solved. But it was his type theory that finally "metamorphosed into the structural theory of carbon compounds."[21]

Laurent and Gerhardt's ideas faced serious challenge, partly for personal reasons. Nevertheless, the acerbity of the debate is still amazing in light of the epistemological status both sides accorded to chemical theories. In an article

---

[20] John H. Brooke, "Laurent, Gerhardt and the Philosophy of Chemistry," *Historical Studies in the Physical Sciences*, 6 (1975), 405–29, (repr. in Brooke, *Thinking about Matter,* chap. 7).

[21] Brock, *The Fontana History of Chemistry,* p. 237. For an excellent introduction to the road to structural chemistry, cf. O. Theodor Benfey, *From Vital Force to Structural Formulas* (Philadelphia: Houghton Mifflin, 1964).

attacking Laurent's theories, Liebig wrote: "Our theories are the expression of contemporary views: in this respect, only the facts are true, whereas the explanation of the relationship of the facts to one another merely more or less approaches truth."[22]

This was Laurent's opinion, too, but it seems to be all too human to "ontologize" in the heat of the debate the ideas one has brought up or tries to refute. Furthermore, not knowing the "positive" reasons for the chemical characteristics of substances does not mean that there cannot be sound arguments for preferring one classification over another. Williamson, for instance, postulated the existence of monobasic acid anhydrides of the water type analogous to the ethers, and when Gerhardt experimentally prepared acetic anhydride, this was a strong vindication of the water type. However, classification according to types often proved arbitrary when chemists had to decide which hydrogen of which type should be considered to have been replaced by fragments of other molecules.

Besides, even in the 1840s, a fundamental question had not been answered: Is it a characteristic part of the molecule that is responsible for the close relationship within chemically similar substances, or is it the arrangement of the atoms of the whole molecule? Trying to extend Berzelius's theory of radicals, Hermann Kolbe (1818–1884), who considered his formulas to be a reflection of molecular reality, broke up chemical molecules and parts of molecules into ever-smaller hierarchically ordered fragments.[23] In the 1850s, chemists such as Kekulé focused on the single atom and its position within the molecule. In 1854 Kekulé offered an example of how the predictive power of a taxonomic assumption may lead to discoveries that, in turn, enlarge the scope of the very classification from which they came. By treating acetic anhydride with phosphorus pentasulfide, he could show that mercaptans belong to the water type even though they contain no oxygen. The realization that no specific atom is in itself decisive and indispensable and, therefore, that there is no hierarchy of atoms in a molecule was a big step toward a theory of chemical structure.

One prerequisite for such a theory was new attention to the question of affinity, which was raised in connection with the focus on the single atom within the molecule. Research on radicals via organo-metallic compounds drew Edward Frankland's (1825–1899) attention to the power of combination of the metals on one side and their organic reaction partners on the other side. In 1852 he postulated that elements may have different, but always definite, combining powers.[24] At the same time, he showed that there is a strict analogy between the organic and the inorganic compounds of metals.

---

[22] Justus Liebig, "Über Laurent's Theorie der organischen Verbindungen," *Annalen der Pharmacie,* 25 (1838), 1–31, at p. 1.

[23] Alan J. Rocke, *The Quiet Revolution: Hermann Kolbe and the Science of Organic Chemistry* (Berkeley: University of California Press, 1993), pp. 243–64.

[24] Colin A. Russell, *Edward Frankland: Chemistry, Controversy and Conspiracy in Victorian England* (Cambridge: Cambridge University Press, 1996), pp. 118–46.

Equivalent weight now became the ratio of atomic weight to valence.[25] In 1857/8 Kekulé and Archibald Scott Couper (1831–1892) independently and without any justification in physical theories stated that all theory of structure must be based on the assumption that carbon atoms are tetravalent and linked together in chains.[26] Alexandr Butlerov (1828–1886) popularized the term "chemical structure" as the basis of the molecule's properties, but so long as there was no knowledge of the inner dynamism of molecules and of reaction mechanisms, the envisaged "structure" of a molecule could not be taken as a true image of reality.

ISOMERS AND STEREOCHEMISTRY

The structural theory proved its worth in the field of isomerism. The theory not only had prognostic value by predicting, for example, primary, secondary, and tertiary alcohols, but it could also restrict the number of isomers to be expected for a given substance. For instance, while the type theory allowed for isomers in which the hydrogen atoms have different functions in respect to their positions in the type, the structural theory was able to demonstrate, for example, that there are not two different ethanes of the hydrogen type, namely, C_2H_5, H and CH_3, CH_3. This instance and others proved that all valencies are equal.

The biggest challenge to structural theory, that is, the problem of aromatic compounds, turned out to be its biggest success. In 1864 Kekulé solved the riddle of why there are exactly three isomers of disubstituted benzene by suggesting that the six carbon atoms of benzene form a ring. So began the successful efforts to classify all substances within this large and distinct family of aromatic substances.[27]

But even structural chemistry could not explain the strange behavior of compounds like tartaric acid and its close relative. The question of optical isomers, but also the question of isomorphism in relation to isomerisms, had to be tackled both from the side of chemistry and the side of crystallography. In 1848 Louis Pasteur (1822–1895) found the phenomenon of enantiomorphism, that is, of mirror-isomerism, and in 1860 he supposed that all optically active molecules must be asymmetrical.[28] One may speculate whether it was a general reluctance among chemists to enter into discussion about the reality

---

[25] Colin A. Russell, *The History of Valency* (Leicester: Leicester University Press, 1971), pp. 34–43.

[26] Cf. O. Theodor Benfey, ed., *Classics in the Theory of Chemical Combination*, Classics series vol. 1 (New York: Dover Publication, 1963); see also Alan J. Rocke's contribution to this volume.

[27] Alan J. Rocke, "Hypothesis and Experiment in the Early Development of Kekulé's Benzene Theory," *Annals of Science*, 42 (1985), 355–81; Hans-Werner Schütt, "Der Wandel des Begriffs 'aromatisch' in der Chemie," in Friedrich Rapp and Hans-Werner Schütt, eds. *Begriffswandel und Erkenntnisfortschritt in den Erfahrungswissenschaften* (Berlin: Technische Universität Berlin, 1987), pp. 255–72.

[28] Hans-Werner Schütt, "Louis Pasteur und das Rätsel der Traubensäure," *Deutsches Museum Wissenschaftliches Jahrbuch 1989* (Munich: R. Oldenbourg, 1990), pp. 175–88.

of atoms possessing a fixed position in space that hindered further development, until in 1874 van't Hoff assumed that the four bonds of a central carbon atom are located at the summits of a tetrahedron.[29] Taking Louis Pasteur's discovery of enantiomorphism as a starting point, Joseph-Achille Le Bel (1847–1930) reached the same conclusion at about the same time.[30] The notion of the asymmetrical carbon atom also gave insight into the stereometry of compounds with double and triple bonds. This facilitated chemical classification enormously, as it helped chemists to "see on paper" what they were doing. For instance, stereochemical insights proved indispensable in the prognostication and classification of the carbohydrates.

Stereochemistry also gave an approximate answer to the question of why double bonds are more reactive than single bonds. The strain theory put forward by Adolf von Baeyer (1835–1917) in 1885 stated that the stability of double bonds is related to the strain to which the valencies of carbon atoms are subjected when they are bent out of their usual directions at the tetrahedron angle of 109° 28′. The question of why benzene as a cyclohexene shows a much greater stability than, for example, cyclopentene was tentatively answered by Ulrich Sachse (1854–1911), who in 1890 put forward the hypothesis that as the six carbon atoms of the benzene ring try to keep their tetrahedron structure, the ring cannot be planar. The assumption that there are two isomers of benzene, the "boat form" and the "chair form," was the first step in the direction of conformational analysis.

Admittedly, structural theory, crystallography, and stereochemistry did not solve all problems of chemical classification. One conundrum was tautomerism – acetoacetic ester investigated after 1866 being the best example – as it suggested that atoms may freely change their position. Nevertheless, stereochemistry proved successful even in the field of inorganic chemistry. After (against Kekulé's resistance) it became clear that elements like phosphorus can display variable valencies, the idea that the stereometric arrangements of atoms in complex inorganic compounds can be inferred from chemical behavior and optical isomerism also gained ground. Pivotal in this respect was Alfred Werner's (1866–1919) coordination chemistry, which he developed after 1893.

Aided by structural theory and by stereochemistry, the clarification of strucure (*Strukturaufklärung*) by analysis and synthesis became the catchword of chemistry during the second half of the nineteenth century – both in inorganic and organic chemistry. In this context, the chemistry of terpenes (Berthelot, William Henry Perkin, Jr. [1860–1929], Otto Wallach [1847–1931], et al.), of carbohydrates (Emil Fischer [1852–1919] et al.), and

[29] Cf. Trevor H. Levere, "Arrangement and Structure: A Distinction and a Difference," in *Van't Hoff-Le Bel Centennial*, ed. O. Bertrand Ramsay (Washington, D.C.: American Chemical Society, 1975), pp. 18–32.
[30] Cf. H. A. M. Snelders, "J. A. Le Bel's Stereochemical Ideas Compared with Those of J. H. van't Hoff (1874)," in *Van't Hoff-Le Bel Centennial*, pp. 66–73; O. Bertrand Ramsay, *Stereochemistry* (London: Heyden, 1981), pp. 81–97.

of heterocyclic compounds such as indigo (Baeyer et al.) should at least be mentioned.

FORMULAS AND MODELS

A few words may be said here about formulas and models, as both played a large role in the development of structural chemistry and of stereochemistry. Analogies of function and supposed analogies of function could be well demonstrated by formulas, as was done in the type theories. But those formulas were not designed to give any references to a "real" spatial arrangement of the atoms. This approach changed when structural chemists, like Butlerov in 1861, stressed that the particular arrangement of atoms within a molecule was the cause of its properties. In the 1860s, graphic formulas indicating valencies, by straight lines – as propagated by Alexander Crum Brown (1938–1922) et al. – came into use. Those formulas were able to visualize certain isomeric relations.

Models, which also were used from the beginning of the nineteenth century, were more problematic than formulas, as the former could not be adequately represented in two-dimensional print. On the other hand, they showed the possible arrangement of particles in space. Dalton had already built wooden models of conglomerates of atoms to demonstrate certain consequences of his theory, such as the relative position of atoms of the element B around a central atom of the element A. Wollaston and Mitscherlich used models to illustrate crystal forms, and it was in the crystallographic tradition that, with the aid of the model of a hypothetical hydrocarbon, Laurent demonstrated that the structure of the molecule is of utmost importance, and that for this reason, substances that do not radically alter their carbon skeleton during substitution retain most of their chemical properties after the reaction. Laurent's "models on paper" were used in important handbooks, accustoming chemists to the principle of minimum structural change. Around 1845, Leopold Gmelin (1788–1853) also used models to explain isomerism.[31] In the 1850s and 1860s, Kekulé, Brown, Hofmann, and Frankland built models to be used in teaching. Those models, however, were not intended to show "real" atomic arrangements in space, as it was not at all clear whether the atoms really had a static configuration within the molecule. When in 1867 James Dewar (1842–1923) proposed a model based on the notion of a tetrahedral carbon, it was disregarded, as it did not have any heuristic and prognostic value. On the other hand, van't Hoff after 1874, with the help of models, was able to demonstrate that the specific rotation power of malic acid salts is not anomalous, as might have been suspected.[32]

---

[31] In his *Handbuch der Chemie*, 4th ed., 6 vols. (Heidelberg, 1843–55).
[32] H. A. M. Snelders, "Practical and Theoretical Objections to J. H. van't Hoff's 1874 Stereochemical Ideas," in *Van't Hoff-Le Bel Centennial*, pp. 55–65.

The example of van't Hoff demonstrates that the relation between the models and the atoms and molecules they depicted was not one of formal analogy. Even the chemists who held an agnostic position with respect to atoms when interpreting models could not avoid implying that if atoms are somewhere situated in space, their existence is proven. Van't Hoff talked of "material points," and one of his earliest supporters, Johannes Wislicenus (1835–1902), in 1887 published a long paper on the "Spatial Arrangement of Atoms in Organic Molecules. . . ."[33] As the configuration of material points of atoms could explain a lot of chemical phenomena, they became more and more "real," regardless of the question of whether or not they can be subdivided.

THE PERIODIC SYSTEM AND STANDARDIZATION IN CHEMISTRY

The basis of all classification in chemistry is the periodic system. The search for such a system goes back to the times of physicotheology, when attempts at classifying chemical substances were made in order to show that God has ordered everything "according to number, measure, and weight" (The Wisdom of Solomon 11, 21). With the number of known elements having grown to about sixty by the mid-nineteenth century, those attempts multiplied, and a significant number of chemists tried to arrange elements.[34] To name but a few: From 1817 onward, Johann Wolfgang Döbereiner (1780–1849) had found several cases in which the equivalent weights of three chemically related elements, Ca, Sr, and Ba, increase in arithmetic progression. Many such efforts by other chemists followed. In 1857 Odling drew attention to the series carbon, nitrogen, oxygen, and fluorine by showing a regular increase in weight and a decrease in the number of valencies from four in the case of carbon to one in the case of fluorine. He also tried to establish a comprehensive periodic system. In 1862 Alexandre Emile Béguyer de Chancourtois (1820–1886) arranged all known elements according to their atomic weights on a spiral he had drawn on a cylinder. Every sixteen units an element appeared above another element, which was often closely related. In 1869 John Alexander Reina Newlands (1837–1898) postulated that, in general, if elements are arranged according to weight, those elements that are eight positions apart from one another are chemically related (law of octaves). Gustavus Detlev

[33] Peter J. Ramberg, "Commentary: Johannes Wislicenus, Atomism, and the Philosophy of Chemistry," *Bulletin for the History of Chemistry*, 15/16 (1994), 45–51; cf. note 39.
[34] Johannes W. van Spronsen, *The Periodic System of Chemical Elements: A History of the First Hundred Years* (Amsterdam: Elsevier, 1969), esp. pp. 97–146; Heinz Cassebaum and George B. Kauffman, "The Periodic System of the Chemical Elements: The Search for its Discoverer," *Isis*, 62 (1971), 314–27; Don C. Rawson, "The Process of Discovery: Mendeleev and the Periodic Law," *Annals of Science*, 31 (1974), 181–204; Bernadette Bensaude-Vincent, "Mendeleev's Periodic System of Chemical Elements," *British Journal for the History of Science*, 19 (1986), 3–17.

Hinrich (1836–1923) should be mentioned in this context, too, as he had another approach, ordering the elements on an Archimedean spiral.

Credited with having introduced "true" and comprehensive periodic tables are Lothar Meyer (1830–1895), who published his proposals in 1868 and 1870, and Dimitry Ivanovich Mendeleyev (1834–1907), who independently put forward his system in 1869. It is worth noting that both Meyer and Mendeleyev developed their concepts while working on books about the theoretical foundations of chemistry. Writing about foundations always involves writing about the principles of classification of the empirical information available. Both chemists had attended the Karlsruhe Congress, both adhered to Avogadro's hypothesis as a means of calculating correct atomic weights, and both proposed "natural systems," in that their taxonomies depended not only on an "educated feeling" for what belongs together chemically, but also on sets of independent and measurable empirical data – be they atomic volumes, atomic weights, aspects of isomorphism, the rule of Dulong and Petit, or valencies. While Meyer in 1870 was publishing an atomic volume curve in relation to atomic weight, Mendeleyev in 1869 was stressing the importance of the number of valencies: "The arrangement of the elements, in the order of their atomic weights, corresponds with their so-called valencies."[35]

It was Mendeleyev's proposals that really proved fruitful; not only did he state – as Meyer also had done – that there are some as yet undetected elements, but he deduced many properties of those unknown elements, such as atomic weight, specific weight, atomic volume and boiling points, and specific weights of expected compounds from their position in the periodic table. Thus, his table presented itself not only as a taxonomic system but also as a theory with prognostic value. The first success of Mendeleyev's table was Paul-Emile Lecoq de Boisbaudran's (1838–1912) discovery of gallium (1875), which was characterized as eka-aluminum in the table. In a fine chauvinistic sequence, scandium (1879, eka-boron) and germanium (1886, eka-silicon) followed suit (eka = one, in Sanskrit).

But there were other problems with the taxonomy of chemical elements, especially in the realm of elements with closely related chemical properties, like the transition elements and the rare earth elements. The addition of the noble gases to the list of known elements – in 1894 by Lord Rayleigh (John William Strutt) (1842–1919) and William Ramsay (1852–1916) – led in 1900 to the insertion of a new zero group in the system between the group of halogens and the group of the alkali metals.

Not only did Mendeleyev's research program give a fresh boost to analytic inorganic chemistry, but it also led to a renewed discussion of Prout's hypothesis and to steps toward standardization in chemistry.[36] Chemists like

---

[35] *J. Russ. Chem. Soc.*, 99 (1896), 60–77, referenced in Partington, vol. 4, p. 894.
[36] Britta Görs, "Chemie und Atomismus im deutschsprachigen Raum (1860–1910)," *Mitteilungen der Fachgruppe Geschichte der Chemie der Gesellschaft Deutscher Chemiker*, 13 (1997), 100–14.

Meyer and Wislicenus voiced the opinion that "the periodicity of the relations between the properties and the weights of the atoms of the elements" suggests that those atoms are somehow systematically composed of smaller particles.[37] Those particles, being simple, could be imagined as consisting of proto hyle. Meyer assumed that if the smallest particle of matter is really the hydrogen atom, one could account for the fact that the other atomic weights are not multiples of the weight of hydrogen by assuming that the smallest particles are surrounded by ponderable ether. Mendeleyev, who tried to stay clear of mere hypotheses, remained skeptical in regard to Prout.

The renewed efforts to determine atomic weights also led to a discussion on how to standardize them, since standard weights and standard substances greatly facilitate chemical calculations. So in the early 1860s, Hofmann had already proposed that the chemists should relate all gas volumes and their weights to the volume of one liter of hydrogen weighing 0.896 grams.

Much of the confusion even in late-nineteenth-century chemistry resulted from the fact that there was no reference atomic weight accepted by all chemists. At the Karlsruhe Congress, participants proposed the calculation of chemical compounds not in equivalents but in atomic weights, which was widely accepted. But the proposal to take $O = 16$ – instead of $H = 1$, $O = 1$, $O = 10$, $O = 100$ – as standard weight apparently met with resistance, as many chemists continued to use $H = 1$ as reference weight.[38] This meant that (as determined by Stas) the value of oxygen had to be taken as 15.96. After long debates between 1895 and 1906 and several votes in national and international commissions, $O = 16$ was finally accepted as the reference basis.

As already mentioned, the debate on atomic weights coincided with the search for elements to fill the positions left open in the periodic table. But research in this field could not solve the riddle of several cases in which the atomic weight of elements did not match their chemical properties as suggested by their place in the periodic table. To give a reasonable picture of those properties, iodine (at. wt. 126.8) and tellurium (127.6) had to swap places, and the same went for nickel (58.69) and cobalt (58.95). The pairs argon-potassium, thorium-protactinium, and neodymium-prasaeodynium also posed problems. All this suggested that the very criteria of taxonomy on which the most important chemical classification system rested had flaws. The riddle of the "wrong" positions of certain elements was not solved until 1913, when in the context of research on radioelements, Henry Gwynn Moseley (1887–1915) found a constant relationship between the relative position of the shortest wavelength x-ray line of an element in the spectrum and its atomic number, which in turn indicates the true position of the element in the context of the periodic system. At about the same time, Frederick

---

[37] Johannes Wislicenus, "Über die Lage der Atome im Raume: Antwort auf W. Lossens Frage," *Berichte der Deutschen Chemischen Gesellschaft*, 21 (1888), 581–5, p. 581f.

[38] Mary Jo Nye, "The Nineteenth-Century Atomic Debates and the Dilemma of the 'Indifferent Hypothesis,'" *Studies in the History and Philosophy of Science*, 7 (1976), 245–68.

Soddy (1877–1956) and Kasimir Fajans (1887–1975) introduced the notion of isotopy.

It now became clear that the chemical identity of a specific element cannot be correctly deduced from its weight. But this was exactly what Dalton and all chemists of the nineteenth century had believed, regardless of whether they were atomist or antiatomist, regardless of whether they relied on atomic weights or on equivalent weights. It turned out that not only the classification of elements in the periodic system, but, to a greater or lesser extent, *all* chemical classifications rested on a false assumption. But all those classifications also rested on good luck, as atomic numbers and atomic weights usually correlate.

TWO TYPES OF BONDS

By the end of the century, not only the "dark forest" of organic chemistry but also that of inorganic chemistry had been transformed into park landscapes. A ditch that seemed to separate these parks was bridged, too. In 1916 Gilbert Newton Lewis (1875–1946) and, independently, Walter Kossel (1888–1956) stated that bonding results either from electron transfer (electrovalency) or electron sharing (covalency). As intermediate states between pure electron transfer and pure electron sharing may occur, one cannot say that in respect to valency, inorganic chemistry, where electrovalency is predominant, is totally different from organic chemistry, where covalency is predominant.

With this completely new concept of the interrelation of the structure of the atom and its chemical behavior, chemical atomism and physical atomism, which for an entire century had developed along different paths, merged again in the concept of a complex atom as a focal point of the quantum physics and quantum chemistry of the twentieth century.

13

THE THEORY OF CHEMICAL STRUCTURE AND ITS APPLICATIONS

Alan J. Rocke

The theory of chemical structure was developed in the 1850s and 1860s, a product of the efforts of a number of leading European chemists.[1] By the late 1860s it was regarded as a mature and powerful conceptual scheme that not only gave important insight into the details of molecular architecture in an invisibly small realm of nature, but also furnished heuristic guidance in the technological manipulation of those molecules, providing assistance in the creation of an important fine chemicals industry. The theory continued to develop in its power and subtlety throughout the following decades, until by the end of the century, it was by all measures the reigning doctrine of the science of chemistry, dominating investigations in both academic and industrial laboratories. Consequently, the story of the rise of this theory is an important component of the history of basic science, and also of the manner in which scientific ideas are applied to industry.

EARLY STRUCTURALIST NOTIONS

Speculations concerning geometrical groupings of the imperceptible particles that make up sensible bodies go back to the pre-Socratics. However, for our purposes, it is expedient to begin the story with the rise of chemical atomism, since structural ideas presuppose atoms in the modern chemical (post-Lavoisien) sense. The founder of the chemical atomic theory was John Dalton (1766–1844), and it is suggestive that immediately following the proposal of chemical atoms, Dalton and others began to speculate how they

---

[1] Overviews of the rise of the theory of chemical structure are provided by G. V. Bykov, *Istoriia klassicheskoi teorii khimicheskogo stroeniia* (Moscow, Akademiia Nauk, 1960); O. T. Benfey, *From Vital Force to Structural Formulas* (Boston: Houghton Mifflin, 1964); J. R. Partington, *A History of Chemistry*, vol. 4 (London: Macmillan, 1964); C. A. Russell, *The History of Valency* (Leicester, England: Leicester University Press, 1971); and A. J. Rocke, *The Quiet Revolution: Hermann Kolbe and the Science of Organic Chemistry* (Berkeley: University of California Press, 1993).

might be arranged into molecules (often then called "compound atoms"). As early as 1808 – about the time Dalton's ideas first began to be known in the chemical community – William Wollaston was "inclined to think... that we shall be obliged to acquire a geometric conception of [the] relative arrangement [of the elementary atoms] in all the three dimensions of solid extension." Four years later, Humphry Davy made a similar suggestion.[2]

These were mere conjectures. What made structuralist hypotheses more compelling was the discovery of isomerism and similar phenomena, such as allotropy and polymorphism, very early in the history of the atomic theory. One could imagine, for instance, that the various species of sugar, all having the same elemental composition, differed in their properties because of differences in the arrangement of the like numbers and kinds of atoms of which their molecules were composed. In 1815, the Swedish chemist Jacob Berzelius (1779–1848) wrote:

> We may then form the idea that the organic atoms [i.e., molecules] have a certain mechanical structure.... It is only by such a structure that we can explain the different products... composed of the same elements, and in proportions (stated in per cents) but little different from each other. I am persuaded that an attempt to study the probabilities of the construction of organic atoms... would be of the greatest importance, and might be even capable of correcting analysis.[3]

Indeed, a number of examples of what was later called isomerism were discovered in the 1810s, and this was just the beginning. In 1826 J. L. Gay-Lussac (1778–1850) distinguished racemic acid from the identically constituted tartaric acid; the same year, Gay-Lussac's German protégé, the Giessen chemist Justus Liebig (1803–1873), confirmed that fulminic and cyanic acids shared the same empirical formula; and two years later, Liebig's new friend Friedrich Wöhler (1800–1882) demonstrated that urea had the same composition as ammonium cyanate. In 1830 Berzelius discussed the now-well-established phenomenon of compounds having identical composition but differing properties, coined the term "isomerism," and suggested a structuralist cause for the phenomenon.[4]

---

[2] Wollaston, "On Super-Acid and Sub-Acid Salts," *Philosophical Transactions of the Royal Society*, 98 (1808), 96–102, p. 101; Davy, *Elements of Chemical Philosophy* (London, 1812), pp. 181–2, 488–9; W. V. Farrar, "Dalton and Structural Chemistry," in D. Cardwell, ed., *John Dalton and the Progress of Science* (Manchester: Manchester University Press, 1968), pp. 290–9.

[3] Berzelius, "Experiments to Determine the Definite Proportions in Which the Elements of Organic Nature are Combined," *Annals of Philosophy*, 5 (1815), 260–75, p. 274. However, Berzelius also toyed with an electrochemical explanation for isomerism: See Farrar, "Dalton," and especially John Brooke, "Berzelius, the Dualistic Hypothesis, and the Rise of Organic Chemistry," in *Enlightenment Science in the Romantic Era*, ed. E. Melhado and T. Frängsmyr (Cambridge: Cambridge University Press, 1992), pp. 180–221.

[4] Berzelius, "Ueber die Zusammensetzung der Weinsäure und Traubensäure," *Annalen der Physik*, 2d ser., 19 (1830), 305–35. On the early history of isomerism, see John Brooke, "Wöhler's Urea, and its Vital Force? – A Verdict from the Chemists," *Ambix*, 15 (1968), 84–114, and A. J. Rocke, *Chemical Atomism in the Nineteenth Century* (Columbus: Ohio State University Press, 1984), pp. 167–74.

ELECTROCHEMICAL DUALISM AND ORGANIC RADICALS

At this time, Berzelius had been enjoying a nearly twenty-year-long reign as the leading theorist of organic chemistry. He was strongly disposed toward electrochemical models, for electrolysis had been the key to many of his chemical investigations, including some of his earliest. It was logical to suppose that the chemical constituents that migrate to the two poles of an electrolytic cell did so because they possessed coulombic charges opposite of those of the respective electrodes; reasonable, as well, to believe that these components possessed those charges *before* electrolysis, that is to say, in the stable molecule before it was torn apart by electricity; and, finally, sensible to conclude that these opposite polarities in the parts of the molecule were the *cause* of cohesion (i.e., stability) of the molecule as a whole. The resulting theory of electrochemical dualism worked exceedingly well in the inorganic realm, and there seemed to be every reason to adopt a similar approach for organic compounds, since organic salts also underwent electrolysis.

Berzelius further argued that electrochemical dualism offered a ready accounting of isomerism in organic chemistry.[5] After all, compounds that have identical overall formulas may have differing *proximate* components; the formula A_4B_4, for instance, could characterize either of two different substances having the more highly resolved formulas $A_2B_2 \cdot A_2B_2$ and $AB_3 \cdot A_3B$, respectively. Berzelius thus distinguished between "empirical" and "rational" (i.e., theoretical) formulas, the former simply summarizing empirical elemental analysis, the latter reproducing the chemist's ideas about how the atoms are grouped within the molecule. Even in the absence of isomerism, rational formulas were interesting for the details they revealed. One could inquire, for example, whether pure grain alcohol, whose Berzelian empirical formula was (and is for modern chemists as well) C_2H_6O, should best be represented as $C_2H_4 \cdot H_2O$, or $C_2H_6 \cdot O$, or $C_2H_5O \cdot H$, or $C_2H_5 \cdot OH$, or by some other pattern.[6] Such rational formulas could be inferred from chemical reactions. For instance, the first of these more resolved formulas appeared to be supported by the fact that one could dehydrate alcohol; the last seemed justified by the fact that alcohol, in condensing with any acid, contributes the elements of ethyl to the resulting ester.

Such considerations gave immediate impetus to a program of elucidating the "constitutions" (rational construction, in the Berzelian sense) of organic compounds, which several leading chemists began to pursue in the early

---

[5] On Berzelius and organic chemistry, see Melhado and Frängsmyr, *Enlightenment Science*; Melhado, *Jacob Berzelius: The Emergence of His Chemical System* (Madison: University of Wisconsin Press, 1981); and essays V, VI, VII, and VIII in John Brooke, *Thinking about Matter: Studies in the History of Chemical Philosophy* (Brookfield, Vt.: Ashgate, 1995).

[6] Berzelius, "Zusammensetzung"; see also *Jahresbericht über die Fortschritte der physischen Wissenschaften*, 11 (1832), 44–8; 12 (1833), 63–4; and 13 (1834), 186–8. Berzelian formula conventions differed slightly, but unsubstantially, from the modernized formulas reproduced here, and the words also carry slightly different meanings.

1830s. In a classic collaborative work of 1832, Liebig and Wöhler found oil of bitter almonds to be the hydride of an entity of the composition $C_{14}H_{10}O_2$ (expressed in a "four-volume" formula, as was then customary). They called this the benzoyl "radical," because they found that it could enter unaltered into the composition of a wide variety of substances (including benzoin and benzoic acid, whence the name).[7] The following years brought ethyl, methyl, acetyl, cacodyl, and other organic radicals to the fore. Radicals were supposed to be integral electropositive pieces of organic molecules that operated constitutively as elements did in the inorganic realm. The electrochemical-dualist-radical program of investigating the constitutions of organic molecules, pursued by such workers as Berzelius, Liebig, Wöhler, and Robert Bunsen, was potentially very powerful, and was regarded in the mid-1830s as promising indeed.

However, this program never fulfilled its promise, even in those early optimistic days. Three problems conspired against it, from the beginning. One was a continuing uncertainty over which atomic groupings to count as a "radical." Liebig's prominent French rival, Jean-Baptiste Dumas (1800–1884), for instance, argued that the ready dehydration of alcohol to ethylene indicated that the latter (or "etherin," as Berzelius called it) must be taken to be the constituent radical of alcohol, rather than Liebig's ethyl. A second problem was the continuing uncertainty over what standards to use for atomic weights and molecular formulas. It was difficult to reason about molecular constituents when agreement could not be reached over how to represent the entities that one was manipulating: Did alcohol have nine atoms per molecule, as Berzelius believed, or eighteen, according to Liebig, or twenty-two, as Dumas thought? One could respond, with justice, that such formula variations had only to do with notational conventions, not substantive distinctions; all agreed on the elemental composition of alcohol, disagreeing only on the atomic weights being used, and each man's notions could readily be translated into any of the others. However, more substantively and more fatally, there was also disagreement over molecular magnitudes. For instance, Berzelius thought that ether was a doubled alcohol molecule, less H_2O, whereas Liebig and Dumas both considered ether to be produced by simple abstraction of water from alcohol.[8]

The third problem was a result of the discovery that chlorine could substitute for the hydrogen of organic compounds. As far back as 1815, Gay-Lussac had shown that chlorine could replace the hydrogen of prussic (hydrocyanic) acid, commenting that "it is quite remarkable that two bodies with such

---

[7] Wöhler and Liebig, "Untersuchungen über das Radikal der Benzoesäure," *Annalen der Pharmacie*, 3 (1832), 249–87.

[8] Berzelius's formula corresponds to the modern one; Liebig, preferring "four-volume" organic formulas, used twice the number of atoms as Berzelius (or, halved atomic weights); Dumas preferred four-volume formulas but used an atomic weight for carbon that was half that preferred by Berzelius and Liebig. For details, see, for example Partington, *History*, chaps. 8–10, or Rocke, *Chemical Atomism*, chap. 6.

different properties play the same role in combining with cyanogen."⁹ Remarkable indeed, because electrochemistry put chlorine and hydrogen at opposite ends of the electronegativity scale. Five years later, Michael Faraday discovered that chlorine could replace the hydrogen of "Dutch oil" (ethylene chloride), and in the late 1820s, both Gay-Lussac and Dumas found that oils and waxes could be similarly chlorinated. In the 1830s and 1840s, chlorine became the organic chemist's reagent par excellence, especially in France. The very existence of chlorinated organic materials was anomalous for electrochemical dualism, for the modified substances were usually little altered in their properties. Highly electronegative chlorine truly appeared to be playing the same chemical role as highly electropositive hydrogen; electrochemical composition no longer provided a reliable predictor of chemical properties.¹⁰

This development turns out to be connected historically with the replacement of Lavoisier's oxygen theory of acidity by a novel hydrogen theory of acids. Gay-Lussac's chlorination of prussic acid, a hydracid, found a parallel in his German student's later work on benzoyl. Like Gay-Lussac's cyanogen, Liebig's and Wöhler's benzoyl radical could combine indifferently with hydrogen (to form benzaldehyde) or chlorine (to form benzoyl chloride). Six years later (1838), Liebig developed these ideas into a thoroughgoing theory of hydrogen-acids, which had much in common with emerging French antidualist chemical theories. Liebig posited that acids were not oxygenated radicals but, rather, substances with replaceable hydrogen.¹¹

THEORIES OF CHEMICAL TYPES

The phenomenon of chlorine substitution, reinforced by an incipient hydracid theory that postulated substitution of the hydrogen of acids by metals, worked against the electrochemical model in general, and cast doubt on Berzelius's explanation for isomerism. Perhaps, some chemists began to think, the properties of substances depended far more on the physical arrangements of atoms within molecules than on the electrochemical character of either atoms or radicals. Inspired by the work of Gay-Lussac and especially Dumas, in the mid-1830s the young Auguste Laurent (1807–1853) developed a theory of "derived radicals," later renamed the "nucleus" theory. Laurent depicted the chemical molecule as a small crystal, where the most important factor influencing the properties of the compound was not the identities of the

---

[9] Gay-Lussac, "Recherches sur l'acide prussique," *Annales de chimie*, 95 (1815), 136–231, at pp. 155, 210. Translated by the author unless otherwise noted.
[10] Recent studies concerning the history of organic radicals and chlorine substitution include, from a cognitive perspective, John Brooke's articles cited in notes 3 and 4, and Ursula Klein, *Nineteenth-Century Chemistry: Its Experiments, Paper-Tools, and Epistemological Characteristics* (Berlin: Max-Planck-Institut für Wissenschaftsgeschichte, 1997); from a more rhetorical and sociological angle, see Mi Gyung Kim, "The Layers of Chemical Language II," *History of Science*, 30 (1992), 397–437; and Kim, "Constructing Symbolic Spaces," *Ambix*, 43 (1996), 1–31.
[11] Liebig, "Ueber die Constitution der organischen Säuren," *Annalen der Pharmacie*, 26 (1838), 113–89.

atoms but their position in the array. Laurent derived these ideas not only from organic chemistry but also from crystallography.[12]

Dumas was at first opposed to what his former student was suggesting. However, when he discovered in 1838 that fully chlorinated acetic acid still possessed all the essential properties of the unchlorinated substance, he too abandoned electrochemical dualism and sought a more holistic and unitary viewpoint, based on substitution. According to Dumas's new "type" theory – the term apparently borrowed from Georges Cuvier's biological notions – organic compounds that are closely interrelated by chemical reactions must all be considered to be based on a single "type" formed from the same number of atoms combined in the same way. As long as the arrangement is conserved, substitution of one atom by another, be their electrochemical properties ever so distinct, does not alter the type, hence, does not alter the essential properties of the substance.[13]

Dumas and Liebig, youthful leaders of the chemical communities in their respective countries, vacillated between close collaboration and intense rivalry. Much of the scientific work described here was made possible only by Liebig's novel modification (1831) of Gay-Lussac's and Berzelius's method for analyzing the carbon and hydrogen content of organic compounds, an innovation that made the process at once fast, simple, and precise; Dumas and everyone else adopted it nearly immediately. Dumas, for his part, devised methods for determining vapor densities (1826) and organic nitrogen (1833) that were nearly as influential. Dumas also attempted to reproduce, in Paris, essential aspects of Liebig's extraordinarily successful method of organizing scientific research and pedagogy – routine laboratory instruction combined with group research – but here he had less success.[14]

Liebig actively participated in the research leading to type theories and, like Dumas, drifted considerably from the Berzelian dualist-radical orthodoxy. However, Liebig grew frustrated and ultimately repelled by the constant theoretical shifts, and by the distressingly contentious disputes. In 1840 he resolved to leave the field of theory to pursue applied chemistry. Dumas underwent a similar epiphany, about the same time. Indeed, it would seem that just at this time, there was a European-wide shift to a more positivistic stance toward questions of atoms, molecules, radicals, and structuralist hypotheses. Laurent was one of the few who resisted this trend.[15] In his aversion to dualistic radicals he was joined by a fellow rebel, Charles Gerhardt

---

[12] On Laurent and the crystallographic traditions from which he borrowed, see S. Kapoor, "The Origin of Laurent's Organic Classification," *Isis*, 60 (1969), 477–527, and especially Seymour Mauskopf, *Crystals and Compounds: Molecular Structure and Composition in Nineteenth Century French Science* (Philadelphia: American Philosophical Society, 1976).

[13] J. B. Dumas, "Mémoire sur la loi des substitutions et la théorie des types," *Comptes Rendus*, 10 (1840), 149–78; S. Kapoor, "Dumas and Organic Classification," *Ambix*, 16 (1969), 1–65.

[14] Leo Klosterman, "A Research School of Chemistry in the Nineteenth Century: Jean-Baptiste Dumas and His Research Students," *Annals of Science*, 42 (1985), 1–80.

[15] Marya Novitsky, *Auguste Laurent and the Prehistory of Valence* (Chur, Switzerland: Harwood, 1992); Clara deMilt, "Auguste Laurent, Founder of Modern Organic Chemistry," *Chymia*, 4 (1953), 85–114.

(1816–1856) – though Gerhardt consistently denied the epistemological accessibility of atomic arrangements. Laurent and Gerhardt were both brilliant chemists, but they did not know how (or refused) to play careerist games, fought with Dumas and the other Parisian leaders, and were given positions only in the provinces (Laurent at Bordeaux, and Gerhardt at Montpellier).[16] Meanwhile, there were still signs of life in dualistic organic chemistry. From the mid-1840s, Edward Frankland (1825–1899) in England and Hermann Kolbe (1818–1884) in Marburg and Braunschweig – students of Liebig, Wöhler, and Bunsen – "stalked" the organic radicals, and had considerable success, as they thought, in isolating several of them.[17]

Laurent and Gerhardt, however, interpreted the Frankland-Kolbe reactions not as extractions of radicals but, rather, as substitution reactions, and the putative isolated radicals as dimers. For instance, what Kolbe regarded as the splitting off and isolation of "methyl" by electrolysis of acetic acid, Laurent and Gerhardt interpreted as a replacement of carboxyl by a second methyl radical (in situ and in the nascent state), to form dimethyl (ethane). Once again, the crucial issue was that of molecular magnitudes, for what was ultimately in dispute was the molecular size of the products vis-à-vis that of the reactants. In the late 1840s, neither side had conclusive evidence for its point of view; both camps were arguing on such criteria as coherence and analogy.

This situation changed suddenly in 1850. Alexander Williamson (1824–1904), recently installed at University College London, announced an elegant new synthesis for ether; this reaction allowed the chemist not only to make conventional ether but also to select the two principal pieces of the product molecule in advance and then join them together, to design new ethers at will.[18] The reaction provided the key to resolving the disputes over molecular magnitudes. Williamson created a novel asymmetric ether (one in which the two radicals were not the same) that was consistent only with the larger formula for ether – Berzelius's old formula, later championed by Laurent. The smaller formula preferred by Liebig and Dumas would have required the product of the reaction to have been a mixture of two different symmetrical ethers. Williamson, who had studied in Paris in the late 1840s and had been converted to Laurent's and Gerhardt's views, had succeeded in finding important evidence, by purely chemical means, to support his elder French friends' theories.[19]

The impact of this work was profound. Williamson's "asymmetric synthesis argument" was applied to different molecular systems several times in the next

---

[16] E. Grimaux and C. Gerhardt, Jr., *Charles Gerhardt: Sa vie, son oeuvre, sa correspondance* (Paris: Masson, 1900).

[17] Russell, *History of Valency*; Rocke, *Quiet Revolution*. On the conflict between dualism and types, see J. Brooke, "Laurent, Gerhardt, and the Philosophy of Chemistry," *Historical Studies in the Physical Sciences*, 6 (1975), 405–29.

[18] Williamson, "Theory of Etherification," *Philosophical Magazine,* 3d ser., 37 (1850), 350–6.

[19] J. Harris and W. Brock, "From Giessen to Gower Street: Towards a Biography of Alexander Williamson," *Annals of Science,* 31 (1974), 95–130.

five years: repeatedly by Williamson himself to various molecular systems, to the organic acid anhydrides by Gerhardt, and to the organic radicals themselves by Adolphe Wurtz (1817–1884). The entire chemical world saw the justice of the argument, and Laurent's and Gerhardt's views finally began to prevail. (Tragically, both died young, just at this time – Laurent in 1853 and Gerhardt three years later.)

Connected with this development was the rise of a new sort of type theory. Wurtz and A. W. Hofmann (1818–1892) – a German chemist then resident in London – explored novel organic bases in the years 1849 to 1851, which suggested Laurent/Gerhardt-style replacements of the hydrogen in ammonia with organic radicals to form primary, secondary, and tertiary amines. Williamson's nearly simultaneous ether work suggested similar substitutions of the two hydrogens of water. Organic compounds began to be interpreted ever more generally in the 1850s as schematically produced by substitutions of the hydrogen of simple inorganic compounds with organic radicals. Thus was born the "newer type theory," pursued especially by Gerhardt and members of his camp. This theory of types led toward an emerging theory of valence and structure.

THE EMERGENCE OF VALENCE AND STRUCTURE

As a result of his pathbreaking work with novel organometallic compounds at the end of the 1840s, Frankland began first to accommodate to, then to adopt, the new type-theoretical viewpoint being advocated by Gerhardt, Wurtz, Hofmann, and Williamson. In a classic paper of 1852, Frankland argued from the reactions of organometallic substances that metal atoms have a maximum combining capacity with other atoms or radicals, and he specified these limits by many examples.[20] This was the first explicit statement of the phenomenon of valence. Others were making similar suggestions. In a paper of 1851, for instance, Williamson intimated that the oxygen atom provided a material connection to exactly two other atoms or groups, providing thereby "an actual image of what we rationally suppose to be the arrangement of constituent atoms" in compounds of oxygen.[21]

Influenced by Williamson, the youthful August Kekulé (1829–1896) stated in 1854 that it was "an actual fact," not merely notational convention, that sulfur and oxygen were both "dibasic," that is, equivalent to two atoms of hydrogen. The same year, an associate of Williamson named William Odling explored the "replaceable, or representative, or substitution value"

---

[20] Frankland, "On a New Series of Organic Bodies Containing Metals," *Philosophical Transactions of the Royal Society*, 142 (1852), 417–44; Colin Russell, *Edward Frankland: Chemistry, Controversy, and Conspiracy in Victorian London* (Cambridge: Cambridge University Press, 1996).

[21] Williamson, "On the Constitution of Salts," *Chemical Gazette*, 9 (1851), 334–9; see Harris and Brock, "Giessen to Gower Street."

of atoms of a variety of metallic and nonmetallic elements. As Odling implied, valence intrinsically promoted types and weakened dualism, since the constancy of valence suggested that substitutions could occur independently of electrochemical properties. Wurtz proclaimed the "tribasic" character of nitrogen in 1855. Even the most consistent opponent of substitutionist type theory, Hermann Kolbe, developed in the late 1850s (under the influence of his friend Frankland and partly collaboratively) a type-theoretical schematization of all organic compounds as derived from substitution in carbon dioxide.[22]

The phenomenon of valence gave insight into certain proximate structural details of molecules, and by the late 1850s, chemists' success in investigating this subject – and the unanimity regarding that success – may have helped to lessen the antistructuralist positivism so characteristic of the preceding twenty years. However, until 1857, valence ideas had not yet been systematically applied to carbon, and details regarding the atomic arrangements within hydrocarbon radicals were still nearly completely inaccessible. Attention was moving in that direction, however, as work published in the mid-1850s by Odling, Wurtz, Kolbe, and Frankland all demonstrate.

Under the probable proximate influence of a paper by Wurtz published in 1855, Kekulé achieved an important breakthrough, enunciating the essentials of the theory of chemical structure in two papers published in the autumn of 1857 and the spring of 1858, respectively.[23] (Less than a month after Kekulé's second paper appeared, A. S. Couper's largely equivalent and entirely independent structure theory was published; compared to Kekulé's work it was not influential, and Couper himself, a Scottish chemist who had studied with Wurtz, vanished shortly thereafter.) In the second article, Kekulé proclaimed it possible to "go back to the elements themselves," that is, to resolve organic molecules all the way down to their individual atoms, and to show how each of those atoms is connected one to another. To do this, it was necessary to conceive of carbon as a "tetratomic" (tetravalent) element, to regard carbon atoms as capable of using valences to bond to *one* another, and consequently to depict organic compounds as composed of "skeletons" in which the backbone was a "chain" of carbon atoms. Heteroatoms, such as oxygen and nitrogen, served linking functions in alcohols, acids, amines, and so on, and hydrogen atoms filled in all the unused atomic valences.[24]

[22] Kekulé, "On a New Series of Sulphuretted Acids," *Proceedings of the Royal Society,* 7 (1854), 37–40; Odling, "On the Constitution of Acids and Salts," *Journal of the Chemical Society,* 7 (1854), 1–22; Wurtz, "Théorie des combinations glycériques," *Annales de Chimie,* 3d ser., 43 (1855), 492–6; Kolbe, "Ueber den natürlichen Zusammenhang der organischen mit den unorganischen Verbindungen," *Annalen der Chemie,* 113 (1860), 292–332. For historical discussions, see Partington, *History*; Russell, *History of Valency*; and Rocke, *Quiet Revolution.*
[23] The influence of Wurtz on Kekulé is asserted in my article "Subatomic Speculations and the Origin of Structure Theory," *Ambix,* 30 (1983), 1–18.
[24] Kekulé, "Ueber die Constitution und die Metamorphosen der chemischen Verbindungen und über die chemische Natur des Kohlenstoffs," *Annalen der Chemie,* 106 (1858), 129–59; Couper, "Sur une nouvelle théorie chimique," *Comptes rendus,* 46 (1858), 1157–60. On Kekulé, the work by his student Richard Anschütz has never really been superseded: *August Kekulé,* 2 vols. (Berlin: Verlag Chemie,

A year later (June 1859), Kekulé published the first portion of his soon-famous *Lehrbuch der organischen Chemie*, containing a short history of organic-chemical theory over the previous thirty years and a revised version of his structure theory articles.[25] This textbook served as a highly effective means of propagating the new ideas. Many leading theorists of the period – Frankland, Williamson, Hofmann, Wurtz, the British Alexander Crum Brown and Henry Roscoe, the Germans Emil Erlenmeyer and Adolf Baeyer, and many others – were profoundly influenced by it; the older generation – Liebig, Wöhler, Bunsen, and Dumas – paid little attention, as they had paid little attention to all structuralist theories for twenty years or more. Indeed, it is a remarkable circumstance that nearly all active organic chemists who were forty years of age or younger in 1858 became structural chemists soon thereafter, whereas all chemists older than forty virtually ignored the theory.

One of the most avid apostles of the new theory in its early years was the Russian chemist Aleksandr Mikhailovich Butlerov (1828–1886). A mature chemist when the possibility of foreign travel for Russian scientists first arose, Butlerov spent 1857–8 in Western Europe, including two visits with Kekulé in Heidelberg and several months in Wurtz's Paris laboratory as a bench mate to Couper. Influenced by the thinking of both Couper and Kekulé, Butlerov became one of the earliest and finest synthetic structural chemists. On a second trip to the West in 1861, Butlerov delivered an important paper, "On the Chemical Structure of Compounds," at the *Naturforscherversammlung* (Congress of German Scientists and Physicians) in Speyer, in which he urged his colleagues to apply the new ideas more consistently, to adopt his coinage "chemical structure," and to eliminate remaining vestiges of Gerhardt's type theory from the new doctrines. Butlerov later complained – with justice – that some of his ideas were not sufficiently appreciated in Western Europe. Soviet historians in the Stalin and Khrushchev periods, along with a few Westerners, have argued that the theory of chemical structure was first stated by Butlerov in 1861, but this position has since been challenged.[26]

Enough has been said here to confirm that the emergence of the structure theory was complex; it occurred in several stages, and many chemists played essential roles in the story, including Berzelius, Liebig, Dumas, Gerhardt, Laurent, Frankland, Kolbe, Williamson, Odling, Wurtz, Butlerov, Couper,

1929); see also O. T. Benfey, ed., *Kekulé Centennial* (Washington, D.C.: American Chemical Society, 1966), and my *Quiet Revolution*.

[25] Kekulé, *Lehrbuch der organischen Chemie*, 2 vols. (Erlangen: Enke, 1861–6). Despite the title page imprint on the first volume, the first fascicle of that volume (pp. 1–240) was published in June 1859.

[26] A. J. Rocke, "Kekulé, Butlerov, and the Historiography of the Theory of Chemical Structure," *British Journal for the History of Science*, 14 (1981), 27–57; a perceptive and helpful response by G. V. Bykov is "K istoriografii teorii khimicheskogo stroeniia," *Voprosy istorii estestvoznaniia i tekhniki*, 4 (1982), 121–30, which was the last article published by this fine historian before his death. See also Nathan Brooks, "Alexander Butlerov and the Professionalization of Science in Russia," *Russian Review*, 57 (1998), 10–24.

and Kekulé. For this reason, priority issues in this matter have been contentious and difficult to resolve, from their day to ours. I would argue, however, that the crucial postulate was stated clearly first by Kekulé in May 1858: the self-linking of carbon atoms. This concept was difficult for many chemists to accept. Neither of the two available macroscopic physical models, coulombic or gravitational attraction, appeared to be a reasonable way of visualizing the phenomenon, and chemists were reduced either to arrant speculation or to positivism as to the cause of these interatomic attractions.

However, it would appear that physics did provide an important impetus for structure theory in another manner. The kinetic theory of gases was being formulated simultaneously with the structure theory, and it provided support for Amedeo Avogadro's gas hypotheses. Avogadro had posited elemental molecules consisting of two or more identical atoms, and so his posthumous victory among the physicists (ca. 1856 to 1859) provided a confirmed precedent that must have made the notion of carbon–carbon combinations more attractive among chemists. By the time of the international chemical Karlsruhe Conference of 1860 – a brainchild of Kekulé, Wurtz, and Karl Weltzien – kinetic theory was making a nice package with the reformed (Gerhardtian) atomic weights and molecular formulas, *and* the new theory of structure. Although the results of the conference were somewhat unsatisfying to the reformers, their success was fuller than it may have appeared at the time.

FURTHER DEVELOPMENT OF STRUCTURAL IDEAS

Despite optimism among some reformers, structure theory got off to a slow start. Even for those who accepted the basic principles, there were innumerable questions of detail and of method to sort through. Was valence necessarily constant? If so, how could one account for the structures of olefins and other "unsaturated" organic compounds? Why were certain predicted compounds (such as methylene oxide) never found? Were the four valences of carbon chemically equivalent? How were they arrayed spatially? Could one even think of investigating the actual spatial arrangements indicated by organic structures? What guidelines could one establish for inferring structural details from chemical reactions? And so on.[27]

As far as olefins were concerned, a number of structuralist notions were explored in the early 1860s by Kekulé, Erlenmeyer, Butlerov, Wurtz, and Crum Brown, among others.[28] By the middle of the decade, a tentative consensus was forming that doubled bonds between carbon atoms provided the best explanation for the apparently reduced total valence of the compounds; the

---

[27] Russell's *History of Valency* provides an excellent guide to these later developments.
[28] A. A. Baker, *Unsaturation in Organic Chemistry* (Boston: Houghton Mifflin, 1968); Russell, *History of Valency*.

high reactivity of the double bonds suggested that the hydrogen-"saturated" state was preferred. Additional empirical experience suggested about this time that the four valences (at that time usually called "affinity units" or "affinities") of carbon were all chemically equivalent. The nature of certain functional groups, such as carboxyl, ester, and hydroxy, became increasingly clear. Wurtz, Marcellin Berthelot, and others fruitfully explored polyfunctional organic compounds. There were, of course, plenty of puzzles remaining, among them "absolute isomerism," which was defined as any isomerism that *could not* be explained by current structure-theoretical ideas.

The saga of structure theory in the 1860s is epitomized by August Kekulé's theory of the benzene molecule.[29] The mythic status of this event was not created by any difficulty in arriving at candidate structures for the molecule whose empirical formula is C_6H_6 (for others had already suggested possible structures), nor in the circumstance that the hexagonal "ring" structure he proposed is substantially identical to what we accept today. Rather, it was the challenge of arriving at a structure that could legitimately be defended from empirical evidence that was then available, and that could guide future work. Empirical experience with aromatic substances was sparse at the time of the first formulation of structure theory, and in the 1850s, Kekulé and most other chemists avoided the question of how the benzene molecule or the phenyl radical was constituted. By early 1864, the time when Kekulé later stated that he privately formulated the benzene ring hypothesis, the field was sufficiently matured. For instance, it was clear by that year (and not much earlier) that the minimum number of carbon atoms in aromatic substances was six, and that substitution and not addition could occur in the benzene nucleus. It was also becoming clear by that year (and not earlier) that every aromatic formula produced by substitution of *one* radical for a hydrogen atom of benzene had only *one* isomer, but substitution of *two* radicals for hydrogen of benzene resulted in exactly *three* isomeric variations, no more and no less. How could one explain these puzzling facts?

In a short paper published in French in January 1865, Kekulé, who was a professor in Ghent at that time, posited a closed chain of six carbon atoms for benzene, with alternating single and double bonds. In a more detailed German article and in the sixth fascicle of his textbook (both published in 1866), he provided many more details, including a full theoretical justification for the isomer numbers that had been empirically noted.[30] From the start, Kekulé's benzene theory was extraordinarily successful, as measured

---

[29] The following discussion is taken from my articles "Hypothesis and Experiment in the Early Development of Kekulé's Benzene Theory," *Annals of Science*, 42 (1985), 355–81, and "Kekulé's Benzene Theory and the Appraisal of Scientific Theories," in *Scrutinizing Science: Empirical Studies of Scientific Change*, ed. A. Donovan, L. Laudan, and R. Laudan (Boston: Kluwer, 1988), pp. 145–61; some material also comes from my *Quiet Revolution*, chap. 12.

[30] Kekulé, "Sur la constitution des substances aromatiques," *Bulletin de la Société Chimique*, 2d ser., 3 (1865), 98–110; "Untersuchungen über aromatische Verbindungen," *Annalen der Chemie*, 137 (1866), 129–96; *Lehrbuch der organischen Chemie*, vol. 2 (1866), pp. 493–744.

by acceptance in the community, by its demonstrated scientific power, and by its technological applications. As early as April 1865, Kekulé wrote his former student Baeyer: "[My] plans are unlimited, for the aromatic theory is an inexhaustible treasure-trove. Now when German youths need dissertation topics, they will find plenty of them here."[31]

A sage prediction. Throughout the 1860s and 1870s, European academic chemistry expanded at an astounding rate. The country where the growth was most explosive was Germany, and the field of growth within chemistry was organic — especially that of aromatic derivatives. Huge new academic laboratories were built throughout the Germanic lands, competition heated up for the top professorial stars, students flooded to the universities and Technische Hochschulen, and even the job market for graduates expanded greatly. There can be little doubt that a principal reason for this growth was the extraordinary intellectual power of structure theory; in any case, the correlation holds, for the country where structure theory most flourished was Germany. Liebig's prescription of routine laboratory education allied with group research had paid off, especially when the subject matter was suitable to the pedagogical and research style.[32]

In Paris, Adolphe Wurtz led an extremely successful group in structural organic chemistry, but on the whole, the theory failed to flourish in France until close to the end of the century, because of a combination of political and intellectual factors that need further study. Given these circumstances, it is not surprising that organic chemistry as a whole stagnated in France.[33] Edward Frankland, Faraday's successor at the Royal Institution and Hofmann's at the Royal College of Chemistry, led the most significant structural chemical laboratory in Britain. Other important British structuralists included Crum Brown in Edinburgh (to whom we owe the sort of letter-and-dash structural formulas to which chemists quickly became accustomed); Henry Roscoe at Manchester; and slightly later, such figures as Henry Armstrong and W. H. Perkin, Jr. In Russia, Butlerov built an excellent school of structural chemistry in Kazan and then St. Petersburg.

Two examples of the sorts of projects typical of structural "organikers" were studies of positional isomerism in the aromatic series, and stereochemistry. If Kekulé's theory were right, there ought to be three series of diderivatives of benzene; but which series represented the 1,2-, which the 1,3-, and which the 1,4- compounds? The first tentative efforts toward the determination of positional isomers in the aromatic realm were made by Wilhelm Körner, who

---

[31] Kekulé to Baeyer, 10 April 1865, August-Kekulé-Sammlung, Institut für Organische Chemie, Technische Hochschule, Darmstadt; cited in Rocke, "Hypothesis," p. 370.

[32] Jeffrey Johnson, "Academic Chemistry in Imperial Germany," *Isis*, 76 (1985), 500–24, and "Hierarchy and Creativity in Chemistry, 1871–1914," *Osiris*, 2d ser., 5 (1989), 214–40; Frederick L. Holmes, "The Complementarity of Teaching and Research in Liebig's Laboratory," *Osiris*, 2d ser., 5 (1989), 121–64.

[33] Robert Fox, "Scientific Enterprise and the Patronage of Research in France," *Minerva*, 11 (1973), 442–73; Mary J. Nye, "Berthelot's Anti-Atomism: A 'Matter of Taste'?" *Annals of Science*, 31 (1981), 585–90, and *Science in the Provinces* (Berkeley: University of California Press, 1986).

worked directly with Kekulé from 1864 to 1867, and by Baeyer, in Berlin. Further steps were taken in Baeyer's lab by Carl Graebe, who used several different approaches to attempt structural assignments. However, the problem was an extraordinarily difficult one. This chaotic situation was clarified by a classic investigation by the young and brilliant Victor Meyer, who devised new reactions that related all the known diderivatives to the one case in which the positional determination was secure, that of the three isomeric dicarboxylic acids. In 1874 Körner then devised a method that could be applied for the general case. The "absolute isomerism" of aromatic positional isomers had thus been subsumed under classical structural theory.[34]

Stereochemistry owed its origin to a young Dutch chemist named J. H. van't Hoff, who studied with both Wurtz and Kekulé, and independently to the Frenchman J. A. LeBel, a student of Wurtz. Van't Hoff outlined the theory in a twelve-page pamphlet in 1874, publishing a more detailed French account the following year. The four valence bonds of carbon had long been established as chemically equivalent. Van't Hoff's idea was that if they were considered *spatially* equivalent in three dimensions (directed toward the four vertices of a tetrahedron), a number of additional cases of absolute isomerism could be understood structurally – especially the curious property of certain substances to rotate the plane of polarized light passing through a solution of the compound. Optical activity, a familiar empirical effect, was thus successfully related to geometrical asymmetries in molecular structures. Van't Hoff's idea was especially championed by the well-known structural chemist Johannes Wislicenus, who in the next decade expanded stereochemical considerations to include compounds possessing double bonds. The name "stereochemistry" was coined by one of the most skilled practitioners in the 1880s and 1890s, Victor Meyer (who became Wöhler's successor at Göttingen in 1884 and Bunsen's at Heidelberg in 1889).[35]

The elucidation of positional aromatic isomerism and stereoisomerism are case studies that reveal the extraordinary power of structural chemistry. That even this degree of success could not compel assent from determined opponents is indicated by the example of Hermann Kolbe. One of the finest organic chemists of his day, Kolbe was intensely skeptical about claims of access to the details of molecular architecture. That this attitude was not completely unreasonable – at least early on – is indicated by the fact that most chemists older than Kolbe, including Liebig, Wöhler, Bunsen, and many others, felt the same way. Using his own type-theoretical version of valence theory (carefully adapted from the older radical theories), Kolbe was

---

[34] W. Schütt, "Guglielmo Koerner und sein Beitrag zur Chemie isomerer Benzolderivate," *Physis*, 17 (1995), 113–25.

[35] Overviews of the subject are provided by O. B. Ramsay, *Stereochemistry* (London: Heyden, 1981), and Ramsay, ed., *Van't Hoff-LeBel Centennial* (Washington, D.C.: American Chemical Society, 1975). On Wislicenus, see Peter Ramberg, "Arthur Michael's Critique of Stereochemistry," *Historical Studies in the Physical Sciences*, 22 (1995), 89–138, and "Johannes Wislicenus, Atomism, and the Philosophy of Chemistry," *Bulletin for the History of Chemistry*, 15/16 (1994), 45–54.

able to contribute substantially to the early phases of what became known as structural chemistry. More sophisticated later developments (such as those just described) were, however, beyond the power of his theory. Kolbe argued aggressively and tenaciously against structure theory, and against its most visible success, Kekulé's benzene ring. Kolbe's own benzene theory was adopted by no one, not even his own students, and in 1874 he conceded that "the great majority of chemists" preferred Kekulé's ideas on the subject. By this date, German organic chemistry (and, increasingly, European organic chemistry) had become fully structuralized.[36]

APPLICATIONS OF THE STRUCTURE THEORY

Many good recent historical studies have been done on the fine chemicals industry in the nineteenth century, and its relations with academic chemistry.[37] A brief synopsis will, therefore, suffice here. Chemical industry in the first half of the nineteenth century was primarily oriented to bulk inorganic substances, such as soda, sulfuric acid, salt, and alum, along with a few organic materials, such as soap and wax, all of which were produced largely by empirically derived manufacturing methods. Despite a good deal of pious contemporary rhetoric to the contrary, the high chemical theory of the first part of the century was not very relevant to the affairs of industrialists.[38]

Gradually, chemists began to acquire a repertoire of synthetic techniques that allowed them both to build larger organic molecules out of smaller pieces and also to produce naturally occurring organic substances at the lab bench. Leading actors in this story were Kolbe, Frankland, Hofmann, Wurtz, Berthelot, Kekulé, Erlenmeyer, Butlerov, Baeyer, and Graebe; the years of the most dramatic transformation of the field of organic synthesis were the 1850s, 1860s, and 1870s. These trends were vastly accelerated by the rise and development of the structural theory the history of which has just been traced.[39]

[36] Rocke, *Quiet Revolution*.
[37] Anthony Travis, *The Rainbow Makers* (Bethlehem, Pa.: Lehigh University Press, 1993); W. J. Hornix, "A. W. Hofmann and the Dyestuffs Industry," in *Die Allianz von Wissenschaft und Industry*, ed. C. Meinel and H. Scholz (Weinheim: VCH Verlag, 1992), pp. 151–65; J. A. Johnson, "Hofmann's Role in Reshaping the Academic-Industrial Alliance in German Chemistry," in ibid., pp. 167–82; A. S. Travis, W. J. Hornix, and R. Bud, eds., *Organic Chemistry and High Technology, 1850–1950* (special issue of *British Journal for the History of Science*, March 1992); Walter Wetzel, *Naturwissenschaft und chemische Industrie in Deutschland* (Stuttgart: Steiner, 1991); R. Fox, "Science, Industry, and the Social Order in Mulhouse," *British Journal for the History of Science*, 17 (1984), 127–68; G. Meyer-Thurow, "The Industrialization of Invention: A Case Study from the German Chemical Industry," *Isis*, 73 (1982), 363–81; F. Leprieur and P. Papon, "Synthetic Dyestuffs: The Relations between Academic Chemistry and the Chemical Industry in Nineteenth-Century France," *Minerva*, 17 (1979), 197–224; Y. Rabkin, "La chimie et le pétrole: Les débuts d'une liaison," *Revue d'Histoire des Sciences*, 30 (1977), 303–36; J. J. Beer, *The Emergence of the German Dye Industry* (Urbana: University of Illinois Press, 1959); and L. F. Haber, *The Chemical Industry During the Nineteenth Century* (Oxford: Clarendon Press, 1958).
[38] R. Bud and G. Roberts, *Science versus Practice: Chemistry in Victorian Britain* (Manchester: Manchester University Press, 1984).
[39] John Brooke, "Organic Synthesis and the Unification of Chemistry: A Reappraisal," *British Journal*

As regards applications of new structural organic-chemical knowledge, the products that led the way were dyes. The textile industry was a leading sector of industrialization, and the chemical arts provided an indispensable adjunct to clothing production. The dye industry was ancient and well established, but there was plenty of room for useful innovation in such qualities as range of colors, fastness, and price. The classic story of the rise of synthetic organic-chemical dyes involves the study of coal tar, a then-useless by-product of coke manufacture. Hofmann provided one of the earliest competent analyses of coal tar as his first major scientific project, as a student of Liebig in 1843. Two years later, he was hired as the first director of the new Royal College of Chemistry, in London. Hofmann was immensely successful, both as a teacher and as a research chemist. He continued his studies of substances derived from coal tar and petroleum, concentrating especially on nitrogen-containing organic compounds.[40]

In 1856 a student of Hofmann's named William Henry Perkin prepared a new purple color with excellent dye properties by oxidizing impure aniline, derived directly from coal tar. Perkin discovered the process by simple trial and error; he was not aware of the details of the constitution of the new compound – the structure theory had not yet been formulated. Perkin patented the material and, against Hofmann's advice, built a factory to produce the dye; full production began in 1858. "Mauve" immediately caught on, especially among the arbiters of fashion in Paris. Perkin became very rich, and the coal tar dye industry had begun.[41]

In 1858 Hofmann noted the production of a crimson color when aniline reacted with carbon tetrachloride. This dye was developed in France by F. E. Verguin, who sold the process to the Renard Frères firm of Lyon. From 1859, new French and English firms marketed this red dye, named "magenta" or "fuchsine," in what Hofmann soon thereafter termed "colossal proportions." Production of mauve soon faded, but magenta proved to be a lasting success. Hofmann's scientific studies of this material in 1862 and 1863 led to the production of alkylated derivatives, which provided different shades of the basic dye. Hofmann patented these "rosaniline" colors, subsequently named "Hofmann violets," and another former student of his, Edward Nicholson, produced them at the London firm of Simpson, Maule, and Nicholson. The growth of the European coal tar dye industry in the early 1860s was nothing short of spectacular; this growth continued throughout the decade, led by French and English firms.

for the History of Science, 5 (1971), 363–92; C. Russell, "The Changing Role of Synthesis in Organic Chemistry," *Ambix*, 34 (1987), 169–80.

[40] A useful recent compendium on Hofmann is Meinel and Scholz, eds., *Allianz*.

[41] Picric acid, whose production began in the mid-1840s, was actually the first coal tar dye. However, the large-scale marketing of synthetic organic products began only with Perkin. See Travis, *Rainbow Makers*, pp. 40–3.

An important turning point was the artificial synthesis of the natural product alizarin, which is the bright red coloring principle of the madder plant, and commercially the most important traditional dye next to indigo. The synthesis was achieved by Carl Graebe and Carl Liebermann in Baeyer's laboratory at the Berlin Gewerbeakademie, in 1868–9. This event transformed the coal tar dye industry. First, it was the occasion of a gradual shift from French and English to German leadership in the new industry; second, this was the first important natural dye to yield to the chemical arts; third, many future large chemical firms established themselves with this dye; and finally, this event marked a shift from more-or-less empirically driven innovation to product development that owed a great deal to chemical theory. In particular, structural chemistry, and especially Kekulé's benzene theory, proved indispensable to future growth in the industry.

In succeeding decades, the coal tar dye trade provided the leadership for other branches of the fine chemicals industry: pharmaceuticals, food and agricultural chemicals, photochemicals, medical supplies, and so on. Corporate research labs began to appear, staffed by chemists educated not only as chemical engineers but also in basic research. In this way, the high theory of molecular and structural chemistry had come to play a defining role in the birth of the modern age of industrial research.

14

THEORIES AND EXPERIMENTS ON RADIATION FROM THOMAS YOUNG TO X RAYS

Sungook Hong

Four different, but related, topics will be examined in this chapter: first, the debate between the emission theory and the undulatory theory of light; second, the discovery of new kinds of radiation, such as heat (infrared) and chemical (ultraviolet) rays at the beginning of the nineteenth century, and the gradual emergence of the consensus that heat, light, and chemical rays constituted the same continuous spectrum; third, the development of spectroscopy and spectrum analysis; and finally, the emergence of the electromagnetic theory of light and the subsequent laboratory creation of electromagnetic waves, as well as the discovery of x rays at the end of the nineteenth century.

The account given here is based on current scholarship in the history of nineteenth-century physics and, in particular, optics and radiation. However, as the current status of historical research on each of these topics is not homogeneous, different historical and historiographical points will be stressed for each topic. The first subject will stress historiographical issues in interpreting the optical revolution, with reference to Thomas Kuhn's scheme of the scientific revolution. The second and third subjects, which have not yet been thoroughly examined by historians, will stress the interplay among theory, experiment, and instruments in the discovery of new rays and the formation of the idea of the continuous spectrum. The fourth subject, Maxwell's electromagnetic theory of light, is rather well known, but the account here concentrates on the transformation of a theoretical concept into a laboratory effect, and then on the transformation of the laboratory effect into a technological artifact.

THE RISE OF THE WAVE THEORY OF LIGHT

The emission theory and what it is best to call "medium" theories of light had competed with each other since the late seventeenth century. Medium theories viewed light as a disturbance of some sort in an all-pervading ether;

the emission theory considered light in terms of particles and the Newtonian forces acting upon them. The emission theory was derived preeminently from Newton's optical work, in particular his 1704 *Opticks,* wherein light particles and forces were called upon to explain, among other phenomena, refraction and possibly dispersion. Medium theories were rooted in Descartes's conception of light as an (instantaneously propagating) pulse. Christiaan Huygens (1629–1695) suggested a novel principle – called Huygens's principle – according to which every point on a front acts as an emitter for secondary wavelets, which combine to form the (finitely propagating) front. Huygens also applied geometrical considerations of the undulation theory to explain a strange effect displayed by the Iceland crystal, an effect called double refraction. He obtained laws of double refraction in some particular cases and provided an experimental confirmation for these cases, but the general confirmation of his law was beyond the scope of experimental physics in the eighteenth century.[1]

It would be Whiggish to classify eighteenth-century optical works solely into the emission and the undulatory theories. G. N. Cantor has suggested a threefold division: the projectile theory, which conceived of light as a projection of material particles; the fluid theory, in which light was viewed on the analogy of the translational motion of hypothetical fluids; and the vibration theory, in which light was regarded as the vibrational motion of pulses in an all-pervading ether. Cantor distinguishes the vibration theory of the eighteenth century from the wave theory of Augustin Fresnel (1788–1827). The latter was characterized by a highly developed mathematics that made experimental predictions possible on the basis of the continuation of Huygens's principle with the principle of interference. By contrast, the vibration theory of Leonhard Euler (1707–1783), who for the first time explicitly introduced the notion of periodicity in considering light pulses, remained qualitative. Although Thomas Young (1773–1829) first produced an undulation theory of light on the basis of the principle of interference, he did not use Huygens's principle and, therefore, did not move completely outside the orbit of previous vibration conceptions. For this reason, Cantor considers Young to be the culmination of the eighteenth-century vibration theory, rather than the beginning of the nineteenth-century wave theory of light.[2]

At the end of the eighteenth and during the early nineteenth century, French optics was dominated by the emission theory. Pierre-Simon Laplace (1749–1827), a staunch Newtonian, had explained atmospheric refraction by analyzing mathematically the interaction between light particles and air.

---

[1] "Medium" theories of light are well examined in Casper Hakfoort, *Optics in the Age of Euler: Conceptions of the Nature of Light, 1700–1795* (Cambridge: Cambridge University Press, 1995). For the history of double refraction, see Jed Z. Buchwald, "Experimental Investigations of Double Refraction from Huygens to Malus," *Archive for History of Exact Sciences,* 21 (1980), 311–73.

[2] Geoffrey Cantor, *Optics after Newton: Theories of Light in Britain and Ireland, 1704–1840* (Manchester: Manchester University Press, 1983).

In 1802, a British natural philosopher, W. H. Wollaston (1766–1828), confirmed Huygens's construction of double refraction. This posed a challenge to Laplace and the Laplacians. Laplace's protégé, Etienne Malus (1775–1812), successfully explained double refraction in terms of the emission theory. Malus also discovered and then explained polarization and partial reflection. Jean-Baptiste Biot (1774–1862), another emission theorist, explained a new phenomenon, that of chromatic polarization.[3] The British wave theory of light suggested by Thomas Young was simply ignored in Paris. However, the emissionist successes in Paris were short-lived. The virtually unknown provincial engineer Augustin Fresnel, armed with mathematics and François Arago's (1786–1853) support, revived the wave theory of light in 1815. He then won an Académie des Sciences prize on theory of diffraction in 1819. Three out of five members of the prize committee – Laplace, Biot, and Siméon-Denis Poisson (1781–1840) – were either emission theorists or ardent Laplacians or both, but they nevertheless awarded the prize to Fresnel. This striking event has often been interpreted as evidence that Fresnel's theory was finally regarded as superior to the emission theory even by emission theorists.[4]

The history, as well as the historiography, of the debate between the wave and the particle theory of light, as outlined here, has for long been conditioned by William Whewell's (1794–1866) earliest description of the two theories. In this influential *History of the Inductive Sciences* (1837), Whewell remarked:

> When we look at the history of the emission theory of light, we see exactly what we may consider as the natural course of things in the career of a false theory. Such a theory may, to a certain extent, explain the phenomena which it was at first contrived to meet; but every new class of facts requires a new supposition, an addition to the machinery; and as observation goes on, these incoherent appendages accumulate, till they overwhelm and upset the original framework. Such has been the history of the hypothesis of the material emission of light.... In the undulatory theory, on the other hand, all tends to unity and simplicity.... It makes not a single new physical hypothesis; but out of its original stock of principles it educes the counterpart of all that observation shows. It accounts for, explains, simplifies, the most entangled cases; corrects known laws and facts; predicts and discloses unknown ones.[5]

To a new generation of wave partisans like Whewell, the result of the debate was, in a sense, predetermined, since light was a wave.

---

[3] For the Laplacian context, see M. Crosland, *The Society of Arcueil* (London: Heinemann, 1967); Robert Fox, "The Rise and Fall of Laplacian Physics," *Historical Studies in the Physical Sciences*, 4 (1974), 81–136; Eugene Frankel, "The Search for a Corpuscular Theory of Double Refraction: Malus, Laplace and the Prize Competition of 1808," *Centaurus*, 18 (1974), 223–45.

[4] After Fresnel's wave theory of light became successful, Thomas Young contended that he had planted the tree and Fresnel had picked up the apples. However, Fresnel, who generally agreed on Young's priority over undulation conceptions, denied Young's influence on him. See Edgar W. Morse, "Thomas Young," *Dictionary of Scientific Biography*, XIV, 568.

[5] William Whewell, *History of Inductive Sciences from the Earliest to the Present Time*, 3 vols. (London: John W. Parker, 1837), 2: 464–6.

A more sophisticated history of the debate between the wave and the emission theory of light was inaugurated with the recognition that in the 1810s and even early 1820s, the emission theory was quite successful in explaining optical phenomena. Eugene Frankel has described in detail the success and the strength of the emission theory of light in this period.[6] Generally speaking, the emission theory was more successful in explaining polarization and related phenomena, while the wave theory explained various aspects of diffraction. Laplacians regarded diffraction as a less important phenomenon than polarization, since they thought diffraction was a secondary phenomenon caused by the interaction between light and material objects. In the early 1820s, Fresnel introduced the idea of transverse waves to explain polarization, but, to do this, he had to accept that the ether must be highly elastic like a solid, a hypothesis that even Fresnel himself found hard to accept. The elastic solid ether model was later elaborated by Augustin-Louis Cauchy (1789–1857), James MacCullagh (1809–1847), and George Green (1793–1841), although it constantly posed hard questions for wave theorists.[7]

The triumph of the wave over the emission theory, according to Frankel, cannot be properly evaluated without considering the wider context in which this shift occurred: A series of battles between the Laplacian "short-range-force" program and its opponents was taking place in almost every field of the physical sciences during this period, including heat, electricity, and chemistry. Frankel drew two significant implications from his study. First, anomalies in existing sciences were detected by people distanced from the center of the main scientific enterprise. Laplacians in Paris, who tried to perfect the emission theory, found no anomalies in it, while Fresnel – far removed from the strong influence of Laplace – was able to suggest an altogether novel hypothesis. Second, Frankel proposed that social and political contexts not only influenced the resolution of the battle between the two different theories of light but also were deeply implicated in the battle's very origin. He proposed that these two conclusions could supplement Thomas Kuhn's scheme of the way in which scientific revolutions should proceed.[8]

Although Frankel's consideration of the social and political contexts in which the debate took place is illuminating, there was one question that he neither asked nor answered: Why, in the 1810s, was the emission theory successful? In other words, how did Malus, for example, obtain his "sine-squared law"? Or, how did Biot formulate equations for chromatic polarization? It

---

[6] Eugene Frankel, "Corpuscular Optics and the Wave Theory of Light: The Science and Politics of a Revolution in Physics," *Social Studies of Science*, 6 (1976), 141–84.

[7] For Fresnel's hypothesis of transverse waves and the subsequent ether models, see Frank A. J. L. James, "The Physical Interpretation of the Wave Theory of Light," *British Journal for the History of Science*, 17 (1984), 47–60; David B. Wilson, "George Gabriel Stokes on Stellar Aberration and the Luminiferous Ether," *British Journal for the History of Science*, 6 (1972), 57–72; Jed Z. Buchwald, "Optics and the Theory of the Punctiform Ether," *Archive for History of Exact Sciences*, 21 (1980), 245–78.

[8] Frankel, "Corpuscular Optics and the Wave Theory of Light"; Thomas S. Kuhn, *The Structure of Scientific Revolutions* (Chicago: University of Chicago Press, 1962).

has usually been assumed by wave theorists and by later historians, including Frankel, that their formulas were "empirical" – that they had been obtained somehow directly from experimental data. On the other hand, Fresnel's formula, such as his integral for interference, was said to be founded directly on theory. According to this traditional way of thinking, as Whewell himself long ago noted, although the emission theory succeeded in explaining some phenomena, "as observation goes on, these incoherent appendages accumulate, till they overwhelm and upset the original framework."[9]

Jed Buchwald, through a painstaking reconstruction of lost theories and their meanings in emissionist optics, has convincingly shown that this simple story is far from correct. He argues that the Laplacian paradigm, according to which forces acting upon moving particles were employed to account for microscopic and even some macroscopic phenomena, was not very successful for optical phenomena. The short-range-force principle did explain (in part) refraction and, in an indirect way, double refraction, but it did not work well for partial reflection, polarization, or chromatic polarization – phenomena that were intensively discussed in the 1810s and 1820s. Emission theorists, such as Biot and Malus, no more used Newtonian forces to explain these phenomena than undulation theorists later used ether mechanics to explain interference patterns – which is to say, hardly at all.

Yet, something was at work here. The central principle of emission theorists, according to Buchwald, lay in their assumption – and alternative practice – that rays of light exist as physical and objective entities and, as such, that they can be counted. To explain polarization, each ray was given an asymmetry orthogonal to its axis, but it was essentially meaningless to say that one ray is or is not polarized, because every ray always is just as asymmetrical as it can ever be. Polarization, properly speaking, refers to a structure in which a group of rays (or a beam) were aligned in the same direction. Counting how many rays are aligned in the same direction (which amounts to a sort of "ray statistics") constituted the emission theorists' actual practice. Polarization, partial reflection, and chromatic polarization were explained on this basis. On the other hand, to wave theorists, a ray of light was thought to be a geometrical line connecting the source of light and a point in the wave front. It is not, accordingly, a physical entity but, rather, a mathematical line, and thus cannot be counted. However, the ray was said to be or not to be polarized, as its asymmetry was specified as a direction at the point of the front intersected by the ray.[10]

The debate between the wave and the emission theory of light, therefore, neither involved an abrupt victory of the one over the other (symbolized by

---

[9] See note 5.
[10] Jed Z. Buchwald, *The Rise of the Wave Theory of Light: Optical Theory and Experiment in the Early Nineteenth Century* (Chicago: University of Chicago Press, 1989). His argument on the ray conception of polarization is well summarized in Jed Z. Buchwald, "The Invention of Polarization," in *New Trends in the History of Science*, ed. R. P. W. Visser, H. J. M. Bos, L. C. Palm, and H. A. M. Snelders (Amsterdam: Rodopi, 1989), pp. 3–22.

Fresnel's winning of the 1819 prize), nor a smooth transition from the latter to the former; rather, it was a difficult and prolonged process of confusion and misunderstandings – in short, of partial incommensurability.[11] For example, according to Buchwald, Fresnel's winning of the 1819 prize was due to the fact that his mathematical formula, which nicely fit experimental data, did not seem to involve any significant physical hypothesis on the nature of light that would have threatened the established status of ray statistics. In other words, although Fresnel's beautiful formula cast much doubt on emissionist principles of light, it did not similarly affect the underlying principles of ray theory. During the initial phase of the debate, emission partisans found it difficult to understand that wave theorists like Fresnel could use the "ray of light" without also assuming the apparatus of ray statistics. Once Fresnel became a fully fledged wave theorist, he criticized Biot for inconsistently employing Newtonian forces in the latter's ray optics. For Biot, however, ray statistics could be distinguished from the Newtonian hypothesis on forces and particles and remained untouched by Fresnel's critique.[12]

NEW KINDS OF RADIATION AND THE IDEA OF THE CONTINUOUS SPECTRUM

Throughout the eighteenth century, the spectrum referred to something visible and colored. In 1800, Frederick William Herschel (1738–1822) discovered an invisible ray. He had accidentally noticed that glasses of different colors, used in the telescope, had different heating effects. This led him to examine the heating action of various parts of the colored spectrum. With a prism and thermometers, he detected a rise in heating effect beyond the red end of the solar spectrum, where no visible light existed, but none beyond the violet end. He named the invisible rays to which he ascribed the effect "heat" or "caloric" rays. He went on to demonstrate that these new rays could be reflected and refracted, which raised the question concerning the identity between the rays and light. Herschel discovered that while an uncolored glass might be perfectly transparent to visible light, it nevertheless absorbed about 70 percent of the heat rays. He performed several different kinds of experiments, including what he called the "crucial experiment," in which the two absorptions by the same colored glass of, say, the visible spectrum of red light and the invisible spectrum of heat rays in the red-light range were compared. The results always pointed to a difference between light and heat rays. Herschel, who held to both the caloric theory of heat and

---

[11] See, for instance, John Worrall, "Fresnel, Poisson, and the White Spot: The Role of Successful Predictions in the Acceptance of Scientific Theories," in *The Uses of Experiment*, ed. D. Gooding, T. Pinch, and S. Schaffer (Cambridge: Cambridge University Press, 1989), pp. 135–57.

[12] Buchwald, *Rise of the Wave Theory of Light*, pp. 237–51. The incommensurability issue is further analyzed in Jed Z. Buchwald, "Kinds and the Wave Theory of Light," *Studies in History and Philosophy of Science*, 23 (1992), 39–74.

the corpuscular theory of light, concluded that two independent spectra existed. Light belonged to the "spectrum of light," whereas the invisible rays belonged to the "spectrum of heat." To him, the only commonality between them lay in the fact that both sorts of rays were refrangible (though to different degrees).[13]

Herschel's discovery of invisible heat rays was much doubted initially. John Leslie, a Scottish natural philosopher, argued that Herschel's observation of the heating effect outside the solar spectrum was due to a rise of room temperature caused by the reflection of light from the stand. Leslie reported that a careful experiment he had himself performed had not revealed any evidence for invisible heat rays beyond the red end of the spectrum. A few people, such as C. E. Wünsch in Germany, confirmed Leslie's experiment. On the other hand, Thomas Young and others were able to confirm Herschel's results. The reason for the discrepancy between them was sought in the different prisms that they used. Johann Wilhelm Ritter (1776–1810) in Germany suggested that Herschel and Wünsch used different prisms, and that Wünsch's result was true for the kind of prism that he used. Thomas J. Seebeck (1770–1831) in 1806 (though not published until 1820) also argued that the different results could be attributed to prisms with different dispersive powers, as well as to the use of different materials with different absorption powers. Seebeck himself demonstrated the existence of a heating effect outside the solar spectrum, confirming Herschel. By the time Seebeck's research was published in 1820, the existence of invisible rays outside the solar spectrum had been generally accepted.[14]

Meanwhile in 1801, Ritter discovered the chemical effect of invisible rays lying outside of the violet side of the solar spectrum. As a follower of German *Naturphilosophie,* Ritter had discovered what he called "deoxidizing rays" while performing his research under the strong conviction that polarity in nature should reveal the cold counterpart of heat rays, with the cold rays lying beyond the violet spectrum. In 1777, K. W. Scheele (1742–1786) had discovered that a paper treated with silver chloride became blackened far sooner in violet light than in other colors. Ritter, who had been aware of Scheele's experiment, employed paper treated with silver chloride as a detector for his invisible radiation. He succeeded in showing that the maximum blackening of the paper occurred beyond the violet. Three years after Ritter's discovery, Thomas Young produced an interference pattern for the ultraviolet rays by using paper treated with silver chloride. After this, techniques of

---

[13] Herschel's discovery of heat rays has been mentioned in many historical and scientific works on infrared spectroscopy. See, for example, D. J. Lovell, "Herschel's Dilemma in the Interpretation of Thermal Radiation," *Isis,* 59 (1968), 46–60; E. Scott Barr, "Historical Survey of the Early Development of the Infrared Spectral Region," *American Journal of Physics,* 28 (1986), 42–54.

[14] Published in 1938, the debate between Herschel and Leslie, as well as Seebeck's contribution, was clearly analyzed during the same year in E. S. Cornell, "The Radiant Heat Spectrum from Herschel to Melloni – The Work of Herschel and his Contemporaries," *Annals of Science,* 3 (1938), 119–37.

detecting ultraviolet rays developed along with improvements in photographic techniques.[15]

Other instrumental developments proved essential to the later emergence of the idea of the continuous spectrum. When Herschel discovered his invisible radiation, he had used mercury thermometers, which remained in common use until Leopoldo Nobili (1784–1835) in Italy devised the thermopile in 1829, which was much more sensitive. Another important advancement occurred with the discovery of the substances that are transparent to infrared radiation (as glass is almost transparent to light). Macedonio Melloni (1798–1854) in Italy found that rock salt was much more transparent or, in his own term, "diathermanous," to infrared radiation than was glass, which allowed him to make prisms out of rock salt. These prisms, with his improved thermopile, made infrared rays much more controllable and manipulable.[16]

After Herschel's and Ritter's discoveries, three different rays – heat, light, and chemical – had been identified. The point of controversy remained whether the heat and chemical rays were extensions of the visible spectrum into the invisible regions, or whether they were utterly different from rays of light. As we have seen, on the basis of his experiments, Herschel thought that heat and light rays were, though similar in nature, distinct in kind. In 1813/14, French physicists Biot, C. L. Berthollet (1748–1822), and J. A. C. Chaptal, who discussed and compared these two hypotheses, concluded, contra Herschel, that heat, light, and chemical rays were all of one kind, the difference among them being solely that of refrangibility. In 1812, Humphry Davy (1778–1829) also rejected Herschel's notion that heat rays were distinct from the light rays. In 1832, with the undulatory theory of light as the basis, André-Marie Ampère (1775–1836) maintained that radiant heat could not be distinguished from light. However, Melloni, who made an enormous contribution to the later research on infrared radiation, argued in the 1830s that since heat and light had different absorption ratios in various materials, they must be due to two distinct agents or to two *essentially distinct* modifications of the same agent. Even in the 1830s, the status of the invisible rays remained uncertain.

Several factors contributed to the emergence of the idea that heat, light, and chemical rays belong to one and the same spectrum, distinguished solely by wavelength – a consensus that became dominant in the 1850s. Robert James

---

[15] For the connection between Ritter's discovery of chemical rays and *Naturphilosophie*, see Kenneth L. Caneva, "Physics and Naturphilosophie: A Reconnaissance," *History of Science*, 35 (1997), 35–106, esp. pp. 42–8. On Ritter in general, see Stuart W. Strickland, "Circumscribing Science: Johann Wilhelm Ritter and the Physics of Sidereal Man," PhD diss., Harvard University, 1992; Walter D. Wetzels, "Johann Wilhelm Ritter: Romantic Physics in Germany," in *Romanticism and the Sciences*, ed. Andrew Cunningham and Nicholas Jardine (Cambridge: Cambridge University Press, 1990), pp. 199–212.

[16] For Melloni's improvement of Nobili's thermopile, see Edvige Schettino, "A New Instrument for Infrared Radiation Measurements: The Thermopile of Macedonio Melloni," *Annals of Science*, 46 (1989), 511–17. For Melloni's contribution to investigations of infrared radiation, see E. S. Cornell, "The Radiant Heat Spectrum from Herschel to Melloni – II: The Work of Melloni and his Contemporaries," *Annals of Science*, 3 (1938), 402–16.

McRae, who examined this issue in great detail, notes that the formation of the consensus was a long process, in which theoretical, experimental, and instrumental factors were interwoven.[17] No single experiment was crucial. Theoretical factors, such as the formulation of the first law of thermodynamics, in which heat was identified with mechanical energy, helped scientists to look at the heating effect of rays in new ways. Technological and instrumental factors, such as the discovery of rock salt as a diathermanous material and the invention of the precise thermocouple, were central.

The rise and spread of the wave theory of light made a positive contribution to this process, as the idea of wavelength became meaningful and useful. Thomas Young said, in 1802, that "it seems highly probable that light differs from heat only in the frequency of its undulations or vibrations."[18] Young's speculation was supported and extended by later experiments. James Forbes (1809–1868), for instance, demonstrated the circular polarization of heat rays and measured their wavelength in 1836. The case of Melloni is even more striking. He had been a strong believer in the difference between heat and light rays. But after he converted to the wave theory of light in 1842, he considered the heat, light, and chemical rays to be identical in nature, the only real difference being wavelength. The experimental establishment of the similarity among these rays in reflection, refraction, polarization, partial transmissibility, and interference, some of which had already been established by Herschel in 1800, was reinterpreted in a new way once the wave theory of light was accepted. In this sense, Herschel had provided a set of tools with which later scientists were to investigate new rays.

THE DEVELOPMENT OF SPECTROSCOPY AND SPECTRUM ANALYSIS

Spectroscopy began in 1802 with the discovery by W. H. Wollaston of several dark lines in the solar spectrum, which he thought to be an instrumental anomaly. It was Joseph Fraunhofer (1787–1826), who was engaged in manufacturing glass for telescopes and prisms, who transformed anomaly into natural phenomenon, and eventually into an extremely influential instrument. By utilizing resources, such as the skilled artisans in a local Dominican monastery, he was able to produce superb prisms and achromatic lenses. With them, he discovered more than 500 fine dark lines in the solar spectrum. Fraunhofer immediately utilized the lines for the calibration of these new lenses and prisms, since the dark lines were good benchmarks for distinguishing the hitherto rather obscure boundaries between different colors.

---

[17] Robert James McRae, "The Origin of the Conception of the Continuous Spectrum of Heat and Light," PhD diss., University of Wisconsin, 1969.
[18] Thomas Young, "On the Theory of Light and Colors," *Philosophical Transactions*, 92 (1802), 12–48, at p. 47.

With them, he was able to measure the refractive indices of glasses for the production of achromatic lenses. However, he did not take further theoretical steps. He did note, for instance, that two very close yellow lines, obtained in the spectrum of a lamp, agreed in position with two dark lines in the solar spectrum (which he named "D"), but did not speculate about the reason for the coincidence. In the 1830s, in connection with the conflict between the wave and emission theories of light, a heated debate arose among David Brewster (1781–1868), George Biddell Airy (1801–1892), and John Herschel (1792–1871) over what caused the dark lines in the spectrum. But Fraunhofer was far removed from such matters.[19]

The idea that line spectra might be related to the structure of the atoms or molecules of the light-emitting or light-absorbing substance was suggested by several scientists. In 1827, John Herschel interpreted dark and bright lines as indicating that the capacity of a body to absorb a particular ray is associated with the body's inability to emit the same ray when heated. L. Foucault (1819–1868) in 1849 and George Gabriel Stokes (1819–1903) in 1852 conjectured the mechanism of dark and bright lines, and William Swan in 1856 attributed the D lines to the presence of sodium in the medium or in light sources. However, it was not until Gustav Kirchhoff (1824–1887) – at the request of R. Bunsen (1811–1899) – in 1859 proposed the law of the identity of emission and absorption spectra under the same physical conditions that spectroscopic investigations of light-emitting or absorbing substances became widespread. Balfour Stewart (1828–1887) in England, nearly simultaneously, suggested a similar idea. The difference between Kirchhoff and Stewart lay in the fact that Kirchhoff's idea was based on general principles of thermodynamics and rigorous demonstration, whereas Stewart's concept was rooted in Pierre Prévost's much older, and looser, theory of exchanges. The priority dispute was fought not only by themselves but also by their followers until the end of the nineteenth century. Their achievements represented, in a sense, features of German and British scientific styles.[20]

Before Kirchhoff, spectral lines had been discussed in the context of physical theories. Wave theorists were concerned with the origins of spectral lines, because emission theorists claimed that the wave theory could not account for the absorption of specific frequencies of light by matter. To explain these absorption lines, wave theorists, such as John Herschel, developed an elaborate resonance model on the analogy of the mechanism of the tuning fork. However, Herschel did not associate spectral lines with the chemical properties

---

[19] For Fraunhofer, see Myles W. Jackson, "Illuminating the Opacity of Achromatic Lens Production: Joseph Fraunhofer's Use of Monastic Architecture and Space as a Laboratory," in *Architecture of Science*, ed. Peter Galison and Emily Thompson (Cambridge, Mass.: MIT Press, 1999); Jackson, *Spectrum of Belief: Joseph von Fraunhofer and the Craft of Precision Optics* (Cambridge, Mass.: MIT Press, 2000).

[20] The priority dispute between Kirchhoff and Stewart is examined in Daniel M. Siegel, "Balfour Stewart and Gustav Robert Kirchhoff: Two Independent Approaches to 'Kirchhoff's Radiation Law,'" *Isis*, 67 (1976), 565–600.

of substances.[21] After Kirchhoff proposed the law of the identity of emission and absorption spectra, emission and absorption lines were soon used to examine chemical properties. This principle was embodied in the spectroscope. Kirchhoff and Bunsen constructed the first spectroscope in 1860, and the word "spectroscopy," or spectrum analysis, began to be widely used in the late 1860s. Various kinds of spectroscopes were constructed. In the mid-1860s, for example, W. Huggins (1824–1910) combined spectroscopy with a stellar telescope for the purpose of examining the stellar spectral lines. This marked the beginning of astrospectroscopy, which made astronomy into a sort of laboratory science.[22]

How did scientists understand the origins of spectral lines? At first, it was commonly believed that the banded spectrum represented the effects of a molecule, whereas the line spectrum represented the effects of an atom. This belief was not unchallenged. J. Norman Lockyer (1836–1920), who had noticed changes in line spectra under certain conditions, in 1873 proposed a scheme involving the "dissociation of an atom," founded on the notion that an atom is a grouping of more elementary constituents. Line spectra, according to Lockyer, were caused by these elementary constituents. Lockyer's hypothesis was not seriously considered by his contemporaries, mainly because an atom had long been considered not further divisible. During these years, the wavelength of various spectra were more exactly determined when, in 1868, A. J. Ångström (1814–1874) published his measurements of approximately 1,000 solar spectral lines, done with diffraction gratings. His wavelengths replaced Kirchhoff's arbitrary units and served as the standard until Henry A. Rowland (1848–1901) set a new one by using his improved gratings.[23]

In the 1870s and 1880s, several scientists tried to find mathematical regularities among the various line spectra of a given substance. The Irish physicist G. J. Stoney (1826–1911), in 1871, thought the hydrogen spectrum to be due to the splitting of the original wave by the medium into several different parts. He suggested that this splitting could be analyzed by employing Fourier's

---

[21] The important role played by Herschel for the development of spectroscopic ideas is stressed in M. A. Sutton, "Sir John Herschel and the Development of Spectroscopy in Britain," *British Journal for the History of Science*, 7 (1974), 42–60. This view was criticized by Frank James, who considered it "a Victorian myth." See Frank A. J. L. James, "The Creation of a Victorian Myth: The Historiography of Spectroscopy," *History of Science*, 23 (1985), 1–22.

[22] For the history of early spectroscopy, see J. A. Bennett, *The Celebrated Phaenomena of Colours: The Early History of the Spectroscope* (Cambridge: Whipple Museum of the History of Science, 1984). For stellar spectroscopy, see Simon Schaffer, "Where Experiments End: Tabletop Trials in Victorian Astronomy," in *Scientific Practice: Theories and Stories of Doing Physics*, ed. Jed Z. Buchwald (Chicago: University of Chicago Press, 1995), pp. 257–99.

[23] For Lockyer, see A. J. Meadows, *Science and Controversy: A Biography of Sir Norman Lockyer* (London: Macmillan, 1972). Rowland's gratings are nicely examined in Klaus Hentschel, "The Discovery of the Redshift of Solar Fraunhofer Lines by Rowland and Jewell in Baltimore around 1890," *Historical Studies in the Physical Sciences*, 23 (1993), 219–77.

theorem and by matching the harmonics that appeared in the theorem with the observed spectrum lines. He noted three hydrogen lines, 4102.37, 4862.11, and 6563.93 Å, and found their ratios to be approximately 20, 27, and 32. In 1881, Arthur Schuster (1851–1934), who claimed that Stoney's ratio could not be considered to be a mathematical regularity, cast strong doubt on the harmonics hypothesis. Schuster, however, could not suggest a plausible alternative theory. In 1884, Johann K. Balmer (1825–1898), a virtually unknown Swiss mathematician, examined four hydrogen lines and formulated the series now named after him:

$$\lambda_n = \left[\frac{n^2}{(n^2 - 2^2)} \right] \lambda_o$$

where $\lambda_o = 3645.6$ Angstrom and $n = 3, 4, 5, \ldots$

The Balmer series was not at all similar to the simple harmonic ratio that had been proposed by Stoney. Although Balmer's formula beautifully linked the four known hydrogen spectra and turned out to be valid for the newly discovered ultraviolet and infrared spectra of hydrogen, it provided more problems than solutions, as scientists failed to agree on any explanation of the regularity. Later, Niels Bohr's quantum model of the hydrogen atom, in which an emission of radiation was said to be caused by the jump of an electron from a higher to a lower energy level, was able to yield the Balmer series.[24]

Spectroscopy of invisible rays was much more difficult than spectroscopy of visible rays. To draw the spectrum of infrared rays, sensitive detectors were crucial. When A. Fizeau (1819–1896) and Foucault had established the description of infrared interference in 1847, they had used a tiny alcohol thermometer, read by a microscope, and had shown that the temperatures at different points followed the alternations of intensity in an interference pattern. Sensitive detectors to replace the thermometer were invented only in the 1880s. Samuel P. Langley (1834–1906) in the United States developed a new detector, the bolometer, in 1881. This device, which utilized the dependence of electrical resistance of metal on temperature, could detect a difference of 0.00001°C and enabled Langley to map the infrared spectrum with unprecedented accuracy.

The first noticeable advancement in investigations of ultraviolet radiation occurred when Stokes, having discovered the transparency of quartz to ultraviolet radiation in 1862, examined ultraviolet spectra of various arcs

---

[24] The search for regularities in spectral lines in the nineteenth century is described in William McGucken, *Nineteenth-Century Spectroscopy: Development of the Understanding of Spectra, 1802–1897* (Baltimore: Johns Hopkins University Press, 1969). See also Leo Banet, "Balmer's Manuscripts and the Construction of His Series," *American Journal of Physics*, 28 (1970), 821–8; J. MacLean, "On Harmonic Ratios in Spectra," *Annals of Science*, 28 (1972), 121–37.

and sparks by means of a phosphate fluorescent screen and a quartz prism. He thereby extended the ultraviolet spectrum down to 2,000 Å and photographed spectral lines in this region. From then until 1890, no spectral line below 2,000 Å was observed, and most scientists tended to believe that this was the natural low limit for the range. V. Schumann (1841–1913), who did not believe this to be so, thought instead that the apparent limit was due to absorption, and he tried to find alternative materials that were more transparent to ultraviolet rays. He noted that three absorbers were present in most experiments: the quartz prism, air, and a photographic plate. Accordingly he inserted the entire apparatus into a vacuum, used fluorite (which he thought to be more transparent to short waves) instead of quartz, and used a photographic plate with a minimum amount of gelatin. In 1893, he was thereby able to extend the ultraviolet spectrum below 2,000 Å, but he could not determine the wavelength of this new region precisely, since there was no available standard. T. Lyman (1874–1954), who later explored the ultraviolet spectrum down to 500 Å in 1917, found that Schumann's investigation had been made on waves of 2,000–1200 Å.[25]

THE ELECTROMAGNETIC THEORY OF LIGHT AND THE DISCOVERY OF X RAYS

Return now to the 1860s, by which time the wave theory of light had long been established. On the basis of an ingenious "idle wheel" model of the electromagnetic ether, James Clerk Maxwell (1831–1879) in 1861 suggested that light itself is a species of electromagnetic disturbance. Maxwell's suggestion did not undermine the status of the established wave theory, since his electromagnetic disturbances had all the standard properties, and more. In 1865, Maxwell formulated his electromagnetic theory of light in a tighter mathematical form without relying on the debated mechanism of his ether. Maxwell's theory implied that the ratio of electrostatic to electromagnetic unit of electricity should be equal to the velocity of light. Although controversial in the 1860s and 1870s, Maxwell's claim became more widely accepted in the late 1870s, although the identity per se did not prove persuasive to those who, like William Thomson (Lord Kelvin, 1824–1907), had not accepted Maxwell's system.[26]

[25] F. Fraunberger, "Victor Schumann," in *Dictionary of Scientific Biography*, XII, 235–6; Ralph A. Sawyer, "Theodore Lyman," *Dictionary of Scientific Biography*, VIII, 578–9.
[26] For Maxwell's electromagnetic theory of light, see Daniel M. Siegel, *Innovation in Maxwell's Electromagnetic Theory: Molecular Vortices, Displacement Current, and Light* (Cambridge: Cambridge University Press, 1991). Measurements of the value of the ratio of electrostatic to electromagnetic unit of electricity is examined by Simon Schaffer, "Accurate Measurements is an English Science," in *The Values of Precision*, ed. M. Norton Wise (Princeton, N.J.: Princeton University Press, 1995), pp. 135–72.

Maxwell himself never attempted to generate or to detect electromagnetic waves lengthier than those of light. Maxwell seemed far less concerned with producing and detecting electromagnetic waves other than light than with revealing the electromagnetic properties of light. Nevertheless, Maxwell's electromagnetic theory of light did naturally suggest that it might be possible to create such disturbances – to produce, as it were, something that could be truly called an electromagnetic wave. The spectrum would then be extended far below the infrared, to centimeter and even meter wavelengths. In the early 1880s, such Maxwellians as George FitzGerald (1851–1901), who had had some doubts about the possibility, and J. J. Thomson (1856–1940) suggested ways to generate such waves by purely electrical methods. In particular, FitzGerald specified rapid electrical oscillations in a closed circuit (caused by condenser discharges) as a proper way to do this and calculated the wavelength that would thereby be generated. But he did not know how to detect such waves. In 1887/88, Oliver Lodge (1851–1940) experimented with Leyden jar discharge, but he did not produce or detect fully propagating waves.[27]

Heinrich Hertz (1857–1894), who had been exposed, via Hermann von Helmholtz (1821–1894), to both the German (Weberian) electrodynamics – in which electric charge and current were considered to be real, and electric potential was though to propagate at a finite speed – and the Maxwellian electrodynamics – in which action-at-a-distance was denied, and electromagnetic fields were considered to be real – observed a curious effect displayed by secondary sparks from a pair of metallic coils, called Riess coils, in 1887. He first tried to abolish the sparks, but failed to do so. Then, he tried to control and manipulate the effect. He fabricated a spark detector, which eventually became a means of probing the propagation of electric forces. Neither Maxwell's nor Helmholtz's theory entirely guided the laboratory practice that led him to the discovery of the electromagnetic wave. The interplay between his local devices and his theories on instruments led to the stabilization of the strange effect. Hertz eventually concluded, after extensive investigations, that he had produced and detected Maxwell's electromagnetic waves. The induction coil and the condensers with which he produced the primary sparks became the generating oscillator, and his spark-based resonator became the first detector of electromagnetic waves. Hertz measured the length of his waves to be 66 centimeters.[28]

Hertz discovered what we now call the microwave spectrum, extending the radiation into a thoroughly new region. This new spectrum was, however, generated by means totally different from those that had been used for the production of infrared radiation, as electromagnetic waves were generated from a rapid electrical oscillation, such as condenser discharge. Following

[27] Bruce J. Hunt, *The Maxwellians* (Ithaca, N.Y.: Cornell University Press, 1991), pp. 33–47, 146–51.
[28] Jed Z. Buchwald, *The Creation of Scientific Effects: Heinrich Hertz and Electric Waves* (Chicago: University of Chicago Press, 1994).

Hertz's experiments, Lodge in Britain, A. Righi (1850–1920) in Italy, and J. Chandra Bose (1858–1937) in Calcutta pushed to shorter wavelengths. For this experiment, spherical oscillators replaced Hertz's linear ones. As for detectors, the coherer, invented by E. Branly and improved by Lodge, replaced Hertz's spark-gap resonators. With these, Bose successfully generated waves with centimeter wavelengths. Experiments on the diffraction, refraction, polarization, and interference of microwaves followed. It is important to note here that because of the physical nature of contemporary electrical circuitry, the waves thus generated were highly damped. Damping produced puzzling effects, such as multiple resonance, which generated much debate in the 1890s and early 1900s.

Practical applications for Hertzian waves were not at first obvious. In 1895/6, Guglielmo Marconi (1874–1937) opened a new field by applying Hertzian waves to telegraphy. What is notable in Marconi is his movement against the mainstream of physics: He tried to increase, rather than to decrease, the wavelength. He erected a tall vertical antenna, and connected one end of the discharge circuit to it and the other end to the ground. The antenna and the ground connections increased the capacitance of the discharge circuit considerably, and this lengthened the wavelength and increased the power that could be stored in the system. When he succeeded in the first transatlantic wireless telegraphy in 1901, the transmitter used 20 kW power, and the estimated wavelength was of the order of one thousand meters. Long waves were the only possible way of combining power and communication.[29]

Near the time when Marconi first succeeded in demonstrating the feasibility of commercial Hertzian-wave telegraphy, W. K. Röntgen (1845–1923) discovered x rays while experimenting with cathode rays. He noticed phosphorescent effects on a screen of barium platino-cyanide placed 2 meters away from the cathode-ray tube. While examining the phenomenon, he discovered the existence of a ray with astounding power to penetrate ordinary matter. The x-ray photograph of his hand created a worldwide sensation, but the nature of the new rays escaped plausible explanation for some time. They were different from cathode rays, because they were not bent in magnetic fields. They were different from Lenard rays, which were believed to exist within the short distance outside the cathode tube, because they were able to travel a long distance in the air. Experiments seemed to indicate that they were neither charged matter nor uncharged particles. The hypothesis that x rays are very short waves (shorter than ultraviolet) was considered at this time, but the nonexistence of refraction, interference, diffraction, and

---

[29] For the early history of the application of Hertzian waves to telegraphy, see Hugh G. J. Aitken, *Syntony and Spark: The Origins of Radio* (New York: Wiley, 1976); Sungook Hong, "Marconi and the Maxwellians: The Origins of Wireless Telegraphy Revisited," *Technology and Culture*, 35 (1994), 717–49; Hong, *Wireless: From Marconi's Black-Box to the Audion* (Cambridge, Mass.: MIT Press, 2001).

polarization of x rays made the wave hypothesis difficult to accept, and wavelength measurement was accordingly impossible.

Among various hypotheses, the notion that x rays were transverse impulses caused by the collision of electrons with the metallic plate or glass in a cathode-ray tube became dominant. It is interesting to note that particle-like properties of x rays emerged from this pulse model. The discovery of x-ray diffraction and interference in 1912/13 by Max von Laue (1879–1960) and others in Germany, and by the Braggs in Britain, made the claim that x rays were waves of extremely short wavelengths plausible to many. Then, previous particle-like properties of x rays were used to justify and consolidate particle-like properties of ordinary light, properties that began to be discovered after the beginning of the twentieth century. As for x-ray diffraction, crystals, which had a lattice structure, were used as gratings, leading to the precise measurement of the wavelength of x rays. At the same time, it opened an entirely new field of x-ray crystallography. In the 1910s, Henry G. J. Moseley (1887–1915) also made an important contribution to the development of x-ray spectroscopy.[30]

THEORY, EXPERIMENT, INSTRUMENTS IN OPTICS

During the nineteenth century, the spectrum of radiation was transformed from the finite spectrum of visible light into a nearly infinite one, including not only invisible (infrared and ultraviolet) rays adjacent to the light spectrum, but also much longer electromagnetic waves and much shorter x rays. The physics of radiation was also shifted from a pure curiosity to a commercially important business. Throughout these transformations, one can find interactions between theory, experiment, and instruments.

In Malus's and Fresnel's optical research, one can see the emergence of an intimate linkage between precise measurement and mathematical theories producing experimentally testable formulas. This feature came to characterize "physics" in the nineteenth century. In the case of Herschel's discovery of infrared rays and the subsequent controversy over them, the difference in prisms and thermometers used by scientists made it difficult for them to reach a consensus. In 1800, Herschel himself was convinced of the existence of invisible thermal rays, but on the basis of his "crucial" experiment, he discarded the possibility that these invisible rays and visible light are of

---

[30] For the discovery of x rays, see Alexi Assmus, "Early History of X-Rays," *Beam Line,* 25 (Summer 1995), 10–24. The history of the pulse model is well probed in Bruce R. Wheaton, "Impulse X-Rays and Radiant Intensity: The Double Edge of Analogy," *Historical Studies in the Physical Sciences,* 11 (1981), 367–90; Wheaton, *The Tiger and the Shark: Empirical Roots of Wave-Particle Dualism* (Cambridge: Cambridge University Press, 1983). For Moseley, see John L. Heilbron, *H. G. J. Moseley: The Life and the Letters of an English Physicist, 1887–1915* (Berkeley: University of California Press, 1974).

the same nature. In the 1830s, Melloni thought exactly the same way on the same grounds. The subsequent acceptance of the idea of the continuous spectrum of infrared, visible, and ultraviolet light was made possible by the formation of the triad consisting of a new and encompassing theory, striking experiments, and more reliable instruments. The combination of the wave theory of light, the establishment of an interference effect of invisible rays, and diathermanous prisms and precise thermometers convinced most physicists to accept the theory of the continuous spectrum.

One can also find rich interactions among theory, experiment, and instruments in the development of spectroscopy, as well as in Maxwell's electromagnetic theory of light. Maxwell's theory produced testable formulas, one of which was that the ratio of the electrostatic to electromagnetic unit of electricity should be equal to the velocity of light. The experimental evidence for this identity, which Maxwell thought crucial for his theory, failed to convince those who were skeptical of Maxwell's theory. Some Maxwellians tried to generate electromagnetic waves through rapid oscillations, but they did not know how to detect them. Hertz's new way of using old instruments (the *Riess* coils), which previously had been used for making sparks, created detectable electromagnetic waves. Electromagnetic waves had existed before their artificial production, but with Hertz, these waves became the subject, and the instrument, of research in physicists' laboratories.

15

FORCE, ENERGY, AND THERMODYNAMICS

Crosbie Smith

Surveying the history of nineteenth-century science in his magisterial *A History of European Thought in the Nineteenth Century* (1904–12), John Theodore Merz concluded that one "of the principal performances of the second half of the nineteenth century has been to find...the greatest of all exact generalisations – the conception of energy."[1] In a similar vein, Sir Joseph Larmor, heir to the Lucasian Chair of Mathematics at Cambridge once occupied by Newton, wrote in the obituary notice of Lord Kelvin (1824–1907) for the Royal Society of London in 1908 that the doctrine of energy "has not only furnished a standard of industrial values which has enabled mechanical power...to be measured with scientific precision as a commercial asset; it has also, in its other aspect of the continual dissipation of energy, created the doctrine of inorganic evolution and changed our conceptions of the material universe."[2] These bold claims stand at the close of a remarkable era for European physical science, which saw, in the context of British and German industrialization, the replacement of earlier Continental (notably French) action-at-a-distance force physics with the new physics of energy.

This chapter traces the construction of the distinctively nineteenth-century sciences of energy and thermodynamics. Modern historical studies of energy physics have usually taken as their starting point Thomas Kuhn's paper on energy conservation as a case of simultaneous discovery. Kuhn's basic claim was that twelve European men of science and engineering, working more or less in isolation from one another, "grasped for themselves essential parts of the concept of energy and its conservation" during the period between 1830 and 1850. Kuhn then offered an account of this phenomenon of "simultaneous discovery" in terms of shared preoccupations, in varying degrees across the twelve protagonists, with experimental conversion processes, engine

---

[1] J. T. Merz, *A History of European Thought in the Nineteenth Century*, 4 vols. (Edinburgh: Blackwood, 1904–12), 2: 95–6.
[2] Joseph Larmor, "Lord Kelvin," *Proceedings of the Royal Society*, 81 (1908), iii–lxxvi, at p. xxix.

performance, and the unity of nature. Kuhn's critics, identifying the extent to which individuals from the original list diverged from such preoccupations, have not, for the most part, offered a substitute for "simultaneous discovery."[3]

Challenging in the light of social constructivist accounts of science Kuhn's assumption that the elements of energy conservation were there to be *discovered in nature,* I employ a contextualist methodology whereby scientific practitioners construct concepts, such as "energy," within specific local contexts and in relation to particular audiences. By employing such terms as "force," "energy," and "thermodynamics" as historical actors' categories, and by focusing on an interacting and self-conscious group of Scottish natural philosophers who promoted a new "science of energy," I offer an account of energy and thermodynamics that avoids the ahistoricism of earlier models, such as "simultaneous discovery."

THE MECHANICAL VALUE OF HEAT

The formation of the British Association for the Advancement of Science (BAAS) in the 1830s was a major attempt by British gentlemen of science to reform the organization and practice of natural knowledge production during a period characterized by industrial change and social instability. First-generation BAAS reformers had long admired the preeminence of French mathematical physics exemplified in Pierre Simon de Laplace's *Mécanique céleste*. Equally, however, they had become increasingly dissatisfied with the basis of the Laplacian doctrines, which assumed action between point atoms over empty space as the explanatory framework for *all* natural phenomena, from light to electricity and from astronomy to cohesion. A second generation of younger and more radical reformers, associated with the *Cambridge Mathematical Journal,* became enamored with the macroscopic and nonhypothetical flow equations of Joseph Fourier in opposition to the microscopic and hypothetical action-at-a-distance physics of Laplace and his disciples, such as S. D. Poisson. By 1840, the very young Glasgow-based William Thomson (later Lord Kelvin) had committed himself to the Fourier cause and begun a lifelong opposition to Laplacian doctrines. Within a short time, Thomson would find common cause with the respected Michael Faraday (1791–1867), whose own electrical doctrines also contrasted with those of the action-at-a-distance school.[4]

---

[3] T. S. Kuhn, "Energy Conservation as an Example of Simultaneous Discovery," in *Critical Problems in the History of Science,* ed. M. Clagett (Madison: University of Wisconsin Press, 1959), pp. 321–56. For criticism see, for example, P. M. Heimann, "Conversion of Forces and the Conservation of Energy," *Centaurus,* 18 (1974), 147–61; "Helmholtz and Kant: The Metaphysical Foundations of Ueber die Erhaltung der Kraft," *Studies in History and Philosophy of Science,* 5 (1974), 205–38.

[4] Jack Morrell and Arnold Thackray, *Gentlemen of Science: Early Years of the British Association for the Advancement of Science* (Oxford: Clarendon Press, 1981); Robert Fox, "The Rise and Fall of Laplacian Physics," *Historical Studies in the Physical Sciences,* 4 (1974), 89–136; Crosbie Smith and M. Norton

In 1840 the BAAS held its annual meeting in Glasgow. William Thomson and his elder brother James played active supporting roles on behalf of the engineering section. Glasgow's links with the legendary James Watt, and more recently the development of the Clyde as a major site for the construction of cross-channel and ocean steamships, lent the hitherto rather lowly section much-needed status. Soon after, James Thomson commenced a series of apprenticeships in engineering, which eventually took him to the Thames iron shipbuilding works of the famous Manchester engineer William Fairbairn. While there, James avidly studied the theoretical and practical problems of economy in relation to long-distance steam navigation. By August 1844 he had written to his younger brother, now nearing the end of training as a Cambridge mathematician, asking if he knew who had offered an account of the motive power of heat in terms of the mechanical effect (or work done) by the "fall" of a quantity of heat from a state of intensity (high temperature as in a steam-engine boiler) to a state of diffusion (low temperature as in the condenser), analogous to the fall of a quantity of water from a high to a low level in the case of waterwheels.[5]

While in Paris the following spring, William located Emile Clapeyron's memoir (1834) on the subject but failed to locate a copy of the original source in the form of a little-known treatise (1824) by Sadi Carnot (son of the celebrated French engineer Lazare Carnot). At the same time, William began to consider solutions to problems in the mathematical theory of electricity (notably those of two electrified spherical conductors, the complexity of which had defied Poisson's attempts to obtain a general mathematical solution) in terms of mechanical effect given out or taken in, analogous to the work done or absorbed by a waterwheel or heat engine. He therefore recognized that measurements both of electrical phenomena and of steam could be treated in absolute, mechanical, and, above all, engineering terms. The contrast to the action-at-a-distance approach of Laplace and Poisson was striking.[6]

During his first session (1846–7) as Glasgow College professor of natural philosophy, William Thomson rediscovered a model air engine, presented to the college classroom in the late 1820s by its designer, Robert Stirling, but long since clogged with dust and oil. Having joined his elder brother as a member of the Glasgow Philosophical Society in December 1846, Thomson addressed the Society the following April on issues raised by the engine when considered as a material embodiment of the Carnot-Clapeyron account of

---

Wise, *Energy and Empire: A Biographical Study of Lord Kelvin* (Cambridge: Cambridge University Press, 1989), pp. 149–68, 203–28. Fox's "Laplacian Physics" offers a compelling historicist model for French force physics in the early nineteenth century.

[5] Smith and Wise, *Energy and Empire*, pp. 52–5, 288–92.

[6] Sadi Carnot, *Reflexions on the Motive Power of Fire: A Critical Edition with the Surviving Scientific Manuscripts*, trans. and ed. Robert Fox (Manchester: Manchester University Press, 1986); M. Norton Wise and Crosbie Smith, "Measurement, Work and Industry in Lord Kelvin's Britain," *Historical Studies in the Physical and Biological Sciences*, 17 (1986), 147–73, esp. 152–9; Smith and Wise, *Energy and Empire*, pp. 240–50.

the motive power of heat. If, he suggested, the upper part of the engine were maintained at the freezing point of water by a stream of water, and if the lower part were held in a basin of water also at the freezing point, the engine could be cranked forward without the expenditure of mechanical effect (other than to overcome friction) because there existed no temperature difference. The result, however, would be the transference of heat from the basin to the stream and the gradual conversion of all the water in the basin into ice.[7]

Such considerations raised two fundamental puzzles. First, the setup would lead to the production of seemingly unlimited quantities of ice without work. Second, heat was required to melt ice, and yet such heat might instead have been deployed to perform useful work. As he explained the second puzzle to J. D. Forbes: "It seems very mysterious how power can be lost in such a way [by the conduction of heat from hot to cold], but perhaps not more so than that power should be lost in the friction of fluids (a plumb line with the weight in water for instance) by which there does not seem to be any heat generated, nor any physical change effected."[8]

At the close of the session, Thomson attended the BAAS Oxford meeting. Although well known in Cambridge and other mathematical circles for a string of avant-garde articles on electricity, he was making his first appearance at the BAAS as a professor of natural philosophy. The event also marked his first encounter with James Prescott Joule (1818–1889), who had been arguing since 1843 for the mutual convertibility of work and heat according to an exact mechanical equivalence.[9]

Joule's earliest publications, directed at a readership of practical electricians through William Sturgeon's *Annals of Electricity*, had focused on the possibilities opened up by electromagnetic engines for the production of motive power. The *Annals*, indeed, placed great emphasis on "the rise and progress of electro-magnetic engines for propelling machinery." Unlike James Thomson with the Carnot-Clapeyron theory, however, Joule had entered a veritable battlefield of competing theories and practices, in which elite experimental philosophers, such as Faraday and Charles Wheatstone, contended with practical electricians whose livelihood depended upon the shocks and sparks of the new science. With aspirations to gentlemanly, elite status, Joule soon began to emulate not Sturgeon but Faraday as he attempted to fashion himself as an experimental philosopher, rather than an ingenious inventor.[10]

Initial concerns with practical electromagnetic engines provided Joule with the engineering measure of engine performance known as "economical

---

[7] Smith and Wise, *Energy and Empire*, pp. 296–8; Crosbie Smith, *The Science of Energy: A Cultural History of Energy Physics in Victorian Britain* (Chicago: University of Chicago Press, 1998), pp. 47–50.

[8] William Thomson to J. D. Forbes, 1 March 1847, Forbes Papers, St. Andrews University Library; Smith and Wise, *Energy and Empire*, p. 294; Smith, *Science of Energy*, p. 48.

[9] Smith and Wise, *Energy and Empire*, pp. 302–3; Smith, *Science of Energy*, pp. 78–9.

[10] Smith, *Science of Energy*, pp. 57–8; Iwan Morus, "Different Experimental Lives: Michael Faraday and William Sturgeon," *History of Science*, 30 (1992), 1–28.

duty," understood to be the load (in pounds weight) raised to a height of one foot by a pound of fuel such as coal (steam engine) or zinc (electromagnetic engine). Recognizing the serious shortcomings of the latter engine compared to the former, Joule directed his investigations to the sources of resistance in electromagnetic engines. Having already established for himself a relationship for the heating effect in a current-carrying wire as proportional to the square of the current multiplied by the resistance, by 1842–3 he turned his attention to other sources of resistance to economical performance, including the "resistances" of the battery and of the electromagnet. This framework provided him with considerable philosophical authority to pronounce upon the limitations of electromagnetic engines invented by various "ingenious gentlemen."[11]

In early 1843, Joule told the Manchester Literary and Philosophical Society, to which he had been elected twelve months previously, that whatever the arrangement of voltaic apparatus in an electrical circuit, "the whole of the caloric of the circuit is exactly accounted for by the whole of the chemical changes." That is, he sought to persuade himself and his audience that he had traced the heat produced or absorbed in every part of the circuit (including that "latent" in the chemicals of the battery) and had found that the gains and losses were all balanced. But with no gain or loss of work, the conclusion was perfectly consistent with a caloric or material theory of heat, whereby heat was simply transferred from one part of the circuit to another without net production or annihilation.[12]

Presented to the Cork meeting of the BAAS a few months later, Joule's "On the Calorific Effects of Magneto-electricity, and on the Mechanical Value of Heat" reported on an experimental arrangement that introduced the means of producing or requiring mechanical work. The key feature was the deployment of a small electromagnet immersed in water between the poles of a powerful magnet. Joule's main conclusion was that when the electromagnet was used as a magnetoelectric machine (generator), the electricity yielded heat over and above that due to the chemical changes in the battery. Thus, the extra heat was not merely transferred from one part of the arrangement to another, as might be expected from a material theory of heat. Already firmly committed to a mechanical view of nature's agents (including heat and electricity), Joule further argued for a constant ratio between the heat and "mechanical power gained or lost," that is, a "mechanical value of heat."[13]

Adopting the mean result of thirteen experiments, Joule claimed that the "quantity of heat capable of increasing the temperature of a pound of water by one degree of Fahrenheit's scale is equal to, and may be converted into,

---

[11] Smith, *Science of Energy*, pp. 57–63; R. L. Hills, *Power from Steam: A History of the Stationary Steam Engine* (Cambridge: Cambridge University Press, 1989), pp. 36–7, 107–8 (on "duty").

[12] Smith, *Science of Energy*, p. 64; D. S. L. Cardwell, *James Joule: A Biography* (Manchester: Manchester University Press, 1989), p. 45.

[13] J. P. Joule, "On the Calorific Effects of Magneto-electricity, and on the Mechanical Value of Heat," *Philosophical Magazine*, 23 (1843), 263–76, 347–55, 435–43, esp. 435; Smith, *Science of Energy*, pp. 64–5; Cardwell, *Joule*, pp. 53–9.

a mechanical force capable of raising 838 lb. to the perpendicular height of one foot." He admitted that there was a considerable difference among some of these results for the mechanical value of heat (which ranged from 587 to 1040), but the differences were not, he asserted, "greater than may be referred with propriety to mere errors of experiment." But Joule's experimental results hardly spoke for themselves, requiring instead a trustworthy experimenter to assure his uneasy readers that the errors were indeed due to mere errors of experiment and not to some more fundamental cause.[14]

Joule's chosen phrase "mechanical value of heat" was significant. If the meaning of "value" was understood not simply in the *numerical* but also in the *economic* sense, then it is easy to see that Joule's investigations were being shaped by a continuing search for the causes of the failure of his electromagnetic engine to match the economy of heat engines. Earlier concerns with "economical duty" were linked directly to the "mechanical value of heat," that is, to the amount of work obtainable from a given quantity of heat, which in turn derived from chemical or mechanical sources. Thus, his primary concern was not with the conversion of work into heat as in frictional cases – the "waste of useful work," which was of most interest to the Thomson brothers – but with maximizing the conversion of heat from fuel into useful work in various kinds of engine. Joule was, therefore, engaged in constructing a new theory of heat, not as an abstract and speculative set of doctrines, but as a means of understanding the principles that govern the operation and economy of electrical and heat engines of all kinds. Only in retrospect can Joule be represented as a "discoverer" of the conservation of energy and a "pioneer" of the science of energy.

Although Joule aspired to the status of a gentleman of science with its concomitant credibility, he had not attained that status in the mid-1840s. Undeterred by the Royal Society's rejection of his 1840 paper on the "i^2r" law, he submitted a second major paper on the mechanical value, deploying data derived from the condensation and rarefaction of gases, for publication in the Society's prestigious *Philosophical Transactions*.[15] To have succeeded would have given Joule that coveted gentlemanly status.

"It is the opinion of many philosophers," wrote Joule in this 1844 paper, "that the mechanical power of the steam-engine arises simply from the passage of heat from a hot to a cold body, no heat being necessarily lost during the transfer." In the course of its passage, the caloric developed vis viva. Joule, however, asserted that "this theory, however ingenious, is opposed to the recognized principles of philosophy, because it leads to the conclusion that *vis viva* may be destroyed by an improper disposition of the apparatus." Aiming his criticism at Clapeyron for a cleverly contrived theory, Joule explained that the French engineer had inferred that the fall of heat from the

---

[14] Joule, "Calorific Effects," p. 441; Smith, *Science of Energy*, p. 66.
[15] Smith, *Science of Energy*, p. 68; Cardwell, *Joule*, p. 35.

temperature of the fire to that of the boiler leads to an enormous loss of vis viva. Invoking a shared belief with two eminent Royal Society Fellows, Joule countered: "Believing that the power to destroy belongs to the Creator alone, I entirely coincide with Roget and Faraday in the opinion that any theory which, when carried out, demands the annihilation of force, is necessarily erroneous." His own theory, then, substituted the straightforward conversion into mechanical power of an equivalent portion of the heat contained in the steam expanding in the cylinder of a steam engine.[16]

Summarizing for the Royal Society's *Proceedings,* the Society's secretary (P. M. Roget) noted that Joule's experimental method relied upon the accurate measurement of the heat produced by work done in compressing a gas. Conversely, Joule was claiming that the expansion of a gas against a piston would result in a loss of heat equivalent to the work done. On the other hand, the argument that no work was done by a gas expanding into a vacuum rested on the contentious claim that no change in temperature had been or could be detected. Much depended upon the audience's trust in the accuracy of the thermometers employed.[17] As Otto Sibum has argued, Joule's own exacting thermometric skills can be located in the context of the family brewing business.[18] Such personal skills, however, initially carried little authority with Joule's peers.

Returning to the BAAS in 1845, Joule presented to the Chemistry Section a further method for determination of the mechanical equivalent. The apparatus consisted of a paddle wheel placed in a can filled with water and driven by strings attached over pulleys to weights that descended vertically. Once again his peers seemed indifferent to his conclusions. Two years later, he addressed the Mathematics and Physics Section and was apparently told to keep his remarks brief on account of pressure of business. In the official BAAS *Report,* the synopsis of his paper was printed under the less-prestigious Chemistry Section. But William Thomson's attention had been attracted by Joule's focus on the conversion of mechanical effect into heat in fluid friction, the very problem of "loss" or "waste" that had been puzzling the Thomson brothers. Other savants present, notably Faraday and G. G. Stokes, offered suggestions for similar experiments with liquids, such as mercury. Before long, Thomson himself was employing assistants, and even considering the use of a steam engine, to demonstrate in dramatic fashion the heating effects of fluid friction. Joule was at last receiving the credibility he had long craved.[19]

---

[16] J. P. Joule, "On the Changes of Temperature Produced by the Rarefaction and Condensation of Air," *Philosophical Magazine,* 26 (1844), 369–83, esp. pp. 381–2; Smith, *Science of Energy,* p. 69; Cardwell, *Joule,* pp. 67–8.
[17] Smith, *Science of Energy,* pp. 68–9.
[18] Otto Sibum, "Reworking the Mechanical Value of Heat: Instruments of Precision and Gestures of Accuracy in Early Victorian England," *Studies in History and Philosophy of Science,* 26 (1994), 73–106.
[19] Smith, *Science of Energy,* pp. 70–3, 79–81; Cardwell, *Joule,* p. 87.

In 1848 a German physician, Julius Robert Mayer (1814–1878), became acquainted with Joule's papers on the mechanical equivalent of heat. Seizing this opportunity to impress upon the scientific establishments the importance of his own contributions during the 1840s, Mayer wrote to the French Academy of Sciences pointing out his claims to priority. Published in the *Comptes Rendu* (the Academy's official reports), his letter drew a rapid defense from Joule. Joule's tactics, agreed upon in consultation with his new advocate, William Thomson, were to acknowledge Mayer's priority with respect to the idea of a mechanical equivalent, but to claim that he (Joule) had established it by experiment.[20]

Mayer's papers, unorthodox and unconvincing to contemporary men of science, had been rejected by most German and French scientific authorities, leaving him to fall back upon the last resort of private publication. Outside the dominant schools of European mathematical and experimental science, Mayer's work nevertheless shared with that of his Prussian contemporary, Hermann von Helmholtz (1821–1894), a straddling of the complementary fields of German physics and physiology. From about the mid-1820s, German physiologists had been reacting strongly against the "speculative" and "unscientific" doctrines of *Naturphilosophie,* with its account of unity and organization in Nature in terms of an immanent mind or *Geist.* In very different local contexts, both Mayer and Helmholtz deployed physics to launch aggressive attacks on the notion that living matter depended on a special vital force, *Lebenskraft.*[21] But only as the priority dispute with Joule developed in the late 1840s and beyond did the writings of Mayer begin to be reread as "pioneering contributions" toward the doctrines of energy physics.

A SCIENCE OF ENERGY

From 1847, Thomson recognized in Joule's claim for the conversion of work into heat an answer to the puzzle (highlighted by the Stirling engine) of what happened to the seeming "loss" of the useful work that might have been done, but that was instead "wasted" in conduction and fluid friction. Unconvinced, however, by Joule's complementary claim that such heat could in principle be converted into work, Thomson remained deeply perplexed by what seemed to him the irrecoverable nature of that heat. Furthermore, he could not accept

---

[20] Smith, *Science of Energy,* pp. 73–6.
[21] Timothy Lenoir, *The Strategy of Life: Teleology and Mechanics in Nineteenth Century German Biology* (Dordrecht: Reidel, 1982), pp. 103–11; M. Norton Wise, "German Concepts of Force, Energy, and the Electromagnetic Ether: 1845–1880," in *Conceptions of Ether: Studies in the History of Ether Theories 1740–1900,* ed. G. N. Cantor and M. J. S. Hodge (Cambridge: Cambridge University Press, 1981), pp. 269–307 esp. 271–5. On Mayer's contexts see K. L. Caneva, *Robert Mayer and the Conservation of Energy* (Princeton, N.J.: Princeton University Press, 1993).

Joule's rejection of the Carnot-Clapeyron theory, with its "fall" of heat from high to low temperature, in favor of mutual convertibility.[22]

With regard to the first puzzle raised by the Stirling engine, James Thomson soon pointed out the implication that since ice expands on freezing, it could be made to do useful work: In other words, the arrangement would function as a perpetual source of power, long held to be impossible by almost all orthodox engineers and natural philosophers. In order to avoid such an inference, he therefore predicted that the freezing point would be found to be lowered with increase of pressure. His prediction, and its subsequent experimental confirmation in William's laboratory, did much to persuade the brothers of the value of the Carnot-Clapeyron theory.[23]

Within a year, William had added another feature to the Carnot-Clapeyron construction, namely, an absolute scale of temperature. In presentations to the Glasgow and Cambridge Philosophical Societies (1848), William explained that the air-thermometer scale provided "an arbitrary series of numbered points of reference sufficiently close for the requirements of practical thermometry." In an absolute thermometric scale, "a unit of heat descending from a body A at the temperature $T°$ of this scale, to a body B at the temperature $(T-1)°$, would give out the same mechanical effect [motive power or work], whatever be the number T." Its absolute character derived from its being "quite independent of the physical properties of any specific substance." In other words, unlike the air thermometer, which depended on a particular gas, he deployed the waterfall analogy to establish a scale of temperature independent of the working substance.[24]

When Thomson acquired from his colleague Lewis Gordon (professor of civil engineering and mechanics at Glasgow since 1840) a copy of the very rare Carnot treatise, he presented an "Account of Carnot's Theory," written in the light of the issues raised by Joule, to the Royal Society of Edinburgh, for publication in its *Proceedings* and *Transactions*. In particular, Thomson read Carnot as claiming that any work obtained from a cyclical process can only derive from transfer of heat from high to low temperature. From this claim, grounded on a denial of perpetual motion, Thomson inferred that no engine could be more efficient than a perfectly reversible engine ("Carnot's criterion" for a perfect engine). It further followed that the maximum efficiency obtainable from any engine operating between heat reservoirs at

---

[22] Smith and Wise, *Energy and Empire*, pp. 294, 296, 310–11; Smith, *Science of Energy*, pp. 82–6.
[23] Smith, *Science of Energy*, pp. 50–1, 95–7; Crosbie Smith, " 'No Where but in a Great Town': William Thomson's Spiral of Class-room Credibility," in *Making Space for Science: Territorial Themes in the Shaping of Knowledge*, ed. Crosbie Smith and Jon Agar (Basingstoke, England: Macmillan, 1998), pp. 118–46.
[24] William Thomson, "On an Absolute Thermometric Scale, Founded on Carnot's Theory of the Motive Power of Heat, and Calculated from the Results of Regnault's Experiments on the Pressure and Latent Heat of Steam," *Philosophical Magazine*, 33 (1848), 313–17; Smith, *Science of Energy*, pp. 51–2; Smith and Wise, *Energy and Empire*, p. 249.

different temperatures would be a function of those temperatures (Carnot's function).[25]

Acquainted with the issues through a reading of Thomson's "Account," the German theoretical physicist Rudolf Clausius (1822–1888) produced in 1850 the first reconciliation of Joule and Carnot. Accepting a general mechanical theory of heat (that heat was vis viva) and, hence, Joule's claim for the mutual convertibility of heat and work, Clausius retained the part of Carnot's theory that required a transfer of heat from high to low temperature for the production of work. Under the new theory, then, a portion of the initial heat was converted into work according to the mechanical equivalent of heat, and the remainder descended to the lower temperature. In order to demonstrate that no engine could be more efficient than a perfectly reversible one, Clausius reasoned that if such an engine did exist, "it would be possible, without any expenditure of force or any other change, to transfer as much heat as we please from a cold to a hot body, and this is not in accord with the other relations of heat, since it always shows a tendency to equalise temperature differences and therefore to pass from hotter to colder bodies."[26]

At the same time, a young Scottish engineer, Macquorn Rankine (1820–1872), had been turning his attention to the question of the motive power of heat from the perspective of a molecular vortex hypothesis. Far more specific than Clausius's very general claims for heat as vis viva of some kind, and far more mathematical than Joule's recent speculations linking heat, electricity, and vis viva at a molecular level, Rankine's hypothesis nevertheless shared with its competitors the view that heat was mechanical in nature. Brought into contact by their mutual acquaintance with J. D. Forbes, Edinburgh professor of natural philosophy, Thomson and Rankine began evaluating in 1850 the claims of Clausius for a reconciliation of Joule and Carnot, and especially the new foundation that Clausius appeared to have offered for the theory of the motive power of heat.[27]

Prompted by these discussions, Thomson finally laid down two propositions early in 1851, the first a statement of Joule's mutual equivalence of work and heat, and the second a statement of Carnot's criterion (as modified by Clausius) for a perfect engine. His long-delayed acceptance of Joule's proposition rested on a resolution of the problem of the irrecoverability of mechanical effect lost as heat. He now privately believed that work "is *lost to man* irrecoverably though *not lost in the material world.*" Thus, although "no destruction of energy can take place in the material world without an act of power possessed only by the supreme ruler, yet transformations take

---

[25] Smith, *Science of Energy*, pp. 86–95; Smith and Wise, *Energy and Empire*, pp. 323–4.
[26] Rudolf Clausius, "On the Moving Force of Heat, and the Laws Regarding the Nature of Heat itself which are Deducible Therefrom," *Philosophical Magazine*, 2 (1851), 1–21, 102–19; Smith, *Science of Energy*, pp. 97–9.
[27] Smith, *Science of Energy*, pp. 102–7; Smith and Wise, *Energy and Empire*, pp. 318–27.

place which remove irrecoverably from the control of man sources of power which ... might have been rendered available." In other words, God alone could create or destroy energy (i.e., energy was conserved in total quantity), but human beings could make use of transformations of energy, for example, in waterwheels or heat engines.[28]

In his private draft, Thomson grounded these transformations on a universal statement that "everything in the material world is progressive." On the one hand, this statement expressed the geological directionalism of Cambridge academics, such as William Hopkins (Thomson's former mathematical coach) and Adam Sedgwick (professor of geology), in opposition to the steady-state uniformitarianism of Charles Lyell. But on the other hand, it could be read as agreeing with the radical evolutionary doctrines of the subversive *Vestiges of Creation* (1844). In his published statement (1852), Thomson opted instead for universal dissipation of energy, a directionalist (and thus "progressive") doctrine that reflected the Presbyterian (Calvinist) views of a transitory visible creation, rather than a universe of ever-upward progression. Work dissipated as heat would be irrecoverable to human beings, for to deny this principle would be to imply that we could produce mechanical effect by cooling the material world with no limit except the total loss of heat from the world.[29]

This reasoning crystallized in what later became the canonical Kelvin statement of the second law of thermodynamics, first enunciated by Thomson in 1851: "[I]t is impossible, by means of inanimate material agency, to derive mechanical effect from any portion of matter by cooling it below the temperature of the coldest of the surrounding objects." This statement provided Thomson with a new demonstration of Carnot's criterion of a perfect engine. Having resolved the recoverability issue, he also quickly adopted a dynamical theory of heat, making it the basis of Joule's proposition of mutual equivalence and abandoning the Carnot-Clapeyron notion of heat as a state function (i.e., that in any cyclic process the change in heat content is zero).[30]

In January 1852, William Thomson saw for the first time Helmholtz's "admirable treatise on the principle of mechanical effect," published nearly five years earlier as *Über die Erhaltung der Kraft*. Far from seeing Helmholtz as a threat to British priorities, however, Thomson rapidly appropriated the German physiologist's essay to the British cause, deploying it ultimately as a means of enhancing the international credibility of the new "epoch of energy." For his part, Helmholtz derived dramatic gains in credibility from Thomson's enthusiastic recognition of the value of the 1847 essay, which

---

[28] Smith and Wise, *Energy and Empire*, pp. 327–32; Smith, *Science of Energy*, p. 110.
[29] Crosbie Smith, "Geologists and Mathematicians: The Rise of Physical Geology," in *Wranglers and Physicists: Studies on Cambridge Physics in the Nineteenth Century*, ed. P. M. Harman (Manchester: Manchester University Press, 1985), pp. 49–83; James A. Secord, "Behind the Veil: Robert Chambers and *Vestiges*," in *History, Humanity and Evolution*, ed. J. R. Moore (Cambridge: Cambridge University Press, 1989), pp. 165–94; Smith, *Science of Energy*, pp. 110–20.
[30] Smith and Wise, *Energy and Empire*, p. 329.

had hitherto received rather mixed reactions from German physicists. John Tyndall, whom Helmholtz first met in August 1853, translated the essay in the same year for *Scientific Memoirs. Natural Philosophy* (edited by Tyndall and the publisher William Francis). Also in 1853, Helmholtz traveled to England for the Hull meeting of the British Association where he met Hopkins, whose presidential address did much to promote, especially on Thomson's behalf, the new doctrines of heat. He became acquainted with other members of Thomson's circle, notably Stokes and the Belfast chemist Thomas Andrews, though it was not until 1855 that he met Thomson in person. By 1853 he could write that his *Erhaltung der Kraft* was "better known here [in England] than in Germany, and more than my other works."[31]

In a draft for his "On a Universal Tendency in Nature to the Dissipation of Mechanical Energy" (1852), Thomson reworked Helmholtz's arguments in the light of his own fundamental convictions. At first sight, the analyses appear identical. But Helmholtz's commitment to a basic physics of attractive and repulsive forces acting at a distance contrasted strikingly with Thomson's early preference for continuum approaches to physical agencies, such as electricity and magnetism. *Erhaltung der Kraft* as conservation of force, whose quantity is measured in terms of vis viva and whose intensity is expressed in terms of attractive or repulsive forces acting at a distance, was now being read as an independent "Universal Truth," "conservation of mechanical energy," whose quantity is measured as mechanical effect and whose intensity is understood in terms of a potential gradient.[32]

Thomson's "On a Universal Tendency" took the new "energy" perspective to a wide audience. In this short paper for the *Philosophical Magazine*, the term "energy" achieved public prominence for the first time, and the dual principles of conservation and dissipation of energy were made explicit: "[A]s it is most certain that Creative Power alone can either call into existence or annihilate mechanical energy, the 'waste' referred to cannot be annihilation, but must be some transformation of energy." Now the dynamical theory of heat, and with it a whole program of dynamical (matter-in-motion) explanation, went unquestioned. And now, too, the universal primacy of the energy laws opened up fresh questions about the origins, progress, and destiny of the solar system and its inhabitants. Two years later, Thomson told the Liverpool meeting of the British Association that Joule's discovery of the conversion of work into heat by fluid friction, the experimental foundation of the new energy physics, had "led to the greatest reform that physical science has experienced since the days of Newton."[33]

---

[31] Leo Koenigsberger, *Hermann von Helmholtz*, trans. F. A. Welby (Oxford: Clarendon Press, 1906), pp. 109–13, 144–6; Smith, *Science of Energy*, chap. 7. See also Fabio Bevilacqua, "Helmholtz's *Ueber die Erhaltung der Kraft*," in *Hermann von Helmholtz and the Foundations of Nineteenth-Century Science*, ed. David Cahan (Berkeley: University of California Press, 1993), pp. 291–333; Smith, *Science of Energy*, pp. 126–7.

[32] Smith and Wise, *Energy and Empire*, p. 384.

[33] William Thomson, "On the Mechanical Antecedents of Motion, Heat, and Light," *Report of the British Association for the Advancement of Science*, 24 (1854), 59–63.

From the early 1850s, the Glasgow professor and his new ally in engineering science, Macquorn Rankine, began replacing an older language of mechanics with such terms as "actual energy" ("kinetic" from 1862) and "potential energy." By 1853, Rankine had formally restyled the "principle of mechanical effect" as "the law of the conservation of energy," that "the sum of the actual and potential energies in the universe is unchangeable." The new language, developed by Thomson and Rankine, signified their concern not merely to avoid ambiguities in speaking about "force" and "energy" in physics and engineering, but also to reinforce a whole new way of thinking about and doing science.[34]

Within a few years, Thomson and Rankine had been joined by like-minded scientific reformers, most notably the Scottish natural philosophers James Clerk Maxwell (1831–1879), Peter Guthrie Tait (1831–1901), and Balfour Stewart (1828–1887), together with the engineer Fleeming Jenkin (1833–1885). With strong links to the British Association, this informal grouping of "North British" physicists and engineers was primarily responsible for the construction and promotion of the "science of energy," inclusive of nothing less than the whole of physical science. Natural philosophy or physics was redefined as the study of energy and its transformations. As William Garnett (Maxwell's assistant at the Cavendish and later one of his biographers) put the issue in the *Encyclopaedia Britannica* (9th edition) in 1879: "A complete account of our knowledge of energy and its transformations would require an exhaustive treatise on every branch of physical science, for natural philosophy is simply the science of energy."[35]

With respect to the material world, Thomson and Rankine had adapted Carnot's theory and set up an ideal of a perfect thermodynamic engine against which existing and future engines could be assessed. All such engines were liable to some incomplete restoration if run in reverse. Friction, spillage, and conduction produced "waste," ensuring that a working engine would fall short of the ideal. No human engineers could ever hope to construct such a perfect engine, but Rankine and his Glasgow friend James Robert Napier (son of the famous Clyde shipbuilder and marine engine builder) collaborated on a new design of air engine which, unlike previous attempts, would embody the new energy principles. Reworking the concept of an indicator diagram as a "diagram of energy" to express the useful work delivered by a prime mover, Rankine did much to promote the new theory of the motive power of heat, restyling the science of thermodynamics by Thomson and Rankine from 1854.[36]

---

[34] W. J. M. Rankine, "On the General Law of the Transformation of Energy," *Philosophical Magazine*, 5 (1853), 106–17; Smith, *Science of Energy*, pp. 139–40.

[35] William Garnett, "Energy," *Encyclopaedia Britannica* [9th ed.], vol. 8, pp. 205–11; Smith, *Science of Energy*, p. 2.

[36] Ben Marsden, "Engineering Science in Glasgow. W. J. M. Rankine and the Motive Power of Air," PhD diss., University of Kent at Canterbury, 1992; "Blowing Hot and Cold: Reports and Retorts on the Status of the Air-Engine as Success or Failure, 1830–1855," *History of Science*, 36 (1998), 373–420.

Alongside the question of imperfect prime movers went the complementary question of nature's ultimate perfection. As the older Calvinist views of a fallen state of man and nature gave way in mid-nineteenth-century Scotland to more liberal Presbyterian doctrines of Christ as perfect humanity, so did the older views of a nature inherently imperfect and decayed yield to nature as a perfect creation with man alone as the fallen creature. Rankine's 1852 speculation regarding nature's reconcentration of energy suggested that the universe as a whole might function as a perfectly reversible thermodynamic engine, thereby limiting "dissipation" to the visible portion only and asserting that the creation did not, in the Rev. Thomas Chalmers's earlier Calvinistic language, contain within itself the seeds of its own destruction. Thomson, on the other hand, preferred to point to an infinite universe of energy with an "endless progress, through endless space," in which the "dissipation of energy" was characterized not as imperfection in nature but as an irreversible stream of energy from concentration to diffusion. Stewart and Tait took this perspective much further, locating the visible and transitory universe within an unseen universe, in which the law of dissipation of energy might not hold as an ultimate principle.[37]

Whatever the ultimate condition of the universe, however, all members of the North British group agreed that the directionality of energy flow (whether expressed as "progression" or "dissipation" in the material world) characterized the visible creation, and that this doctrine was the strongest weapon in the armory against anti-Christian materialists and naturalists. By his direct involvement with the history and meaning of energy physics, John Tyndall (1820–1893) had rapidly assumed the status of bête noire for the scientists of energy. Tyndall's elevation of Mayer gave Tait a golden opportunity to caricature the German physician as the embodiment of speculative and amateurish metaphysics, and to set him against the trustworthy and gentlemanly producer of experimental knowledge from Manchester. But Tyndall's associations with other scientific naturalists, such as Thomas Henry Huxley and Herbert Spencer, made him especially dangerous. However much Tyndall might profess views above those of rank materialism, his opposition to dogmatic Christianity and his seeming commitment to scientific determinism throughout both inanimate and living nature made him a ready, if subtle, embodiment of materialism.[38]

For the North British group, and especially for Thomson and Maxwell, the core doctrine of "materialism" was reversibility. In a purely dynamical system, there was no difference between running forward or backward. If, then, the visible world were a purely dynamical system, we could in principle have a cyclical world that would run in either direction. But the doctrine of irreversibility killed all such cyclical cosmologies stone dead. The ramifications

---

[37] Developed in Smith, *Science of Energy*, esp. pp. 15–30, 110–20, 307–14.
[38] Ibid., pp. 170–91, 253–5.

of the doctrine of irreversibility were indeed manifold. At one level, Thomson and his allies deployed it to construct estimates of the past ages of Earth and Sun that would police geological and biological theorizing, in general, and undermine Charles Darwin's doctrine of natural selection, in particular. At another level, they would use it to reinforce, as Maxwell put it, "the doctrine of a beginning."[39]

To these ends of demonstrating the limits to the mechanical effect available for past, present, and future life on Earth, Thomson examined the principal source of this energy, namely the Sun. Arguing that the Sun's energy was too great to be supplied by chemical means or by a mere molten mass cooling, he at first suggested that the Sun's heat was provided by vast quantities of meteors orbiting around the Sun but inside the Earth's orbit. Retarded in their orbits by an ethereal medium, the meteors would progressively spiral toward the Sun's surface in a cosmic vortex analogous to his brother's vortex turbines (horizontal waterwheels). As the meteors vaporized by friction, they would generate immense quantities of heat. In the early 1860s, however, he adopted Helmholtz's version of the Sun's heat, whereby contraction of the body of the Sun released heat over long periods. Either way, the Sun's energy was finite and calculable, making possible order-of-magnitude estimates of the limited past and future duration of the Sun. In response to Darwin's demand for a much longer time for evolution by natural selection, and in opposition to Lyell's uniformitarian geology upon which Darwin's claims were grounded, Thomson deployed Fourier's conduction law to make similar estimates for the Earth's age. The limited timescale of about 100 million years (later reduced) approximated estimates for the Sun's age. But the new cosmogony was itself evolutionary, offering little or no comfort to strict biblical literalists within the Scottish churches (especially the recently founded Free Church of Scotland whose clergy reaffirmed traditional readings of the Old and New Testaments).[40]

Parallel North British concerns about the importance of free will as a directing agency in a universe of mechanical energy provided a principal context for Maxwell's statistical interpretation of the second law of thermodynamics in 1867. Framing his insight in terms of a microscopic creature possessed of free will to direct the sorting of molecules, he interpreted the second law's meaning relative to human beings, who were imperfect in their ability to know molecular motions and to devise tools to control them. Available energy, then, became "energy which we can direct into any desired channel," whereas dissipated energy was "energy which we cannot lay hold of and direct at pleasure." The notion of dissipation would not, therefore, occur either to

---

[39] James Clerk Maxwell to Mark Pattison, 7 April 1868, in *The Scientific Letters and Papers of James Clerk Maxwell*, ed. P. M. Harman, 2 vols. published (Cambridge: Cambridge University Press, 1990–), 2: 358–61; Smith, *The Science of Energy*, pp. 239–40.
[40] Smith and Wise, *Energy and Empire*, pp. 497–611; J. D. Burchfield, *Lord Kelvin and the Age of the Earth* (London: Macmillan, 1975); Smith, *Science of Energy*, esp. pp. 110–25, 140–9.

a creature unable to "turn any of the energies of nature to his own account" or to one, such as Maxwell's imaginary demon, who "could trace the motion of every molecule and seize it at the right moment." Only to human beings, then, did energy appear to be passing "inevitably from the available to the dissipated state."[41]

Maxwell's "demon" purported to illustrate the statistical character of the second law of thermodynamics. Maxwell and his colleagues, therefore, disapproved strongly of Continental attempts by Clausius and others to deduce the law from purely mechanical principles. More generally, such Continental approaches contrasted strikingly with the North British emphasis on visualizable processes and experimentally grounded concepts. Maxwell could admire the thermodynamics of the American Josiah Willard Gibbs with its graphical representations, but condemn the mathematical complexities of Ludwig Boltzmann. As Maxwell told Tait in 1873: "By the study of Boltzmann I have become unable to understand him. He could not understand me on account of my shortness and his length was and is an equal stumbling block to me."[42]

The new science of thermodynamics was embodied in successive textbooks by Rankine (1859), Tait (1868), and Maxwell (1871). The most celebrated text for the "science of energy," however, was Thomson and Tait's *Treatise on Natural Philosophy* (1867). Originally intended to treat all branches of natural philosophy, the *Treatise* was limited to volume one only, comprising dynamical foundations. Taking statics to be derived from dynamics, Thomson and Tait reinterpreted Newton's third law (action–reaction) as conservation of energy, with action viewed as rate of working. Fundamental to this work-based physics was the move to make extremum conditions, rather than point forces, the theoretical foundation of dynamics. The tendency of an entire system to move from one place to another in the most economical way would determine the forces and motions of the various parts of the system. Variational principles (especially least action) played a central role in the new dynamics.[43]

THE ENERGY OF THE ELECTROMAGNETIC FIELD

The delay of the *Treatise*, due in large part to Thomson's preference for practical projects over literary ones, made space for Maxwell to produce a complementary *Treatise on Electricity and Magnetism*. As Tait wrote in a

---

[41] James Clerk Maxwell, *The Scientific Papers of James Clerk Maxwell*, ed. W. D. Niven, 2 vols. (Cambridge: Cambridge University Press, 1890), 2: 646; Smith and Wise, *Energy and Empire*, p. 623; Smith, *Science of Energy*, pp. 240–1, 247–52.

[42] James Clerk Maxwell to P. G. Tait, ca. August 1873, in *The Scientific Letters and Papers of James Clerk Maxwell*, 2: 915–16; Smith, *Science of Energy*, pp. 255–67.

[43] Smith and Wise, *Energy and Empire*, pp. 348–95; M. Norton Wise, "Mediating Machines," *Science in Context*, 2 (1988), 77–113.

contemporary review for *Nature,* the main object of Maxwell's *Treatise* (1873), "besides teaching the experimental facts of electricity and magnetism... is simply to upset completely the notion of *action at a distance.*" In the mid-1840s, Wilhelm Weber had constructed a major new unifying theory of electricity based on the interaction of electric charge at a distance. But between 1854 and his death a quarter of a century later, James Clerk Maxwell made relentless efforts to depose Weber's theory from its preeminent position as the most powerful and persuasive interpretation yet on offer.[44]

Locating Maxwell in opposition to Continental action-at-a-distance theories and in alignment with Faraday's "field" theories, however, reveals only part of the historical picture. Shaped by a distinctive Presbyterian culture, Maxwell's deeply Christian perspective on nature and society became inseparable from his central commitment to the science of energy. Yet the science of energy was in a state of *construction,* rather than a finished edifice. It provided the cultural and conceptual framework within which Maxwell would build credibility for himself and for his controversial electromagnetic theory. To that end, he would depend heavily on private, critical discussions with his closest scientific colleagues, Thomson and Tait, and would attempt to tailor his successive investigations to specific public audiences.[45]

Written by a young Trinity College don for an audience representing (since the foundation of the Cambridge Philosophical Society in 1819) the university's mathematical and scientific establishment, Maxwell's first electrical paper, "On Faraday's Lines of Force" (1856), was designed to appeal to an older generation of Cambridge mathematical reformers, notably William Whewell, who had advocated geometrical reasoning over analytical subtleties as the pedagogical core of the university's "liberal education." This career-making paper belonged to a strong Cambridge "kinematical" research tradition (exemplified by the hydrodynamical and optical papers of Stokes and the physical geology of Hopkins), which regarded the formulation of geometrical laws as the prerequisite to mathematical dynamical theory.[46]

Maxwell's second paper, "On Physical Lines of Force" (1860–1), addressed instead the wider readership of the *Philosophical Magazine*. Published in four installments (1861–2), "On Physical Lines" aimed "to clear the way for speculation" in the direction of understanding the *physical* nature (rather than simply the geometrical form) of lines of magnetic force. Although in Part I (magnetism), Maxwell employed the language of "mechanical effect" and "work done," rather than energy, it was in Part II (electric current) that he began introducing Rankine's "actual" and "potential energy." Emphasizing

---

[44] P. G. Tait, "Clerk-Maxwell's Electricity and Magnetism," *Nature,* 7 (1878), 478–80; Smith, *Science of Energy,* pp. 211, 232–8.

[45] Smith, *Science of Energy,* p. 211.

[46] James Clerk Maxwell, "On Faraday's Lines of Force," *Transactions of the Cambridge Philosophical Society,* 10 (1856), 27–83; Smith and Wise, *Energy and Empire,* 61–5; Smith, *Science of Energy,* pp. 218–22.

that he had there attempted to imitate electromagnetic phenomena "by an imaginary system of molecular vortices," he issued a subtle challenge to his opponents: "Those who look in a different direction for the explanation of the facts, may be able to compare this theory with that of the existence of currents flowing freely through bodies, and with that which supposes electricity to act at a distance with a force depending on its velocity, and therefore not subject to the law of conservation of energy." Weber especially would have to answer for his theory's seeming violation of energy conservation.[47]

At the same time, however, Maxwell admitted to the idle-wheel hypothesis, introduced to represent electric current, being "somewhat awkward" and of "provisional and temporary character." While he emphasized that he had not brought it forward as "a mode of connexion existing in nature," it was "a mode of connexion which is mechanically conceivable, and easily investigated." Concerned to offer a *possible* explanation in terms of a continuous mechanism in opposition to action-at-a-distance force models, he later explained to Tait that the vortex theory "is built up to shew that the phenomena are such as can be explained by mechanism. The nature of this mechanism is to the true mechanism what an orrery is to the solar system."[48]

From his extended molecular vortex model, Maxwell in Part III deduced energy expressions for the magnitude of the forces, as an inverse square law, acting between two charged bodies. Comparing this "force law" with its familiar counterpart in electrostatic measure (Coulomb's law) enabled a direct relation to be established between "the statical and dynamical measures of electricity." He then made the dramatic assertion that he had shown, "by a comparison of the electro-magnetic experiments of MM. Kohlrausch and Weber with the velocity of light as found by M. Fizeau, that the elasticity of the magnetic medium in air is the same as that of the luminiferous medium, if these two coexistent, coextensive, and equally elastic media are not rather one medium." In other words, Maxwell had calculated a theoretical velocity of transverse undulations in the "magnetic medium." This velocity, he reiterated, "agrees so exactly with the [experimentally measured] velocity of light . . . that we can scarcely avoid the inference that *light consists in the transverse undulations of the same medium which is the cause of electric and magnetic phenomena.*"[49] Such a radical claim would form the core of Maxwell's "electromagnetic theory of light." But convincing his scientific peers was going to require a more credible formulation than one based upon artificial mechanisms.

Concern about credibility formed a key motivation for "A Dynamical Theory of the Electromagnetic Field," published in the Royal Society's *Phil.*

---

[47] James Clerk Maxwell, "On Physical Lines of Force," *Philosophical Magazine*, 21 (1861), 161–75, 281–91, 338–48; 23 (1862), 12–24, 85–95; Daniel Siegel, *Innovation in Maxwell's Electromagnetic Theory: Molecular Vortices, Displacement Current, and Light* (Cambridge: Cambridge University Press, 1991), pp. 35–41.

[48] James Clerk Maxwell to P. G. Tait, 23 December 1867, in *The Scientific Letters and Papers of James Clerk Maxwell*, 2: 176–81.

[49] Maxwell, "On Physical Lines," pp. 20–4; Siegel, *Innovation*, pp. 81–3.

Trans. (1865) as Maxwell's third substantial paper on electricity and magnetism. As he told Tait in 1867, the paper departed from the style of "Physical Lines": It was "built on Lagrange's Dynamical Equation and is not wise about vortices." Seeking once again to go beyond a kinematical, geometrical description of electromagnetic phenomena, Maxwell turned to a distinctive style of "dynamical" theory that had found recent exposition in the optical and hydrodynamical investigations of the Lucasian professor at Cambridge, Stokes, and in which specific mechanisms yielded to very general assumptions of matter in motion. In this case, the ethereal medium, made credible by the (by now) highly reputable undulatory theory of light and by recent energy cosmology, was to be the means by which energy was transmitted between gross bodies.[50]

Exploring the electromagnetic field through the phenomena of induction and attraction of currents, and mapping the distribution of magnetic fields, Maxwell sought to express the results in the form of "the General Equations of the Electromagnetic Field," requiring at this stage some twenty equations in total, involving twenty variable quantities: electric currents by conduction, electric displacements, total currents, magnetic forces, electromotive forces, electromagnetic momenta (each with three components), free electricity, and electric potential. Maxwell attempted to express in terms of these quantities what he now named "the intrinsic energy of the Electromagnetic Field as depending partly on its magnetic and partly on its electric polarization at every point." He also made clear that he wanted his readers to view "energy" as a literal, real entity and not simply a concept for dynamical illustration:

> In speaking of the Energy of the field... I wish to be understood literally. All energy is the same as mechanical energy, whether it exists in the form of motion or in that of elasticity, or in any other form. The energy in electromagnetic phenomena is mechanical energy. The only question is, Where does it reside? On the old theories it resides in the electrified bodies, conducting circuits, and magnets, in the form of an unknown quality called potential energy, or the power of producing certain effects at a distance. On our theory it resides in the electromagnetic field, in the space surrounding the electrified and magnetic bodies, as well as in those bodies themselves, and is in two different forms, which may be described without hypothesis as magnetic polarization and electric polarization, or, according to a very probable hypothesis, as the motion and the strain of one and the same medium.[51]

Refined further, this energy approach to electromagnetism would find full conceptual expression in the *Treatise*.[52] But the science of energy had its

---

[50] James Clerk Maxwell, "A Dynamical Theory of the Electromagnetic Field," *Phil. Trans.*, 155 (1865), 459–512; Smith, *Science of Energy*, pp. 228–32.
[51] Maxwell, "Dynamical Theory"; Maxwell, *Scientific Papers*, 1: 564.
[52] James Clerk Maxwell, *A Treatise on Electricity and Magnetism*, 2 vols. (Oxford: Clarendon Press, 1873); Smith, *Science of Energy*, pp. 232–8.

focal point as much in the physical laboratory as in mathematical treatises. Ever since his participation in Henri-Victor Regnault's laboratory practice in 1845, Thomson had resolved to make physical measurements in absolute or mechanical measures. This commitment derived from a realization that electricity could be measured simply in terms of the work done by the fall of a quantity of electricity through a potential, just as in the way that work was done by the fall of a mass of water through a height. His absolute scale of temperature utilized the same notion of absolute measurement in the case of heat, but his first public commitment to a system of absolute units for electrical measurement coincided both with his reading of Wilhelm Weber's contribution "On the Measurement of Electric Resistance According to an Absolute Standard" to Poggendorff's *Annalen* (1851), and with his own "Dynamical Theory of Heat" series. In contrast to Weber's system founded on absolute measures of electromotive forces and intensities, Thomson's approach continued to be grounded on measurements of mechanical effect or work. His 1851 paper on the subject, therefore, deployed Joule's mechanical equivalent to calculate the heat produced by the work done in an electrical circuit. Further applying Joule's earlier relationship of heat to current and resistance squared yielded an expression for resistance in absolute measurement.[53]

Unable to attend the 1861 Manchester meeting of the British Association in person, Thomson nevertheless worked vigorously to secure the appointment of a Committee "On Standards of Electrical Resistance." Fleeming Jenkin, only recently introduced to Thomson, handled on his behalf the delicate negotiations among practical electricians and natural philosophers. The outcome was a committee, already heavily weighted toward scientific men, which eventually included most members of the North British energy group: Thomson, Jenkin, Joule, Balfour Stewart, and Maxwell. Throughout the 1860s, Thomson played a leading role both in shaping the design of measuring apparatus and in promoting the adoption of an absolute system of physical measurement, such that all the units (including resistance) of the system should bear a definite relation to the unit of work, "the great connecting link between all physical measurements."[54]

RECASTING ENERGY PHYSICS

By the 1880s, the science of energy was fast slipping from the control of its original British promoters. Rankine and Maxwell had already gone from

---

[53] Smith and Wise, *Energy and Empire,* pp. 684–98.
[54] "Provisional Report of the Committee Appointed by the British Association on Standards of Electrical Resistance," *Report of the British Association for the Advancement of Science,* 32 (1862), 126; Smith and Wise, *Energy and Empire,* p. 687; Bruce Hunt, "The Ohm is Where the Art Is: British Telegraph Engineers and the Development of Electrical Standards," *Osiris,* 9 (1994), 48–63.

the scene. During the coming decade, death would exact a further toll with the passing of Jenkin, Stewart, and Joule. Thomson and Tait alone would continue to assert their authority over physics in Britain. But against the new generations of physical scientists – theoretical and experimental physicists, as well as physical chemists – Thomson especially began to look increasingly conservative, a survivor from a past era of natural philosophy.

In contrast, the rising generations began to recast the energy doctrines for their own purposes. A self-styled British group of "Maxwellians," comprising G. F. FitzGerald (1851–1901), Oliver Heaviside (1850–1925), and Oliver Lodge (1851–1940), reinterpreted Maxwell's *Treatise* for their own ends and in accordance with energy principles. But for them, "Maxwell was only half a Maxwellian," as Heaviside noted wryly in 1895, after he and his associates had wrought a transformation in Maxwell's original perspective. Later "Maxwellians" increasingly located energy in the field around an electrical conductor, tended to carry mechanical model building to extremes, and began to reify energy, rather than regard it as mechanical energy or the capacity to do work.[55]

It was, above all, this fundamental link between matter and energy, whereby all energy was ultimately regarded as mechanical energy measured in terms of work done, that had characterized the scientists of energy. Any remaining link between matter and energy was to be decisively severed by the rise of the so-called Energeticist school in Germany. This school marked a far more radical departure from the "science of energy." Led by the physical chemist Wilhelm Ostwald (1853–1922), the Energeticists rejected atomistic and other matter theories in favor of a universe of "energy" extending from physics to society.[56]

Such late-nineteenth-century recastings of energy physics highlight the contingent character of the "science of energy" as it was constructed in the period of 1850 to 1880. That construction was the product of an identifiable, though informal, network of scientific practitioners located mainly in Scotland and sharing a culture characterized by the twin features of engineering and Presbyterianism. On the one hand, their reshaping of Carnot's theory into thermodynamics offered an ideal standard by which the economy of all actual heat engines, especially marine engines, could be assessed. On the other hand, Carnot's theory was now grounded upon a "directional" tendency in visible nature, which reflected the traditional Presbyterian doctrine that God alone could "regenerate" a fallen man and a fallen nature. Whether expressed as "progression" or "dissipation," directionality became the strongest weapon in the North British armory against metropolitan, anti-Christian materialists and naturalists.

[55] Bruce Hunt, *The Maxwellians* (Ithaca, N.Y.: Cornell University Press, 1991).
[56] See, for example, Erwin N. Hiebert, "The Energetics Controversy and the New Thermodynamics," in *Perspectives in the History of Science and Technology*, ed. D. H. D. Roller (Norman: University of Oklahoma Press, 1971), pp. 67–86.

I have argued in this chapter that the construction of energy physics was not the inevitable consequence of the "discovery" of a principle of energy conservation in midcentury, but the product of a North British group concerned with the reform of physical science and with the rapid enhancement of its own scientific credibility. As a result of careful dissemination of the energy principles through well-chosen forums, such as the British Association, the energy proponents succeeded in redrawing the disciplinary map of physics and in carrying forward a reform program for the whole range of physical and even life sciences. "Energy," therefore, became the basic intellectual property of these elite men of science, a construct rooted in industrial culture, but now transcending that relatively local culture to form the core of a science claiming to have universal character and universal marketability.

16

ELECTRICAL THEORY AND PRACTICE IN THE NINETEENTH CENTURY

Bruce J. Hunt

The nineteenth century saw enormous advances in electrical science, culminating in the formulation of Maxwellian field theory and the discovery of the electron. It also witnessed the emergence of electrical power and communications technologies that have transformed modern life. That these developments in both science and technology occurred in the same period and often in the same places was no coincidence, nor was it just a matter of purely scientific discoveries being applied, after some delay, to practical purposes. Influences ran both ways, and several important scientific advances, including the adoption of a unified system of units and of Maxwellian field theory itself, were deeply shaped by the demands and opportunities presented by electrical technologies. As we shall see, electrical theory and practice were tightly intertwined throughout the century.

EARLY CURRENTS

Before the nineteenth century, electrical science was limited to electrostatics; magnetism was regarded as fundamentally distinct. In the 1780s, careful measurements by the French engineer Charles Coulomb established an inverse-square law of attraction and repulsion for electric charges, and electrostatics occupied a prominent place in the Laplacian program, based on laws of force between hypothetical particles, then beginning to take hold in France. The situation was soon complicated, however, by Alessandro Volta's invention in 1799 of his "pile," particularly as attention shifted from the pile itself to the electric currents it produced.[1] Much of the history of electrical science in the nineteenth century can be read as a series of attempts to come to terms with

---

[1] Theodore M. Brown, "The Electric Current in Early Nineteenth-Century French Physics," *Historical Studies in the Physical Sciences*, 1 (1969), 61–103. For a thorough history of nineteenth-century electrical science, see Olivier Darrigol, *Electrodynamics from Ampère to Einstein* (Oxford: Oxford University Press, 2000).

the puzzles posed, and the opportunities presented, by currents like those generated by Volta's pile.

In 1820 the Danish physicist H. C. Oersted, influenced in part by *Naturphilosophie* and its doctrine of the unity of forces, sought a connection between magnetism and electric currents. He found that a magnetized needle placed near a current-carrying wire would turn across the direction of the wire. News of his surprising discovery spread rapidly, and researchers struggled to understand the peculiar twisting force and the mixing of electric and magnetic effects. In France, André-Marie Ampère (1775–1836) soon showed that parallel currents attract one another and argued that the dualism between electricity and magnetism could be eliminated by treating all magnets as composed of myriad molecular electrical currents.[2] In 1826 he formulated an inverse-square law for forces between current elements that fully accounted for Oersted's effect, as well as much else.

Oersted's discovery led to the invention of the galvanometer and the electromagnet, which were soon put to use in the first practical electric telegraphs. In 1833 the German scientists C. F. Gauss and Wilhelm Weber exchanged signals over a double wire strung through Göttingen. In 1837 the English entrepreneur W. F. Cooke teamed with the physicist Charles Wheatstone to patent the first commercially viable electric telegraph, and in 1844 the Americans S. F. B. Morse and Alfred Vail brought their own system into use.[3] Electricity had moved into the practical realm, and experience with currents, coils, and magnets would no longer be confined to the narrow circle of laboratory researchers.

THE AGE OF FARADAY AND WEBER

Two of the leading figures in electrical science from the 1830s through the 1850s were Michael Faraday (1791–1867) and Wilhelm Weber (1804–1891). Both were active experimentalists, but in other ways they followed very different scientific paths, Faraday propounding a radically new field theory of electricity and magnetism, while Weber pursued the more orthodox task of formulating laws of electric force. Their contrasting approaches set the stage for the striking national differences – field theory in Britain, action-at-a distance theories in Germany – that were to mark electrical science later in the century.

Faraday began his scientific career as a chemical assistant to Sir Humphry Davy of the Royal Institution. As David Gooding has emphasized, Faraday's

---

[2] James Hofmann, *André-Marie Ampère* (Oxford: Blackwell, 1995); R. A. R. Tricker, *Early Electrodynamics: The First Law of Circulation* (Oxford: Pergamon Press, 1965).
[3] There is no satisfactory history of early telegraph technology, but see Jeffrey Kieve, *The Electric Telegraph: A Social and Economic History* (Newton Abbot: David and Charles, 1973), and Robert Thompson, *Wiring a Continent: The History of the Telegraph Industry in the United States, 1832–1866* (Princeton, N.J.: Princeton University Press, 1947).

background as a chemist led him to value direct experience over mathematical theorizing, a tendency reinforced by his literalist religious views.[4] His first electrical discovery came in 1821, when he found that a current-carrying wire could be made to rotate around the pole of a magnet, an effect that later became the basis of virtually all electric motors. In 1831, while trying to produce the converse of Oersted's effect – that is, to use a magnet to generate an electric current – he found that moving a magnet rapidly near a coil of wire produced a brief jolt of current, a process he called electromagnetic induction. He could now generate a current by simply turning a coil between the poles of a magnet, a discovery that later led to the invention of the dynamo.

Other electrical researchers revered Faraday as a discoverer – not just of electromagnetic induction but also of specific inductive capacity (1837), magneto-optic rotation (1845), and a host of other phenomena – but had less regard for his theoretical ideas, at least at first. Eschewing mathematical laws of attraction and repulsion, Faraday pictured electric and magnetic phenomena in terms of curved lines of force spreading out from charges or poles, in patterns like those revealed when one sprinkles iron filings around a magnet. In the 1840s and 1850s, he generalized these views into a theory of electric and magnetic fields, treating space not as empty and inert but as the locus of power and activity. Electrified or magnetized bodies do not act directly across empty space, Faraday said, but only by altering the state of the field around them, so that apparent actions at a distance are, in fact, the result of contiguous actions through an intervening medium. But while Faraday came to base his thinking more and more on the notion of a field, most mathematically trained physicists looked on it as little more than a mental crutch, suited to one who could not handle the more elegant and rigorous force law approach. When in 1845 the young William Thomson (1824–1907), later Lord Kelvin, showed mathematically that Faraday's approach led to the same results as Coulomb's action-at-a-distance law, it served as much to protect the orthodox force laws from apparent conflict with Faraday's experiments as to advance acceptance of Faraday's notion of contiguous action.[5] In 1855 the English Astronomer Royal, G. B. Airy, declared that no one who really understood the inverse-square law would "hesitate an instant in the choice between this simple and precise action, on the one hand, and anything so vague and varying as lines of force, on the other hand."[6]

By the time Weber took it up in the 1830s, the task of devising a law of electric force was not as simple as it had been in Coulomb's day. A comprehensive

---

[4] David Gooding, *Experiment and the Making of Meaning* (Dordrecht: Kluwer, 1990); Geoffrey Cantor, *Michael Faraday, Sandemanian and Scientist: A Study of Science and Religion in the Nineteenth Century* (London: Macmillan, 1991).
[5] William Thomson, "On the Mathematical Theory of Electricity in Equilibrium" (1845), in Thomson, *Reprint of Papers on Electrostatics and Magnetism* (London: Macmillan, 1872), pp. 15–37.
[6] G. B. Airy to John Barlow, 7 February 1855, quoted in L. P. Williams, *Michael Faraday: A Biography* (New York: Basic Books, 1965), p. 508.

law now had to account not only for electrostatic attraction and repulsion (Coulomb's law), but also for Oersted's electromagnetic effect (Ampère's law) and Faraday's electromagnetic induction. Weber not only managed to combine all three into a single law but also devised ways to test it experimentally. By the 1850s he had built the model of forces acting directly between particles into a formidable theoretical edifice.

After becoming professor of physics at Göttingen in 1831, Weber worked with Gauss to formulate an "absolute" system for expressing magnetic and electrodynamic measurements in terms of length, time, and mass, and also developed his electrodynamometer, a delicate moving-coil device for measuring electromagnetic forces. Political troubles interrupted Weber's work in 1837, but on taking it up again in the 1840s, he soon achieved remarkable success.[7]

Following G. T. Fechner, Weber pictured an electric current as a double stream of tiny positively and negatively charged particles flowing in opposite directions through a conductor. His task, as he saw it, was to determine the forces between these particles. Coulomb's law required no revision, but Weber transformed Ampère's law for current elements into one depending on the relative velocities of electric particles, and added a third term depending on their relative accelerations to account for electromagnetic induction. In 1846 he published a long paper laying out this fundamental force law, or *Grundgesetz*, and the experimental evidence supporting it. At about the same time, Franz Neumann of Königsberg formulated a parallel set of laws based on current elements and potential functions, but Weber's more comprehensive theory won wider acceptance in Germany. Yet the theory remained troublingly speculative. Weber could point to no evidence on the actual size, charge, or mass of his hypothetical particles, nor could he even demonstrate directly that they existed. Even worse, Hermann von Helmholtz (1821–1894) argued that the dependence of Weber's force law on velocities led to violations of the conservation of energy. Weber was able to parry Helmholtz's objections for a time, but in the 1870s they returned and began to eat away at physicists' acceptance of forces acting directly at a distance.

TELEGRAPHS AND CABLES

The rapid spread of telegraph lines in the late 1840s and 1850s forever transformed everything from the dissemination of news to the operation of world markets. Science, too, soon felt the effects, as telegraphy generated both new demands for electrical knowledge and new means for obtaining it. This was especially true of submarine telegraphy, a field British firms dominated from the time the first successful cable was laid across the English Channel in 1851. Undersea cables presented more complex electrical conditions than did the

---

[7] Christa Jungnickel and Russell McCormmach, *Intellectual Mastery of Nature: Theoretical Physics from Ohm to Einstein*, 2 vols. (Chicago: University of Chicago Press, 1986), 1: 70–7, 130–7, 146–8.

overhead lines used on the Continent and in America, and the task of wrestling with the peculiarities of submarine telegraphy gave British electrical science much of its distinctive flavor in the second half of the nineteenth century.

The chief such peculiarity came to light in the early 1850s, when the British engineer Latimer Clark noticed that initially distinct signals sent into a submarine cable or long underground line emerged at the far end slightly delayed and badly blurred. Clark soon demonstrated such "retardation" effects to Faraday, who brought them to wider notice in a lecture at the Royal Institution in January 1854. Although he recognized the threat it posed to rapid signaling, Faraday welcomed Clark's discovery as confirmation for his own long-held (and long-ignored) view that conduction could not occur in a wire until the surrounding insulation (or "dielectric") had been thrown into a state of electrostatic strain, with the consequent storage of a quantity of charge.[8] This happened so quickly in ordinary wires that it usually passed unnoticed, but the inductive capacity of a long cable was so great that it took an appreciable time for the strain to be set up or decay away, resulting in the retardation Clark had observed.

Faraday's lecture was of keen interest to British telegraphers, and its publication drew new attention to his theoretical ideas. It also helped prompt William Thomson to work out a mathematical theory of telegraphic transmission, which indicated that the retardation on a cable would increase with the square of its length – the "law of squares." That same year, Thomson took out the first of what would become a lucrative series of telegraph patents, a clear sign of the growing convergence of electrical science and cable telegraphy in Britain.[9]

The interaction between electrical science and telegraphic technology was raised to a new level in 1856 when the American entrepreneur Cyrus Field, backed by British capital and expertise, launched an ambitious effort to span the Atlantic with a 2,000-mile-long cable. To calm fears that retardation would make signaling through such a long cable too slow to be profitable, Field brought in Wildman Whitehouse, a former Brighton surgeon who claimed to have shown experimentally that Thomson's law of squares was a mere "fiction of the schools," and that retardation would pose no real obstacle to the operation of the cable.[10] Although Thomson protested that

---

[8] Michael Faraday, "On electric induction – associated cases of current and static effects" (1854), in Faraday, *Experimental Researches in Electricity*, 3 vols. (London: Taylor and Francis, 1839–55), 3: 508–20; Bruce J. Hunt, "Michael Faraday, Cable Telegraphy and the Rise of Field Theory," *History of Technology*, 13 (1991), 1–19.

[9] Crosbie Smith and M. Norton Wise, *Energy and Empire: A Biographical Study of Lord Kelvin* (Cambridge: Cambridge University Press, 1989), pp. 452–3, 701–5.

[10] Wildman Whitehouse, "The Law of Squares – is it applicable or not to the Transmission of Signals in Submarine Circuits?" *British Association Report* (1856), 21–3; Bruce J. Hunt, "Scientists, Engineers and Wildman Whitehouse: Measurement and Credibility in Early Cable Telegraphy," *British Journal for the History of Science*, 29 (1996), 155–70.

Whitehouse had misapplied his theory, he was sufficiently intrigued by the project to sign on as a director of Field's company.

Field rushed the work along, and the cable was hastily manufactured and inadequately tested. When, after five abortive laying attempts in August 1857 and June 1858, it was finally completed from Ireland to Newfoundland on 5 August 1858, the rejoicing was rapturous – but short-lived. Whitehouse's receiving instruments worked only haltingly, and the huge jolts of currents he sent rippling along the cable further weakened its already fragile insulation. Amid mounting recriminations, Whitehouse was soon dismissed. Thomson took over and gently nursed the cable along for several weeks, using only weak currents and receiving signals on his sensitive mirror galvanometer, but the damage had already been done; by mid-September the cable was dead.[11]

The failure of the first Atlantic cable prompted a sharp reassessment of practices in the industry and helped convince both engineers and industrialists that accurate electrical measurement would be crucial to the future success of submarine telegraphy. Whitehouse's idiosyncratic methods were roundly discredited, both in the press and by an official investigating committee, and Thomson's more scientific approach was heaped with praise.[12] By the mid-1860s, the demands of the cable industry had produced a revolution in the practice of electrical measurement – with enormous effects on science as well as technology. Looking back in 1883, Thomson remarked that "resistance coils and ohms, and standard condensers and microfarads, had been for ten years familiar to the electricians of the submarine-cable factories and testing-stations, before anything that could be called electric measurement had come to be regularly practised in almost any of the scientific laboratories of the world."[13] Thomson had helped launch this revolution himself, calling on the British Association for the Advancement of Science in 1861 to provide a standard of electrical resistance (soon dubbed the ohm) for both practical and scientific use.[14] Clark and other cable engineers soon joined in, and the British Association Committee on Electrical Standards began the work that produced substantially the system of ohms, volts, and amperes still used today. Much of its experimental work was carried out by a young physicist then just beginning to make his name in electrical science: James Clerk Maxwell.

[11] Charles Bright, *Submarine Telegraphs, Their History, Construction, and Working* (London: Lockwood, 1898), pp. 38–54; Silvanus P. Thompson, *The Life of William Thomson, Baron Kelvin of Largs*, 2 vols. (London: Macmillan, 1911), 1: 325–96; Vary T. Coates and Bernard S. Finn, *A Retrospective Technology Assessment: Submarine Telegraphy – The Transatlantic Cable of 1866* (San Francisco: San Francisco Press, 1979), pp. 2–17.

[12] Smith and Wise, *Energy and Empire*, pp. 674–9.

[13] William Thomson, "Electrical Units of Measurement" (1883), in Thomson, *Popular Lectures and Addresses*, 3 vols. (London: Macmillan, 1891–4), 1: 73–136, quotation at pp. 82–3.

[14] Bruce J. Hunt, "The Ohm is Where the Art Is: British Telegraph Engineers and the Development of Electrical Standards," *Osiris*, 9 (1994), 48–63. On work in Germany, see Kathryn M. Olesko, "Precision, Tolerance, and Consensus: Local Cultures in German and British Resistance Standards," in *Scientific Credibility and Technical Standards in 19th and early 20th century Germany and Britain*, ed. Jed Z. Buchwald (Archimedes: New Studies in the History and Philosophy of Science and Technology, 1) (Dordrecht: Kluwer, 1996), pp. 117–56.

MAXWELL

James Clerk Maxwell (1831–1879) was one of the towering figures of nineteenth-century physics. Although his work ranged from the kinetic theory of gases to color vision, his greatest contributions came in electromagnetic theory. A small but active "Maxwell industry" has produced several editions of his writings and many studies of his work, but attempts to plumb his personality have been hampered by the loss of most of his personal papers after his death. No modern biography has yet superseded the *Life* published in 1882 by Lewis Campbell and William Garnett. Recent studies have, however, shed much light on the roots and context of Maxwell's work.[15]

Maxwell was born in Edinburgh and grew up there and at his family's estate in southwestern Scotland. After attending Edinburgh University, he went on to Cambridge, where he finished as second Wrangler in the mathematics tripos in January 1854. Casting about for a research topic, he wrote the next month to Thomson, saying he meant to take up electricity and asking what he should read of Faraday's.[16] We do not know why Maxwell chose electricity, which had been excluded from the Cambridge curriculum, or why he was drawn to Faraday's unorthodox approach, but it is worth noting that his letter to Thomson came amid the flurry of interest stirred up by Faraday's Royal Institution lecture on cable retardation.

The first fruit of Maxwell's electrical researches was a paper, "On Faraday's Lines of Force," completed early in 1856.[17] Drawing an analogy between lines of electric and magnetic force and streamlines in a fluid, he cast many of Faraday's allegedly vague ideas into strict mathematical form. In 1861–2 he followed with an ambitious paper, "On Physical Lines of Force," built around an elaborate mechanical model of the ether composed of tiny vortices and idle wheels. This model led Maxwell not only to equations linking the main electromagnetic quantities, and to the notion that changing electric forces produced "displacement currents," but also to the surprising conclusion (the italics are his) that *"light consists in the transverse undulations of the same medium which is the cause of electric and magnetic phenomena."*[18]

The electromagnetic theory of light stands as one of the grandest unifications in all of physics, and Maxwell's route to it has been closely studied. Historians and philosophers have particularly debated just how realistic Maxwell meant his "Physical Lines" model to be. Those who have studied

---

[15] Lewis Campbell and William Garnett, *The Life of James Clerk Maxwell* (London: Macmillan, 1882). The best brief biography is C. W. F. Everitt, *James Clerk Maxwell: Physicist and Natural Philosopher* (New York: Scribners, 1975). Maxwell's writings have been collected in *The Scientific Papers of James Clerk Maxwell*, ed. W. D. Niven, 2 vols. (Cambridge: Cambridge University Press, 1890), and *The Scientific Letters and Papers of James Clerk Maxwell*, ed. P. M. Harman, 2 vols. (Cambridge: Cambridge University Press, 1990–). See also M. Norton Wise, "The Maxwell Literature and British Dynamical Theory," *Historical Studies in the Physical Sciences*, 13 (1982), 175–205.

[16] Maxwell to Thomson, 20 February 1854, in Harman, ed., *Maxwell Papers*, 1: 237–8.

[17] J. C. Maxwell, "On Faraday's Lines of Force" (1855–6), in Niven, ed., *Scientific Papers*, 1: 155–229.

[18] J. C. Maxwell, "On Physical Lines of Force" (1861–2), in Niven, ed., *Scientific Papers*, 1: 451–513, quotation at p. 500.

the question most closely generally agree that Maxwell thought there really were whirling vortices in a magnetic field, but that he regarded the interstitial idle wheel particles as little more than convenient fictions, probably quite unlike the real connecting mechanism.[19]

Maxwell's best evidence for his electromagnetic theory of light was the coincidence between the speed of light, as measured by Hippolyte Fizeau and others, and the ratio between the electrostatic and electromagnetic systems of units, as measured by Weber and Rudolph Kohlrausch. A better measurement of the ratio of units would provide a sharp test of his theory, and it was in hope of securing one that he joined the British Association Committee on Electrical Standards in 1862. By then a professor of physics at Kings College London, Maxwell worked closely with the cable engineer Fleeming Jenkin over the next two years to determine the value of the ohm, using a spinning coil apparatus devised by Thomson. He then measured the ratio of units in terms of the ohm, and despite some discrepancies, found close enough agreement with the speed of light to bolster his own confidence in his theory, though Thomson remained unconvinced.[20]

In December 1864, Maxwell presented to the Royal Society his "Dynamical Theory of the Electromagnetic Field," in which he derived the electromagnetic equations not from a specific mechanical model, as in "Physical Lines," but from the general dynamics of a connected system.[21] Building on Faraday's view of charges and currents as merely epiphenomenal reflections of the state of the surrounding field, Maxwell now expressed that state chiefly in terms of variations in what he called the "electromagnetic momentum" (later renamed the vector potential), and began to delineate how energy is distributed in the field. He boasted to Charles Hockin, his assistant in the ohm experiments, that he had "cleared the electromagnetic theory of light from all unwarrantable assumption" and could now safely determine the speed of light from purely electromagnetic measurements.[22] Maxwell still believed in a mechanical ether, but until more light could be shed on the details of its substructure, he thought it best to formulate his theory with a minimum of conjecture.

Maxwell left Kings College in 1865 and spent the next few years at his Scottish estate writing his great *Treatise on Electricity and Magnetism* (1873).[23] Although filled with ideas, it was rambling and notoriously hard to follow.

---

[19] Daniel Siegel, *Innovation in Maxwell's Electromagnetic Theory: Molecular Vortices, Displacement Current, and Light* (Cambridge: Cambridge University Press, 1991); Ole Knudsen, "The Faraday Effect and Physical Theory, 1845–1873," *Archive for History of Exact Sciences*, 15 (1976), 235–81; for a contrary view, see T. K. Simpson, *Maxwell on the Electromagnetic Field* (New Brunswick, N.J.: Rutgers University Press, 1997), p. 140.

[20] Simon Schaffer, "Accurate Measurement Is an English Science," in *The Values of Precision*, ed. M. Norton Wise (Princeton, N.J.: Princeton University Press, 1995), pp. 135–72.

[21] J. C. Maxwell, "A Dynamical Theory of the Electromagnetic Field" (1865), in Niven, ed., *Scientific Papers*, 1: 562–4.

[22] Maxwell to Hockin, 7 September 1864, in Harman, ed., *Maxwell Papers*, 2: 164.

[23] J. C. Maxwell, *Treatise on Electricity and Magnetism*, 2 vols. (Oxford: Clarendon Press, 1873).

By the time it appeared, Maxwell had taken up the Cavendish professorship of experimental physics at Cambridge, created in 1871, and was building up the new Cavendish Laboratory, where he installed precision electrical measurement as a prime activity.[24] While revising his *Treatise* for a new edition, he was stricken with cancer. He died on 5 November 1879, at age 48. His theory of the electromagnetic field was as yet neither well understood nor widely accepted; indeed, it was in a sense still unfinished.

CABLES, DYNAMOS, AND LIGHT BULBS

In a striking passage in the preface to his *Treatise on Electricity and Magnetism*, Maxwell remarked that

> [t]he important applications of electromagnetism to telegraphy have... reacted on pure science by giving a commercial value to accurate electrical measurements, and by affording to electricians the use of apparatus on a scale which greatly transcends that of any ordinary laboratory. The consequences of this demand for electrical knowledge, and of these experimental opportunities for acquiring it, have already been very great, both in stimulating the energies of advanced electricians, and in diffusing among practical men a degree of accurate knowledge which is likely to conduce to the general scientific progress of the whole engineering profession.[25]

Maxwell no doubt had in mind his own labors for the Standards Committee and, even more, Thomson's work on the second Atlantic cable, finally completed in 1866.

After the 1858 debacle, Field and his partners took several years to regroup for another try. They followed Thomson's scientific advice more closely this time, ordering a thicker cable and testing it carefully. The *Great Eastern* – the only ship afloat capable of carrying the entire length – began laying the cable from Ireland in July 1865, only to have it snap 1,200 miles out. The project seemed cursed, but backed by John Pender, a wealthy Manchester cotton merchant, Field ordered yet another cable and set out again the next summer. This time all went smoothly and the cable was landed at Newfoundland on 27 July 1866. The *Great Eastern* then grappled up the 1865 cable, spliced it, and completed it, too, so that from September 1866, the Atlantic was spanned by two working cables.[26]

The successes of 1866 set off a global boom in cable laying. By 1875, British firms (most of them controlled by Pender) had laid cables to India, Australia,

---

[24] Simon Schaffer, "Late Victorian Metrology and Its Instrumentation: A Manufactory of Ohms," in *Invisible Connections: Instruments, Institutions and Science*, ed. Robert Bud and Susan Cozzens (Bellingham, Wash.: SPIE, 1992), pp. 23–56.
[25] Maxwell, *Treatise*, 1: vii–viii.
[26] On the 1865–6 cables, see Bright, *Submarine Telegraphs*, pp. 78–105; Thompson, *Kelvin*, 1: 481–508; Coates and Finn, *Transatlantic Cable*, pp. 21–5.

and Hong Kong; by 1890, cables ringed the coasts of South America and Africa and were reaching out around the world. Cable telegraphy became a large and lucrative industry and one of the strategic bulwarks of the British Empire. It also provided the chief market for advanced electrical expertise from the 1850s until the 1880s. This market was overwhelmingly British – British firms owned over two-thirds of the cable mileage in place in the 1880s and had built and laid all but a tiny fraction of the rest – and its growth reinforced the focus of British electrical science on precision measurement and electromagnetic propagation.

In the late 1870s, a new market for electrical knowledge began to appear with the emergence of the electric power industry. Current from voltaic batteries, or from magnetos using permanent magnets, had been too weak and expensive for any but a few specialized uses. Around 1866–7, however, several inventors hit on the idea of feeding some of the current from a magneto back into coils around its field magnets, thus turning them into electromagnets that would grow stronger as the magneto was cranked harder.[27] As the design of self-excited dynamos was refined over the next decade, electric power became, for the first time, relatively cheap and abundant. The first major use of dynamo current was for arc lights, but they were too bright to use indoors. Thomas Edison (1816–1890) responded by inventing the incandescent light bulb in 1879, and went on to develop a system for generating and distributing electric power; he installed the first central power station at Pearl Street in New York in 1882. The story of the subsequent growth of electric "networks of power" has been well told by Thomas P. Hughes.[28]

Edison's original system used direct current (DC), which worked well over short distances but lost so much power in its transmission lines that it could serve customers only within a few miles of the generating station. Alternating current (AC) systems avoided this limitation by using transformers to step currents up to high voltages for efficient long-distance transmission and then down to low voltages for safe local distribution. After an epic "battle of the systems," AC won the upper hand in the 1890s.

Early DC systems were fairly simple; the electric current in their wires could, for most purposes, be treated like water flowing in a pipe. AC systems were more complex, particularly as polyphase transmission was adopted, and were subject to various field effects. The rapid growth of the power industry in the 1880s and 1890s generated an unprecedented demand for trained electrical engineers, particularly those competent to handle AC. This demand was initially met, especially in the United States, by university physics departments, which added staff and facilities to handle the influx of students.[29]

[27] Silvanus P. Thompson, *Dynamo-Electric Machinery*, 4th ed. (London: Spon, 1892), pp. 6–21.
[28] Thomas P. Hughes, *Networks of Power: Electrification in Western Society, 1880–1930* (Baltimore: Johns Hopkins University Press, 1983).
[29] Robert Rosenberg, "American Physics and the Origins of Electrical Engineering," *Physics Today*, 36 (October 1983), 48–54.

Indeed, the growth of physics as a discipline throughout the second half of the nineteenth century owed much to the demands and opportunities presented by electrical technology.

THE MAXWELLIANS

At the time of Maxwell's death in 1879, his theory of electromagnetism was only one among several and by no means the clear leader. By 1890 it had swept its rivals from the field and was taking its place as one of the most successful and fundamental theories in all of physics. This transformation was largely the work of a group of younger "Maxwellians," especially G. F. FitzGerald (1851–1901), Oliver Lodge (1851–1940), Oliver Heaviside (1850–1925), and Heinrich Hertz (1857–1894), who during the 1880s recast the theory into a clearer and more compact form, confirmed it experimentally, and extended it in directions that Maxwell himself would have scarcely anticipated.[30] In the process, the theory acquired ever closer ties to existing electrical technologies, particularly through the work of Heaviside, while also, through the work of Hertz and Lodge, spawning the new technology of radio.

The first link in the formation of the Maxwellian group was forged in 1878 when Lodge and FitzGerald met and found they shared an enthusiasm for Maxwell's *Treatise*, although neither yet claimed to understand it very well. Indeed, FitzGerald initially thought that Maxwell's theory ruled out the direct production of electromagnetic waves, and in 1879 he talked Lodge out of trying to generate them experimentally. FitzGerald soon found his error, and in 1883 described in print how waves a few meters in length could be produced by discharging a capacitor through a small resistance. But neither he nor Lodge could think how to detect such rapid oscillations, and by the mid-1880s they had put the problem aside.[31]

Late in 1883, the Cambridge-trained physicist J. H. Poynting discovered that, on Maxwell's theory, energy should pass through the field along paths perpendicular to the lines of both electric and magnetic force. Poynting's theorem had some surprising consequences, including the finding that the energy of an electric current does not flow along within the wire itself, as everyone had always assumed, but instead passes through the seemingly empty space around it. FitzGerald and Lodge soon came to regard the energy flux theorem as one of the keys to understanding Maxwell's theory – although Maxwell himself had had no inkling of it. In 1885 FitzGerald built a model out of brass wheels and rubber bands to illustrate how energy flowed through

---

[30] Jed Z. Buchwald, *From Maxwell to Microphysics: Aspects of Electromagnetic Theory in the Last Quarter of the Nineteenth Century* (Chicago: University of Chicago Press, 1985); Bruce J. Hunt, *The Maxwellians* (Ithaca, N.Y.: Cornell University Press, 1991).

[31] Hunt, *Maxwellians*, pp. 30–47.

the ether, while Lodge described the workings of a similar model at length in his influential *Modern Views of Electricity* (1889).[32]

Heaviside, an eccentric former cable engineer who had "retired" at age 24 to devote himself to electrical theory, hit on the energy flux theorem independently a few months after Poynting.[33] He too saw it as central to Maxwell's theory. Indeed, convinced that Maxwell's use of the vector potential had obscured the true distribution and flow of energy in the field, Heaviside was inspired to recast the long list of equations Maxwell had given in his *Treatise* into the compact set of four vector equations now universally known as "Maxwell's equations":

$$\text{div}\, \varepsilon\boldsymbol{E} = \rho \qquad \text{curl}\, \boldsymbol{H} = k\boldsymbol{E} + \varepsilon\, d\boldsymbol{E}/dt$$
$$\text{div}\, \mu\boldsymbol{H} = 0 \qquad -\text{curl}\, \boldsymbol{E} = \mu\, d\boldsymbol{H}/dt.$$

These equations led to the energy flux formula ($\boldsymbol{S} = \boldsymbol{E} \times \boldsymbol{H}$) in a simple and direct way and clarified many other aspects of Maxwell's theory.

In the mid-1880s, Heaviside used his new set of field equations, along with a refined version of the line equations relating voltage and current, to analyze the propagation of electromagnetic waves along wires – the basic problem of telegraphy. According to Heaviside, signals did not travel within a wire, but slipped along through the space around it. In 1886 he found theoretically that retardation (or "distortion") could be radically reduced, even eliminated, by loading a line with extra inductance, such as properly spaced coils.[34] Inductive loading later proved of great value to the telephone and cable industries, but in the short run, Heaviside's discovery led only to a bitter fight with W. H. Preece, the powerful chief electrical engineer of the British Post Office telegraph system. Preece had publicly declared inductance to be detrimental to clear signaling (as indeed it is if not properly applied), and he took steps to block publication of Heaviside's contrary findings. Heaviside's important work on Maxwell's theory and the propagation of electromagnetic waves had as yet attracted little notice, and now Preece threatened to choke it off altogether.

To Heaviside, it thus seemed "a sort of special Providence" when, beginning early in 1888, experimental work by Lodge in England and Hertz in Germany suddenly drew physicists' attention to his writings.[35] While trying to mimic the effects of lightning by discharging large capacitors into wires, Lodge had found signs that electromagnetic waves were sent skittering along in the

---

[32] G. F. FitzGerald, "On a Model Illustrating some Properties of the Ether" (1885), in FitzGerald, *The Scientific Writings of the Late George Francis FitzGerald*, ed. J. Larmor (Dublin: Hodges and Figgis, 1902), pp. 142–56; Oliver Lodge, *Modern Views of Electricity* (London: Macmillan, 1889).

[33] Hunt, *Maxwellians*, pp. 48–72, 109–14; Paul J. Nahin, *Oliver Heaviside: Sage in Solitude* (New York: IEEE Press, 1988).

[34] James E. Brittain, "The Introduction of the Loading Coil: George A. Campbell and Michael Pupin," *Technology and Culture*, 11 (1970), 36–57; Hunt, *Maxwellians*, pp. 137–46; Ido Yavetz, *From Obscurity to Enigma: The Work of Oliver Heaviside, 1872–1889* (Basel: Birkhäuser, 1995).

[35] Heaviside to Lodge, 24 September 1888, quoted in Hunt, *Maxwellians*, p. 149.

surrounding space, as they did in Heaviside's theory of telegraph signals. He and Heaviside soon struck up a correspondence and became close allies. Then in mid-1888, news reached Britain of Hertz's even more striking experiments on electromagnetic waves in air. Having been taught a hybrid version of Maxwell's theory by Helmholtz at Berlin, Hertz was intent on testing it against Weber and Neumann's action-at-a-distance theories. Following up a chance observation of sparks produced by rapid electrical oscillations, he set up patterns of interfering waves several meters long in his Karlsruhe lecture hall and measured their properties.[36]

Hertz's experiments were warmly welcomed by the British Maxwellians, who identified them as the long-sought confirmation of their own theoretical predictions. At the September 1888 meeting of the British Association, FitzGerald declared that Hertz's "splendid" experiments had finally proven that electromagnetic forces acted through a medium, rather than directly at a distance. Lodge, who had come so close to the discovery himself, joined in the praise, as did Heaviside – while noting the irony that such dramatic confirmation of Maxwell's field theory had come from Germany, the great home of action-at-a-distance theories.[37] Continental physicists soon took up Maxwell's theory, or at least the pared-down set of "Maxwell's equations," very similar to Heaviside's, that Hertz published in 1890. Greatly strengthened, Maxwellian theory entered the 1890s poised to absorb not just optics but a whole string of other subjects. In Lodge's words, electricity had become "an imperial science."[38]

Electricity was, of course, already an imperial science in a more direct sense: Submarine cables formed the "nervous system" of the British Empire. It is ironic that in light of its long-standing ties to the British cable industry, Maxwellian theory helped spawn the new technology – radio – that eventually broke the British monopoly on global telecommunications. Hertz had focused on studying the properties of electromagnetic waves, not using them to convey messages. But by 1894, Lodge, without really intending it, had developed the means to send wireless signals in Morse code, and the young Italian Guglielmo Marconi (1874–1937) soon turned a similar combination of spark gaps and coherers into a practical signaling system.[39] Wireless telegraphy came into wide use after 1900, supplementing and eventually competing with the cable network. The advent of continuous wave transmission and vacuum tube amplification further transformed radio communications, and

---

[36] Jed Z. Buchwald, *The Creation of Scientific Effects: Heinrich Hertz and Electric Waves* (Chicago: University of Chicago Press, 1994); Heinrich Hertz, *Electric Waves,* trans. D. E. Jones (London: Macmillan, 1893); Joseph Mulligan, ed., *Heinrich Rudolf Hertz (1857–1894): A Collection of Articles and Addresses* (New York: Garland, 1994).
[37] Hunt, *Maxwellians,* pp. 153–4, 160–1.
[38] Lodge, *Modern Views,* p. 309.
[39] Hugh G. J. Aitken, *Syntony and Spark: The Origins of Radio* (New York: Wiley, 1976); Sungook Hong, "Marconi and the Maxwellians: The Origins of Wireless Telegraphy Revisited," *Technology and Culture,* 35 (1994), 717–49.

the introduction of broadcasting in the 1920s carried the interaction between electrical science and technology into even wider realms.[40]

ELECTRONS, ETHER, AND RELATIVITY

By the time it passed into general circulation around 1890, "Maxwell's theory" had already been considerably modified. It would change even more in the years to come. Although most Maxwellians continued to share Maxwell's belief in a mechanical ether, they had little luck in unlocking its underlying structure. Nor could they say exactly how the field was connected to matter. Maxwell had sidestepped the question by treating material bodies simply as regions of the field with different electric and magnetic constants, but accumulating evidence, particularly from magneto-optics and the study of conduction in rarefied gases, pointed to the need to take the microstructure of matter explicitly into account. The motion of matter through the ether posed even deeper puzzles and eventually drove physicists to rethink some very basic issues. The 1890s were a time of triumph for Maxwellian theory, but also one in which it underwent a major change of direction.

The characteristic Maxwellian conception of charge as a superficial discontinuity in dielectric displacement, and of current as the breakdown of electrical strain within a conductor, had always been hard for most Continental physicists to grasp, and in the 1890s several of them sought to meld the more concrete and familiar Weberian notion of charged particles with a Maxwellian treatment of the field. The first and most successful was the Dutch physicist H. A. Lorentz (1853–1928), who in 1892 began to develop the hypothesis that matter contains enormous numbers of tiny charges (soon dubbed electrons) that are able to move freely within conductors but are bound in place in material dielectrics. Lorentz's theory involved none of the puzzling breakdown of strain that Maxwellians had identified as true conductivity; instead, electric currents were simply flows of electrons, and macroscopic charges were simply local accumulations of positive or negative electrons. Joseph Larmor began developing a similar theory in England in 1894, though he treated electrons as singularities in a rotational ether, rather than simply as small electrified bodies, and in Germany, Emil Wiechert and others formulated electron theories of their own.[41] Such theories were found to account well not only for the ordinary phenomena of electrodynamics, but also for various optical phenomena, and even for the magnetic splitting of spectral lines that Pieter Zeeman discovered in 1896.

---

[40] Hugh G. J. Aitken, *The Continuous Wave: Technology and American Radio, 1900–1932* (Princeton, N.J.: Princeton University Press, 1985).
[41] Buchwald, *Maxwell to Microphysics*; Russell McCormmach, "H. A. Lorentz and the Electromagnetic View of Nature," *Isis*, 61 (1970), 459–97.

By the late 1890s, the theoretical projects of Lorentz, Larmor, and Wiechert had begun to merge and to link up with a growing body of experimental evidence. J. J. Thomson's work on cathode rays at the Cavendish Laboratory proved particularly important. Availing himself of the improved vacuum pumps that had recently been developed to meet the demands of the electric light bulb industry, Thomson by 1899 had produced convincing evidence that electrons – or at least negatively charged ones – were real, measureable components of the physical world.[42] With the advent of the electron, electrical science turned its focus, as so much of physics would in the twentieth century, toward the microstructure of matter.

Electron theory also answered the puzzle of motion through the ether, but in a way that raised new questions of its own. In 1881 the American physicist A. A. Michelson devised a delicate interferometer that, by splitting a beam of light into two parts that were sent off in different directions and then recombined, should have been able to detect Earth's motion through the ether, although the effect depended on the very small square of the ratio of Earth's speed to that of light (v^2/c^2). Yet neither in 1881, nor when he repeated the experiment more carefully with E. W. Morley in 1887, did Michelson see any sign of such motion – flatly contradicting the electromagnetic theory of light, or any theory that assumed a motionless ether.[43] In 1889 FitzGerald proposed a startling solution. Electromagnetic forces were known to be affected by motion through the ether; indeed, Heaviside had just shown theoretically that such motion altered the force between two electric charges by an amount depending on v^2/c^2. FitzGerald now suggested that all matter was held together by similar forces, so that motion through the ether made the interferometer – or any other object – change in size by just enough to compensate for the effect Michelson had sought to detect. Lorentz hit on the same idea independently in 1892 and later showed that the "FitzGerald-Lorentz contraction" followed rigorously from his electron theory.[44]

From one perspective, the contraction hypothesis was a triumph for electron theory, converting Michelson's result from a troubling anomaly into striking evidence of the electronic constitution of matter.[45] But from another,

---

[42] G. P. Thomson, *J. J. Thomson: Discoverer of the Electron* (New York: Doubleday, 1964), p. 57; see also Edward A. Davis and Isobel Falconer, *J. J. Thomson and the Discovery of the Electron* (London: Taylor and Francis, 1997).

[43] Stanley Goldberg and Roger Stuewer, eds., *The Michelson Era in American Science, 1870–1930* (New York: American Institute of Physics, 1988); Loyd S. Swenson, Jr., *The Ethereal Aether: A History of the Michelson-Morley-Miller Aether-Drift Experiments, 1880–1930* (Austin: University of Texas Press, 1972).

[44] Bruce J. Hunt, "The Origins of the FitzGerald Contraction," *British Journal for the History of Science*, 21 (1988), 67–76.

[45] Andrew Warwick, "On the Role of the FitzGerald-Lorentz Contraction Hypothesis in the Development of Joseph Larmor's Electronic Theory of Matter," *Archive for History of Exact Sciences*, 43 (1991), 29–91.

it all seemed too neat, as if nature were somehow conspiring to hide from us the effects of our motion through the ether. By 1900, the French mathematician Henri Poincaré and others were asking what "motion through the ether" really meant, and suggesting that its undetectability reflected not a conspiracy of forces but the operation of a far more general principle of the relativity of motion.[46]

It was, of course, Albert Einstein (1879–1955), then an obscure Swiss patent examiner, who put this principle to remarkable use in his pathbreaking 1905 paper, "On the Electrodynamics of Moving Bodies."[47] After studying at the Federal Polytechnic in Zurich, Einstein had devoted close attention to Maxwell's and Lorentz's theories and to the problems in applying them consistently to moving bodies. At the patent office, he often examined designs for electric motors and dynamos, and it was in this technological context that he was struck by an asymmetry in the way electromagnetic induction was explained: If the magnet moved, it was said to generate a new electric field, which then produced a current in the stationary coil, but if only the coil moved, no electric field was generated, the current arising instead from the "Lorentz force" acting on the electrons in coil as they moved across the magnetic field. Yet the measured current depended solely on the relative motion of the magnet and coil.

To eliminate what he called "asymmetries which do not appear to be inherent in the phenomena," Einstein proposed to start with two basic principles: that only relative motion counts, in optics and electrodynamics as well as in ordinary mechanics, and that the speed of light in empty space is the same in any inertially moving reference frame. To combine these principles consistently, he first had to clarify the meaning of such basic notions as "time," "length," and "simultaneity." Once he had done so, everything fell into place, but in surprising ways. Time and space were now strangely mixed together, so that clocks appeared to run more slowly, and rods to contract along their line of motion, when viewed by observers moving relatively to them. The notion of "absolute rest" lost its meaning, and the ether, the foundation of so much of nineteenth-century physics, became, Einstein said, "superfluous."[48]

It took some time for Einstein's radical new approach to win over other physicists, but by 1911 it was rapidly gaining ground, especially in Germany, as a way out of the impasse into which attempts to found electrodynamics on

---

[46] Olivier Darrigol, "The Electrodynamic Roots of Relativity Theory," *Historical Studies in the Physical and Biological Sciences*, 26 (1996), 241–312.

[47] Albert Einstein, "On the Electrodynamics of Moving Bodies" (1905), in H. A. Lorentz, A. Einstein, H. Minkowski, and H. Weyl, *The Principle of Relativity*, trans. W. Perrett and G. B. Jeffery (1923; reprint, New York: Dover, 1952), pp. 37–65. David Cassidy, "Understanding the Special Theory of Relativity," *Historical Studies in the Physical and Biological Sciences*, 16 (1986), 177–95, reviews the large historical literature on Einstein and relativity.

[48] Einstein, "Electrodynamics," pp. 37–8. On the technological context of Einstein's work, see Peter Galison, "Three Laboratories," *Social Research*, 64 (1997), 1127–55, esp. pp. 1134–6.

a physical ether had been led.[49] Though rooted in electrodynamics, Einstein's theory moved beyond it, invoking far more general principles that applied to any type of physical interaction.

Throughout most of the nineteenth century, electrical theory had been one of the hottest areas in physics. In the twentieth century it became, if not a backwater, mainly a resource on which those in other scientific and technological fields could draw; except for the development of quantum electrodynamics in the 1940s, electromagnetic theory was not itself the source of much scientific excitement. This was, in fact, testimony to how solidly Maxwell and the other nineteenth-century masters of electrical science had built: Though stripped of its ether and cast into relativistic form, their theory turned out to stand up remarkably well in both the scientific and technological realms.

---

[49] Arthur I. Miller, *Albert Einstein's Special Theory of Relativity: Emergence (1905) and Early Interpretation (1905–1911)* (Reading, Mass.: Addison-Wesley, 1981).

Part IV

ATOMIC AND MOLECULAR SCIENCES IN THE TWENTIETH CENTURY

17

QUANTUM THEORY AND ATOMIC STRUCTURE, 1900–1927

Olivier Darrigol

Quantum mechanics is a most intriguing theory, the empirical success of which is as great as its departure from the basic intuitions of previous theories. Its history has attracted much attention. In the 1960s, three leading contributors to this history, Thomas Kuhn, Paul Forman, and John Heilbron, put together the Archive for the History of Quantum Physics (AHQP), which contains manuscripts, correspondence, and interviews of early quantum physicists.[1] In the same period, Martin Klein wrote clear and penetrating essays on Planck, Einstein, and early quantum theory; and Max Jammer published *The Conceptual Development of Quantum Mechanics*, still the best available synthesis.[2]

Since then, this subfield of the history of science has grown considerably, as demonstrated in accounts such as the five-volume compilation by Jagdish Mehra and Helmut Rechenberg; the philosophically sensitive studies by Edward MacKinnon, John Hendry, and Sandro Petruccioli; Bruce Wheaton's work on the empirical roots of wave-particle dualism; my own book on the classical analogy in the history of quantum theory; and a number of biographies.[3]

---

[1] For a description, cf. T. Kuhn, P. Forman, and L. Allen, *Sources of Quantum Physics: An Inventory and Report* (Philadelphia: American Philosophical Society, 1967).

[2] Martin Klein, "Max Planck and the Beginning of the Quantum Theory," *Archive for the History of Exact Sciences* (hereafter *AHES*), 1 (1962), 459–79, and "Planck, Entropy, and Quanta," *Natural Philosopher*, 1 (1963), 83–108; and Max Jammer, *The Conceptual Development of Quantum Mechanics* (New York: McGraw Hill, 1966).

[3] Jagdish Mehra and Helmut Rechenberg, *The Historical Development of Quantum Theory*, 5 vols. (hereafter *HD*) (New York: Springer-Verlag, 1982–7); Edward MacKinnon, *Scientific Explanation and Atomic Physics* (Chicago: University of Chicago Press, 1982); John Hendry, *The Creation of Quantum Mechanics and the Bohr-Pauli Dialogue* (Dordrecht: Reidel, 1984); Sandro Petruccioli, *Atoms, Metaphors and Paradoxes: Niels Bohr and the Construction of a New Physics* (Cambridge: Cambridge University Press, 1993); Bruce Wheaton, *The Tiger and the Shark: Empirical Roots of Wave-Particle Dualism* (Cambridge: Cambridge University Press, 1983); Olivier Darrigol, *From c-numbers to q-numbers: The Classical Analogy in the History of Quantum Theory* (hereafter *CQ*) (Berkeley: University of California Press, 1992); John Heilbron, *The Dilemmas of an Upright Man: Max Planck as Spokesman for German Science* (Berkeley: University of California Press, 1986); Martin Klein,

These sources nicely complement one another. There have been, however, a couple of bitter controversies. Historians notoriously disagree about the nature of Planck's quantum work around 1900. Whereas Klein sees in it a sharp departure from classical electrodynamics, Kuhn denies that Planck introduced any quantum discontinuity before Einstein. Here I take Kuhn's side, although it follows from Allan Needell's insightful dissertation that neither Klein nor Kuhn fully identified Planck's goals.[4]

Another controversial issue is the extent to which the introduction of non-causal assumptions in quantum theory depended on the broader cultural context in which quantum physicists lived. In 1971 Forman proposed that the antirational ideology of the Weimar republic was essential. His critics, including Hendry, argued that internal considerations by themselves required the renunciation of causality. In recent years, the conflict and the underlying internal/external divide have lost much of their sharpness. According to the present wisdom, the inventors of quantum probability were discussing highly constrained issues of physics, but in a language that bore various subtexts, including Weimar politics, neo-Kantian philosophy, and forms of probabilistic thinking.[5]

THE QUANTUM OF ACTION

The first quantum concepts emerged from the study of a problem at the border between two major achievements of nineteenth-century physics, thermodynamics and electrodynamics. The problem was the thermal equilibrium of electromagnetic radiation, which was known to occur in a "blackbody,"

Paul Ehrenfest, vol. 1: *The Making of a Theoretical Physicist* (Amsterdam: North Holland, 1970); Abraham Pais, *"Subtle is the Lord—": The Science and Life of Albert Einstein* (Oxford: Oxford University Press, 1982), and *Niels Bohr's Times, in Physics, Philosophy, and Polity* (Oxford: Clarendon Press, 1991); Max Dresden, *H. A. Kramers: Between Tradition and Revolution* (Berlin: Springer, 1987); Michael Eckert, *Die Atomphysiker: Eine Geschichte der Theoretischen Physik am Beispiel der Sommerfeldschen Schule* (Brauchschweig: Wieweg, 1993); David Cassidy, *Uncertainty: The Life and Science of Werner Heisenberg* (New York: Freeman, 1992); Helge Kragh, *Dirac: A Scientific Biography* (Cambridge: Cambridge University Press, 1990); Walter Moore, *Erwin Schrödinger: Life and Thought* (Cambridge: Cambridge University Press, 1987).

[4] Thomas S. Kuhn, *Black-body Theory and the Quantum Discontinuity, 1894–1912* (New York: Oxford University Press, 1967); Allan A. Needell, "Irreversibility and the Failure of Classical Dynamics: Max Planck's Work on the Quantum Theory, 1900–1915," PhD diss., University of Michigan, Ann Arbor, 1980. Cf. Klein, A. Shimony, and T. Pinch, "Paradigm Lost, A Review Symposium," *Isis*, 70 (1979), 429–40; Darrigol, *CQ*, part A.

[5] Paul Forman, "Weimar Culture, Causality, and Quantum Theory, 1918–1927: Adaptation by German Physicists and Mathematicians to a Hostile Intellectual Environment," *Historical Studies in the Physical (and Biological) Sciences* (hereafter *HSPS*), 3 (1971), 1–116; John Hendry, "Weimar Culture and Quantum Causality," *History of Science*, 18 (1980), 155–80; Catherine Chevalley, "Le dessin et la couleur," introduction to Bohr, *Physique atomique et connaissance humaine* (Paris: Gallimard, 1991), 19–140; "Niels Bohr and the Atlantis of Kantianism," in *Niels Bohr and Contemporary Philosophy*, ed. J. Faye and H. J. Folse (Dordrecht: Kluwer, 1994), pp. 33–56; Norton Wise, "How Do Sums Count? On the Cultural Origins of Statistical Causality," in *The Probabilistic Revolution*, ed. L. Krüger et al., vol. 1: *Ideas in History* (Cambridge, Mass.: MIT Press, 1987), pp. 395–425.

concretely, a cavity (enclosure) whose walls have a definite temperature and absorb any incoming radiation. According to a theorem proved by Gustav Kirchhoff in 1859 on the basis of the second principle of thermodynamics, the blackbody spectrum has a very remarkable property: It is a universal function of temperature only. In the 1880s, Ludwig Boltzmann and Willy Wien restricted the form of this function by combining electromagnetism and thermodynamics. In the 1890s, spectroscopists working at Berlin's newly founded Physikalisch-Technische Reichsanstalt (PTR), measured it with the aim of determining an absolute standard for high-temperature measurement. At the same time, the Berlin theorist Max Planck (1858–1947) attempted a complete theoretical determination of the blackbody spectrum.[6]

Having studied in Berlin under Helmholtz and Kirchhoff, Planck shared the former's focus on general, organizing principles and the latter's preference for macroscopic phenomenology over atomistic speculation. He ascribed an absolute validity to the principles of his preferred science, thermodynamics, and he denigrated Boltzmann's kinetic-theoretical approach for lacking original results and limiting the second principle of thermodynamics to a statistical validity. In Planck's opinion, a deeper explanation of thermodynamic irreversibility could only be found in the dynamics of a continuum, such as the electromagnetic ether. In 1895 he embarked on an ambitious quest for an electromagnetic foundation of thermodynamics. He hoped to deduce the form of the universal blackbody spectrum as a side result.[7]

The development of this program turned out to be polemical and intricate. Planck originally saw irreversibility in the scattering of waves by small electric resonators. Soon he conceded to Boltzmann that irreversibility required the notion of disordered state. The derivation of the equilibrium spectrum was even more problematic. Planck's first guess was quickly contradicted by new measurements of the infrared end of the spectrum. In October 1900, seeking the simplest form of the entropy compatible with the low- and high-frequency limits of the spectrum, he obtained a new blackbody law that fitted excellently the data from the PTR. Before the end of the year, he announced a more fundamental proof of this law, based on a formula by Boltzmann that makes the entropy S of a set of identical oscillators proportional to the logarithm of the number W of distributions of equal energy-elements among them. The success of the derivation requires a definite value of the energy-elements: h times the frequency v of the resonators, where h is a universal constant with the dimension of action.[8]

---

[6] Cf. Hans Kangro, *Vorgeschichte des Planckschen Strahlungsgesetzes* (Wiesbaden: Steiner, 1970); Kuhn, *Black-body*; David Cahan, *An Institute for an Empire: The Physikalisch-Technische Reichsanstalt, 1871–1918* (Cambridge: Cambridge University Press, 1989), chap. 4; Max Jammer, *The Conceptual Development of Quantum Mechanics* (hereafter *CD*) (New York: McGraw Hill, 1966), chap. 1.1; Mehra and Rechenberg, *HD*, vol. 1, chap. 1.1.

[7] Cf. Klein, Kuhn, and Needell, "Irreversibility"; Darrigol, *CQ*, part A.

[8] M. Planck, "Über eine Verbesserung des Wienschen Spektralgleichung," in Deutsche Physikalische Gesellschaft, *Verhandlungen* (hereafter *PGV*), 2 (1900), 202–4, and "Zur Theorie des Gesetzes der

If Planck had strictly followed Boltzmann's methods, he could not have fixed the energy-elements in this manner, and he would have reached an absurd blackbody law (the so-called Rayleigh-Jeans law, for which the total energy of radiation is infinite). However, Planck did not accept Boltzmann's statistical interpretation of irreversiblity until the 1910s. For reasons too complex to be detailed here, his nonstatistical reinterpretion of Boltzmann's entropy formula permitted the elements $h\nu$ without yet contradicting the continuous dynamics of the resonators. Planck needed this continuity in his derivation of the equilibrium between radiation and resonators.[9]

Planck never claimed to have restricted the energies of his resonators or of radiation to discrete values. Instead, he regarded the introduction of the new fundamental constant h as his major innovation. He also emphasized that the constant k and the "Boltzmann formula" $S = k \ln W$ had general significance for thermodynamics and should constitute an important bridge between gas theory and radiation theory. His intention was to unify physics, not to turn it upside down. The revolutionary view of discontinuous energy exchanges came from Albert Einstein (1879–1955).[10]

QUANTUM DISCONTINUITY

Unlike Planck, the young Einstein was highly impressed by Boltzmann's statistical thermodynamics and tried to improve on it. He regarded the entropy decrease of an isolated system as only improbable, and he accepted the possibility of fluctuations of the macroscopic state around equilibrium. He even took entropy to be a measure of the fluctuations and expected them to be observable in certain systems, such as Brownian suspensions.

In 1905, Einstein computed the entropy of dilute thermal radiation from the high-frequency limit of Planck's law. Accordingly, the probability of a fluctuation in which the radiation of frequency ν is confined to a fraction of the available volume has the same form as that expected for a gas of quanta $h\nu$. Taking the analogy seriously, Einstein assumed that the emission and absorption of radiation always occurred by means of such quanta, and he thus derived the frequency-dependence of the photoelectric effect. At the beginning of the same paper, he argued, implicitly against Planck, that a proper application of statistical thermodynamics to classically interacting matter and radiation led to an absurd blackbody law. The following year, in order to save Planck's law he quantized Planck's resonators, in conformity with their absorbing or emitting light by discrete quanta. Extending the

---

Energeivertheilung im Normalspektrum," *PGV*, 2 (1900), 237–45. Cf. Kuhn, *Black-body*; Klein, "Max Planck"; Needell, "Irreversibility,"; Darrigol, *CQ*, part A.

[9] Cf. Needell, "Irreversibility"; Darrigol, *CQ*, part A.
[10] Cf. Kuhn, *Black-body*.

procedure to the oscillations of atoms in a crystal, he was able to explain a long-known anomaly: the decrease of specific heats at low temperature.[11]

No major theorist approved of Einstein's light quantum. Maxwell's field equations were the best confirmed part of physics, and Einstein's fluctuation argument could be criticized in various ways. A few experts on high-frequency radiation, for example Johannes Stark (1874–1957) and William Henry Bragg (1862–1942), were about the only physicists to welcome the light quantum: It easily explained the strange fact that the energy of electrons produced by x-ray ionization was of the same order as that of the electrons in the x-ray tube, no matter how far the target was from the source. In the following years, Einstein provided increasingly strong arguments in favor of the light quantum, all based on statistical fluctuations, but he also deplored his inability to explain the wave-aspects of light simultaneously. Until the early 1920s, the few theorists who appreciated the full force of his considerations preferred to drop one of the premises, energy conservation, rather than accept the light quantum.[12]

However, Einstein's opinion that ordinary electrodynamics and statistical thermodynamics necessarily led to an absurd blackbody law gradually gained ground. The supreme authority on the matter, the Dutch theorist Hendrik Lorentz (1853–1928), admitted in 1908 that statistical mechanics (in Gibbs's form) confirmed Einstein's conclusion. Even Planck ended up admitting the necessity of some discontinuity, if only for the emission of light. The most successful aspect of Einstein's quantum ideas was by far his theory of specific heats. The energetic Walther Nernst (1864–1941) perceived a link with his cherished third principle of thermodynamics (which implies that the entropy of pure substances should vanish at zero temperature, and relates chemical equilibrium constants to heat measurements) and launched an extensive program for studying specific heats at low temperature in the light of Einstein's theory.[13]

---

[11] A. Einstein, "Über eine die Erzeugung und die Verwandlung des Lichtes betreffenden heuristischen Gesichtspunkts," *Annalen der Physik* (hereafter *AP*), 17 (1905), 132–48; "Zur Theorie der Lichterzeugung und Lichtabsorption," *AP*, 20 (1906), 199–206; and "Die Plancksche Theorie der Strahlung und die Theorie der spezificischen Wärme," *AP*, 22 (1907), 180–90. Cf. Klein, "Einstein's First Paper on Quanta," *Natural Philosopher*, 2 (1963), 59–86; "Thermodynamics in Einstein's Thought," *Science*, 157 (1967), 509–16; and "Einstein, Specific Heats, and the Early Quantum Theory," *Science*, 148 (1965), 173–80; Darrigol, "Statistics and Combinatorics in Early Quantum Theory," *HSPS*, 19 (1988), 17–80; Kuhn, *Black-body*, chap. 7; Jammer, *CD*, chap. 1.3; Mehra and Rechenberg, *HD*, vol. 1.

[12] Cf. Klein, "Thermodynamics" and "Einstein and the Wave-Particle Duality," *Natural Philosopher*, 3 (1964): 1–49; Kuhn, *Black-body*, chap. 9; Darrigol, "Statistics"; Wheaton, *Tiger*; John Stachel, "Einstein and the Quantum: Fifty Years of Struggle," in *From Quarks to Quasars*, ed. R. Colodny (Pittsburgh, Pa.: University of Pittsburgh Press, 1983).

[13] On conversions to the quantum discontinuity, cf. Kuhn, *Black-body*, chap. 8. On Nernst, cf. Erwin N. Hiebert, "Nernst, Hermann Walther," *Dictionary of Scientific Biography*, supp. 1 (1978), 432–53; Diana Barkan, *Walther Nernst and the Transition to Modern Physical Science* (Cambridge: Cambridge University Press, 1999); Klein, "Einstein, Specific Heats," pp. 176–80; Kuhn, *Black-body*, pp. 210–20. On theoretical works regarding specific heats, cf. Richard Staley, "Max Born and the German Physics Community: The Education of a Physicist," PhD diss., Cambridge University, 1990, chap. 6.

Nernst was also largely responsible, with the sponsorship of the industrial chemist Ernest Solvay, for the international conference on the new quantum ideas held in 1911 in Brussels. The gathered elite agreed on the necessity of quantum discontinuity in physics, although opinions varied on the manner of introducing the discontinuity and on the possibility of reducing it to previously known laws of physics.[14]

In the following years, the quantum was used roughly in two different manners. The first approach, dominated by the Berlin physicists Einstein, Planck, and Nernst and by the Leyden professor Paul Ehrenfest (1881–1933), was an extension of the original context of the quantum and involved the thermodynamic properties of matter and the nature of radiation; it was often purely theoretical but received experimental support from thermal measurements and x-ray physics. The other trend, centered socially in Copenhagen, Munich, and Göttingen, consisted of the application of the quantum to individual atoms and molecules; it relied heavily on the wealth of spectroscopic data from many different sources, Berlin and Tübingen being the most important German ones. The following analysis describes the course of this second type of quantum physics, as well as its modifications through interactions with the first.

FROM EARLY ATOMIC MODELS TO THE BOHR ATOM

The easiest nonthermodynamic application of the new quantum theory was to the rotation of individual molecules and the vibration of their atoms: The quantization of such simple motions could easily be guessed, and conclusions drawn about the emitted spectra. The Danish physicist Niels Bjerrum (1879–1958) inaugurated this fruitful path after a suggestion by Nernst. Several Americans, especially Edwin Kemble (1889–1984) and Robert Mulliken (1896–1986), used the theory as a link between molecular structure and spectra, and they reached some of the principal results of modern molecular spectroscopy, well before quantum mechanics was born.[15]

The internal structure of atoms, however, was the main interest of the more ambitious atom builders. Since the late nineteenth century, evidence in favor of a composite atom had accumulated in the study of spectra, cathode rays, and radioactivity. A growing number of physicists assumed that the electron

---

[14] Cf. Barkan, "The Witches' Sabbath: The First International Solvay Congress in Physics," *Science in Context*, 6 (1993), 59–82; *Les conseils Solvay et les débuts de la physique moderne*, ed. P. Marage and G. Wallenborn (Brussels: Université libre de Bruxelles, 1995).

[15] Cf. Gerald Holton, "On the Hesitant Rise of Quantum Physics Research in the United States," in *Thematic Origins of Scientific Thought: Kepler to Einstein*, 2d ed. (Cambridge, Mass.: Harvard University Press, 1987), pp. 147–87; Alexi Assmus, "The Molecular Tradition in Early Quantum Theory," *HSPS*, 22 (1992), 209–31, and "The Americanization of Molecular Physics," *HSPS*, 23 (1992), 1–34.

was a universal constituent of matter, and a few British physicists tried to construct atoms with electrons only. The most successful atom builder in the early years of this century was Joseph John Thomson (1856–1940), director of the Cavendish Laboratory from 1884 to 1919, and codiscoverer of the electron in 1897. Benefiting from the British familiarity with mechanical models, Thomson sought to integrate, in the same model, wide-ranging experimental data on ionization, chemical regularities, spectra, and the production and scattering of various radiations. In contrast, the Germans usually avoided detailed mechanical pictures and confined their models to the representation of spectral properties.[16]

Thomson's model of 1903 involved a uniform and immaterial positively charged sphere in which a large number of electrons orbited in circles. The symmetrical arrangement of the electrons on the circles prevented radiative collapse. Mechanical stability implied restrictions on the number of electrons on successive rings, in which Thomson recognized the chemical periodicity of elements (successive elements differing by the addition of one electron). By 1906, x-ray and β-ray scattering experiments forced Thomson to reduce the number of electrons to the order of the atomic number. This made the nature of the positive charge and the radiation drain more problematic. Also, there were not enough electrons to explain all spectral lines in terms of perturbations of the normal configuration of the atom. Anticipating Bohr's notion of excited states, J. J. Thomson admitted, with Philipp Lenard (1862–1947) and Johannes Stark (1874–1957), that the various spectral series were emitted during the recombination of electrons with ionized atoms.[17]

In 1911, from the large-angle scattering of α-rays observed in 1908 by his Manchester colleagues Hans Geiger (1882–1945) and Ernest Marsden, Ernest Rutherford (1871–1937) concluded that the positive charge of the atom was concentrated in a very small "nucleus." Ignoring all problems of stability, he proposed that atoms were made of electrons revolving around the nucleus. The model originally attracted little attention, except from the Manchester physicists. Among them was a young Danish visitor, Niels Bohr (1885–1962), who soon combined the new model with Frederick Soddy's (1877–1956) recent doctrine of isotopes to enunciate the modern function of the nucleus: The chemical identity of the atom depended only on the nuclear charge, and radioactivity was a decay of the nucleus. Bohr also tried to improve the existing scattering theories – which played a crucial role in Thomson's and

---

[16] Cf. Heilbron, "A History of the Problem of Atomic Structure from the Discovery of the Electron to the Beginning of Quantum Mechanics," PhD diss., University of California at Berkeley, 1964, chap. 3.

[17] Cf. Heilbron, "Thomson, Joseph John," in *Dictionary of Scientific Biography*, XIII, 362–72; "J. J. Thomson and the Bohr Atom," *Physics Today*, 30 (1977), 23–30; and "Lectures on the History of Atomic Physics," in *History of Twentieth Century Physics*, ed. C. Weiner (New York: Academic Press, 1977), pp. 40–108.

Rutherford's interpretations of scattering data – and thus faced the mechanical instability of the Rutherford atom. Following the Oxford physicist John Nicholson, Bohr decided to postulate an unmechanical kind of stability, based on Planck's quantum of action. Comparing the energies of various quantized ring configurations of the electrons, he found similarities with Mendeleyev's chart of elements, as Thomson had done with his purely mechanical model.[18]

Bohr reached his most important ideas in 1913, while trying to account for Balmer's simple formula for the hydrogen spectrum. He then assumed that every atom or molecule could exist only in a discrete series of stationary states, the stability of which eluded ordinary mechanics and electrodynamics. The emission or absorption of radiation involved sudden jumps between such states. For the hydrogen atom, Bohr further assumed that ordinary mechanics applied to the motion of the electron around the nucleus, and he selected the stationary states by the "quantum rule" $T = nh\omega/2$, where T is the kinetic energy of the electron, and ω its rotation frequency; he determined the frequency v of the radiation emitted during a transition between two states of energy E and E' by the "frequency rule" $E - E' = hv$. The latter rule represented a radical break from ordinary electrodynamics: The frequency of radiation no longer reflected the frequency of the emitter. Yet Bohr could show that his "horrid" assumptions agreed with the classical radiation spectrum for highly excited states.[19]

Within a year or so, Bohr's theory had attracted considerable attention, thanks to several experimental successes. Most spectacularly, Bohr's prediction that Edward Pickering's stellar hydrogen series actually belonged to the ion He^+ was confirmed. The identification by Bohr and Antonius van den Broek of the atomic number with the nuclear charge inspired Henry Moseley's (1887–1915) tracking of chemical elements through x-ray spectral analysis. In 1914, Sommerfeld's student Walther Kossel (1888–1956) successfully applied Bohr's stationary states and frequency rule to explain x-ray absorption and emission. Also, in 1915 Bohr interpreted the experiments by James Franck (1882–1964) and Gustav Hertz (1887–1975) on the stopping power of mercury vapor for an electron beam in terms of quantum jumps of the mercury atoms.[20]

Yet Bohr remained unsure about the generality of his assumptions. He trusted them only for strictly periodic systems, and he expected the frequency rule and ordinary mechanics to fail for more complex systems. This was

---

[18] Cf. Heilbron and Kuhn, "The Genesis of the Bohr Atom," *HSPS*, 1 (1969), 211–90; Pais, *Niels Bohr*, chap. 8.

[19] N. Bohr, "On the Constitution of Atoms and Molecules," *Philosophical Magazine* (hereafter *PM*), 26 (1913), 1–25, 476–502, 857–75.

[20] Cf. Jammer, *CD*, chap. 2; Mehra and Rechenberg, *HD*, vol. 1, chap. 2; Pais, *Niels Bohr*, chap. 8; J. Heilbron, *H. G. J. Moseley: The Life and Letters of an English Physicist, 1887–1915* (Berkeley: University of California Press, 1974); Heilbron and Kuhn, "The Genesis"; Heilbron, "The Kossel-Sommerfeld Theory and the Ring Atom," *Isis*, 58 (1967), 451–82; Wheaton, *Tiger*, part 8; Giora Hon, "Franck and Hertz versus Townsend: A Study of Two Types of Experimental Error," *HSPS*, 20 (1989), 79–106.

exaggeratedly pessimistic, as Einstein's and Sommerfeld's contributions to Bohr's theory soon revealed.

EINSTEIN AND SOMMERFELD ON BOHR'S THEORY

Einstein's reaction to Bohr's theory was enthusiastic. In 1916 he applied it to a new discussion of the thermodynamic equilibrium between matter and radiation. To Bohr's quantum transitions he ascribed specific probabilities represented by three coefficients (in Bohr's terminology: spontaneous emission, stimulated emission, and absorption). Kinetic equilibrium between Bohr atoms and radiation could only occur if Bohr's frequency rule held generally; then the radiation had to obey Planck's law; and the fluctuation in its emission could only be explained in terms of light quanta, as long as energy and momentum were conserved. Whereas Einstein was mostly interested in the latter point, Bohr emphasized the general validity of the frequency rule, which he soon called the "second postulate," the first being the existence of stationary states.[21]

Another essential contribution to Bohr's theory came in the same year from the Munich professor Arnold Sommerfeld (1868–1951), a Göttingen-trained master of applied mathematics. Sommerfeld quantized the relativistic Kepler motion with two quantum numbers (principal and azimuthal), and found a splitting of the hydrogen spectrum in conformity with the known fine structure. Following his lead, Karl Schwarzschild (1878–1916) and Paul Epstein (1883–1966) treated the Stark effect, which is the splitting of the spectrum in an electric field; Sommerfeld and Peter Debye (1884–1966) dealt with the magnetic splitting to which Pieter Zeeman's name is attached. In general, quantum rules could be given for the so-called multiperiodic systems, and they could be shown to be compatible with the conditions established by Paul Ehrenfest for the slow deformations of quantum systems (the "adiabatic principle"). The frequency rule applied to any transition, against Bohr's original intuition. Bohr quickly admitted these conclusions, and soon completed them with a new principle that played a central role in guiding future researches in Copenhagen.[22]

---

[21] Einstein, "Strahlungs-emission und -absorption nach der Quantentheorie," *PGV,* 18 (1916), 47–62, and "Zur Quantentheorie der Strahlung," *Physikalische Zeitschrift* (hereafter *PZ*), 18, 121–8. Cf. Klein, "Einstein and the Development of Quantum Physics," in *Einstein: A Centenary Volume,* ed. A. P. French (Cambridge, Mass.: Harvard University Press, 1979); Mehra and Rechenberg, *HD,* vol. 1, chap. 5.1; Darrigol, *CQ,* pp. 118–20.

[22] A. Sommerfeld, "Zur Quantentheorie der Spektrallinien," *AP,* 51 (1916) 1–94, 125–67; K. Schwarzschild, "Zur Quantenhypothese," Akademie der Wissenschaften zu Berlin, Physikalisch-mathematische Klasse, *Sitzungsberichte* (hereafter *BB*) (1916), 548–68. Cf. Jammer, *CD,* chap. 3.1; Sigeko Nisio, "The Formation of the Sommerfeld Quantum Theory of 1916," *Japanese Studies in the History of Science (JSHS),* 12 (1973), 39–78; Helge Kragh, "The Fine Structure of Hydrogen and the Gross Structure of the Physics Community, 1916–1926," *HSPS,* 15 (1985), 67–125; Mehra and Rechenberg, *HD,* vol. 1, chap. 2.4; Darrigol, *CQ,* chap. 6; Ulrich Benz, *Arnold Sommerfeld: Lehrer und Forscher an der Schwelle zur Atomzeitalter, 1868–1951* (Stuttgart: Wissenschaftliche Verlagsgesellschaft, 1975); Eckert, *Die Atomphysiker;* Klein, *Paul Ehrenfest.*

BOHR'S CORRESPONDENCE PRINCIPLE VERSUS MUNICH MODELS

In order to obtain agreement with observed spectra, Sommerfeld had introduced ad hoc "selection rules" that permitted only certain transitions between quantum states. For example, in the case of the Zeeman effect, the magnetic quantum number could only vary by 0, +1, or −1. Bohr noticed that this rule reflected a similarity between the classical and quantum spectrum: To the three frequencies of the perturbed motion (ω_0, $\omega_0 + \omega_L$, $\omega_0 - \omega_L$ if ω_0 is the original frequency and ω_L the Larmor frequency) corresponded to the possible changes of the magnetic quantum number. More generally, Bohr assumed a close connection between the harmonics of the motion in a stationary state and the transitions from this state. This relation between atomic motion and radiation was the "correspondence principle," which Bohr regarded as an essential heuristic tool in the construction of the quantum theory.[23]

The most immediate application of this principle was, of course, the derivation of selection rules. But there was much more to it. Bohr and his skillful Dutch assistant Hendrik Kramers (1894–1952) gave approximate values for the intensities of spectral lines; they sketched the theory of perturbed multiperiodic systems; they obtained general properties of the spectra of higher atoms; and they discussed the construction of atoms. In this last application, the correspondence principle served to select possible histories for the formation of a given atom through the radiative capture of successive electrons by a bare nucleus. Bohr reasoned in part inductively, from the known optical and x-ray spectra to the atomic structure; in part deductively, from an a priori discussion of possible motions. In this more ambitious register, he relied on analogy with the helium atom, the only one for which Kramers could manage the required calculations. The resulting classification of elements, published in 1921, made a great impression and won Bohr the Nobel Prize for the following year. The year 1921 also saw the coronation of Bohr's efforts to create an institute for theoretical physics in Copenhagen. He thus secured the means to diffuse his methods and develop international collaborations, especially with Sommerfeld's students.[24]

Around that time, Sommerfeld and his followers were developing the atomic theory with different, but equally powerful, methods. Whenever a

---

[23] Bohr, "On the Quantum Theory of Line Spectra, Part I: On the General Theory; Part II: On the Hydrogen Spectrum," Det Kongelige Danske Videnskabernes Selskab, *Matematisk-fysiske Meddelser*, 4 (1918), 1–36, 36–100. Cf. Jammer, *CD*, chap. 3.2; Mehra and Rechenberg, *HD*, vol. 1, chap. 2.5; Darrigol, *CQ*, chap. 6; Klaus Meyer-Abich, *Korrespondenz, Individualität und Komplementarität: Eine Studie zur Geistesgeschichte der Quantentheorie in den Beiträgen Niels Bohrs* (Wiesbaden: Franz Steiner, 1965); Petruccioli, *Atoms*.

[24] Cf. Heilbron, "A History," chap. 6.5; Helge Kragh, "Niels Bohr's Second Atomic Theory," *HSPS*, 10 (1979), 123–86; Darrigol, *CQ*, chap. 7; Dresden, *H. A. Kramers*. On the Copenhagen Institute, cf. Pais, *Niels Bohr*, chap. 9, and A. Kozhevnikov, "Niels Bohr and the Copenhagen Network in Physics," forthcoming.

system was not multiperiodic (helium already was not), Bohr tried to guess the main features of its electronic motion through subtle applications of the correspondence principle. By contrast, Sommerfeld, his wunderkind students Werner Heisenberg (1901–1978) and Wolfgang Pauli (1900–1958), and the imaginative Tübingen professor Alfred Landé (1888–1975) replaced the orbital atoms with simpler multiperiodic systems to which the known quantum rules could be applied. For example, an alkali atom was reduced to a rigid core coupled with the outer electron. When the simplified model did not work, some liberty could be taken with the quantum rules – for example, integral quantum numbers were replaced with half-integral ones; or retreat was made to a pure phenomenology of quantum numbers, seeking the simplest classification of the spectral terms in accordance with the observed spectra. These strategies worked well for the helium spectrum, and for the anomalous Zeeman effect, on which Bohr had relatively little to say. They were clear and easily exportable, whereas Bohr's subtle harmonies of motion could be learned *only directly from Bohr* in Copenhagen. Moreover, Sommerfeld's outstanding teaching skills and his reference book *Atombau und Spektrallinien* facilitated the spread of his methods.[25]

A CRISIS, AND QUANTUM MECHANICS

In 1922, the Göttingen professor Max Born (1882–1970), with the help of Pauli and Heisenberg, used sophisticated celestial mechanics to perfect the Bohr-Kramers perturbation theory. In principle, they could compute any atom with this method and test Bohr's intuitions on the harmonies of atomic motions. By 1923, the rigorous treatment of the helium case flatly contradicted Bohr and Kramers's description and thereby threatened Bohr's explanation of chemical periods. The crisis was amplified by Bohr and Pauli's failure to make sense of the Munich theory of the anomalous Zeeman effect. From an improved version of Landé's and Heisenberg's models for this effect, and from the doublet structure of x-ray spectra, Pauli concluded that the azimuthal quantum number of the atomic electrons was intrinsically ambiguous.[26]

Having thus doubled the number of accessible quantum states, Pauli noted a correspondence with Edmund Stoner's new numbers for the populations of electron groups in atoms, and he inferred the "exclusion principle": Two electrons could never occupy the same quantum state in a given atom. Pauli

---

[25] Cf. Jammer *CD*, chap. 3.3; Forman, "Alfred Landé and the Anomalous Zeeman Effect, 1919–1921," *HSPS*, 2: 153–261; Cassidy, "Heisenberg's First Core Model of the Atom," *HSPS*, 10 (1979), 187–224. On Sommerfeld's school, cf. Eckert, *Die Atomphysiker*.

[26] On helium, cf. Mehra and Rechenberg, *HD*, vol. 1, chap. 4.2; Darrigol, *CQ*, 175–9. On the anomalous Zeeman effect and related puzzles, cf. Jammer, *CD*, chaps. 3.3, 3.4; Forman, "The Doublet Riddle and the Atomic Physics *circa* 1924," *Isis*, 59 (1968), 156–74; Daniel Serwer, "*Unmechanischer Zwang*: Pauli, Heisenberg, and the Rejection of the Mechanical Atom," *HSPS*, 8 (1977), 189–256.

and Sommerfeld opposed this result to Bohr's claim that chemical periods could be deduced from the correspondence principle. Pauli further argued that the electronic orbits had to be abandoned in favor of a new, nonvisualizable kinematics. A few months later, he objected to Ralph Kronig's (b. 1904) private explanation of the azimuthal ambiguity by means of a spinning electron. George Uhlenbeck (1900–1988) and Samuel Goudsmit (1902–1978) reached the same idea independently in late 1925 and had the better luck of meeting with Ehrenfest's sympathy and support.[27]

We return to 1923. Confronted with the Göttingen calculations of helium, Bohr admitted that ordinary mechanics failed to describe the interactions between electrons in stationary states, even approximately. But he maintained that the correspondence principle should guide the construction of the theory. For further inspiration, he turned to the paradoxes of radiation, which recent "proofs" of Einstein's light quantum had considerably sharpened. According to Maurice de Broglie (1875–1960), the usual subterfuges to avoid the light quantum theory of the photoelectric effect failed in the case of x rays, for which the atoms of the target acted simply and individually. Most strikingly, in 1923 Arthur Compton (1892–1962) interpreted the angular dependence of the frequency of x rays scattered by quasi-free electrons in terms of a light quantum collision.[28]

Bohr still refused to admit the light quantum. With John Slater (1900–1976) and Kramers, he managed to explain both the discontinuous and the continuous aspects of radiation without giving up the Maxwell equations for free radiation. The basic idea was to divorce the quantum jumps of atoms from the emission and absorption of radiation. In the Bohr-Kramers-Slater (BKS) picture, the electromagnetic field interacts continuously with atoms during their sojourn in stationary states, in harmony with the idea of a close "correspondence" between atomic motion and radiation. This interaction depends on "virtual oscillations" at the spectral frequencies "corresponding" to the harmonics of the electronic motion. The field is itself virtual since it cannot carry any energy. Its only observable effect is to control the probability of quantum jumps. Energy is only conserved statistically.[29]

The BKS theory was short-lived. In the spring of 1925, Walther Bothe (1891–1957) and Geiger obtained evidence of energy conservation in

---

[27] Hendry, in *The Creation*, discusses Pauli relativity-theory background and philosophy; B. L. van der Waerden, "Exclusion Principle and Spin," in *Theoretical Physics in the Twentieth Century*, ed. M. Fierz and V. Weisskopf (New York: Interscience, 1960), pp. 199–244; Heilbron, "The Origins of the Exclusion Principle," *HSPS*, 13 (1983), 261–310.

[28] Cf. Wheaton, *Tiger*, part IV; Roger Stuewer, *The Compton Effect: Turning Point in Physics* (Canton, Mass.: Science History Publications, 1975).

[29] Bohr, Kramers, and Slater, "The Quantum Theory of Radiation," *PM*, 47 (1924), 785–822. Cf. Klein, "The First Phase of the Bohr-Einstein Dialogue," *HSPS*, 2 (1970), 1–39; Klaus Stolzenburg, introduction to Bohr, *Collected Works*, ed. L. Rosenfeld and E. Rüdinger, 10 vols. (Amsterdam: North Holland, 1972–99), vol. 5 (1984), 1–96; Dresden, *H. A. Kramers*, chaps. 6, 8; Jammer, *CD*, chap. 4.3; Hendry, *The Creation*, chap. 5; Darrigol, *CQ*, chap. 9.

individual Compton processes. Even before that, Bohr had been disturbed by Pauli's criticism and by the difficulties he encountered in the extension of his views to collision processes. He now decided to erase all visual elements of the quantum theory, the orbits, the waving fields, and the light quanta, and he recommended a further exploration of "symbolic analogies" between classical and quantum theory. By the latter phrase, Bohr meant a formal method of translation of well-chosen classical relations into formulas that involved quantum concepts no longer reminiscent of the electronic orbits, such as quantum states, quantum numbers, transition probabilities, and spectral frequencies.[30]

This method, inaugurated by Kramers in 1924 with a new dispersion formula, had quickly been extended by Born and Heisenberg, who hoped to find hints of a future quantum mechanics. Even though it was originally interpreted in terms of the BKS picture and virtual oscillators, the procedure only required the formal kernel of the correspondence principle: the connection between Fourier components and quantum jumps. In the early summer of 1925, Heisenberg realized that the classical equations of motion could themselves be written in terms of Fourier components and translated into a complete system of equations for the quantum amplitudes whose square gives the spectral intensities. Thus were born the rudiments of a new quantum mechanics.[31]

The Göttingen mathematical physicists Born and Pascual Jordan (1902–1980) soon joined Heisenberg in his efforts. By the end of the year, the three men had widely generalized the initial scheme, thanks to the methods of advanced matrix calculus, and Pauli had solved the hydrogen atom. At the same time, a young Cambridge theorist, Paul Dirac (1902–1984), obtained similar results with more elegant mathematics. Being familiar with noncommutative algebras and their geometric interpretations, Dirac focused on Heisenberg's noncommutative product and discovered that the new quantum rule ($pq - qp = h/2\pi i$) had a classical counterpart in the Poisson brackets of Hamiltonian mechanics. Dirac systematically exploited this analogy to develop quantum mechanics, unlike his Göttingen competitors, who insisted on the novelty and self-sufficiency of the new theory. Dirac was also original in his introduction of an abstract algebra of "q-numbers," to be developed according to the needs of the theory. In his theory, matrices came last, in the identification of the physical content of some algebraic relations.[32]

---

[30] Bohr to Born, 1 May 1925, in Bohr, *Collected Works*, vol. 5, 310–11.

[31] H. Kramers, "The Quantum Theory of Dispersion," *Nature*, 114 (1924), 310–11; W. Heisenberg, "Über die quantentheoretische Umdeutung kinematischer und mechanischer Beziehungen," *Zeitschrift für Physik* (hereafter *ZP*), 33 (1925), 879–93. Cf. Jammer, *CD*, chap. 5.1; Mehra and Rechenberg, *HD*, vol. 2, chap. 3.5; Dresden, *H. A. Kramers*, chap. 8; Darrigol, *CQ*, chaps. 9–10; Hiroyuki Konno, "Kramers' Negative Dispersion, the Virtual Oscillator Model, and the Correspondence Principle," *Centaurus*, 36 (1993), 117–66; Cassidy, *Uncertainty*.

[32] Born, Heisenberg, and Jordan, "Zur Quantenmechanik II," *ZP*, 35 (1926), 557–615; P. Dirac, "The Fundamental Equations of Quantum Mechanics," Royal Society of London, Series A, *Proceedings*

Heisenberg, Dirac, and the Göttingen theorists convinced their peers that the long-sought mechanics of atoms had been found.[33] Yet, with their infinite matrices or q-numbers, they achieved little more than the old quantum theory had given. Even the helium problem seemed a formidable challenge. Before they had time to improve their mathematics, they became aware of a new, concurrent mechanics.

QUANTUM GAS, RADIATION, AND WAVE MECHANICS

Wave mechanics emerged from the other kind of quantum theory, which concerned itself with statistical thermodynamics and radiation properties. Whereas in the atomic theory the main heuristic principle was its analogy with ordinary electrodynamics, in the quantum statistical theory it was the analogy between matter and radiation. Planck's original theory of radiation relied on an analogy with Boltzmann's gas theory and Einstein's light quantum on a more daring use of the same analogy; the quantum theory of specific heats rested on the reverse analogy, from Planck's radiators to the vibrations of solids.

In 1911/12, Otto Sackur and Hugo Tetrode extended the procedure to gases and determined their "chemical constant" (the additive constant in the entropy, on which chemical reactions among gases depend) as a function of Planck's quantum. Planck applauded the new bridging of gas and radiation theory, and the increased definiteness of the entropy function. In 1916 he generalized the quantization of gas molecules (the division of phase-space into quantum cells) to obtain quantum rules equivalent to Sommerfeld's, as well as a theory of gas "degeneracy." According to Nernst's theorem, there had to be a degenerate low-temperature behavior of ideal gases, with a gradual vanishing of the specific heat. Even though the prospect of observing the phenomenon seemed small, most quantum theorists agreed that it ought to be part of any consistent theory of gases.[34]

The few marginal supporters of the light quantum pursued the reverse analogy, from gas to radiation. Somewhat naively, they tried to derive Planck's radiation law from the equilibrium of a gas of light quanta. The most successful of these attempts was that of an obscure Indian physicist, Satyendra

---

(hereafter *PRS*), 109 (1925), 642–3. Cf. Jammer, *CD*, chap. 5.2; Mehra and Rechenberg, *HD*, vol. 3, chaps. 1–3, and vol. 4, part 1; Kragh, *Dirac: A Scientific Biography* (Cambridge: Cambridge University Press, 1990); Darrigol, *CQ*, part C; R. H. Dalitz and R. Peierls, "Paul Adrien Maurice Dirac, 1902–1984," *Biographical Memoirs of Fellows of the Royal Society*, 32 (1986), 138–85.

[33] Cf. Alexei Khozhevnikov and Olga Novik, *Analysis of Information Ties Dynamics in Early Quantum Mechanics (1925–1927)* (Moscow: Academia Nauk, 1987); Mehra and Rechenberg, *HD*, vol. 4, part 2.

[34] Cf. Darrigol, "Statistics and Combinatorics in Early Quantum Theory, II: Early Symptoms of Indistinguishability and Holism," *HSPS*, 21 (1991), 237–98; A. Desalvo, "From the Chemical Constant to Quantum Statistics: A Thermodynamical Route to Quantum Mechanics," *Physis*, 29 (1992), 465–538.

Nath Bose (1894–1974), published in 1924 with Einstein's support. Bose distributed the light quanta over quantum cells according to a special statistics, which he seems to have mistaken for Boltzmann's. Adopting the new procedure without further justification, Einstein transposed it to ordinary gas molecules, and obtained a new theory of degeneracy now called the Bose-Einstein theory. As was usual for him, Einstein studied the corresponding density fluctuation and found a wavelike term besides the usual corpuscular term. He immediately perceived a connection with the speculations of a younger brother of Maurice de Broglie.[35]

As the house theorist of his brother's laboratory, Louis de Broglie (1892–1987) developed a double competence in the two varieties of quantum theory, atomic and statistical. Maurice's x-ray spectroscopy concerned simultaneously the structure of atoms and the nature of radiation. On the latter issue, Louis was struck by his brother's arguments in favor of a dual nature of x rays and radiation in general. In a corpuscular derivation of Planck's law proposed in 1922, Louis gave the light quantum a small finite mass, so that the light/matter analogy would be more perfect. This led him to reflect on the way to associate an oscillatory phenomenon with a relativistic particle. In 1923 he offered the covariant picture of corpuscles internally oscillating at the proper frequency $m_0 c^2 / h$ (m_0 being the rest mass) and sliding on a plane monochromatic wave; the internal oscillations were synchronous with the wave if the velocity V of the wave and that v of the corpuscles were related according to $Vv = c^2$.[36]

Quite daringly, Broglie assumed that such fictitious, supraluminal waves were also associated with *material* corpuscles, and suggested that electron beams could perhaps be diffracted. Hair-raising though it was, the speculation provided him an explanation of the Bohr-Sommerfeld quantum rules, as the global synchronicity condition for the electron and associated-wave motions around the nucleus. It also provided a new basis for the quantum gas and thermal radiation: the statistics of a system of waves with a varying number of associated corpuscles.[37]

Broglie's matter waves enjoyed publicity from Einstein and caught the attention of a number of physicists. Among them was Walter Elsasser (b. 1904) of Göttingen, who perceived a connection with the anomalous

---

[35] S. N. Bose, "Plancks Gesetz und Lichtquantenhypothese," *ZP*, 26 (1924), 178–81, and Einstein, "Quantentheorie des einatomigen idealen Gases," *BB*, (1924), 261–7. Cf. Jammer, *CD*, pp. 248–9; Mehra and Rechenberg, *HD*, vol. 1, chap. 5.3; Darrigol, "Statistics, II."

[36] Cf. Jammer, *CD*, chap. 5.3; F. Kubli, "Louis de Broglie und die Entdeckung der Materiewellen," *AHES*, 7 (1970), 26–68; Mehra and Rechenberg, *HD*, vol. 1, chap. 5.4; Wheaton, *Tiger*, part 5; Darrigol, "Strangeness and Soundness in Louis de Broglie's Early Works," *Physis*, 30 (1993), 303–72; Mary Jo Nye, "Aristocratic Culture and the Pursuit of Science: The de Broglies in Modern France," *Isis*, 88 (1997), 397–421.

[37] L. de Broglie, "Ondes et quanta," Académie des Sciences, *Comptes-rendus Hebdomadaires des Séances*, 177 (1923), 507–10; "Quanta de lumière, diffraction et interférences," *Comptes-rendus*, 548–50; "Les quanta, la théorie cinétique des gaz et le principe de Fermat," *Comptes-rendus*, 630–2; and *Recherches sur la théorie des quanta* (Paris: Masson, 1924).

low-energy scattering of electrons observed by Carl Ramsauer (1879–1955).[38] Most important, the Austrian theorist Erwin Schrödinger (1887–1961), who had also been interested in gas degeneracy, read Broglie's dissertation. Schrödinger justified the Bose-Einstein statistics by the natural statistics of a system of quantized Broglie waves, and turned to a generalization of Broglie's derivation of the quantum rules. In early 1926, dropping the relativistic aspect of Broglie's waves, Schrödinger hit upon a wave equation the single-valued solutions of which corresponded to the energy spectrum of the hydrogen atom. Thus he hoped to represent all atoms by systems of stationary waves, and the corpuscular aspect of electrons by wave packets.[39]

In the following months, he realized that such mechanistic interpretation of the waves was impossible, because the wave function for several electrons belonged to the more abstract configuration-space and because of the unavoidable spread of wave packets. However, he maintained that the wave motion correctly represented the electromagnetic properties of the atom.

THE FINAL SYNTHESIS

Schrödinger's equation created a sensation among quantum physicists, if only because they were better equipped for solving wave equations than for diagonalizing infinite matrices. Within a couple of months, several proofs appeared that wave mechanics was formally equivalent to matrix mechanics. In a first adaptive step, Heisenberg and Dirac kept the general framework of their quantum mechanics, but used the Schrödinger equation to help solve the matrix or q-number equations. They were soon able to conquer the helium atom, to derive the two kinds of quantum statistics (Fermi-Dirac and Bose-Einstein), to devise a perturbation theory (as did Schrödinger), and to calculate Einstein's transition probabilities.[40]

---

[38] Cf. Jammer, CD, p. 249–51; Mehra and Rechenberg, HD, vol. 1, pp. 624–6; H. A. Medicus, "Fifty Years of Matter Waves," Physics Today, 27 (1974), 38–47; Arturo Russo, "Fundamental Research at Bell Laboratories: The Discovery of Electron Diffraction," HSPS, 12 (1981), 117–60.

[39] E. Schrödinger, "Quantisierung als Eigenwertproblem," AP, 79 (1926), 361–76, 489–527, 734–56; AP, 80 (1926), 437–90, 109–39. Cf. Jammer, CD, chap. 5.3; Paul Hanle, "The Coming of Age of Erwin Schrödinger: His Quantum Statistics of Ideal Gases," AHES, 17 (1977), 165–92; V. V. Raman and P. Forman, "Why Was It Schrödinger Who Developed de Broglie's Ideas," HSPS, 1 (1969), 291–314; Darrigol, "Schrödinger's Statistical Physics and Some Related Themes," in Erwin Schrödinger: Philosophy and the Birth of Quantum Mechanics, ed. M. Bitbol and O. Darrigol (Gif-sur-Yvette: Frontières, 1992), pp. 237–76; Linda Wessels, "Schrödinger's Route to Wave-Mechanics," Studies in History and Philosophy of Science (SHPS), 10 (1979), 311–40; Kragh, "Erwin Schrödinger and the Wave Equation: The Crucial Phase," Centaurus, 26 (1982), 154–97; Mehra and Rechenberg, HD, vol. 5.

[40] Heisenberg, "Über die Spektra von Atomsystemen mit zwei Elektronen," ZP, 29 (1926), 499–518; Dirac, "On the Theory of Quantum Mechanics," PRS, 112 (1926), 661–77; M. Born, "Zur Quantentheorie der Stossvorgänge," ZP, 37 (1926), 863–7; "Quantenmechaniker Stossvorgänge," ZP, 38 (1926), 803–27. Cf. Jammer, CD, chap. 6.1, 8.1; Mehra and Rechenberg, HD, vol. 3, chap. 5.6.

More inclined to seek a physical meaning for Schrödinger's waves, Max Born analyzed electron collisions in terms of scattered waves. There he introduced the first germs of the statistical interpretation of the wave function and expressed his view that determinism no longer held in the atomic domain. Several physicists had already argued for a breakdown of causality during the quantum jumps of the older quantum theory. Possibly, the antirational ideology of the Weimar republic eased German physicists' admission of acausality. However, internal arguments for this radical step were quite strong, and the strongest ones often came from physicists who did not belong to the Weimar culture, like Bohr and Dirac.[41]

Even though the protagonists of the two new mechanics easily borrowed each other's mathematical tools, they strongly disagreed on the deeper nature of quantum phenomena. For the Göttingen physicists, quantum discontinuity was fundamental, and the visualization of atomic processes had to be given up definitively. For Schrödinger, discontinuity was only an appearance, and atomic physics had to be reduced to the continuous play of more intuitable waves. Opposition between the wave and the quantum camps was violent, the more so because the shares in the discovery of quantum mechanics were at stake. Yet a solid synthesis emerged in the winter of 1926/7.[42]

Through a global study of the transformation properties of the equations of quantum mechanics, Dirac and Jordan reached general interpretive rules that seemed to answer any possible question about the empirical behavior of atomic systems. Later elaborations by the Göttingen mathematicians John von Neumann (1903–1957) and Hermann Weyl (1885–1955) consolidated the theory but brought little of physical interest, except for the notion of state vector. The Dirac-Jordan interpretation, which is similar to the modern form of quantum mechanics, reduced the wave-function to a symbolic tool for calculating the probabilities of measurable quantities, and implied the impossibility of simultaneously ascribing sharply defined values to conjugate variables like position and momentum. Although these limitations seemed mathematically necessary, their intuitive meaning remained unclear.[43]

In early 1927, Heisenberg argued that the theoretical indetermination of conjugate variables exactly corresponded to the perturbing action of measurement, as judged from simple thought experiments. For example, defining the

---

[41] Hiroyuki Konno, "The Historical Roots of Born's Probability Interpretation," *Japanese Studies in History of Science,* 17 (1978), 129–45; Mara Beller, "Born's Probabilistic Interpretation: A Case Study of 'Concepts in Flux,'" *SHPS,* 21 (1990), 563–88; Im Gyeong Soon, "Experimental Constraints on Formal Quantum Mechanics: The Emergence of Born's Quantum Theory of Collision Processes in Göttingen, 1924–1927," *AHES,* 90 (1996), 73–101. On acausality, cf. Forman, "Weimar"; Hendry, "Weimar"; Wise, "How Do Sums Count?"; Yemima Ben-Menahem, "Struggling with Causality: Schrödinger's Case," *SHPS,* 20 (1989), 307–34.

[42] Cf. Beller, "Matrix Theory Before Schrödinger: Philosophy, Problems, Consequences," *Isis,* 74 (1983), 469–91; Jammer, *CD,* chap. 6.2 (transformation theory), chap. 6.3 (von Neumann); Mehra and Rechenberg, *HD,* vol. 4, part 1. On Dirac, cf. Kragh, *Dirac,* chap. 2; Darrigol, *CQ,* chap. 13.

[43] Dirac, "The Physical Interpretation of Quantum Dynamics," *PRS,* 113 (1927), 621–41; P. Jordan, "Über eine neue Begründung der Quantenmechanik," *ZP,* 40 (1927), 809–38, and 44 (1927), 1–25.

position of an electron by its image through a γ-ray microscope, he made the quanta of the γ radiation responsible for the indefiniteness of the conjugate momentum. Bohr immediately detected flaws in the reasoning. More fundamentally, he condemned the instrumentalist outlook and the privilege given to the particle picture. In his view, both waves and particles were necessary to understand the behavior of the γ rays in Heisenberg's microscope. The wave aspect of the electron, which Clinton Davisson (1881–1950) and Lester Germer had recently exhibited by crystal diffraction at Bell Laboratories, could not be silenced.[44]

According to Bohr, the true foundation of the uncertainty relation was the mutual incompatibility of the wave and particle pictures: following a well-known property of the Fourier transform, a wave packet of width Δx has a wave-number spread Δk_x at least equal to $1/\Delta x$; therefore, the associated momentum $p_x = hk_x/2\pi$ satisfies the uncertainty relation $\Delta x \Delta p_x \gtrsim h$. This duality excludes any control of the discontinuous interaction occurring in a measurement, and implies the physical incompatibility of devices that would simultaneously measure conjugate variables. Bohr regarded the various properties of a quantum object as "complementary": Each of them represents a possible type of prediction for the behavior of the object, but they can never be all simultaneously determined.[45]

Bohr offered this solution to the various paradoxes of quantum theory in his Como lecture of September 1927, after a broad epistemological introduction of complementarity. He asserted the necessity of classical concepts in experimental reports but declared the failure of the classical type of description in the quantum domain. The quantum postulate, he argued, implied an uncontrollable perturbation of the state of an object during any attempt to locate it in space and time, so that the classical demand of space-time coordination became complementary to that of causality.[46]

Bohr's difficult utterances failed to convince some of the founders of quantum theory, notably Planck, Einstein, and Schrödinger. But the rising generation of physicists agreed that quantum mechanics, strange as it looked, was a basically consistent theory. They quickly accumulated successes in its applications to atomic structure, chemistry, magnetism, and the solid state of matter. After a long period of cooperative and competitive efforts, the crisis

---

[44] Heisenberg, "Über den anschaulichen Inhalt der quantentheoretischen Kinematik un Mechanik," *ZP,* 43 (1927), 172–98; Bohr, "The Quantum Postulate and the Recent Development of Atomic Theory," *Nature,* 121 (1928), 580–90. Cf. Jammer, *CD,* chap. 7; MacKinnon, *Scientific Explanation*; Henry Folse, *The Philosophy of Niels Bohr: The Framework of Complementarity* (Amsterdam: North Holland, 1985); Dugald Murdoch, *Niels Bohr's Philosophy of Physics* (Cambridge: Cambridge University Press, 1987); Beller, "The Birth of Bohr's Complementarity: The Context and the Dialogues," *SHPS,* 23 (1992), 147–80.
[45] Bohr, "The Quantum Postulate."
[46] Ibid.

of physics announced by Einstein in 1905 had finally been resolved to most physicists' satisfaction.[47]

It would be a mistake, however, to believe that quantum mechanics was cast in a final, inalterable form in 1927. Although its basic principles and mathematical structure have remained the same, its formulation and notations have evolved considerably. The earlier-mentioned contributions of such mathematicians as Hermann Weyl brought the rigor and power of Hilbert spaces and group theory. Dirac's lectures and *Principles* brought physical clarity and operational efficiency. Also, the applications of quantum mechanics have diversified its formulations and its canonical methods of resolution and approximation. In one case – the application to the atomic nucleus and to relativistic field theories – the general validity of quantum mechanics was questioned for some time. The Copenhagen interpretation of the theory has occasionally been challenged, for instance by David Bohm's hidden-variable theory of 1952. Theories of the measurement process have been proposed. The relationship between quantum and classical behavior is better understood, and we have more precise ways to characterize the strangeness of the quantum world, for instance, with the nonlocality expressed in the violation of John Bell's inequalities. Quantum mechanics still is a subject of wonder for physicists, philosophers, and historians of science.[48]

---

[47] On the origins of Bohr's philosophy, cf. Jan Faye, *Niels Bohr, His Heritage and Legacy* (Dordrecht: Kluwer, 1991); Wise, "How Do Sums Count?"; Chevalley, "Le dessin." On the diffusion of quantum mechanics, cf. Jammer, *CD,* chap. 8; Mehra and Rechenberg, *HD*, vol. 4, part 2, and vol. 5, chap. 4.

[48] Cf. Jammer, *CD*, chaps. 8, 9, and *The Philosophy of Quantum Mechanics* (New York: Wiley, 1974); Michael Eckert's and Silvan Schweber's contributions to this volume in chapters 21 and 19, respectively.

18

RADIOACTIVITY AND NUCLEAR PHYSICS

Jeff Hughes

Few branches of the physical sciences have had more of an impact on the twentieth-century world than radioactivity and nuclear physics. From its origins in the last years of the nineteenth century, the science of radioactivity spawned the discovery of hitherto unsuspected properties of matter and of numerous new elements. Its practitioners charted a novel kind of understanding of the structure and properties of matter, their achievements gradually winning wide acceptance. With its emphasis on the internal electrical structure of matter and its explanation of atomic and molecular properties by subatomic particles and forces, radioactivity transformed both physics and chemistry. Its offspring, nuclear physics and cosmic ray physics, consolidated and extended the reductionist approach to matter, ultimately giving rise to high energy physics, the form of physical inquiry that became characteristic of late-twentieth-century science: large, expensive machines designed to produce ever smaller particles to support ever more complex and comprehensive theories of the fundamental structure of matter.

The significance of nuclear physics extends far beyond the laboratory and even science itself, however. Practiced in only a handful of places in the 1930s, nuclear physics boomed during World War II, when it provided the scientific basis for the development of nuclear weapons. During the Cold War, nuclear and thermonuclear weapons were the key elements in the precarious military standoff between the superpowers. At the same time, the development of the civil nuclear power industry, of nuclear medicine, and of many other applications brought nuclear phenomena to the attention of a large public. Nuclear physicists came to enjoy enormous prestige and to command enormous resources for their science in the context of the nuclear state. From the 1960s to the 1980s, however, with increasing public fears of nuclear catastrophe, the dumping of radioactive waste from the nuclear industry, and other deleterious effects, nuclear science came to lose its aura, again with significant consequences for its practitioners in terms of both research funding and public and professional status.

The shadow of the nuclear atom loomed large over the century just ended. It is this very visibility, ironically, that complicates historical analysis of the subjects of radioactivity and nuclear physics. In the years after World War II, scientists and their historians looked back to the prewar period to give an account of the development of the subject that preferentially emphasized those elements that would become important in the making of nuclear weapons. Thus, from a postwar perspective, the successive discoveries of radioactivity itself, the disintegration theory of radioactivity, and the notion of half-life, the nucleus, isotopes, the neutron, the liquid-drop model of the nucleus, artificial transmutation, and ultimately nuclear fission formed a linear, teleological, "internalist" sequence of theoretical developments and associated "significant" experimental discoveries through which nuclear history could be given shape and meaning. Complemented by an "externalist" history that portrayed the disciplines largely as exemplars of the Mertonian ideals of internationalism and intellectual freedom; buttressed by a strong strand of biography, autobiography, reminiscence, and uncritical popular history; and supported by enduring scientific and public fascination with those connected to nuclear science, this canonical account dominated the historiography of radioactivity and nuclear physics from the 1940s to the 1980s. Moreover, it has secured remarkable consensus among scientists and historians alike, so that there has been very little historical debate over the fundamentals of interpretation or of this "bomb historiography."[1]

More recently, however, historians have begun to rethink the history of radioactivity and nuclear physics in terms of the material and social practices involved in establishing, maintaining, and extending the disciplines. Cautious of the teleological judgments and values inherent in earlier histories, scholars are increasingly focusing on the instruments, materials, conceptual tools, and standards of evidence brought to bear in the creation of new facts in these domains of inquiry. They seek to understand how the boundaries within and between these and other fields were established and enforced, and how they changed over time. And they attempt to understand the dynamic interrelationships between the various individuals and collectives involved in the elaboration and maintenance of credible knowledge concerning the atom. The products of radioactivity and nuclear physics took shape within a dynamic disciplinary network of individuals and institutions, shaped by complex local circumstances, but held

---

[1] For exemplary instances of the canonical "internalist" history, see L. M. Brown and H. Rechenberg, *The Origin of the Concept of Nuclear Forces* (Bristol, England: Institute of Physics Publishing, 1996); M. Mladjenovic, *The History of Early Nuclear Physics (1896–1931)* (Singapore: World Scientific, 1992). For the "externalist" history, see C. Weiner, "Institutional Settings for Scientific Change: Episodes from the History of Nuclear Physics," in *Science and Values: Patterns of Tradition and Change*, ed. A. Thackray and E. Mendelsohn (New York: Humanities Press, 1974), pp. 187–212; C. Weiner, ed., *Exploring the History of Nuclear Physics* (New York: American Institute of Physics, 1972).

together by shared material and conceptual practices. Far from being a history of theoretical developments, the emerging history locates radioactivity and nuclear physics at the intersection of academia, industry, and the modern state. Rather than taking them as self-evidently significant, it sees their practitioners as having actively to justify their own work to one another and their collective efforts to other scientists and to the wider polity. The nuclear age is only now coming to be understood as a contingent accomplishment, rather than an inevitable outcome, of scientific activity.[2]

RADIOACTIVITY AND THE "POLITICAL ECONOMY" OF RADIUM

The origins of radioactivity lie in the work of the Parisian physicist Antoine Henri Becquerel (1852–1908), whose exploration of the relationship between phosphorescence and the "x rays" discovered by Wilhelm Conrad Röntgen in 1895 led him to observe in 1896 that uranium-bearing minerals produce a radiation capable of darkening photographic plates and discharging electroscopes. A burst of work on uranium radiation followed, although interest soon waned as it became clear that Becquerel's radiation (which most workers regarded as a form of phosphorescence) was very feeble in comparison with x rays and the other types of radiative emissions for which claims were made in the 1890s.

Becquerel's work was taken up in 1897 by Marie Sklodowska Curie (1867–1934) and her husband Pierre Curie (1859–1906). Using a piezoelectric method devised by the latter, the couple were able to show that uranium radiation rendered air electrically conductive and to establish a method for quantifying the new radiation. This work also led to the discovery of thorium's radioactivity. The finding that the uranium-bearing ore pitchblende was more "active" than uranium itself led to the isolation of the new elements polonium and radium from pitchblende (1898) and the stabilization of a new phenomenon characteristic of certain heavy elements ("active matter"), for which Marie Curie coined the term "radioactivity" (from the Latin $radius$ = ray). While Curie herself devoted the next several years to the manufacture of tangible quantities of the new substances for the spectroscopic work needed to identify them as new elements, other researchers began to explore the properties of the new "radioelements" and their radiations.[3]

---

[2] For an introduction to these issues, see J. Golinski, *Making Natural Knowledge: Constructivism and the History of Science* (Cambridge: Cambridge University Press, 1998).

[3] Lawrence Badash, "Radioactivity Before the Curies," *American Journal of Physics*, 33 (1965), 128–35; "Radium, Radioactivity, and the Popularity of Scientific Discovery," *Proceedings of the American Philosophical Society*, 122 (1978), 145–54, and "The Discovery of Thorium's Radioactivity," *Journal*

With differing combinations of physical and chemical skills, a small number of researchers – the Curies in Paris; Ernest Rutherford (1871–1937) and Frederick Soddy (1877–1956) at McGill University, Montreal; Julius Elster (1854–1920) and Hans Geitel (1855–1923) in Wolfenbüttel; Stefan Meyer (1872–1949) and Egon von Schweidler (1873–1948) in Vienna; and a few others in Europe and America – carried out systematic work on the properties of the new radioactive substances in the period 1898 to 1902. As Lawrence Badash and Thaddeus Trenn have shown, this work sought to characterize the radiations emitted by the new substances through their behavior in electric and magnetic fields and their electrical effects as measured by electroscopes and electrometers. It also deployed the tools of analytical chemistry to investigate the new elements and their interrelationships. Among its outcomes were the identification of alpha, beta, and gamma radiation and the disclosure of further radioelements, such as actinium. The discovery of gaseous radioactive "emanations" from some of the radioelements and the elaboration of genetic sequences of elemental transformation led to Rutherford and Soddy's disintegration theory of radioactivity (1902). With its notion that radioactive elements "transmute" one to another with characteristic "half-lives" by the emission of radiations, this theory did much to consolidate the nascent discipline by acting as an intellectual focus for discussion. Though many chemists argued forcefully for several years against the new theory, associating it with the ionic theory being promoted by Svante Arrhenius (1859–1927) and others, those experienced in radioactivity research quickly adopted the theory as a cogent organizing principle.[4]

The small, close-knit group of researchers in the field obtained their raw materials from various sources, often from the Curies themselves. Eager both to promote and to control research in the field in which she and her husband were leaders, Marie Curie quickly developed an industrial extraction process for radium, farmed it out to the Société Centrale de Produits Chimiques, and maintained strong oversight of the production and distribution of the product through her collaborator André Debierne (1874–1949). In Germany, Friedrich Oskar Giesel (1852–1927) of the Braunschweig chemical company Buchler & Co. and E. de Haën of List, near Hannover, also began producing radioactive materials for sale beginning in 1903. With material thus readily available, a radium market quickly developed in which researchers sought to amass as much active material as possible in order to obtain new and more stable results. It was with commercially available radium bromide, for example, that Soddy and the London chemist William Ramsay (1852–1916) were able to show spectroscopically that radium emanation transformed itself

---

of Chemical Education, 43 (1966), 219–20; R. Pflaum, *Grand Obsession: Marie Curie and Her World* (New York: Doubleday, 1989).
[4] T. J. Trenn, *The Self-Splitting Atom: A History of the Rutherford-Soddy Collaboration* (London: Taylor and Francis, 1977).

into helium. Within this radium economy, the ability to produce stable, reliable results was deemed to depend heavily on both the quantity and quality of radioactive matter at the experimentalists' disposal and on the skill that they were able to bring to bear in the laboratory.[5]

The demonstration of the heating effect of radium and the award of the 1903 Nobel Prize in Physics to Becquerel and the Curies brought radioactivity to a new prominence, both through its therapeutic applications – a number of medically oriented radium institutes were opened between 1903 and 1914 – and through the intepretation of the heating effect in terms of the release of internal atomic energy, the source of much satirical comment and utopian speculation. As public and scientific interest soared, the radium market, driven largely by increased medical demand, boomed. Popular books, such as William Hampson's *Radium Explained* (1905), many of which included practical experiments to be performed with small quantities of radium salts, brought radioactivity to a wide audience. With increased public visibility came an increase in the numbers of those wishing to pursue research in the field. The concentration of radioactive resources and skills in relatively few laboratories led to a high degree of mobility among fledgling radioactivity researchers and consolidated the tightly knit research community that had already taken shape. Paris and Montreal, in particular, quickly became centers for training in radioactivity technique – the young German chemist Otto Hahn (1879–1968), for example, traveled to Canada to learn the new science from Rutherford. Among the growing numbers of researchers in the field were an unusually high proportion of women, indicating to some historians both the relatively marginal status of the field and the sympathies of its gatekeepers.[6]

Pedagogically, the emergent discipline was well provided for in Marie Curie's monumental treatise *Radioactivité* (1903), the first systematic survey of empirical data in the subject, while Soddy's *Radioactivity from the Standpoint of the Disintegration Theory* (1904) and Rutherford's *Radioactivity* (1905) both sought to establish the legitimacy of the disintegration theory. By 1907, the theory had achieved wide acceptance among specialists, not least because it provided a coherent interpretation of the decay series of the various radioelements and a scheme of work for the chemical elaboration of those series. As the number of researchers and the quantity of research increased, specialist journals also began to mark the boundaries of the emergent discipline. Yet those boundaries are markedly different than the canonical history assumes. While radioactivity researchers were well represented in established

---

[5] S. Boudia and X. Roque, eds., *Science, Medicine and Industry: The Curie and Joliot-Curie Laboratories*, special issue of *History and Technology*, 13 (1997), 241–354; L. Badash, *Radioactivity in America: Growth and Decay of a Science* (Baltimore: Johns Hopkins University Press, 1979), 135–6.

[6] W. Hampson, *Radium Explained: A Popular Account of the Relations of Radium to the Natural World, to Scientific Thought, and to Human Life* (London: T. C. & E. C. Jack, 1905); M. F. Rayner-Canham and W. Rayner-Canham, eds., *A Devotion to Their Science: Pioneer Women of Radioactivity* (Montreal: McGill-Queen's University Press, 1997).

national scientific journals, such as the *Philosophical Magazine* in Britain and the *Comptes Rendus* of the French Académie des Sciences, they were also increasingly involved, both as editors and as contributors, with new journals, such as *Le Radium* (1903), the *Jahrbuch der Radioaktivität und Elektronik* (1904), and the short-lived English periodical *Ion: A Journal of Electronics, Atomistics, Ionology, Radioactivity and Raumchemistry* (1908). These indicate contemporary assessments of the intellectual place of the field on the one hand, as being, *between* orthodox physics and chemistry but also, on the other, as being part of a new cluster of analytical practices associated with electrons, ionism, and physical and spatial chemistry.

Popular lectures and books, too, were crucial tools in establishing both the identity of radioactivity and the meaning and legitimacy of its products. Most of the leading workers were zealous proselytes for their subject, publishing frequent nontechnical, explanatory, and interpretative articles in popular magazines and journals. While some like Curie chose to stay close to their experimental data and exercised caution with respect to theoretical interpretation, others, like Soddy, consistently emphasized the potential cosmological implications of radioactivity. Although Soddy's utopian speculations on the possible uses of atomic energy by no means represented the views of all researchers, they did enroll a large and interested public for the subject from which all benefited. Indeed, Soddy's book *The Interpretation of Radium* (1909) was the inspiration for H. G. Wells's dystopic scientific romance *The World Set Free* (1914), like his *Tono-Bungay* (1909), an example of the wider cultural appropriation of images from radioactivity. Some historians have seen here evidence of the existence of a "nuclear culture," a set of popular understandings of atomic science that provided the cultural grounds for the "reception" of later nuclear phenomena (including nuclear weapons). Others have argued, however, that such a view is ahistorical since, like the canonical account that it both draws upon and supports, it retrospectively reifies a set of social and natural categories that would only later come to be put in place.[7]

INSTITUTIONALIZATION, CONCENTRATION, AND SPECIALIZATION: THE EMERGENCE OF A DISCIPLINE, 1905–1914

Over the period 1905 to 1910, a distinct disciplinary topography of radioactivity emerged as researchers and radioactive materials became concentrated at a few research centers – principally Paris, Montreal, Berlin, and Vienna.

---

[7] S. Weart, *Nuclear Fear: A History of Images* (Cambridge, Mass.: Harvard University Press, 1988); K. Willis, "The Origins of British Nuclear Culture, 1895–1939," *Journal of British Studies*, 34 (1995), 59–89; R. Ward, "Before and After the Bomb: Some Literary Speculations on the Use of the Atomic Bomb," *Ambix*, 44 (1997), 85–95.

Radioactivity's practitioners came from a variety of backgrounds, the ways in which they situated themselves within the emergent discipline and in which they chose to develop radioactivity practically and intellectually depending heavily on their earlier intellectual formation and practical training. Thus Rutherford, trained in Cambridge ionist physics, promoted a theoretically reductionist approach to the subject, while Curie and her heirs, inheritors of a strongly positivist tradition, worked along predominantly chemical lines, eschewing theoretical abstraction. In several places, collaboration between chemists and physicists was important – as with Rutherford and Soddy at Montreal, or Otto Hahn and physicist Lise Meitner (1878–1968) at the Kaiser Wilhelm Institute for Chemistry in Berlin.

A significant shift in the geography of radioactivity came in 1907 when Rutherford returned to Britain to become professor of physics at Manchester University. With improved resources now at his disposal, Rutherford established a sizable research school of students and visitors. He and his co-workers institutionalized a training regime for researchers in radioactivity, codified in Walter Makower (1879–1945) and Hans Geiger's (1882–1945) influential textbook *Practical Measurements in Radioactivity* (1912). The award of the 1908 Nobel Prize for Chemistry to Rutherford for his work on the alpha particle both ratified his place as a leader in radioactivity research and tellingly indicated how leading scientists outside the subject placed it in relation to established disciplinary categories. When Rutherford obtained access to a large quantity of radioactive material supplied through Stefan Meyer by the Vienna Academy of Sciences, the Manchester group obtained a number of new results, including Rutherford and Royds's spectroscopic identification of the alpha particle with an ionized helium nucleus.

In their quest for a physicalist, reductionist understanding, the Manchester group also explored the forces inside the atom. On the basis of the results of a series of experiments undertaken in his laboratory by Geiger and Ernest Marsden (1889–1970), in which energetic alpha particles were fired into various substances and the products observed and analyzed by the scintillations they produced on fluorescent zinc sulfide screens, Rutherford developed a new atomic model in opposition to the "plum pudding" model of his mentor, Joseph John Thomson (1856–1940). In Rutherford's model (1910–11), the mass of the atom was held to be concentrated in an intensely electrically charged central core, or *nucleus*.[8]

Later commentators have tended to assume both that Rutherford's nuclear atomic model was unambiguously articulated and that it had an immediate impact. In fact, it was only after an initial period of ambiguity that the nucleus came to be regarded as positively charged, with the atom's complement

---

[8] W. Makower and Hans Geiger, *Practical Measurements in Radioactivity* (London: Longmans, Green, 1912); D. Wilson, *Rutherford: Simple Genius* (London: Hodder & Stoughton, 1983), pp. 216–405; J. L. Heilbron, "The Scattering of α and β Particles and Rutherford's Atom," *Archive for History of Exact Sciences*, 4 (1968), 247–307.

of electrons then deemed to be circulating in quasi-planetary orbits some distance from the central core, maintaining the atom's electrical neutrality. Similarly, the new hypothesis received no mention at all at the 1911 Solvay Congress, and only a cursory discussion (largely by Rutherford himself) at the second Congress two years later. Even in the new round of textbooks reflecting the state of the mature discipline, Rutherford's *Radioactive Substances and their Radiations* (1913) and Stefan Meyer and Egon von Schweidler's *Radioaktivität* (1916), the nuclear model was presented as one of several possible alternatives. It is clear, then, that there was no firm consensus as to the "best" atomic model in this period; indeed, models continued to proliferate for several years, and not until the 1920s did any form of unanimity begin to emerge, largely through the mathematization of the nuclear model by Niels Bohr (1885–1962), Arnold Sommerfeld (1868–1951), and others in efforts to "explain" the phenomena of spectroscopy.[9]

Meanwhile, research elsewhere continued along both chemical and physical lines. In 1910, a new Institut für Radiumforschung opened in Vienna. Headed by Stefan Meyer, the institute supported a variety of researches, including work on atmospheric radioactivity by Victor Hess (1883–1964) that disclosed the existence of penetrating radiation in the atmosphere – what would later be called "cosmic rays." By this time, radioactivity had become a recognized, mature, if not a well-established discipline situated *between* physics and chemistry, with significant links to medicine, geology, oceanography, and meteorology. The relatively large quantities of radioactive matter now required to make significant contributions to research meant that only well-resourced institutions were able to remain effective players, with the laboratories in Manchester (Rutherford), Paris (Curie), Berlin (Hahn-Meitner), Vienna (Meyer), London (Ramsay), and Glasgow (Soddy) being the principal research centers. The disciplinary network was consolidated by a number of specialist international conferences and congresses at which business of mutual interest was transacted and through which close informal ties developed among members of the research community. Far from being instances of a benign internationalism, as many subsequent commentators have assumed, the numerous meetings of this sort were essential to the creation of a shared material culture and cognitive world among radioactivity's practitioners, not least through the negotiation of measurement standards and units that allowed the commercial, medical, and academic apects of the subject to operate between laboratories and across national boundaries.[10]

Intellectually, the inaugural Solvay Congress of 1911 vaulted Curie and Rutherford onto another emergent international platform, that of mathematical-theoretical physics, and gave radioactivity a place at the heart of

---

[9] J. Mehra, *The Solvay Conferences on Physics: Aspects of the Development of Physics since 1911* (Dordrecht: Reidel, 1975).
[10] E. Crawford, *Nationalism and Internationalism in Science, 1880–1939: Four Studies of the Nobel Population* (Cambridge: Cambridge University Press, 1992).

the theoretical, discontinuous microphysics of electrons, ions, and quanta being promoted by Walther Nernst (1864–1941) and others. While this conjunction indicates the negotiation of shared interests between radioactivity and the rarefied world of European mathematical physics, it also raises an important historical question concerning the relationship between "experiment" and "theory" in the new sciences of matter in the early twentieth century. Scientists and historians alike have tended to see mathematical theory as the quintessential defining feature of modern physics, yet this was far from self-evident to the radioactivists. As with the distinction between chemical and physical approaches to experimental radioactivity, it is becoming clear that a variety of conceptual practices contributed to "theoretical" understanding of atomic phenomena, from the qualitative modeling characteristic of work on the radioactive decay sequences, through the low-level mathematical elaboration employed in the formulation of the nuclear model, to the more complex treatments advanced by those with mathematical training, as in the advanced mathematics underlying Thomson's atomic models. For many radioactivity researchers, "theory" in its many forms was a useful adjunct to experiment, either as an organizing principle, as a source suggestive of new experiments, or even as a source of legitimation, rather than as an epistemologically privileged form of explanation.[11]

With institutional and disciplinary stability, radioactivity flourished from 1910 to 1914. At the leading centers, significant accomplishments included work on the physical and chemical properties of radioactive substances and their radiations, the continuing elaboration of the decay sequences through ever-more-careful measurements of half-lives, the articulation of generalized principles for the transmutation of one element to another in radioactive decay (the "displacement laws"), work on the x-ray spectra of the elements, the use of Charles Thomas Rees Wilson's (1869–1959) cloud chamber to make visible the paths of ionizing radiations, and the introduction of the concept of isotopy. Elsewhere, a large number of studies of terrestrial and atmospheric radiation contributed to the burgeoning literature, and explorations of the use of radioactivity in medicine continued.

None of this is to suggest, of course, that the field was without controversy. In a domain in which many of the experimental phenomena under investigation were liminal, each new observational claim and speculative assertion was scrutinized and the credentials of its author assessed, especially by those (like Rutherford) with strong theoretical agendas of their own. In a field increasingly characterized by competition between laboratories at the research frontier, personal and professional affiliations and antipathies helped shape the politics of the discipline. Curie was sometimes criticized for her

---

[11] J. Hughes, "'Modernists With a Vengeance': Changing Cultures of Theory in Nuclear Science, 1920–1930," *Studies in History and Philosophy of Modern Physics*, 29 (1998), 339–67.

autocratic and proprietorial attitude, for example, while Ramsay's hubris, his controversy with Rutherford over control of English radium supplies, his unverified claims to have "transmuted" copper to produce lithium, and his death in 1916 combined to exclude him from the historical record of radioactivity by those who constructed it in the 1920s and afterward.[12]

By 1914, then, a diverse array of radioactive phenomena had been elucidated experimentally and elaborated theoretically. The subject was institutionalized in several universities and research institutes, with moderate numbers of researchers and an apparatus of textbooks, journals, and training courses for its reproduction. Standards reposed at national institutions, underpinning both academic and commercial aspects of the subject. International contacts through conferences, visits, and student exchanges were at a high level, and although the discipline still uneasily straddled the boundaries between chemistry and physics, its practitioners continued to promote it tirelessly in popular articles and lectures. Intellectual and social exchanges and student mobility among radioactivity's key institutions came to an abrupt end in 1914, however, with the outbreak of war and the drafting of many researchers into military service of one kind or another. Research in radioactivity (as in all other civil science) was drastically curtailed, and an international congress to have been held in Vienna in 1916 was canceled. However, it is testimony to the strength of the professional and personal bonds holding the radioactivity community together that when they were trapped in Europe by the outbreak of war in 1914, James Chadwick (1891–1974) in Berlin and Robert Lawson (1890–1960) in Vienna received financial and material support from Geiger and Meyer, respectively, even during the height of hostilities. It is equally noteworthy that during World War I, no serious attention seems to have been given to possible military applications of radioactivity, other than their established medical uses and the use of radioactive substances in luminous watch dials and various optical munitions.[13]

Some limited research was carried out during the war: In Berlin, for example, Lise Meitner completed a sequence of experiments begun with Hahn, culminating in 1918 in the discovery of protactinium. In Manchester in 1917, between stints of war work for the British Admiralty, Rutherford and his assistant William Kay obtained surprising evidence from a difficult series of experiments, apparently indicating that the nucleus of nitrogen could be disintegrated by energetic alpha particles to yield hydrogen nuclei – "protons," as Rutherford would christen them in 1920. And in Copenhagen, Bohr created a new synthesis of three independent sets of work: that on apparently inseparable radioactive substances (labeled

---

[12] T. J. Trenn, "The Justification of Transmutation: Speculations of Ramsay and Experiments of Rutherford," *Ambix*, 21 (1974), 53–77.
[13] C. H. Viol and G. D. Kammer, "The Application of Radium in Warfare," *Transactions of the American Electrochemical Society*, 32 (1917), 381–8.

"isotopes" by Soddy); results from the Cavendish Laboratory suggesting that neon existed in two forms separable only with the greatest of difficulty; and the nuclear model of the atom. The implication of Bohr's synthesis was that the individual chemical elements could exist in forms with different masses, and that these differences could be explained by differences in the composition and structure of their nuclei. This synthesis would have a significant impact on postwar matter theory, the relations between physics and chemistry, and, through the ascendance of reductionism, the broader development of science itself in the middle of the twentieth century.

"AN OBSCURE ODDITY"? RADIOACTIVITY RECONSTITUTED, 1919–1925

After a devastating war in which science and technology had demonstrated their capacity to multiply the destructive capacities of nations manyfold, radioactivity research was slow to recommence. An English technical journal summed up the position of the discipline in 1919:

> Radioactivity, discovered in 1896, came of age during the war, but it was hardly due to the war that the event passed almost unnoticed, though the war interfered with radioactive researches as with all philosophical study. Radioactivity never enjoyed real popularity even in its infancy. There is too little in the radioactive phenomena to catch the popular fancy.... [Even] the most striking phenomena, the scintillation visible in the spinthariscope, can only be watched by one person at a time.... [Yet] the band of workers in the new field swelled, order was established in the apparent chaos, radioactive phenomena were found to occur with the regularity of astronomical events, and at present radioactivity is generally accepted, though as an obscure oddity rather than perhaps as anything likely to play a part in matters technical and general.[14]

When the academic network of radioactivity workers began to reconstitute itself in 1918, the political and economic changes wrought by the war meant that there were profound changes in the conditions of work at each laboratory, and in the relationships between individual laboratories. Against the general exclusion of Germany from international scientific circles (which lasted until 1926), direct personal and scientific communication between members of the radioactivity community resumed almost immediately after the Armistice – Rutherford, for example, corresponded intensively with Stefan Meyer and was able to negotiate the purchase of radium, which previously had been loaned him by the Vienna Academy of Sciences, in order to give financial assistance to the impoverished Institut für Radiumforschung. The French adopted a

---

[14] "Radioactivity," *The Electrician*, 107 (1919), 673–4, at 673.

rather more isolationist attitude. In war-worn Paris, the largest effort was at Curie's long-planned Institut du Radium, which opened in 1919. Oriented primarily toward the chemical aspects of radioactivity, the Laboratoire Curie accommodated gradually increasing numbers of researchers during the 1920s, among them Curie's daughter Irène (1897–1956), who quickly mastered the techniques of radiochemistry. Though Marie Curie remained abreast of developments elsewhere, much of the work of the Laboratoire Curie for the next decade involved the elaboration of the chemical properties of the radioelements and of ways of preparing and manipulating them, as well as oversight of radioactive standards and French commercial production of radioactive substances.[15]

In Germany and Austria, economic conditions made research next to impossible for some time. In Vienna, Meyer's attempts to promote research met with little success, while in Berlin, Hahn and Meitner gradually resumed their work, though on a less closely collaborative basis than hitherto. With a small number of students and guest researchers, their laboratory continued to contribute to both physical and chemical aspects of the subject. At the nearby Physikalisch-Technische Reichsanstalt (PTR), Geiger and a succession of students and co-workers, among them Walther Bothe (1891–1957), oversaw German radioactivity standards and developed a program of research into instrumentation, particularly electronic methods of detecting and measuring radioactive phenomena. Overall, however, the quantity of research in the war-torn countries of Europe remained low and would take several years to reach its prewar levels.

In Britain, conditions for scientific work were rather better, although there were significant changes in the distribution and nature of radioactivity research. The death of William Ramsay in 1916 led to the suspension of radioactivity research in London as his co-workers moved on to new pastures. The appointment of Soddy to the professorship of inorganic chemistry at Oxford in 1919 led many to expect the development of a major school of radioactivity research there, especially after Soddy dramatically obtained a large quantity of Czech radium. Although he continued his own research and supported a small number of co-workers, institutional difficulties prevented Soddy from establishing a research group comparable to that at prewar Manchester or postwar Cambridge, where Rutherford succeeded Thomson as professor of experimental physics in 1919. During the 1920s, Rutherford and a growing group of colleagues and research students marshaled the considerable resources of an elite university to establish a comprehensive program of research into the structure and properties of the atomic nucleus, as well as a series of more traditional studies of the radiations emitted by radioactive bodies and their effect on matter.

[15] Pflaum, *Grand Obsession*.

In the canonical history of nuclear physics, the Cavendish Laboratory of the early 1920s has usually been seen as important for Rutherford's prediction of the existence of the neutron, an uncharged nuclear constituent. Yet Rutherford's "prediction" was one speculative remark among several made during the course of a general lecture on "The Nuclear Constitution of Atoms" at the Royal Society in 1920. Far more significant, historians now argue, were the numerous instruments and techniques developed at the Cavendish during this period, which were widely copied by those seeking to enter the field of nuclear research. For example, Rutherford's theorizing concerning the nucleus was bolstered by the work of Francis William Aston (1877–1945), whose new mass spectrograph provided evidence of the existence of Soddy's isotopes among the light, as well as the heavy, elements (it is highly indicative that Soddy and Aston received the 1921 and 1922 Nobel Prizes for Chemistry at the same Stockholm ceremony during which Einstein and Bohr received the corresponding Physics awards). Similarly, the Wilson cloud chamber, deployed to brilliant effect at the Cavendish in the early 1920s by Patrick Maynard Stuart Blackett (1897–1974) and others, was a key source of evidence on the processes involved in nuclear disintegration. Taken together, Bohr's tripartite synthesis, Rutherford and Chadwick's ongoing nuclear disintegration experiments, and photographic evidence from the mass spectrograph and the cloud chamber provided the bases for a comprehensive new picture of nuclear structure, which Rutherford tirelessly (and largely successfully) promoted to other scientists and the public in the early 1920s.[16]

INSTRUMENTS, TECHNIQUES, AND DISCIPLINES: CONTROVERSY, 1924–1932

The role of controversy in shaping science is too often underestimated. From 1919 to 1923, Rutherford and Chadwick were alone in their work on nuclear structure. In 1923, however, Cambridge's domination of nuclear disintegration experiments was challenged by two researchers at the Institut für Radiumforschung in Vienna, Hans Pettersson (1888–1966) and Gerard Kirsch (1890–1956). Deploying ostensibly the "same" experimental methods, the Vienna workers systematically repeated and extended the Cambridge experiments. As Roger Stuewer has shown, Pettersson and Kirsch challenged not just the experimental results of the Cavendish researchers, but also Rutherford's theory of the nucleus. Over the next five years, the Vienna workers doggedly pursued their opposition to the Cambridge results, developing state-of-the-art techniques of nuclear research in order to do so. With neither side able to establish an independent basis for determining the "correct" outcome to the experiments, and both claiming the legitimacy of their own

[16] R. H. Stuewer, "Rutherford's Satellite Model of the Nucleus," *Historical Studies in the Physical Sciences*, 16 (1986), 321–52.

methods and interpretations, by 1927 the two groups had reached an impasse, with each side countering the experimental and conceptual claims of the other in an excellent example of what Harry Collins has called the "experimenters' regress."[17]

The situation was exacerbated by a parallel controversy between Cambridge's Charles Ellis (1895–1980) and Meitner in Berlin concerning the nature and interpretation of the beta-ray spectrum, in which issues of experimental technique and interpretation were again at stake, and in which there were also significant implications for Rutherford's nuclear model. Rutherford sought to contain the damaging disputes within the private sphere of personal communication characteristic of the radioactivity network. Yet with journal publications of claim and counterclaim, the controversies could not but attract attention from the wider scientific public. While the beta-ray controversy was ultimately resolved amicably in Meitner's favor, the Cambridge-Vienna controversy came to a head in December 1927, when Chadwick visited Vienna. By running a series of control experiments, he was able to demonstrate crucial differences between the protocols of the Vienna and the Cambridge scintillation-counting experiments. Historians have seen this episode as representing a decisive closure to the controversy, with Cambridge largely vindicated. There is substantial evidence to suggest, however, that the Viennese workers continued to press their claims against the Cavendish Laboratory for several more years by using new techniques, especially electronic methods of particle counting. In a canonical history structured by retrospective assessments, the controversy has also been seen as being responsible for the lack of "progress" in nuclear science in the 1920s. It is now becoming clear, however, that far from holding back the development of nuclear physics, the controversy and its aftermath significantly shaped it.[18]

In response to the controversies of the 1920s, the Cambridge group organized a conference in the summer of 1928 to discuss problems in radioactivity. Most of the key workers in the subject were invited, and as a result of the discussions at the meeting, several reoriented their research toward the contested artificial disintegration experiments. In so doing, they were able to draw on new repertoires of technique. Since 1926, workers in several laboratories – notably Geiger, now working with Otto Klemperer (b. 1899) and Walther Müller (1905–1979) at Kiel University, his erstwhile student Bothe, still at PTR, and Eryl Wynn-Williams (1903–1978) at the Cavendish Laboratory – had been trying to develop viable electrical detectors and counters in an attempt to establish alternative evidential grounds for the disintegration

---

[17] R. H. Stuewer, "Artificial Disintegration and the Cambridge-Vienna Controversy," in *Observation, Experiment and Hypothesis in Modern Physical Science*, ed. P. Achinstein and O. Hannaway (Cambridge, Mass.: MIT Press, 1985), 239–307; H. M. Collins, *Changing Order: Replication and Induction in Scientific Practice* (London: Sage, 1985).

[18] C. Jensen, *A History of the Beta Spectrum and its Interpretation, 1911–1934* (Basel: Birkhäuser, 2000); J. Hughes, "The Radioactivists: Community, Controversy and the Rise of Nuclear Physics," (PhD diss., Cambridge University, 1993).

experiments. With the development of reliable and inexpensive electronic components in the booming radio industry, and with the newly available skills of young wireless enthusiasts in the late 1920s, it became possible to construct stable electronic counting equipment for use in the laboratory. In 1928, Geiger and Müller unveiled an electrical counter capable of counting particles under a variety of conditions. Much effort was expended in making and calibrating reliable amplifiers and complex counting circuits, so that by 1930, electronic particle counters were in regular use in disintegration experiments in several laboratories, including those in Cambridge, Berlin, Paris (laboratoires Curie and de Broglie), Vienna, Halle, Kiel, and Giessen, with a number of others beginning to develop the technology.[19]

This expansion of the number of laboratories engaged in active work on the disintegration experiments fundamentally reshaped radioactivity research around a new set of tools and a new set of questions. The shift was reinforced by a simultaneous set of changes in the relationship between experimentalists and mathematical theoreticians. The 1920s saw significant development in theoretical atomic physics, centered on Bohr's Institute for Theoretical Physics in Copenhagen. With the support of the Rockefeller Foundation and other philanthropic bodies, a new generation of students turned their attention to atomic theory, forming a small and highly mobile international community. The articulation of wave mechanics by Werner Heisenberg (1901–1976) and Erwin Schrödinger (1887–1961) from 1926, and its application by George Gamow (1904–1968) and others to the nucleus from 1928, offered new resources for the understanding of nuclear phenomena. In particular, the development of the notions of nuclear energy levels, quantum tunneling, and resonance nuclear penetration allowed experimentalists to focus on novel kinds of phenomena, using the new electrical techniques now increasingly at their disposal. In the wake of experimentalists' crisis of certitude, their adoption of wave mechanics simultaneously legitimated the new mathematics and its practitioners and restructured the relationship between laboratory researchers and the growing community of mathematical theorists, creating a new and mutually reinforcing dialogue.[20]

The confluence of new interpretative strategies and novel experimental techniques defining the new nuclear research community led to the elaboration of a wide range of new phenomena, some of which became reified as experimental objects. It was out of a 1930 series of experiments on the nuclear energy levels of light nuclei, for example, that Bothe observed the emission of an unusually penetrating gamma radiation by beryllium under bombardment by polonium alpha particles. The experiments were quickly

---

[19] T. J. Trenn, "The Geiger-Müller Counter of 1928," *Annals of Science*, 43 (1986), 111–35.
[20] R. H. Stuewer, "Gamow's Theory of Alpha Decay," in *The Kaleidoscope of Science*, ed. E. Ullmann-Margalit (Dordrecht: D. Reidel, 1986), pp. 147–86; Hughes, "'Modernists With a Vengeance.'"

taken up by the Joliots in Paris and others, and resulted in Chadwick's identification of the *neutron* in February 1932. With half a dozen or more laboratories equipped with the materials, instruments, and skills to repeat Chadwick's work, neutrons rapidly appeared in Paris, Berlin, and elsewhere, and Chadwick's interpretation was quickly accepted. Moreover, with extensive publicity in the scientific and popular press (led by science journalist J. G. Crowther, a social and political ally of several Cavendish researchers), what quickly became reified as the "discovery of the neutron" promoted the new physics of the nucleus as the most exciting branch of contemporary science.

The rapid acceptance of Chadwick's neutron has typically been seen as self-evident, so that a literature has developed explaining why Bothe and the Joliots *failed* to make the "correct" interpretation of their work. It is noteworthy, however, that while experimentalists and theoreticians alike accepted Chadwick's neutron because it helped save the energy conservation laws that the Joliots' interpretation violated, they fiercely debated the nature and properties of the putative new particle. Indeed, the neutron could be regarded as having being accepted so quickly *because* it was understood in several different ways, thereby serving as a fruitful new research object for both experimentalists and mathematical theorists. In broader terms, the neutron helped consolidate and focus the new network of institutions interested in nuclear questions, so that by the summer of 1932, laboratories and institutes in Berkeley, Berlin, Cambridge, Copenhagen, Halle, London, New York, Paris, Rome, Vienna, Washington, D. C., and numerous other places were equipped or equipping themselves with cloud chambers, electronic counters, valve amplifiers, polonium, and the other paraphernalia of nuclear science to take part in what was increasingly seen as one of the most exciting and productive areas of physics: experimental neutron research.[21]

Other new forms of instrumentation and experimentation, too, were becoming significant. A second line of development was related to the appropriation of wave mechanics by experimentalists. Gamow's work raised the possibility that fast protons might be able to penetrate light nuclei, potentially leading to disintegration. Long confined to the use of the fast particles available from naturally occurring radioactive materials, experimentalists at the Cavendish Laboratory used Gamow's calculations as a resource to reorient existing programs of electron accelerator research toward the acceleration of protons. In May 1932, Ernest Walton (1903–1995) and John Cockcroft (1897–1967) succeeded in disintegrating lithium nuclei by using electrically accelerated protons. A former electrical engineer, Cockcroft had links with

---

[21] J. Six, *La découverte du neutron (1920–1936)* (Paris: Editions du CNRS, 1987); R. H. Stuewer, "Mass-Energy and the Neutron in the Early Thirties," *Science in Context,* 6 (1993), 195–238; J. Hughes, "The French Connection: The Joliot-Curies and Nuclear Research in Paris, 1925–1933," *History and Technology,* 13 (1997), 325–43.

the Manchester electrical company Metropolitan-Vickers that were crucial in obtaining equipment and materials for the acceleration work. With their emphasis on the development of its theories, historians have typically ignored the role of industry in the development of nuclear physics. Just as the radio industry was crucial to the development of the electrical counting methods that changed bench practice in the laboratory, so the electrical engineering industry played a key part in the construction of the large particle acelerators that came to refashion and redefine the scale and scope of the laboratory itself in the 1930s. Indeed, the role of industry, not merely in providing materials but in justifying what to many was a marginal subject, is only now beginning to attract historical attention.[22]

Similar developments were in progress elsewhere. At Berkeley, California, Ernest Lawrence (1901–1958) and his students built ever-larger cyclotrons – machines using electrical and magnetic fields to accelerate particles in gently spiraling orbits. At MIT, and later Princeton, Robert J. van de Graaff (1901–1967) constructed an electrostatic particle accelerator. The identification of a heavy isotope of hydrogen by Harold Urey (1893–1981) and others in 1931 and the production of tangible quantities of "heavy water" added another particle – the "deuteron" – to the laboratory toolkit of experimentalists. With the increased control available over particle energies and experimental conditions, many laboratories invested in one or more accelerators in the 1930s, beginning a race to ever-higher machine energies and a research regime of "atom smashing" that would persist to the end of the century.

Funding such developments created new difficulties, and appeals to philanthropies and private donors often stressed the potential *medical* applications of the big machines, recasting a link with medicine that had existed since the early days of radioactivity. The new machines also brought with them the problem of organizing science on a large scale, including the division of labor between physicists and engineers, the hierarchical organization and time scheduling of work, and the creation of new forms of laboratory space and practice. While historians have seen here the origins of the "Big Science" taken to be characteristic of postwar physics, they are only beginning to explore the impact of these new forms of organization on the practices and values of physicists and on the shape of physics – and its historiography.[23]

---

[22] J. Hughes, "Plasticine and Valves: Industry, Instrumentation and the Emergence of Nuclear Physics," in *The Invisible Industrialist: Manufactures and the Construction of Scientific Knowledge*, ed. J. P. Gaudillère and I. Löwy (London: Macmillan, 1998), pp. 58–101.

[23] J. L. Heilbron and R. W. Seidel, *Lawrence and His Laboratory: A History of the Lawrence Berkeley Laboratory*, vol. 1 (Berkeley: University of California Press, 1989); F. Aaserud, *Redirecting Science: Niels Bohr, Philanthropy, and the Rise of Nuclear Physics* (Cambridge: Cambridge University Press, 1990); P. Galison and B. Hevly, eds., *Big Science: The Growth of Large-Scale Research* (Stanford, Calif.: Stanford University Press, 1992); P. Galison, *Image and Logic: A Material Culture of Microphysics* (Chicago: University of Chicago Press, 1997); J. Hughes, "1932: Une 'annus mirabilis' pour la physique nucléaire?" *La Recherche*, 309 (May 1998), 66–70.

The identification of the positive electron ("positron") in cloud chamber photographs in 1932 by Carl D. Anderson (1905–1991) opened up further avenues of inquiry into the nature of the penetrating radiations in the earth's atmosphere – what Robert Millikan (1868–1953) had labeled "cosmic rays." Many experimentalists now used their electronic counters, magnetic fields, and cloud chambers to study and map cosmic rays and the "fundamental particles" of nature. Nuclear disintegration experiments also continued apace. Early in 1934, the Joliot-Curies announced their production of new, positron-emitting isotopes ("artificial radioactivity") in alpha-bombardment experiments, opening up yet more possibilities for experimental work. The production and manipulation of neutrons in the laboratory allowed them to be used as projectiles in nuclear disintegration experiments; following up the Joliots' work, for example, a group led by Enrico Fermi (1901–1954) used neutrons as projectiles to produce a series of new isotopes, including what they believed to be transuranic elements. The Italians also realized the efficacy of slow neutrons (neutrons filtered through paraffin) in producing nuclear reactions. A spate of new experimental studies in "transmutation" – labelled the "Newer Alchemy" by Rutherford – followed, each set of claims typically being repeated, checked, and extended by collaborating physicists and chemists in the rapidly extending network of laboratories.[24]

Theoretical work, too, developed quickly in the period 1932 to 1935. For theoreticians, as for experimentalists, the new particles and processes disclosed in the early 1930s provided significant opportunities for new work, new theories based on proton-neutron nuclei being rapidly developed by Heisenberg and others. Although historians question the immediate impact of the neutron in challenging the prevailing proton-electron model of the nucleus, the variety of theoretical responses to the new results of the early 1930s mirrored that of the experimentalists, as theoreticians sought to make sense of the burgeoning mass of experimental data. Building on the notion of the neutrino postulated in 1930 by Wolfgang Pauli (1900–1958) to preserve the conservation of energy in nuclear processes, Fermi developed a widely accepted theory of beta decay (1934), which used the idea of the creation and annihilation of material particles. In the mid-1930s, Bohr and others developed Gamow's earlier work on the liquid-drop model of the nucleus, ultimately arriving at the notion of the compound nucleus, which sought to account both for the properties of nuclei and for the increasing number of nuclear reactions and excitations observed in the laboratory. This model dominated nuclear theory for the next two decades.[25]

---

[24] E. Rutherford, *The Newer Alchemy* (Cambridge: Cambridge University Press, 1937).
[25] R. H. Stuewer, "The Nuclear Electron Hypothesis," in *Otto Hahn and the Rise of Nuclear Physics*, ed. W. R. Shea (Amsterdam: Reidel, 1983), pp. 19–67; L. M. Brown and H. Rechenberg, "Field Theories of Nuclear Forces in the 1930s: The Fermi Field Theory," *Historical Studies in the Physical Sciences*, 25 (1994), 1–24; R. H. Stuewer, "The Origin of the Liquid-Drop Model and the Interpretation of Nuclear Fission," *Perspectives on Science*, 2 (1994), 76–129.

FROM "RADIOACTIVITY" TO "NUCLEAR PHYSICS": A DISCIPLINE TRANSFORMED, 1932–1940

Over the period 1928 to 1933, then, diversification in experimental and theoretical practice allowed several new groups to enter the field of nuclear research. Unlike those who had been in the field since the 1900s and 1910s, the newcomers did not identify with the radioactivity tradition. Instead, they labeled their field "nuclear physics." In 1931, a Rome conference hosted by Fermi's novitiate research group was devoted to "Il Fisica Nucleare," denoting an emergent sense of disciplinary identity. Similarly, the 1933 Solvay Congress on "The Structure and Properties of Atomic Nuclei" and a 1934 conference in London on "Nuclear Physics" seemed to ratify a new disciplinary space, while a series of Nobel prizes in physics in the 1930s – Heisenberg in 1932, Chadwick in 1935, Fermi in 1938, Lawrence in 1939 – signaled the growing importance of the field within important parts of the scientific community. Whereas the influential 1930 monograph by Rutherford, Chadwick, and Ellis had been titled *Radiations from Radioactive Substances,* by the mid-1930s, the next generation were producing textbooks on nuclear physics. In 1936–7, a key series of articles in *Reviews of Modern Physics* by Hans Bethe (b. 1906), with Robert Bacher (b. 1905) and Stanley Livingston (1905–1986), synthesized all current theoretical and experimental work on the nucleus. With three parts covering the properties of nuclei; nuclear forces; alpha, beta, and gamma radiations; neutrons and deuterons; the statistical theory of heavy nuclei; nuclear moments; nuclear processes as many-body problems; scattering; and experimental methods and data, the articles ran to almost 500 pages. The "Bethe Bible" remained the standard reference work in nuclear physics until the 1950s.[26]

The subject's social and intellectual geographies were simultaneously undergoing significant transformation. While the Bohr Institute in Copenhagen and Max Born's institute at Göttingen University were important centers for the dissemination of the new mathematical physics of wave mechanics, an increased emphasis on research by American universities in the late 1920s took a number of European nuclear scientists, particularly theoreticians, to the United States. This shift was reinforced in the mid-to-late 1930s by the flight of Jewish scientists from the fascist regimes in Germany and Italy to the Soviet Union, Britain, and, especially, the United States. The diaspora helped establish or consolidate nuclear physics at a number of institutions, further reifying the geographical transformation that had taken place since 1930. Meanwhile, the subject matter of nuclear physics was itself multivalent, with

---

[26] N. Feather, *An Introduction to Nuclear Physics* (Cambridge: Cambridge University Press, 1936); F. Rasetti, *Elements of Nuclear Physics* (London: Blackie, 1937); H. A. Bethe and R. F. Bacher, "Nuclear Physics: A. Stationary States of Nuclei," *Reviews of Modern Physics,* 8 (1936), 82–229; H. A. Bethe, "Nuclear Physics: B. Nuclear Dynamics, Theoretical," *Reviews of Modern Physics,* 9 (1937), 69–244; M. S. Livingston and H. A. Bethe, "Nuclear Physics: C. Nuclear Dynamics, Experimental," *Reviews of Modern Physics,* 9 (1937), 245–390.

studies of the characteristics of individual nuclei, transmutation processes (explored both with neutrons and with artificially accelerated particles), and cosmic rays forming differentiated but related experimental clusters, whose products (such as the mesotron, disclosed in 1937) the theoreticians sought to integrate within a phenomenological and mathematical framework.[27]

Despite the rapid institutional growth of nuclear physics in the late 1930s, as increasing numbers of institutions felt it important to participate in the most "modern" branch of the subject, historians have only recently begun to explore this modernist imperative and the ways in which the boundaries between the different traditions of experiment and between experiment and theory shifted in this period. Typically, they have focused instead on the discovery of nuclear fission in December 1938 and its consequences. Hahn, Meitner, and Fritz Strassmann (1902–1980) in Berlin had been competing with groups in Paris and Rome in their work on neutron-induced nuclear transmutation. The work continued after Meitner's flight from the Nazis in 1938, including the announcement by Hahn and Strassmann in December of that year that they had apparently produced barium by slow-neutron bombardment of uranium, was assimilated quickly. As news of the experiment was disseminated among American physicists by Bohr, leading to rapid replication of the result at several suitably equipped institutions early in 1939, the exiled Meitner and Otto Frisch (1904–1979) were elaborating a theory of uranium transmutation, coining the term "fission" for the process. Detailed exploration of the process of fission followed, including attempts to establish the number of neutrons released and investigations of the possibility of a self-sustaining "chain reaction."[28]

The mobilization of European scientists for war led to the cancellation of the planned 1939 Solvay Congress on particle physics. More significantly, it sparked fears (particularly among émigré Jewish physicists) that the Nazis might exploit fission for military purposes. This context fundamentally shaped wartime research in nuclear physics in Britain and the United States, as those with the relevant skills were recruited for state-supported programs of research into the physics and chemistry of fission (although it is important to note that many of those who had worked in nuclear physics in the 1930s were actually mobilized into the development of radar and code-breaking technology because of their electronics expertise). The outbreak of war and the establishment of national programs to explore the military possibilities of fission in Germany, Russia, Japan, Britain, and the United States saw the enrollment of nuclear physics by the state on a massive scale. Ultimately, the Manhattan Engineering District in the United States, arguably the largest

---

[27] P. K. Hoch, "The Reception of Central European Refugee Physicists of the 1930s: U.S.S.R., U.K., U.S.A.," *Annals of Science*, 40 (1983), 217–46.

[28] W. R. Shea, ed., *Otto Hahn and the Rise of Nuclear Physics* (Dordrecht: Reidel, 1983); L. Badash, E. Hodes, and A. Tiddens, "Nuclear Fission: Reaction to the Discovery in 1939," *Proceedings of the American Philosophical Society*, 130 (1986), 196–231.

planned human project since the building of the pyramids, a secret state-within-a-state with a $2 billion budget, produced the weapons dropped on Hiroshima and Nagasaki in August 1945.

This is not the place to discuss the wartime projects in detail; a large literature exists on the subject, documenting in detail nuclear weapons research in America, Britain, Germany, Russia, and elsewhere. In terms of the nuclear physics underlying the production of fission weapons, the wartime projects produced copious amounts of data concerning the properties and behavior of nuclei and about the chemistry and metallurgy of radioelements (including plutonium and the other transuranics produced in nuclear piles). They also led to the rapid development and mass manufacture of new forms of instrumentation and the explosive growth of the field of nuclear electronics – *nucleonics* – which supported the development of the postwar civil and military nuclear industry, medical physics, and academic nuclear physics. They transformed nuclear physicists' horizons of expectation of what was possible through the proper organization of scientific and technical work. And, finally, they brought nuclear physics and nuclear physicists forcibly to the attention of the public, giving them – and their history – a new and powerful credibility.[29]

NUCLEAR PHYSICS AND PARTICLE PHYSICS: POSTWAR DIFFERENTIATION, 1945–1960

Nuclear physicists emerged from the war with a new prestige and authority. While historians endlessly debate the development and motivation for the use of nuclear weapons and their role in the origins and development of the Cold War, little historical attention has been paid either to the postwar construction of a legitimating "history of nuclear physics" or to the development of postwar nuclear physics itself. Physicists and historians alike have instead focused almost exclusively on the development of the domain known as high energy physics, or particle physics, to which prewar radioactivity and nuclear physics are usually portrayed as direct precursors. Yet, just as with the distinction between radioactivity and nuclear physics, the development of boundaries between different intellectual subfields and groups of practitioners in the postwar period is crucially important in understanding the way nuclear science has developed since 1945.

As in 1939, nuclear physics in 1945 was an umbrella term for research on the particles and forces forming the underlying structure of matter. Thus,

---

[29] L. Hoddeson, P. W. Henriksen, R. A. Meade, and C. Westfall, *Critical Assembly: A Technical History of Los Alamos during the Oppenheimer Years, 1943–1945* (Cambridge: Cambridge University Press, 1993); M. Gowing, *Britain and Atomic Energy, 1939–1945* (London: Macmillan, 1964); M. Walker, *German National Socialism and the Quest for Nuclear Power, 1939–1949* (Cambridge: Cambridge University Press, 1989); D. Holloway, *Stalin and the Bomb: The Soviet Union and Atomic Energy, 1939–1956* (New Haven, Conn.: Yale University Press, 1994).

scattering, disintegration and nuclear reaction research, neutronics, accelerator physics, field theory, and cosmic rays, as well as the new physics of nuclear weapons, reactors, and isotopes (the latter with many new industrial and medical applications), all belonged to the broad domain of nuclear physics. The institutional geography of nuclear physics was considerably expanded during and after the war with the creation of a number of new governmental and military establishments devoted to nuclear science (often including the increasingly differentiated categories of "classical" radioactivity, "nuclear chemistry," and "nuclear medicine"), especially those aspects with industrial, medical, or military applications. With the increasing demand for nuclear physicists came a rapid increase in the size of the academic nuclear physics research community and a proliferation of conferences and other gatherings, which helped reconstitute national and international communities of practitioners.

Postwar nuclear physicists sought to meet the expectations placed on them by applying technical and organizational lessons learned in the war to the problems left suspended in 1939. The rapid development of electronics during the war and the consolidation of links with industry allowed the development of new and more sophisticated forms of instrumentation, both for detectors and for the accelerators that now formed an increasingly central part of the nuclear physicist's armamentarium. At the same time, research on cosmic rays continued, using the traditional small-scale methods of cloud chambers and photographic plates. Even here, however, lessons learned from the war and new industrial and political contacts had a considerable impact: For example, the use of new and more sensitive photographic materials in collaboration with the photographic industry and larger-scale analytical practices elicited novel phenomena, like the π-meson in 1948, creating much work for both experimentalists and theoreticians.[30]

Following the production of π-mesons and other cosmic-ray particles by accelerators in the early 1950s, particle physics became increasingly associated with large accelerators and increasingly distinct both intellectually and organizationally from nuclear physics. With the articulation of the principle of phase stability (1945) and, later, the alternating-gradient focusing principle (1952), machines of ever-greater energy, complexity, and diversity were constructed as the technical bases of ambitiously reductionist scientific programs. As Peter Galison has demonstrated, large bubble chambers and industrial modes of organization began to displace cloud chambers and photographic emulsions, while electronic detectors increased in scale and competed with bubble chambers for epistemic authority and funds. Ultimately, this increasingly costly regime led to the creation and expansion of large regional, national, and international laboratories, such as the Brookhaven facility in the United States and the Centre Européen pour la Recherche Nucléaire

---

[30] Galison, *Image and Logic*, pp. 160–218.

(CERN) at Geneva, constituting what came to be known as "high energy physics." It thus became increasingly clearly differentiated from work on nuclear structure, which explored nuclear forces in greater depth within a much more restricted range of energies (typically, several MeV up to 200–300 MeV, the threshold for production of π-mesons). This shift was mirrored among theoreticians, as a distinction became apparent between those theoreticians who sought an understanding of the properties of the nucleus as a structural entity and those who sought "a more and more refined ultimate analysis of matter" – the latter eventually becoming identified as elementary-particle physicists.[31]

While particle physicists investigated the underlying constituents of nucleons, then, nuclear physicists in the 1950s and 1960s sought to understand the nucleons' collective behavior – the "feel of nuclear matter."[32] Using relatively low-energy accelerators coupled with increasingly sophisticated detection equipment, such as scintillation counters and high-resolution semiconductor detectors, experimentalists explored proton, deuteron, and neutron scattering reactions, nuclear moments, orientations, spins and energy levels, stripping reactions, and heavy-ion induced reactions. With magnetic beta spectrometers and the neutron-rich isotopes made available by nuclear reactors, they sought to study the decay schemes of various nuclei and other nuclear properties, as well as the merits of collective versus single-particle models of the nucleus (particularly in the light of the independent particle "shell" model of nuclear structure put forward in 1948 by Maria Goeppert-Mayer [1906–1972] and Hans Daniel Jensen [1907–1973]). As yet, however, historians have not explored this terrain in any detail, having been seduced instead by the development of the more "glamorous" and better-publicized fields of particle and high energy physics.

In many respects, the forms of work organization and collaboration developed at the large facilities of the high energy physicists served as a model for nuclear physicists in the emerging domain of intermediate energy nuclear physics. As in high energy physics, for example, the much-debated trend to larger machines, computerization, increasingly complex forms of organization, and increasing cost of nuclear physics research encouraged the development of regional, national, and eventually supranational research facilities in nuclear physics also. More recently, however, high energy physics has begun to follow a pattern established in the 1980s by its lower-energy partner toward declining funding and lower prestige in the scientific rankings – in the United Kingdom, for example, the percentage of the total research council budget allocated to particle and nuclear physics fell from

---

[31] Ibid., pp. 313–552; A. Hermann, J. Krige, U. Mersits, and D. Pestre, *History of CERN*, vol. 1: *Launching the European Organization for Nuclear Research* (Amsterdam: North Holland, 1987), pp. 3–52, esp. p. 9.

[32] S. S. Schweber, "From 'Elementary' to 'Fundamental' Particles," in *Science in the Twentieth Century*, ed. J. Krige and D. Pestre (Amsterdam: Harwood Academic Publishers, 1998), pp. 599–616, at 607.

46.1 percent in 1966 to 21.5 percent in 1986, while in the United States, the cancellation of the Superconducting Super Collider in 1993 is seen by some as the last act in the drama of high energy physics. These seismic, and perhaps conclusive, shifts will undoubtedly have significant implications for our understanding of the history both of nuclear physics and of the reductionist project in twentieth-century physics.[33]

At the end of the twentieth century, historians of science were only beginning to reassess one of its defining features: nuclear science. Its historiography long dominated by nuclear physicists themselves, the principal contours of the canonical scientists' history of radioactivity and nuclear physics have, until recently, been implicitly accepted by historians of science who, in many cases, merely repeated or finessed the details of the "standard" account. With the emergence of a generation of historians perhaps less closely allied to the values of physics and physicists, and the adoption of properly historical questions and contextual approaches, the way in which nuclear history should be written and the relationship between nuclear history and its contemporary conditions of intellectual and social production are all open issues for discussion.

For the period up to 1939, it is becoming clear that radioactivity and nuclear physics were not self-evident or inevitable enterprises whose boundaries and goals were predefined by nature. Through studies of experimental and theoretical practice and the flux of institutional and disciplinary politics, historians are starting to understand the social and epistemological dimensions of the interplay between "theory" and "experiment," which underwrote scientists' claims about the subatomic world. They are only beginning to comprehend the roles of industry, medicine, the military, gender, the media, and the public in constituting radioactivity and nuclear physics and the dynamics of their development in relation to science as a whole. While a considerable literature exists on the development of nuclear physics during World War II, or at least of those elements of it related to the production of nuclear weapons, the history of postwar nuclear physics remains almost completely unexplored in relation to its more culturally visible cognate, particle or high energy physics. Little historiographical effort has been devoted to understanding the construction of the canonical history of nuclear physics in the postwar period, with its linear, teleological narrative and, more importantly, its implicit naturalistic justification of the creation of nuclear weapons. Whether these topics will receive the historical attention they properly deserve remains to be seen, however. Where once the history of nuclear physics occupied the commanding heights of the history of physics, and perhaps even the history of science more broadly, it now appears to be mirroring the

---

[33] W. E. Burcham, "Nuclear Physics in the United Kingdom, 1911–1986," *Reports on Progress in Physics*, 52 (1989), 823–79; D. Kevles, *The Physicists: The History of a Scientific Community in Modern America*, 2d ed. (Cambridge, Mass.: Harvard University Press, 1995), pp. ix–xlii.

decline in prestige of its object, with fewer graduate students choosing to work in this field and fewer papers being published.

Against this picture of apparent decline, however, must be pitched the intellectual challenge of rewriting the history of radioactivity and nuclear physics, and of exploring *historically* the ideology of reductionism, which has been so influential in the wider culture of twentieth-century science. Many questions remain to be explored in the history of prewar nuclear science, and the postwar period has barely been touched. There remain links to be forged with the political, diplomatic, and cultural historians of the nuclear era, who are busily reassessing the history of nuclear weapons and nuclear politics, and a new frame to be defined for an integrated historical understanding of both the social and the technical aspects of nuclear science. At a time of such far-reaching intellectual ferment, the history of radioactivity and nuclear physics is a field ripe for further research. And it is perhaps only now, as the scientific community adjusts itself to the end of the nuclear Cold War and as wider publics seek an understanding of the century that saw the birth of the nuclear age, that such research can properly be undertaken.

19

QUANTUM FIELD THEORY
From QED to the Standard Model

Silvan S. Schweber

Until the 1980s, it was usual to tell the story of the developments in physics during the twentieth century as "inward bound" – from atoms, to nuclei and electrons, to nucleons and mesons, and then to quarks – and to focus on conceptual advances. The typical exposition was a narrative beginning with Max Planck (1858–1947) and the quantum hypothesis and Albert Einstein (1879–1955) and the special theory of relativity, and culminating with the formulation of the standard model of the electroweak and strong interactions during the 1970s. Theoretical understanding took pride of place, and commitment to reductionism and unification was seen as the most important factor in explaining the success of the program. The Kuhnian model of the growth of scientific *knowledge,* with its revolutionary paradigm shifts, buttressed the primacy of theory and the view that experimentation and instrumentation were subordinate to and entrained by theory.[1]

The situation changed after Ian Hacking, Peter Galison, Bruno Latour, Simon Schaffer, and other historians, philosophers, and sociologists of science reanalyzed and reassessed the practices and roles of experimentation. It has become clear that accounting for the growth of knowledge in the physical sciences during the twentieth century is a complex story. Advances in physics were driven and secured by a host of factors, including contingent ones. Furthermore, it is often difficult to separate the social, sociological, and political factors from the technical and intellectual ones.

In an important and influential book, *Image and Logic,* published in 1997, Peter Galison offered a framework for understanding what physics was about in the twentieth century. Galison makes a convincing case for regarding experimentation, instrumentation, computational modeling, and theory as

---

[1] See A. Pais, *Inward Bound: Of Matter and Forces in the Physical World* (Oxford: Clarendon Press, 1986); T. Yu Cao, *Conceptual Developments of 20th-Century Field Theories* (Cambridge: Cambridge University Press, 1997); Paul Davies, ed., *The New Physics* (Cambridge: Cambridge University Press, 1989); and R. E. Marshak, *Conceptual Foundations of Modern Particle Physics* (Singapore: World Scientific, 1993).

quasi-autonomous subcultures with languages and practices that are distinct, yet linked and coordinated. Experimental, theoretical, and instrumental practices do not all change of a piece – each has its own periodization; and their relation to one another varies with the specific historical situation in which each is embedded. There is, in fact, continuity of experimental practices across theoretical and instrumental breaks.[2]

Image and Logic is a brief for "mesoscopic history" – for history written at a level between macroscopic, universalizing history and microscopic, nominalistic history. Galison proposes treating the movement of ideas, objects, and practices as one of *local* coordination – both social and epistemic – and their interconnections and linkages are made possible through the establishment of pidgin and creole languages. He sees the separate, but correlated, subcultures of physics as bound and stabilized by such interlanguages. These suggestions are attractive and valuable. However, limited as I am in this chapter to choices for presentation, most of the following account lies squarely within the history of ideas. The reader is referred to recent books by Andrew Pickering, Gerard 't Hooft, Lillian Hoddeson, and others for more mesoscopic accounts.[3]

I have not tried to fit my presentation of the history of quantum field theory (QFT) from QED (quantum electrodynamics) to QCD (quantum chromodynamics) into a preconceived pattern – whether that of Thomas Kuhn or that of Imre Lakatos. My concern has been with the telling of the story. One could easily cast the history into a Lakatosian mold of research programs – with S-matrix and field theory the two competing modes.[4] Similarly, one could pick from that history examples that would instantiate both of Kuhn's notions of paradigm; namely, paradigm as achievement – the body of work that emerges from a scientific crisis and sets the standard for addressing problems in the subsequent period of normal science – and paradigm as a set of shared values – the methods and standards shared by the core of workers who decide what are interesting problems and what counts as solutions, and determine who shall be admitted to the discipline and what shall be taught to them.

Furthermore, one could readily give examples of Kuhnian revolutions. Renormalization theory as formulated in the period from 1947 to 1949,

---

[2] Peter Galison, *Image and Logic: A Material Culture of Microphysics* (Chicago: University of Chicago Press, 1997); see also Galison, *How Experiments End* (Chicago: University of Chicago Press, 1987).

[3] L. M. Brown and Lillian Hoddeson, *The Birth of Particle Physics* (Cambridge: Cambridge University Press, 1983); Laurie M. Brown, Max Dresden, and Lillian Hoddeson, eds., *Pions to Quarks: Particle Physics in the 1950s* (Cambridge: Cambridge University Press, 1989); L. M. Brown and H. Rechenberg, "Quantum Field Theories, Nuclear Forces, and the Cosmic Rays (1934–1938)," *American Journal of Physics*, 59 (1991), 595–605; Gerard 't Hooft, *In Search of the Ultimate Building Blocks* (Cambridge: Cambridge University Press, 1997); Andrew Pickering, *Constructing Quarks: A Sociological History of Particle Physics* (Edinburgh: Edinburgh University Press, 1984); Michael Riordan, *The Hunting of the Quark: A True Story of Modern Physics* (New York: Simon and Schuster, 1987).

[4] Steve Weinberg, *Dreams of a Final Theory: The Search for the Fundamental Laws of Nature* (New York: Pantheon, 1992); J. T. Cushing, *Theory Construction and Selection in Modern Physics: The S Matrix* (Cambridge: Cambridge University Press, 1990).

culminating with the work of Freeman Dyson (b. 1923), is surely one such revolution; broken symmetry, as formulated by Jeffrey Goldstone (b. 1933) and Yoichiro Nambu (b. 1921), in the early 1960s another. One probably could constrain the history of quantum field theory into a Kuhnian mold. But I believe that much would be lost in so doing, in particular, a perspective on the cumulative and continuous, yet novel, components of the developments. It seems to me that the later Kuhn's emphasis on "lexicons" – the learnable languages, algorithms, laws, and facts of a given tribe of scientific workers – constitutes a more useful approach to the growth of our knowledge in high energy physics.

Equally helpful, I believe, is Ian Hacking's notion of a style of scientific reasoning: "A style of reasoning makes it possible to reason toward certain kinds of propositions, but does not of itself determine their truth value."[5] A style determines what may be true or false. Similarly, it indicates what has the status of evidence. Styles of reasoning tend to be slow in evolution and are vastly more widespread than paradigms. Furthermore, they are not the exclusive property of a single disciplinary matrix. Thus, Feynman's space-time approach to nonrelativistic quantum mechanics encapsulates a new style of reasoning: All physical measurements and interactions can be considered as scattering processes. I believe Hacking's notion of a style of reasoning captures something right about the history of quantum field theory.

The use of symmetry is another example of a style of reasoning. The fact that such styles of reasoning are useful in both particle physics and condensed matter physics – and, in point of fact, cross-fertilize these fields – illustrates the (nonlinear) additive properties of styles of reasoning. Since a style of reasoning can accommodate many different paradigms, it is not surprising that one should discern Kuhnian revolutionary episodes within its evolution. The delineations of such revolutions are helpful guidelines and periodizations of the history of the field. But it is the identification of the different styles of reasoning that is, I believe, the important task for the intellectual historian attempting to relate that history.[6]

QUANTUM FIELD THEORY IN THE 1930s

The history of theoretical elementary particle physics from the 1930s until the mid-1970s can be narrated in terms of oscillations between the particle and field viewpoints epitomized by Paul Dirac (1902–1984) and by Pascual

---

[5] Ian Hacking, "Styles of Scientific Reasoning," in *Post-Analytic Philosophy*, ed. John Rajchman and Cornell West (New York: Columbia University Press, 1986), pp. 145–65.
[6] For other accounts, and in greater detail, Silvan S. Schweber, "From 'Elementary' to 'Fundamental' Particles," in *Science in the Twentieth Century*, ed. John Krige and Dominique Pestre (Amsterdam: Harwood, 1997), pp. 599–616, and Schweber, *QED and the Men Who Made It* (Princeton, N.J.: Princeton University Press, 1994).

Jordan (1902–1980), as noted by Olivier Darrigol in Chapter 17.[7] That the *field* approach was richer in potentialities and possibilities than the *particle* one is made evident by the quantum field theoretic developments of the 1930s (QFT). All these advances took as their point of departure insights gained from the quantum theory of the electromagnetic field and, in particular, from the centrality of the concept of emission and absorption of quanta.

Enrico Fermi's (1901–1954) theory of beta decay was an important landmark in the field theoretic developments of the 1930s. It had been recognized since 1915 that the nucleus was the site of all radioactive processes, including β-radioactivity in which a nucleus ejects an electron. It was, therefore, natural to believe that electrons existed in the nucleus. Already in 1914, Ernest Rutherford (1871–1937) had assumed that the hydrogen nucleus is the positive electron – he called it the H-particle – and he conjectured that nuclei were made of H-particles and electrons. During the 1920s, the generally accepted model of a nucleus was that it consisted of the two elementary particles then known: protons and electrons. Rutherford in his Bakerian Lecture of 1920 had suggested that a proton and an electron could bind and create a neutral particle, which he believed was necessary for the building up of the heavy elements. However, if nuclei were assumed to be composed of protons and electrons, the Pauli principle made it difficult to understand the spin of certain nuclei, such as N^{14}. Similarly, should there be electrons in the nucleus, their magnetic moment – as determined by the hyperfine structure of atoms – ought to be much larger than the values determined experimentally, which are three orders of magnitude smaller than atomic moments. Confusion reigned, compounded by the difficulty in understanding β-decay.

The process of β-decay – wherein a radioactive nucleus emits an electron (β-ray) and increases its electric charge from Z to $Z+1$ – had been studied extensively during the first decade of the century. If the process is assumed to be a two-body decay, that is, if the decay consists in a nucleus undergoing the process $A^Z \to A^{Z+1} + e^-$, then energy and momentum conservation requires the electron to have a definite energy. However, in 1914, James Chadwick had found that the energy of the emitted electrons had a continuous energy spectrum – up to some maximum energy. At the maximum electron energy, energy conservation was found to hold – to the accuracy of the measurements in the experiment.

By the end of the 1920s, no explanation of the continuous β-spectrum had proven satisfactory, and some physicists, in particular Niels Bohr, were ready to give up energy conservation in β-decay processes. In December 1930, Wolfgang Pauli (1900–1958), in a letter addressed to the participants of

---

[7] See Helge Kragh, *Dirac: A Scientific Biography* (Cambridge: Cambridge University Press, 1990); Steven Weinberg, *The Quantum Theory of Fields*, 2 vols. (Cambridge: Cambridge University Press, 1995–6).

a conference on radioactivity, proposed saving energy conservation with "a desperate remedy," suggesting that

> there could exist in the nuclei electrically neutral particles that I wish to call neutrons [later renamed neutrinos by Fermi], which have spin $1/2$ and obey the exclusion principle.... The continuous β-spectrum would then become understandable by the assumption that in β-decay a [neutrino] is emitted together with the electron, in such a way that the sum of the energies of the [neutrino] and electron is constant.[8]

Fermi took Pauli's hypothesis seriously and in 1933 formulated his theory of β-decay. It marked a change in the conceptualization of "elementary" processes. In the introduction to his paper, Fermi indicated that the simplest model of a theory of β-decay assumes that electrons *do not* exist as such in nuclei before β-emission occurs

> but that they, so to say, acquire their existence at the very moment when they are emitted; in the same manner as a quantum of light, emitted by an atom in a quantum jump, can in no way be considered as pre-existing in the atom prior to the emission process. In this theory, then, the total number of the electrons and of the neutrinos (like the total number of light quanta in the theory of radiation) will not necessarily be constant, since there might be processes of creation or destruction of these light particles.[9]

Fermi's theory made clear the power of a quantum field theoretical description.

For nuclear physics, 1932 was the annus mirabilis. The discovery of the neutron by James Chadwick (1891–1974) working at the Cavendish Laboratory led quickly to the view that a nucleus of mass number A is a composite system built out of Z protons and (A − Z) neutrons. The neutron, which was assumed to be an electrically neutral, spin $1/2$ particle with a mass roughly equal to that of the proton, made possible the application of quantum mechanics to the elucidation of the structure of the nucleus, as was shortly done in a series of papers by Heisenberg, based on short-range (static) two-body nucleon–nucleon interactions.

After the discovery of the neutron and of the positron, matter was thought to consist of two sets of entities: electrons and neutrinos (and their antiparticles) and neutrons and protons (and their antiparticles). The charged members of the two groups could interact with one another electromagnetically. Electrons and neutrinos interacted with neutrons and protons through the Fermi interaction; neutrons and protons interacted "strongly" through nuclear forces. The neutron and the protons were recognized as being very

---

[8] Wolfgang Pauli, letter to the "Dear Radioactive Ladies and Gentlemen," 4 December 1930, in K. von Meyenn, *Wolfgang Pauli: Wissenschaftlicher Briefwechsel,* vol. 2 (New York: Springer-Verlag, 1979).
[9] Enrico Fermi, "Versuch einer Theorie der β-Strahlen. I," *Zeitschrift für Physik,* 88 (1934), 161–71.

similar, yet also different. They have different electric charges and electromagnetic interactions, but interact very similarly in their "strong" (nuclear) interactions.

The indifference of the nuclear force to the nucleons involved became expressed formally by considering the neutron and the proton as having a new "internal" quantum property, called isotopic spin. Neutrons and protons differ merely in the value of the z-component of their "isotopic" spin. This attribution of an isotopic spin to nucleons by Heisenberg was the first example of the two kinds of internal quantum numbers eventually used to classify particles, namely: (1) (conserved or approximately conserved) additive quantum numbers, like electric charge, strangeness, baryon, and lepton numbers; and (2) "non-abelian" quantum numbers, such as isotopic spin, that label families of particles.[10]

In 1935, Hideki Yukawa (1907–1981) published a paper in which he proposed a field theoretical model to account for the nuclear forces. In Yukawa's theory, the neutron-proton force was mediated by the exchange of a scalar particle between the neutron and proton, with the mass of the scalar particle – called a meson – so adjusted as to yield a reasonable range for the nuclear forces. Yukawa had writ large what had been known in QED, namely that the electromagnetic force between charged particles could be conceptualized as arising from the exchange of "virtual" photons – called virtual because these photons did not obey the relation $E = h\nu$, which is valid for free photons. The masslessness of photons implies that the range of electromagnetic forces is infinite. In Yukawa's theory, the exchanged quanta are massive, and the range, R, of the resulting interaction is related to the mass, μ, of the quanta by $R = h/\mu c$. This association of interactions with exchanges of quanta is a general feature of all quantum field theories.

Shortly after the Caltech cosmic ray physicists Carl Anderson (1905–1991) and Seth Neddermeyer (1907–1989) had given evidence for the existence of a new type of particle in the penetrating component of cosmic rays, Robert Oppenheimer (1904–1967) and Robert Serber (1909–1996) in 1937 published a short note in the *Physical Review* in which they pointed out that the mass of the newly discovered particle specified a length that they connected with the range of the nuclear forces, as had been suggested by Yukawa. Oppenheimer and Serber's note was responsible for drawing the attention of American physicists to the meson theories of nuclear forces that Yukawa, Ernest Stückelberg (1905–1984), and Gregor Wentzel (1898–1978) had advanced. The existence of this "heavy electron" – which existed in both a positive and a negative variety – was authenticated by its direct observation in a cloud chamber by Curry Street (1906–1981) and Edward C. Stevenson

---

[10] In 1953 Gell-Mann, and independently Nakano and Nijishima, proposed the property of matter called "strangeness." The quantum numbers which are associated with operators that do not commute with the electric charge operators are called "non-abelian." See M. Gell-Mann and Y. Ne'eman, *The Eightfold Way* (New York: Benjamin, 1964).

(b. 1907), who also determined its mass (150–220 electron masses) from measurements of the ionization it produced and from the curvature of its track in a magnetic field. Its lifetime was estimated to be about 10^{-6} sec. By 1939 Hans Bethe (b. 1906) could assert that "it was natural to identify these cosmic ray particles with the particles in Yukawa's theory of nuclear forces."[11]

QED, Fermi's theory of β-decay and Yukawa's theory of nuclear forces established the model upon which all subsequent developments were based.[12] The model postulated new "impermanent" particles to account for interactions and assumed that relativistic QFT was the natural framework in which to attempt the representation of phenomena at ever smaller distances, that is, at higher and higher energies. It led to a description of nature in terms of a sequence of families of elementary constituents of matter with fewer and fewer members.

By the late 1930s, the formalism of quantum field theory was fairly well understood. However, it was recognized that all relativistic QFTs are beset by divergence difficulties that manifest themselves in perturbative calculations beyond the lowest order. These problems impeded progress throughout the 1930s, and most of the workers in the field doubted the correctness of QFT in view of these divergence difficulties. Numerous proposals to overcome these problems were advanced during the 1930s, but all ended in failure.[13]

The pessimism of the leaders of the discipline – Bohr, Pauli, Heisenberg, Dirac, Oppenheimer – was partly responsible for the lack of progress. They had witnessed the overthrow of the classical concepts of space-time and were responsible for the rejection of the classical concept of determinism in the description of atomic phenomena. They had brought about the quantum mechanical revolution, and they were convinced that only further conceptual revolutions would solve the divergence problem in quantum field theory.

Heisenberg in 1938 noted that the revolutions of special relativity and of quantum mechanics were associated with fundamental *dimensional* parameters: the speed of light, c, and Planck's constant, h. These delineated the domain of classical physics. He proposed that the next revolution be associated with the introduction of a fundamental unit of length, which would delineate the domain in which the concept of fields and local interactions would be applicable.

The S-matrix theory, which Heisenberg developed in the early 1940s, was an attempt to make this approach concrete. He observed that all experiments can be viewed as scattering experiments. In the initial configuration, the system is prepared in a definite state. The system then evolves and the final

---

[11] Hans Bethe, "The Meson Theory of Nuclear Forces," *Physical Review*, 57 (1940), 260–72.
[12] L. M. Brown, "How Yukawa Arrived at the Meson Theory," *Progress of Theoretical Physics*, suppl. 85 (1985), 13–19; Olivier Darrigol, "The Origin of Quantized Matter Waves," *Historical Studies in the Physical Sciences*, 16:2 (1986), 198–253.
[13] Steven Weinberg, "The Search for Unity: Notes for a History of Quantum Field Theory," *Daedalus* (Fall, 1977); Pais, *Inward Bound*.

configuration is observed after a time that is long compared with the characteristic times pertaining in the interactions. The S matrix is the operator that relates initial and final states. Its knowledge allows the computation of scattering cross-sections and other observable quantities. By again suggesting that only variables referring to experimentally ascertainable quantities should enter theoretical descriptions, Heisenberg opened a new chapter in the development of quantum field theories.[14]

FROM PIONS TO THE STANDARD MODEL: CONCEPTUAL DEVELOPMENTS IN PARTICLE PHYSICS

Modern particle physics can be said to have begun with the end of World War II. Peace and the Cold War ushered in an era of new accelerators of ever-increasing energy and intensity that were able artificially to produce the particles that populate the subnuclear world. Simultaneously, there developed the expertise to construct particle detectors of ever-increasing complexity and sensitivity that allowed the recording of the imprints of high energy subnuclear collisions. Challenges, opportunities, and resources attracted practitioners: The number of "high energy" physicists worldwide was to grow from a few hundred after World War II to some 8,000 in the early 1990s.

John Archibald Wheeler (b. 1911) summarized the state of affairs in elementary particle physics in the fall of 1945 by observing that the experimental and theoretical researches of the 1930s had made it possible to identify four fundamental interactions: (a) gravitation, (b) electromagnetism, (c) nuclear (strong) forces, and (d) weak-decay interactions. Wheeler believed that the interesting and exciting areas of research were the investigations of the electromagnetic, the strong, and the weak interactions, and these, indeed, became the traditional domain of high energy physics.[15]

Two important developments in 1947 shaped the further evolution of particle physics. Both were the result of intense discussions that followed experimental findings presented to the Shelter Island Conference. This was the first of three meetings sponsored by the National Academy of Sciences, which assembled the young American theorists who had made important contributions to the wartime weapons research in order to discuss foundational problems in physics. These conferences were the precursors of the (international) Rochester conferences begun in 1950 that brought together high energy physicists – experimentalists and theorists – biannually.

---

[14] See Cushing, *Theory Construction*.
[15] John A. Wheeler, "Problems and Prospects in Elementary Particle Research," *Proceedings of the American Philosophical Society*, 90 (1946), 36–52.

At the 1947 Shelter Island Conference, the curious results that Marcello Conversi (b. 1917), Ettore Pancini (1915–1981), and Oreste Piccioni (b. 1915) had obtained regarding the decay of mesons observed at sea level led Robert Marshak (1916–1992) to formulate the "two-meson" hypothesis. He suggested that there existed two kinds of mesons. The heavier one, the π-meson, which was identified with the Yukawa meson responsible for the nuclear forces, is produced copiously in the upper atmosphere in nuclear collisions of cosmic ray particles with atmospheric atoms. The lighter one, the μ-meson observed at sea level, is the decay product of a π-meson and interacts but weakly with matter. A similar suggestion had been made earlier by Shoichi Sakata (1911–1970) in Japan.

Within a year, Cecil Powell (1903–1969), using nuclear emulsions sent aloft in high altitude balloons, corroborated the two-meson hypothesis by exhibiting $\pi \to \mu$ decays. During the early 1950s, the data pouring out of the plethora of π-meson-producing accelerators led to the rapid determination of the characteristic properties of the pion or π-meson, which occurs in three varieties: positively charged, negatively charged, and neutral.

The two-meson hypothesis also suggested that the list of the particles comprising the two distinct kinds of matter had to be amended. There were particles like the electron, the muon, and the neutrino that do not experience the strong nuclear forces; these were called leptons. Then there were the particles like the neutron, proton, and π-mesons that do interact strongly with one another and were named hadrons.

In January 1949, Jack Steinberger (b. 1921) gave evidence that the μ-meson decays into three light particles

$$\mu^+ \to e^+ + \nu + \nu$$

and shortly thereafter, Gianpietro Puppi (b. 1917), Oskar Klein (1894–1977), Jaime Tiomno (b. 1924) and Wheeler, and Tsung Dao Lee (b. 1926), Marshall Rosenbluth (b. 1930), and Ning Yang (b. 1922) indicated that this process could be described by a Fermi-like interaction, as in the case of ordinary β-decay. Moreover, they pointed out that the coupling constant describing this interaction was of the same magnitude as the one occurring in nuclear beta decay. Thus, the pre-1947 period can be characterized as that of *classical beta decay*, while the postwar period initiated the modern period of *universal Fermi interactions*.

The second important development in the immediate post–World War II period was a theoretical advance. It stemmed from the attempt to explain quantitatively the discrepancies between the empirical data and the predictions of the relativistic Dirac equation for the level structure of the hydrogen atom and the value it ascribed to the magnetic moment of the electron. These deviations had been observed in reliable and precise molecular beam experiments carried out by Willis Lamb (b. 1913), and by Isidor Rabi (1898–1988) and co-workers at Columbia, and were reported at the Shelter Island Conference. Shortly after the conference, it was shown by Bethe that the Lamb shift (the deviation of the $2s$ and $2p$ levels of hydrogen from the values given by the Dirac equation) was of quantum electrodynamical origin, and that the effect could be computed by making use of what became known as "mass renormalization," an idea that had been put forward by Hendrik Kramers (1894–1952).

The parameters for the mass, m_o, and for the charge, e_o, that appear in the Lagrangian-defining QED, are not the observed charge and mass of an electron. The observed mass, m, of the electron is to be introduced in the theory by the requirement that the energy of the physical state corresponding to an electron moving with momentum p be equal to $(p^2 + m^2)^{1/2}$. Similarly, the observed charge should be defined by the requirement that the force between two electrons (at rest), separated by a large distance r, be described by Coulomb's law, e^2/r^2, with e the observed charge of an electron.

It was shown by Julian Schwinger (1918–1994) and by Richard Feynman (1918–1988) that the divergences encountered in the low orders of perturbation theory could be eliminated by reexpressing the parameters m_o and e_o in terms of the observed values m and e, a procedure that became known as mass and charge renormalization. In 1948, Freeman Dyson (b. 1923), working at the Institute for Advanced Study in Princeton, was able to show that charge and mass renormalization were sufficient to absorb all the divergences of the scattering matrix (S-matrix) in QED to all orders of perturbation theory. More generally, Dyson demonstrated that only for certain kinds of quantum field theories is it possible to absorb *all* the infinities by redefinition of a *finite* number of parameters. He called such theories renormalizable. Renormalizability thereafter became a criterion for theory selection.[16]

The ideas of mass and charge renormalization, implemented through a judicious exploitation of the symmetry properties of QED – that is, the Lorentz invariance and the gauge invariance of the theory – made it possible to formulate and to give physical justifications for algorithmic rules to eliminate all the ultraviolet divergences that had plagued the theory and to secure unique finite answers. The success of renormalized QED in accounting for the Lamb shift, the anomalous magnetic moment of the electron and of the muon, and the radiative corrections to the scattering of photons by electrons, to pair production, and to bremsstrahlung, was spectacular.

Perhaps the most important theoretical accomplishment of the 1947–52 period was providing a firm foundation for believing that local quantum field theory was the framework best suited for the unification of quantum theory and special relativity. The most perspicacious theorists, for example, Murray Gell-Mann (b. 1929), also noted the ease with which symmetries – both space-time and internal symmetries – could be incorporated into the

---

[16] Schweber, *QED*.

framework of local quantum field theory. Local QFTs were thus advanced for the description of the "elementary particles" and their internal symmetries. Photons, pions, nucleons, electrons, muons, and neutrinos (the elementary particles as perceived in the early fifties) corresponded to localized excitations of the underlying, "fundamental" local fields.

Although experiments with cosmic rays during the 1940s and 1950s had indicated the presence of new "strange" particles, high energy physics during most of the 1950s was dominated by pion physics. The success of QED rested on the validity of perturbative expansions in powers of the coupling constant, $e^2/\hbar c$, which is small, $\approx 1/137$. However, the pseudoscalar meson theory of the pion-nucleon interaction required the coupling constant to be large – of the order of 15 – for the theory to yield nuclear potentials that would bind the deuteron. No valid method was found to deal with such strong couplings. It also became clear that meson theories were woefully inadequate to account for the properties of all the new hadrons being discovered. The importance of the *tempo* of new experimental findings by the particle accelerators that were coming on-line cannot be overemphasized. The plethora of new experimental discoveries quelled any hope for a rapid, neat, and systematic transition from QED to the formulation of a dynamics for the strong interaction.

To Sam Treiman (1925–1999), an important contributor to the development in particle physics from the 1950s to the 1980s and the teacher and mentor at Princeton University of many of the best young theorists coming of age during that period, "the prospect of finding the *right* quantum field theory, if indeed there were a right one, or even recognizing it if it were presented seemed remote [from 1955 to 1965]."[17]

Thus, at the end of the 1950s, QFT faced a crisis because of its inability to describe the strong interactions and the impossibility of solving any of the realistic models that had been proposed to explain the dynamics of hadrons. Efforts to develop a theory of the strong interactions along the model of QED were generally abandoned, although a local gauge theory of isotopic spin symmetry, advanced by Yang and Robert Mills (b. 1927) in 1954, was to prove influential later on.

There were several responses to the crisis that developed in theoretical particle physics at the end of the 1950s. For some theorists, the failure of quantum field theory and the superabundance of experimental results was, in fact, emancipating. It led to explorations of the generic properties of QFT when only such general principles as causality, the conservation of probability (unitarity), and relativistic invariance are invoked and no specific assumptions are made regarding the form of the interactions.

Geoffrey Chew's (b. 1924) S-matrix program, which rejected QFT and attempted to formulate a theory that made use only of observables embodied

[17] S. Treiman, "A Life in Particle Physics," *Annual Reviews of Nuclear and Particle Science,* 46 (1996), 1–30, at p. 6.

in the S-matrix, was more radical. Physical consequences were to be extracted without recourse to any dynamical field equations, by making use of general properties of the S-matrix, such as unitarity and Lorentz invariance, and certain assumptions (analyticity) regarding its dependence on the variables describing the initial and final energies and momenta of the particles involved.[18]

Another response to the crisis was to make symmetry concepts central. Symmetry considerations were first applied to the weak and the electromagnetic interactions of the hadrons, and they were later extended to encompass low energy strong interactions. Phenomenologically, the strong interactions seemed to be well modeled by (effective) Hamiltonians, the physical variables of which were hadron *current* operators. No dynamical assumptions were made on how these hadron current operators were to be constructed from hadron field operators, but commutation relations were imposed on them, reflecting the underlying symmetries that were assumed to be independent of dynamic details and to be universally valid. These symmetries and their group structure were derived from the exact or approximate regularities that manifested themselves in the experimental data. This research program, known as current algebra, took shape during the late 1950s and early 1960s. The program reached its limit around 1967 because some of its predictions were in direct conflict with experiments.

In fact, during the 1950s and 1960s, progress in classifying and understanding the phenomenology of the ever-increasing number of hadrons was not made by virtue of a fundamental theory. It was accomplished by shunning dynamical assumptions and, instead, making use of symmetry principles (and their associated group theoretical methods) and exploiting kinematical principles that embodied the essential features of a relativistic quantum mechanical description.

Symmetry thus became one of the fundamental concepts of modern particle physics. It is used both as a classificatory and organizing tool and as a foundational principle to describe dynamics. The notion of symmetry was enriched by two developments in the second half of the 1950s: (1) the realization by Lee and Yang that parity is not conserved in the weak interactions; and (2) the extension by Yang and Mills in 1955 of the global isotopic spin symmetry of nucleons to a local symmetry, in analogy with gauge invariance in QED.

This local symmetry, or local gauge invariance, demands that the photon be massless. The requirements of relativistic invariance, gauge invariance, and the absence of dimensionality in the coupling constant scaling the strength of the interaction determine the form of the Lagrangian describing the interaction between charged particle fields and the electromagnetic field.

[18] Cushing, *Theory Construction*.

A Lagrangian that is invariant under some global transformation of the form

$$\psi(x) \to e^{ie\Lambda} \psi(x)$$

with Λ constant, can be made locally invariant under such a transformation, that is, with $\Lambda = \Lambda(x)$, by introducing appropriate gauge fields. Yang and Mills made use of this observation to introduce a gauge theory of the strong interactions, by extending to a local symmetry the invariance of the nucleon fields under global isotopic rotations:

$$\psi(x) \to e^{ig\tau \cdot \phi} \psi(x)$$

Local gauge invariance, however, implies that the gauge bosons are massless. This is not the case for the pion, and thus Yang and Mills's theory was considered an interesting model but without relevance for understanding the strong interactions.

Interest in field theories, and in particular in gauge theories, was revived after the notion of spontaneous symmetry breaking (SSB) became fully appreciated in the early 1960s. Jeffrey Goldstone (b. 1933) and Yoichiro Nambu (b. 1921) noted that in quantum field theories, symmetries could be realized differently: It was possible to have the Lagrangian invariant under some symmetry, yet have this symmetry not respected by the vacuum (that is, by the ground state of the theory). Such symmetries are known as spontaneously broken (SBS). It turns out that if the symmetry that is spontaneously broken is a global one, there will be massless (Goldstone) spin zero bosons in the theory. If the (broken) symmetry is a local gauge symmetry, then the Goldstone bosons disappear from the theory, but each of the gauge Bosons associated with broken symmetries acquires a mass. This is the Higgs mechanism.[19]

In 1967 Steven Weinberg (b. 1933), and somewhat later in 1968, Abdus Salam (1926–1996) independently proposed a gauge theory of the weak interactions that unified the electromagnetic and the weak interactions and made use of the Higgs mechanism. Their model incorporated previous suggestions that Sheldon Glashow had advanced in 1961 on how to formulate a gauge theory of the weak interactions, in which the weak forces were mediated by gauge bosons. The original Glashow theory had been set aside because the consistency of gauge theories with *massive* gauge bosons was doubted, and by the fact that such theories were nonrenormalizable.

SBS offered the possibility of giving masses to the gauge bosons, but whether such theories with spontaneously broken symmetries via a Higgs mechanism were renormalizable was not known. The renormalizability of

---

[19] For an overview of the mechanisms which implement the broken symmetry, see the presentation by L. M. Brown and the subsequent discussion in Hoddeson et al., *Birth of Particle Physics*, pp. 478–522.

such theories was proved by Gerard 't Hooft (b. 1946) in his dissertation in 1972 at Utrecht University under the supervision of Martinus Veltman (b. 1931). The status of the Glashow-Weinberg-Salam theory changed dramatically thereafter. As Sidney Coleman noted, " 't Hooft's kiss transformed Weinberg's frog into an enchanted prince."[20]

Gauge theory, the mathematical framework for generating dynamics incorporating symmetries into a QFT, has played a crucial role in the further development of QFT. It can rightly be said that symmetry, gauge theories, and spontaneous symmetry breaking have been the three pegs upon which modern particle physics rests.

QUARKS

All the phenomenological theorizing of the 1960s led to the view that the elementary constituents of matter at the smallest distances, or equivalently at the highest energies, are quarks and leptons. In 1961, Gell-Mann and Yuval Ne'eman (b. 1925) independently proposed classifying the hadrons into families on the basis of a symmetry that became known as the "eightfold way." They realized that the mesons grouped naturally into octets, the baryons into octets, and decuplets. The mathematical expression of the "eightfold way" symmetry was the group of (unitary) transformations $SU(3)$, the generalization to hadrons of the symmetry group $SU(2)$ that had been used to express mathematically the charge independence of the nuclear forces between neutrons and protons. A fundamental representation of $SU(3)$ is three-dimensional, which led Gell-Mann, and independently George Zweig (b. 1937), to suggest that hadrons were composed of three elementary constituents, which Gell-Mann named quarks (from a passage in James Joyce's *Finnigan's Wake:* "Three quarks for Master Mark!") and Zweig called aces.

To account for the observed spectrum of hadrons, Gell-Mann and Zweig assumed that there were three "flavors" of quarks (generically indicated by q), called up (u), down (d), and strange (s), that had spin $½$, isotopic spin $½$ for the u and d and isotopic spin 0 for the s, and strangeness 0 for the u and d and -1 for the s quark. Ordinary matter contains only u and d quarks; "strange" hadrons contain strange quarks or antiquarks. The three quarks were to carry baryonic charge of $1/3$ and an electrical charge that is $2/3$ for the u, and $-1/3$ for the d and s, that of the proton's charge.[21]

This was a rather startling assumption since there is no experimental evidence for any macroscopic object carrying a positive charge smaller than that of a proton or a negative charge smaller than an electron. Since a relativistic quantum mechanical description implies that for every charged particle

---

[20] See Weinberg, *Quantum Theory of Fields.*
[21] M. Gell-Mann and Y. Ne'eman, *The Eightfold Way*; and Gell-Mann, "Quarks, Color and QCD," in P. M. Zerwas and H. A. Kastrup, *QCD 20 Years Later* (Singapore: World Scientific, 1993), pp. 3–15.

there exists an "antiparticle" with the opposite charge, it was assumed that there are likewise antiquarks (generically denoted by $\bar{q}$), having the opposite electric charge and opposite sign of strangeness. Quarks were assumed to interact with one another and to form bound states, giving rise to the observed hadrons. Thus a π^+ meson was assumed to be a bound state of an up and antidown quark. Similarly, a proton was "made up" of two up quarks (that contributed $4/3\ e$ to the electrical charge) and a down quark (of electrical charge $-1/3\ e$), giving rise to an entity with an electrical charge of $+1\ e$. In fact, all baryons could be made up of three quarks, all mesons with one quark and one antiquark.

However, in order to satisfy the Pauli principle in a structure like the Ω^-, which is presumably constituted of three *identical* spin $^1/_2$ strange quarks, all in s states, quarks had to be given a new attribute, a new form of charge, called color, in order to distinguish otherwise identical quarks. Color is a "three-dimensional" analog of electric charge: It occurs in three varieties (sometimes taken to be red, yellow, and blue). Thus, there are positive and negative red, yellow, and blue colors. Quarks carry positive color charges and antiquarks carry the corresponding negative charge. The observed hadrons are required to be color singlets, that is, to have zero net color charge.[22]

If the SU(3) symmetry were exact, all the quarks, and all the baryons in a given octet or decuplet, would have the same mass. Since they do not, the symmetry must be broken; this comes about by virtue of the three flavors of quarks having different masses, with the s quark assumed to have a greater mass than the u and the d quarks.

An entire phenomenology grew out of this classificatory scheme. In the early 1960s, the flavor SU(3) quark model, in which the u, d, and s quarks are considered the building blocks, could classify all the then-known hadrons into three families: an octet of spin 0 mesons (that included the ρ and K mesons); an octet of spin $^1/_2$ baryons (that included the neutron and the proton, the Λ and the Σ), and a decuplet of spin 3/2 baryons.[23]

In the late 1960s, experiments in which high energy electrons were inelastically scattered off protons were carried out at the Stanford Linear Accelerator (SLAC).[24] Since the early 1950s, it had been known that protons had an internal structure. By 1968 electrons were being accelerated at SLAC to 20Gev, at which energy their wavelength was such that they could resolve entities appreciably smaller than the size of the proton. Such electrons were thus ideal probes for investigating the internal structure of the proton. If charge were uniformly distributed within the proton, high energy electrons would

---

[22] See O. W. Greenberg, "Color: From Baryon Spectroscopy to QCD," in M. Gai, ed., *Baryons '92: International Conference on the Structure of Baryons and Related Mesons* (Singapore: World Scientific, 1993), pp. 130–9.

[23] J. J. J. Kokkedee, *The Quark Model* (New York: Benjamin, 1969); K. Gottfried and V. Weisskopf, *Concepts of Particle Physics* (New York: Oxford University Press, 1986).

[24] See the presentations by Jerome Friedman and James Bjorken in Hoddeson et al., *Birth of Particle Physics*, pp. 566–600.

tend to go through the proton without being appreciably deflected. If on the other hand – in analogy with Rutherford's interpretation of Geiger and Marsden's experiment on the scattering of α-particles by gold atoms – the charge within the proton were localized on internal constituents, then an electron, if it were to pass close to one of these concentrations of charge, would be strongly deflected.

Just such large-angle scatterings were observed at SLAC. Upon hearing these experimental findings, Feynman suggested that the proton is composed of pointlike particles that he called "partons," and from the angular distribution of the scattered electrons he inferred that partons had spin $\frac{1}{2}$. The partons were soon recognized as identical with the quarks of Gell-Mann's and Zweig's model. There were, however, paradoxical aspects with this identification of the proton constituents. First, the partons/quarks appeared to be very light, much less than one-third of the mass of the proton; and second, they appeared to move almost freely inside the proton – difficulties that were addressed and resolved only later.

The discovery in November 1974 of the J/ψ, a spin 1 meson, gave further evidence for the correctness of the quark picture and gave credence to the existence of a fourth quark with a new flavor, called charm. (The charmed quark was denoted by c). The existence of such a quark had been suggested by James Bjorken and Glashow in 1964, and the proposal had been elaborated further by Glashow, John Iliopoulos (b. 1940) and Luciano Maiani (b. 1941) in 1970. It was immediately conjectured that the J/ψ was a bound state of a c and $\bar{c}$. The subsequent detection of the ψ', a "particle" related to the J/ψ by its decay, made the notion of quarks in general, and of charmed quarks in particular, compelling. The discoveries of November 1974 revolutionized high energy physics. With the November revolution, the conceptualization of hadrons as

> quark composites was put beyond dispute, and gauge theory received a tremendous boost – the Weinberg-Salam plus Glashow-Iliopoulos-Miaini Model became the basis of a new hadron spectroscopy. At the heart of these developments was charm.... The triumph of charm was simultaneously a triumph for gauge theory.[25]

With the discovery of the charmed quark, and subsequently of the third family of particles – the τ lepton and its neutrino – and of the "bottom" (or "beauty") b quark in 1977, and of the "top" (t) quark in 1994, six different "flavors" of quarks were needed to account for the observed hadron spectroscopy. Each successively discovered quark is more massive than its predecessors: The u and d quarks have (effective) masses of 5 and 10 Mev/c^2, respectively; the s an (effective) mass of 180 Mev/c^2; the c has a mass 1.6 Gev/c^2; the b of 4.8 Gev/c^2 and the t of 174 Gev/c^2. All are spin $\frac{1}{2}$ particles

---

[25] Pickering, *Constructing Quarks*, p. 254.

that partake in the strong, electromagnetic, and weak interactions, and all come in pairs: up and down (u, d), charm and strange (c, s), and top and bottom (t, b). The first member of each pair has electric charge 2/3 and the second −1/3. Each flavor comes in three colors.

From the time they were introduced as "hypothetical" particles, an important problem connected with quarks loomed large: If indeed all hadrons are made up of fractionally charged quarks, why is it that one does not eventually reach an energy high enough to liberate the constituent quarks in a collision process and thus allow a fractionally charged hadron to be observed? This is the so-called confinement problem. And even were one able to provide a mechanism that accounts for the confinement of quarks, what meaning is to be attached to the reality of quarks as constituents of hadrons if they can never be observed empirically?

GAUGE THEORIES AND THE STANDARD MODEL

As currently described, a common mechanism underlies the strong, weak, and electromagnetic interactions. Each is mediated by the exchange of a spin 1 gauge boson. In the case of the strong interactions, the gauge bosons are called gluons; in the case of the weak interactions, $W^{\pm}$ and Z bosons; and in the electromagnetic case, photons. A general chromatic terminology has become popular, and one often refers to the charges as "colors." Thus, one speaks of QED – the paradigmatic gauge theory – as a theory of a single gauge boson, the photon, coupled to a single "color," namely, the electric charge. The gauge bosons of the strong interactions carry a 3-valued color; those mediating the weak interactions carry a "two-dimensional" weak color charge. Weak gauge bosons interact with quarks and leptons, and in the act of being emitted or absorbed, some of them can transform one kind of quark or lepton into another. When these gauge bosons are exchanged between leptons and quarks, they are responsible for the force between them. They can also be emitted as radiation when the quarks or leptons are accelerated.

Quantum chromodynamics (QCD) describes the strong interactions between the six quarks. Quarks carry electrical charge and, in addition, carry a ("three-dimensional") strong color charge. Each of the six quarks carries this color charge and can be in any of three colors states. QCD is a gauge theory with three colors and involves eight massless gluons, the color-carrying gauge bosons, six that alter color, and two that merely react to them. QCD possesses a gauge invariance: The theory is invariant under the addition to the gluon field potentials of a set of gradients and a simultaneous change of the phases of the quark fields. A quark's color is changed when it absorbs or emits a color-changing gluon. However, a quark's flavor is not changed by the absorption or emission of a gluon – nor by the emission or absorption of a photon.

The GWS (Glashow-Weinberg-Salam) gauge theory of the weak interactions is a gauge theory involving two colors. Each of the quarks thus carries an additional weak color (or weak charge). There are four gauge bosons that mediate the weak interactions between the quarks. Three of them (the W^+, W^-, and W^o) change the flavor of the quark when absorbed or emitted; the fourth, the B^o boson, reacts but does not alter the weak color charges.

As just described, the standard model, although aesthetically beautiful, does not accord with the known characteristics of the weak interactions nor with the properties of quarks as envisaged in their phenomenological descriptions. Local gauge invariance requires that the gauge bosons be massless and, therefore, that the range of the forces they generate be long range. Yet it is known that the weak force is of very short range (less than 10^{-16} cm), and that the mass of the W boson is 80Gev and that of the Z, 91Gev. Nor can it accommodate the masses of the quarks. A Higgs mechanism for spontaneously breaking symmetries – accomplished by introducing a (complex) doublet of scalar fields – is the mechanism most commonly invoked to overcome these difficulties. Establishing the reality of such Higgs particles became an important reason for justifying the building of the Superconducting Super Collider (SSC).[26]

The past two decades have seen a large number of successful explanations of high energy phenomena using QCD. The substantiation in 1973 at the Centre Européen pour la Recherche Nucléaire (CERN) of the process $\nu_\mu + e^- \rightarrow \nu_\mu + e^-$ corroborated the existence of weak neutral currents as embodied in the Glashow-Weinberg-Salam electroweak theory. The detection and identification of the $W^\pm$ and of the Z_0 in 1983 by Carlo Rubbia and co-workers at CERN gave further important confirmation of that theory. Similarly, the empirical data obtained in lepton and photon deep inelastic scattering, and in the study of jets in high energy collisions can be accounted for quantitatively by QCD. Furthermore, computer simulations have presented convincing evidence that QCD does produce quark and gluon confinement inside hadrons.[27] Frank Wilczek, one of the important contributors to the field, in his opening remarks at a conference in 1992 devoted to an assessment of QCD since its initial formulation, could assert that "QCD is now a mature theory, and it is now possible to begin to view its place in the conceptual universe with appropriate perspective."[28]

The empirical data that can be accounted for quantitatively are indeed impressive.[29] As Guido Altarelli remarked in his review of "QCD and

---

[26] Daniel J. Kevles, "Preface 1995: The Death of the Superconducting Super Collider in the Life of American Physics," in *The Physicists: The History of a Scientific Community in Modern America*, 2d ed. (Cambridge, Mass.: Harvard University Press, 1995).
[27] M. Creutz, *Quarks, Gluons and Lattices* (Cambridge: Cambridge University Press, 1983).
[28] Wilczek in P. M. Zerwas and H. A. Kastrup, eds., *QCD 20 Years Later* (Singapore: World Scientific, 1993), p. 16.
[29] See Frank Wilczek, "The Future of Particle Physics as a Natural Science," in *Critical Problems in Physics*, ed. V. A. Fitch, D. R. Marlow, and M. A. E. Dementi (Princeton, N.J.: Princeton University Press, 1997), pp. 281–308.

Experiment" at that same conference:

> [Since the late 80s] [m]any relevant calculations, often of unprecedented complexity, have been performed. As a result of 3 years of really remarkable progress, our confidence in QCD has been further consolidated, ... and a lot of additional checks from many different processes have become possible.[30]

QCD is the accepted framework for describing the interactions of leptons, quarks and gluons below 1 TeV.

The standard model is one of the great achievements of the human intellect. It will be remembered – together with general relativity, quantum mechanics, and the unraveling of the genetic code – as one of the outstanding intellectual advances of the twentieth century. But the standard model is not the "final theory," for too many parameters that have to be empirically determined enter the description, for example, the masses of the quarks and the various coupling constants.

Very shortly after the realization that the strong and the electroweak interactions could be described by gauge theories that had similar mathematical structures, a new phase in the unification of the different forces of nature began. The similarity between the transformation properties of gluon and quark fields under the (three) color gauge transformations and those of the quark and lepton fields under the (two) weak color gauge transformations immediately suggested the possibility that a larger (five-dimensional) gauge group, $SU(5)$, might encompass both the strong and the electroweak interactions. Howard Georgi (b. 1947) and Glashow advanced such a grand unified (gauge) theory (GUT) as soon as QCD was recognized as the likely theory of the strong interactions.[31]

The greatest immediate impact of GUTs has been in cosmology and in the description of the physics of the early universe. Asymptotic freedom implies that matter at extreme temperatures and densities becomes weakly interacting and, therefore, that its equation of state is rather simply calculable. GUTs made it possible to calculate the consequences of various unification scenarios for cosmology with some confidence. It also offered an explanation for how the observed asymmetry between matter and antimatter could have developed from a symmetric starting condition. In fact, probably the most consequential unification during the past twenty years has been the "unification" of particle physics and astrophysics. The early universe, the immediate aftermath of the big bang, has become the laboratory in which to explore the implications of foundational theories (such as GUTs and string theory) at temperatures and energies that are and will remain inaccessible in terrestrial laboratories.

---

[30] Altarelli in Zerwas and Kastrup, *QCD 20 Years Later*.
[31] H. Georgi and S. L. Glashow, "Unified Theory of Elementary Particle Forces," *Physics Today*, 33, no. 9 (1980), 30–9; and S. Dimopoulous, S. A. Raby, and F. Wilczek, "Unification of Couplings," *Physics Today*, 44, no. 10 (1991), 25–33.

20

CHEMICAL PHYSICS AND QUANTUM CHEMISTRY IN THE TWENTIETH CENTURY

Ana Simões

In 1967, Per-Olov Löwdin introduced the new *International Journal of Quantum Chemistry* in the following manner:

> Quantum chemistry deals with the theory of the electronic structure of matter: atoms, molecules, and crystals. It describes this structure in terms of wave patterns, and it uses physical and chemical experience, deep-going mathematical analysis, and high-speed electronic computers to achieve its results. Quantum mechanics has rendered a new conceptual framework for physics and chemistry, and it has led to a unification of the natural sciences which was previously inconceivable; the recent development of molecular biology shows also that the life sciences are now approaching the same basis.
>
> Quantum chemistry is a young field which falls between the historically developed areas of mathematics, physics, chemistry, and biology.[1]

In this chapter I address the emergence and establishment of a scientific discipline that has been called at times quantum chemistry, chemical physics, or theoretical chemistry. Understanding why and how atoms combine to form molecules is an intrinsically chemical problem, but it is also a many-body problem, which is handled by means of the integration of Schrödinger's equation. The heart of the difficulty is that the equation cannot be integrated exactly for even the simplest of all molecules. Devising semiempirical approximate methods became, therefore, a constitutive feature of quantum chemistry, at least in its formative years.

The first section presents a brief outline of the traditional narrative of the history of quantum chemistry, as generally offered by chemists and built around the conflict between the two alternative methods of dealing with

---

[1] Per-Olov Löwdin, "Program," *International Journal of Quantum Chemistry*, 1 (1967), 1–6, at p. 1.
I wish to thank Stephen G. Brush, Andreas Karachalios, Helge Kragh, and Mary Jo Nye for making available preprints of some of their works, as well as the volume *The Pauling Symposium;* A. Stadler for helping with the translation of Karachalios's paper; Kostas Gavroglu for his always useful comments on an earlier version of this paper; and the anonymous referees for suggestions.

valence problems – the *valence bond method* (VB) and the *molecular orbital method* (MO). The following sections present alternative schemes of historical analysis centered on methodological issues related to foundationalism and the question of reductionism, then proceed to address the importance of national contexts in theoretical chemistry, and, finally, discuss discipline building in theoretical chemistry and chemical physics.

PERIODS AND CONCEPTS IN THE HISTORY OF QUANTUM CHEMISTRY

In the case of recent disciplines such as quantum chemistry, the writing of memoirs, autobiographies, and other recollections are often produced by its first practitioners, who therefore become its first historians-scientists. This is a common move to highlight the identity of the discipline through the creation of its own history. In the case of quantum chemistry, Robert Sanderson Mulliken (1896–1986), Linus Pauling (1901–1994), John Clarke Slater (1900–1976), Walter Heitler (1904–1981), Erich Hückel (1896–1980), and Charles Alfred Coulson (1910–1974), to name a few of the most representative, contributed to this sort of literature.

William Shakespeare, following a rich medieval and Renaissance tradition, once spoke of the seven ages of man. In 1971, in an after-dinner speech, Coulson offered an analysis in which he followed Shakespeare's lead and looked for periods in the history of what he called theoretical chemistry, and what we may call quantum chemistry.[2] He distinguished five periods in the past history of quantum chemistry: the age of birth, Pauling era, Mulliken era, the many-electron era, and the computer age. He considered that a new age was starting, yet did not make predictions too far into the future.

The age of birth arrived with the formulation of quantum mechanics, in the period from 1926 to 1928. In 1926 in a paper about the helium atom, Werner Heisenberg (1901–1976) assumed that the two electrons are indistinguishable and so should be interchanged in writing the atomic wave function. Heisenberg called the phenomenon a resonance effect. In 1927, Heitler and Fritz London (1900–1954) extended the former notion to the two electrons, one belonging to each hydrogen atom, which come together in the formation of the hydrogen molecule.[3] Pairing between the two electrons occurred when the electrons had opposed spins, and thus, covalent bonds were shown to be purely quantum mechanical effects. In cases involving more than two electrons, London proceeded to a formulation of the Pauli principle, which proved convenient for his later work in group theory: The wave function

---

[2] Coulson Papers, Ms. Coulson 40, B.20.9, Bodleian Library, Department of Western Manuscripts, Oxford, After-dinner Speech, 16 August 1971, Faculty Club of the University of British Columbia, The Fourth Canadian Symposium on Theoretical Chemistry.

[3] Cathryn Carson, "The Peculiar Notion of Exchange Force. I: Origins in Quantum Mechanics, 1926–1928," *Studies in the History and Philosophy of Modern Physics*, 27 (1996), 23–45.

can contain terms symmetric in pairs, and electron pairs on which the wave function depends symmetrically have antiparallel spins. Spin was, therefore, taken to be a constitutive feature of quantum chemistry, and one of the most significant indicators of valence behavior.

The next five years spanned what Coulson called the Pauling era. This period provided the conceptual apparatus by means of which chemical explanations were tied to the explanatory framework of quantum mechanics. Then, Coulson said, "it became clear that what had started as an extra bit of physics was going to become a central part of chemistry."[4] Elsewhere, Coulson referred to the period as that in which the pioneers of quantum chemistry started to escape from the "thought forms of the physicist."[5]

In "The Nature of the Chemical Bond" series (1931–3), Pauling outlined a chemical theory based on the concept of resonance. This concept played a fundamental role in the discovery of the hybridization of bond orbitals, the one-electron and the three-electron bonds, and the discussion of the partial ionic character of covalent bonds in heteropolar molecules. To explain the tetravalency of carbon and the directionality of its bonds, Pauling introduced the idea of "changed quantization" of bond orbitals. In the *Nature of the Chemical Bond* (1939), the same idea was called hybridization, a name that illustrates the interplay in the genesis of Pauling's ideas of what Mary Jo Nye calls "physical and biological modes of thought."[6] For example, to account for the tetravalency of carbon (supposed to have two s electrons and two p electrons, and therefore expected to have on quantum theoretical grounds a valence of two), Pauling suggested that one of the s electrons is promoted to a p electron, thus giving carbon four unpaired electrons available for bond formation. By calculating the wave functions in both situations, he was able to show that the energy required to promote the s electron was more than compensated for by its allowing carbon to form four bonds instead of two. He then showed that the four bonds were tetrahedrally oriented (sp^3 hybridization).

Furthermore, Pauling suggested that in certain aromatic compounds, such as benzene, the wave function should be written as a superposition of wave functions associated with the different valence bond structures that chemists had introduced to account for all its chemical and physical properties. Therefore, the idea of resonance among several hypothetical bond structures explained in "an almost magical way" the many puzzles that had plagued organic chemistry and established the connecting link between Pauling's new valence theory and the classical structure theory, which had been developed

---

[4] Coulson, After-dinner Speech, p. 3.
[5] Charles A. Coulson, "Recent Developments in Valence Theory," *Pure and Applied Chemistry*, 24 (1970), 257–87, at 259.
[6] Mary Jo Nye, "Physical and Biological Modes of Thought in the Chemistry of Lines Pauling," *Studies in the History and Philosophy of Modern Physics*, 31 (2000), 475–89.

throughout the second half of the nineteenth century by such chemists as August Kekulé, Archibald Scott Couper, Aleksandr Butlerov, and Jacobus van't Hoff.[7]

Pauling's program was an extension of Gilbert Newton Lewis's contribution to the explanation of the covalent bond developed in the context of the work of the community of physical chemists. By contrast, Mulliken's work on band spectra structure can be seen as an instantiation of the research agenda carried out by the American molecular physics community.[8]

The clarification of the relations between electronic states and band spectra structure led Mulliken to dispense altogether with classical valence theory. He rejected the accepted notion of chemical structure, which viewed atoms as the combining units in molecule formation, and which he considered as an instantiation of the "ideology of chemistry."[9] He further proposed to analyze the phenomena of molecule formation in terms of the electronic structure of molecules. Reasoning by analogy with Niels Bohr's building-up principle for atoms, Mulliken considered molecules to be formed by the feeding of electrons into molecular orbitals, that is, into orbitals that encircled two or more nuclei. Electrons were delocalized in the sense that there was a nonzero probability of finding them near more than one nucleus.

The assignment of quantum numbers to electrons in molecules and the classification of molecular orbitals were achieved by exploring the relations to the united-atom description and the separated-atom description. New auxiliary concepts were introduced, such as promoted and unpromoted electrons, bonding, nonbonding and antibonding electrons, and the varying bonding power of electrons. In 1929, John Edward Lennard-Jones (1894–1954) introduced the physical simplification of representing molecular orbitals as the linear combination of atomic orbitals (LCAO), a step that proved to be fundamental to the mathematical development of MO theory.

From 1933 to the end of the Second World War, in what Coulson named the Mulliken era, a debate was staged between the two different approximations to chemical bonding – the molecular-orbital and valence-bond methods. Although these two approximations started from very different presuppositions, their equivalence was then proved. Simultaneously, new concepts (fractional bond orders), new techniques (UV spectra), and new methods were introduced. Mulliken successfully used group theory in the classification of the

[7] Ava Helen and Linus Pauling Papers, Kerr [now, Valley] Library Special Collections, Oregon State University, Box 242, Popular Scientific Lectures 1925–1955, "Resonance and Organic Chemistry," 1941. [References to the Pauling Papers follow a cataloging system that has been under revision.]
[8] Alexi Assmus, "The Molecular Tradition in Early Quantum Theory," *Historical Studies in the Physical and Biological Sciences*, 22 (1992), 209–31; Alexi Assmus, "The Americanization of Molecular Physics," *Historical Studies in the Physical and Biological Sciences*, 23 (1993), 1–33; Ana Simões and Kostas Gavroglu, "Different Legacies and Common Aims," in *Conceptual Perspectives in Quantum Chemistry*, ed. J.-L. Calais and E. S. Kryachko (Dordrecht: Kluwer, 1997), pp. 383–413.
[9] Robert Sanderson Mulliken, "Electronic Structures of Polyatomic Molecules and Valence: VI. On the Method of Molecular Orbitals," *Journal of Chemical Physics*, 3 (1935), 375–8.

symmetries of the electronic states of polyatomic molecules, and strove to convince chemists and physicists of the usefulness of this mathematical theory.

To the period that extended from 1945 to 1960 Coulson gave the name the many-electron era. This was the period of the chemical understanding of the many-electron character of a molecule, but it was concomitantly the period of the study of increasingly sophisticated approximations to Schrödinger's equation for the molecule with the introduction of small terms into the Hamiltonian operator (electron-spin resonance and nuclear-magnetic resonance).

Throughout a period extending to the 1950s, the VB method dominated quantum chemistry for reasons that were not due to its superiority. In sharp contrast with Mulliken, who was not a persuasive writer or an eloquent teacher, Pauling had a consummate ability to present the theory of resonance as an extension of former chemical theories and to show its power in providing explanations for a broad range of chemical phenomena, especially in small molecules, thus accounting for the popularity of the VB method. In addition, Pauling's emphasis on model building and on visualizability as constitutive features of his chemical theory of the bond were decisive for its adoption. In the other camp, few visual representations existed to compete with Pauling's. In the case of Hückel's work, the *problématique* in which it developed was, on the contrary, that of the unvisualizability heralded by the new quantum mechanics.

Discussions and controversies over the meaning of resonance contributed to the downfall of the VB approach. The ontological status of resonance was the object of a dispute between Pauling and George Wheland (1907–1972), Pauling's former student and longtime collaborator, who worked hard for the extension of resonance theory into organic chemistry. To Pauling, resonance held the same status – the same "man-made" character – as chemical notions, such as double bonds, bond lengths, or bond angles, and therefore the theory of resonance involved, in his view, "the same amounts of idealization and arbitrariness as the classical valence-bond theory." Wheland, on the contrary, believed that resonance was more man-made in the sense that

> the statement that benzene is a hybrid of the two Kekulé structures does not describe the properties of a molecule so much as the mental processes of the person who makes the statement.... [R]esonance is not something that the hybrid *does,* or that could be "seen" with sensitive apparatus, but is instead a description of the way that the physicist or chemist has arbitrarily chosen for the approximate specification of the true state of affairs.[10]

This view was also advocated by Coulson, who declared in a semipopular magazine that "resonance is a "calculus,"... a method of calculation; but it

---

[10] Ava Helen and Linus Pauling Papers, Box 115, Pauling to Wheland, 26 January, 8 February 1956; Wheland to Pauling, 20 January 1956.

has no physical reality."[11] On the other hand, the theory of resonance was attacked in the Soviet Union during the 1950s under the methodological objection that one could not obtain meaningful results by starting from conditions and structures that did not correspond to reality.[12]

The ascendancy of the MO theory accompanied the downfall of the VB theory. At long last, the MO camp found an advocate – Coulson himself – with as much rhetorical and pedagogical skill as Pauling. On the other hand, MO theory proved better adapted to the classification of the excited states of molecules – one of the realms of molecular spectroscopy – and above all, better adapted to computer programs. The successful utilization of digital computers in quantum chemistry to compute wave functions and energy levels was prepared by a program discussed and agreed upon at the Shelter Island Conference in 1951.[13] The program aimed at obtaining formulas for the "troublesome" integrals needed for the integration of Schrödinger's equation and making them available to the community of quantum chemists in standardized tables.

The 1960s saw, in fact, the coming of age of the computer, the emergence of high-speed digital computers. At the computational level, a shift of emphasis from semiempirical approximations toward wholly theoretical ('ab initio') calculations occurred. In semiempirical calculations, the computation of molecular properties was carried out by setting up a theoretical framework; then, at certain points, integrals that were difficult to compute were substituted by experimentally determined quantities. The use of computers to calculate the time-consuming integrals of the increasingly sophisticated versions of the MO method also opened the way to the investigation of molecules that were otherwise inaccessible to experimentation. At the experimental level, computers in some instances replaced laboratory experiments as sources of new data. However, Coulson warned that, at the conceptual level without new unifying ideas, computers were basically useless.

Coulson could only make some predictions as to the characteristics of the coming stage in the history of theoretical chemistry. He considered it probable that the concern with molecular structure (molecular architecture) or chemical statics would give way in the 1970s to chemical dynamics and

---

[11] Charles A. Coulson, "The Meaning of Resonance in Quantum Chemistry," *Endeavour*, 6 (1947), 42–7, at 47.

[12] D. N. Kursanov et al., "The Present State of the Chemical Structural Theory," *Journal of Chemical Education* (January 1952), 2–13; V. M. Tatevskii and M. I. Shakhparanov, "About a Machist Theory in Chemistry and its Propangadists," *Journal of Chemical Education* (January 1952), 13–14; I. Moyer Hunsberger, "Theoretical Chemistry in Russia," *Journal of Chemical Education* (October 1954), 504–14; Loren R. Graham, *Science, Philosophy and Human Behavior in the Soviet Union* (New York: Columbia University Press, 1987).

[13] These conferences were started in 1947, and the first has often been compared with the Solvay Congress of 1911 in the sense of having played for quantum field theory a role equivalent to the Solvay Congress for quantum theory. Silvan S. Schweber, "Shelter Island, Pocono, and Oldstone: The Emergence of American Quantum Electrodynamics after World War II," *Osiris*, 2 (1986), 265–302.

chemical reactivity, and that the extension of theoretical chemistry into biology would play an increasingly dominant role. He was, therefore, in syntony with Löwdin, who, in the remark quoted at the beginning of this chapter, also foresaw the extension of quantum chemistry into biology.

THE EMERGENCE OF QUANTUM CHEMISTRY AND THE PROBLEM OF REDUCTIONISM

A new scientific discipline is accepted as such only if it manages to establish a conceptual, methodological, and institutional identity through definition of problems to be solved, values shared by its practitioners, integration into academic curricula, and so on. The claim is often found in the chemical literature that quantum chemistry began in 1927 with the famous paper written by the German physicists Heitler and London.[14] Scientists and historians of science are eager to find landmarks in their field, thereby revealing a willingness to highlight features that are deemed to confer singularity on an event – almost as if disclosing its "essence." In choosing the 1927 paper as the birth date of quantum chemistry, one is putting emphasis on the first quantum mechanical explanation of the formation of a molecule – the molecule of hydrogen, the simplest of all molecules.

The hydrogen molecule plays a relatively minor role in chemistry, however, and so it has been suggested that the pivotal date be shifted to 1931, the year in which Pauling and Slater, independently from each other, explained in quantum mechanical terms the reason that bonds are directed toward privileged directions – the reason that carbon, in general, may form four bonds tetrahedrally oriented.[15] However, both birth dates place too much emphasis on the role played by quantum mechanics in the genesis of quantum chemistry. Although quantum mechanics provided the mathematical apparatus necessary for quantum chemistry to embark on a computational phase, one should note that some of the more important concepts of quantum chemistry were arrived at quite independently of quantum mechanics – I have in mind the concept of "electron promotion" and of molecular orbitals suggested in the context of Mulliken's interpretation of molecular electronic spectra, which took place in the framework of the old quantum theory. Even Pauling liked to stress that it had been an accident in the history of physics and chemistry that the resonance theory – Pauling's hallmark – was not completely formulated before quantum mechanics. On the other hand, a comparative analysis of the genesis of quantum chemistry in the United States and in Germany, which

---

[14] Walter Heitler and Fritz London, "Wechselwirkung neutraler Atome und homöpolare Bindung nach Quantenmechanik," *Zeitschrift für Physik*, 44 (1927), 455–72.
[15] Mary Jo Nye, *From Chemical Philosophy to Theoretical Chemistry* (Berkeley: University of California Press, 1993), p. 228.

focused on the contributions of Mulliken, Pauling, Slater, John Hasbrouck Van Vleck (1890–1980), Heitler, London, and Friedrich Hund (1896–1996), showed that the genesis of quantum chemistry involved the explicit or implicit discussion of issues that transcended the question of the application of quantum mechanics to chemical problems.[16]

In the opening paragraph of the first paper on "The Nature of the Chemical Bond," Pauling contrasted his methodology with the one adopted by Heitler and London. His applications provided many more results that could be obtained in the form of rules, the usefulness of which served as an empirical criterion for their acceptance.[17] Coming back full circle, in a paper published in the *Foundations of Physics* two years before his death, Pauling once more reiterated his lifetime belief in the usefulness of "rough quantum mechanical calculations" in producing new insights, and suggested that nuclear physicists look at the origins of quantum chemistry for methodological guidance:

> In thinking about the history of science in the period around 60 years ago, I have come to the conclusion that much progress was the result of carrying out approximate quantum mechanical calculations. It is my impression that in recent years the effort has been made to carry out quantum mechanical calculations that are as quantitatively accurate as possible. Instead of making calculations of energy levels and other properties of a system with use of a simple approximate wave function corresponding to some simple model, the effort of many physicists is to formulate as complicated a wave function as can be handled by the computers.... With a complicated wave function, however, it is essentially impossible to formulate an interpretation in terms of a model of the system.[18]

The key to Pauling's success and to the Americans' success was their ability to develop, to use, and to transmit to their audience a "feeling for chemistry" as a prerequisite for doing quantum chemistry.[19] Coulson also called the attention of the reader to the idiosyncrasies of chemical thinking in the preface to *Valence*: "[T]he theoretical chemist is not a mathematician, thinking mathematically, but a chemist, thinking chemically."[20]

---

[16] Ana Simões, *Converging Trajectories, Diverging Traditions: Chemical Bond, Valence, Quantum Mechanics and Chemistry, 1927–1937*, PhD thesis, University of Maryland, 1993 (University Microfilms Inc., Publication # 9327498); Kostas Gavroglu and Ana Simões, "The Americans, the Germans and the Beginnings of Quantum Chemistry: The Confluence of Diverging Traditions," *Historical Studies in the Physical and Biological Sciences*, 25 (1994), 47–110; Kostas Gavroglu, *Fritz London: A Scientific Biography* (Cambridge: Cambridge University Press, 1995).

[17] Linus Pauling, "The Nature of the Chemical Bond: Application of Results Obtained from the Quantum Mechanics and from a Theory of Paramagnetic Susceptibility to the Structure of Molecules," *Journal of the American Chemical Society*, 53 (1931), 1367–1400.

[18] Linus Pauling, "The Value of Rough Quantum Mechanical Calculations," *Foundations of Physics*, 22 (1992), 829–38, at 834–5.

[19] Ava Helen and Linus Pauling Papers, Box 157, California Institute of Technology General Files 1922–64, letter Pauling to A. A. Noyes, 18 November 1930.

[20] Charles A. Coulson, *Valence* (Oxford: Clarendon Press, 1952), p. v.

In the 1930s, quantum chemistry developed an autonomous language relative to physics, and many of the problems debated were about ontological priorities and methodological commitments. Disputes and disagreements were as much about getting the correct solution to a problem as they were part of a rhetoric about how to go about solving similar kinds of problems. It has therefore been suggested that the usual division between the Heitler-London-Slater-Pauling valence bond method (VB) and the Hund-Mulliken method of molecular orbitals (MO) should give way to another – the Mulliken-Pauling versus the Heitler-London-Hund.[21] This new dichotomy puts the emphasis on the contrasting methodological options that oppose the Americans to the Germans.

The pragmatic Americans acknowledged the importance of quantum mechanics but aimed at developing semiempirical methods, in which shortcut rules, based on a sort of induction from the available data (which, in many instances, they gathered themselves) and dependent only partially on quantum mechanics, were to play a prominent role. For the Germans, the theories of the chemical bond, as any physical theory, should be derived from first principles firmly based on the postulates of quantum mechanics. Whereas the Americans were contributing to the consolidation of a chemical culture, the Germans saw themselves as contributing to the development of yet another branch of applied physics. Without ever contrasting the Americans to the Germans, Van Vleck and Albert Sherman, in their 1935 joint paper, considered that two diverging attitudes had emerged in developing a "quantum theory of valence": One could be content to adopt the mental attitude and procedure of the optimist or that of the pessimist.[22]

In the months following the appearance of Heitler and London's paper, several physicists thought that chemistry was now at their mercy, reducible to a field of applied physics, and therefore on the verge of losing its status as an autonomous scientific discipline. In 1929, P. A. M. Dirac was the first to give voice to this reductionist dream in a paper, asserting that "the underlying physical laws for the mathematical theory of a large part of physics and the whole of chemistry are thus completely known."[23] But before Dirac, Pauling had already stated, in lectures given at the regional meeting of the American Chemical Society in Pasadena and at the American Association for the Advancement of Science meeting in Pomona that:

> science itself progresses ... toward a goal: the reduction of phenomena included within its domain to the simplest possible form; that is the description of these phenomena in the most economical and esthetically satisfying

---

[21] Gavroglu and Simões, "The Americans and the Germans."
[22] J. H. Van Vleck and Albert Sherman, "A Quantum Theory of Valence," *Reviews of Modern Physics*, 7 (1935), 167–227, at 169.
[23] P. A. M. Dirac, "Quantum Mechanics of Many-Electron Systems," *Proceedings of the Royal Society*, A123 (1929), 714–33, at 714.

terms.... Theoretical chemistry, being more complicated and extensive [than physics], must necessarily follow theoretical physics in its development.... [W]e can now predict with considerable measure of confidence the general nature of the future advances. We can say and partially vindicate the assertion, that the whole of chemistry depends essentially upon two fundamental phenomena: these are 1) the one described in the *Pauli Exclusion Principle;* and 2) the *Heisenberg-Dirac Resonance Phenomena.*[24]

Pauling's statements about the question of reduction changed dramatically in time to a nonreductionist stance, however. In 1936, he gave another talk at the meeting of the Southern California section of the American Chemical Society at Pasadena, the same place where he had spoken eight years before. Now he was sure that "there is more to chemistry than an understanding of general principles. The chemist is also, perhaps even more, interested in the characteristics of individual substances – that is, of individual molecules."[25]

Pauling was then in the process of implementing a lifelong agenda aimed at reforming the science of chemistry from the point of view of quantum chemistry. In this context, he came to reevaluate the status of chemistry within the hierarchy of the sciences. He believed in the desirability of the "integration" of the sciences, which he considered to be achieved through the transfer of tools and methods from one science to the other.[26] The most important kind of transfer was, however, what he called the "technique of thinking." It is in this respect that he considered chemistry, and specifically the structure theory of chemistry, to play a central role within the physical and biological sciences. Pauling's work on quantum chemistry in the 1930s, and its extension to molecules of biological relevance during the 1940s and 1950s, justified his claim of a central place for chemistry, a place formerly occupied by physics.

The culture and tradition of physics played a fundamental role in indoctrinating physicists in the avenues of reductionism. With few exceptions, reductionist statements were made by physicists. In different circumstances and degrees they were made by Slater, Heitler, London, and Hückel. As Heitler confessed in an interview, his and London's perhaps too-ambitious initial goal had been "to understand the whole of chemistry," a statement that much amused Eugene Wigner, who confessed to be a little skeptical about Heitler's overarching aim.[27]

---

[24] Ava Helen and Linus Pauling Papers, Box 242, Popular Scientific Lectures 1925–1955, "The Nature of the Chemical Bond," 6 April, 14 June 1928.
[25] Ibid., "Recent Work on the Structure of Molecules," part II, "Pauling's Response to the Heitler and London Paper."
[26] Linus Pauling, "The Place of Chemistry in the Integration of the Sciences," *Main Currents in Modern Thought,* 7 (1950). Selections from this paper in Barbara Marinacci, ed., *Linus Pauling in His Own Words* (New York: Simon and Schuster, 1995), pp. 107–11.
[27] Archives for the History of Quantum Physics, interview with Walter Heitler conducted by J. L. Heilbron, Zurich, March 1963; interview with E. P. Wigner conducted by T. S. Kuhn, Rockefeller Institute, November 1963.

Not all physicists partook of similar opinions, however. At the 1931 Centenary Meeting of the British Association for the Advancement of Science, Ralph H. Fowler introduced his talk with the following comment:

> One may say now that the chemical theory of valency is no longer an independent theory in a category unrelated to general physical theory, but just a part – one of the most gloriously beautiful parts – of a simple self-consistent whole, that is of non-relativistic quantum mechanics. I have at least sufficient chemical appreciation to say rather that quantum mechanics is glorified by this success than that now "there is some sense in valencies," which would be the attitude, I think, of some of my friends.[28]

Heisenberg also wondered in his address "whether the quantum theory would have found or would have been able to derive the chemical results about valency, if it had not known them before."[29] In the development of quantum chemistry, physics provided part of the tools, but at the same time, the explanation of empirical chemical facts, such as rules of valence and stereochemistry, guided the development of the theory.

THE EMERGENCE OF QUANTUM CHEMISTRY IN NATIONAL CONTEXT

The identity of an emerging discipline may disclose characteristics peculiar to different times and places. If this is the case, the adequacy and utility of speaking about national styles is a problem to be discussed historically. Given that the mobility of Americans, Germans, and British to study, lecture, or do research in foreign institutions might have enhanced a common approach to the new discipline, it is rather surprising to find out that this was not in fact the case.

It has been claimed that the Americans' ability to move between chemistry and physics, and between theory and experiment, in the context of a congenial institutional atmosphere, made them particularly apt to make significant contributions to, and therefore shape, the new discipline.[30] A particular kind of institutional atmosphere accounted for the appearance of this new type of scientist, whose definition as a chemist or physicist was in many instances a matter of chance, individual preference, or institutional affiliation.

The institutional ties between chemistry and physics were stronger in the United States than in Europe. In universities like Berkeley and Caltech, chemistry students were often learning as much physics as chemistry, and

---

[28] R. H. Fowler, "A Report on Homopolar Valency and its Quantum Mechanical Interpretation," in *Chemistry at the (1931) Centenary Meeting of the British Association for the Advancement of Science* (Cambridge: W. Heffer & Sons, 1932), pp. 226–46, at 226.

[29] Werner Heisenberg, "Contribution to the Discussion on the Structure of Simple Molecules," in ibid., pp. 247–8, at 247.

[30] Ana Simões, PhD thesis; Simões and Gavroglu, "Different Legacies and Common Aims."

they thus were more apt to learn and accept quantum mechanics than were their European counterparts. Harvard, MIT, Princeton, Chicago, Michigan, Minnesota, and Wisconsin also promoted cooperation between physics and chemistry departments.

In Germany by the early 1930s, chemistry and physics were well-established disciplines, entertaining few disciplinary, methodological, or institutional ties to each other. Therefore, scientists whose profile could favor an attack on chemical problems using the tools of the newly developed quantum mechanics were hard to find. Several German physicists, not chemists, were interested in applications to chemistry, and they contributed initially to the field but were unable in the long run to carry out their research programs. Such were the cases of Heitler, London, Hund, and Max Born.

An exceptional case in the German context was the physicist Erich Hückel. He was able to overcome his deficient chemical background by taking advantage of his brother Walter Hückel's expertise in organic chemistry, which probably helped him to ask pertinent questions in organic chemistry to be answered within the framework of quantum mechanics.[31] Erich Hückel developed a reductionist program in which the facts of organic chemistry were to be interpreted by taking seriously the peculiar theoretical features of quantum mechanics. Its nonvisualizability was thought to forbid Pauling's description of the structure of benzene by means of resonance among several fictitious valence bond structures.

In his groundbreaking papers, Hückel explained the aromatic properties of benzene within the framework of MO theory and then proceeded to extend its theory to any conjugated systems (rings or chains). Although one might, as the chemist Jerome Berson has recently suggested, characterize Hückel's scientific style as "pragmatic," in the sense of producing a theory based on bold, simplified assumptions, not totally justified at the time, Hückel was unable to formulate his original results as "rules." On the contrary, they appeared in highly technical papers full of mathematical formulas, a fact that may have driven away potential readers. By 1937, he had abandoned the field, unable to challenge a scientific establishment in which German physicists were not yet ready to accept research on the quantum mechanical properties of the chemical bond as a topic of research for physicists, just as German chemists did not consider quantum chemistry a field of chemistry, because they defined chemistry as what they did, and they did not do quantum chemistry.

---

[31] Walter Hückel's *Theoretische Grundlagen der organischen Chemie* (1931) included quantum interpretations and was very influential when eventually translated into English. Helge Kragh, "The Young Erich Hückel: His Scientific Work until 1925," invited paper given at the Erich Hückel Festkolloquium at the Philipps-Universität, Marburg, 28 October 1996; Jerome A. Berson, "Erich Hückel, Pioneer of Organic Quantum Chemistry: Reflections on Theory and Experiment," *Angew. Chem. Int. Ed. Engl.*, 35 (1996), 2750–64; Andreas Karachalios, "Die Entstehung und Entwicklung der Quantenchemie in Deutschland," *Mitt. Ges. Deut. Chem. Fachgr. Gesch. Chem.*, 13 (1997), 163–79; Andreas Karachalios, "On the Making of Quantum Chemistry in Germany," *Studies in the History and Philosophy of Modern Physics*, 31 (2000), 493–510.

During the 1930s many scientists migrated from Germany, including Heitler, London, and Hans Hellman, who was preparing a book on quantum chemistry, when he was offered a job at the Karpow Institute of Physical Chemistry in Moscow. After the Second World War, the new discipline experienced a boost mainly with the creation of two research centers, one in Frankfurt headed by Hermann Hartmann, the other in Göttingen under Heinz-Werner Preuss. This period marks a shift from concern with conceptual and methodological questions to institutional establishment and recognition of the discipline.

In nineteenth-century France, an antiatomistic climate had been largely responsible for the neglect of theoretical questions by organic chemists, and simultaneously, physical chemists were to privilege the use of thermodynamics in the study of chemical equilibrium. The main contributor to the revival of atomism was Jean Perrin, an outstanding physical chemist who dominated the French chemical scene up to the 1940s. Still, there was not a positive response to the quantum theory of the chemical bond, neither on the part of the French physical chemists, such as Jean Perrin, nor on the part of quantum physicists, such as Louis de Broglie.[32] The two world wars and the rigidity of the French civil service system for obtaining permanent university positions, which hindered the accommodation of scientists leaving Hitler's Germany, could hardly have reversed this trend: Research in physical chemistry slowed down considerably, keeping its mainly experimental profile. Therefore, the beginning of quantum chemistry on French soil was to be postponed until the last years of the Second World War. A group of theorists, including Alberte Pullman, gathered around Raymond Daudel and started to work on molecules of biological interest using the MO approximation.

In the same period during which Mulliken, Pauling, and their research groups were steadily building the foundations of quantum chemistry, the discipline was to make a good start in Great Britain. At first inspection, the situation in Great Britain looks quite different from that of the United States. On the one hand, the separation between the two disciplines of physics and chemistry was marked and, on the other, according to Douglas Hartree's own recollections, in the late 1920s and early 1930s physics really meant experimental physics, and therefore, theoretical physics was usually the realm of mathematicians.[33]

On a second look, however, two characteristics need to be emphasized. On the one hand, the Cambridge physicist Ralph Fowler was very successful in calling attention to quantum physics, promoting it, and supervising or starting to supervise some of the British pioneers in quantum chemistry – Lennard-Jones, Hartree, and Coulson. On the other hand, N. V. Sidgwick

---

[32] Jules Guéron and Michel Magat, "A History of Physical Chemistry in France," *Ann. Rev. Phys. Chem.*, 22 (1971), 1–25; Mary Jo Nye, *Science in the Provinces* (Berkeley: University of California Press, 1986).

[33] Fritz London Archives, Duke University, Hartree to London, 16 September 1928.

played a decisive role in creating a milieu particularly receptive to quantum chemistry. By means of his book *Some Physical Properties of the Covalent Link in Chemistry* (1933), his annual reports, and his presidential addresses, he became one of the most effective propagandists of the immense usefulness of resonance for chemistry. Lennard-Jones, Hartree, and Coulson contributed to what might be characterized as the "British approach" to quantum chemistry. The British quantum chemists perceived the problems of quantum chemistry first and foremost as calculational problems and, by devising novel calculational methods, they tried to bring quantum chemistry within the realm of applied mathematics. In that specific context, the demand for more rigor was a demand not primarily for rethinking the conceptual framework but, rather, for developing, as well as legitimizing, formal (mathematical) techniques and methods to be used in solving chemical problems.[34]

QUANTUM CHEMISTRY AS A DISCIPLINE

In addition to the conceptual and methodological features that converge in the creation of an identity for every scientific subject, institutional and sociological features also contribute in varying degrees. In the case of quantum chemistry, defined in the formative years as an interdisciplinary subject in the borderland between physics and chemistry, its identity was built through the negotiation of an autonomous space within chemistry, as well as in relation to physics.

Demarcation within chemistry was reflected in the concern about securing a name for the new discipline, the introduction of new outlets for its publications, the establishment of its own language, the concomitant standardization of notation, the creation of chairs and professorships in quantum or theoretical chemistry, the organization of meetings, conferences, and summer schools, and the writing of textbooks.

The editorial written by the young physical chemist H. C. Urey, then at Columbia University, introducing the first issue of the new *Journal of Chemical Physics* in 1933, presented the new discipline as a natural outcome of recent developments in the chemical and physical sciences. A transition was taking place from a physical chemistry concerned with the properties of bulk matter, and founded on the physical theory of thermodynamics, to a "chemical physics" concerned with individual atoms and molecules and founded on quantum mechanics. The decision process that preceded

---

[34] Ana Simões and Kostas Gavroglu, "Quantum Chemistry *qua* Applied Mathematics: The Contributions of Charles Alfred Coulson (1910–1974)," *Historical Studies in the Physical and Biological Sciences*, 29(1999), 363–406; Simões and Gavroglu, "Quantum Chemistry in Great Britain: Developing a Mathematical Framework for Quantum Chemistry," *Studies in the History and Philosophy of Modern Physics*, 31 (2000), 511–48.

the publication of this new journal by the American Institute of Physics made clear the confrontation between two diverging attitudes concerning the interaction of chemistry and physics.[35] Some chemists, like Wilder D. Bancroft, thought that the identity of chemistry could only be secured by its isolation from physics and by a methodological emphasis on the qualitative and nonmathematical aspects of chemistry. In contrast, a growing fraction of the community of American physical chemists greeted the appearance of a journal that could house papers that were "too mathematical for the *Journal of Physical Chemistry,* too physical for the *Journal of the American Chemical Society,* or too chemical for the *Physical Review.*"[36] The use of appropriate channels of communication was therefore perceived as a fundamental step toward the consolidation of the new discipline and the capture of as large an audience as possible. In 1967, more than thirty years later, the *International Journal of Quantum Chemistry* was created to meet the growing internationalization of the discipline: "Today quantum chemistry is undergoing such a fast development that many scientists feel that it deserves its own journal on an international basis," wrote Löwdin.[37]

Starting in the 1920s, meetings and congresses were organized in Great Britain, in the United States, and in Germany, giving us an indicator of reactions by the chemical community to the new events. The receptivity of British organic chemists to physical considerations dated back to the 1923 meeting of the Faraday Society held in Cambridge that was organized by T. Lowry and J. J. Thomson. That meeting was concerned with the discussion of "The Electronic Theory of Valence" and, notably, with Lewis's electronic model for the covalent bond.[38] The next meeting organized by the Faraday Society on similar topics was held in 1929 and was devoted to "Molecular Spectra and Molecular Structure." Lennard-Jones, Mulliken, and Raymond Thayer Birge were present. Hund and Rudolf Mecke were among the Germans attending the meeting. The problems under discussion already involved a first attempt at comparing the VB and the MO methods. This was also the meeting where the young quantum chemist Lennard-Jones presented a paper in which he introduced the LCAO method. Two years later, the 1931 Centenary Meeting of the British Association for the Advancement of Science also provided a forum for discussion of the quantum theory of valence.

In 1928, the American Chemical Society organized a meeting in St. Louis on "Atomic Structure and Valence." The papers delivered were published in the journal *Chemical Reviews,* and they reveal a deep awareness by the

---

[35] John Servos, *Physical Chemistry from Ostwald to Pauling: The Making of a Science in America* (Princeton, N.J.: Princeton University Press, 1990).

[36] Lewis Correspondence, Bancroft Library, CU-30, Box 2, folder on K. T. Compton, Compton to Lewis, 6 August 1932.

[37] Löwdin, "Program," p. 1.

[38] The papers delivered were published in *Transactions of the Faraday Society,* 19 (1923), 450–558.

American chemical community of the radical changes to be effected in their culture and tradition. In Germany, two meetings of the Bunsen Gesellschaft were organized in 1928 and 1930, and the topics addressed were, respectively, "The mode of chemical binding and the structure of atoms" and "Spectroscopy and the structure of molecules."

From the 1930s to the first years after the end of the Second World War, meetings had the participation of scientists from an increasing number of countries. They served the function of consolidating quantum chemistry as an international discipline, as well as an autonomous subdiscipline within chemistry. The Colloque de la Liaison Chimique, the first important meeting convened after the war, took place in Paris in 1948 and launched quantum chemistry in France. Starting in the 1950s, quantum chemistry meetings focused on the assessment of the impact of computers in reshaping the aims of the discipline. Initiated by the 1951 Shelter Island Conference, this topic of discussion was again addressed during the Conference on Molecular Quantum Mechanics held at Boulder, Colorado, in 1959, as well as in other international meetings, such as the Sanibel Island Conferences and the Gordon Research Conferences, which became obligatory meeting points for quantum chemists from all over the world.

The creation of professorships in theoretical or quantum chemistry extended throughout a fifty-year period, varying from place to place even within the same nation. In America, Pauling in 1927 became assistant professor of theoretical chemistry and started immediately to give graduate lectures on the nature of the chemical bond. Mulliken held professorships in physics and chemistry, but only in the 1960s did he get a professorship in chemical physics. In Great Britain, Lennard-Jones was the first to hold a chair in theoretical chemistry – the Plummer Chair of Theoretical Chemistry at Cambridge University – starting in 1932. His lectures on quantum chemistry were considered by his former student Coulson to be the first at the undergraduate level to be delivered worldwide. Coulson became the first occupant of the first chair of theoretical chemistry in the new department of theoretical chemistry at Oxford in 1973.

Demarcation across disciplines, that is, demarcation of quantum chemistry from physics, at the outset involved discussions over the fruitfulness or lack of fruitfulness of strengthening ties with physics. It involved, simultaneously, an evaluation of the distinctive role played by quantum mechanics in the new context and the challenges it posed to the autonomy that the new discipline was claiming for itself. The obstacle posed by chemists' lack of knowledge of sophisticated mathematics or of quantum theory, together with the intrinsic "strangeness" of quantum mechanics as a physical theory, had to be overcome, and the strategies followed assumed many faces.

In 1928, two articles, one by Pauling and the other by Van Vleck, appeared in *Chemical Reviews* with the explicit aim of "educating" chemists in the ways of the new quantum mechanics.³⁹ Seven years later, the publication of Pauling and E. B. Wilson's *Introduction to Quantum Mechanics with Applications to Chemistry* (1935) appeared as part of a much more ambitious strategy. With this and two other textbooks, *The Nature of the Chemical Bond* (1939) and *General Chemistry* (1947), Pauling was trying to reform chemistry from the standpoint of quantum chemistry.

In *The Nature of the Chemical Bond,* Pauling presented the major questions that he had discussed in his papers in a language more appropriate for a larger professional audience, and made clear what he considered the methodological approach to be followed in dealing with quantum chemistry. The book was to have a tremendous impact not only on research but also on the teaching of chemistry. Joseph Mayer published a review of the book in which he considered it to be

> unfortunate that this treatise will almost certainly tend to fix, even more than has been done by the author's excellent papers, the viewpoint of most chemists on this, and only this one, approach to the problem of the chemical bond. It appears likely that the Heitler-London-Slater-Pauling method will entirely eclipse, in the minds of chemists, the single electron molecular orbital picture, not primarily by virtue of its greater applicability or usefulness, but solely by the brilliance of its presentation.⁴⁰

It was only in 1952, with the publication of Coulson's *Valence,* that a book was to have, at last, an impact equivalent to Pauling's *The Nature of the Chemical Bond* in shaping the new discipline.

Many textbooks on quantum chemistry were written, starting in the 1940s. Intended to serve mainly an educational purpose, they aimed at presenting quantum mechanics to chemistry students by the adoption of different strategies. In some cases, they presented quantum mechanics with full consideration of its mathematical methods and different degrees of emphasis on topics of chemical interest. In other cases, they provided qualitative discussions of the applications of quantum theory to chemistry, particularly to chemical bonds, avoiding as much as possible the mathematical structure of the theory. In still other cases, they attempted to combine the advantages of the two other classes of books. In some instances, these different strategies reflected implicit or explicit views about the autonomy of quantum chemistry, that is, about the hypothetical reduction of chemistry to physics.⁴¹

---

[39] Linus Pauling, "The Application of Quantum Mechanics to the Structure of the Hydrogen Molecule and Hydrogen-Molecule Ion and to Related Problems," *Chemical Reviews,* 5 (1928), 173–213; J. H. Van Vleck, "The New Quantum Mechanics," *Chemical Reviews,* 5 (1928), 467–507.

[40] Ava Helen and Linus Pauling Papers, Box 399, The Nature of the Chemical Bond 1932–59.

[41] Kostas Gavroglu and Ana Simões, "One Face or Many? The Role of Textbooks in Building the New Discipline of Quantum Chemistry," in *Communicating Chemistry: Textbooks and Their Audiences 1789–1939,* ed. Anders Lundgren and Bernardette Bensaude-Vincent (Canton, Mass.: Science History Publications, 2000), pp. 415–49; Buhm Soon Park, "Chemical Translators: Pauling, Wheland and their Strategies for Teaching the Theory of Resonance," *British Journal for the History of Science,* 32 (1999), 21–46.

THE USES OF QUANTUM CHEMISTRY FOR THE HISTORY AND PHILOSOPHY OF THE SCIENCES

In 1985, to stress the lack of attention given to the history of chemistry in America, relative to the more glamorous disciplines of physics and biology, John Servos designated chemistry as the "dismal science."[42] Physical chemistry, and even more so quantum chemistry, had not been the topic of many historical and philosophical studies. The last decade has witnessed a positive reversal in this state of affairs with the first set of works addressing quantum chemistry per se, or as the culmination of other case studies. Different methodologies have been followed. Discipline formation and disciplinary identity have been addressed through the point of view of biographical studies, discipline genealogy, and comparative case studies involving different national contexts.[43] In some cases, problems pertinent for philosophers of science have been addressed from a historical perspective.[44]

Quantum chemistry may also help in the clarification of such topics as the language of chemistry, representation in chemistry, and reduction of chemistry to physics. These subjects have been recently at the forefront of research carried out by philosophers of science and are evidence of the growing interest in the philosophy of chemistry.[45]

Historians have concentrated mainly on the 1930s, and on the American and German cases. We need more comparative case studies in order to be able to form a comprehensive view of the genesis and development of quantum chemistry; on the other hand, the period after the Second World War has not yet been systematically studied. Is it the case that, with the advent of computational quantum chemistry, the problématique involved in the genesis

---

[42] J. W. Servos, "History of Chemistry," in *Historical Writing in American Science, Osiris*, 1 (1985), 132–46.

[43] R. S. Krishnamurthy, ed., *The Pauling Symposium: A Discourse on the Art of Biography* (Corvallis: Special Collections, Oregon State University Libraries, 1996) and references therein; Kostas Gavroglu, *London*; S. S. Schweber, "The Young John Clarke Slater and the Development of Quantum Chemistry," *Historical Studies in the Physical and Biological Sciences*, 20 (1990), 339–406. For discipline genealogy, see Servos, *Physical Chemistry*; Nye, *Chemical Philosophy*; Assmus, "The Molecular Tradition in Early Quantum Theory," and "The Americanization of Molecular Physics." For comparative case studies, see Gavroglu and Simões, "The Americans and the Germans."

[44] Mary Jo Nye, "Physics and Chemistry: Commensurate or Incommensurate Sciences?" in *The Invention of Physical Science* (Dordrecht: Kluwer, 1992); Nye, *Chemical Philosophy*; S. G. Brush, "Dynamics of Theory Change in Chemistry: Part I. The Benzene Problem 1865–1945," and "Dynamics of Theory Change in Chemistry: Part II. The Benzene Problem 1945–1980," *Studies in the History and Philosophy of Science*, 30 (1999), 21–79, 263–302.

[45] Volume III (1997) of *Synthese* has been entirely dedicated to the philosophy of chemistry and includes a bibliography on the philosophy of chemistry. The journals *Hyle*, edited by Joachin Schummer, and *Foundations of Chemistry*, edited by Eric Scerri, include papers on the philosophy of chemistry. For a review of the topic, see Jeff Ramsey, "Recent Work in the History and Philosophy of Chemistry," *Perspectives on Science*, 6 (1998), 409–27. Kostas Gavroglu edited a volume in *Studies in the History and Philosophy of Modern Physics*, 31 (2000), which is dedicated to the history and philosophy of quantum chemistry. See also Ana Simões and Kostas Gavroglu, "Issues in the History of Theoretical and Quantum Chemistry," in *Chemical Sciences in the 20th Century: Bridging Boundaries* (Weinheim: Wiley-VCH, 2001), pp. 57–74.

of the discipline has changed dramatically, from a concern with conceptual and methodological issues to exclusively technical ones? Is it the case that the coexistence of semiempirical and accurate calculations has brought about a new quantitative model of the chemical bond? If the quantum chemist R. G. Parr could proclaim that "accurate descriptions of the electronic structure of molecules are upon us,"[46] is it the case that, after all, Dirac's 1929 prediction has been fulfilled to a significant degree? These problems still wait to be addressed.

[46] R. G. Parr, *Quantum Theory of Molecular Electronic Structure* (New York: Benjamin, 1963), p. 123.

21

PLASMAS AND SOLID-STATE SCIENCE

Michael Eckert

There is a common trait in reviews by plasma- and solid-state physicists of their specialties: an emphasis on the ubiquity of the subject matter with which they are concerned. "As we now know, planet Earth is but a small non-plasma island in a vast sea of plasma. Though tenuous in outer space, plasma is dense and omnipresent in the stars and in their coronas. In fact this 'fourth state of matter' – plasma – is seen as the dominant form of matter in the Universe," a pioneer of plasma physics writes in a review about his discipline.[1] With the same zeal but a slightly different emphasis, a German solid-state protagonist has created a link between the omnipresent solid matter and human culture: "Solid substances have given their names to the great historical epochs of mankind. Stone, bronze, and iron have caused epochal changes," he begins in a book on the history of solid-state electronics. Now, at the end of the iron age, we are entering a new era. "Probably this epoch will be given the name of the crystal.... This new epoch perhaps will be called silicon age."[2]

What is the message behind such a "ubiquity" rhetoric? Beyond the pleading for recognition, funds, and other means of furthering plasma and solid-state physics, are we supposed to consider research in those omnipresent substances as a cultural obligation? In contrast to specialties like elementary particle physics, which appear to the public as more fundamental, the study of plasmas is regarded as a corollary to the quest for controlled thermonuclear fusion. Similarly, solid-state physics seems to derive its importance more from technological applications than from intellectual curiosity. Although the conceptual roots of both specialties may be traced back to the nineteenth century and earlier, they acquired their disciplinary identities only in the second half of the twentieth century.

[1] Richard F. Post, "Plasma Physics in the Twentieth Century," in *Physics in the Twentieth Century*, vol. 3, ed. L. Brown et al. (Bristol: Institute of Physics Publishing, 1995), pp. 1617–90, at p. 1618.
[2] Hans Queisser, *Kristallene Krisen: Mikroelektronik – Wege der Forschung, Kampf um Märkte* (Munich: Piper, 1985), pp. 7–8. Translated by the author unless otherwise noted.

414   *Michael Eckert*

PREHISTORY: CONTEXTUAL VERSUS CONCEPTUAL

The sheer quantity of subfields comprised by modern plasma and solid-state physics makes elusive any attempts to trace their historical developments concept by concept. Except from a Whiggish selection of "Who named the -on's," a mere conceptual approach is inappropriate to explain the factors that finally gave those specialties their identity.[3] Instead of presenting a list of the conceptual "beginnings" of plasma and solid-state physics, it is more rewarding to consider the variety of contexts for early plasma and solid-state researches.

Before the 1920s, a host of solid-state know-how was accumulated under such diverse umbrellas as mineralogy, crystallography, or metallurgy. Institutionally, this research was performed in universities, as well as in national establishments (like the Physikalisch-Technische Reichsanstalt in Berlin) and in new industrial laboratories of firms such as Siemens or General Electric [GE]. Sometimes research was well organized (such as in the long-term and systematic efforts of materials testing), but sometimes it was haphazard (like the discovery of x-ray diffraction) or a combination of luck and long-term research strategies (like the discovery of superconductivity). It would be hard to isolate a single and predominant feature of the emergence of our knowledge about solid matter – apart from this very heterogeneity of contexts, backgrounds, and approaches.[4]

During the 1920s and early 1930s, new contexts shaped this pattern. The new quantum mechanics formed a backbone for solid-state theory – both intellectually and socially. Solid-state topics now became a major target of opportunity for ambitious young theorists, who happened to experience the tremendous advances of atomic theory and quantum mechanics at that time. John Clarke Slater (1900–1976), one of this elite group of "maybe 50 to 100 well-known names" from the "classical decade of 1923–1932," recalled: "During all this time, almost countless momentous discoveries had been made in quantum theory and its application to molecular and solid-state problems."[5] The names of those who shared Slater's experience read like a who's who of solid-state theory: Hans Bethe (b. 1906), Felix Bloch (1905–1983), Léon Brillouin (1889–1969), Herbert Fröhlich (b. 1905), Werner Heisenberg (1901–1976), Walter Heitler (1904–1981), Ralph

---

[3] Charles T. Walker and Glen A. Slack, "Who Named the -ON's," *American Journal of Physics*, 38 (1970), 1380–9.

[4] Michael Eckert et al., "The Roots of Solid-State Physics before Quantum Mechanics," in *Out of the Crystal Maze: Chapters from the History of Solid-State Physics*, ed. Lillian Hoddeson et al. (New York: Oxford University Press, 1992), pp. 3–87; Paul Forman, "The Discovery of the Diffraction of X-Rays by Crystals: A Critique of the Myths," *Archive for History of Exact Sciences*, 6 (1969), 38–71; Peter Paul Ewald, "The Myth of the Myths: Comments on P. Forman's Paper," ibid., 72–81.

[5] John Clarke Slater, *Solid-State and Molecular Theory: A Scientific Biography* (New York: Wiley, 1975), pp. 3–7. Most of these solid-state pioneers did not consider themselves as "solid-state theorists," and they became known for fundamental work in other fields as well.

Kronig (b. 1904), Fritz London (1900–1954), Nevill Mott (1905–1996), Wolfgang Pauli (1900–1958), Linus Pauling (1904–1994), Rudolph Peierls (1907–1996), and Edward Teller (b. 1908), to name only the most prominent.

Solid-state concepts like Bloch wall, Brillouin zone, or London-Heitler method are textbook reminders of these pioneers. After they had established their own careers in a university chair for theoretical physics or as head of a physics department, a few of them chose solid-state theory as a major research area. Heisenberg's institute at Leipzig University in Germany, Slater's department at MIT in the United States, and Mott's institute at the University of Bristol in England became renowned during the 1930s for solid-state theory. By 1933, another transition was changing the social and intellectual environment of solid-state theory: Many of those who had started their careers in Germany with applications of quantum mechanics were forced to emigrate as a consequence of the rise of the Nazi regime. By that time, Eugene Wigner (1902–1995), a Hungarian emigrant at Princeton University, and Frederick Seitz (b. 1911) had developed a model to compute the electronic energy bands of a particular metal, sodium. With this method, the theory became applicable to real solids, whereas between 1926 and 1933 the focus had been on ideal crystals.[6]

While the focus of this quantum mechanical application mainly addressed the electronic properties of solids, other origins of modern solid-state physics were unrelated to quantum mechanics – and consequently emerged and grew up in quite different contexts. Research on the mechanical properties of solids, for example, had its main roots in the engineering environment of metallurgy. Empirical observations of mechanical failures of aircraft and engine structures in the British Royal Aircraft Factory in Farnborough during World War I, for example, gave rise to a theory of crack propagation as a mode of plastic deformation, which in turn became an immediate predecessor of modern dislocation theory.[7] The engineering context of metallurgy met eventually with the interests of physical chemistry, crystallography, and mineralogy, which were rooted both in academic and industrial environments. The landmark discovery in 1912 of x-ray diffraction in crystals, which had been made without a primary interest in materials, was followed by a rise in applications in basic science and in industrial material research, providing a common conceptual background to a vast variety of scientific activities in university institutes the world over, as well as in the I. G. Farben laboratories in Ludwigshafen and in the newly created Kaiser Wilhelm Institutes in Germany, in the research laboratory of the General Electric Company in Schenectady, New York, and in the Tokyo Institute of Physical and Chemical Research in Japan. The materials investigated under the x-ray (and later electron) beams

---

[6] Lillian Hoddeson et al., "The Development of the Quantum Mechanical Electron Theory of Metals, 1926–1933," in *Out of the Crystal Maze*, pp. 88–181; Paul Hoch, "The Development of the Band Theory of Solids, 1933–1960," in *Out of the Crystal Maze*, pp. 182–235.

[7] Ernest Braun, "Mechanical Properties of Solids," in *Out of the Crystal Maze*, pp. 317–58.

in those laboratories were metals and alloys, as well as organic matter. The physical, chemical, and biological research topics ranged through the study of phase transitions, crystal growth, chemical fibers, and magnetic, electrical, and optical properties of any solid matter to the quest for the genetic code.[8]

On a smaller scale, this heterogeneity is also observed in the prehistory of plasma physics. It has important roots in academic astronomical research and in the industrial investigation of electronic tubes. The very notion of plasma was born in an industrial research laboratory in the late 1920s. The study of electrical discharges in gases was vital to firms like GE and AT&T in the United States, Telefunken and Siemens in Germany, or Philips in the Netherlands. The names of Irving Langmuir (1881–1957) (GE) and Gustav Hertz (1887–1975) (Siemens, Philips), both Nobel Prize winners, illustrate this heritage of plasma physics most prominently. We could add work performed in the 1930s by Walter Schottky (1886–1976), Max Steenbeck (1904–1981), and others at Siemens, who achieved theoretical and experimental advances in such diverse areas as electric boundary layers, magnetic plasma properties, and short time measurements, or research in thermionic resistance and thermal noise in vacuum tubes at the Bell Laboratories.[9]

Another context for early plasma physics was ionospheric research, which arose as a side effect of investigations of the propagation of radio waves. In the 1920s it was demonstrated that the so-called Kennelly-Heaviside layer (postulated twenty years earlier) reflected radio waves like a mirror and, therefore, played a decisive role in the propagation of radio waves over the curved surface of the earth. Henceforth, the study of the ionospheric plasma became part of the growing radio technology. In addition to industrial and state laboratories, such as the Bell Labs or the U.S. Naval Research Laboratory, this research was also considered important for national organizations like the British Radio Research Board, set up after World War I in order to suggest standards for the ever-spreading "wireless" technology. Edward Appleton's (1892–1965) magneto-ionic theory, awarded the Nobel Prize in 1947, emerged from this context – to name just another concept that contributed to the growing knowledge of plasma physics in the 1920s and 1930s.[10] "Probing the Ionosphere" was also a major effort of the National Bureau of Standards in Washington, which was proud to describe, in a review entitled "Achievement in Radio," accounts of a potpourri of ionospheric research programs. The spectrum ranges from pioneering experiments of long-distance transmission of radio signals, performed by Louis W. Austin (1867–1932) before

---

[8] Peter Paul Ewald, ed., *Fifty Years of X-Ray Diffraction* (Utrecht: International Union of Crystallography, N. V. A. Oosthoek's Uitgeversmaatschapij, 1962).

[9] Lewi Tonks, "The Birth of 'Plasma,'" *American Journal of Physics*, 35 (1967), 857–8; Ferdinand Trendelenburg, "Aus der Geschichte der Forschung im Hause Siemens," *Technikgeschichte in Einzeldarstellungen*, 31 (1975), 1–279.

[10] C. Steward Gillmor, "The History of the Term 'Ionosphere,'" *Nature*, 262 (1976), 347–8, and "Wilhelm Altar, Edward Appleton, and the Magneto-Ionic Theory," *Proceedings of the American Philosophical Society*, 126, no. 5 (1982), 395–440.

World War I aboard navy vessels, to "investigating the ionosphere for the FBI," which led the National Bureau of Standards in 1935 to "a series of tests over a period of one year to determine the feasibility of voice transmission from a transmitter at Washington, D.C. to cover the entire United States."[11] We may doubt whether the FBI's request contributed to the advancement of plasma theory, but it underlines the importance of the radio context, which was certainly not without consequences for the development of techniques of measurements.

At the other extreme, early plasma research had also been undertaken in a purely academic context and with completely different orientations. As with the solid-state, the new quantum mechanics proved to be fertile ground for young theorists to win laurels in widely scattered areas. Albrecht Unsöld's (1905–1995) investigation of stellar atmospheres, for example, became a landmark for astro- and plasma physics. Even before quantum mechanics was available, the stars offered topics of research that today are considered within the framework of plasma physics. In 1920, the astronomer Arthur Stanley Eddington (1882–1944) calculated the amount of energy liberated in the stars when helium is made out of hydrogen by nuclear fusion. Drawing upon measurements of the tiny mass differences of isotopes, which Francis Aston (1877–1945) had obtained with a newly designed mass spectrometer, and in view of Ernest Rutherford's (1871–1937) recent experiments at the Cavendish Laboratory, where an artificial transformation of atomic nuclei had been obtained, Eddington speculated that "what is possible in the Cavendish laboratory may not be too difficult in the Sun."[12] The source of the stars' energy production remained a major riddle at the junction of astronomy and nuclear physics, until it finally became understood, at the end of the 1930s, as a cascade of nuclear fusion processes. Most of this research was performed by pure theorists (Bethe, Carl Friedrich von Weizsäcker [b. 1912], Charles L. Critchfield [b. 1910]) and considered highly academic at the time. From the perspective of modern plasma physics, it only became pertinent when controlled thermonuclear fusion was declared as the ultimate aim.

WORLD WAR II: A CRITICAL CHANGE

Even before World War II, therefore, plasma and solid-state research had been practiced for decades. However, its practitioners worked in diverse environments and generally did not share a common self-understanding as solid-state or plasma physicists. Retrospectively, it turns out that something

---

[11] Wilbert F. Snyder and Charles L. Bragaw, eds., *Achievement in Radio: Seventy Years of Radio Science, Technology, Standards, and Measurement at the National Bureau of Standards* (Boulder, Colo.: National Bureau of Standards, 1986), pp. 171–242, at p. 232.

[12] Quoted in John Hendry, "The Scientific Origins of Controlled Fusion Technology," *Annals of Science*, 44 (1987), 143–68.

was still missing that could have served the purposes of a communal conscience in both specialties. Reducing a complicated story to its most prominent features, World War II may be considered the critical point for the transition of "prehistorical" solid-state and plasma investigations to modern solid-state and plasma physics.

Historiographically, it is important again to make a distinction between a conceptual and a contextual perspective: Conceptually, World War II did not further solid-state and plasma knowledge to a considerable extent. From a retrospective view, the two most important conceptual pillars of solid-state physics had been firmly established before the 1940s. Since its inception in 1912, x-ray crystallography had turned the study of crystal lattices into a worldwide research area; and by 1940 the quantum mechanical study of electrons in solids was advanced to the point that its basic principles had already been incorporated into textbooks.[13] Plasma concepts had not yet found such comprehensive presentation, but nevertheless had reached the status of review articles.[14] Conceptually, the outbreak of the war interrupted ongoing research in solid-state and plasma phenomena by reorienting the work of its pioneers from basic science toward more practical ends. Alluding to Heraclitus's famous dictum ("War is father of all, ruler of all"), the war was *not* the "father" of modern solid-state and plasma physics, as far as the advancement of its concepts is concerned.

The same reorientation of research toward military purposes, however, created those new contexts from which the postwar scientific communities of both specialties derived their identities. Plasma research became tied to the study of thermonuclear fusion, solid-state physics to semiconductor electronics. In the former case, this new context had its roots in the quest for atomic bombs, in the latter, in the development of radar detectors. In the United States, both research efforts assumed a scale of heretofore unparalleled dimensions, and henceforth, the country would become the leader in plasma and solid-state research. At the same time, national security aspects began to play a role where previously only scientific or industrial interests had determined the pace of progress.

Although the Manhattan Project's role in thermonuclear fusion research was not as prominent as in the case of nuclear fission, there is more in it than Teller's early ideas about a hydrogen bomb. "Work on the deuterium bomb or Super project [the hydrogen bomb] was secondary, but continued throughout the course of the project," the official history says of this research. It was affirmed "that although the Super might not be needed as a weapon for the war, the Laboratory had a long-range obligation to continue this

---

[13] E.g., Nevill F. Mott and R. W. Gurney, *Electronic Processes in Ionic Crystals* (Oxford: Oxford University Press, 1940); Frederick Seitz, *The Modern Theory of Solids* (New York: McGraw Hill, 1940).

[14] E.g., Lewi Tonks, "Theory of Magnetic Effect in the Plasma of an Arc," *Physical Review*, 56 (1939), 360–73.

investigation."[15] In addition, plasma became an important topic of research in other bomb-related efforts. Ernest Lawrence's (1901–1958) Radiation Laboratory in Berkeley, for example, developed an alternative approach to isotope separation by bending ion beams in strong magnetic fields (the "calutron").[16] While trying to increase the output of such devices, a theorist on Lawrence's team, David Bohm (1917–1992), discovered an unstable behavior between the ions and their accompanying electrons; this "Bohm diffusion" would become a famous riddle for future efforts on the magnetic confinement of plasmas.[17]

Similar to the Manhattan Project's role in plasma physics, the Radiation Laboratory (RadLab) at MIT became a pacesetter for solid-state physics. Its major task, the development of microwave radar, involved the investigation of the rectifying properties of semiconductors for the detection of centimeter waves, which was beyond the capabilities of the traditional electronic valves. Bethe, a pioneer of the early quantum mechanical electron theory of solids, was one of the consultants for the RadLab's detector project, before he became the head of the theoretical division of the bomb project at Los Alamos; in 1942, he worked out a theory of the rectifying contact between a metal wire and a silicon crystal, which served as a lead for the development of point-contact detectors throughout the war. Specific investigations were farmed out by the RadLab to university laboratories. Karl Lark-Horovitz (1893–1958) and his group of graduate students at Purdue University, for example, performed experimental investigations on germanium as contractual work for the RadLab. Seitz did similar work concerning the properties of silicon at the University of Pennsylvania. Other important research on radar detectors was done at the Bell Laboratories, where a few years later the transistor was invented. One of the inventors, John Bardeen (1908–1991), denied that "the war had any direct effect on that," and concluded that for himself, "[t]he work during the war was a diversion from my earlier interest in fundamental semiconductor research." Nevertheless, the quest for radar detecting materials provided the context for the first coordinated research in those semiconductors which would become the catchwords for postwar semiconductor and solid-state physics: germanium and silicon.[18]

The war also had a considerable effect on the development of specific techniques and instruments for solid-state and plasma research. Microwave

---

[15] David Hawkins, *Project Y: The Los Alamos Story.* Part I: *Toward Trinity* (Los Angeles: Tomash Publishers, 1983), pp. 86–7; originally published as *Manhattan District History LAMS 2532*, Los Alamos Scientific Laboratory, 1961.

[16] John L. Heilbron and Robert W. Seidel, *Lawrence and His Laboratory: A History of the Lawrence Berkeley Laboratory*, vol. 1 (Berkeley: University of California Press, 1989), pp. 516–17.

[17] Richard F. Post, "Plasma Physics in the Twentieth Century," in *Physics in the Twentieth Century*, p. 1630.

[18] Ernest Braun, "Selected Topics from the History of Semiconductor Physics and Its Applications," in *Out of the Crystal Maze*, pp. 443–88, at 454–63.

radiation, for example, became an important probe for solids and plasmas in the arsenal of postwar laboratories. The same is true for another technological offspring of World War II, the nuclear reactor; the intensive neutron beams from research reactors were used in the most versatile technique of neutron diffraction, with applications to research on the structure of all kinds of materials, from polymer investigations in chemistry to the study of biomolecules or the analysis of the magnetic properties of solids.[19]

Although this technological offspring of the war should not be underrated, its social effects probably exerted the most pervasive influence on postwar science. In the United States, the experience of coordinated research, organized and funded by the military or the government through organizations such as the Office of Scientific Research and Development (OSRD), changed general awareness about the conduct of science. The "mutual embrace of science and the military" and the "crystallization of a strategic alliance," as was observed by historians of modern physics, "altered the character of science in a fundamental and irreversible way."[20]

FORMATIVE YEARS, 1945–1960

From this perspective, the growth in the fostering of research in plasmas and solid materials after World War II is just an illustration of a broader trend in the modern history of science. Comparing the military expenditures for research and development in the United States before and after World War II, Paul Forman found an increase from about $23 million in 1938, amounting to less than one-third of all federal expenditure for research and development (R&D), to more than $1,600 million in 1945, comprising about 90 percent of all federal R&D. He quotes RadLab scientist Jerrold Zacharias, who called World War II "a watershed for American science and scientists. It changed the nature of what it means to do science and radically altered the relationship between science and government[,] the military . . . and industry." This marked the beginning of the "megabuck era" for American science. Federal funds for R&D in the electronics industry between 1945 and 1960 increased tenfold. Similar growth rates can be observed in the number of physicists and other personnel engaged in electronic R&D during this period.[21] Radar,

---

[19] Stephen T. Keith and Pierre Quédec, "Magnetism and Magnetic Materials," in *Out of the Crystal Maze*, pp. 359–442; G. E. Bacon, ed., *Fifty Years of Neutron Diffraction: The Advent of Neutron Scattering* (Bristol, England: Adam Hilger, 1986).

[20] Silvan S. Schweber, "The Mutual Embrace of Science and the Military: ONR and the Growth of Physics in the United States after World War II," in *Science, Technology and the Military*, ed. Everett Mendelsohn et al. (Dordrecht: Kluwer, 1988), pp. 3–46; Paul K. Hoch, "The Crystallization of a Strategic Alliance: The American Physics Elite and the Military in the 1940s," ibid., pp. 87–118.

[21] Paul Forman, "Behind Quantum Electronics: National Security as Basis for Physical Research in the United States, 1940–1960," *Historical Studies in the Physical and Biological Sciences*, 18, no. 1 (1987), 149–229, here n. 5 and figs. 1 and 3.

in particular, proved to be a technology that found a host of applications in physical research after the war.[22]

The rise of solid-state physics, material science, plasma research, and possibly a couple of other specialties after World War II, therefore, was not merely a consequence of intrinsic developments in these areas. The United States had emerged from this war as the world's leading power, scientifically as well as militarily, and the dynamics of international competition during the Cold War provided enough impetus for this growth to take place in other countries as well. The growth of academic programs both in America and elsewhere created a host of new opportunities across all scientific disciplines. Cross-fertilization among various specialties happened here and there. Bohm's work on plasmas, for example, led to a many-body theory that helped to establish the theory of superconductivity. By such cross-fertilization the new and thriving field of cooperative phenomena emerged.[23] In addition, there were specific events that facilitated the formation of solid-state physics and plasma physics as distinct subdisciplines. In 1947, the invention of the point contact transistor opened a new era of solid-state electronics. The expansion of plasma physics was sparked in the early 1950s, when premature expectations of controlled thermonuclear fusion as an endless source of energy were raised. The processes by which solid-state and plasma physics became self-contained specialties were quite different and deserve separate analysis in detail.

Neither the timing nor the discovery of the point-contact transistor at the Bell Laboratories was accidental. The development of a microscopic theory for solid-state phenomena relevant to communication devices had been part of the Bell Labs research plans since the late 1930s. A host of military projects during World War II, connected with solid-state components for radar and communication systems, stimulated the rapid growth of technological semiconductor know-how. In 1945, the Physical Research Department of the Bell Labs was reorganized, and the exploration of solid-state properties became the task of a new Solid State Department, with the purpose of obtaining "new knowledge that can be used in the development of completely new and improved components and apparatus elements of communication systems." The new department was divided into several subgroups, each composed of a balanced mixture of specialists with quite different backgrounds. As the responsible manager observed, the efficiency of such an organization was proven by the big wartime laboratories at Los Alamos and MIT. Out of this plan for multidisciplinary teamwork and because of the strong focus on basic research in solid-state properties, a semiconductor subgroup was shaped, which represented the pertinent range of specialties for

---

[22] Paul Forman, "Swords into Ploughshares: Breaking New Ground with Radar Hardware and Technique in Physical Research after World War II," *Reviews of Modern Physics*, 67, no. 2 (1995), 397–455.
[23] Lillian Hoddeson et al., "Collective Phenomena," in *Out of the Crystal Maze*, pp. 489–616.

future discoveries: theoretical solid-state knowledge (William Shockley, John Bardeen), experimental physics (Walter Brattain, Gerald Pearson), physical chemistry (Robert Gibney), electronics engineering (Hilbert Moore), and general technical assistance (Thomas Griffith, Philip Foy). By the end of 1947, this group had assembled enough theoretical and technological expertise to repeat earlier efforts aimed at a "field effect amplifier," and on 16 December 1947, they were able to demonstrate, for the first time, a small amplification with a device in which a metallic conductor (gold) and a semiconductor (germanium) formed a suitable junction in order to produce the desired effect. The device was called the "point-contact transistor" and, after being kept "laboratory confidential" for seven months, it was reported by the press to be a "revolution in the electronics industry."[24]

This discovery was followed by a mushrooming of solid-state research, in both industrial laboratories and university departments. By June 1949, the military started to fund Bell Labs transistor research. Within the following ten years, this amounted to $8.5 million, representing about one-fourth of total Bell Labs expenditure on material development during this period. Within a few years after the discovery of the point-contact transistor, the basis for technological applications had been considerably broadened. New methods like zone refining (W. G. Pfann) and techniques of doping soon enabled the production of planar transistors, which in contrast to the clumsy device of the point-contact transistors offered the prospects of further miniaturization. In September 1951, a course on transistor physics and technology was held at the Bell Labs, which was attended by 121 military personnel, 41 university scientists, and 139 industrial researchers. In April 1952, a second transistor symposium was organized for an industrial audience. Twenty-six domestic and fourteen foreign firms sent delegates. Siemens in Germany and Philips in the Netherlands, for example, entered the "new semiconductor era" on this occasion.[25]

Plasma physics, too, had its key events by this time, although these were of a totally different character. During the early postwar years at Los Alamos (and probably various other secret weapon laboratories in the East and the West), the possibility of controlled thermonuclear fusion was being studied as a corollary to the development of the hydrogen bomb. A number of schemes for the magnetic confinement of hot plasmas were discussed, such as the proposals as early as 1945 of George Paget Thomson (1892–1975) and Peter

[24] Lillian Hoddeson, "The Discovery of the Point-Contact Transistor," *Historical Studies in the Physical and Biological Sciences*, 12, no. 1 (1981), 41–76.
[25] Ernest Braun, "Selected Topics from the History of Semiconductor Physics and Its Applications," *Out of the Crystal Maze*, pp. 443–88, here, 474–6; Ernest Braun and E. MacDonald, *Revolution in Miniature*, 2d. ed. (Cambridge: Cambridge University Press, 1982); Joop Schopman, "Philips' Antwort auf die neue Halbleiterära Germanium und Silicium (1947–1957)," *Technikgeschichte*, 50 (1983), 146–61; Michael Eckert and Helmut Schubert, *Kristalle, Elektronen, Transistoren: Von der Gelehrtenstube zur Industrieforschung* (Reinbek: Rowohlt, 1986; Am. translation: New York: American Institute of Physics, 1990).

Thoneman for "toroidal solenoids," resulting in experimental investigations of the "pinch effect" at Great Britain's central nuclear research facility in Harwell by 1949. These efforts led to the setting up of the first coherent program of controlled fusion research in Great Britain. At the same time, similar schemes seem to have been debated in Teller's "wild ideas" seminars at Los Alamos, and by Andrei Sakharov (1921–1989) and Igor Tamm in Russia (1895–1971).[26]

In 1951, the secret fusion research in the East and the West was stirred up by news from Argentina that a German physicist had achieved controlled nuclear fusion in a secret island laboratory for the dictator Juan Peron.[27] Although the news was received with skepticism (as it turned out, Peron had been the victim of a quack), it became the "proximate cause" of the U.S. fusion program, as historian Joan L. Bromberg acknowledged. At least it sparked the astrophysicist Lyman Spitzer, Jr.'s (1914–1997) interest in fusion research; he subsequently developed the so-called stellarator scheme of magnetic plasma containment. For some planners in the U.S. Atomic Energy Commission, the funding of this and other controlled fusion efforts fitted quite well into a program that could turn the deadly outlook of the thermonuclear research for the hydrogen bomb into the benefaction of endless energy for mankind. The AEC's fusion program ("Project Sherwood"), starting with a contract for $50,000 for Spitzer's stellarator scheme at Princeton University, expanded within the next few years into a large-scale effort. By 1957, more than $10 million had been spent in a half-dozen laboratories at Los Alamos, Livermore, Oak Ridge, Princeton, and Washington, where a variety of magnetic confinement techniques began to take shape.

By this time, the field as a whole had become reshaped. Nurtured in the secrecy of the weapons laboratories and within a framework of national security and the Cold War, fusion was now the catchphrase of the "peaceful atom." In his opening speech at the first International Conference on Peaceful Uses of Atomic Energy in Geneva in 1955, Homi Bhabha (1909–1966) ventured a prediction that a method would be found for liberating fusion energy in a controlled manner within twenty years. At this time, even the very existence of Project Sherwood was secret. The second Atoms-for-Peace Conference, held in Geneva from 1 to 13 September 1958, was used as the propagandistic vehicle to disclose its results to the public. Delegates from the Sherwood Project rivaled their colleagues from Russia, Great Britain, and other nations in explaining schemes to deal with plasma. Stellarator, pinch,

---

[26] John Hendry, "The Scientific Origins of Controlled Fusion Technology," *Annals of Science*, 44 (1987), 143–68; Joan Lisa Bromberg, *Fusion: Science, Politics, and the Invention of a New Energy Source* (Cambridge, Mass.: MIT Press 1982); I. N. Golowin and W. D. Schafranow, "Die Anfänge der kontrollierten Kernfusion," in *Andrej D. Sacharow: Leben und Werk eines Physikers in der Retrospektive seiner Kollegen und Freunde in der Akademie der Wissenschaften* (Heidelberg: Spektrum Akad. Verlag, 1991; Russian original Moscow: Nauka/Priroda, 1990), pp. 45–59.

[27] Mario Mariscotti, *El Secreto Atomico de Huemul* (Buenos Aires: Sudamericana/Planeta, 1985).

and magnetic-mirror machines were presented in museum-like exhibits. (The Russian tokamak scheme, a variant of the pinch, already existed but broke into prominence only ten years later.) The "watershed" year for international fusion research was 1958.[28]

Controlled fusion fueled plasma research, much as the transistor had rekindled the interest in solid-state physics a few years earlier. By the late 1950s, a solid-state community had emerged, with its special conferences, journals, and prizes. In 1952, for example, the Bell Laboratories endowed a solid-state physics prize to be named after the the lab's retired president, Oliver E. Buckley (awarded in 1953 to William Shockley [1910–1989]). Many industrial firms followed Bell's example of establishing solid-state physics divisions. In Germany, the Siemens-Schuckert company reorganized its research laboratory in 1949, and shortly afterward set up a department of solid-state physics under the direction of Heinrich Welker (1912–1981), where a new type of semiconductor (III-V compounds) was developed. Welker, like many of his American colleagues, had been involved in radar research during World War II.[29]

The emerging solid-state community, however, did not focus on semiconductors exclusively. In America, efforts to found a "metal physicists" organization eventually were combined with broader academic interests, embracing all solid matter. A most influential group for the integration of solid-state physics into a coherent framework, the Solid State Physics Panel of the Office of Naval Research (ONR), even endorsed plans for a concentrated funding of materials science. In 1960, backed by the post-Sputnik burst of funding research with military applications, the newly created Advanced Research Projects Agency (ARPA) of the Department of Defense, together with the Atomic Energy Commission and other agencies, took over these initiatives, which finally led to the foundation of a dozen laboratories for interdisciplinary materials research around the country, such as the Center for Materials Science and Engineering at MIT.[30] In plasma physics, the Geneva event in 1958 was followed in rapid succession by further international meetings and conferences, such as the "International Summer Course in Plasma Physics 1960" in Risø, Denmark, and the first of a series of fusion conferences organized by the International Atomic Energy Agency (IAEA) in 1961 in Salzburg, where the growing community of plasma physicists confirmed that the "classified years" were over.[31]

---

[28] Joan Lisa Bromberg, *Fusion: Science, Politics, and the Invention of a New Energy Source*, pp. 89–105.

[29] Michael Eckert, "Theoretical Physicists at War: Sommerfeld Students in Germany and as Emigrants," in *National Military Establishments and the Advancement of Science and Technology: Studies in 20th Century History*, ed. Paul Forman and Jose M. Sanchez-Ron (Dordrecht: Kluwer 1996), pp. 69–86.

[30] Spencer R. Weart, "The Solid Community," in *Out of the Crystal Maze*, pp. 617–69.

[31] Richard F. Post, "Plasma Physics in the Twentieth Century," pp. 1617–90.

CONSOLIDATION AND RAMIFICATIONS

The post-1958 period of controlled fusion research started like gold fever. The expectation of rapid success is evident in the hasty transformation of small-scale experimental devices into large machines, such as the "Model C" stellarator, with which the newly founded Princeton University Plasma Physics Laboratory (PPL) in 1961 entered its first decade. After physicists had explored the fundamentals of the stellarator plasma containment with tabletop models A and B under the code name "Matterhorn," the Model C was designed as a scale model of a reactor. It involved industrial contractors, a cooling tower with the capability of evaporating 200,000 gallons of water per day, the first computers that could run Fortran programs, special "shops" for dealing with microwaves, reactor engineering, and so on, including a variety of personnel with expert knowledge of half a dozen specialties. In short, the Model C stellarator represented a model for the new type of project-oriented research, to which plasma physics has become addicted throughout the world since then.[32]

Other countries were eager to climb on the bandwagon. In Germany, for example, a new Institute für Plasmaphysik (IPP) near Munich was founded in 1960 under the umbrella of the Max-Planck-Gesellschaft. With the benefit of hindsight resulting from the failures of others, stellarator research at the IPP did not immediately rush into a Model-C-like decade but focused on small-scale basic research. By the early 1960s it had become evident that the so-called Bohm diffusion, an unacceptably high loss of plasma ("pump out") from the magnetically confined toroidal volume, could not be overcome by enlarging the dimensions as in the Princeton Model C. The plasma physicists at the IPP, therefore, decided to analyze more carefully the pump out and other instabilities, for example, by experimenting with "quiet" alkali-model plasmas (instead of the hot hydrogen plasmas suitable for thermonuclear fusion) before entering the phase of large-scale projects. Theoretical progress (for example, the "Pfirsch-Schlüter theory" on the conditions for magnetohydrodynamic equilibrium in stellarators) and the demonstration of tolerable (that is, classical, non-Bohmian) loss rates of plasma in a small tabletop device (famous in the stellarator community as the "Munich mystery") rewarded this approach. Nevertheless, this only delayed the beginning of the big projects by a few years; when the IPP finally entered this phase by the late 1960s, it did so with more confidence. Whereas its Princeton rival gave up stellarator research when Russian experiments promised better results with a so-called tokamak device, the Munich "Wendelstein" stellarators came of age.[33]

---

[32] Earl C. Tanner, *The Model C Decade: An Informal History*, rev. ed. (Princeton, N.J.: Princeton University Plasma Physics Laboratory, 1982).

[33] Michael Eckert, "Vom 'Matterhorn' zum 'Wendelstein': Internationale Anstösse zur nationalen Grossforschung in der Kernfusion," in Michael Eckert and Maria Osietzki, *Wissenschaft für Macht*

The rush to controlled thermonuclear fusion research tended to push nonfusion-related plasma physics to the background. The physics Nobel Prize in 1970 for Hannes Alfvén's (1908–1995) "fundamental work in magnetohydrodynamics with fruitful applications in different parts of plasma physics" is a reminder of research unrelated to the post-1958 fusion fever. Alfvén's work dealt with astrophysical and geophysical plasma phenomena, such as sunspots, the aurora, and magnetic storms. In 1942, he had predicted a new type of "electromagnetic-hydrodynamic waves" in electrically conducting fluids, now known as "Alfvén waves." Such waves were observed in the ionospheric plasma after nuclear bomb tests in 1958 and were measured in space by the *Pioneer* and *Explorer* satellites. By the mid-1960s, Alfvén waves had become just another item among the various subspecialties of plasma physics. From an epistemological point of view, their existence apparently never was a matter of dispute – these waves were just a logical consequence of Maxwell's equations and hydrodynamics. Alfvén's "discovery," as well as his other predictions about space plasma phenomena, therefore, have been analyzed as a test case for philosophical theses about science.[34] From the perspective of the disciplinary growth of plasma physics, Alfvén's work illustrates another postwar context: With bomb tests, rockets, and satellites, Earth's ionosphere and the interplanetary plasma became the subject of experimental investigation, with hitherto unseen efforts. Here, too, an internationaliziation of originally military research took place after the late 1950s (the International Geophysical Year, 1957–8, may be taken as the turning point). As with other areas of large-scale military funding after World War II, geophysical and astrophysical plasma research went through an epoch of enormous growth and restructuring, and finally became consolidated under such new umbrellas as "geospace" and "space science."[35]

Solid-state physics, too, gathered renown, as is illustrated by the other half of the 1970 physics Nobel Prize, awarded to Louis Néel (b. 1904) "for his pioneering studies of the magnetic properties of solids." Néel's entrepreneurial activities for the installation of a nuclear research center at Grenoble (Centre d'Etudes Nucléaires de Grenoble, CENG) make evident that solid-state physicists did not content themselves with small-scale research. Like their American colleagues at the Oak Ridge or Argonne National Laboratories, by the 1960s physicists in France, Germany, and elsewhere had become beneficiaries of the nuclear boom, by constructing

---

und Markt. Kernforschung und Mikroelektronik in der Bundesrepublik Deutschland (Munich: C. H. Beck, 1989), pp. 115–37.

[34] Stephen G. Brush, "Prediction and Theory Evaluation: Alfvén on Space Plasma Phenomena," *Eos*, 71, no. 2 (1990), 19–33.

[35] C. Stewart Gillmor, "Geospace and its Uses: The Restructuring of Ionospheric Physics Following World War II," in *The Restructuring of Physical Sciences in Europe and the United States 1945–1960*, ed. Michelangelo De Maria et al. (Singapore: World Scientific, 1989), pp. 75–84; Paul Hanle and V. D. Chamberlain, eds., *Space Science Comes of Age* (Washington, D.C.: Smithsonian Institution Press, 1981).

ever-more-powerful high-flux neutron sources for the study of materials.[36] But solid-state physicists did not content themselves with the crumbs of bread falling from the abundant table of nuclear big science. With the zeal of latecomers, they initiated the foundation of their own centers for solid-state research. In 1961 in Germany, for example, it was argued in a report to the Deutsche Forschungsgemeinschaft "that for the industrial development of a country, solid-state physics plays at least a role comparable to that allotted recently to nuclear physics." Such was the beginning of a major effort that in 1969 was crowned with the foundation of the Max-Planck-Institut für Festkörperforschung in Stuttgart.[37]

Although solid-state physics, material research, and plasma physics had become firmly consolidated institutionally by the last third of the twentieth century, ramification into a host of subspecialties was giving these disciplines an appearance of patchwork. This outcome reflects historical roots in quite diverse environments, which have preserved disciplines' autonomy in some cases (magnetism, for example, did not become entirely subordinated to solid-state physics but is still considered an independent specialty). On the other side, the emergence of new specialties, such as nonlinear dynamics ("chaos theory"), tends to change conceptual borderlines. Pioneering contributions to nonlinear dynamics originated from the study of plasmas as well as solids. From a modern perspective, order–disorder phenomena, phase transitions, and other collective phenomena in matter are often better identified with specialties like self-organization, chaos, synergetics, or, more generally, nonlinearity, instead of being distinguished according to the solid, fluid, or plasma state of matter.

MODELS OF SCIENTIFIC GROWTH

The time is not yet ripe for a historical appraisal of how nonlinear physics has restructured solid-state and plasma physics during the last decades of the twentieth century. However, in view of the various crucial events, watersheds, consolidations, and ramifications in the development of these disciplines, this new trend once more brings to the fore some doubts about the established historiographical views on the growth of science. Usually, disciplinary or subdisciplinary growth is considered a process of specialization. Plasma and solid-state physics, however, acquired their identities less by differentiation than by integration. Furthermore, the Kuhnian model of scientific growth does not seem appropriate here. Neither in solid-state nor in plasma physics

[36] Dominique Pestre, *Louis Néel, le Magnétisme et Grenoble* (Paris: CNRS, 1990); Michael Eckert, "Neutrons and Politics: Maier-Leibnitz and the Emergence of Pile Neutron Research in the FRG," *Historical Studies in the Physical Sciences*, 19, no. 1 (1988), 81–113.
[37] Michael Eckert, "Grosses für Kleines – Die Gründung des Max-Planck-Instituts für Festkörperforschung," in *Wissenschaft für Macht und Markt*, pp. 181–99.

does the notion of scientific revolutions make sense; there were no switches of incommensurable paradigms – or too many of them on a smaller scale. Even less appropriate would be an evaluation in terms of corroboration and falsification in a Popperian framework or any other epistemological scheme that does not account for political and social environments. At the beginning of a new century, a retrospective view on the development and organization of our knowledge about matter (solid or otherwise) displays a patchwork-like image, which has been restructured almost permanently in response to the varying needs of our (post)modern society.

22

MACROMOLECULES
Their Structures and Functions

Yasu Furukawa

The concept of the macromolecule was formed and evolved within the framework of two sciences that emerged in the twentieth century: polymer chemistry (or macromolecular chemistry) and molecular biology. Over the past three decades, a large number of books have been published on the history of these two fields. While practicing scientists have provided their personal reminiscences as well as technical reviews, historians have shed light on intellectual, institutional, and industrial aspects of the history of these sciences.[1]

---

[1] On the history of polymer chemistry, see Hermann Staudinger, *From Organic Chemistry to Macromolecules: A Scientific Autobiography Based on My Original Papers,* trans. Jerome Fock and Michael Fried (New York: Wiley-Interscience, 1970); Frank M. McMillan, *The Chain Straighteners: The Fruitful Innovation: the Discovery of Linear and Stereoregular Synthetic Polymers* (London: Macmillan, 1979); Claus Priesner, *H. Staudinger, H. Mark und K. H. Meyer: Thesen zur Grösse und Struktur der Makromoleküle* (Weinheim: Verlag Chemie, 1980); Allan G. Stahl, ed., *Polymer Science Overview: A Tribute to Herman F. Mark* (Washington, D.C.: American Chemical Society, 1981); Raymond B. Seymour, ed., *History of Polymer Science and Technology* (New York: Marcel Dekker, 1982); Herbert Morawetz, *Polymers: Origins and Growth of a Science* (New York: Wiley, 1985); Peter J. T. Morris, *Polymer Pioneers: A Popular History of the Science and Technology of Large Molecules* (Philadelphia: Center for History of Chemistry, 1986); Raymond B. Seymour, ed., *Pioneers in Polymer Science* (Dordrecht: Kluwer, 1989); Herman F. Mark, *From Small Organic Molecules to Large: A Century of Progress* (Washington, D.C.: American Chemical Society, 1993); Yasu Furukawa, *Inventing Polymer Science: Staudinger, Carothers, and the Emergence of Macromolecular Chemistry* (Philadelphia: University of Pennsylvania Press, 1998).

On the industrial and technological aspects of polymer chemistry, see Raymond B. Seymour and Tai Cheng, eds., *History of Polyolefins: The World's Most Widely Used Polymers* (Dordrecht: Reidel, 1986); David A. Hounshell and John K. Smith, *Science and Corporate Strategy: Du Pont R&D, 1902–1980* (Cambridge: Cambridge University Press, 1988); S. T. I. Mossman and Peter J. T. Morris, eds., *The Development of Plastics* (Cambridge: Royal Society of Chemistry, 1994); Jeffrey I. Meikle, *American Plastic: A Cultural History* (New Brunswick, N.J.: Rutgers University Press, 1995).

On the history of molecular biology, see John Cairns, Gunther S. Stent, and James D. Watson, eds., *Phage and the Origins of Molecular Biology* (Cold Spring Harbor, N.Y.: Cold Spring Harbor Laboratory of Quantitative Biology, 1966); James D. Watson, *The Double Helix: A Personal Account of the Discovery of the Structure of DNA* (London: Atheneum, 1968); François Jacob, *The Logic of Life: A History of Heredity,* trans. Betty E. Spillmann (Princeton, N.J.: Princeton University Press, 1973); Robert Olby, *The Path to the Double Helix* (Seattle: University of Washington Press, 1974); Franklin H. Portugal and Jack S. Cohen, *A Century of DNA: A History of the Discovery of the Structure and Function of the Genetic Substance* (Cambridge, Mass.: MIT Press, 1977); Horace F. Judson,

The existing literature, however, has rarely covered both fields simultaneously. Just as polymer chemistry and molecular biology are separate disciplines that have demarcated the communities and goals of the practitioners, in like manner have their histories been compiled and treated in isolation. While historians have been eager to look into the origins of molecular biology, they tend to pay little attention, if any, to polymer chemistry.[2] The historical link between polymer chemistry and molecular biology is a subject yet to be explored. The macromolecule was a common conceptual ground that sustained the intellectual framework of the two sciences: Scientists of both fields sought a causal chain of evidence from macromolecular structures to their properties and functions. The development of the two sciences may well be seen as a process of elaboration of the macromolecular concept, as well as the "molecularization" of the physical and life sciences. In keeping with this perspective, this chapter will focus on the "science of macromolecules" from the 1920s to the 1950s.

FROM ORGANIC CHEMISTRY TO MACROMOLECULES

The term "macromolecule" (*Makromolekül*) was coined in 1922 by the German organic chemist Hermann Staudinger (1881–1965), professor at the Federal Polytechnic Institute in Zurich. He introduced this term to designate long-chain molecules, which constitute a class of substances with colloidal nature, exemplified by rubber, cellulose, starch, proteins, and plastics. They were known as "polymers" – so named after the Greek word for "many parts," as introduced in 1832 by the Swedish chemist Jöns J. Berzelius (1779–1848) – that is, bodies having molecules in which the same atomic groups are arranged repeatedly, without regard for the size of molecules. Staudinger's coinage of

---

The Eighth Day of Creation: The Makers of the Revolution in Biology (New York: Simon and Schuster, 1979); Salvador E. Luria, *A Slot Machine, A Broken Test Tube: An Autobiography* (New York: Basic Books, 1984); Francis Crick, *What Mad Pursuit: A Personal View of Scientific Discovery* (New York: Basic Books, 1988); Lily E. Kay, *The Molecular Vision of Life: Caltech, The Rockefeller Foundation, and the Rise of the New Biology* (Oxford: Oxford University Press, 1993); Max F. Perutz, *I Wish I'd Made You Angry Earlier: Essays on Science, Scientists, and Humanity* (New York: Cold Spring Harbor Laboratory Press, 1998); Michael Morange, *A History of Molecular Biology*, trans. Matthew Cobb (Cambridge, Mass.: Harvard University Press, 1998).
  Besides books, there are numerous papers and articles on the history of these fields. For brief surveys of the history of polymer chemistry, see Yasu Furukawa, "Polymer Chemistry," in *Science in the Twentieth Century*, ed. John Krige and Dominique Pestre (London: Harwood, 1997), pp. 547–63, and "Polymer Science: From Organic Chemistry to an Interdisciplinary Science," in *Chemical Sciences in the 20th Century: Bridging Boundaries*, ed. Carsten Reinhardt (Weinheim: Wiley-VCH, 2001) pp. 228–45. For historiographical overviews of molecular biology, see Robert Olby, "The Molecular Revolution in Biology," in *Companion to the History of Science*, ed. R. Olby, G. N. Cantor, J. R. R. Christie, and M. J. S. Hodge (London: Routledge, 1990), pp. 503–20; Pnina G. Abir-Am, "'New' Trends in the History of Molecular Biology," *Historical Studies of the Physical and Biological Sciences*, 26 (1995), 167–96.

[2] Robert Olby is perhaps the only exception: "The Macromolecular Concept and the Origins of Molecular Biology," *Journal of Chemical Education*, 47 (1970), 168–71 and "The Significance of the Macromolecules in the Historiography of Molecular Biology," *History and Philosophy of the Life Sciences*, 1 (1979), 185–98.

the term macromolecule was intended to depart from the current notion of polymers, which were taken to be composed of relatively small molecules. This is why Staudinger preferred to call his field of activity "macromolecular chemistry" (*makromolekulare Chemie*), rather than "polymer chemistry," although the latter expression is used more often in English-speaking countries today.[3]

When Staudinger proposed the macromolecular concept, there were two related views supporting polymers as small molecules. One was developed by colloid chemists, among them Wolfgang Ostwald (1883–1943) at Leipzig. The son of the celebrated physical chemist Wilhelm Ostwald, Wolfgang defined colloids as dispersed systems consisting of particles that are too small to be seen and too large to be called molecules – a world he called "neglected dimensions." A colloid was a physical state of matter into which any substance might be brought; and under appropriate conditions, any substance could form a colloidal solution. There existed, he claimed, no definite connection between molecular structure and a colloidal state. Properties of colloids were determined by the physical state outside the molecule, for example, the degree of dispersion. Influential colloid chemists, such as Richard A. Zsigmondy (1865–1929) and Herbert Freundlich (1880–1941) in Germany and Wilder D. Bancroft (1867–1953) in the United States, shared with Ostwald the idea of colloid as a state of matter.

Complementary to the colloidalist views was the so-called aggregate theory developed by such notable organic chemists as Carl D. Harries (1866–1923), Hans Pringsheim (1876–1940), Rudolf Pummerer (1882–1973), Max Bergmann (1886–1944), Kurt Hess (1888–1935), and Paul Karrer (1889–1971). According to this theory, colloidal substances, such as cellulose, rubber, starch, proteins, resins, and synthetic polymers, were the physical aggregates of small cyclic molecules held together by certain intermolecular forces. Rubber, for example, was believed to be composed of an eight-membered cyclic molecule, consisting of two isoprene units. Colloid particles were seen as the aggregates of these rubber molecules, held together by the weak "partial valences" that were derived from the carbon–carbon double bonds in the cyclic molecule. The "molecular weight" meant the weight of a colloidal particle. The apparent high molecular weights of polymers, measured by such available means as freezing-point depressions and osmotic pressures, were not taken literally as the weights of the real chemical molecules but as those of their physical aggregates.[4]

Staudinger was uncomfortable with the upsurge of colloidalist views that appeared to be invading his territory, organic chemistry. Nor was he convinced by the aggregate theory proposed by his fellow organic chemists. He

---

[3] Hermann Staudinger and Jakob Fritschi, "Über Isopren und Kautschuk: Über die Hydrierung des Kautschuks und über seine Konstitution," *Helvetica Chimica Acta*, 5 (1922), 785–806. On the origins of the macromolecular theory, see Furukawa, *Inventing Polymer Science*, chap. 2.

[4] For example, Carl D. Harries, "Zur Kenntnis der Kautschukarten: Ueber Abbau und Constitution des Parakautschuks," *Berichte der deutschen chemischen Gesellschaft*, 38 (1905), 1195–203.

was a firm believer in classical organic structural chemistry, developed by August Kekulé (1829–1896) and others in the middle of the previous century. Properties of matter, Staudinger thought, stemmed not only from the kinds of constituent elements but also from the topological arrangement of atoms in the molecule, namely, the molecular structure. Polymers were not exceptional. Colloid particles were, in many cases, themselves macromolecules that were composed of between 10^3 and 10^9 atoms linked together by the Kekulé valence bonds. Colloidal properties of polymers were determined by the structure and large size of these molecules. Colloidal phenomena should not be interpreted on principles of colloid chemistry but on those of organic chemistry. Ostwald's "neglected dimensions" were, he claimed, not colloid chemists' territory but the new world that organic chemists must explore.

Unlike ordinary organic compounds, colloids were not susceptible to established methods of purification and isolation, as they could hardly be crystallized from a solution or distilled without decomposition. Contemptuously dubbed "grease chemistry," the study of these gluey colloid materials did not attract many organic chemists. Adolf von Baeyer (1835–1917), the German authority in classical organic chemistry, once lamented at the turn of the century that now that studies of most organic compounds were completed, "the field of organic chemistry is exhausted... and then all that remains is the chemistry of grease."[5] Challenging Baeyer's premature judgment of the approaching decline of organic chemistry, Staudinger declared in 1926: "Despite the large number of organic substances which we know today, we are only standing at the beginning of the chemistry of true organic compounds and have not reached anywhere near a conclusion."[6]

Upon moving to the University of Freiburg in the summer of 1926, Staudinger met considerable opposition to his theory. One blow came from x-ray crystallography. Physical chemists and physicists had begun applying the x-ray method to the investigation of polymers. This type of research was especially advanced at the Kaiser Wilhelm Institute for Fiber Chemistry, founded in 1920 in Berlin-Dahlem. Silk and a part of cellulose became known to exhibit a crystalline form to which x-ray analysis was applicable. It was found that rubber, when stretched, showed a crystalline form exhibiting a fiber diagram similar to those of silk and cellulose. The result indicated that rubber, silk, and cellulose had a common structure, although it remained an open question why stretching brought about crystallization in rubber. Meanwhile, x-ray experts observed that unit cells – the recurring atomic groups in the crystalline lattice – of polymers were as small in size as ordinary molecules. There was a common assumption among crystallographers

---

[5] Quoted in Joseph S. Fruton, *Contrasts in Scientific Style: Research Groups in the Chemical and Biochemical Sciences* (Philadelphia: American Philosophical Society, 1990), p. 162.

[6] Hermann Staudinger, "Die Chemie der hochmolekularen organischen Stoffe im Sinne der Kekuléschen Strukturlehre," *Berichte der deutschen chemischen Gesellschaft*, 59 (1926), 3019–43, at p. 3043. Translated by the author unless otherwise noted.

that the molecule could not be larger than the unit cell. Thus, Reginald O. Herzog (1878–1935), director of the Institute for Fiber Chemistry, and his followers concluded that the molecular size of the polymer must be small. This conclusion was in contradiction to Staudinger's concept.

Staudinger had to accumulate a mass of evidence for his theory in the face of vigorous criticism. The conservative organic chemist disdained physical chemistry and distrusted physical methods, such as x-ray crystallography. Instead, Staudinger's methods for demonstration were purely organic-chemical, mobilizing his skills in organic analysis and synthesis. To take one example, the aggregate theory of rubber indicated that hydrogenation of rubber would yield a normal small-molecule substance, because saturation of the double bonds in the cyclic rubber molecule would occur and destroy the partial valences between the molecules. Staudinger performed this experiment but obtained a contradictory result. The properties of the hydro-rubber were similar to those of the original natural rubber; the hydro-rubber did not crystallize but produced a colloidal solution like rubber. Therefore, colloidal particles of rubber could not be the aggregates of small molecules held together by partial valences, but were themselves big molecules.[7]

In the early 1930s, the climate of opinion was shifting to Staudinger's side. The causes of this shift were manifold. The ultracentrifuge, developed in the mid-1920s by the Swedish physical chemist The(odor) Svedberg (1884–1971) at Uppsala, made it possible to estimate the molecular weight of some proteins even in the range of several millions.[8] Other significant support came from the other side of the Atlantic. The American organic chemist Wallace H. Carothers (1896–1937) at the DuPont Company demonstrated the macromolecularity of synthetic polymers through his intensive research on the mechanism of polymerization reactions. His study of macromolecular syntheses, which resulted in the synthetic fiber nylon and the synthetic rubber neoprene, paved the way for the birth of the science-based polymer industry that ushered in the "plastics age" in the postwar period.[9]

The role of the Austrian chemist Herman F. Mark (1895–1992) was also seminal to the triumph of the macromolecular concept. Mark was among the earliest who rejected as a misconception the assumption that polymer molecules must be smaller than the x-ray elementary cell of crystals. Trained in organic chemistry, Mark became a physical chemist when he studied the x-ray diffraction of polymers at Herzog's Institute. Upon moving to the

---

[7] Staudinger and Fritschi, "Über Isopren und Kautschuk."
[8] On Svedberg's ultracentrifugal study, see Boelie Elzen, "The Failure of a Successful Artifact: The Svedberg Ultracentrifuge," in *Center on the Periphery: Historical Aspects of 20th-Century Swedish Physics*, ed. Svante Lindqvist (Canton, Mass.: Science History Publications, 1993), pp. 347–77.
[9] On Carothers and his work at DuPont, see Hounshell and Smith, *Science and Corporate Strategy*, chaps. 12 and 13; Matthew E. Hermes, *Enough for One Lifetime: Wallace Carothers, Inventor of Nylon* (Washington, D.C.: American Chemical Society and Chemical Heritage Foundation, 1996); Furukawa, *Inventing Polymer Science*, chaps. 3 and 4.

I. G. Farben Industrie in 1927, Mark set up an improved x-ray apparatus and continued his studies of structures of cellulose and other fibers.[10]

From 1928 to 1930, he and his colleague, Kurt H. Meyer (1883–1952), developed the so-called new micelle theory, which made a compromise between Staudinger's macromolecular concept and the aggregate theory. According to Mark and Meyer, colloid particles in solution were not themselves macromolecules but "micelles," which were the aggregates of long-chain molecules held together by the Van der Waals–type micellar forces. Although Staudinger criticized their eclecticism (and, indeed, the proposed molecular sizes would prove to be too small), the Mark-Meyer theory played a significant role in disseminating the concept of long-chain molecules, drawing favorable attention even from the advocates of the aggregate theory. After all, the Faraday Society General Discussion meeting on "The Phenomena of Polymerisation and Condensation," held at Cambridge in 1935, which Staudinger, Carothers, Mark, and Meyer attended as guest speakers, illustrated a wide acceptance of the macromolecular concept in the chemical community.

As one of his former disputants put it when Staudinger won the Nobel Prize in 1953, "Staudinger succeeded where others failed because he knew and believed in organic chemistry."[11] Staudinger clung to the classical notion of molecule, regarding the molecule as the only entity from which all properties of the substance arise. He maintained that properties originate in the molecular structure.

The triumph of Staudinger's macromolecular theory cannot, however, be sufficiently explained simply in terms of his return to classical organic chemistry. Whereas organic chemists had long believed that a pure substance consisted of a single and definite molecular component, Staudinger gave up this belief. Macromolecular substances were composed of molecules of various, nonidentical sizes. Their molecular weights, therefore, could only be expressed by average values, rather than by precise numbers. In this respect, the macromolecular concept represented a sharp break from the classical notion of chemical compounds.

Staudinger also realized that the shapes of macromolecules determined the properties of polymers, such as fibrousness, elasticity, tensile strength, viscosity, and swelling phenomena. He thought that cellulose and many other macromolecular compounds were linear. Because of this shape, they were fibrous and tough and dissolved with considerable swelling to give gel solutions of high viscosity. The shape of macromolecules, he said, affected the physical and chemical properties of the substances considerably more strongly than was the case with ordinary compounds. With his emphasis

---

[10] See Stahl, *Polymer Science Overview*; Herman F. Mark, "Polymer Chemistry in Europe and America – How It All Began," *Journal of Chemical Education*, 58 (1981), 527–34, and *From Small Organic Molecules to Large*.

[11] Quoted in "Nobel Prize to German, Hollander," *Chemical and Engineering News*, 31 (1953), 4760–1, at p. 4761.

on molecular size and shape in explaining properties, Staudinger departed from classical structural chemists, who confined their studies to the atomic arrangement inside the molecules.

Staudinger claimed that the macromolecule – when viewed as a whole – exhibited its own unique properties:

> Molecules as well as macromolecules can be compared with buildings which are built essentially from a few types of building stones: carbon, hydrogen, oxygen, and nitrogen atoms. If only 12 or 100 building units are available, then only small molecules or relatively primitive buildings can be constructed. With 10,000 or 100,000 building units an infinite variety of buildings can be made: apartment houses, factories, skyscrapers, palaces, and so on. Constructions, the possibilities of which cannot even be imagined, can be realized. The same holds for macromolecules. It is understandable that new properties will therefore be found which are not possible in low molecular [weight] materials.[12]

Qualities came from the whole, rather than the isolated parts. Thus, macromolecular compounds exhibited properties that could not be predicted even by a thorough study of small-molecular substances. With this conviction, Staudinger considered macromolecular chemistry a new field of organic chemistry, rather than a part of classical organic chemistry.

PHYSICALIZING MACROMOLECULES

Staudinger explored macromolecules only from the standpoint of organic chemistry, and he continued to insist that this field was a new branch of organic chemistry. By the late 1930s, however, it became clear that organic chemistry alone could not solve the whole problem of polymers, and that the structural approach had its limits. For example, macromolecules move. The dynamics of macromolecules would prove to be a substantial aspect that the organic chemists' vision of static molecules had failed to take into account. Once the macromolecular concept received a wide acceptance by the mid-1930s, physical chemists and physicists began to take up this subject, finding ample room for application of their methods and theories to polymers. The physics and physical chemistry of polymers were beginning to flower.

The viscosity law, which Staudinger had introduced in 1930, turned out to trigger the rise of the physics and physical chemistry of macromolecules. Viscometry did not remain merely a technique for measuring molecular weight, but also posed the question of what macromolecules were really like. According to Staudinger's law, viscosity was in direct proportion to molecular size. The implication was clear to him: Linear macromolecules

---

[12] Staudinger, *From Organic Chemistry to Macromolecules*, p. 92.

have a thin, rigid rod shape like a glass fiber in solution. Long rigid rods of macromolecules would move across a flowing liquid, rotating as they moved in a disklike plane. The intrinsic viscosity was directly proportional to the volume of disklike cylinders swept out by linear macromolecules.

Staudinger's concept of rigid macromolecules soon ignited strong opposition from physical chemists and physicists. Mark claimed that Staudinger's concept of sticklike molecules contradicted all the basic requirements of physical chemistry. The bent form of macromolecules, Mark said at the 1935 Faraday Society meeting, was "in fairly good agreement with all experimental evidence... and remains at the same time in concordance with fundamental statistical considerations and with the principle of the free rotation round the single carbon bond."[13]

The Swiss physicist-turned-physical chemist Werner Kuhn (1899–1963) developed views similar to Mark's. He assumed that the macromolecule was partially rigid but, as a whole, flexible; that is, the separate rigid links of the molecular chain could rotate freely around single chemical bonds in relation to each other. He envisaged the macromolecule as a flexible, coiled chain-molecule analogous to a pearl necklace.

Once physicists and physical chemists had adopted the concept of macromolecules, they were fascinated by the very complexities peculiar to polymers. Here they found a great opportunity to exploit their mathematical and physical methods. The study of the distribution of molecular weights, sizes, and shapes and of the behavior of macromolecules seemed amenable to probability, statistics, kinetics, thermodynamics, and hydrodynamics. The period between the late 1930s and 1940s marked an exciting formative stage "when polymer science looked easy," as American physical chemists recalled.[14]

The concept of flexible molecular chains, as well as the statistical and kinetic approaches, led to new interpretations of the origin of physical properties of polymers. For example, by the early 1940s, the cause of rubber elasticity was thermodynamically and statistically explained by Meyer, Mark, Kuhn, and others. The kinetic units of macromolecules were regarded as segments, rather than whole molecules, which could only move sluggishly because of their great sizes. Owing to the frequent, free movements of many thousands of individual segments ("microbrownian motions"), the macromolecule could take a wide variety of irregularly contorted configurations. The probability of the random motions and shapes of the macromolecule was statistically calculated. Probability of the stretched state of the macromolecule would be nearly zero, while a thread-ball-like state would be highly probable.

Rubber, when relaxed, tended to be restored to its original state, because the stretched macromolecule tended to return to its innate thread-ball shape.

---

[13] Herman F. Mark, discussion in *Transactions of the Faraday Society*, 32 (1936), pt. 1, p. 312.
[14] Walter H. Stockmayer and Bruno H. Zimm, "When Polymer Science Looked Easy," *Annual Review of Physical Chemistry*, 35 (1984), 1–21.

This was in harmony with the second law of thermodynamics: In nature, entropy increased from the molecularly ordered system (stretched state) to the molecularly disordered system (rolled state). Rubber elasticity increased as the temperature increased, because the higher the temperature, the greater the microbrownian motions. The fact that rubber, when stretched and kept cool, exhibited a fibrous x-ray crystallographic pattern could now be explained as a phenomenon in which the stretched macromolecules were arranged in one direction in an orderly manner and held by an intermacromolecular force in a crystalline form like a fiber. When heated, rubber immediately lost its crystalline form and shrank because heat increased the microbrownian motions enough to break the intermacromolecular force, and the macromolecules shrank to take the original thread-ball-like shape.

The American government's wartime synthetic rubber research program, launched in 1942 in the face of Japan's occupation of the Pacific area, provided chemists and physicists with unique opportunities to work closely together on polymers.[15] The case of Peter J. W. Debye (1884–1964) illustrates how a physicist became involved in polymer research. When he joined the rubber program in 1943, it was urgent for the members to develop a new way to obtain the accurate molecular weight of rubber. Debye was quick to recall a 1910 paper by Albert Einstein (1879–1955), which suggested that fluctuations of the solute concentration in a solution causes light scattering. The relationship between the concentration and the refractive index of the solution could be known empirically; the molecular weight of the solute could then be obtained by measuring the solution's turbidity at different concentrations. In cooperation with industrial chemists, Debye could develop mathematical equations on light scattering to calculate molecular weights of various polymers, a method that allowed a quick and accurate molecular weight determination of polymers.

By the 1950s, physics and physical chemistry had proven not to be inveterate foes of organic chemistry but its essential complementary partners in perfecting the science of polymers. At the same time, the polymer community was able to expand the size and scope of its professional activity by winning over to its side those physical chemists and physicists who now could share a disciplinary identity as polymer scientists.

EXPLORING BIOLOGICAL MACROMOLECULES

At mid-century, a number of chemists and physicists, often not trained in biology, had migrated into the life sciences, as biology looked like the new frontier of science. In 1944, the Viennese physicist Erwin Schrödinger

---

[15] On this program, see Peter J. T. Morris, *The American Synthetic Rubber Research Program* (Philadelphia: University of Pennsylvania Press, 1989).

(1887–1961) had published an influential work, *What Is Life?*, in which he suggested that the new physics of quantum mechanics could explain the new results in genetics. He also noted:

> Organic chemistry, indeed, in investigating more and more complicated molecules, has come very much nearer to that "aperiodic crystal" [chromosome fiber] which, in my opinion, is the material carrier of life. And therefore it is small wonder that the organic chemist has already made large and important contributions to the problem of life, whereas the physicist has made next to none.[16]

What role this little book, and physics more generally, actually played in the emergence of molecular biology has been controversial among biologists and historians of science.[17] Whatever the outcome, Schrödinger intended to draw physicists' attention to the problem of life to which, according to him, organic chemists had already made significant contributions. The book at least inspired such young physicists as Francis H. C. Crick (b. 1916) and Maurice H. F. Wilkins (b. 1916) to convert to molecular biology.

Three years after the appearance of the Schrödinger book, Staudinger wrote an ambitious book, *Makromolekulare Chemie und Biologie,* in which the organic chemist did not fail to cite *What Is Life?* The idea of the application of physics to biology was by no means Staudinger's concern, but he argued that the new science of macromolecular chemistry was the key to understanding biological phenomena. It was essential, he claimed, to understand life processes on the basis of the macromolecular concept. He explained the biological implications of macromolecules by utilizing the isomer concept; that is, compounds of the same composition having the same molecular weight could exhibit different properties in accordance with their structural differences. Due to the great size of macromolecules, there was an infinite number of structural possibilities for molecules. This was especially true for protein molecules; even slight differences in the structure could yield different biological properties. Staudinger also discussed deoxyribonucleic acid (DNA) as a biologically important macromolecular substance.[18] Neither Schrödinger nor Staudinger ever practiced molecular biology, but as a physicist and a chemist, respectively, they publicized a new vision of biology in which the physical sciences could play a fundamental role in the life sciences.

---

[16] Erwin Schrödinger, *What Is Life? The Physical Aspect of the Living Cell* (Cambridge: Cambridge University Press, 1944), p. 5.

[17] Robert Olby, "Schrödinger's Problem: What Is Life?" *Journal of the History of Biology,* 4 (1971), 119–48; E. J. Yoxen, "Where Does Schrödinger's 'What Is Life?' Belong in the History of Molecular Biology?" *History of Science,* 17 (1979), 17–52; Evelyn F. Keller, "Physics and the Emergence of Molecular Biology: A History of Cognitive and Political Synergy," *Journal of the History of Biology,* 23 (1990), 389–409; Morange, *A History of Molecular Biology,* pp. 67–78, 99–101.

[18] Hermann Staudinger, *Makromolekulare Chemie und Biologie* (Basel: Wepf Verlag, 1947), pp. 1–11, 48.

Proteins and DNA were known to be the major constituents of the chromosome in the nuclei of all living cells. They were the major focus of interest in early molecular biology, which was interdisciplinary in origin, advanced by chemists, physicists, and biologists. It was a grand fusion of the methods, techniques, and concepts of organic chemistry, polymer chemistry, biochemistry, physical chemistry, x-ray crystallography, genetics, and bacteriology. In the 1930s and 1940s, there were several research schools with varied approaches and concerns, though they were often isolated from one another. Prominent among them were the structuralist schools that saw the knowledge of precise three-dimensional molecular structure, drawn particularly from x-ray data, as the key to understanding biological functions, an approach that was inherited from organic structural chemistry and x-ray crystallography. The definition given by William T. Astbury (1898–1961), a British propagandizer for the field, represents the stance of this school: "Molecular biology is predominantly three-dimensional and structural.... It must at the same time inquire into genesis and function."[19] To this tradition belonged Linus C. Pauling (1901–1994), Max F. Perutz (b. 1914), and John C. Kendrew (b. 1917), who gained enormous success in elucidating the three-dimensional structures of proteins.

As early as the 1910s, the German organic chemist Emil Fischer (1852–1919) had shown that proteins were made of polypeptides in which many different amino acids were linked together. While considering proteins to be "giant molecules" (*Riesenmoleküle*), Fischer never conceived of organic compounds of a molecular weight greater than 5,000.[20] During the 1920s, polymer chemists, such as Staudinger, Mark, Meyer, and Carothers, had no difficulty extending their views on naturally occurring organic polymers like rubber and cellulose to proteins. They argued that proteins were made up of macromolecules, far larger molecules than Fischer had imagined.

During the 1920s there was lack of communication between polymer scientists and protein researchers (then consisting mainly of biochemists and colloid chemists). Svedberg, for example, was totally unaware of Staudinger's work on macromolecules when he embarked on his protein research in the mid-1920s. It is striking that articles and textbooks on biochemistry written around 1930 rarely cited polymer chemists' views of protein molecules. Although some biochemists independently conceived proteins to be big molecules from available means of molecular weight measurement, they rarely addressed the details of the large molecular structures and their relations to properties and functions. After all, few of them knew how heatedly these issues were debated in general terms by polymer organic chemists and physical chemists. Only in the late 1930s did polymer researchers' issues become

---

[19] William T. Astbury, "Molecular Biology or Ultrastructural Biology?" *Nature*, 190 (1961), 1124.
[20] Emil Fischer, "Synthesis of Depsides, Lichen-Substances, and Tannins," *Journal of the American Chemical Society*, 36 (1914), 1170–1201, at p. 1201.

known to many protein specialists through personal communications and publications and via interdisciplinary scientific meetings (such as the Faraday Society General Discussions and the Royal Society symposiums), which brought together scientists from various countries and different disciplines.[21]

THE STRUCTURE OF PROTEINS: THE MARK CONNECTION

Most of the first generation of molecular biologists embarked on their researches after a fair recognition of the macromolecular concept had taken hold in the scientific community. From the outset of their studies, they took for granted the macromolecularity of proteins and were immune to the colloidalist views. It is important to emphasize some early contacts between polymer chemists and future trailblazers of molecular biology, which resulted in remarkable consequences in the history of molecular biology.

Mark inspired the young American physical chemist Pauling to study biological macromolecules when the latter visited his laboratory in Germany in 1930. Already known for his application of quantum mechanics to the chemical bond, Pauling learned from Mark the x-ray analysis of polymers (such as crystallized rubber and fibrous proteins), as well as recent developments in the chemistry of macromolecules. Mark explained to Pauling, much to the latter's edification, his opinion that the elasticity of rubber was due to a spiral shape of its macromolecules. Pauling was also fascinated by Mark's theory of the structure of fibrous proteins, a theory based on x-ray fiber diagrams, as well as considerations of the length of chemical bonds, bond angles, and intermolecular forces. In a 1932 paper, Mark, with Meyer, suggested that the proteins were large molecules with polypeptide chains attracted to one another by forces between the C=O groups and the NH groups of adjunct chains. His conclusions, as well as techniques, triggered Pauling's research on protein structure that began in the mid-1930s at California Institute of Technology under a grant from the Rockefeller Foundation, a principal patron of molecular biology from the 1930s to the 1950s.[22]

---

[21] For a discussion of the historical significance of the macromolecular concept in protein research, see John T. Edsall, "Proteins as Macromolecules: An Essay on the Development of the Macromolecule Concept and Some of Its Vicissitudes," *Archives of Biochemistry and Biophysics*, Supp. 1, (1962), 12–20. On the history of biochemistry, see Joseph S. Fruton, *Molecules and Life: Historical Essays on the Interplay of Chemistry and Biology* (New York: Wiley-Interscience, 1972). Cf. Charles Transford and Jacqueline Reynolds, "Protein Chemists Bypass the Colloid/Macromolecule Debate," *Ambix*, 46 (1999), 33–51.

[22] Linus Pauling, "Herman F. Mark and the Structure of Crystals," in *Polymer Science Overview*, pp. 93–9. The impact of the Rockefeller Foundation's grants on the rise of molecular biology has been an issue of heated debate among historians: Robert E. Kohler, "The Management of Science: Warren Weaver and the Rockefeller Foundation Programme in Molecular Biology," *Minerva*, 14 (1976), 279–306; Pnina Abir-Am, "The Discourse of Physical Power and Biological Knowledge in the 1930s: A Reappraisal of the Rockefeller Foundation's 'Policy' in Molecular Biology,"

Using x-ray diffraction to determine the bond length and angles between atoms, and building molecular models, Pauling and his colleagues were able to develop a picture of the three-dimensional structure of protein molecules, an achievement that culminated in their landmark 1951 papers. They demonstrated that the long polypeptide chain of proteins was folded into a coiled configuration as the secondary structure, which they called "alpha helix," and they showed that the folds were maintained by the weak interactions, hydrogen bonds, between the amino-acid groups. The alpha-helix concept satisfactorily explained the behavior of proteins, including denaturation and renaturation – phenomena that had captured Pauling's interest from an early period. Heat and acidity would cause denaturation because they broke the hydrogen bond to unfold the polypeptide chain, and it thus lost its biological activity. Renaturation was the reverse process, in which the chain was folded again into its original functional shape. Pauling's method of model building, as well as his concept of the alpha helix as a basic configuration of protein macromolecules, provided a basis for the structurist approach to proteins and DNA.

Pauling prided himself on his dual capacity as a structural chemist and an x-ray crystallographer. He sarcastically saw the rival British structurist school as a group of physicists who knew little about structural chemistry.[23] Actually, the British school had a few eminently trained chemists. Stimulated by the German x-ray work at the Kaiser Wilhelm Institute for Fiber Chemistry, the British school of x-ray crystallography of biological materials had been initiated in the 1930s by Astbury, John D. Bernal (1901–1971), and their pupils. Financed by the Rockefeller Foundation, Astbury, a textile physicist at the University of Leeds, conducted an x-ray study on the structure of keratin, a protein that forms the major constituent of wool and hair. He attributed the elasticity and shrinkability of wool and hair to the folded configuration of the keratin molecule.

The Cambridge physicist Bernal was among the first to obtain clear images of x-ray diffraction by protein crystals. His picture of the enzyme pepsin impressed Mark, then professor at the University of Vienna, who met Bernal at the 1935 Faraday Society meeting at Cambridge. It was Mark who persuaded his chemistry student, Perutz, to join Bernal's laboratory. Perutz went to Cambridge the following year and stayed there during and after the war, eventually working with the x-ray physicist Lawrence Bragg (1890–1971) at the Cavendish Laboratory.[24]

Kendrew had been trained in physical chemistry and became Perutz's collaborator at the Cavendish after the war. The two chemists embarked on painstaking investigations of the structure of two related proteins,

---

Social Studies of Science, 12 (1982), 341–82, and "Responses and Replies," *Social Studies of Science,* 14 (1984), 225–63; Kay, *The Molecular Vision of Life.*
[23] Linus Pauling, "How My Interest in Proteins Developed," *Protein Science,* 2 (1993), 1060–3, at p. 1063.
[24] Olby, *The Path to the Double Helix,* p. 263.

hemoglobin (the oxygen carrier present in red blood cells) and myoglobin (a protein in muscle that stores oxygen within the cell). Using molecular models as well as computers for mathematical analysis of x-ray photographs, they were finally able to elucidate the immensely complex, precise, three-dimensional structures by 1960.[25] The myoglobin molecule, one-fourth the size of the hemoglobin molecule, contained one oxygen-binding iron atom. In both proteins the iron atom was located in a "heme group," an active site of the molecule that could bind and release oxygen. The peculiar physical shape and its configurational change facilitated the binding of oxygen with the heme group. The cylindrical parts of the myoglobin molecule proved to correspond in structure to Pauling's alpha helix.

Many scientists had sought to find some simple, regular pattern in the way protein molecules are arranged. From the ultracentrifugal data, Svedberg, for instance, found "an unmistakable regularity" that all proteins were multiples of a "subunit" with a molecular weight of about 34,500 (later altered to 17,600). Svedberg's subunit theory (which would later prove to be an illusion) inspired Dorothy M. Wrinch (1894–1976), a Cambridge mathematician who was fascinated by the problem of protein structure, to propose that proteins were composed of compact, globule-shaped, honeycomb-like molecules, consisting of cycled polypeptides based on hexagons. After all, the Wrinch model, based on her "geometrical instincts and deductions," possessed a degree of elegance and symmetry marvelous enough to capture the attention of eminent scientists such as Irving Langmuir (1881–1957).[26]

Now Kendrew and Perutz's work showed that this was not the case. Although the whole molecular arrangement of myoglobin, for example, looked highly compact and globular, the "most striking features" of the molecule, Kendrew stated, were "its irregularity and its total lack of symmetry."[27] As more protein structures were disclosed, they proved to be equally irregular and asymmetrical, though differing in detailed pattern from myoglobin. They were far from elegant; or as John Edsall and David Bearman put it more politely, "these strange molecular shapes seemed like nothing except perhaps some of the abstract creations of modern art."[28]

---

[25] See Max F. Perutz, "Molecular Biology in Cambridge," in *Cambridge Minds*, ed. Richard Mason (Cambridge: Cambridge University Press, 1994), pp. 193–203, and *I Wish I'd Made You Angry Earlier*, pp. 255–77; Harmke Kamminga, "Biochemistry, Molecules and Macromolecules," in *Science in the Twentieth Century*, ed. Krige and Pestre, pp. 525–46.

[26] Quotation from Dorothy C. Hodgkin's obituary of Wrinch, *Nature*, 260 (1976), 564. On Wrinch, see M. M. Julian, "Dorothy Wrinch and the Search for the Structure of Proteins," *Journal of Chemical Education*, 61 (1984), 890–2; Pnina G. Abir-Am, "Synergy or Clash: Disciplinary and Marital Strategies in the Career of Mathematical Biologist Dorothy Wrinch," in *Uneasy Careers and Intimate Lives: Women in Science, 1789–1979*, ed. Pnina G. Abir-Am and Dorinda Outram (New Brunswick, N.J.: Rutgers University Press, 1987), pp. 239–80.

[27] John C. Kendrew, "Myoglobin and the Structure of Proteins," *Science*, 139 (1963), 1259–66, at p. 1261.

[28] John T. Edsall and David Bearman, "Historical Records of Scientific Activity: The Survey of Sources for the History of Biochemistry and Molecular Biology," in *Archival Sources for the History of Biochemistry and Molecular Biology*, ed. David Bearman and John T. Edsall (Philadelphia: American Philosophical Society, 1980), pp. 3–16, at p. 9.

THE PATH TO THE DOUBLE HELIX: THE SIGNER CONNECTION

The molecular biologist Gunther S. Stent (b. 1924) stated in retrospect that the influence of the structuralist school on biology was "not revolutionary," as that school was preoccupied with only structure, rather than information. As a member of the so-called phage group of Max Delbrück (1906–1981), he believed that what revolutionized biology was the achievement of his group in pursuing biological information as its central theme.[29] Genes were known to be located in chromosomes. Were genes, then, proteins or DNA? In the 1940s, there was general acceptance that genes were special types of protein molecules. Whereas the DNA molecule had only four kinds of nucleotide bases, the protein molecule was made of twenty kinds of amino acids. The almost infinite variety of possible amino acid sequences in the protein macromolecule seemed to carry far more coded genetic information. However biased Stent's appraisal was, it was the phage group, and not the structurist school, that played a cardinal role in refuting experimentally this point of view. A 1944 paper on bacterial transformation by the geneticist Oswald T. Avery (1877–1955) and his colleagues at the Rockefeller Institute indicated that DNA, not proteins, was the primary carrier of genetic information. Eight years later, this suggestion was confirmed by the experiments conducted by phage-group members Alfred D. Hershey (1908–1997) and Martha Chase (b. 1927), using radioactive tracers to study the process of bacteriophage infection.

Stent's historical assessment of molecular biology ignores the role of the macromolecular concept. The recognition of the macromolecularity of DNA was gained later than that of proteins. Until the late 1930s, many researchers had thought that DNA consisted of aggregates of relatively small molecules with a molecular weight of about 1,500. The polymer chemist Rudolf Signer (b. 1903) at Bern, Staudinger's faithful student, was among the first to demonstrate the macromolecularity of DNA. While working on DNA at the request of the Swedish biochemists Einar Hammarsten (1889–1968) and Torbjoerr O. Caspersson (b. 1910), Signer was quick to consider DNA to be "one of the biologically most important polymers."[30] In 1938, by means of flow birefringence, he estimated that the molecular weight of DNA lay between 500,000 and 1,000,000.[31] The enormous size of DNA molecules convinced scientists that DNA could store genetic information, a conviction prerequisite to the phage group's results.

---

[29] Gunther S. Stent, "That Was the Molecular Biology That Was," *Science,* 160 (1968), 390–5, at p. 391. Cf. John C. Kendrew, "How Molecular Biology Started," *Scientific American,* 216 (1967), 141–4.

[30] Transcript of interview with Rudolf Signer by Tonja Koeppel, 30 September 1986, Chemical Heritage Foundation, p. 17.

[31] R. Signer, T. Caspersson, and E. Hammarsten, "Molecular Shape and Size of Thymonucleic Acid," *Nature,* 141 (1938), 122.

Throughout the 1940s, Signer spent considerable time on the improvement of DNA preparation. When in May 1950 he delivered a paper at the Faraday Society General Discussion, "Physico-chemical Properties and Behavior of the Nucleic Acids," he brought a bottle of DNA finely prepared from calf thymus, and distributed it. One of the participants, Maurice Wilkins at Kings College, London, was fortunate to receive one part of the sample. The specimen was, in turn, handed down to the physical chemist Rosalind Franklin (1920–1958) and her graduate student Raymond G. Gosling (b. 1926) at Kings. One of the most intriguing tasks for the molecular biologists, then, was to find the precise three-dimensional structure of the DNA macromolecule and to clarify the mechanism of replication and transmission of genetic information. Although there were other DNA preparations, such as those from wheat germ and from pig thymus, the Signer DNA was evaluated as "the best DNA preparation."[32] With this DNA, Franklin and Gosling were able to discover that high humidity led DNA to transform from its crystalline form (A-DNA) into the paracrystalline form (B-DNA). Wilkins, while in the midst of a confrontation with Franklin, showed the young American biologist James D. Watson (b. 1928) the print of a fine B-DNA x-ray photograph produced by them. Inspired by this picture, Watson and Crick at the Cavendish built their famous double-helix model of DNA in the spring of 1953. After all, the basic molecular configuration was shown to have great simplicity.

The Watson-Crick model allowed scientists to grasp the molecular mechanism of heredity. The sequence of four kinds of nucleotide bases in the DNA macromolecule contained coded genetic information. DNA itself could replicate by each strand serving as a template for constructing a new partner strand; when the two strands unwind and separate, each directs the synthesis of its complementary partner. Portions of the coded genetic information in DNA could also be transcribed into ribonucleic acid (RNA). RNA could then carry that information to the cell, where it could be translated into a specific sequence of amino acids to synthesize a protein. This hereditary mechanism at the molecular level, later elaborated and called the "central dogma," opened an era of unprecedented growth for research in molecular biology in the 1960s, followed by the deciphering of the genetic code, the synthesis of artificial DNA and RNA, the elucidation of the structure of the transfer RNA, and the development of a technique to determine quickly the sequence of bases in DNA macromolecules.

As we have seen, conceptually, polymer chemistry and molecular biology were linked by their common focus upon macromolecules. Methodologically,

---

[32] H. R. Wilson, "The Double Helix and All That," *Trends in Biochemical Sciences*, 13 (1988), 275–8, at p. 275.

the two fields were linked by their common reliance on such tools as x-ray crystallography and, more generally, by their common interest in three-dimensional structures and functions. Historically, they were linked through such figures as Staudinger and, most especially Mark and Signer.

From the birth of classical organic chemistry through the rise of polymer chemistry to the maturation of molecular biology, understanding of molecular structure and function underwent considerable development. With a conviction that everything could be explained in terms of molecules and that the knowledge of the molecular structures could lead to an understanding of functions, scientists extended their pursuits from delineation of internal atomic arrangements within the molecule to elucidation of the shapes, dynamic behavior, and precise three-dimensional structures of macromolecules. Although the molecular-structural approach originated in chemistry, physicists and biologists, as well as chemists, elaborated it by supplementing their concepts and techniques. Whether this molecular-reductionist bent has its limits, as some contemporaries fear, the molecularization of the physical and biological sciences certainly shaped styles of research, subjects of study, and communities of professional scientists in the twentieth century.

Part V

MATHEMATICS, ASTRONOMY, AND COSMOLOGY SINCE THE EIGHTEENTH CENTURY

23

THE GEOMETRICAL TRADITION
Mathematics, Space, and Reason in the Nineteenth Century

Joan L. Richards

Until the end of the nineteenth century, geometry was the study of space. As such, geometrical knowledge can be found in virtually all civilizations. Ancient Sumerians, Babylonians, Chinese, Indians, Aztecs, and Egyptians surveyed their lands, constructed their pyramids, and knew the relation among the sides of a right triangle. The Western geometrical tradition dates from Euclid's (fl. 295 B.C.E) *Elements*. What marks this work as seminal lies not so much in its content per se as in how that content was known.

Two tightly interwoven characteristics marked Euclidean geometrical knowledge. First, the objective characteristic was the strict correspondence between the terms of the geometry and the objects to which those terms referred. Euclid's geometry dealt with something that we would call space.[1] For example, the Euclidean definition "a point is that which has no part" neither explains the concept of point nor shows how to use it nor establishes its existence. It does, however, indicate what a point is. The definition has meaning; it refers to an aspect of space that we already know.

Euclidean axioms are self-evident truths; the postulates are obvious statements that must be accepted before the rest can follow. Like the definitions, the axioms and postulates are statements about space that make explicit what we already know. Euclid's axioms and postulates do more, however. They support and structure all of the subsequent argument; all of the rest of the subject is drawn out of or built upon these basics. The adequacy of this axiomatic structure to support all legitimate geometrical conclusions is the second, rational, characteristic of Euclidean knowledge.

The value of knowledge with these characteristics has been differently assessed in different times and places. Its position in nineteenth-century

[1] Hans Freudenthal, "The Main Trends in the Foundations of Geometry in the 19th Century," in *Logic, Methodology and Philosophy of Science*, ed. Ernst Nagel, Patrick Suppes, and Alfred Tarski (Stanford, Calif.: Stanford University Press, 1962). For other overviews, see Felix Klein, *Vorlesungen über die Entwicklung der Mathematik im 19. Jahrhundert* (1926–27, reprint; New York: Chelsea, 1967), and Morris Kline, *Mathematical Thought from Ancient to Modern Times* (New York: Oxford University Press, 1972).

European thought was laid down by Isaac Newton (1642–1727), in his *Principia* of 1687, and refined by his eighteenth-century heirs. Geometry was an essential part of Newton's understanding of the physical universe, which meant that it held a central position in post-Newtonian epistemological thinking. Within mathematics, however, its pride of place was not so clear. Enthralled with the power of the calculus, most eighteenth-century mathematicians paid little attention to the spatial and rational demands of geometrical rigor.

In geometry, as in so much else, the nineteenth century began with the French Revolution. In the eventful decades that followed the cataclysm of 1789, geometry was viewed and pursued somewhat differently in different parts of Europe. In France, a combination of practical and ideological interests supported a flowering of interest in geometry at the Ecole Polytechnique in the first decades of the century. In Germany, a romantic interest in intuition supported geometrical study within the research university. In England, the subject was pursued as part of a neo-Newtonian natural theological tradition that found institutional expression in the curriculum of Cambridge University. In all of these contexts, the special value of geometry lay in its powerful validity, which rested on the twin pillars of its spatial referent and its rational structure.

Even as its unique validity supported the pursuit of geometry in various contexts throughout Europe, a small number of scattered investigators questioned that validity in investigations of Euclid's fifth, or parallel, postulate. In the 1860s, the ideas of these mathematical mavericks burst into the European consciousness as non-Euclidean geometries, mathematical descriptions of spaces that were essentially different from Euclid's. The existence of these alternative geometries was seen as a fundamental challenge to the self-evidence on which geometrical validity had rested for so long. In the second half of the nineteenth century, mathematicians throughout Europe struggled to respond to the challenge of non-Euclidean ideas. The effort was creative, and geometry was a thriving research field by the end of the century. At the same time, it was conservative in that all new developments were seen as taking place on the spatial ground that had for so long defined the limits of geometry.

A critical turning point came with the publication in 1899 of David Hilbert's (1862–1943) *Grundlagen der Geometrie*, in which the German resolutely broke the connection between geometry and space that had characterized the subject since the time of Euclid. Following his lead, geometry in the twentieth century became a formal subject with only the most tenuous ties to the space it had so long described.

THE EIGHTEENTH-CENTURY BACKGROUND

The central importance of Euclidean knowledge was reinforced for the nineteenth century by the celestial mechanics of Newton's 1687 *Philosophiae*

Naturalis Principia Mathematica. As its title attests, Newton's goal in this book was to present the mathematical principles of natural philosophy. His mathematical model was Euclid, and he carefully structured his book on the Greek's axiomatic example, beginning with definitions and axioms, which were then used in proofs of more complex relations.

At the same time as he was developing a mathematics, however, Newton was writing a natural philosophy. His axioms were laws of motion, drawn from the natural world. His challenge was to draw together these two modes of explanation, the mathematical and the natural. In the first explanatory "Scholium," appended to his definitions, he carefully explained the basis on which he claimed to have done so.

Like Euclid's, Newton's mathematics was rooted in things that were already known: "I do not define time, space, place, and motion, as being well known to all." But, he continued, people were often unclear about the ways in which their common ideas of these concepts related to the mathematical ones he was developing in the *Principia*. So, Newton explicitly laid out the relationship that he saw between the mathematical space of his *Principia* and the relative space of everyday life.

> Absolute space, in its own nature, without relation to anything external, remains always similar and immovable. Relative space is some movable dimension or measure of the absolute spaces; which our senses determine by its position to bodies; and which is commonly taken for immovable space.... Absolute and relative space are the same in figure and magnitude; but they do not remain always numerically the same. For if the earth, for instance, moves, a space of our air, which relatively and in respect of the earth remains always the same, will at one time be one part of the absolute space into which the air passes; at another time it will be another part of the same, and so, absolutely understood, it will be continually changed.[2]

Thus, Newton's universe was embedded in absolute, infinite, mathematical space. This distinguished it from the classical universe of Aristotle (384–322 B.C.E.) or Ptolemy (100–170). For classical astronomers, places were real and significant; that the earth stood at the center of the universe explained the motion of the heavenly bodies around it. Geometrical space, on the other hand, is homologous. The absolute position of geometrical figures – whether or not they are in the center of the universe, for example – is not relevant to geometrical proofs. This marked a significant difference between physical and mathematical explanation. In Newton's universe, however, both geometrical and physical spaces were infinite. This meant that for both of them, places are merely relative and "positions properly have no quantity."[3] This identification

---

[2] Isaac Newton, *Mathematical Principles of Natural Philosophy,* trans. Andrew Motte, rev. Florian Cajori (Berkeley: University of California Press, 1962), p. 6.
[3] Ibid., p. 7.

of physical and geometrical space allowed Newton to use a mathematical approach to physical problems.[4]

That Newton followed Euclidean modes of argument in the *Principia* represented a decided move away from the analytic approach of his mathematical predecessor, René Descartes (1596–1650). The Frenchman had, in his *Géometrie* of 1637, developed a method for solving geometrical problems that did not rely on reasoning from geometrical figures. Descartes had argued that two-dimensional geometrical curves and equations of two variables were equivalent, that a circle, for example, could also be represented by an equation. This recognition enabled him to solve conceptually complicated geometrical problems by means of relatively straightforward algebraic manipulations.[5]

Newton was well aware of the power of the Cartesian approach; he first developed his calculus of motion using Cartesian symbolic manipulations. In the *Principia,* his masterwork, however, he worked geometrically. This was because Newton believed Euclidean reasoning to be philosophically unimpeachable in a way that analytic reasoning was not. Manipulating algebraic symbols could be a very powerful method for solving problems, but the solidity of Euclid's self-evidence was lost in the process. Reasoning directly on geometrical figures, as Euclid had done, kept the focus clearly directed at the spatial objects themselves.

This was of critical importance, because for Newton, space shared the essential attributes of God. God, Newton wrote in the "General Scholium" to the *Principia,* "is eternal and infinite, omnipotent and omniscient; that is, his duration reaches from eternity to eternity; his presence from infinity.... He is not eternity and infinity, but eternal and infinite; he is not duration or space, but he endures and is present. He endures forever, and is everywhere present; and, by existing always and everywhere, he constitutes duration and space." In his Queries to the *Opticks,* Newton described a God "who in infinite Space, as it were in his Sensory, sees the things themselves intimately... by their immediate presence to himself."[6]

Newton's picture of a God who was literally mindful of his universe was not widely accepted, but many among his compatriots found in his physics a true insight into the workings of the divine mind. For them, the new science promised a clear understanding of the world that would supersede the mysteries of religion. The strength of the threat to traditional theological authority is clearly reflected in George Berkeley's (1685–1753) 1734 attack on Newton's calculus in *The Analyst or a discourse addressed to an infidel mathematician.*

---

[4] Alexandre Koyré, *From the Closed World to the Infinite Universe* (Baltimore: Johns Hopkins University Press, 1957); I. Bernard Cohen, *The Newtonian Revolution* (Cambridge: Cambridge University Press, 1980), esp. pp. 61–8.

[5] H. J. M. Bos, "The Structure of Descartes' *Géometrie*" *Il Metodo e I Saggi: Atti del Convegno per il 350 Anniversario della Publicazione del Discours de la Methode e degli Essais* (Rome: Istituto della Enciclopedia Italiana, 1990).

[6] Newton, *Mathematical Principles,* 2: 389–90, and *Opticks* (New York: Dover, 1952), p. 370, based on the fourth edition, London, 1730.

"The Method of Fluxions," Berkeley wrote, "is the key by help whereof the modern mathematicians unlock the secrets of Geometry and consequently of Nature." He then went on to inquire into the validity of the work, "whether this method be clear or obscure." The grounds on which he based his judgment were conceptual; the question was whether we could clearly "conceive" Newtonian mathematical ideas. For Berkeley, "conceiving" would entail sensing, perceiving, or even imagining them. Putting Newton's mathematical ideas to this test, Berkeley found:

> Now, as our sense is strained and puzzled with the perception of objects extremely minute, even so the imagination, which faculty derives from sense, is very much strained and puzzled to frame clear ideas of the least particles of time.... The further the mind analyseth and pursueth these fugitive ideas the more it is lost and bewildered;... take it in what light you please, the clear conception of it will, if I mistake not, be found impossible.[7]

Hence, he concluded, the Newtonian calculus was not legitimately grounded.

In his 1742 *Treatise of Fluxions,* Colin Maclaurin (1698–1746) took up Berkeley's challenge to free Newton's mathematics of its conceptual obscurity. He did so by proving Newton's results using only the universally respected "geometry of the antients."[8] The result was as exhausting as it was exhaustive, but it was embraced by Maclaurin's compatriots as the answer to Berkeley's objections.

Maclaurin's work was not embraced on the Continent, however. The dispute over whether Newton or Gottfried Wilhelm Leibniz (1646–1716) deserved the credit for having invented the calculus was so vituperative that it served to separate English from Continental mathematics throughout the eighteenth century. The battle line between the two groups can be clearly seen in a symbolical distinction. In England, Newton's symbols, in which derivatives were indicated by raised dots ($y = t^3$: $\dot{y} = 3t^2$: $\ddot{y} = 6t$) were used. This symbology arose from Newton's dynamic focus; the dependent variable, y, was a fluxion, a moving quantity plotted against the independent variable of time. On the Continent, Leibniz's symbols in which derivatives were indicated as if they were infinitesimal fractions were used ($y = x^3$: $dy/dx = 3x^2$: $dy^2/d^2x = 6x$). Leibniz's symbols were more flexible than Newton's; the independent variable did not always have to be time. When it came to problems that necessitated differentiating the functions of more than one variable, for example, Leibniz's symbols were vastly superior to Newton's fluxions.[9]

By the middle of the century, few on the Continent would question the power of Newton's physics, but many challenged his geometrical approach.

---

[7] D. J. Struik, *A Source Book in Mathematics, 1200–1800* (Cambridge, Mass.: Harvard University Press, 1969), pp. 334–5.

[8] Quoted in Niccolò Guicciardini, *The Development of Newtonian Calculus in Britain 1700–1800* (Cambridge: Cambridge University Press, 1989), p. 47.

[9] H. J. M. Bos, "Differentials and Derivatives in Leibniz's Calculus," *Archive for History of Exact Sciences,* 14 (1974), 1–90.

The proofs of the *Principia* were often convoluted and made difficult certain problems that could be rather easily solved by using analytic techniques. Berkeley's theological concerns carried very little weight outside of England, and although it was widely acknowledged that Maclaurin had answered them, there was considerable doubt about whether his ponderous efforts had been worth the trouble. "The axioms of geometry are rigorous," wrote the anonymous author of the article "Rigueur" in the *Encyclopédie,* but, he also noted, "Genius does not tolerate rigor."[10]

As these phrases suggest, there was actually a move away from geometrical rigor in the middle of the Enlightenment, a move that can be seen from the sides both of axiomatic structures and of self-evident meaning. A number of elementary texts rejected Euclidean rigor in an attempt to make geometry "natural." "All reasoning, which applied to that which good sense knows in advance, is a pure loss and serves only to obscure truth and disgust the reader," Alexis-Claude Clairaut (1713–1765) wrote in his *Elémens de géométrie.*[11]

As elementary texts challenged the value of Euclidean axiomatic rationality, more advanced works challenged the importance of spatial meanings. Many artifacts of eighteenth-century analysis, including negative numbers, imaginary numbers, and divergent series, did not have clearly understood spatial or other interpretations. Eighteenth-century mathematicians recognized this, but usually they proceeded nonetheless in the spirit of Jean LeRond d'Alembert's (1717–1783) oft-quoted dictum *"Allez en avance, la foi vous viendra!"* (go ahead, the faith will come to you!).

GEOMETRY AND THE FRENCH REVOLUTION

The period immediately following the French Revolution of 1789 saw considerable change in mathematics. Empowered by the powerful postrevolutionary conviction that science could change the world, a new system of secondary and postsecondary schools was created to educate the new French citizenry. They drew their faculties from the top of the French scientific community, who were challenged to teach their subjects from their foundations, and to present them in such a way that they would be accessible to all. Geometry, the quintessentially reasonable study of universally known space, had a central role to play in educating a rational populace.[12]

On the elementary level, Adrien-Marie Legendre's (1752–1833) *Eléments de géométrie* challenged the eighteenth-century tradition of naturalistic geometry, and trumpeted a return to strict standards of rational proof. "[T]here is no need to fear seeming long and detailed," Legendre wrote. "Length ... is a small sacrifice for ... exactitude."[13]

---

[10] *L'Encyclopédie ou Dictionnaire raisonné des sciences des arts et des métiers* (Paris, 1851; New York: Readex Microprint Corporation, 1969), s.v. "Rigueur."
[11] [Alexis Claude] Clairaut, *Elémens de géométrie* (Paris: Lambert & Durand, 1741), pp. x–xi.
[12] Judith V. Grabiner, *The Origins of Cauchy's Rigorous Calculus* (Cambridge, Mass.: MIT Press, 1981).
[13] Adrien-Marie Legendre, *Eléments de géométrie* (Paris: F. Didot, 1794), pp. v–vi.

Beyond the elementary level, geometry was developed most strikingly in the context of the Ecole Polytechnique. Founded in 1794, this school brought together the best mathematicians in France to teach a new generation of French engineers. The attempt to teach their subject focused attention on foundations.

Notable among the mathematicians at the Ecole, was Gaspard Monge (1746–1818), who taught descriptive geometry there from its founding. Descriptive geometry was essentially the mathematical theory behind mechanical drawing. Narrowly understood, the subject focused on techniques for projecting the essential aspects of three-dimensional objects onto perpendicular planes. But Monge did not approach it narrowly.

Monge took a fluid view of geometry in which the boundary that separated the subject from dynamic processes was essentially blurred; he presented figures as generated from one another by continuous processeses. Lines formed families as they rotated around fixed points. Ellipses became circles as their foci moved together; ellipses became parabolas as one focus moved away to infinity. As Charles Dupin (1784–1873), one of Monge's students, put it: "[In the study of descriptive geometry] the mind [*esprit*] learns to see internally and with perfect clarity, the individual lines and surfaces, [and the] families of lines and surfaces; it acquires a sense of the character of these families and individuals;... it compares them, combines them and predicts the results of their intersections and their more or less intimate contacts."[14]

For Monge and his disciples, the methods of descriptive geometry held the promise of mathematical rewards far beyond engineering drawing. In particular, they suggested an answer to those who questioned the foundations of analysis. Echoing a neo-Lockean French tradition in which "all our knowledge and all our faculties are derived from the senses," a number of "physicalist" mathematicians viewed mathematical symbols as simple signs for sensed objects.[15] For them, the validity of mathematical arguments depended on the tight fit between the signs and the things symbolized. The strength of this fit was attested by the ease with which one could move from the one to the other, by the clarity with which one could visualize mathematical objects. For physicalist geometers, analysis was merely a way of talking about the really "moving geometrical spectacle."[16]

In the first decades of the nineteenth century, students of Monge worked to develop a "modern" geometry that would be as powerfully flexible as analysis, even as it remained unimpeachably grounded in spatial sensation. The most

---

[14] Charles Dupin, *Essai historique sur les services et les travaux scientifiques de Gaspard Monge* (Paris, 1819), p. 177. Translated by the author unless otherwise noted.

[15] The phrase is from Condillac, quoted in L. Pearce Williams, "Science, Education and the French Revolution," *Isis*, 44 (1953), 311–29, at p. 313. The term "physicalist" is from Lorraine J. Daston, "The Physicalist Tradition in Early Nineteenth Century French Geometry," *Studies in History and Philosophy of Science*, 17 (1986), 269–95.

[16] The phrase is from Monge, quoted in Eduard Glas, "On the Dynamics of Mathematical Change in the Case of Monge and the French Revolution," *Studies in History and Philosophy of Science*, 17 (1986), 249–68, at p. 257.

articulate spokesman for this group in the 1820s was Jean Victor Poncelet (1788–1867), who in 1822 published his *Traité des propriétés projectives des figures*. As the title suggests, Poncelet here focused on the transformations of figures through central projection, and his work marks the beginning of the nineteenth-century study of projective geometry.

A key result in Poncelet's book is the principle of duality. This principle made explicit a curious fact that many of the modern geometers had noticed: A true geometrical statement will remain true if the words "line" and "point" are interchanged. Thus, for example, the dual of the statement "Two points determine a line" is "Two lines determine a point" (their intersection). Poncelet elevated this observation to a principle in his *Traité*, but only for particular cases. Joseph Diez Gergonne (1771–1859) saw it as more generally applicable, and began the practice of listing theorems and their duals in parallel columns.

Poncelet constructed his geometry on a principle of continuity: "If one figure is derived from another by a continuous change, and the latter is as general as the former, then any property of the first figure can be asserted at once for the second figure."[17] With its inclusion of motion in the heart of geometry and its quasi-inductive emphasis on general properties, this principle reflected the fluid approach to geometry that Poncelet had learned from Monge.

By the 1820s, however, Monge was dead, the Bourbons were again on the throne, and a new wind was blowing in French mathematics. The most powerful voice for a new approach was that of Augustin-Louis Cauchy (1789–1857). In 1822, Cauchy published his *Cours d'analyse*, in which he declared his intent to proceed with "all the rigor that one demands in geometry." But Cauchy's rigor was not the rigor of the physicalist geometers. Their generalizing approach had involved "inductions which may sometimes lead to truth, but which accord little with the vaunted exactitude of the mathematical sciences."[18] It had blurred the lines between legitimate and illegitimate objects, between convergent and divergent series, between real and imaginary quantities. In order for mathematics to be exact, these lines had to be precise and exactly delineated.

Cauchy's insistence on exactitude meant, however, that mathematics had to be removed from the accessible but poorly delineated world of the everyday. It is not because they adequately capture the meanings of terms like "limit" that Cauchy's definitions are valued, but because they fix the meaning of terms in ways that can be used precisely. They do not appeal to familiar experience, but they do permit precise judgments about the results of mathematical calculations.

The abstract rigor of Cauchy's *Cours d'analyse* was highly controversial. His austere distance from the common-experience approach so alienated him from his students that they rebelled against his classes. The issues were

---

[17] Quoted in Kline, *Mathematical Thought*, p. 842.
[18] Augustin Cauchy, *Cours d'analyse de l'école royale polytechnique* (Paris, 1821), pp. ii, iii.

not merely pedagogical. The question of whether mathematics should remain true to spatial experience fueled a battle between synthetic and analytic mathematics that raged among France's mathematicians throughout the 1820s. By the end of the decade, Cauchy's approach had essentially won, however, and his interpretation of rigor pointed the direction for most European mathematics during the nineteenth century.[19]

The triumph of analysis over synthesis in French mathematics was both supported by and reflective of the larger cultural and institutional framework within which mathematics was pursued. Just as the attempt to teach the subject had focused attention on foundational issues, so too were the goals of that teaching reflected in the kinds of interpretation that were given to the study of mathematics as a whole. These goals changed considerably during the first three decades of the century, and the triumph of analytic over synthetic views of mathematics can be seen as a mirror of these larger changes.

In the postrevolutionary and early Napoleonic period, Monge's descriptive geometry was seen not just as a practical subject but also as a central part of an educational program intended to reach and unify a wide variety of interests. Minds trained in descriptive geometry were both highly versatile and eminently practical. A subject accessible to all, it served as a common thread among the competing specialties at the Ecole Polytechnique. Military engineers who knew descriptive geometry could judge terrain at a glance and formulate appropriate strategies; builders of canals and bridges could see suitable sites; naval architects could visualize the most useful and efficient boats. Studying it joined together all of the various interests at the school.[20]

As the years of Napoléon's reign wore on, however, the universal accessibility of geometry became less clearly a value. More and more clearly protected within the institutional confines of the *grandes écoles,* French mathematicians no longer felt a need to maintain its claims to universality and its ties to other subjects. As their professional identity was increasingly tied to their particular mathematical knowledge, the value of that knowledge came to lie more in its separateness and inaccessibility than in its universal appeal. As early as 1803, Sylvestre François Lacroix (1765–1843) was arguing that the widely accessible geometrical approach to the subject is appropriate for children and elementary instruction. For more advanced work, however, constant references to spatial examples are cumbersome and confining. The advanced mathematical thinker had to be able to calculate freely, without being bound to spatial meanings. As Lacroix put it, one should not "borrow from appearances and sensations those things that can be drawn from judgment alone."[21]

---

[19] For mathematics in France during this period, see Bruno Belhoste, *Cauchy: Un Mathématicien Légitimiste au XIX Siècle*, preface by Jean Dhombres (Paris: Belin, 1985); Jesper Lützen, *Joseph Liouville 1809–1882: Master of Pure and Applied Mathematics* (New York: Springer-Verlag, 1990).

[20] Dupin, *Essai*, p. 177.

[21] Sylvestre François Lacroix, *Essais sur l'enseignement en général et sur celui des mathématiques en particulier* (Paris, 1816), p. 174.

Lacroix acknowledged that abstract mathematics was difficult, that many people could not understand it. This meant that it was not a good model for democratic, populist thought. On the other hand, it had other, perhaps more important, social uses. The over-subscribed Ecole Polytechnique needed a way to gauge merit, both for admission to the school and for ranking students within it. Lacroix argued that judgments based on examinations of abstract mathematical prowess was a fair way to judge merit in a postaristocratic society.

GEOMETRY AND THE GERMAN UNIVERSITY

The French Revolution was not just a national, but also an international, cataclysm. In Germany, the effects were felt most directly in the 1806 defeat of the Prussian army by Napoléon. This event was politically demoralizing, but in other ways highly constructive. It led to a veritable *Geistesrevolution,* an intellectual revolution that radically changed the shape of German intellectual life. In the first decades of the century, the Prussian university system was radically restructured as part of an attempt to reform and revivify German society through education.

In the eighteenth century, German universities had been primarily teaching institutions; any research was carried out in separate research institutions. The major innovation of the early-nineteenth-century reforms was to bring these two functions together in the newly reformed universities. The rise of the German research university had profound consequences for the study of mathematics in Germany.

In the eighteenth-century universities, mathematics was studied within the philosophy faculty, which laid the educational ground for students interested in going on to the more prestigious faculties of law, theology, or medicine. Interest in the subject reflected this hierarchy of value. In 1800, the only major German mathematician was Carl Friedrich Gauss (1777–1855), at the University of Göttingen. Gauss was a towering mathematical intellect; his work was central to nineteenth-century mathematics, but he was a solitary researcher. The growth of a community of German mathematicians in the early nineteenth century was due to others.

In the era of post-Napoleonic reform, the philosophy faculty, in which mathematics was housed, was given the job of educating Prussia's secondary school teachers. From this position, it expanded to become the center of the nineteenth-century Prussian research university. The study of mathematics was one of the mainstays of the secondary liberal education, and benefited accordingly. In 1826, August Leopold Crelle's (1780–1855) *Journal für die reine und angewandte Mathematick* was established with the backing of the Prussian ministry of education. In 1834, the mathematician Carl Jacobi (1804–1851) and the theoretical physicist Franz Neumann (1798–1895) set up the combined mathematics-physics seminar at Königsberg. From these beginnings grew a research tradition that by the middle of the century made Prussia the undisputed center of European mathematics.

The defining characteristic of the mathematics supported by these developments was its freedom from practical applications. In his inaugural lecture at Königsberg in 1832, Jacobi contrasted the purity of the program he intended to follow with the applied interests that characterized mathematical work at the Ecole Polytechnique: "While they seek to obtain the only salvation for mathematics in physical problems, they desert that true and natural path of the discipline, which . . . has brought the analytical art to the importance that it now enjoys."[22]

Two tightly intertwined kinds of concerns, one philosophical, the other institutional, supported this determined focus on pure mathematics. From a philosophical point of view, the Germans approached mathematics very differently than did the French. The French tradition was a neo-Lockean one, in which the neonatal mind is a blank slate: We build our notion of space from experience. For the German philosopher Immanuel Kant (1724–1804), on the other hand, the slate had significant shape to begin with; space was an integral part of the perceptual and cognitive apparatus through which we have our experience. This was a significant difference. Within the French tradition, it was always more or less artificial to separate the objects from the mathematics of those objects. From the Kantian point of view, the objects of the two studies were essentially different, and the order of study moved most naturally from the pure to the applied.

Basic institutional factors also supported the German interest in pure mathematics. In the context of the research university, the impetus was to develop an independently grounded, autonomous discipline that could take its place as an equal among the other specialties in the philosophy faculty. In Germany, as in France, the growth of a professionally supported group able to focus primarily, if not exclusively, on mathematics removed the impetus to keep mathematics accessible or to relate the subject to other areas of thinking. Thus, in Germany as in France, mathematics became ever more abstract and abstruse as the century progressed.

The German interest in pure mathematics did not particularly encourage the pursuit of physicalist geometry. Nonetheless, Jakob Steiner (1796–1863), extraordinary professor of mathematics at the University of Berlin, enthusiastically pursued projective geometry. A passionate disciple of Johann Heinrich Pestalozzi (1746–1827), Steiner held a view of his work that reads like a German romantic translation of Dupin's Monge. His goal was to discover the

---

[22] Quoted in Gerd Schubring, "The German Mathematical Community," *Möbius and his Band*, ed. John Flauvel, Raymond Flood, and Robin Wilson (Oxford: Oxford University Press, 1993), p. 29. See also R. S. Turner, "The Growth of Professorial Research in Prussia," *Historical Studies in the Physical Sciences*, 3 (1971), 137–82; D. Rowe, "Klein, Hilbert and the Göttingen Tradition," in Kathryn M. Olesko, ed., "Science in Germany," 2d ser., *Osiris*, 5 (1989), 186–213.

organism [*Organismus*] through which the different appearances in the spatial world are tied to each other.... [T]he heart of the matter... consists in the dependence of the figures upon each other and will be discovered in the forms and ways that their properties grow from the simpler to the more complex. This relation and transition is the essential source of all the other individual propositions of geometry.[23]

Like Monge, Steiner was an inspired teacher, but there was a difference between French physicalist and German intuitionist geometry. Whereas Monge was known for vivid illustrations, Steiner was known to teach in the dark in order to maintain a focus on intuitive knowing.

In the 1830s, Steiner's intuitive approach to geometry embroiled him in conflicts that mirrored the issues that fueled the French analytic-synthetic debates. His compatriot August Möbius (1790–1868) had in the 1820s interpreted algebraically some of the major insights of the French geometers. Möbius developed new barycentric and projective coordinate systems that made the basic equivalence relations and dualities of the projective plane algebraically evident. In the following decade, Julius Plücker (1801–1868) used an algebraic approach to resolve some disturbing problems that had arisen from the principle of duality. Möbius and Plücker's algebra led to significant geometrical insights, but their move away from direct spatial reasoning so infuriated Steiner that he threatened not to publish in Crelle's journal if their work was also included.

Thus, German mathematics thrived in the early nineteenth century. The basis for its strength lay in a professionally defined group within the universities, however, which meant that the readily accessible study of space was not a major focus of interest. Nonetheless, the research tradition was strong enough that German mathematicians made major contributions to geometry, as well as to other areas of mathematics.

GEOMETRY AND ENGLISH LIBERAL EDUCATION

The French did not conquer England, but the Revolution and its aftermath profoundly affected the island across the channel. In politics, there was a powerful conservative backlash, but by the second decade of the nineteenth century, it was clear that there was much to be admired in French mathematics. In the 1810s, a small but influential group of young men at Cambridge University formed the Analytical Society with the stated goal of raising Cambridge mathematics to the level of the French. In their view, Newton's geometrical calculus was outmoded, and the Newtonian cast at Cambridge was making it harder for them to work as effectively as their Continental counterparts. Therefore, they wanted to replace Newtonian fluxional symbols with more easily manipulated Leibnizian ones. In the oft-quoted words of their

---

[23] Jacob Steiner, *Gesammelte Werke*, 2 vols. (New York: Chelsea Publishing Company, 1971), p. 233–4.

leader, Charles Babbage (1791–1871), they wanted to replace the "dot"age of the university with the "dei"sm of the Continent.

On the surface, the young Analytics succeeded; by the end of the decade, Babbage, George Peacock (1791–1858), and John Herschel (1792–1871) had translated Lacroix's *Traité élémentaire du calcul* into English and eliminated Newton's fluxional notation from the Cambridge examination. But a closer look reveals that the Analytics did not give up the conceptual view of mathematics that had for so long supported the fluxional approach. In those places where Lacroix's text moved toward a more abstract view of the calculus, they tempered his message with critical notes that emphasized its conceptual grounding. Even as they claimed to be bringing French mathematics to England, Newton's heirs firmly defended its conceptual base.

Through the 1820s and 1830s, the Analytics moved from Cambridge into a variety of positions connected to English science. However, William Whewell (1794–1866), who was a close satellite of the original group, remained at Cambridge throughout his adult life, defining and defending the mathematical education there. As Whewell saw it, mathematics at Cambridge was the heart of a liberal education designed to teach young men to think effectively. Mathematical study was not pursued as narrow training, nor was it a way to develop specialized or professional skills. Instead, it was advocated as a broad education, as a way to educe from students their full human potential.

A neo-Newtonian natural theology lay behind the nineteenth-century geometrical education at Cambridge. Classical geometry was valued because it was known with the same absolute certainty with which God was known. The evidences of science could lead to contingent truth, but mathematics was known necessarily. Cambridge students studied the geometry in order to experience this kind of necessary truth directly. As Whewell, put it: "[O]ne of the most important lessons which we learn from our mathematical studies is a knowledge that there are such truths, and a familiarity with their form and function."[24]

Although Whewell resolutely held classical geometry at the heart of the Cambridge curriculum, the mathematics pursued there went far beyond Euclid. Throughout the 1830s, competition for top honors on the exit examination, or Tripos, became fierce. When it came to ranking students, geometry's commonsensical base and eminent reasonableness were not helpful, and the advanced parts of the Tripos blossomed into ever-more-difficult and abstract problems. So, after demonstrating their geometrical powers, the first rank of Cambridge students came to pursue mathematics at levels as abstruse as those to be found on the Continent.[25]

---

[24] William Whewell, *Of a Liberal Education in General* (London: J. W. Parker, 1845), p. 163.
[25] On William Whewell and the Cambridge Tripos, see Harvey Becher, "William Whewell and Cambridge Mathematics," *Historical Studies in the Physical Sciences*, 11 (1980), 1–48; Menachem Fisch and Simon Schaffer, eds., *William Whewell: A Composite Portrait* (Oxford: Clarendon Press, 1991); Andrew Warwick, *Masters of Theory: The Pursuit of Mathematical Physics in Victorian Cambridge* (Cambridge: Cambridge University Press, forthcoming).

Cambridge was neither the Ecole Polytechnique nor a German research university, however. Its natural theological focus meant that, ultimately, all of its mathematics had to be conceptually grounded. Although by the 1840s England's mathematical adepts had followed Cauchy's lead and based the calculus on the limit, they never accepted his abstract definition of rigor nor translated his *Cours d'analyse* into English. Though extraordinarily proficient in symbolic manipulations, Cambridge students were always kept aware of the potential "evil effect of this in giving rise to vagueness of conception," and their teachers insisted on "the continual interpretation and translation of [mathematical symbols] into the language of the subject."[26]

Well into the nineteenth century, there was no place to pursue mathematics in any English institutional setting beyond Cambridge. The amateur Royal Society was no match for the French Académie des Sciences, and in the British Association for the Advancement of Science, mathematics was a poor stepsister to physics throughout the century. In the words of a Parliamentary commission evaluating the Cambridge education in 1852, "There were danger that Mathematics would vanish from the face of the earth if not made an especial part of a liberal education."[27]

Despite all of these constraints, there were some moves to support mathematical research in England. The *Cambridge Mathematics Journal*, founded in 1837 by Duncan Gregory (1813–1844), was an attempt to move past the fixed boundaries of the Cambridge curriculum and publish original research. During the 1840s and 1850s, a small but prolific group of British mathematicians, including Gregory, Robert Leslie Ellis (1817–1859), Arthur Cayley (1821–1895), James Joseph Sylvester (1814–1897), and George Salmon (1819–1904) published mathematical researches in the *Journal*'s pages. Their work often appears relentlessly abstract, but a closer look shows a continued respect for the importance of geometrical interpretations for their algebraic results. Therefore, from the middle of the century there was a considerable English interest in and development of the "modern" or projective geometry.[28]

EUCLIDEAN AND NON-EUCLIDEAN GEOMETRY

Even before nineteenth-century European thinkers were struggling to incorporate conceptually grounded classical geometry into their rapidly changing worlds, little-noticed developments within the subject itself were raising

---

[26] Great Britain, Parliament, 1852, *Parliamentary Papers*, 1852–3, vol. 44, "Report of Her Majesty's Commissioners Appointed to Inquire into the State, Discipline, Studies, and Revenues of the University and College of Cambridge," p. 113.

[27] Ibid., p. 105.

[28] Joan L. Richards, "Projective Geometry and Mathematical Progress in Mid-Victorian Britain," *Studies in the History and Philosophy of Science*, 17 (1986), 297–325.

questions about its absolute validity. In the eighteenth century, scattered individuals interested in the certainty of Euclid's system scrutinized its foundations. By the early nineteenth century, some were concluding that although consistent, Euclid's was not the only possible mathematical description of space, and that there were non-Euclidean geometries that were equally viable.[29]

Initially, the route to non-Euclidean geometries led through Euclid's fifth, or paralled postulate, which states: "That, if a straight line falling on two straight lines make the interior angles on the same side less than two right angles, the two straight lines, if produced indefinitely, meet on that side on which are the angles less than the two right angles."[30] This statement fits Aristotle's criteria for a postulate: It is the kind of statement that would be proved as a theorem were it not assumed, and it is not necessarily self-evident. Nonetheless, Euclidean commentators since antiquity had tried either to prove it or to construct the geometry without it.

The Italian Jesuit Girolamo Saccheri (1667–1733) is notable as the earliest voice of a modern revival of interest in the fifth postulate. When he took up the question early in the eighteenth century, he found the historical landscape littered with failed attempts to prove it outright. So, Saccheri developed an alternative approach. He set out to prove the fifth postulate indirectly by assuming that it was false and generating an internal contradiction.

Saccheri's approach required that he be precise about the alternatives to the postulate. To do this, he constructed a quadrilateral, in which he assumed that the base angles were right angles. In Euclidean space, the nature of parallels would guarantee that the two remaining angles would be right angles as well. Saccheri's goal was to prove their equality without using the parallel postulate.

Saccheri was able to prove quite easily that the two remaining angles were equal to each other. From there he went on to show that if they were both right, Euclid's parallel postulate would hold. In order to prove the necessity of Euclidean geometry, however, Saccheri had to demonstrate the impossibility of the acute and the obtuse angle hypotheses. He dismissed the acute angle hypothesis quite quickly, but proved a number of theorems that would be true were the obtuse angle hypothesis assumed, before he concluded that he had been lead to a "manifest falsity."[31]

Saccheri apparently believed he had succeeded in proving the fifth postulate, but others, such as Johann Heinrich Lambert (1728–1777), were more

---

[29] For the history of non-Euclidean geometry, see Roberto Bonola, *Non-Euclidean Geometry: A Critical and Historical Study of its Developments*, trans. with appendices by H. S. Carslaw (New York: Dover, 1955); Jeremy Gray, *Ideas of Space* (Oxford: Clarendon Press, 1979); Joan L. Richards, *Mathematical Visions: The Reception of Non-Euclidean Geometry in Victorian England* (Boston: Academic Press, 1988).

[30] *The Thirteen Books of Euclid's Elements*, 3 vols., trans. and ed. Sir Thomas Heath (New York: Dover, 1956), p. 155.

[31] Giorlamo Saccheri, *Euclides vindicatus*, trans. by G. B. Halsted (Chicago: Open Court Publishing, 1920), p. 14.

ambivalent. By the early decades of the nineteenth century, a number of individuals were beginning to argue that the theorems developed from assuming the parallel postulate false were legitimate and not inconsistent. Gauss and Ferdinand Schweikart (1780–1859), in private letters, and Nicholai Ivanovich Lobachevsky (1793–1856) and János Bolyai (1802–1860), in published works, all developed the implications of Saccheri's obtuse angle hypothesis. These men saw the theorems that they generated as an alternative geometry, as the description of non-Euclidean space. They also saw that their success in creating alternatives to Euclid's geometrical system raised significant questions about the absolute truth that Newton and his posterity ascribed to that space. For these thinkers, it was no longer necessarily the case that space was Euclidean; there was an alternative possibility. This meant that epistemologies that rested on absolute knowledge of Euclidean space were no longer tenable.

In the first half of the nineteenth century, the notion of geometry's absolute truth was so deeply rooted in European consciousness that non-Euclidean objections were simply not heard. Gauss's decision not to publish his ideas because he feared "the chatter of the Boetians" is legendary, but both Lobachevsky and Bolyai did publish and there was no chatter to speak of. When he read Lobachevsky's work in 1865, Cayley was simply confused by the claims that the formulas constituted a non-Euclidean trigonometry. "I do not understand this," the Englishman wrote, "but it would be very interesting to find a *real* [that is Euclidean] geometrical interpretation of the last-mentioned system of equations."[32]

GEOMETRY IN TRANSITION: 1850–1900

The situation changed dramatically in the 1860s, when an energetic group of mathematicians in France, England, and Italy took up and publicized non-Euclidean ideas. By 1866, Gauss's correspondence on matters of non-Euclidean geometry, a French translation of Lobachevsky's book, and Bernhard Riemann's (1826–1866) *Habilitationsvortrag* had all been published, and non-Euclidean geometry had arrived.

When non-Euclidean geometry burst into European consciousness, there were two major approaches to the subject. In addition to the "synthetic" approach of the geometers who were concerned with the fifth postulate, there was a newer "metric" approach, which had been pioneered by Riemann. In 1854 the young German mathematician had presented *"Über die Hypothesen welche der Geometrie zu Grunde liegen"* to the faculty at Göttingen. In it he tried to move past the concept of space to the basic hypotheses that undergirded it.

---

[32] Arthur Cayley, "Note on Lobatchewsky's Imaginary Geometry," *Philosophical Magazine*, 29 (1865), 231–3, at p. 233.

Riemann's starting point was analytic. He constructed the concept of space on a highly abstract notion of multiply extended magnitude. He showed that from the mass of possible ways that number triplets could be structured, the identifying mark of space was its measure relations or metric. According to Riemann, metric relations distinguished Euclidean from non-Euclidean spaces. Euclidean space was characterized by the particular distance function, $ds = \sqrt{dx^2 + dy^2}$; other distance functions generated alternative spaces. What is more, Riemann speculated, space might not have a single, constant distance function, but its measure might be different for the infinitely large or small.

All of these possibilities led Riemann, like the synthetic non-Euclidean geometers before him, to conclude that the ultimate choice among possible geometries could not be decided by mathematical criteria. "Hence," Riemann wrote, it "flows as a necessary consequence that the propositions of [Euclidean] geometry cannot be derived from general notions of magnitude, but that the properties which distinguish [Euclidean] space from other conceivable triply extended magnitudes are only to be deduced from experience." He did not try to specify what those experiences might be because he thought this was "a problem which from the nature of the case is not completely determinate."[33] Thus, Riemann left open the question of why we think we live in Euclidean space.

When Riemann's lecture was published in 1866, his ideas were picked up almost immediately by Hermann von Helmholtz (1821–1894). As an empirical physiologist, Helmholtz believed that we learn everything through experience. At the time he read Riemann's paper, he was embroiled in a major debate with another group of physiologists, the nativists, who believed that some concepts are innate. Crucial to the discussion was the concept of space. Helmholtz thought we learned about space through infant experience; the nativists thought it developed from within.

Since newborns cannot talk about their experiences, Helmholtz was unable to establish directly how much they know about space. He could, however, construct plausible grounds for his hypothesis by showing that there are experiences all infants have had that are sufficient to generate the concept of Euclidean space. He first argued that all infants, even blind ones, directly experience the motions of rigid bodies as they manipulate objects that are given to them. He then showed how all of the basic structures of Euclidean space could be produced from rigid-body motions.

When Helmholtz read Riemann's paper, he recognized an essential link between Riemann's mathematical approach and his empirical one. The experience of his infants began in the amorphous manifolds Riemann described; it was by moving rigid bodies around that they learned the essential properties

---

[33] Bernhard Riemann, "On the Hypotheses which Lie at the Bases of Geometry," trans. W. K. Clifford, *Nature*, 8 (1873), 14–15.

of distance. Helmholtz wrote a series of papers in German and in English that detailed what the experience of living in a non-Euclidean space would be like – what we would see, what we would feel were we to live in such spaces. In this way, he transformed the often dense mathematics of non-Euclidean geometry into a set of widely accessible ideas. In the final decades of the nineteenth century, his conceivable non-Euclidean geometry caught the attention of a rapidly growing reading public, which speculated about space with undisciplined exuberance.

Of particular popular interest were geometries of more than three dimensions. This was a departure. Non-Euclidean geometers were interested in curvature; even the ever-colorful Helmholtz refused to speculate about worlds of more than three dimensions. But in 1882, an English schoolmaster, Edwin Abbott Abbott (1838–1926), published a book in which spaces of any number of dimensions are explored by a two-dimensional hero. *Flatland* was an immediate popular success, and at the end of the century, Charles H. Hinton (1853–1907), H. G. Wells (1866–1946), and others followed Abbott's lead with science fiction explorations of four-dimensional worlds. The exploration was not just literary. In an age in which spiritualism was being taken very seriously, the fourth dimension served many as a convenient way to understand the place of the spirit world. Abbott had clearly described how three-dimensional objects would appear to fade in and out of a two-dimensional world. Many saw that four-dimensional spirits could in the same way fade in and out of our three-dimensional one. The mathematical community disapproved of such unbridled speculations, but so long as geometry was grounded in space, the boundary between legitimate and illegitimate geometrical thinking was hard to enforce.

More troubling were the philosophical meanings that could be attached to non-Euclidean geometries. Even with no empirical evidence that they were real, the mere possibility of clearly conceiving non-Euclidean alternatives threatened the necessary truth of Euclidean geometry. Newton had taken for granted that absolute space was Euclidean, but he did not know of the alternatives. Non-Euclidean geometries raised questions about whether or not he was right.

By the 1870s, many mathematicians believed they had found a resolution of this challenge in projective geometry. This interpretation emphasized that metric relations are not preserved in projections; a person close to the viewer may be seen as larger than a faraway building. This means that the concept of distance, which Riemann had shown to be what distinguished Euclidean from non-Euclidean geometries, is not an essential part of the space described by projective geometry.

For many nineteenth-century mathematicians, projective geometry, thus, was perfectly situated to deal with the non-Euclidean challenge. It penetrated into a spatial bedrock that lay deeper than the notion of distance or of parallel lines. In this way, projective geometers were able to defend the necessity of geometrical knowledge, albeit one rooted in a projective, rather than Euclidean, view of space. The choice among metric geometries might be contingent, but knowledge of the projective space in which they were embedded was not.

Cayley naively pioneered this approach in his "Sixth Memoir upon Quantics," published in 1859. Here, he developed a function from projective properties that displayed the defining characteristics of a Euclidean distance function. Having shown how metric geometry could be generated from the more amorphous structure of projective space, Cayley concluded: "Metrical geometry is thus a part of [projective] geometry, and [projective] geometry is *all* geometry."[34]

When Cayley referred to metrical geometry, he meant Euclidean geometry. He completed his work before he knew of non-Euclidean geometry, and it is doubtful that he ever accepted the legitimacy of non-Euclidean ideas. However, later in the century, the German mathematician Felix Klein (1849–1925) showed that Cayley's insight about metrics in projective space did not need to be confined to Euclidean metrics, that non-Euclidean metrics could also be generated within projective spaces. In 1872, Klein developed this insight into the "Erlanger Program." Here, Klein developed the projective approach to defining non-Euclidean geometries into a powerful research program in which geometries, whether Euclidean or not, are defined by the invariants of different algebraic groups of transformations.

It is often difficult to see, but ultimately, space lay behind the highly abstract algebra of Klein's school. Bertrand Russell (1872–1970) was a committed student and disciple of Klein, when in 1897 he wrote that "none but a madman... would throw doubt on [geometry's rational] validity, and none but a fool would deny its objective [spatial] reference."[35] Thus, as the nineteenth century drew to a close, geometry was still the study of space.

Within less than five years, however, a twentieth-century Russell would be pursuing the route his nineteenth-century self had reserved for fools and madmen, by embracing the geometrical point of view set forth by Hilbert in his 1899 *Der Grundlagen der Geometrie*. Before it, many had quibbled with details of Euclid's argument, but in this highly influential work, Hilbert radically redefined the subject. His goal was to create a geometry the strength of which lay in the integrity of its internal structure, not in its description of space. Hilbert began with three undefined objects – point, line, and plane – but, as he later put it, he could equally easily have used the words "tables, chairs and beermugs."[36] Geometry was no longer the study of space.

[34] Arthur Cayley, "A Sixth Memoir upon Quantics," in *The Collected Works of Arthur Cayley*, 11 vols. (Cambridge: Cambridge University Press, 1889–97), 2: 92.
[35] Bertrand A. W. Russell, *An Essay on the Foundations of Geometry* (New York: Dover, 1956), p. 1.
[36] Jeremy Gray, "The Revolution in Mathematical Ontology," in *Revolutions in Mathematics*, ed. Donald Gillies (Oxford: Clarendon Press, 1992), pp. 226–48, at p. 240.

24

BETWEEN RIGOR AND APPLICATIONS
Developments in the Concept of Function in Mathematical Analysis

Jesper Lützen

In this chapter I shall illustrate some of the general trends in the development of mathematical analysis by considering its most basic element: the concept of function. I shall show that its development was shaped both by applications in various domains, such as mechanics, electrical engineering, and quantum mechanics, and by foundational issues in pure mathematics, such as the striving for rigor in nineteenth-century analysis and the structural movement of the twentieth century. In particular, I shall concentrate on two great changes in the concept of function: first, the change from analytic-algebraic expressions to Dirichlet's concept of a variable depending on another variable in an arbitrary way, and second, the invention of the theory of distributions.[1] We shall see that it is characteristic of both of the new concepts that they were initiated in a nonrigorous way in connection with various applications, and that they were generally accepted and widely used only after a new basic trend in the foundation of mathematics had made them natural and rigorous. However, the two conceptual transformations differ in one important respect: The first change had a revolutionary character in that Dirichlet's concept of function completely replaced the earlier one. Furthermore, some of the analytic expressions, such as divergent power series, which eighteenth-century mathematicians considered as functions, were considered as meaningless by their nineteenth-century successors. The concept of distributions, on the other hand, is a generalization of the concept of function in the sense that most functions (the locally integrable functions) can be considered distributions. Moreover, the theory of distributions builds upon the ordinary theory of functions, so that the theory of functions is neither superfluous nor meaningless.

---

[1] For a more extensive account of the history of the concepts of function and generalized function, see A. P. Youschkevich, "The Concept of Function up to the Middle of the 19th Century," *Archive for History of Exact Sciences*, 16 (1976), 37–85; Jesper Lützen, *The Prehistory of the Theory of Distributions* (Studies in the History of Mathematics and Physical Sciences 7) (New York: Springer Verlag, 1982).

EULER'S CONCEPT OF FUNCTION

With Leonhard Euler (1707–1783), calculus and, more generally, mathematical analysis became a science of functions. However, to Euler and his contemporaries, a function was not a mapping but "an analytic expression composed in any manner from a variable quantity and numbers or constant quantities."[2] In other words, a function was defined as a formula expressed in mathematical notation. The Eulerian concept of function is typical of the eighteenth-century style of analysis, in which conceptual analysis played only a minor role, whereas algebraical manipulations were at the center.

This so-called algebraic analysis was also characterized by its global nature. According to Euler, "a variable quantity is an indeterminate universal quantity which comprises in itself all determinate values without exception."[3] This means that a function was considered defined for all values of its variable, and all identities between functions were supposed to have general validity. For example, by the usual rule for dividing two polynomials, Euler found the identity:

$$\frac{1}{1-x} = 1 + x + x^2 + x^3 + \cdots \qquad (1)$$

He was well aware of the fact that the right-hand side was only convergent for $|x| < 1$, but he maintained the general validity of the identity, defining the sum of a (diverging) series as the value of the expression whose series expansion gives rise to the series.[4]

Euler's search for the value of the logarithm of negative and complex numbers was also characteristic of this global philosophy. He did not consider the problem as one of *defining* the function in places where it had not been defined before, but as a platonic search for the "true" value. In a more general way, Euler believed that all analytic operations and formulas were universally valid: "The differential calculus operates on variable quantities, i.e. on quantities considered generally. Therefore, if it were not generally true that $dlx = dx/x$ whatever value one attributes to x, one would never be able to use this rule, because the truth of the differential calculus is based on the generality of the rules it includes."[5]

---

[2] Leonhard Euler, *Introductio in analysin infinitorum*, vol. 1 (Lausanne, 1748), *Leonardi Euleri Opera Omnia* (Leipzig: Teubner, 1911–) (hereafter *LEOO*), ser. 1, vol. 8, p. 18. Translation in Diether Rüthing, "Some Definitions of the Concept of Function from Joh. Bernoulli to N. Bourbaki," *The Mathematical Intelligencer*, 6 (1984), 72–7.

[3] Euler, *Introductio*, 17.

[4] Leonhard Euler, *Institutiones calculi differentialis* (St. Petersburg, 1755), *LEOO*, ser. 1, vol. 10, § 111. Translated by the author unless otherwise noted.

[5] Leonhard Euler, "De la controverse entre Mrs. Leibniz et Bernoulli sur les logarithmes des nombres négatifs et imaginaires," *Mém. Acad. Sci. Berlin*, 5 (1749), 139–79. (*LEOO*, ser. 1, vol. 17, pp. 195–232.)

Thus, according to Euler, generality was the fundamental property of analysis.[6]

NEW FUNCTION CONCEPTS DICTATED BY PHYSICS

In the year he published his analytical definition of function, Euler was already advocating a generalization of it in connection with a debate on the mathematical description of vibrating strings.[7] In 1747, Jean le Rond d'Alembert (1717–1783) set up the wave equation

$$\frac{\partial^2 y}{\partial t^2} = \frac{\partial^2 y}{\partial x^2}, \qquad (2)$$

where y is the oscillation of the string as a function of time t and distance x along the string, and he argued that the general solution was of the form

$$y = \varphi(x+t) + \psi(x-t), \qquad (3)$$

where φ and ψ are "arbitrary" functions.[8] The following year, Euler published his own lucid account of d'Alembert's solution and pointed out that φ and ψ need not be given by one analytic expression.[9] For example, in order to describe the motion of a plucked violin string, one would need functions that are piecewise linear. More generally, Euler would allow functions given by various analytic expressions in various intervals, or even given by arbitrary hand-drawn curves for which the analytic expression according to Euler changes from point to point. Euler called such functions discontinuous.

D'Alembert sharply disagreed with Euler on this point. He insisted that φ and ψ in (3) must be given by one analytic expression: "In all other cases the problem cannot be solved, at least not with... the powers of the known analysis."[10] Expressing the second derivative geometrically in a way that corresponds to

$$\frac{d^2 f(z)}{dz^2} = \frac{f(z) + f(z+2\varepsilon) - 2f(z+\varepsilon)}{\varepsilon^2} \quad (\varepsilon \text{ infinitely small}), \qquad (4)$$

---

[6] For a deeper analysis of the algebraic analysis of the eighteenth century, see Hans Niels Jahnke, "Die algebraische Analysis des 18. Jahrhunderts," in *Geschichte der Analysis*, ed. H. N. Jahnke (Heidelberg: Spektrum, 1999), pp. 131–70. English edition: "The Algebraic Analysis of the 18th Century," in *A History of Analysis* (Providence, R.I.: American Mathematical Society, 2002).

[7] The debate on the vibrating string has been discussed in many works. See, e.g., C. Truesdell, *The Rational Mechanics of Flexible or Elastic Bodies, 1638–1788, LEOO*, ser. 2, vol. 11, part 2; J. R. Ravetz, "Vibrating Strings and Arbitrary Functions," *Logic of Personal Knowledge, Essays Presented to M. Polanyi on His 70th Birthday* (London: Routledge and Kegan Paul, 1961), pp. 71–88.

[8] Jean le Rond d'Alembert, "Recherches sur la courbe que forme une corde tendue mise en vibration," *Mém. Acad. Sci. Berlin*, 3 (1747), 214–19.

[9] Leonhard, Euler, "Sur la vibration des cordes," *Mém. Acad. Sci. Berlin*, 4 (1748; pub. 1750), 69–85. (*LEOO*, ser. 2, vol. 10, pp. 63–77.)

[10] Jean de Rond d'Alembert, "Addition au mémoire sur la courbe que forme une corde tendue mise en vibration," *Mém. Acad. Sci. Berlin*, 6 (1750), 355–66.

he argued that if ψ has a radius of curvature that jumps at particular values of $x - t$ then $\psi(x - t)$ would not satisfy the wave equation. Indeed, according to (4) the second partial derivative with respect to x would correspond to the radius of curvature to the right of $x - t$, whereas the derivative with respect to t would correspond to the radius of curvature to the left of $x - t$. In his later papers, d'Alembert realized that this does not necessitate that ψ be one analytic expression, and he came close to formulating the classical requirement of a solution ψ, namely, that it be twice differentiable.

Euler agreed with d'Alembert that the known analysis did not deal with functions that are not given by one analytic expression, but he insisted that it was the job of the mathematical community to generalize analysis to such functions, and he gave various arguments to support his claim that $\psi(x - t)$ is a solution of the wave equation, whether or not ψ is a single analytic expression.[11]

The debate between Euler and d'Alembert on the vibrating string was a discussion based on two very different attitudes toward mathematics. D'Alembert (at least here) valued rigor so highly that he was willing to limit radically the range of his own mathematical discovery. Euler, on the other hand, insisted that mathematics must be made general enough to deal with all situations in physics, and he was willing to extend his own concept of function and to use somewhat questionable arguments to attain this goal.

The discussion also involved Daniel Bernoulli (1700–1782), whose music theoretic approach made him suggest that arbitrary functions, or at least those that could describe vibrating strings, could be written on the form

$$y = a \sin x + b \sin 2x + c \sin 3x + \cdots \tag{5}$$

However, on this issue Euler and d'Alembert joined forces with a third combatant, Joseph Louis de Lagrange (1736–1813), and argued that only very special functions would be represented as such a trigonometric series.

The discussion of the vibrating string focused on two related problems: the correct concept of function and the concept of a solution of a (partial) differential equation. I shall now pursue the first point and come back to the second later.

DIRICHLET'S CONCEPT OF FUNCTION

In his *Institutiones Calculi Differentialis* of 1755, Euler offered a new definition of a function: "Thus when x denotes a variable quantity, then all quantities that depend on x in any manner whatever or are determined by it are called functions of x."[12]

---

[11] Jesper Lützen, "Euler's Vision of a General Partial Differential Calculus for a Generalized Kind of Function," *Mathematics Magazine*, 56, 299–306.
[12] Euler, *Institutiones*, p. 4.

Thus, any correspondence between two variables is called a function. It is possible that this definition was inspired by the discussion of the vibrating string. However, it is conspicuous that even after 1755, Euler never appealed to this definition in his continued contributions to the dispute, but referred to discontinuous functions given by changing analytic expressions. Variations of Euler's new definition were repeated in textbooks by Lagrange (1801), Sylvestre François Lacroix (1810–19), and Augustin Louis Cauchy (1821).[13] For example, Lacroix wrote: "Every quantity whose value depends on one or several other quantities is called a function of these quantities, whether one knows or does not know the operations one has to employ to get from the latter quantities to the former."[14]

Yet following Hermann Hankel (1870), the concept of a function as an arbitrary dependence between variables is usually named after Johann Peter Gustav Lejeune Dirichlet (1805–1859), who in his important paper of 1837 on the convergence of Fourier series defined (continuous) functions as follows:

> If every x gives a unique y in such a way that when x runs continuously through the interval from a to b then $y = f(x)$ varies little by little, then y is called a continuous function of x in this interval. It is not necessary that y depends on x according to the same law in the entire interval. One does not even need to think of a dependence that can be expressed through mathematical operations.[15]

In his earlier French paper of 1829 on the same subject, Dirichlet had even given the function

$$\varphi(x) = \begin{cases} c & \text{for } x \in \mathbb{Q} \\ d & \text{for } x \notin \mathbb{Q} \end{cases} \qquad (6)$$

as an example of a function that cannot be integrated.[16] This was the first explicitly stated function that was not given through one or several analytic expressions.

It may seem unjust to call this new concept of function after Dirichlet, rather than after Euler, when the latter formulated it three-quarters of a

---

[13] Translations of many definitions of functions can be found in Diether Rüthing, "Some Definitions of the Concept of Function from Joh. Bernoulli to N. Bourbaki," pp. 72–7.

[14] Sylvestre François Lacroix, *Traité du calcul différentiel et du calcul intégral. Seconde édition, revue et augmentée*, 3 vols. (Paris, 1810–19), p. 1.

[15] Hermann Hankel, *Untersuchungen über die unendlich oft oscillirenden und unstetigen Funktionen* (Tübingen, 1870), *Mathematische Annalen*, 20 (1882), 63–112. (*Ostwalds Klassiker der exakten Wissenschaften* [Leipzig: Akademische Verlagsgesellschaft Geest und Portig, 1889–] [hereafter *Ostwalds Klassiker*], vol. 153, pp. 44–112.) Johann Peter Gustav Lejeune Dirichlet, "Über die Darstellung ganz willkürlicher Funktionen durch sinus- und cosinus-Reihen," *Repertorium der Physik*, 1 (1837), 157–74 (*Werke*, vol. 1, pp. 133–60) (*Ostwalds Klassiker*, vol. 116, pp. 3–34).

[16] Dirichlet, "Sur la convergence des séries trigonométriques qui servent à représenter une fonction arbitraire entre les limites données," *Journal für die reine und angewandte Mathematik*, 4 (1829), 157–69. (*Werke*, vol. 1, pp. 117–32).

century earlier. It is, however, justified if one takes into consideration how the new concept was used. Dirichlet used it consistently in his proof, whereas Euler does not seem to have been aware that his new concept was different from the earlier one. In fact, the whole machinery built up in Euler's *Introductio* of 1748 is used unconditionally in his book of 1755. A similar remark holds for Euler's successor, Lacroix, as well as Lagrange, who explicitly mentioned that any function is an analytic expression.[17] Thus, before the 1820s there was a gulf between the general definition of a function and its use.

Before Dirichlet, one mathematician consistently insisted that functions were given as a dependence between two variables, namely Dirichlet's teacher, Joseph Fourier (1768–1830). In his *Theorie analytique de la chaleur* (1822), he defined a function as follows:

> In general, the function $f(x)$ represents a succession of values or ordinates each of which is arbitrary. An infinity of values being given to the abscissa x, there are an equal number of ordinates $f(x)$. All have actual numerical values, either positive or negative or nul. We do not suppose these ordinates to be subject to a common law; they succeed each other in any manner whatever, and each of them is given as it were a single quantity.[18]

Like Euler, Fourier needed such a definition of a function in order to be able to deal with a physical situation in all its generality. In Fourier's case, it was heat conduction in a solid. He set up the partial differential equation describing the situation and succeeded in solving it in special cases using separation of variables. This led him to the conclusion that any arbitrary function could be represented by a trigonometric series

$$\pi f(x) = a_0 + \sum_{n=1}^{\infty} (a_n \cos nx + b_n \sin nx), \tag{7}$$

where

$$a_0 = \frac{1}{2} \int_{-\pi}^{\pi} f(x) dx, \quad a_i = \int_{-\pi}^{\pi} f(\alpha) \cos \alpha \, d\alpha, \tag{8}$$

$$b_i = \int_{-\pi}^{\pi} f(\alpha) \sin \alpha \, d\alpha, \quad i = 1, 2, \ldots$$

I shall return to his attempt to prove the convergence of the so-called Fourier series (7). Here it suffices to mention that he realized that the infinite series in (7) only represents $\pi f(x)$ in the interval $(-\pi, \pi)$. This went against the eighteenth century belief in the generality of analytic formulas and explains many of Euler's arguments against Daniel Bernoulli's claims.

---

[17] Joseph Louis Lagrange, *Leçons sur le calcul des fonctions* (Paris, 1801), *Oeuvres*, vol. 10, Leçons 1, def. 2.
[18] Joseph Fourier, *Théorie analytique de la chaleur* (Paris, 1822), *Oeuvres*, vol. 1, in particular p. 430. Translation in Rüthing, "Some Definitions of the Concept of Function," pp. 72–7.

EXIT THE GENERALITY OF ALGEBRA – ENTER RIGOR

The reason that the general concept of function took so long to enter into the core of mathematical analytical reasoning was probably that it sat very uncomfortably with the prevalent algebraic formal ideas on foundations. Indeed, Euler's algebraic and general formalistic style was reinforced toward the end of the eighteenth century when Lagrange argued that any function could be expanded in the form

$$f(x + i) = f(x) + p(x)i + q(x)i^2 + r(x)i^3 + \cdots \qquad (9)$$

and then *defined* "the derived function" $f'(x)$ to be equal to $p(x)$.[19] In this way, he believed he had based analysis on algebra (after all, power series are just polynomials of infinite degree), circumventing earlier problematic notions of infinitesimals, fluxions, or limits. This algebraic notion of analysis prevailed until 1821 and with it also the first Eulerian concept of function.

It was a total reorientation of the foundation of analysis that made the new general concept of function become a natural and integrated part of mathematics. This reorientation was suggested independently by at least three mathematicians, Bernhard Boltzano (1781–1848), Carl Friedrich Gauss (1777–1855), and Cauchy (1789–1857).[20] The latter, whose works were by far the most influential, developed his ideas in connection with his teaching of analysis to the students at the Ecole Polytechnique in Paris. In the introduction to his textbook *Cours d'analyse* (1821), he wrote:

> As for methods, I have sought to give them all the rigor that one demands in geometry, in such a way as never to revert to reasoning drawn from the generality of algebra. Reasoning of this kind, although commonly admitted, particularly in the passage from convergent to divergent series and from real quantities to imaginary expressions, can, it seems to me, only occasionally be considered as inductions suitable for presenting the truth, since they accord so little with the precision so esteemed in the mathematical sciences. We must at the same time observe that they tend to attribute an indefinite extension to algebraic formulas, whereas in reality the larger part of these formulas exist only under certain conditions and for certain values of the quantities that they contain.[21]

---

[19] Joseph Louis Lagrange, *Théorie des Fonctions Analytiques* (Paris, 1797), 2d ed. 1813 (*Oeuvres*, vol. 9).

[20] The history of the foundation of analysis in the nineteenth century is described in Umberto Bottazzini, *The Higher Calculus* (New York: Springer Verlag, 1986) and in Jesper Lützen, "Grundlagen der Analysis im 19. Jahrhundert," *Geschichte der Analysis*, pp. 191–244. To appear as "The Foundation of Analysis in the 19th Century," in the English translation of the book.

[21] Augustin Louis Cauchy, *Cours d'analyse de l'Ecole Royale Polytechnique, 1re Partie. Analyse Algebrique* (Paris, 1821), in a new edition with an interesting introduction by Umberto Bottazzini (Bologna: Editrice CLUEB, 1990); translation from Bottazzini's *The Higher Calculus*. Another analysis of Cauchy's new calculus and its roots in earlier works, in particular in those of Lagrange, is in Judith V. Grabiner, *The Origins of Cauchy's Rigorous Calculus* (Cambridge, Mass.: MIT Press, 1981).

He admitted that "in order to remain constantly true to these principles I have been forced to admit certain propositions that may seem a bit severe at first sight. For example in chapter VI I announce that a divergent series has no sum."[22]

Thus, Cauchy insisted that analytic expressions (or theorems) are not necessarily generally true. For example, the formula

$$\frac{1}{1-x} = 1 + x + x^2 + x^3 + \cdots, \qquad (10)$$

which Euler had given universal validity, only held true for Cauchy when $x \in (-1, 1)$. Outside of this interval the series is divergent and therefore has no sum. Moreover, he maintained that functions such as $\sin x$ were only defined where we have defined them, and if we want to extend them, for example, from the real axis to the complex numbers, we must explicitly give a separate definition valid for nonreal complex numbers. One cannot appeal to the generality of algebra as Euler had done. Cauchy also pointed out that Euler's concept of continuity was not well defined. Indeed, the function $|x|$ can be written in various ways:

$$|x| = \begin{cases} x & \text{for } x \geq 0 \\ -x & \text{for } x < 0 \end{cases} = \sqrt{x^2} = \frac{2}{\pi} \int_0^\infty \frac{x^2}{t^2 - x^2} dt \qquad (11)$$

of which the first is clearly discontinuous in Euler's sense, because it consists of *two* analytic expressions, whereas the last two are continuous, that is, given by one analytic expression.[23] Cauchy replaced Euler's obscure definition by a more precise one defining a function to be continuous in an interval if it is one-valued, finite, and the difference

$$f(x + \alpha) - f(x) \qquad (12)$$

"decreases indefinitely with α."[24]

It is characteristic of this new definition, as well as Cauchy's other definitions and theorems, that it is local in character compared with Euler's concept: Cauchy defined continuity in an interval, whereas Euler's concept concerns the global behavior of the function. Since Cauchy, the definition has been localized even further, to continuity in a point. This is, of course, a highly unintuitive concept – what is continued by a function that is only continuous in one point? It is no wonder that Cauchy did not go that far.

Cauchy's definitions were also distinguished by being operational in the sense that they were used explicitly in proofs of theorems. For example,

---

[22] Cauchy, *Cours d'analyse*, p. iv.
[23] Augustin Louis Cauchy, "Mémoire sur les fonctions continuées ou discontinuées," *Comptes Rendus de l'Académie des Sciences, Paris*, 18 (1844), 116–30 (in *Oeuvres*, ser. 1, vol. 8, pp. 145–60, in particular p. 145).
[24] Cauchy, *Cours d'analyse*, p. 43.

the concept of continuity is used in a crucial way in Cauchy's solution of functional equations, in his proof of the binomial theorem, and in his proof of existence of the integral of a continuous function. This may seem a matter of course for a modern mathematician, but in previous works on analysis, definitions had not entered into proofs in a precise way. For example, Euler never used his concept of continuity in the proof of a theorem. One may even argue that Cauchy's definitions were generated by the proofs.

Although Cauchy's new orientation in analysis was generally hailed for its rigor, some problems were detected over time. For example, around 1870 it became clear that one has to distinguish between pointwise continuity and uniform continuity in an interval.[25] This distinction was not made by Cauchy. The argument was also made that one has to distinguish between pointwise and uniform convergence of a series of functions. Cauchy had very carefully defined convergence of a series of numbers, but he had given no special definition of convergence of a series of functions. On the other hand, he had "proved" that a convergent series of continuous functions will have a continuous sum.[26] Intepreting Cauchy to mean pointwise convergence, Niels Henrik Abel pointed out that Fourier series of discontinuous functions yielded counterexamples to this theorem.[27] George Gabriel Stokes and Philipp von Seidel showed that at points where the sum function is discontinuous, the convergence becomes "infinitely" or "arbitrarily" slow. Finally, Karl Weierstrass (1815–1897) and Eduard Heine (1821–1881) showed that Cauchy's theorem holds true if the convergence is uniform in the given interval.[28]

The problems spotted in Cauchy's calculus were mostly due to definitions that turned out to be imprecise. They were made precise by supplying the necessary quantifiers, ε's and δ's. Cauchy had used such techniques in many of his proofs, but it was mainly Weierstrass who showed how to supply rigor to Cauchy's *definitions* in this way. This happened in Weierstrass's lectures at Berlin University from 1857 to 1887. Although he did not publish these introductory parts of his lectures, his ideas quickly became known through his many German and foreign students.

Another major problem that was gradually spotted in Cauchy's calculus was his unsatisfactory basis for the real numbers. To Cauchy, numbers arise from the line by choosing a unit. However, with such a definition,

---

[25] Eduard Heine, "Die Elemente der Funktionenlehre," *Journal für die reine und angewandte Mathematik*, 74 (1872), 172–88.

[26] Cauchy, *Cours d'analyse*, chap. VI, Theorem 1.

[27] Niels Henrik Abel, "Untersuchung über die Reihe $1 + \frac{m}{1}x + \frac{m(m-1)}{1 \cdot 2}x^2 + \frac{m(m-1)(m-2)}{1 \cdot 2 \cdot 3}x^3 + \cdots,$" *Journal für die reine und angewandte Mathematik*, 1 (1826), 311–39, in *Oeuvres complètes*, vol. 1, pp. 219–50 (*Ostwalds Klassiker*, vol. 71, Leipzig, 1921).

[28] See references in Bottazzini, *The Higher Calculus*, and Lützen, "The Foundation of Analysis." Lakatos has emphasized this cause of events as a typical example of a development through "proofs and refutations"; see Imre Lakatos, *Proofs and Refutations: The Logic of Mathematical Discovery*, ed. J. Worrall and E. Zahar (Cambridge: Cambridge University Press, 1976).

it is in fact impossible to give rigorous proofs of three key theorems in Cauchy's *Cours d'analyse:* the convergence of Cauchy series (or fundamental sequences); the existence of the definite integral of a continuous function; and the intermediate value theorem, which states that if a continuous function attains positive as well as negative values, then it attains the value zero. These problems were discovered by Richard Dedekind (1831–1916) and Weierstrass independently, and both found that the way out was to *construct* the real numbers from the rational (and thereby the natural) numbers, instead of relying on geometry.[29] Another construction, based on Weierstrass's ideas, was published by Georg Cantor (1872) and Heine (1872). In this way, analysis was emancipated: Geometric intuition was banished, and the basic "space" in which analysis worked was defined through purely arithmetical notions. "Arithmetization of analysis" became a catchword.

THE DREADFUL GENERALITY OF FUNCTIONS

Cauchy's flat rejection of the generality of algebra, as well as the subsequent move toward rigor, made Euler's concept of function obsolete and even meaningless. However, it was more than half a century before Dirichlet's concept of function was generally accepted. As late as 1870 Hankel wrote: "One mathematician defines the functions essentially in Euler's sense; another requires that y must change with x according to a law, but fails to explain this vague concept; a third defines it in Dirichlet's way; a fourth does not define it at all: but all of them deduce from their concept consequences that are not contained in it."[30]

It is indeed a fact that even mathematicians who publicly accepted Dirichlet's concept of function were often of the opinion that functions generally behave nicely. For example, in his textbooks Cauchy defined the derivative of f as the limit of $(f(x + \alpha) - f(x))/\alpha$ for α tending towards zero, and he was careful to state that the limit is denoted $f'(x)$ *when it exists*.[31] However, later in the book, whenever he differentiated a function, he only assumed it to be continuous. In this way, he clearly gave the impression that a continuous function was differentiable (at least almost everywhere). His colleague at the Ecole Polytechnique, André Marie Ampère (1775–1836), even provided a "proof" of this theorem.

In general, one can say that Cauchy's textbooks gave the impression that analysis was generally valid in the domain of continuous functions. Though many mathematicians continued to dream about such a general domain for

---

[29] Richard Dedekind, *Stetigkeit und irrationale Zahlen* (Braunschweig, 1872) (*Werke*, vol. 3, 315–34). English trans. *Essays on the Theory of Numbers* (New York: Dover, 1963).
[30] Hankel, *Untersuchungen*, "Einleitung."
[31] Augustin Louis Cauchy, *Résumé des leçons données a l'école royale polytechnique sur le calcul infinitesimal. Tome premier* (Paris, 1823) (*Oeuvres*, ser. 2, vol. 4, pp. 5–261, esp. p. 22).

all of analysis, one gradually had to give up the dream owing to the discovery (or construction) of a series of pathological functions. Dirichlet's function (6) may be considered an early pathological function, but the most famous one is Weierstrass's example (1872) of a continuous but nowhere differentiable function, which showed that Ampère had been entirely mistaken.[32] Two connected conclusions were drawn from these pathological functions: (1) General functions do not behave nicely, and (2) Analysis is not generally applicable within a fixed domain of functions. Each theorem of analysis has its own regularity requirements for the functions involved. In the eighteenth century, a typical theorem of analysis looked like this:

$$y(x) \text{ has max/min when } \frac{dy}{dx} = 0 \text{ or } \infty.$$

After 1870 a typical analytic theorem looked like this:

> Let $f(x)$ be a ($C^j/L^1, \ldots$) function defined in a (closed/open/..., bounded...) subset of $\mathbb{R}$. Moreover assume that f' (or ...) be... on ... and assume that ... and .... Then (some formula).

Not all mathematicians welcomed this new style in analysis: For example, the most important mathematician toward the end of the nineteenth century, Henri Poincaré (1854–1912), wrote:

> For half a century we have seen a mass of bizarre functions which appear to be forced to resemble as little as possible honest functions which serve some purpose....
>
> In former times when one invented a new function it was for a practical purpose; today one invents them purposely to show up defects in the reasoning of our fathers, and one will deduce from them only that.[33]

It turned out to be impossible to go back to the more innocent style, however, once the snakes of pathological functions had been let loose in the Eden of Analysis. In fact, neither Poincaré nor Charles Hermite (1822–1901), who agreed with him, had any suggestion of how to avoid the evils of pathological functions. Other mathematicians, however, tried to limit the general concept of function. For example, Weierstrass declared that Dirichlet's general concept of function was "totally untenable and unfruitful. In fact it is impossible to deduce any general properties of functions from it."[34] Instead,

---

[32] Karl Weierstrass, "Über continuirliche Funktionen eines reellen Arguments, die für keinen Werth des letzteren einen bestimmten Differentialquotienten besitzen," in *Werke*, vol. 2, pp. 71–4.

[33] Henri Poincaré, "La logique et l'intuition dans la science mathématique et dans l'enseignement," *L'enseignement mathématique* I (1899), 157–62 (*Oeuvres*, vol. II, pp. 129–33). Translation from Morris Kline: *Mathematical Thought from Ancient to Modern Times* (New York: Oxford University Press, 1972), p. 973.

[34] According to a quotation by Hurwitz, see Pierre Dugac, "Eléments d'analyse de Karl Weierstrass," *Archive for History of Exact Sciences*, 10 (1973), 41–176, esp. p. 116.

in his theory of complex functions, he *defined* an analytic function as a collection of power series

$$\sum_{n=0}^{\infty} a_n(z-z_0)^n, \qquad (13)$$

which are analytic continuations of each other. In this way, he tried to return to an algebraic concept of function similar to that of Euler and Lagrange.

Another attempt at restricting the general concept of function came from the French analysts René Baire, Emile Borel, and Henri Lebesgue. From a logical point of view, they argued that a function is only defined if one has a way to construct $f(x)$ for all x in its domain of definition. They did not quite agree on what a "construction" should mean, but their ideas were taken up and continued in the twentieth century by L. E. J. Brouwer's so-called intuitionistic school in mathematics, which called for a general reform of the foundation of mathematics along constructive lines.

Despite these attempts to restrict Dirichlet's general concept of function, the majority of analysts around 1900 chose to accept it. From the domain of applications some even called out for further generalizations.

THE DELTA "FUNCTION"

Probably the best-known generalized function is Dirac's δ-function. In his classical book on quantum mechanics, Paul A. M. Dirac (1902–1984) defined it as a function that is zero everywhere except at $x = 0$, at which point it is so infinite that $\int_{-\infty}^{\infty} \delta(x)dx = 1$.[35]

He stressed that the δ-function has the fundamental property that

$$f(x) = \int_{-\infty}^{\infty} \delta(x-\alpha) f(\alpha) d\alpha \qquad (14)$$

for "all" functions f. The δ-function rather naturally suggests itself as a description of the density of a unit point mass. Indeed, if $f(x)$ describes the density of a mass distribution, then $\int_a^b f(x)dx$ is the mass contained between the limits a and b. A unit point mass at zero must therefore be represented by a density function that is zero everywhere except at $x = 0$. But if an ordinary function only has a finite value at zero, its integral $\int_{-\infty}^{\infty} f(x)dx$ is equal to zero; thus, in order to represent a unit mass, it must be infinite in the way described by Dirac.

The δ-function was not new with Dirac. Fourier had already introduced it in his "proof" of convergence of the general Fourier series (7). Fourier

---

[35] Paul A. M. Dirac, *The Principles of Quantum Mechanics* (Oxford: Clarendon Press, 1930). The basic ideas were already published in "The Physical Interpretation of the Quantum Dynamics," *Proceedings of the Royal Society*, A, no. 113 (1926), 621–41.

was first confronted with the δ-function when he inserted the expressions (8) into (7) and interchanged summation and integration (this step was not considered problematic until around 1870). Thereby he got

$$F(x) = \frac{1}{\pi} \int_{-\pi}^{\pi} F(\alpha) d\alpha \begin{cases} \frac{1}{2} + \cos x \cos \alpha + \cos 2x \cos 2\alpha + \cdots \\ + \sin x \sin \alpha + \sin 2x \sin 2\alpha + \cdots \end{cases} \quad (15)$$

which by a simple trigonometric formula yields

$$F(x) = \frac{1}{\pi} \int_{-\pi}^{\pi} F(\alpha) \left(\frac{1}{2} + \sum_{n=1}^{\infty} \cos n(x-\alpha) \right) d\alpha. \quad (16)$$

He concluded: "The expression $\frac{1}{2} + \sum \cos i(x-\alpha)$ represents a function of x and α such that if one multiplies it with an arbitrary function $F(\alpha)$, and after writing $d\alpha$ one integrates between the limits $\alpha = -\pi$ and $\alpha = +\pi$, one has changed the given function $F(\alpha)$ to a similar function of x, multiplied by the semicircumference."[36]

If we compare this with Dirac's definition we see that

$$\frac{1}{2} + \sum \cos n(x-\alpha) = \pi \delta(x-\alpha), \quad (17)$$

at least in the interval $[-\pi, \pi]$.

In a footnote in Fourier's collected works from 1888, the editor Gaston Darboux commented: "Since the series $\frac{1}{2} + \cos(x-\alpha) + \cos 2(x-\alpha) + \cdots$ has an indeterminate sum, one cannot attach any sense to the expression $\frac{1}{2} + \sum_{i=1}^{\infty} \cos i(x-\alpha)$."

This is a typical reaction of a rigorous classical analyst of the late nineteenth century. Despite the banishment of the δ-function by rigorous mathematicians, it continued to pop up in connection with applied mathematics. Gustav Robert Kirchhoff (1824–1887) used it in 1882 in connection with a discussion of the fundamental solution to the wave equation, and, more generally, it explicitly or implicitly showed up in considerations of the so-called Green's function.[37] Among electrical engineers, the δ-function became quite popular after it had been introduced by Oliver Heaviside (1850–1925). He developed a rather peculiar version of the operational calculus in which he calculated freely with differential operators. In particular, he was interested in the response of an electrical circuit when a switch (a telegraph key) was suddenly connected. In this case, he represented the resulting voltage by QH where H is the Heaviside function

$$H(t) = \begin{cases} 0 & \text{for } t < 0 \\ 1 & \text{for } t \geq 0. \end{cases}$$

---

[36] Fourier, *Théorie analytique de la chaleur*, § 235.
[37] For a later example, see Richard Courant and David Hilbert, *Methoden der Mathematischen Physik*, vol. 1 (Berlin: Springer Verlag, 1924), p. 274. In English trans., *Methods of Mathematical Physics* (New York: Springer Verlag, 1962).

The current could then in particular cases be represented as $\frac{d}{dt}Q$, or pQ as he would write it. He discussed the result, which is, of course, the δ-function, in his characteristic polemical style:

> Since Q is constant for any finite value of time, the result is zero.... Is this nonsense? Is it an absurd result, indicating the untrustworthy nature of the operational mathematics, or at least indicative of some modifications of treatment being desirable? Not at all.... We have to note that if Q is any function of time, then pQ is its rate of increase. If, then, as in the present case Q is zero before and constant after $t = 0$, pQ is zero except when $t = 0$. It is then infinite. But its total amount is Q. That is to say $p\,1\,[pH(t)]$ means a function of t which is wholly concentrated at the moment $t = 0$, of total amount 1. It is impulsive so to speak.[38]

It is probably not a coincidence that Dirac, who really made the δ-function known among mainstream phycisists and mathematicians, had been trained as an electrical engineer. The δ-function entered, in a fundamental way, into his very influential version of quantum mechanics (1926, 1932). Around this time, some applied mathematicians or electrical engineers tried to bring rigor to Heaviside's operational calculus, including the δ-function, mostly by using the Laplace transform.[39] However, most mathematicians rejected its use. For example, John von Neumann (1903–1957) explicitly declared that it "lies outside the usual mathematical methods," so instead, he put forward a foundation of quantum mechanics based on an axiomatic introduction to Hilbert space and operators on such a space.[40]

The main motivation for the rigorous introduction of the δ-function and other generalized functions came from other corners, namely, from the attempt to generalize (1) the concept of a solution to differential equations, and (2) the Fourier transform.

GENERALIZED SOLUTIONS TO DIFFERENTIAL EQUATIONS

From the time of Euler, various attempts were made to generalize the concept of solution of an nth order differential equation to functions that are not n times differentiable. The overall strategy was to replace the differential equation with another problem having a larger set of solutions. The solutions to the latter problem can then be considered generalized solutions of the differential equation. This replacement could be of a *physical nature*. If the

---

[38] Oliver Heaviside, *Electromagnetic Theory*, vol. 2 (London: Office of "The Electrician," 1899), § 238–42.
[39] Jesper Lützen, "Heaviside's Operational Calculus and the Attempts to Rigorize It," *Archive for History of Exact Sciences*, 21 (1979), 161–200.
[40] John von Neumann's ideas first appeared as a paper in 1927: "Mathematische Begründung der Quantenmechanik," *Göttinger Nachrichten* (1927), 1–57, and then as a book, *Mathematische Grundlagen der Quantenmechanik* (Berlin: Springer Verlag, 1932). The quotation is from § 3 of the book.

differential equation is a model of a physical situation, one can choose another model with more solutions. For example, in the discussion of the vibrating string, Lagrange considered the string as the limiting case of a finite number of point masses distributed equidistantly along a weightless string. Letting the number of point masses tend to infinity while keeping their total mass constant, he concluded that Euler was right when he claimed that even nonanalytic functions φ and ψ could occur in the general expression (3) for the motion of the string. In the nineteenth century, similar ideas were employed by Bernhard Georg Riemann (1826–1866) and E. B. Christoffel (1829–1900) in their treatment of shock waves.[41]

Of more interest here are the different mathematical methods of generalization. One method was to *replace several limit processes with one*. This was also used by Lagrange. Recall that d'Alembert had argued that $\psi(x-t)$ was not a solution of the wave equation in points where its second derivative makes a jump. His argument was based on the asymmetric expression (4). Lagrange, on the other hand, pointed out that if one uses the symmetric expression

$$\frac{d^2 f(z)}{dz^2} = \frac{f(z-\varepsilon) + f(z+\varepsilon) - 2f(z)}{\varepsilon^2} \quad (\varepsilon \text{ infinitely small}) \qquad (18)$$

instead, $\psi(x-t)$ would satisfy the wave equation even in points where the second derivative makes a jump.[42] A similar technique was applied by Riemann in his study of trigonometric series.[43] The second derivative classically involves two limiting procedures, whereas the right-hand side of (18) involves one. In Lagrange's case, one cannot speak of the generalization of a classical concept of solution, because such a classical concept had not yet been formulated, and even Riemann did not put his idea forward as a generalization. However, the method was used explicitly as a method of generalization in 1908 when H. Petrini generalized the Laplace operator.[44]

Another method of generalization can be called the test curve or test surface method. It had its origin in potential theory and was based on Green's theorem:

$$\int_\Omega (u\Delta v - v\Delta u)d\bar{x} = \int_{\partial\Omega} \left(v\frac{\partial u}{\partial n} - u\frac{\partial v}{\partial n} \right) ds. \qquad (19)$$

If we let $u \equiv 0$ we have

$$\int_\Omega \Delta v\, d\bar{x} = \int_{\partial\Omega} \frac{\partial v}{\partial n} ds. \qquad (20)$$

---

[41] Lützen, *The Prehistory*, § 14.
[42] Joseph Louis Lagrange, "Nouvelles recherches sur la nature et la propagation du son," *Miscellanea Taurenencia*, 2 (1760–1) (*Oeuvres*, vol. 1, pp. 151–332).
[43] Bernhard Georg Riemann, "Über die Darstellbarkeit einer Funktion durch eine trigonometrische Reihe," *Anhandlungen der Gesellschaft der Wissenschaft zu Göttingen Mathematische Klasse*, 13 (1867; pub. 1868), 133–52.
[44] H. Petrini, "Les dérivées premiers et secondes du potentiel," *Acta Mathematica*, 31 (1908), 127–332.

This leads to the
Definition. v is a generalized solution to $\Delta v = 0$ if

$$\int_S \frac{\partial v}{\partial n} ds = 0 \qquad (21)$$

for all (suitable) curves S. In this way we do not need to suppose that v is twice differentiable but only that it is once differentiable.

This method was suggested by Maxime Bôcher (1867–1918), who took the curves to be circles.[45] He even showed that any generalized solution to $\Delta v = 0$ is a usual solution. This turns out to be true of all reasonable methods of generalization of Laplace's equation. Yet the method leads to generalized solutions of other equations. In 1913, Hermann Weyl (1885–1955) defined Δv (generalized) to be a function that satisfies (in $\mathbb{R}^3$)

$$\int_\Omega \Delta v \, d\bar{x} = -\int_{\partial\Omega} \frac{\partial v}{\partial n} ds \qquad (22)$$

for all suitable domains Ω.[46] He showed that with this generalized meaning of Δv, the Newtonian potential

$$v(p) = \int \frac{1}{r(p, p')} f(p') \, dp', \qquad (23)$$

where f is any continuous function, is a generalized solution to Poisson's equation

$$\Delta v = -4\pi f, \qquad (24)$$

but it need not be an ordinary solution.

A third method of generalizing the concept of a solution to a differential equation is the *test function method*. If, in Green's theorem (19), we fix Ω but choose various functions u such that $u = \frac{\partial u}{\partial n} = 0$ at the boundary, then we have

$$\int_\Omega u \Delta v \, d\bar{x} = \int_\Omega v \Delta u \, d\bar{x}. \qquad (25)$$

This leads to the
Definition. v is a generalized solution of $\Delta v = f$ in Ω if

$$\int_\Omega v \Delta u \, d\bar{x} = \int_\Omega u f \, d\bar{x} \qquad (26)$$

for all "test functions" $u \in \mathcal{C}_c^2$, that is, twice continuously differentiable functions with compact support in Ω.

---

[45] Maxime Bôcher, "On Harmonic Functions in Two Dimensions," *Proceedings of the American Academy of Science*, 41 (1905–6), 577–83.
[46] Hermann Weyl, "Über die Randwertaufgabe der Strahlungstheorie und asymptotische Spectralgesetze," *Journal für die reine und angewandte Mathematik*, 143 (1913), 177–202.

This method was suggested explicitly first by Norbert Wiener (1894–1964) in a paper of 1926 on the operational calculus.[47] More generally, he remarked that when L is a second-order linear differential operator, there exists an adjoint operator L' such that

$$\int_{\mathbb{R}^n} L(v) u\, d\bar{x} = \int_{\mathbb{R}^n} v L'(u)\, dx \qquad (27)$$

whenever u and v are sufficiently regular and u has compact support. Therefore he could define: v is a generalized solution of $L(v) = 0$ if

$$\int_{\mathbb{R}^n} v L'(u)\, d\bar{x} = 0 \qquad (28)$$

for all sufficiently regular test functions u with compact support (the name "testing function" is due to Salomon Bochner [1945]). Again we see that in the equation (28), v is not differentiated at all, and so we can allow nondifferentiable generalized solutions. The test function method was anticipated by Lagrange (1761) and Axel Harnack (1887) and used by Jean Leray (1934), Sergei Sobolev (1937), Richard Courant and David Hilbert (1937), Kurt Otto Friedrichs (1939), and Weyl (1940). It was mostly used for hyperbolic partial differential equations. For example, with this definition, $\varphi(x+t) - \psi(x-t)$ is a generalized solution of the wave equation for any function φ and ψ. Thus, Euler was finally vindicated.

This sketch of the different methods of generalization does not reflect the historical driving forces behind the development. In general, it was not a wish for further clarification of a specific method that led mathematicians to work on these ideas. Rather, it was problems, often of a physical nature, that drove the development. For example, Leray was interested in hydrodynamics and therefore tried to generalize the concept of a solution of Navier Stokes's equation; although he used the test function method, he did not build on the ideas of Wiener, but he was inspired by C. W. Oseen's completely different generalization of Navier Stokes's equation.

DISTRIBUTIONS: FUNCTIONAL ANALYSIS ENTERS

In 1945, the French mathematician Laurent Schwartz (b. 1915) had the idea that in order to define generalized solutions to a differential equation, it was preferable to generalize the concept of function, such that any (generalized) function had generalized derivatives of any order.[48] He called his generalized functions "distributions" because they generalize the idea of a distribution

---

[47] Norbert Wiener, "The Operational Calculus," *Mathematische Annalen*, 95 (1926), 557–84, esp. § 43.
[48] Laurent Schwartz, "Généralisation de la notion de fonction, de dérivation, de transformation de Fourier, et applications mathématiques et physiques," *Annales de l'Université de Grenoble. Sect. Sci. Math. Phys.*, 21 (1945; pub. 1946), 57–74.

of mass or electricity. His definition was based on the observation that any (locally integrable) function gives rise to a functional T (i.e., a mapping that takes functions φ into real numbers) defined by

$$T(\varphi) = \int_{\mathbb{R}} f(x)\varphi(x)dx \qquad (29)$$

for any test function φ in C_c^{∞}.

He therefore defined a distribution as any functional on C_c^{∞} that is continuous in a certain way. Moreover, he defined the derivative of such a distribution T by the formula

$$\frac{d}{dx}T(\varphi) = -T\left(\frac{d}{dx}\varphi\right). \qquad (30)$$

According to the formula for partial integration, this generalizes the concept of differentiation of ordinary functions. In this way, any (locally integrable) function (and even any distribution) is differentiable infinitely often, but the derivatives are distributions, not necessarily functions.

The definition and theory of distributions rested heavily on the previously developed theories of functionals and, more generally, on functional analysis. This highly abstract branch of analysis emerged during the period 1907 to 1932 by the confluence of many different technical and conceptual developments: Partly as a result of the acceptance of non-Euclidean geometry, the idea of a space was detached from physical space around 1870, and various types of spaces (high-dimensional spaces, Riemannian manifolds, configuration spaces in mechanics, and so forth) were introduced. Also spaces of functions began to appear around 1900. In particular, the Italian school developed abstract theories of operators on such spaces, and Maurice Frechet (1878–1973) introduced topological notions. Moreover, functionals were considered by Vito Volterra and Jacques Hadamard in connection with variational calculus. However, these abstract ideas had little impact before they were combined with Hilbert's technical work on integral equations and with the new concept of integral developed by Lebesgue in 1902.[49] In the hands of Erhard Schmidt, Frigyes Riesz, Ernst Fischer, and others, these ideas were combined (1907–20) into a theory of function spaces and operators and functionals on such spaces. During the following years, functional analytic ideas were developed and generalized from spaces of functions to axiomatically defined spaces, such as Banach spaces or Hilbert spaces, that were defined as sets of arbitrary objects with certain operations, for example, addition, and a norm satisfying certain axioms.[50]

---

[49] David Hilbert, *Grundzüge einer allgemeinen Theorie der linearen Integralgleichungen* (Leipzig: Teubner, 1912). See also Michael Bernkopf, "The Development of Function Spaces with Particular Reference to their Origins in Integral Equation Theory," *Archive for History of Exact Sciences*, 3 (1966), 1–96.

[50] The development of functional analysis is described and analyzed in Jean Dieudonné, *History of Functional Analysis* (Amsterdam: North Holland 1981); Reinhard Siegmund-Schultze, "Die Anfänge

This last development followed a general trend in the history of mathematics during the twentieth century, when such axiomatically defined structures were introduced in various branches of mathematics. The development followed Hilbert's axiomatization of geometry and his more general call for axiomatization of other branches of mathematics and physics. In algebra, for example, groups, rings, and fields had already been abstractly defined at the end of the nineteenth century, and in 1930 Bartel Leendert van der Waerden wrote the first textbook on algebra based on such a structural point of view. Two years later, the first textbooks on axiomatic functional analysis were published, namely Stefan Banach's *Théorie des Operations Linéaires,* and Marshall Stone's *Linear Transformations in Hilbert Spaces,* as well as von Neumann's *Mathematische Grundlagen der Quantenmechanik.*

Laurent Schwartz was well acquainted with these developments.[51] In 1944 he was inspired by a paper by Gustave Choquet and Jaques Deny to try to generalize solutions to the polyharmonic equation $\Delta^n V = 0$. He came up with a "sequence" generalization, defining f as a generalized solution, if there exist ordinary solutions f_i that converge in a certain sense to f. Such a method of generalization had also been suggested by D. C. Lewis (1933) and Friedrichs (1939). Schwartz observed that if f is a generalized solution, the convolution $f * \varphi$ is an ordinary solution for $\varphi \in C_c^\infty$.

During "the most beautiful night of my life," sometime in October or November 1944, Schwartz realized how this observation suggested a generalization of the concept of function as "convolution operators," that is, operators mapping C_c^∞ into C^∞, satisfying certain rules. He developed this idea for a couple of months until he realized that it would be much simpler to define generalized functions as functionals (distributions), rather than as operators. Here it helped that he had himself studied the abstract theory of duality on C^∞, that is, the space of continuous functionals on C^∞.

In fact, Schwartz was not the first to suggest functionals as a generalization of the concept of functions. The Russian mathematician Sergei Sobolev, in connection with his studies of aerodynamics and partial differential equations, had come up with the same idea in 1936.[52] However, there are several reasons for considering Schwartz the father of the *theory* of distributions:

- There are, in fact, technical differences between Sobolev's and Schwartz's functionals, those of the latter being the most convenient ones.

der Funktionalanalysis und ihr Platz im Umwälzungsprozess der Mathematik um 1900," *Archive for History of Exact Sciences,* 26 (1982), 13–71; G. Birkhoff and E. Kreyszig, "The Establishment of Functional Analysis," *Historia Mathematica,* 11 (1984), 258–321.

[51] Laurent Schwartz has described his own route to the theory of distributions in his autobiography: *Un mathématicien aux prises avec le siècle* (Paris: Edition Odile Jacob, 1997).

[52] Sergei Sobolev, "Méthode nouvelle à résoudre le problème de Cauchy pour les équations linéaires hyperboliques normales," *Matematiceskii Sbornik,* 1, no. 43 (1936), 39–71.

- Sobolev used his idea only in one paper, whereas Schwartz wrote a series of papers from 1945 to 1950, presenting the main ideas of his distribution theory, and then in 1950/1 published a two-volume textbook on the subject.[53]
- Sobolev had only one application in mind, namely, the generalization of solutions of partial differential equations. Schwartz, in addition to this application, presented many other important applications. For example he showed how Dirac's δ-function could be given rigorous existence as the distribution

$$\delta(\varphi) = \varphi(0).$$

He also showed how one could generalize the concept of Fourier transformation. Here he linked up to a long development wherein Hans Hahn, Wiener (1924–6), and Bochner (1927–32) had generalized the Fourier transform to functions for which the Fourier integral does not exist in the classical sense. However, Schwartz's generalization was much more general and much more elegant than those of his predecessors.[54] Finally, Schwartz showed how distribution theory provided a rigorous framework for Heaviside's operational calculus.

Schwartz's generalization of the concept of functions earned him the Fields Medal in 1950. Since then, other generalizations of the concept of functions have been suggested, such as Jan Mikusinski's operators (1950–9), Mikio Sato's hyperfunctions (1959–60), and nonstandard functions developed by Detlef Laugwitz and Curt Schmieden (1958) and Abraham Robinson (1961). But none of these have been as influential as Schwartz's distributions.[55]

The development of the concept of function reflects many of the general trends in the history of analysis. Both the change from Euler's concept of analytic expressions to Dirichlet's modern concept, as well as the extension to generalized functions (distributions), were first suggested by applications of analysis to physics. This illustrates the continued strong links between analysis and its applications. On the other hand, we saw that it was a development in the foundations of analysis that made Euler's concept of function obsolete and made its replacement natural. Similarly, the theory of distributions was only made possible through the introduction of axiomatic structural thinking into analysis, which led to the creation of functional analysis. Thus, the modern concept of function and generalized function, as well as modern analysis as a whole, has been shaped by continuous interactions between applications and autonomous developments of a foundational nature.

---

[53] Laurent Schwartz, *Théorie des distributions*, vols. 1 and 2 (Paris: Hermann, 1950–1).
[54] Cf. Lützen, *The Prehistory*, chap. 3.
[55] Ibid., pp. 166–70.

25

STATISTICS AND PHYSICAL THEORIES

Theodore M. Porter

Until about 1840, the theory of probability was used almost exclusively to describe and to manage the imperfections of human observation and reasoning. The introduction of statistical methods to physics, which began in the late 1850s, was part of the process through which the mathematics of chance and variation was deployed to represent objects and processes in the world. If this was a "probabilistic revolution," it was a multifarious and gradual one, the vast scope of which went largely unremarked. Yet it challenged some basic scientific assumptions about explanation, metaphysics, and even morality. For this reason, it sometimes provoked searching reflection and debate within particular fields, including physics, over what, in retrospect, appears as an important new direction in science.

At the most basic level, statistical method meant replacing fundamental laws whose action was universal and deterministic with broad characterizations of heterogeneous collectives. Statistics, whether of human societies or of molecular systems, involved a shift from the individual to the population and from direct causality to mass regularity. In social writings, it was linked to bold claims for scientific naturalism. Statisticians claimed to have uncovered a lawlike social order governing human acts and decisions that had so far been comprehended by Christian moral philosophy in terms of divine intentionality and human will. Their science seemed to devalue moral agency, perhaps even to deny human freedom. In other contexts, and especially in physics, statistical principles appeared, rather, to limit the domain of scientific certainty. They directed attention to merely probabilistic regularities, the truth of which was uncertain and approximate. Statistical physics, the mathematics of molecules, involved a kind of reduction to mechanics, yet its laws assigned to chance an ever-more-fundamental role in nature. With the development of quantum mechanics in the 1920s, physicists would be driven to wonder, as did Einstein, whether God played dice.[1]

---

[1] Gerd Gigerenzer et al., *The Empire of Chance* (Cambridge: Cambridge University Press, 1989); Lorenz Krüger et al., eds., *The Probabilistic Revolution,* 2 vols. (Cambridge, Mass.: MIT Press, 1987).

STATISTICAL THINKING

The "statistics" of "statistical physics" referred originally to a social science, and not a branch of applied mathematics. This is clear, for example, in the use of the term by James Clerk Maxwell (1831–1879), often identified as the founder of statistical physics. He observed in a lecture of 1873 that it is impossible to follow the chaotic motions of countless molecules. "The modern atomists," he explained, "have therefore adopted a method which is, I believe, new in the department of mathematical physics, though it has long been in use in the section of Statistics." That method looks away from the dynamical or historical laws of individuals, and proceeds instead by counting and classifying, "distributing the whole population into groups, according to age, income-tax, religious belief, or criminal convictions." Similarly, the raw data of physicists consist of "sums of large numbers of molecular quantities." Their conclusions, on this account, cannot claim "that character of absolute precision which belongs to the laws of abstract dynamics." Instead, they must be content "with a new kind of regularity, the regularity of averages."[2]

Statistics, etymologically, was a human science, a descriptive science of the state. Early in the nineteenth century, the term came to be attached to census results and other quantitative social investigations. Although demographic numbers had long provided material for mathematical probability, the new science of statistics only gradually became allied to probability theory. As Maxwell's remarks imply, the defining feature of statistical thinking, at its most basic level, was just the regularity of averages, first observed in social investigations. In the eighteenth century, philosophers and public officials noticed a certain stability in the annual numbers of marriages and deaths, and most famously in the ratio of male to female births. These were generally ascribed to Divine Providence, which took care to compensate for the greater mortality of boys with more annual births, so that the sexes would come into balance at the age of marriage. For some decades, beginning about 1829, the most celebrated instances of mass regularity were drawn from the new official criminal statistics. The reading public was amazed and sometimes distressed to learn that suicide, murder, and theft took place in almost constant numbers from year to year.

Here, the providential consequences of divine planning were less manifest. Inquirers, such as the Belgian astronomer and statistician Adolphe Quetelet (1796–1874), soon began to ascribe these uniformities to the social order, or "society," a new object of scientific investigation. Statistical regularities were the prototype of empirical social laws. Suicide, viewed numerically, was not a question of individual choice and personal morality but rather a property of a whole society, the consequence of its laws, educational system, religion, climate, and customs. From the perspective of social science, the individual

---

[2] James Clerk Maxwell, "Molecules" (1873), in Maxwell, *Scientific Papers*, 2 vols. (Cambridge: Cambridge University Press, 1890), 2: 373–4.

will was at best an accidental cause. The proper method of statistics was to put aside the individual and to reason about large numbers. The statistician deployed mean values in order to typify a whole population and compiled tables to learn how crime varied as a function of age, sex, season, religion, town size, and laws. Quetelet doubted if we could ever plumb the depths of the human soul, so as to comprehend how physical suffering, financial loss, shame, dishonor, or disappointment in love might have driven any particular man to take his own life. Yet a straightforward tabulation and aggregation of cases could be used to identify causes at a higher level, the basis of a "social physics" with its own natural laws.

"Moral statistics" was mainly about immoral behavior, about acts of passion and defiance. The large-scale order they displayed was almost wholly unanticipated. But what was shocking in 1830 had become a commonplace by midcentury. The mathematical economist and statistician Francis Edgeworth (1845–1926) expressed the point in full generality and with epigrammatic brilliance in 1884. The paradox of probability, he wrote, "is that our reasoning appears to become more accurate as our ignorance becomes more complete; that when we have embarked upon chaos we seem to drop down into a cosmos." This was also the fundamental proposition of statistical physics. Auguste Krönig (1822–1879), whose paper of 1856 initiated the modern development of the kinetic gas theory, expressed himself in just these terms. The walls of a container are, on an atomic scale, highly uneven, so that "the path of each gas atom must be so irregular that it defies calculation. In accordance with the laws of probability, however, one can suppose, in place of this absolute irregularity, complete regularity."[3]

The kinetic gas theory was a strategy for linking the macroscopic behavior of gases to the motions of countless molecules. Maxwell identified antecedents of the kinetic theory going back all the way to Lucretius. A more focused history of science dates its origin to 1738, when Daniel Bernoulli showed how the laws of gas pressure could be understood as the consequence of collisions involving gas molecules. Up to the mid-nineteenth century, though, this is a story of ineffective precursors, of scientific originals whose work had little or no contemporary effect.[4] Also, these scientific theorists did not develop the kinetic theory as a statistical one. They never formulated the problem of how chaotic molecular motions would give rise to stable averages of pressure or heat flow. None, in fact, clearly viewed the molecular motions as disorderly. Some pictured the molecules as bouncing back and forth against their neighbors in a regular array. Krönig, in contrast, saw the walls of the container as generating randomness. Rudolf Clausius (1822–1888), who was inspired by Krönig's paper to publish his own ideas on the subject,

---

[3] Quotations from Theodore M. Porter, *The Rise of Statistical Thinking, 1820–1900* (Princeton, N.J.: Princeton University Press, 1986), pp. 260, 115.
[4] Stephen Brush, *The Kind of Motion We Call Heat*, 2 vols. (Amsterdam: North Holland, 1976).

emphasized the importance of intermolecular collisions, through which almost any initial arrangement would rapidly become disorderly. After 1857, the explanatory success of the kinetic theory always presupposed the stability of molecular averages.

LAWS OF ERROR AND VARIATION

Maxwell's first paper on the kinetic theory, published in 1860, introduced another form of statistical order and a subtler form of probabilistic reasoning. Clausius had calculated the average distance traveled by a molecule between collisions, or mean free path, on the assumption that a mean value could be used in place of variable molecular velocities without affecting the result. Maxwell recognized that this was mistaken. His first paper on the kinetic gas theory, published in 1860, set out from the proposition that molecular velocities vary according to a particular law, now known in physics as the Maxwell-Boltzmann distribution and in statistics as the Gaussian, the normal, or simply the bell curve. Maxwell referred to it by its customary name, the astronomer's error law. From his correspondence in 1859, it appears that the error curve was his point of entry into the kinetic theory of gases, and indeed, the entire first half of his 1860 paper consists of mathematical reasoning that presumes its validity. His derivation, we may add, is extremely abstract and seems to back up an intuition, rather than to have grounded his belief in the first place. As the editors of a recent collection of Maxwell documents remark, Maxwell's mathematics "gives a strange appearance of having nothing to do with molecules or their collisions."[5]

In fact, the derivation Maxwell used was originally devised for a very different purpose. He encountered it in a review by John Herschel (1792–1871) of Quetelet's 1846 book, *Letters on the Theory of Probabilities,* which Herschel published anonymously in the *Edinburgh Review* of July 1850 and then reprinted seven years later in a collection of essays. Maxwell read the review both times and remarked on it in his letters. It has since come to be seen as a classic, in its way – a hinge between social science and mathematical physics. In 1963, Charles Gillispie remarked on a similarity of temper, a shared empiricism and epistemological modesty, linking Herschel's review to Maxwell's kinetic theory. In a formal comment on his paper, the philosopher Mary Hesse dismissed Gillispie's argument as gratuitous. Probabilistic reasoning was already part of physics, she argued – one need look no further than to Clausius's derivation of mean free path. But when Stephen Brush and others examined Herschel's essay more closely, they found that Maxwell's derivation of the error curve was closely modeled on it. We can now understand more

---

[5] Porter, *Rise of Statistical Thinking,* 117–18; Elizabeth Garber, Stephen G. Brush, and C. W. F. Everitt, eds., *Maxwell on Molecules and Gases* (Cambridge, Mass.: MIT Press, 1986), Introduction, p. 8.

richly and concretely what John Theodore Merz argued as early as 1904, that the kinetic gas theory was joined culturally and intellectually to social investigation and also to Francis Galton's biometry, forming what Merz called "the statistical view of nature." The natural scientists did not imitate, but drew selectively from, a statistical tradition that arose within social science, social reform, and social administration.[6]

Herschel's derivation was framed in terms of the distribution of errors while dropping balls at a target. He made two assumptions: that the density of errors was independent of direction and that errors in the x-direction are wholly uninfluenced by errors in the y-direction. The exponential error law followed readily from these simple, seemingly natural, and highly abstract assumptions of symmetry and independence. The derivation could be applied readily to a host of problems and topics. Herschel urged in his review that such a derivation was valuable because the error law already had so wide a range of applications. These reached well beyond games of chance to include errors in population estimates, astronomical observations, and most other quantitative measurements. He reaffirmed also Quetelet's bold addition to the ambit of the error law. Quetelet had urged, on the basis of some distributions of human measurements, that departures from the mean of human heights, chest sizes, and all manner of physical and even moral quantities were governed by this same formula. His extension of the error law was facilitated, but also confined, by his disposition to confuse variation with error, as if human variety could be reduced to a failure of nature to realize the ideal type of *l'homme moyen,* the average man. Still, however inadvertently, he helped make the error curve into something more than a law of error, a principle of natural variation. This certainly was how it worked for Maxwell, who argued already in 1860 that Clausius had been led to incorrect results by his failure to consider the inequalities of molecular velocities.

The Maxwell distribution remained one of the most important topics of research on the kinetic theory of gases for several decades. In his next major paper on the topic, published in 1867, Maxwell gave a new derivation. Here he proceeded from the assumption that, at equilibrium, the rate of collisions from which molecules of given initial velocities emerge with specified final velocities will equal the rate of collisions that effect the reverse transformation. Once again he was able to reason mathematically to the exponential distribution law. This time his derivation was tied more closely to the physical problem of systems of colliding particles. By this time, too, the whole approach had become much more credible on the empirical side. Maxwell

---

[6] C. C. Gillispie, "Intellectual Factors in the Background to Analysis by Probabilities," in *Scientific Change,* ed. A. C. Crombie (New York: Basic Books, 1963), pp. 431–53, and Mary Hesse's commentary, pp. 471–6; Brush, *The Kind of Motion We Call Heat,* pp. 184–7; J. T. Merz, *A History of European Thought in the Nineteenth Century,* 4 vols. (New York: Dover, 1965), pp. 548–626; Theodore M. Porter, "A Statistical Survey of Gases: Maxwell's Social Physics," *Historical Studies in the Physical Sciences,* 12 (1981), 77–116; Porter, *Rise of Statistical Thinking.*

had argued in 1860 that if his theory were true, gaseous friction should be independent of density. This was in violation of common sense, and the only measurements he could turn up at the time seemed to be against him. Rather self-effacingly, he anticipated that laboratory results would put an end to his mathematical sport. But his own experiments with his wife Katherine in 1865 supported the theory, and this helped greatly to give it credibility with physicists.[7]

Ludwig Boltzmann (1844–1906) wrote his first paper on the kinetic theory in 1866, at a time when he knew no English. Soon afterward his teacher in Vienna, Josef Stefan, gave him an English dictionary and one of Maxwell's papers on electricity and magnetism. Boltzmann read Maxwell on the kinetic theory, and thereafter he was one of Maxwell's greatest admirers. Immediately he placed Maxwell's distribution at the focus of his work. In 1868 he rederived it for the more complicated case of polyatomic molecules in a uniform force field. In 1872 he wrote perhaps his most influential paper, in which he aimed not merely to show what would be the equilibrium distribution of molecular velocities, but also that other initial conditions must converge to it. In this effort, he began as Maxwell had in 1867 by considering the frequency of collisions for molecules within each energy range. From this starting point, he defined a quantity E, now called H, which was minimized by Maxwell's distribution, and whose derivative in time was negative for all others. From any nonequilibrium starting point, the value of H for a system of molecules should diminish steadily until it reached its minimum.

In 1877 Boltzmann worked out yet another formulation of the problem. This time he showed how to define equally probable cases for the allocation of energy to molecules in a system. Deploying once again his formidable mathematical skill, he derived from this a formula for the probability of any given energy distribution. With suitable manipulations, this probability formula was revealed as another guise of the familiar H-function. He concluded that any system commencing in an improbable state must pass progressively through more probable states until stabilizing at the familiar Maxwell distribution.[8] By this time, he had a deep understanding of his most famous equation, which he first formulated in 1871 and which is printed on his gravestone:

$$S = k \log W,$$

linking entropy S to probability W.

---

[7] Garber et al., *Maxwell on Molecules*, p. 12; also, more generally, Elizabeth Garber, Stephen Brush, and C. W. F. Everitt, *Maxwell on Heat and Statistical Mechanics* (Bethlehem, Pa.: Lehigh University Press, 1995), pp. 274–92.

[8] Martin Klein, "The Development of Boltzmann's Statistical Ideas," *Acta Physica Austriaca*, supp. X (1953), 53–106; Thomas S. Kuhn, *Black-Body Problem and the Quantum Discontinuity, 1894–1912* (New York: Oxford University Press, 1978), chap. 2.

Boltzmann's ambition from the beginning had been to reduce the second law of thermodynamics to a mechanical formulation. Entropy, a term coined by Clausius, was in its origins part of the theory of heat engines. When the flow of heat from a hot to a cold object is harnessed by an engine to produce useful work, this is balanced (or overbalanced) by an increase of entropy as hot and cold are mixed. According to the second law of thermodynamics, formulated by Clausius in Germany and by William Thomson in Britain, the entropy of the world is always increasing, and hence the energy available for work is decreasing. Boltzmann aimed from the outset to understand these thermodynamic processes in terms of molecules and motion. His reformulation of the law in the preceding equation did not quite amount to a mechanical reduction, however. The move to mechanics required another crucial concept, probability.

MECHANICAL LAW AND HUMAN FREEDOM

Statistical physics, like statistical social science, was initially about order and not about uncertainty. Maxwell and Boltzmann reasoned mathematically to deduce laws governing the behavior of countless particles whose small size made them inaccessible to human senses. Indeed, the very existence of molecules was seen as hypothetical, and often as doubtful, until the first decade of the twentieth century. Maxwell's subsequent insistence on the uncertainty of statistical laws was entirely missing from his papers of 1860 and 1867, both of which used titles identifying their subject as "the dynamical theory of gases." In these papers, he included variation in his theory, but not uncertainty or chance. In his 1867 derivation of the error law, he did not demonstrate, but assumed, the existence of an equilibrium for which the number of molecules moving within any given range of velocities will be uniform in time. And Boltzmann was still less inclined to make chance a player in his gas theory. His 1866 paper canceled out variation by averaging the velocity of each molecule over an indefinite time. He claimed in this way to reduce the second law of thermodynamics to a rigorous result of analytical mechanics.

Such reasoning was consistent with the social statistical tradition, the claims or pretensions of which reached their peak just as Krönig and Clausius were publishing their first papers on the kinetic theory. The author of the most celebrated, and the most criticized, paean to statistical law was the English historian Henry Thomas Buckle (1821–1862). Volume 1 of the general introduction to his projected but unfinished *History of Civilization in England* was published in 1857. Near the beginning, he recited some of Quetelet's favorite examples of the regularities of moral statistics: murders and suicides with guns, knives, ropes, water, and poison. Like Quetelet, he drew the lesson that there are laws of society, just as there are laws of nature. He concluded that history provided proper material for a science and should not be a mere recital of anecdotes. His history was thoroughly anticlerical and libertarian, much more radical in its politics than the writings of a bureaucratic reformer like Quetelet. It was also more extreme in its scientism; Buckle claimed without qualification that the prevalence of social law leaves no room for human free will or for providential interference in history. Moreover, the work was an enormous literary success, in Germany and Russia as well as England. Many of its admirers were political radicals. Establishment intellectuals in Britain and university professors in Germany were most often critical. For several years after 1857, the respectable press rained down refutations.

Maxwell read the book when it came out and praised it with reservations in a letter to his friend and subsequent biographer Lewis Campbell.[9] Boltzmann, who seemingly had none of Maxwell's conservative instincts, was less restrained: "As is well known, Buckle demonstrated statistically that if only a sufficient number of people is taken into account, then not only is the number of natural events like death, illness, etc., perfectly constant, but also the number of so-called voluntary actions – marriages at a given age, crimes, and suicides. It occurs no differently among molecules."[10] Like Maxwell, he considered the laws of gases as analogous to mass regularities in society, although when he referred to statistical laws, he always emphasized their reliability and never their uncertainty. In the classic and otherwise highly technical 1872 paper where he first derived his H-theorem, he explained that the "wholly determinate laws" of heat are at bottom stable averages:

> For the molecules of a body are indeed so numerous, and their movements are so rapid, that nothing ever becomes perceptible to us except average values. The regularity of these mean values may be compared to the astonishing constancy of the average numbers furnished by statistics, which are also derived from processes in which each individual occurrence is conditioned by a wholly incalculable collaboration of the most diverse external circumstances. The molecules are like so many individuals.[11]

Maxwell began in the late 1860s to interpret the statistical character of gas physics as evidence of its limitations. No more than Boltzmann, though, did he doubt that in gas physics, statistical knowledge was quite good enough for all practical purposes. The second law of thermodynamics, he told Rayleigh in 1870, has the same truth as the assertion that you cannot recover a tumbler of water thrown into the sea. As a basis for deterministic claims, however, he found atomism and its allied method, statistics, wanting.

---

[9] Maxwell to Campbell, Dec. 1857, in Lewis Campbell and William Garnett, *The Life of James Clerk Maxwell* (London: Macmillan, 1882), p. 294.
[10] Ludwig Boltzmann, "Der zweite Hauptsatz der mechanischen Wärmetheorie" (1886), *Populäre Schriften* (Leipzig: J. A. Barth, 1905), p. 34. Translated by the author unless otherwise noted.
[11] Ludwig Boltzmann, "Weitere Studien über das Wärmegleichgewicht unter Gasmolekülen," in Boltzmann, *Wissenschaftliche Abhandlungen*, 3 vols. (1909; reprint, New York: Chelsea, 1968), 1: 316–17.

In a letter Maxwell discussed the implications of lawlike regularity for human freedom just a few months after reading Buckle. He also participated, though not at first in print, in the debates set off by Buckle about statistics and free will. In 1858, the *Edinburgh Review* published a 48-page essay on Buckle's *History of Civilization* by James Fitzjames Stephen (1829–1894), who urged that Buckle was mistaken in supposing that free will must act irregularly. Stephen's point is echoed, perhaps, in a letter Maxwell wrote at very nearly the same time, where he proposed "on my own authority" that "certain men who write books" assume wrongly the incompatibility of conscious action with whatever is "orderly, certain, and capable of being accurately predicted." Stephen went on to argue that the regularities of moral statistics are no greater than those involving large numbers of rolls of the dice. They cannot possibly justify an inference "that unknown causes" such as the human will "do not exist." In 1859, the year Maxwell presented his first paper on the kinetic theory to the British Association for the Advancement of Science, his childhood friend Robert Campbell (1832–1912) delivered a paper on the regularities of statistics to the very same section. Campbell's paper was framed as a critique of Buckle. He presented some basic probability mathematics to show that the regularities of statistics were not at all "remarkable" if judged against our expectations of "purely fortuitous" events. Such regularities could provide no support for an argument against free will.[12]

The arguments of Stephen and Campbell almost certainly came to Maxwell's attention. Other critics wrote in a similar vein, both in Britain and in Germany. They drew a sharp line between collective uniformities and causation at the level of individuals. They urged, as had few before them, the limits of statistical knowledge. Maxwell himself did not begin deploying such arguments in physics until a decade later. Still, there are impressive continuities between these social debates and later discussions in physics. The form of argument was similar. So also was one key issue: During the 1870s, Maxwell was deeply troubled by arguments from mechanics and from thermodynamics against the possibility of human free will. The defense of human freedom was one of the purposes of his 1873 lecture on molecules, quoted here earlier for its remarks on statistical method. In his earliest published discussion of the limitations of statistical knowledge in physics, in 1871, he concluded similarly: "If the actual history of Science had been different, and if the scientific doctrines most familiar to us had been those which must be expressed in this way, it is possible that we might have considered the existence of a certain kind of contingency a self-evident truth, and treated the doctrine of philosophical necessity as a mere sophism."[13]

---

[12] Porter, *Rise of Statistical Thinking*, pp. 201, 166–7, 195, 241–2.
[13] James Clerk Maxwell, "Introductory Lecture on Experimental Physics," *Scientific Papers of James Clerk Maxwell*, 2 vols. (Cambridge: Cambridge University Press, 1890), 2: 253.

Maxwell first implied that the second law of thermodynamics might admit of exceptions in 1867, in a playful letter to his friend, the Scottish physicist P. G. Tait (1893–1901). There he mentioned that entropy increase could be reversed if a small and neat-fingered being were left to guard a tiny hole in the diaphragm separating two vessels of gas at different temperatures. This creature, dubbed "Maxwell's demon" by William Thomson (later Lord Kelvin) (1824–1907) had merely to exploit the wide range of molecular velocities by permitting only the slowest molecules to escape from the hot to the cold chamber, and only the fastest to pass from the cold to the hot. In this way, the hot gas would get still hotter and the cold one still colder, without requiring that any work be done. Such manipulations are not possible for mere humans, and he did not regard this scenario as technologically possible. Still, this act of imagination could be instructive. It showed that the second law of thermodynamics was a matter of high probability, not mechanical necessity. It suggested that this was a law only from the perspective of beings like us, beings who can harness some of the forces of nature but are incapable of working at the level of atoms and molecules.

Maxwell's ruminations about statistics and free will were directed not merely or mainly at Buckle but also at contemporary scientific naturalists. John Tyndall (1820–1893) may be taken as exemplary. Tyndall wanted to rescue natural science from clerics and theologians, just as Buckle had pursued an independent science of history. The passion in his insistence on the determinism of nature arose from his opposition to religious interference in science. He spoke forcefully against natural theology, miracles, and human free will. We should, he urged, derive our scientific creed from the ancient atomists, from Democritus, who preached the sufficiency of material causation and who had no place for soul in nature. We should pay heed to the iron law of energy conservation, which leaves no room for God or an immaterial human mind to alter the course of the world. Like other scientific naturalists, Tyndall had been energized by the reaction to Charles Darwin's theory of evolution, which tended also to limit the role of God in the universe.[14]

Maxwell did not campaign against Darwin, yet he seems to have rejected evolution privately. At least he was deeply religious, and he was outspoken in opposition to mechanical determinism. Indeed, on this territory he did not stay clear of controversy, perhaps because several of the most influential arguments in favor of determinism derived from some of his scientific specialties: thermodynamics, atomism, and statistics. Maxwell turned these arguments on their head. Precisely because the material world is made up of molecules, our sensory information about it is only statistical and, hence, necessarily incomplete. Not the first law of thermodynamics – energy

---

[14] John Tyndall, "The Belfast Address," [1874] in *Victorian Science,* ed. George Basalla et al. (New York: Anchor, 1970), pp. 435–78; Frank M. Turner, "The Victorian Conflict between Science and Religion: A Professional Dimension," *Isis,* 69 (1978), 356–76.

conservation – but the second – entropy increase – should be taken as exemplary of our physical knowledge. For Tyndall's deterministic Democritus, Maxwell substituted Lucretius, with his account of the uncaused atomic swerve that gave birth to the cosmos.

A paper Maxwell read to the Eranus Club at Cambridge, and which was later published in Lewis Campbell's biography, shows how he combined arguments about uncertainty and instability to rescue the possibility of human freedom. Because our physical knowledge is statistical, we cannot exclude the possibility of minute deviations or swerves at the molecular level. In unstable systems like a gun, and perhaps also like the brain, small causes can have massive effects.[15] But Maxwell was enough of a physicist to be uneasy with violations of physical law, even when so minute as to be totally unobservable. Near the end of his life, he became hopeful that some French work on singular solutions of differential equations might provide for nonmaterial causation. Joseph Boussinesq (1842–1929) showed in 1878 how mechanical systems could have genuine points of singularity, where not just one path but an "envelope" of paths was consistent with mechanical laws and with the constraint of energy conservation.[16] In a letter to Francis Galton in February 1879, Maxwell declared this a promising approach to the problem of physical law and human will, "much better than the insinuation that there is something loose about the laws of nature."[17] His indeterminism, if it may be so labeled, operated only within special systems and under rare circumstances. It involved not acausality but, rather, the substitution of mental causes for physical ones. He valued statistics as providing a space of ignorance in which the grip of mechanical law was relaxed.

REGULARITY, AVERAGE, AND ENSEMBLE

Boltzmann's career in physics has often been interpreted as a story of reluctant accommodation to the special problems raised by probability and the second law, in contrast to Maxwell, who actively exploited them. We should observe, though, that Maxwell no more expected to witness deviations from the second law as a result of statistical fluctuations than did Boltzmann. Also, he developed his thoughts on the limits of knowledge in letters, in popular or philosophical writings, and in a textbook, and we find almost no trace of them in his technical memoirs. Still, Boltzmann, too, gave superb popular

---

[15] Maxwell, "Does the Progress of Physical Science Tend to Give any Advantage to the Opinion of Necessity (or Determinism) over that of the Contingency of Events and the Freedom of the Will?" in Campbell and Garnett, *Maxwell*, 434–44.

[16] Mary Jo Nye, "The Moral Freedom of Man and the Determinism of Nature: The Catholic Synthesis of Science and History in the *Revue des questions scientifiques*," *British Journal for the History of Science*, 9 (1976), 274–92.

[17] The letter is printed in Porter, *Rise of Statistical Thinking*, pp. 205–6.

lectures, and these consistently emphasized the power and not the limits of mechanical explanation. His technical memoirs, less divided from his popular writings than were Maxwell's, reflect the same ambitions and tastes. His cosmopolitan faith in liberalism, science, and progress was unswerving, and this was reflected also in his understanding of physics.

Boltzmann cannot be construed, however, as an opponent of probability theory. He recognized from very early that his reliance on it implied an element of uncertainty in human knowledge. To be sure, he invoked probability concepts to confine uncertainty, not to exalt it. Still, his opponents in science were not advocates of chance in nature. They were positivists and energeticists who were skeptical of atomistic mechanism and skeptical equally of the probabilistic conceptions associated with it. In resisting any role for pure chance in physics, Boltzmann was striving to purge his statistical physics of what critics saw as the absurdities and contradictions associated with its reliance on probability.

Boltzmann and Maxwell both used averages to move from ignorance and unpredictability on a microscale to perfect regularity at the level of what can be sensed. Both tended to think the regularities so perfect that exceptions would never be visible. Only in 1896, long after Maxwell's death, did Boltzmann recognize that the kinetic theory might be related to those irregular vibrations of tiny particles called Brownian motion, that, a decade later, would come to be regarded as definitive evidence of molecular reality.[18] Both stressed, instead, the uniformities of statistics. They did so, however, in subtly different ways. Boltzmann relied generally on averages of individual molecules over time. Maxwell, in contrast, used mean values of many molecules at a given time.

One of Maxwell's last papers, in 1881, developed the concept of the ensemble, a strategy that the American Josiah Willard Gibbs (1839–1903) would make foundational for statistical mechanics.[19] Here, one considered not merely the statistical characteristics of molecules within a given gas but also those of the near infinity of possible states of a molecular system having a given energy. Boltzmann, who had previously mentioned this approach, pursued it further in the wake of Maxwell's paper. He was interested especially in the idea of ergodicity, which derived from a theorem of Henri Poincaré (1854–1912). Ergodic theory implies, roughly, that a mechanical system, such as a gas, will, over time, pass arbitrarily close to every possible state, where a "state" is defined by specified positions and velocities of every molecule. Ergodicity is associated with a flawlessly deterministic order, a purely mechanical succession of states, even if it defies all scientific efforts at detailed prediction. Probability in an ergodic system refers to the proportion of time

---

[18] Mary Jo Nye, *Molecular Reality* (New York: American Elsevier, 1972).
[19] Josiah Willard Gibbs, *Elementary Principles in Statistical Mechanics* (New Haven, Conn.: Yale University Press, 1902).

that the gas is in any particular state. Ensembles, too, permit probability to be defined as a ratio, the proportion of molecular systems in any given state, with no implication of acausality. This line of approach to statistical physics easily survived the quantum indeterminism of the 1920s. The two, however, had rather different implications for the shaping of modern probability theory in mathematics and especially in philosophy, where objective probability is often associated exclusively with quantum mechanics.[20]

REVERSIBILITY, RECURRENCE, AND THE DIRECTION OF TIME

At almost the same time that Maxwell first conceived his sorting demon, one of the most troublesome problems for the kinetic theory was posed: the "reversibility paradox." This was formulated independently, it seems, in Britain and Austria, primarily by William Thomson and Josef Loschmidt (1821–1895). The paradox is based on the observation that while the laws of Newtonian mechanics run equally well forward and backward, the flow of heat is always in one direction, tending to the equalization of temperature and not to increased differentiation. But if heat is merely molecular motion, and the flow of heat a communication of energy from particle to particle, then heat could equally well flow from cold to hot as from hot to cold. It is easy to imagine an entropy-reversing process. Suppose, at some fixed time, the direction of motion of every molecule in a gas, or in the universe, were precisely reversed. The gas would then retrace its path backward in time, becoming progressively more ordered. The second law of thermodynamics would no longer be valid, but, indeed, would be unfailingly false.

Joseph Fourier's theory of heat flow, published in book form in 1822, was among the most influential sources of nineteenth-century mathematical physics. It provided a model for Georg Simon Ohm's theory of the electrical circuit, and also for Thomson's pioneering electrical field theory. The theory of heat also was linked to grand issues of efficiency, morality, and eschatology. Thomson and Clausius had, at midcentury, revived and reformulated Sadi Carnot's 1824 theory of heat engines. Thomson, in particular, discussed engines using a moral vocabulary of work and waste. His analysis applied equally to physics and to economics. In economic terms, the work of engines might hold off, at least for a time, the unhappy fate anticipated by T. R. Malthus and David Ricardo, in which rising population would lead to shortages of food and rising grain prices, and in this way to a massive shift of wealth into the hands of aristocratic landowners. Engines might improve the human condition because they could perform the work of many men. In the end, however, waste must have the victory over useful work. Extrapolating

---

[20] Jan Von Plato, *Creating Modern Probability* (Cambridge: Cambridge University Press, 1994).

the second law into the indefinite future, the final destiny of the universe must be a heat death, where the universal uniformity of temperature excludes the possibility of life.[21]

Loschmidt viewed his reversibility paradox as pointing to a possible escape from the iron grip of entropy increase. Others saw this conundrum instead as a fatal shortcoming of molecular theories of heat. Their skepticism had a strong positivistic resonance. Molecules were thoroughly unpositivistic entities, since it was impossible to see them and difficult to conceive that one was even experiencing their effects. The kinetic theory gave force and focus to these doubts. A mechanical theory of heat provides no easy grounding for the second law of thermodynamics. Since every entropy-increasing process corresponds to an entropy-decreasing process, namely, the one with all velocities reversed, there seems to be no reason that nature should prefer one rather than the other. These reversibility questions were seconded in the 1890s by another so-called paradox, proposed by Ernst Zermelo (1871–1953) and related to ergodicity. If, as Poincaré demonstrated, mechanical systems must eventually return very nearly to their original state, then again there can be no mechanical basis for a law of heat diffusion or of any directional change in time. The two paradoxes, reversibility and recurrence, have in common the implication that mechanical systems should display, on occasion, a decrease of entropy, and not always its increase. Yet, in fact, we never see heat flow from cold to warm bodies. The laws of thermodynamics, expressed in terms of heat, temperature, and entropy, seem to describe the relevant phenomena more accurately than this speculative reduction of heat to a mechanics of molecules. What possible advantage can there be in giving up the language of hot and cold, a language of experience, in favor of a hypothesis of innumerable, purely hypothetical molecules, moving chaotically, beyond the reach of our senses?

Boltzmann encountered this skeptical line in various forms over much of his career. Ernst Mach (1838–1916) was among the most forceful critics of the kinetic theory, which he opposed, above all, because of its atomism. Boltzmann's sharpest opposition, however, came in the 1890s from "energeticists," such as the physical chemist Wilhelm Ostwald (1853–1932), who argued in a broadly positivistic fashion that our nervous systems experience only energy, and that molecules (and even matter) are purely hypothetical. From about 1870 until the 1890s, these challenges provided the occasion for most of Boltzmann's advances in technical methods. It is crucial to note that he was not defending mechanical reductionism against probabilism or indeterminism. Rather, he was defending a combination of mechanics and statistics against objections grounded partly in a commonsensical empiricism. If the

---

[21] Crosbie Smith and M. Norton Wise, *Energy and Empire: A Biographical Study of Lord Kelvin* (Cambridge: Cambridge University Press, 1989); Stephen Brush, *Statistical Physics and the Atomic Theory of Matter from Boyle and Newton to Landau and Onsager* (Princeton, N.J.: Princeton University Press, 1981), chap. 2, "Irreversibility and Indeterminism."

second law is merely a statement of probability, then where are the exceptions? And if it is rooted in time-reversible mechanics, how can the physicist explain the directionality of time?

Boltzmann's H-theorem of 1872 was, in a way, a response to this challenge, since it was designed as a demonstration that all systems must converge to an entropy maximum. His conclusions at that time were expressed in nonprobabilistic language. His 1877 paper, in which he used combinatorics to reach a similar result, involved a more fundamental reliance on probability concepts. At this time, he stated plainly that the second law cannot be valid with mechanical necessity. Every distribution of velocities is possible, he wrote, as probability theory itself teaches. Still, the mathematics he developed here provided him with an ostensible solution to the problem of mechanics and time. The second law, he now held, was at bottom a statement of probability. From the improbable states in which they begin, systems move through ever-more-probable ones until they reach equilibrium at maximum probability, meaning maximum entropy. Two decades later, he referred to this progression toward ever-more-probable states as almost tautologous. But it was not; the mathematics of probability contains no historical or even temporal terms. The theory gave directionality in time only by assuming that the initial conditions were highly ordered – that is, involved very low entropy.

This last assumption invited cosmological speculations, in which Boltzmann occasionally indulged. Perhaps the universe simply began in an extremely improbable state. Possibly, however, it is so huge that from time to time vast areas of it fluctuate by chance into highly improbable states. These define the conditions in which life is possible, which may explain why living beings can only witness conditions of increasing entropy. Or perhaps our experience of the direction of time is defined by increasing entropy, so that in entropy-decreasing epochs, we have memories, as it were, of the future, and never of the past.

All this was marked off as speculative, however. In his more sober physics, Boltzmann tried to exclude violations of the second law. Entropy increase fails only under certain, extremely special, conditions. The time-reversed motion discussed by Loschmidt, he argued, is really no more than an artificial product of calculation, deliberately chosen to violate the mathematical laws of probability. It may properly be assumed away. The magnitude of the improbability can be inferred from Boltzmann's answer to the recurrence paradox. Even a small volume of gas would only separate into hydrogen and oxygen about once in $(10^{10})^{10}$ years. Statistics justifies our neglect of such improbabilities:

> One may recognize that this is practically equivalent to *never*, if one recalls that in this length of time, according to the laws of probability, there will have been many years in which every inhabitant of a large country committed suicide, purely by accident, on the same day, or every building burned down

at the same time – yet the insurance companies get along quite well by ignoring the possibility of such events.[22]

CHANCE AT THE *FIN DE SIÈCLE*

At the end of his life, shortly before his suicide (seemingly for reasons unrelated to his work), Boltzmann retreated from his strong defense of the reality of molecules, defending them merely as useful aids to reasoning. He was among the first to promote the modern scientific language of models, a language that connotes structural similarity and utility and is more cautious about truth.[23] His was not a universe of chance, but his statistical physics provided crucial background for the recognition of chance by science. This was, in the first instance, a triumph of knowledge, rather than a defeat. By the late nineteenth century, chance had been tamed. To admit its role in the generation of events was fully consistent with the expectation that the world is orderly.

Boltzmann's combinatorial mathematics was a key tool in Max Planck's solution to the problem of blackbody radiation in 1900, a key source of the new quantum theory. In this and other ways, statistical physics helped to shape the physics of the quantum, which, in the 1920s, would assign to chance a fundamental role in the physics of elementary particles.[24] But it is more fitting to link the gas theory of Maxwell and Boltzmann to the statistical thought of their own time. Karl Pearson (1857–1936), founder of the modern mathematical form of statistics, was trained in physics and mathematics at Cambridge, in part by Maxwell. He defended, in his positivistic philosophy, a view of nature quite different from Mach's, for he made unexplained variation, if not precisely chance, central to a whole conception of nature. Pearson linked social science with Darwinian biology and with statistical physics, insisting that all were properly quantitative studies and that none could furnish perfect accuracy or complete certainty. He held that the methods and ideas of probability were appropriate for investigating and for understanding all of science.

Charles Sanders Peirce (1839–1914), the American pragmatist, gave perhaps a more coherent philosophical expression to a related set of ideas. He regarded chance variation as key to Darwinian evolutionary progress. He used the direction of time and of history not to contradict statistical physics but, in a

[22] Ludwig Boltzmann, *Lectures on Gas Theory* (1896–1898), trans. Stephen Brush (Berkeley: University of California Press, 1964), p. 444.
[23] Theodore M. Porter, "The Death of the Object: Fin-de-siècle Philosophy of Physics," *Modernist Impulses in the Human Sciences*, ed. Dorothy Ross (Baltimore: Johns Hopkins University Press, 1994), pp. 128, 151; John Blackmore, *Ludwig Boltzmann: His Later Life and Philosophy, 1900–1906*, 2 vols. (Dordrecht: Kluwer, 1995), vol. 2 (Boston Studies in the Philosophy of Science, vol. 174).
[24] Gigerenzer et al., *Empire of Chance*, chap. 5.

way, to confirm it. The manifest importance of growth and decay in life and in history proved, in his view, the insufficiency of mechanical law and the centrality of chance. At the end of the nineteenth century, statistics remained a metaphor for the production of order where individual causes remained elusive, but by then it was no longer necessary to assume that such hidden causes suffice to determine the future unambiguously.[25]

[25] Ian Hacking, *The Taming of Chance* (Cambridge: Cambridge University Press, 1990).

26

SOLAR SCIENCE AND ASTROPHYSICS

Joann Eisberg

During the nineteenth and twentieth centuries, astronomy has changed from a relatively homogeneous discipline to one of tremendous diversity. Before this period, the main business of astronomy had been the measurement and prediction of planetary motion and stellar position. Earlier astronomers depended on a limited range of observational equipment – the optical telescope and various instruments for measuring angles and positions against the sky – in order to map the locations of the stars and to track the motions of the planets as they wandered against this fixed background. By the early nineteenth century, astronomers had, as theoretical tools, not only Newtonian gravitation but also the fruits of a century's further refinement of celestial mechanics. Not only could astronomers calculate the orbits of individual planets around the Sun; they could also investigate the mutual perturbations of the various bodies and the stability of the solar system as a whole, far into the future. Within their well-defined realm, early-nineteenth-century astronomers congratulated themselves on possessing a predictive power exceeding that of all other fields of natural science.

Yet astronomers were eventually to trade their sure grasp of their traditional portion of the world for a much less certain hold on broad, new domains: the study not just of position and motion but also of the physical nature of celestial objects of all kinds, from the Sun, stars, and planets to nebulae and galaxies. This expansion of subject was, in significant part, technology driven, and many new observational technologies contributed to making it possible, including the building of telescopes with tremendously increased light-gathering power and finer resolution, and the introduction of photography as an astronomical tool. The single most revolutionary technical development, however, was the introduction of astronomical spectroscopy, which came to be regarded as the hallmark of the new approach. To distinguish it from traditional activities, the new study was, in the last third of the nineteenth century, variously called physical astronomy, astronomical physics, astrophysics (at first, often hyphenated, emphasizing its hybrid quality), astrochemistry, celestial or solar physics, or frequently "The New Astronomy."

The range of theory upon which the new study depended was even wider than the multiplicity of its names suggests. Before these technical advances yielded much detailed information that could be used to probe the makeup of stars, planets, and other objects, natural philosophers speculated freely and sometimes presciently on their origins and evolution. How, then, did they construct meaningful theories about the histories of objects the natures of which they barely knew? Since at least the late eighteenth century, astronomers have had faith that it must be possible to discern developmental sequences among the tremendous variety of objects that can be observed in the sky. Stars of different kinds, for instance, might represent earlier and later stages of one process, while nebulae might be the progenitors of stars and planetary systems, or perhaps their death throes. Astronomers' faith that developmental sequences were a subject for fruitful study has been buttressed by arguments of two kinds: physical arguments that gravitation and thermodynamics must drive change in massive material bodies, and arguments of a much more biological character. As William Herschel (1738–1822) wrote in a passage cited approvingly for generations:

> The heavens . . . resemble a luxuriant garden which contains the greatest variety of productions in different flourishing beds; and one advantage we may at least reap from it is, that we can, as it were, extend the range of our experience to an immense duration. . . . Is it not almost the same thing whether we live successively to witness the germination, blooming, foliage, fecundity, fading, withering, and corruption of a plant, or whether a vast number of specimens, selected from every stage through which the plant passes in the course of its existence, be brought at once to our view?[1]

Herschel's realization was as powerful and practical as it was poetic: Since so many astronomical processes (especially those visible before the twentieth century) happen on a timescale that dwarfs human time, variety observed as if in a snapshot was the main empirical insight astronomers would long have into celestial process. That physics and biology both fed astronomers' interest in developmental theories of stars nicely illustrates a point worth emphasizing: For much of the period discussed here, astrophysics and solar physics contained some elements that clearly belonged to the exact sciences and other elements more closely resembling natural history. In their effort to make sense of the tremendous range of phenomena they observed in the sky, nineteenth- and early-twentieth-century astrophysicists came to recognize cataloging, classification, and taxonomy as central activities, crucial steps in transforming raw data into grist for theory.

---

[1] William Herschel, *Philosophical Transactions*, 79, p. 226, cited in Agnes Clerke, *A Popular History of Astronomy During the Nineteenth Century*, 4th ed. (London: Adam and Charles Black, 1908), pp. 23–4.

In this chapter we will briefly review the early, phenomenological beginnings of solar physics; the developments in spectroscopy that permitted its application to the Sun, stars, and other objects; the development of communities of workers pursuing astrophysics and solar physics; and the flowering of twentieth-century stellar astrophysics, producing models of stars incorporating quantum and nuclear physics. By the middle of the twentieth century, astronomy had moved far beyond the narrow limits that had been its borders only a century and a half before. No longer the dedicated proving ground of Newtonian mechanics, astronomy had blossomed into a study of physical constitution and process in a range of objects more varied and with forms of matter more exotic than ever imagined by its late-eighteenth-century practitioners. Astronomy had also forged close ties to other disciplines within the physical sciences, and it had come increasingly to depend upon, and respond to, technological development.

Before we trace this story, it is important to emphasize just how little late-eighteenth- and early-nineteenth-century astronomers knew of the nature of the astronomical objects whose positions and motions they had so long observed. Before spectroscopy, the astronomers' only way of telling what distant bodies might actually be like was to look at them through a telescope, and even the highest-quality instruments available at the end of the eighteenth century afforded but little information for surprisingly large amounts of effort. The Moon was unique in presenting a wealth of easily visible detail.[2] Sunspots, too, could be observed, and their motion was used to measure the rotation of the Sun. Surface features on planets, however, were ambiguous and so hard to see that it was difficult to measure planets' rotation. The images of the distant stars were perhaps the least revealing of all celestial observables; they remained pointlike at even the highest magnifications.

With so little information available in the images of celestial objects, the infant field of descriptive astronomy was very much the poor stepsister of positional astronomy. As Friedrich Wilhelm Bessel (1784–1846), soon to be among the first to measure the distance to a star, contended in 1832:

> What astronomy must do has always been clear – it must lay down the rules for determining the motions of the heavenly bodies as they appear to us from the earth. Everything else that can be learned about the heavenly bodies, e.g. their appearance and the composition of their surfaces, is certainly not unworthy of attention; but it is not properly of astronomical interest.[3]

Bessel was hardly an extremist; the philosopher Auguste Comte is infamous for having declared a few years later that positive knowledge of the

---

[2] The first true lunar map, by Michael Van Langren, is shown in Michael Hoskin, *The Cambridge Illustrated History of Astronomy* (Cambridge: Cambridge University Press, 1977), p. 143.
[3] Friedrich Wilhelm Bessel, *Populäre Vorlesungen über wissenschaftliche Gegenstände* (Hamburg: Perthes-Besser und Maucke, 1848), pp. 5–6, translated in Karl Hufbauer, *Exploring the Sun: Solar Science Since Galileo* (Baltimore: Johns Hopkins University Press, 1991), p. 43.

physical and chemical natures and temperatures of stars would forever be unattainable.[4] While it is ironic that even as Bessel and Comte wrote, pathbreaking spectroscopic studies of sunlight had already been made by Joseph Fraunhofer (1787–1826), it is understandable that they sought to distance the rigor of positional astronomy and celestial mechanics from the free speculation of early solar physics. As we shall see in the next section, however, even before the introduction of spectroscopy, the study of the Sun was fast becoming rich in observations. This growth in observational knowledge underlay fresh speculation about the structure, temperature, weather, and even the habitability of the Sun.

SOLAR PHYSICS: EARLY PHENOMENOLOGY

Sustained debate on the nature of sunspots began with the 1612 exchange between Galileo and Christoph Scheiner.[5] By the beginning of the nineteenth century, however, the most extensive discussions of the nature of the Sun were those of William Herschel. Though a professional musician and an amateur astronomer, Herschel was the builder of the largest telescopes of his day and became internationally famous for his discovery of the planet Uranus, the first new planet since antiquity. In 1780, Herschel argued that the Sun was not hot and luminous throughout but, instead, was mainly a solid, dark body surrounded by a multilayered atmosphere, of which only the topmost layer was luminous. Sunspots were holes in the top layer that revealed the regions beneath. Herschel went on to speculate that clouds in the Sun's lower atmosphere shielded the inner sun from its fiery upper atmosphere. This supported the appealing idea that the cool body of the Sun could be a suitable abode for life and, Herschel judged, most probably was inhabited. Herschel was also interested in the possibility that the Sun, like many other stars, varies in brightness, and that the variations result from changing numbers of dark sunspots. Fluctuating solar luminosity should directly affect the Earth's weather and crop production, and so Herschel attempted to correlate past sunspot frequencies with the historical series of grain prices given in Adam Smith's *Wealth of Nations* of 1776.

The important point about Herschel's solar speculations is not whether they seemed as far-fetched to his contemporaries as they do today. In fact, the idea that the Sun was inhabited was quite common. Of greater historical significance is how his approach differed from that typical of professional astronomers. As Karl Hufbauer has pointed out, the routine of observing solar eclipses before 1840 shows astronomers' preoccupation with position:

---

[4] Auguste Comte, *Cours de philosophie psitive*, II, nineteenth lesson, 1853, cited in J. B. Hearnshaw, *The Analysis of Starlight: One Hundred and Fifty Years of Astronomical Spectroscopy* (Cambridge: Cambridge University Press, 1986), p. 1.
[5] Much of this section depends upon the analysis in Hufbauer, *Exploring the Sun*.

The measurements of greatest interest were the timing of the moments when the disk of the Moon first and last appears to make contact with the disk of the Sun, and of the beginning and end of totality. These permitted refined estimates of the relative positions of the Earth, Moon, and Sun and could be used in the calculations of celestial mechanics. Absorbed in the timing of contacts and totality, astronomers took little notice of qualitative phenomena, such as the solar corona and prominences, bright extensions that were briefly visible around the edge of the Sun when the light from its disk was blocked by the Moon.

The person most responsible for changing this situation was the amateur astronomer Francis Baily (1774–1844). Observing the eclipse of 1836, Baily was so struck by the sudden appearance of a string of brilliant, irregular, beadlike lights around the circumference of the Moon as it eclipsed the edge of the Sun that he called for their observation during future eclipses. The effect of Baily's dramatic description was to stimulate wide interest in the aspects of eclipses that revealed the physical properties of the Sun and Moon and that had previously been regarded as entirely secondary.[6] During the eclipse of 1842, observers from many nations marveled at the luminous, halolike corona, with an obvious radial structure that, however, was portrayed very differently by astronomers observing from different locations. Also impressive were several prominences, which stood out like immense red flames or peaks. Baily, overwhelmed by the wealth of phenomena to be seen, recommended to members of the Royal Astronomical Society of London that astronomers adopt a formal division of labor so that each observer could give full attention to a single, selected feature. This became common practice.

Multiplying the number and kinds of observations did as much to raise controversy as to settle theoretical interpretations. For instance, the question of whether the corona and prominences belong to the Sun or are artifacts of diffraction of light in the atmosphere of the Earth or the Moon was debated into the 1860s. By then, photography and spectroscopy were added to the list of tasks a typical eclipse team might undertake. Moreover, by then, the physics of the Sun had become the object of much more widespread investigation than in the days of William Herschel, with more detailed studies of sunspots joining astronomical spectroscopy as grist for observers and theorists alike.

The study of sunspots was a main focus of solar observation outside of eclipses, and it attracted growing attention from 1850 onward. The year 1851 saw the publication of the fruits of a quarter century of daily sunspot records made by a retired German pharmacist, Heinrich Schwabe (1789–1875). Tabulating the number of spots visible each day, Schwabe identified a ten-year cycle of maxima and minima. Schwabe's finding was soon refined from ten to eleven years by J. Rudolf Wolf, an astronomer at Bern University, who

---

[6] Clerke, *Popular History of Astronomy*, pp. 61–2.

reviewed historical sunspot counts dating back to Galileo. The sunspot cycle was found to coincide with cyclical occurrences of magnetic storms, episodes during which compass needles could be observed to vibrate suddenly and wildly here on Earth. In part because of the practical connection between magnetism and navigation, variations in terrestrial magnetism had attracted growing interest over the previous several decades, and this terrestrial connection increased interest in solar studies. Richard Carrington (1826–1875), a wealthy British amateur, began mapping daily sunspot locations, announcing in 1858 that they shifted systematically in latitude as the cycle progressed: The first spots of a cycle appeared in zones about 35 degrees to the north and south of the solar equator, whereas the later spots occurred ever nearer the equator. A year later, Carrington announced that the pattern of sunspots rotated not as though attached to a solid body, but differentially: Zones near the equator rotated faster than those near the poles.

Between eclipse and sunspot observations, the phenomenology of the Sun had expanded greatly by 1860. Solar theory, however, had changed little since William Herschel. Some of the observations fit Herschel's model fairly well. Carrington, for instance, was typical in believing that the Sun was a solid body surrounded by a fluid atmosphere, and he explained differential rotation of sunspots as the effect of permanent equatorial trade winds around the sun. The sunspot cycle and the correlated magnetic effects, however, were not easily explained by analogy to terrestrial weather. During the 1860s, a new solar model was formulated. Its central components were spectroscopy, which previously had been pursued both as part of astronomy and as part of analytical chemistry, and thermodynamics, largely an import from outside astronomy. We will briefly turn to a review of astronomical spectroscopy before returning to consider thermodynamics and the solar model that emerged in the remaining decades of the nineteenth century.

ASTRONOMICAL SPECTROSCOPY

Astronomical spectroscopy, like so much else, dates back at least to Newton.[7] He and his contemporaries were well aware that sunlight, passed through a prism, diverges into a rainbow-like spectrum of colors. As Newton argued in 1672, this happens not because the prism creates colors, but because sunlight, though apparently white, is actually made up of many colors, each of which is bent or refracted to a different degree as it passes through the prism. For more than a century, however, the main focus of spectroscopy was the nature of light itself, and the use of spectroscopy to answer astronomical questions was minimal.

---

[7] The most extensive discussion of astronomical spectroscopy is given in Hearnshaw, *The Analysis of Starlight*.

The years at the turn of the nineteenth century saw the beginning of investigations into color that refined spectroscopy into a prime tool of solar physics. In 1802, William Wollaston (1766–1828) noted that the spectrum of sunlight was crossed by a number of dark lines that, he thought, marked the boundaries of each color of the spectrum. The lines in the solar spectrum suggested themselves to the optician Joseph Fraunhofer as possible standards of homogeneous color for measuring the refractive properties of the different kinds of glass to be used in achromatic lenses. Studying the spectrum of sunlight at a very high dispersion, Fraunhofer found and mapped more than five hundred lines between 1814 and 1823. These lines clearly were not boundaries between colors; indeed, on Fraunhofer's close inspection it appeared that the spectral colors had no boundaries but, instead, shaded evenly from one to the next without regard for the positions of the lines. Fraunhofer found some of sunlight's darkest lines in the spectrum of light from the planet Venus, and he found quite different lines in the spectra of stars.

What, then, caused spectral lines? Chemical spectroscopy was still in its infancy, but it was already realized that characteristic bright lines could be seen in the spectra of flames, and that the pattern of lines changed when different substances were introduced into the flames. John Herschel (1792–1871; William's son) argued as early as 1823 that spectra could be used for chemical analysis and that the dark lines in the spectrum of sunlight should reveal the composition of the solar atmosphere. Numerous investigators followed up Herschel's suggestion by measuring thousands more lines throughout the visible part of the spectrum, attempting to identify the chemicals that caused them, and even, by ingenious methods, identifying lines in the ultraviolet and infrared. In 1842 and 1843, Edmond Becquerel and J. W. Draper each succeeded in photographing solar line spectra from the ultraviolet to the near infrared, initiating a technique that would come to dominate astronomical spectroscopy.

Despite this profuse activity, however, decades passed before spectroscopy could be accepted without qualification as a tool for the chemical analysis of celestial bodies. Sorting out which lines belonged to which element was slowed by the near ubiquity of contaminants (especially sodium) in spectroscopists' samples. Moreover, it was difficult to prove that the lines in the spectrum of sunlight actually originated with the Sun, and not as the light passed through the Earth's atmosphere. Attempts to settle this issue began in the 1830s and produced mixed results that were still being debated into the 1860s. For example, David Brewster found that the intensity of some lines depended on the thickness of the terrestrial atmosphere through which they were observed, while the intensity of others did not. Brewster concluded that the former were terrestrial, and the latter, truly solar. James Forbes, by contrast, compared the spectrum of the limb (edge) of the Sun to the spectrum from the center of the Sun's disk. Rays from the limb pass through more of the Sun's atmosphere and should, Forbes reasoned, show broader lines.

Forbes found no difference, and concluded that all spectral lines observed in sunlight were terrestrial.

In 1859, the Heidelberg physicist Gustav Kirchhoff (1824–1887) initiated the work that, over the next half century, would establish spectroscopy as the preeminent astronomical tool. The prominent pair of dark lines in the solar spectrum that Fraunhofer had labeled the D lines had long been known to lie at the same wavelengths as the bright yellow lines in the spectrum of a flame containing sodium. Kirchhoff intended to demonstrate the perfect coincidence of the lines by passing sunlight through a sodium flame, expecting that the bright lines would just fill in the dark lines. However, instead of being continuous, the resulting spectrum showed even darker lines. Two important points distinguished Kirchhoff's explanation: The absorption of light as it passed through the sodium flame must be the same process as the absorption of light by sodium vapor in the solar atmosphere, and, as Kirchhoff subsequently demonstrated in laboratory experiments, such absorption takes place whenever light from a hotter source is passed through a cooler absorbing medium. This neat demonstration that, depending on its temperature relative to a light source, a medium may produce either an emission or an absorption spectrum, was enough to overcome lingering doubt that the lines in the solar spectrum really reflected the chemistry of the solar atmosphere. Implicit in Kirchhoff's explanation was support for a solar model very different from Herschel's – a hot, glowing interior, surrounded by a cooler, absorbing atmosphere.

Kirchhoff and his chemist colleague Robert Bunsen (1811–1899) also made extensive analyses of the chemistry of the Sun, identifying numerous elements as constituents of the solar atmosphere. That few questioned the validity of their results, while a number disputed their priority, is a good measure of the wide and rapid acceptance of Kirchhoff and Bunsen's work, as well as of the fact that some of their observations and line identifications had already been suggested by others. Where they surpassed other work was in Kirchhoff's perceptive demonstration of the relationship between the absorption and emission lines that might be emitted by a medium, in Bunsen's contribution of exceptionally pure chemical samples for analysis, and in the thoroughness of their joint solar analysis.

THEORETICAL APPROACHES TO SOLAR MODELING: THERMODYNAMICS AND THE NEBULAR HYPOTHESIS

Thermodynamics was the second new component to enter solar theory in the decades before 1860. The amount of energy radiated by the Sun had been estimated with increasing accuracy from the early nineteenth century onward, and by 1833 John Herschel had, in his popular *Treatise on Astronomy*, publicized the fact that no known source of chemical energy could possibly fuel the Sun for long.[8] Were the Sun made entirely of coal, for instance, it would have energy to burn for only a few thousand years, shorter even than the traditional timescales of biblical studies, let alone the tremendously extended timescales already under discussion in geology. What other source might be sufficient remained a mystery. In the 1840s and 1850s, two investigators independently suggested variations of one hypothesis: that the light radiated by the Sun was derived in some way from mechanical energy.[9] The German doctor Julius Mayer (1814–1878) repeatedly, but unsuccessfully, sought to publish his theory that solar heat was generated by a rain of interplanetary meteors as they struck the solar surface. The Scottish engineer John James Waterston (1811–1883) was slightly more successful in gaining a public hearing for a version of the same idea, as well as for the alternative suggestion that heat might be mechanically generated as the body of the Sun contracted, and even that a contraction large enough to generate the measured amount of radiation would still be imperceptible from Earth.

Waterston's ideas were adopted and extended by the thermodynamicists Hermann von Helmholtz (1821–1894) and William Thomson (1824–1907; later Lord Kelvin). Both of these men, already versed in the theory of heat, learned of Waterston's work about 1853. Both toyed with the meteoritic hypothesis but adopted contraction as the single significant current source of solar energy. As Peggy Kidwell has pointed out, both sought to harmonize their theory of solar energy not only with the nebular hypotheses (to which we will turn in a moment), but also with observable astronomical data. Thus, for instance, they rejected the meteoritic hypothesis on the grounds that a steady infall of meteorites would add enough mass to the Sun to shorten the earthly year by a measurable but unobserved amount. Thomson was particularly interested in the finiteness of solar energy that the contraction hypothesis implied.[10] In spite of this implication, which would remain a significant problem in astronomy for decades, the thermodynamic model of a contracting Sun remained the main energy model of the Sun (and eventually of other stars) into the early twentieth century.[11]

The past history, or evolution, of the Sun was treated as a question quite separate from the question of modeling its current structure and processes. Throughout the nineteenth century, the dominant theory of the origin of the Sun, as of all stars, was the nebular hypothesis: The Sun and its planets

---

[8] John Herschel, *A Treatise on Astronomy* (London: Longman, Reese, et al., 1833), reprinted and expanded as *Outlines of Astronomy* (Philadelphia: Blanchard and Lea, 1861), p. 212.
[9] Peggy Aldrich Kidwell, "Solar Radiation and Heat from Kepler to Helmholtz (1600–1860)," PhD diss., Yale University, 1979, chap. 8.
[10] Joe D. Burchfield, *Lord Kelvin and the Age of the Earth* (Canton, Mass.: Science History Publications, 1975).
[11] Agnes Clerke, *Modern Cosmogonies* (London: Adam and Charles Black, 1905).

formed out of a huge, diffuse, and slowly rotating nebula of hot, glowing gas.[12] As the nebula contracted under the action of gravity, it collapsed into a flattened disk. A dense mass of matter in the center became the Sun, while planets formed in orbit around it. The same scenario, repeated in miniature, produced satellites orbiting some of the planets, and the rings around Saturn.

Versions of the nebular hypothesis remained a part of stellar and planetary theory from its elaboration by Pierre-Simon de Laplace (1749–1827) in the 1790s into the early twentieth century; even today's theories of stellar and planet formation bear a family resemblance to it. Nonetheless, Laplace's hypothesis predated the development of thermodynamics and was *not* a thermodynamic theory of the generation of stellar energy from gravitational collapse. Laplace's nebular fluid started out hot and luminous; therefore, the star that formed at the center of the nebula was so as well. Planets and their satellites were dark because they were small enough already to have cooled off. That luminous fluid existed and could be seen condensing into stars seemed well supported by the observations of William Herschel, who had, in his indefatigable sweeping of the heavens, found stars embedded in glowing nebulae. Even before Laplace elaborated his theory of the origin of the solar system as a whole, Herschel had captured the essence of what might be regarded as an alternative formulation of the nebular hypothesis, writing that a shining nebula seemed "more fit to produce a star by its condensation, than to depend on the star for its existence."[13]

The nebular hypothesis could be reconciled with a wide range of other physical concerns. Herschel believed it compatible with his theory, as did the thermodynamicists. The makers of stellar models in the early twentieth century also considered themselves to be working in the same tradition, even though they were at least as interested in modeling binary and variable stars as in planetary systems. The common thread is the prominent role played by Newtonian mechanics.

STELLAR SPECTROSCOPY

As we have seen, solar physicists took many routes to the study of the nature of the Sun: sunspots, eclipse phenomena, magnetic observation, spectroscopy, and thermodynamic theory. Because of the great distance and pointlike appearance of the stars, those who would study their nature had to depend largely on a single tool, spectroscopy. Stellar spectroscopy formed the core of a discipline that was, by the turn of the twentieth century, to emerge as astrophysics. Like solar physics, astrophysics has roots in the seventeenth

---

[12] Ronald L. Numbers, *Creation by Natural Law: Laplace's Nebular Hypothesis in American Thought* (Seattle: University of Washington Press, 1977).
[13] William Herschel, *Philosophical Transactions,* 81, p. 72, cited in Clerke, *Popular History of Astronomy,* p. 24.

century, but because the light from a star is so much fainter than the light from the Sun, stellar spectroscopy got a later start than did similar solar work. Moreover, stellar spectra were, at first, best studied not by attempting the meticulous measurement and identification of individual lines, as Kirchhoff and Bunsen had done with high resolution spectra of the Sun but, instead, by observing, at a lower resolution, the gross features of the spectra of many stars, then sorting the spectra into categories. During the 1860s, Angelo Secchi (1818–1878) of the Collegio Romano observed the spectra of more than four thousand stars and concluded that they could be divided into four categories that formed a sequence running from blue or white stars, whose spectra showed only a few absorption lines, to red stars, with plentiful absorption lines. Similar schemes were later devised by a number of workers, among them Hermann Carl Vogel (1841–1907) of Potsdam and the American amateur Lewis M. Rutherfurd. Guided by color change, astrophysicists not infrequently suggested that sequences of spectral types were evolutionary sequences: Stars passed from blue to red as they cooled in the course of their lives.

Stellar spectroscopy was revolutionized by improvements in photography. The dry collodion plates introduced in the 1870s made longer exposure times possible. The plates thus collected much more light than could the eye, so that lines in stellar spectra could be measured and identified, as had already been done for the Sun. The most famous of those who pioneered this work was probably William Huggins (1824–1910); the most infamous is almost certainly Norman Lockyer (1836–1907), British War Office clerk turned solar physicist.[14] Lockyer, after a decade as director of the South Kensington Solar Physics Observatory, became interested in stellar spectroscopy, and between 1890 and 1900 he developed an ill-fated scheme that attempted to integrate detailed stellar chemistry with his own, unorthodox meteoritic hypothesis of stellar evolution and dissociative physics of matter. Lockyer believed that stars originated in nebulae where swarms of meteorites violently collided, heating and vaporizing their material. The compounds of which they were made broke down into elements, and those dissociated further into proto-elements, visible in the spectra of the hottest stars. Lockyer's theory entailed both heating and cooling phases of stellar evolution, manifest in distinct series of spectral types. No part of Lockyer's theories won acceptance, but he deserves credit for appreciating both that stellar evolution might be more complicated than the simple, unidirectional spectral sequences suggested, and that the physics of the stars might involve matter in more exotic forms than those accessible to the terrestrial chemist.

The successful integration of stellar constitution, stellar evolution, and the physics of matter was not to be a single-handed effort like Lockyer's.

---

[14] A. J. Meadows, *Science and Controversy: A Biography of Sir Norman Lockyer* (Cambridge, Mass.: MIT Press, 1972).

Instead, it approached the other extreme. Not only was it a synthesis forged by specialists in several fields; but crucial work also came from the activities of a newly emerged style of astronomical institution, the factory observatory, staffed by large teams, often marked by sharply gendered divisions of labor. In the final section we will trace the story of stellar models further, but first we will pause to look at the communities of solar physics and astrophysics, as they had developed by the end of the nineteenth century.

FROM THE OLD ASTRONOMY TO THE NEW

As astrophysics began to distinguish itself from astronomy, relations between the two were competitive and cooperative by turns.[15] Especially in America, where it received little federal support, astrophysics was an entrepreneurial science. Key to its success was the founding of new observatories, furnished not with transit circles for measuring position but with huge telescopes (often reflectors) sporting cameras and spectroscopes. It sought wealthy patrons eager to see their own names upon great new telescopes that might – each for a brief while – be the biggest in the world. Astrophysics' master promoter, visionary, and statesman was George Ellery Hale (1868–1938), to whom Yerkes, Mount Wilson, and Palomar Observatories owe their existence. The excitement and novelty of astrophysics could be contrasted with the monotonous routine in which practitioners of the old astronomy seemed – some almost perversely – to glory. Believing that the proper goal of astronomy was the measurement and prediction of the position and motion of heavenly bodies, traditional astronomers valued precision, repetition, routine, the pursuit of long-term research programs, and the meticulous analysis of error. Some seemed afraid of the seductions of spectroscopy and decried its obsession with ever larger, ever newer instrumentation, its promise of important science to come even from the first efforts in this new direction, its much more speculative theory, and above all, its unabashed hustling after patrons, money, and notoriety.

Astronomers and astrophysicists recognized common interests in statistical work. It was difficult to distinguish a star's inherent brightness – the physically interesting parameter – from the apparent brightness it owed to the accident of its distance. The most obvious way of determining distance was by measuring the parallax a star exhibited due to the Earth's orbit, but this was such a difficult measurement that by the end of the nineteenth century, only

---

[15] Tensions between the two have been best studied in the American case, for which John Lankford has recently completed a large quantitative history. John Lankford, with the assistance of Ricky L. Slavings, *American Astronomy: Community, Careers, and Power, 1859–1940* (Chicago: University of Chicago Press, 1997). An alternate point of view, emphasizing routine aspects of spectroscopy, is David H. DeVorkin, *Henry Norris Russell: Dean of American Astronomers* (Princeton, N.J.: Princeton University Press, 2000).

a few hundred accurate parallax measurements had been made.[16] To elicit information on distance, astronomers, therefore, turned to proper motion, the apparent motion that nearer stars exhibit compared to more distant stars, due to the Sun's motion through space. To be informative, proper motion had to be studied statistically. The Dutch astronomer Jacobus Cornelius Kapteyn (1851–1922) spearheaded a large, coordinated, international cataloging effort intended to collect enough data for a statistical determination of the distribution of stars in space. In Kapteyn's campaign, spectral type and proper motion took their place beside the traditional astrometric parameters, position and luminosity. These parameters were to prove essential to astrophysicists in the first decades of the twentieth century, as they worked out the first recognizably modern models of the structure of stars. Where Kapteyn and his colleagues had been relatively uninterested in stars for their own sake, and treated them mainly as the subunits of the universe, astrophysicists would later mine the same collection of data for clues to the nature of the stars themselves.

The most important program of stellar classification at the turn of the twentieth century was the Henry Draper Memorial, an immense project under the direction of Edward Pickering (1846–1919) of the Harvard College Observatory. In the decades between 1885 and 1924, using funds donated by the widow of an amateur astronomer who had been the first to photograph a stellar spectrum, Pickering and his staff classified the spectra of nearly a quarter of a million stars. This project has become famous among historians of astronomy not only for the importance of its data to the development of stellar physics but also because it is now widely cited as an example of a factory observatory. The stellar spectra used in the project came in efficient form: large photographic plates, on which each star in the field had been recorded as a small spectrum, instead of a point image. The use of photography permitted a division of labor between observer and plate reader, and the division was strikingly gendered. While the nighttime work of taking the plates was done by men, reading them was assigned to teams of women who, Pickering discovered, would work diligently and consistently, without expectation of high salary, professional status, or career advancement.[17]

Nonetheless, some of the women made significant individual contributions. Annie Cannon (1863–1941) devised a new spectral sequence of ten types (the OBAFGKMRNS sequence) and numerous subtypes. Cannon's sequence was empirical, simple, and all-encompassing. It soon became, and in its general form still remains, the standard scheme of stellar classification. Nor was Cannon's the only taxonomy devised by the Harvard women. Antonia Maury

---

[16] Erich Robert Paul, *The Milky Way Galaxy and Statistical Cosmology, 1890–1924* (Cambridge: Cambridge University Press, 1993), p. 47.

[17] Pamela E. Mack, "Straying from Their Orbits: Women in Astronomy in America," in *Women of Science: Righting the Record*, ed. G. Kass-Simon and Patricia Farnes (Bloomington: Indiana University Press, 1990).

(1866–1952), recognizing subtle but systematic differences between the spectra of stars otherwise grouped in the same class, suggested the existence of two spectral sequences, perhaps representing two different courses of stellar evolution. Maury's work, though published, was not incorporated into the rest of the Draper project. In part, this was because Pickering doubted that the data were good enough for such fine distinctions, but perhaps he also feared that injecting greater complexity into the Harvard classification would jeopardize its universal acceptance.[18]

TWENTIETH-CENTURY STELLAR MODELS

By the turn of the twentieth century, various threads of the story began to fall together. Statistical astronomy had advanced to the point where average distance and brightness of stars could be measured spectral class by spectral class. When this was done, the Dane Ejnar Hertzsprung (1873–1967) and the American Henry Norris Russell (1877–1957) independently found that stars could be divided into two groups: the numerous small, dim, "dwarf" stars and the rarer enormous, brilliant "giants." These differences proved to lie behind Maury's two spectral sequences. The impact of the discovery was increased by its presentation as a diagram of luminosity versus spectral type, now called the Hertzsprung-Russell (H-R) diagram. As more stars were placed upon the diagram, the stellar population seemed increasingly to be distributed in continuous, intersecting "main" and "giant" sequences, lending weight to the widely held belief that the sequences tracked stellar evolution.[19]

Russell's diagram, presented at a meeting of the Royal Astronomical Society when Russell visited London in 1913, was seized upon by the English astronomer Arthur Stanley Eddington (1882–1944), whose work as chief assistant at the Royal Greenwich Observatory had already involved him in statistical surveys of position and motion. For Eddington, the most interesting feature of Russell's work was the way in which it bridged the gap between traditional astronomy and stellar physics. Eddington turned to stellar physics himself, producing, in 1926, *The Internal Constitution of the Stars*, a monumental text offering a physical model that explained why stars fit the parameters of the H-R diagram.[20] Eddington's approach marks a considerable shift in the project of stellar modeling. It is true that old models, including the nebular hypothesis and Lockyer's meteoritic hypothesis,

---

[18] David DeVorkin, "Community and Spectral Classification in Astrophysics: The Acceptance of E. C. Pickering's System in 1910," *Isis*, 72 (1981), 29–49.

[19] David DeVorkin, "Stellar Evolution and the Origin of the Hertzsprung-Russell Diagram," in *Astrophysics and Twentieth-Century Astronomy to 1950*, ed. Owen Gingerich (Cambridge: Cambridge University Press, 1984), vol. 4A: *The General History of Astronomy*, pp. 90–108.

[20] Arthur Stanley Eddington, *The Internal Constitution of the Stars* (Cambridge: Cambridge University Press, 1926).

had focused on dimensions, contraction, temperature, and, in Lockyer's case, color. However, the most significant efforts at stellar modeling in the immediately preceding decades were those of George Darwin and James Jeans. Working on the mechanics of orbits and gravitating systems, they had concentrated on accounting for other observable stellar phenomena: the formation of solar and planetary systems, and binary and variable stars.

In making his models, Eddington drew on earlier mathematical formulations of the equilibrium structure of a self-gravitating gaseous sphere, but with the difference that Eddington's stars were supported against gravitational collapse not just by gas pressure but by pressure due to radiation pouring out from a source concentrated in the center of the star. By focusing on the way in which radiation was transported, Eddington was able to circumvent perhaps the greatest problem that loomed over stellar theory, the source of stellar energy. Most accounts of the stellar energy problem emphasize its solution by the theoretical physicist Hans Bethe (b. 1906), in 1939, when processes of nucleosynthesis were better understood. One might, however, take an alternative view: that what astrophysics needed most in the early twentieth century was a way to make progress while temporarily avoiding this then-intractable puzzle. The problem of stellar energy had not progressed much beyond its original formulation. It was clear that the enormous energy output of the Sun could not come even from gravitational collapse, the most efficient known source, and still produce the energy required to keep the Sun shining over the long periods of time required by terrestrial biology and geology. By basing his model almost entirely on the mathematics of radiative transport, Eddington was able to reproduce many of the features of the purely empirical H-R diagram. He crowned his achievement by deriving an equation describing the relationship between a star's mass and luminosity that held for all stars for which measurements of both quantities were available.[21]

The chemical composition of stars played little role in Eddington's model. Early-twentieth-century advances in the theory of atomic spectra, however, made possible quantitative chemical analyses of stellar spectra. In 1920 the Indian physicist Meghnad Saha (1894–1956) showed that spectral lines reflect not just an element's presence but also its ionization.[22] The sequence of stellar spectra is, thus, a temperature sequence, not a sequence of differing chemistries. In 1925 Cecilia Payne (1900–1979) derived the relative abundances of elements in stellar atmospheres and showed that their distribution was nearly uniform from star to star.[23] One result of Payne's calculation was

---

[21] Joann Eisberg, "Eddington's Stellar Models and Twentieth-Century Astrophysics," PhD diss., Harvard University, 1991.
[22] Meghnad Saha, "Ionization in the Solar Chromosphere," *Philosophical Magazine*, 40 (1920), 479–88.
[23] Cecilia H. Payne, "The Relative Abundances of the Elements," in *Stellar Atmospheres*, Harvard Observatory Monograph no. 1 (Cambridge, Mass.: Harvard University Press, 1925), chap. 13.

that hydrogen dominated stellar composition. Though the rarity of hydrogen on Earth led Payne to dismiss the result as anomalous, several confirming lines of evidence led Henry Norris Russell to conclude in 1929 that hydrogen was the main component of stars' atmospheres.[24] Bengt Strömgren soon confirmed hydrogen's dominance in stars' interiors.[25]

Despite the fact that his models had succeeded, in part, by sidestepping the problem of stellar energy, Eddington lobbied for the possibility that subatomic interactions powered the stars. Until nuclear physics itself developed further, this could only be speculation. Beginning in 1929, Robert Atkinson, Fritz Houtermans, and Carl Friedrich von Weizsäcker investigated the proton–proton reaction and the CNO cycle, processes by which hydrogen nuclei might combine in the hot stellar interior to form helium and release energy.[26] In 1939, Bethe announced that the CNO cycle offers temperature-dependent energy generation in agreement with the luminosity of massive main-sequence stars.[27] Bethe had also investigated the proton–proton reaction with C. L. Critchfield and suggested that it accounted for the luminosity of lighter main-sequence stars.[28]

Energy production is but one aspect of nucleosynthesis; another is the creation of new chemical elements. Though Bethe's resolution of the stellar energy problem called for the transmutation of certain relatively light elements, a very important question remained unanswered: Where did the assortment of elements in stars come from? One suggestion was that heavy elements were generated from light ones in the hot, dense conditions of a primordial big bang, and that the stars' composition reflected the distribution that resulted as the universe cooled. When this idea proved unworkable, Fred Hoyle (who rejected the big bang in favor of a steady state cosmology, in any case) considered the possibility that heavier elements are generated in stellar cores. In 1957 he, together with E. Margaret Burbidge, Geoffrey Burbidge, and William Fowler, published an extensive survey of nuclear processes that contribute to element building in stars. This paper (colloquially referred to as B^2FH, for the authors' initials) is widely recognized as the foundation for future work on stellar nucleosynthesis, although some of the same material was covered independently by Alastair Cameron, who also emphasized the role of supernovae in the formation of new Elements.[29]

---

[24] Henry Norris Russell, "On the Composition of the Sun's Atmosphere," *Astrophysical Journal*, 70 (1929), 11–82.

[25] Bengt Strömgren, "The Opacity of Stellar Matter and the Hydrogen Content of the Stars," *Zeitschrift für Physik*, 4 (1932), 118–53.

[26] Robert d'Escourt Atkinson and F. G. Houtermans, "Zur Frage der Aufbaumöglichkeit der Elemente in Sterne," *Zeitschrift für Physik*, 54 (1929), 656–65; Robert d'Escourt Atkinson, "Atomic Synthesis and Stellar Energy, I, II," *Astrophysical Journal*, 73 (1931), 250–95, 308–47; Robert d'Escourt Atkinson, *Astrophysical Journal*, 84 (1936), 73; Carl Friedrich von Weizsäcker, "Element Transformation Inside Stars, II," *Physikalische Zeitschrift*, 39 (1938), 633–46.

[27] Hans Albrecht Bethe, "Energy Production in Stars," *Physical Review*, 55 (1939), 434–56.

[28] Hans Albrecht Bethe and C. L. Critchfield, *Physical Review*, 54 (1938), 248.

[29] E. Margaret Burbidge, Geoffrey R. Burbidge, William A. Fowler, and Fred Hoyle, "Synthesis of

Let us pause to consider how stellar models have changed since the first model mentioned in this chapter. William Herschel's theory of the sun was a simple, qualitative account of how sunspots, some of the most visible prespectroscopic features of the Sun's surface, might arise. He assumed that beneath the Sun's bright outermost layers, there might lie a habitable planet, much like the Earth. Laplace and Lockyer, though they took very different approaches, were both most interested in questions of origin and evolution. Spectral classification sequences, for Lockyer and most other stellar physicists of the nineteenth and early twentieth century, were assumed to track paths of stellar evolution. Eddington's approach was to model the physical properties of the stellar interior: temperature, pressure, and density at different depths. A consequence of his model was to overturn much of existing stellar evolutionary theory. While his calculations left the question of stellar energy unanswered, Eddington, by exploring the physical conditions at the core of a star, laid a fruitful groundwork for later explorations of stellar nucleosynthesis and other applications of modern physics to the constitution and development of stars.[30]

The sociology of the field paralleled its intellectual development. Except in a few fields like the observation of variables, it was soon nearly impossible for those without formal training in physics to participate in the study of the stars. At the same time (as described in Robert W. Smith's chapter, Remaking Astronomy: "Instruments and Practice in the Nineteenth and Twentieth Centuries"), instrumentation became more sophisticated and expensive. Modern observatories are typically accessible only to card-carrying or diploma-wielding professionals with ties to major institutions. This contrasts sharply with the past century, as evidenced by the tremendous number and variety of important contributions made by amateurs.

We may summarize this short history by saying that the stars, once points whose location might be measured but whose nature seemed unknowable, had become prime subjects of physical science. During the nineteenth century terrestrial physics was applied to the stars, and during the twentieth the study of stars became a laboratory for testing our understanding of the behavior of matter.

---

the Elements in Stars," *Reviews of Modern Physics,* 29 (1957), 547–650. See also A. G. W. Cameron, in *Publications of the Astronomical Society of the Pacific,* 69 (1957), 201–22, and "Nuclear Astrophysics," *Annual Review of Nuclear Science,* 8 (1958), 299–326; and Geoffrey R. Burbidge, "Nuclear Astrophysics," *Annual Review of Nuclear Science,* 12 (1962), 507–76.

[30] See David DeVorkin and Ralph Kenat, "Quantum Physics and the Stars (I), (II)," *Journal for the History of Astronomy,* 15 (1983), 102–32, 180–222.

27

COSMOLOGIES AND COSMOGONIES OF SPACE AND TIME

Helge Kragh

For over three millennia, cosmology had closer connections to myth, religion, and philosophy than to science. Cosmology as a branch of science has essentially been an invention of the twentieth century. Because modern cosmology is such a diverse field and has ties with so many adjacent scientific disciplines and communities (mathematics, physics, chemistry, and astronomy), it is not possible to write its history in a single chapter. Although there is no complete history of modern cosmology, there exist several partial histories that describe and analyze the main developments. The following account draws on these histories and presents some major contributions to the knowledge of the universe that emerged during the twentieth century. The chapter focuses on the scientific aspects of cosmology, rather than on those related to philosophy and theology.

THE NINETEENTH-CENTURY HERITAGE

Cosmology, the study of the structure and evolution of the world at large, scarcely existed as a recognized branch of science in the nineteenth century; and cosmogony, the study of the origin of the world, did even less. Yet there was, throughout the century, an interest, often of a speculative and philosophical kind, in these grand questions. According to the nebular hypothesis of Pierre-Simon de Laplace and William Herschel, some of the observed nebulae were protostellar clouds that would eventually condense and form stars and planets in a manner similar to the way in which the solar system was believed to have been formed. This widely accepted view implied that the world was not a fixed entity, but in a state of evolution.

Evolutionary processes were described by the laws of thermodynamics that emerged in the 1840s and 1850s, and these were applied to cosmology from an early date. The German physicist Rudolf Clausius's famous 1865 formulations were framed cosmologically, namely, that the energy of the universe is constant and the entropy of the universe tends to a maximum. The idea that the second law of thermodynamics would lead to a maximum-entropy state of the universe was popular in the late nineteenth century. The final state was often referred to as the heat-death, a lifeless universe with no further possibility of evolution. However, many scientists and philosophers found the heat-death scenario unacceptable and not a necessary consequence of the second law. As early as 1852, the British physicist William Rankine suggested the existence of counterentropic processes that would lead to an everlasting creative universe. Similar kinds of speculations abounded in the late nineteenth century and early twentieth century. For example, in 1895, Ludwig Boltzmann used his statistical theory of entropy to suggest that although our part of the world is approaching thermal equilibrium, there will always be other parts of the world in evolving, low-entropy states. In 1913, the British geophysicist Arthur Holmes revived Rankine's old idea of thermodynamic reversibility on a cosmic scale in order to explain why the universe has not already reached a state of maximum entropy. The alternative was to "believe in a definite beginning," which Holmes rejected.[1]

Whereas the final ages of Earth and the solar system were well established before the First World War, the cosmogonic notion of the universe being of finite age had no place in physics and astronomy. This is not to say that the notion cannot be found in the nineteenth century, but it figured only rarely and marginally. In 1861, the German astronomer Johann Mädler suggested that "a finite amount of time has passed from the beginning of Creation until our day."[2] Neither Mädler nor other astronomers followed up the suggestion. Much more common were ideas of a cyclical or recurrent universe, that is, that the universe develops cyclically and eternally in such a way that there is no unidirectional evolution on a very long time scale. Such ideas go back to antiquity and were popular in the late nineteenth century when they were discussed both by scientists (including Boltzmann and Henri Poincaré) and by nonscientists (including the philosopher Friedrich Nietzsche). However, they were of no importance to scientific astronomy.

GALAXIES AND NEBULAE UNTIL 1925

Terms such as "cosmology" and "universe" were rarely used by astronomers prior to the 1920s, and when they were, they usually referred to the stars and nebulae making up the Milky Way. Around 1900, most astronomers believed that the nebulae were located inside the Milky Way, rather than being structures apart from it. There were good observational reasons against

---

[1] Arthur Holmes, *The Age of the Earth* (New York: Harper, 1913), p. 121.
[2] Frank J. Tipler, "Olber's Paradox, the Beginning of Creation, and Johann Mädler," *Journal for the History of Astronomy*, 19 (1988), 45–8.

the rival "island universe" theory, according to which the nebulae were huge extragalactic congregations of stars – galaxies as they became known. For example, if the nebulae were independent galaxies, they would be expected to be equally distributed over the sky; but observations showed that they avoided the plane of the Milky Way. Some astronomers speculated that there were perhaps other galaxies or "universes" hidden in infinite space, only these would be forever invisible. However, the majority of astronomers had no patience for such speculations and rejected the island universe theory. Observations and statistical studies of stars seemed to indicate a populated universe of roughly the same size as the Milky Way. About 1912, the influential Dutch astronomer Jacobus C. Kapteyn (1851–1922) argued that the major radius of the ellipsoid-shaped Milky Way was about 50,000 light years. Outside this distance there was space, but no stars.[3]

The main reason for the uncertainty, both with regard to possible island universes and the size of the Milky Way, was lack of reliable methods for determining distances to the nebulae. Two discoveries made in 1912 proved to be important in establishing a new and larger picture of the world. The Harvard astronomer Henrietta Leavitt (1868–1921) found a method for determining the distance of Cepheid variable stars relative to the Magellanic Clouds. The Cepheid method was quickly developed by other astronomers, and by 1918, the existence of a single Cepheid associated with a given nebula was enough to determine the distance of that nebula. In 1912, Vesto Slipher (1875–1969) at the Lowell Observatory found the first Doppler shift for a spiral nebula, the Andromeda galaxy, and subsequent spectroscopic measurements indicated that most of the nebulae receded from Earth. In 1917 Slipher reported measurements of the radial velocities of twenty-five nebulae, of which four were receding with velocities of more than 1,000 km per second. Slipher did not interpret the redshifts cosmologically, and, for a period, the receding nebulae were a puzzle to the astronomers. In 1924 a cosmological interpretation was suggested by the Polish-born physicist Ludwik Silberstein (1878–1942), who argued that the redshifts were proportional to the distances of the nebulae. However, Silberstein's evidence was flawed, and his work did not convince astronomers of a linear relationship between redshift and distance.[4]

Slipher's observations contributed to a revival of the island universe theory. If spiral nebulae receded from the central part of the Milky Way with enormous velocities, it seemed unlikely that they were gravitationally bound. The whole issue was further complicated by Harlow Shapley's (1885–1972) 1917 claim that the Milky Way was much larger than previously assumed, namely, with a diameter of about 300,000 light years. A galaxy of this size was thought to be a strong argument against the island universe theory. In 1920

---

[3] Erich R. Paul, *The Milky Way and Statistical Cosmology 1890–1924* (Cambridge: Cambridge University Press, 1993).

[4] Robert Smith, *The Expanding Universe: Astronomy's 'Great Debate' 1900–1931* (Cambridge: Cambridge University Press, 1982).

the entire subject was discussed in the "Great Debate" between Heber Curtis and Shapley, with Curtis arguing for the island universe view and against Shapley's big galaxy. No consensus was achieved, and for a couple of years the question remained controversial. In 1923 Edwin Hubble (1889–1953) at the Mount Wilson Observatory in California discovered a Cepheid in the Andromeda nebula, and the observation led him to determine its distance at about one million light years. The very large distance strongly indicated that Andromeda was located outside the Milky Way. By the time that the discovery was officially announced on the first day of 1925, it had already been known for some time. It settled the Great Debate in favor of the island universe theory. Now the universe came to be seen as a vast congregation of galaxies, somewhat analogous to a gas made up of molecules. This major transformation of the world picture was independent of changes in theoretical cosmology that occurred in the same period.

COSMOLOGY TRANSFORMED: GENERAL RELATIVITY

In February 1917, Albert Einstein (1879–1955) completed a work that exposed him "to the danger of being confined in a madhouse," as he told his friend Paul Ehrenfest.[5] The work was an application of his new general theory of relativity to the entire universe. It heralded a revolution in theoretical cosmology and is still, eighty-five years later, considered the foundation of the science of the universe. Einstein solved the problem of formulating boundary conditions for an infinite space – a problem first considered by Isaac Newton – by conceiving the universe as a spatially closed continuum, as described by the general theory of relativity. In Einstein's model, time was linear and space "spherical" in four dimensions. His universe was static and spatially finite in spite of having no boundary. The formal core of Einstein's theory was the gravitational field equations of 1915, now modified by adding a term proportional to the metrical tensor.

Einstein's model universe was homogeneously filled with dilute matter and could be ascribed a definite radius, volume, and mass. Temporally it was infinite, the radius of curvature having the same value at all times. Einstein at first believed that this was the only solution compatible with general relativity, but later in 1917, the Dutch astronomer Willem de Sitter (1872–1934) found another solution very different from Einstein's. De Sitter's model included no matter, but was nonetheless spatially closed. Furthermore, if particles (or galaxies) were introduced in the de Sitter universe, light from them would appear redshifted to the receiver. This was later seen as an effect of the exponential expansion of the de Sitter world, but until 1930, the model was

---

[5] Abraham Pais, 'Subtle is the Lord...': *The Science and Life of Albert Einstein* (Oxford: Oxford University Press, 1982), p. 285.

interpreted as being static with the redshifts implying only a "spurious radial velocity."

Einstein found de Sitter's model objectionable, not only because it contained a world horizon beyond which no signals can reach the observer, but also because he found the curved space-time in the absence of matter to disagree with the spirit of general relativity.[6] Of course, the absence of matter was a problem in itself. All the same, de Sitter's solution soon became popular among the few theoretical cosmologists and experts in relativity, not least because of its connection with the galactic redshifts that were reported at the time. In the 1920s, a group of no more than a dozen mathematical physicists and astronomers investigated which of the two relativistic alternatives was the most satisfactory. The main actors in this development were, apart from Einstein and de Sitter, Arthur Eddington (1882–1944), Herman Weyl, Cornelius Lanczos, Georges Lemaître (1894–1966), Howard Robertson, and Richard Tolman. During the course of this work, they gradually recognized that it was not a question of either Einstein's or de Sitter's model. Neither of the two classical solutions seemed to represent the real universe, and consequently, some scientists developed hybrid theories in which the space-time metric depended on the time coordinate in a matter-filled universe. Such nonstatic theories were suggested by Lanczos in 1922, Lemaître in 1925, and Robertson in 1928. In spite of their nonstatical features, these theories were not considered to be evolutionary in a physical sense. With two exceptions, as will be mentioned, the framework of cosmological thinking in the 1920s was constrained by the static universe paradigm.

AN EXPANDING UNIVERSE

The collapse of the static universe paradigm took place by the interaction of two separate approaches, the one observational and the other theoretical. In 1929, Hubble published data on galactic redshifts and related them to galactic distances. The result was a nearly linear relationship between the distances (r) and the "apparent velocities" (v), as inferred from the redshifts. The Hubble relation is $v = Hr$, where H became known as the Hubble parameter and was found by Hubble to be about 500 km per second per megaparsec. (One megaparsec is about 3.26 million light years.) More extended data published in 1931 by Hubble and his assistant Milton Humason confirmed the linear relationship.[7] Hubble's 1929 paper is often identified with the discovery of the expanding universe, but Hubble did not conclude

---

[6] Pierre Kerzberg, *The Invented Universe: The Einstein-de Sitter Controversy (1916–17) and the Rise of Relativistic Cosmology* (Oxford: Clarendon Press, 1992).

[7] Norriss S. Hetherington, "Philosophical Values and Observation in Edwin Hubble's Choice of a Model of the Universe," *Historical Studies in the Physical Sciences*, 13 (1982), 41–68.

that the galaxies are actually receding from us. Even after most astronomers had accepted the expanding universe, the cautious Hubble emphasized the empirical nature of the redshift-distance relationship and the problems connected with the hypothesis of a universe in expansion. Shortly after Hubble's publication, the Swiss astronomer Fritz Zwicky (1898–1974) proposed that the Hubble relation could be explained by a "tired light" mechanism that made the galactic recession hypothesis unnecessary. Although this and other kinds of alternatives attracted some interest in the 1930s, the majority of astronomers accepted the Doppler interpretation of the redshifts and, hence, the universal recession of the galaxies.

Unknown to Hubble, the possibility of an expanding universe had already been formulated by theoreticians, first by Alexander Friedmann (1888–1925) in 1922. Friedmann, a Russian theoretical physicist, gave a complete analysis of the solutions to Einstein's cosmological field equations and showed that the static Einstein and de Sitter solutions were merely two special cases of a more general solution. This included cyclical and ever-expanding models, "a world in which the curvature of space is independent of the three spatial coordinates but does depend on time."[8] Friedmann formulated the fundamental equations governing the time variation of the curvature of the universe, later known as the Friedmann equations. Five years later, in 1927, the Belgian physicist Lemaître found independently the same equations and subjected them to systematic analysis. Whereas Friedmann, whose approach was basically mathematical, was not much concerned with the real universe, Lemaître explicitly argued that the universe is expanding. He connected his theory with current redshift measurements and described the recession of the galaxies as a cosmic effect of the expansion of the universe. He even derived the later Hubble law ($v = Hr$) and found for the H-factor a value of about 625 km per second per megaparsec. According to Lemaître, the universe had gradually evolved from a static Einstein state and was now rapidly expanding.[9]

It is most remarkable that neither Friedmann's nor Lemaître's works made any impact at all. The reasons for the neglect are not entirely clear, but ingrained belief in the static nature of the universe was undoubtedly an important sociopsychological factor. Attitudes about the expanding universe changed dramatically in the early part of 1930, however. As a result of Hubble's data, and also of theoretical work done by Robertson and Tolman, the climate now became receptive to the idea of an evolving universe. Eddington studied Lemaître's old paper and realized that it provided the solution to the cosmologists' dilemma. With the enthusiastic support of Eddington and de Sitter, the expanding universe became quickly accepted by most specialists,

[8] Alexander Friedmann, "Über die Krümmung des Raumes," *Zeitschrift für Physik*, 10 (1922), 377–86. Translated by the author unless otherwise noted.
[9] Odon Godart and Marian Heller, *Cosmology of Lemaître* (Tucson, Ariz.: Pachart Publishing House, 1985).

and cosmology experienced a sudden paradigm shift. It was only now that Hubble's discovery was interpreted as a discovery of the expanding universe.

Who "discovered" the expanding universe? Among the three main candidates, Lemaître was the only one who clearly argued that the universe is expanding and drew on both theoretical and observational arguments. Friedmann showed that the universe might be expanding, but only as one possibility among many; and Hubble, although he provided strong observational evidence, refrained from concluding that the universe is expanding. It is, therefore, reasonable to credit Lemaître with the discovery, possibly the most important ever in the history of cosmology.

Lemaître's 1927 model was expanding but did not have an origin in time. In his paper of 1922, Friedmann discussed finite-age models originating from a space-time singularity and wrote about "the creation of the world." However, he seems to have considered the idea a mathematical curiosity, rather than a possible physical reality. It was only in March 1931 that Lemaître introduced into scientific cosmology the notion of the beginning of the world in a realist sense. He suggested that the universe, including space and time, had started in a kind of explosive radioactive decay of a "primeval atom" in which the entire mass of the universe was concentrated. The original superatom was of finite size and density, and Lemaître's model was thus not a big bang theory in the strict sense of including an initial singularity of infinite density. In works between 1931 and 1934, he developed his suggestion, and he remained faithful to it throughout his life. Most other cosmologists hesitated to consider models with a sudden origin, and for a period, Lemaître was alone in defending the big bang idea. Insofar as relativist cosmologists considered big bang models in the 1930s, they restricted their considerations to the mathematical aspects and were reluctant to endow big bang solutions with physical reality.

There were good reasons that the big bang hypothesis was coolly received in the 1930s. First, the notion of the creation of the world was widely seen as conceptually problematic. After all, a creation needs something to create it, and what (or who) could possibly be the cause of the universe? Second, the hypothesis had no convincing observational support. And, third, the age of the big bang universe as inferred from the Hubble parameter seemed much too low, namely, smaller than the ages of the stars and even Earth. This problem, known as the time-scale difficulty, was much discussed in the 1930s and 1940s. In spite of the initial lack of positive response to Lemaître's theory, big bang ideas were well known in the 1930s. In 1938, the German physicist Friedrich von Weizsäcker (b. 1912) sought to explain energy production in the stars in terms of nuclear processes, and in the course of this work, he was led to his own version of the big bang. Von Weizsäcker speculated that the early universe was extremely hot and of nuclear density, and that the initial nuclear reactions had produced the energy necessary for the expansion. Von Weizsäcker's cosmological hypothesis supplemented

Lemaître's, but had only a little impact on the further development of cosmology.

NONRELATIVISTIC COSMOLOGIES

Theoretical cosmology was far from identical with models based on general relativity. On the contrary, the 1930s witnessed a proliferation of cosmological ideas and models that were opposed to standard relativistic theory. On the whole, cosmology had very little disciplinary and theoretical unity. No theory of the universe was clearly the most favored. Among the more heterodox alternatives to relativistic evolution cosmology were various attempts to picture the universe as being in a steady state, with decay of matter being balanced by formation of new matter. The American astronomer William MacMillan suggested such a world picture in the 1920s, and his ideas were endorsed by Robert Millikan. In the 1920s and 1930s, the German chemist Walther Nernst developed his own version of an eternal, recycling universe. MacMillan and Nernst denied the expansion of the universe and believed that the Hubble law could be explained without this hypothesis.[10] Their ideas were not taken seriously by mathematical cosmologists.

Of more importance was the alternative developed by Edward Milne (1896–1950) in England from 1932 onward. Milne built on the special, but not the general, theory of relativity, and his theory was based on simple kinematic considerations, rather than on field equations. His model belonged to the big bang category insofar as the galaxies receded proportionally with time. Milne's system of "kinematic relativity," as he called it, was very influential in the 1930s when it set the agenda for a large part of cosmological work. It was as much a philosophical as a scientific system, and its deductive nature and ambitious rationalism caused a good deal of controversy.[11] Paul Dirac's cosmological theory of 1937–8 was inspired by the works of Milne and Eddington. He was led to a big bang model in which the universe expands with the cube root of cosmic time. More controversially, the model was based on the hypothesis that the gravitational constant varies in time, in contradiction with general relativity. Dirac's cosmological theory inspired the German physicist Pascual Jordan to develop it further and formulate it in a field-theoretical framework. Although much work was done on Dirac-Jordan cosmology after 1945, most astronomers and physicists considered it speculative and without empirical support.

[10] Helge Kragh, "Cosmology Between the Wars: The Nernst-MacMillan Alternative," *Journal for the History of Astronomy*, 26 (1995), 93–115.
[11] John Urani and George Gale, "E. A. Milne and the Origins of Modern Cosmology: An Essential Presence," in *The Attraction of Gravitation: New Studies in the History of General Relativity*, ed. John Earman, Michel Janssen, and John D. Norton (Boston: Birkhäuser, 1993), pp. 390–419.

GAMOW'S BIG BANG

The new nuclear physics originating about 1930 provided big bang cosmology with a much-needed physical perspective. Why do the stars shine? How was the present distribution of chemical elements formed? In the late 1930s, these questions were addressed by the Russian-American George Gamow (1904–1968) and a few other physicists who believed that the answers had to be framed cosmologically. At a 1942 conference in Washington, D.C., on "the problems of stellar evolution and cosmology," it was agreed that "the elements originated in a process of explosive character, which took place at the 'beginning of time' and resulted in the present expansion of the universe."[12] Developing this conclusion, in 1946 Gamow presented a revised big bang theory that combined nuclear physics in the early universe with the Friedmann equations. He imagined the early, high-density universe to have consisted of relatively low-energy neutrons forming a kind of gigantic neutronic complex. From this starting point he indicated how the chemical elements were formed during the earliest phase of expansion.

Within two years, the theory was substantially modified and improved in collaboration with Ralph Alpher (b. 1921). Their brief 1948 paper described a primordial, hot neutron gas that started to decay into protons and electrons. Gamow and Alpher argued that subsequent nuclear processes would lead to relative abundances of the elements in agreement with those estimated from observations. Moreover, they realized that the early universe was dominated by electromagnetic radiation, not matter. What had happened to this radiation? According to Alpher and his collaborator Robert Herman (b. 1914), the radiation cooled with the expansion of the universe and would now have a temperature of about 5 K. Although Alpher and Herman's prediction of a cosmic microwave bath appeared in print several times between 1948 and 1956, it attracted no attention, and there were no attempts to detect the feeble radiation.[13]

It soon turned out that the original universe could not consist purely of neutrons. In 1950, the Japanese physicist Chushiro Hayashi argued for a primordial universe consisting equally of protons and neutrons, and three years later, Alpher, Herman, and James Follin included electrons, neutrinos, and other elementary particles in the model. The Alpher-Herman-Follin version of Gamow's theory was a sophisticated quantitative theory that followed the evolution of the universe in mathematical details from 10^{-4} seconds after the initial explosion until about 600 seconds. Among other results, the authors calculated that the percentage of helium would be about 32 percent.

---

[12] Helge Kragh, *Cosmology and Controversy: The Historical Development of Two Theories of the Universe* (Princeton, N.J.: Princeton University Press, 1996), p. 105.

[13] Ralph A. Alpher and Robert C. Herman, "Early Work on 'Big-Bang' Cosmology and the Cosmic Blackbody Radiation," in *Modern Cosmology in Retrospect*, ed. B. Bertotti et al. (Cambridge: Cambridge University Press, 1990), pp. 129–58.

Unfortunately, in the early 1950s there was no reliable empirical figure with which the prediction could be compared. With the Alpher-Herman-Follin theory, big bang cosmology came to an almost complete stop. A dozen or so physicists had been engaged in developing Gamow's theory after 1948, but after 1953, interest decreased drastically. Between 1956 and 1964, only a single research paper was devoted to what a few years earlier had looked like a flourishing research program. The reasons for this lack of interest are complex and must be ascribed to both social and scientific factors. In spite of its successes with regard to the lightest elements, Gamow's theory was unable to explain the abundances of the heavier elements, which was widely seen as a serious objection to the theory. Moreover, it shared the time-scale difficulty of most other evolutionary models. And it faced stiff competition from a new cosmological theory, the steady state theory of the universe.

THE STEADY STATE CHALLENGE

In the same year that Gamow and Alpher introduced their big bang theory, the Cambridge physicists Hermann Bondi (b. 1919), Thomas Gold (b. 1920), and Fred Hoyle (b. 1915) proposed a radically different theory of the universe. The steady state universe was expanding but nonetheless stationary, which required that matter be continually created throughout space.[14] The steady state theory was considered controversial from its very beginning, not only because of its unorthodox scientific features but also because Hoyle used it as an argument against religious beliefs. The postulated creation of matter aroused much debate among physicists and philosophers. Matter created from nothing violated the principle of energy conservation, and for this reason, the theory was sometimes accused of being "unscientific romanticizing" or "science-fiction cosmology," as some of its opponents called it.

Among the scientific successes of the steady state theory was that it led to a promising theory of galaxy formation, a problem that the big bang theory seemed incapable of solving. Even more important was the work that Hoyle did in the mid-1950s on nucleosynthesis in a universe without a big bang. In 1957, Hoyle and his collaborators (William Fowler, Margaret and Geoffrey Burbidge) produced a comprehensive and successful theory of stellar element formation in which they calculated the abundances of almost all elements in good agreement with observations.[15] Because the theory did not refer to a hypothetical earlier state of the universe, it was widely seen as a strong argument against the big bang theory. On the other hand, the big bang theory received support when the German-American astronomer

---

[14] Hermann Bondi, *Cosmology* (Cambridge: Cambridge University Press, 1952); Kragh, *Cosmology and Controversy*.

[15] Stephen F. Mason, *Chemical Evolution: Origins of the Elements, Molecules, and Living Systems* (Oxford: Clarendon Press, 1992).

Walter Baade discovered in 1952 that the Hubble parameter was much less than previously thought. With a new Hubble time of 3.6 billion years, soon to increase to about 10 billion years, there was no longer any serious difficulty with the age of the world, compared with the age of Earth.

The steady state theory was developed primarily by British scientists, whereas American astronomers either ignored or rejected it. The theory was also rejected in the Soviet Union, where the notion of continual creation of matter was considered unscientific, as well as ideologically illegitimate. During the 1950s, the official attitude to the rival big bang theory, with its creation of the universe and religious associations, was basically the same. As a consequence, very little cosmological work was done in the Soviet Union until the early 1960s when ideological constraints loosened.[16] For more than a decade, the steady state theory was a strong competitor to relativistic evolution theories. By the late 1950s, it was by no means evident which of the two theories would be victorious. The majority of astronomers were in favor of a universe with a finite age, but their conviction did not rest on incontrovertible observational facts.

RADIO ASTRONOMY AND OTHER OBSERVATIONS

One way of distinguishing between the two rival cosmological theories would be to measure the rate of expansion of space. Attempts to determine the geometry of space by relating the redshifts of galaxies to their brightness went back to the 1930s, and in 1956, Allan Sandage (b. 1926) and his coworkers at the Mount Palomar Observatory announced results that clearly disagreed with the steady state prediction. However, their observations were disputed by other astronomers and turned out to be inconclusive. The observational program continued during the 1960s but failed to produce results that unambiguously ruled out the steady state theory.

Radio astronomy entered the cosmological controversy in 1955 when Martin Ryle (1918–1984) at Cambridge University concluded that "there seems no way in which the observations can be explained in terms of a steady-state theory."[17] Ryle's dislike of the steady state theory may have colored his conclusion and interpretation of the data of the distribution of radio sources. At any rate, his results were contradicted by measurements made in Sydney, and for a while, radio astronomy appeared to share with optical astronomy an inability to deliver a conclusive refutation of the steady state theory. But in 1960–1, new data were produced in Cambridge that were more reliable and in clear disagreement with the steady state prediction.

---

[16] Loren R. Graham, *Science and Philosophy in the Soviet Union* (New York: Knopf, 1972), pp. 139–94.
[17] Woodruff T. Sullivan, III, "The Entry of Radio Astronomy into Cosmology: Radio Stars and Martin Ryle's 2C Survey," in *Modern Cosmology in Retrospect*, ed. Bertotti et al., pp. 309–30.

"These observations do ... appear to provide conclusive evidence against the steady-state theory," Ryle wrote.[18] This time he was supported by the Sydney astronomers. Although none of the steady state cosmologists found Ryle's conclusion to be compelling, most astronomers did. By 1964 the radio astronomical measurements had seriously shattered the reputation of the steady state theory and, contrarywise, strengthened the case for relativistic evolution cosmology.

A decisive moment came in 1965–6. At that time it became clear that counts of quasars contradicted the steady state theory. The evidence from quasars convinced Dennis Sciama, a leading British steady state physicist, to abandon the theory and accept the big bang model. At the same time, nuclear-physical calculations made by Hoyle, Roger Tayler, J. Peebles, and others showed that the abundance of helium in the universe – about 27 percent – could be nicely reproduced from big bang assumptions. The steady state theory, on the other hand, could not account for the percentage in a statisfactory way.

A NEW COSMOLOGICAL PARADIGM

During the years 1953 to 1963, no progress took place within big bang theory, and Alpher and Herman's prediction of a cosmic microwave background radiation was effectively forgotten. In 1964 the Princeton physicist Robert Dicke (1916–1995) reached the same conclusion independently, and in early 1965, his collaborator James Peebles (b. 1935) estimated the present radiation temperature to be about 10 K. Meanwhile, Arno Penzias (b. 1933) and Robert Wilson (b. 1936) at the Bell Laboratories had found an excess antenna temperature in their radiometer of about 3.5 K. Their experiments indicated that the unexplained excess temperature might be of cosmic origin, but Penzias and Wilson had no explanation for it and did not connect the anomaly with cosmological theory. Dicke's group realized that Penzias and Wilson had unknowingly made an important cosmological discovery, namely, of the cosmic background radiation left over from the big bang. In the summer of 1965, the discovery was published and its significance fully understood: The big bang theory predicted the existence of a blackbody-distributed radiation of temperature of about 3 K, while the radiation could not be explained on steady state assumptions. Other experiments quickly confirmed the finding of Penzias and Wilson and verified that the shape of the spectrum was as predicted.

The discovery of the cosmic microwave background radiation was a great triumph for the big bang theory and was generally seen as the last nail in the coffin of the steady state alternative. After 1965, the latter theory was no longer important, and the victorious big bang theory achieved paradigmatic status.

[18] Kragh, *Cosmology and Controversy,* p. 324.

The discovery of the background radiation, together with quasar counts, radio astronomy, and helium calculations, resulted in a new consensus, a "renaissance of observational cosmology," as Sciama called it.

Together with the new observations, theoretical progress in general relativity contributed to the renaissance of big bang cosmology. The theory of general relativity experienced its own renaissance in the early 1960s, when the theory was brilliantly confirmed and came to be regarded as a fundamental and universally true theory. In 1965–6, Roger Penrose (b. 1931), Stephen Hawking (b. 1942), and others reinvestigated the old question of a cosmic singularity at $t = 0$. They proved that under very general conditions, the universe must necessarily have started in a space-time singularity. In other words, not only is the big bang scenario compatible with general relativity; it seems to follow from it.

The takeoff that cosmology experienced in the wake of the discoveries in the mid-1960s manifested itself socially as well as cognitively. The number of students increased, connections between physicists and astronomers strengthened, and new textbooks appeared that defined the content and context of the new cosmology. Important examples were Peebles's *Physical Cosmology* (1971), Steven Weinberg's *Gravitation and Cosmology* (1972), and Yakov Zel'dovich and Igor Novikov's *Relativistic Astrophysics* (1983; Russian original 1975). The number of annual research publications on cosmology increased rapidly, from about 50 in 1962 to 250 in 1972.

DEVELOPMENTS SINCE 1970

The collaboration between nuclear physics and cosmology, which started with Gamow in the 1940s, accelerated in the 1970s when elementary particle physics became an important ingredient of the new cosmology.[19] For example, detailed calculations made in 1977 by the Americans Gary Steigman, David Schramm, and James Gunn showed that the number of different neutrinos could not be larger than three if the hot big bang theory were correct. This prediction was later confirmed by high-energy accelerator experiments in Europe and the United States, and it served to increase confidence in the basic correctness of the big bang model. Particle physicists have also applied grand unified theories (GUTs) in order to understand processes taking place in the universe a fraction of a second after the big bang. In this way, they have been able to explain the observed ratio of photons to protons and neutrons, rather than accepting the ratio as just a contingent fact of nature. These calculations started in the late 1970s and offered additional inspiration for particle physicists to take up cosmological problems. Even more ambitious

---

[19] Norriss S. Hetherington, ed., *Encyclopedia of Cosmology: Historical, Philosophical, and Scientific Foundations of Modern Cosmology* (New York: Garland, 1993).

attempts to apply particle and quantum physics cosmologically have resulted in theories of so-called quantum cosmology, the aim of which is to explain the origin of the world without relying on initial conditions. In 1983, Hawking and James Hartle proposed such a theory of the creation of the universe, and other physicists have developed alternative quantum cosmologies. However, a satisfactory theory of quantum gravity does not yet exist, and for this and other reasons, quantum-cosmological theories are not generally considered to give a proper explanation of why the universe exists or how it was created.

The most important contribution of particle physics to recent cosmology has undoubtedly been the inflationary theory, introduced by the American physicist Alan Guth (b. 1947) in 1981. According to this theory, the very early universe underwent extreme supercooling and expanded suddenly by a gigantic factor. After the initial explosion, the expansion slowed down in agreement with the standard big bang theory. In 1982, Guth's original theory was improved independently by Andrei Linde in the Soviet Union and Andreas Albrecht and Paul Steinhardt in the United States. The inflationary universe model explained, among other things, the large-scale homogeneity of the universe and the near flatness of space, neither of which phenomena could be explained by the standard big bang theory. Although the inflationary model is not unproblematical and has been accused of being "metaphysical," it has been highly successful, causing a major change in cosmological thinking.[20]

The inflationary model requires space to be completely flat, meaning that ordinary Euclidean geometry is valid. However, in that case there must be much more mass in the universe than can be observed. The problem of unseen or dark matter was noticed by Zwicky as early as 1933, but it was taken seriously within the astronomical community only after theoretical analyses made in the mid-1970s by Peebles, Jeremiah Ostriker, Amos Yahil, and others. It is known that most of the matter in the universe must be "dark," that is, acting gravitationally but not emitting light. The precise amount of dark matter is unknown, and so is the nature of this mysterious matter. The dark-matter problem was considered the most exciting unsolved problem in late-twentieth-century cosmology. It will remain a challenge to cosmologists in the twenty-first century.

Observational cosmology experienced a minor revolution with the launching of artificial satellites specially designed for measurements of cosmological significance. The Cosmic Background Explorer (COBE) satellite, launched in 1989, measured the cosmic background radiation much more precisely than did earlier experiments on Earth. Analysis of its data from 1990 to 1992 by George Smoot and others showed a perfect fit of the background radiation with a blackbody spectrum of temperature 2.736 K. It is more interesting to

---

[20] Alan Lightman and Roberta Brawer, *Origins: The Lives and Worlds of Modern Cosmology* (Cambridge, Mass.: Harvard University Press, 1990).

note that the results also showed small departures from isotropy that were interpreted as "wrinkles in space-time" or inhomogeneities in the early universe. These are consistent with inflationary cosmology, where they provide the seeds necessary for the evolution of galactic structures. The COBE observations turned out to be a great triumph for the big bang theory and made it even more difficult to believe that this theory is not essentially correct. Yet not all observations have agreed so nicely with the big bang theory. The Hubble Space Telescope led in 1994 to improved measurements of galactic distances and then also to an improved value of the Hubble time. Much to the consternation of astronomers, the time-scale difficulty of the 1930s and 1940s reappeared, as it turned out that the age of the universe appears to be smaller than the age of certain clusters of galaxies. In spite of this and several other problems, the big bang theory has a paradigmatic status in modern cosmology and is accepted by almost all physicists and astronomers as the best offer of a correct theory of the universe.

Cosmology has made remarkable progress since Einstein's pioneering work of 1917. Together with the theory of relativity and quantum mechanics, it has caused profound changes in the physical worldview. Indeed, it is tempting to speak of the conceptual changes around 1917, 1930, and 1965 as a series of revolutions. But they can hardly be characterized as revolutions in the sense of Thomas Kuhn, according to whom a revolution consists in a theory change in which the new theory is radically different from, and incompatible with, the old one. Einstein's relativistic foundation of 1917 did not replace an older theory, but, rather, created a new science almost from scratch. An important element in the new science was a continuation of the traditional belief in a static universe, which was perhaps the only truly paradigmatic part of the early phase of cosmology. With the introduction of the expanding universe, we can speak of a revolution of a sort, but again not quite in Kuhn's sense. The new theory rested safely on old ground, Einstein's cosmological field equations, and (contrary to the "Planck-Kuhn thesis") most of the pioneers of the old paradigm welcomed the new dynamic picture of the world. Finally, the so-called 1965 revolution was merely a continuation of the tradition founded by Gamow and his collaborators. In general, Kuhn's model is not easily applicable to the development of modern cosmology. Only in the last decades, after the big bang theory has achieved a hegemonic status in cosmology, can one identify a paradigm-ruled phase of normal science comparable to that of many other sciences.[21]

In spite of the awesome greatness of what modern cosmology has accomplished during a period of eighty-five years, its historical development is not well understood and has attracted relatively little interest compared with, for

---

[21] C. M. Copp, "Professional Specialization, Perceived Anomalies, and Rival Cosmologies," *Knowledge: Creation, Diffusion, Utilization*, 7 (1985), 63–95.

example, the theories of microphysics. Since the 1970s, the history of general relativity has been examined in great detail, and we now have a good picture of how this theory was formed and has developed. But the historical interest in general relativity has not extended to one of its foremost applications, cosmology. One of the reasons is undoubtedly cosmology's lack of disciplinary unity. Cosmologists have always been primarily physicists, mathematicians, or astronomers, and each of these disciplines has its own and distinct historiographical traditions. Cosmology is not only a highly technical science; it also evidently relates to deep philosophical and theological questions. Although much less has been written about the philosophy of cosmology than about the philosophy of quantum mechanics, there is a substantial literature dealing with cosmology's philosophical aspects. Some of this literature is historically relevant and makes use of historical sources. But on the whole, the historiography of cosmology is underdeveloped and has, to some extent, been left to the not-always-satisfactory accounts written by physicists, astronomers, and science journalists. What has been written on the history of cosmology is almost exclusively oriented toward scientific and intellectual aspects.

There is a great need for broader studies that take up social, institutional, and technological questions and relate these to the scientific aspects. At present, there are almost no historical works dealing with funding, public appeal and responses, disciplinary interactions and tensions, education and training, the geography of cosmological research, and networks and school building in cosmology. There are enough of these uncultivated areas to keep historians of cosmology busy many years into the twenty-first century.

28

THE PHYSICS AND CHEMISTRY OF THE EARTH

Naomi Oreskes and Ronald E. Doel

Throughout the late eighteenth and nineteenth centuries, there were two distinctly different ways of thinking about the earth – two different evidentiary and epistemic traditions. Such men as Comte Georges de Buffon and Léonce Elie de Beaumont in France, William Hopkins and William Thomson (Lord Kelvin) in the United Kingdom, and James Dwight Dana in the United States tried to understand the history of the earth primarily in terms of the laws of physics and chemistry. Their science was mathematical and deductive, and it was closely aligned with physics, astronomy, mathematics, and, later, chemistry. With some exceptions, they spent little time in the field; to the degree that they made empirical observations, they were likely to be indoors rather than out. In hindsight, this work has come to be known as the *geophysical* tradition. In contrast, such men as Abraham Gottlob Werner in Germany, Georges Cuvier in France, and Charles Lyell in England tried to elucidate earth history primarily from physical evidence contained in the rock record. Their science was observational and inductive, and it was, to a far greater degree than that of their counterparts, intellectually and institutionally autonomous from physics and chemistry. With some exceptions, they spent little time in the laboratory or at the blackboard; the rock record was to be found outside. By the early nineteenth century, students of the rock record called themselves *geologists*. These two traditions – geophysical and geological – together defined the agenda for what would become the modern earth sciences. Geophysicists and geologists addressed themselves to common questions, such as the age and internal structure of the earth, the differentiation of continents and oceans, the formation of mountain belts, and the history of the earth's climate.

Portions of this essay are adapted from *The Rejection of Continental Drift: Theory and Method in American Earth Science* by Naomi Oreskes, copyright 1999 by Naomi Oreskes, used by permission of Oxford University Press, Inc.; and from "Earth Sciences and Geophysics" by Ronald E. Doel, in *Science in the Twentieth Century*, ed. John Krige and Dominque Pestre (Paris: Harwood, 1997), pp. 361–8.

The identification of these two traditions should not be taken to imply that they were necessarily mutually exclusive or insulated from each other, or that a sharp boundary could always be drawn between them. In some institutions, geology and geophysics coexisted, and some individuals attempted to transcend the gap between them and to argue for advancement by unification. But more often than not, geologists and geophysicists approached common questions from divergent perspectives and obtained divergent answers. The history of the physics and chemistry of the earth is thus a history of enduring tensions and occasional open conflict. Geologists and geophysicists frequently clashed in their interpretations and sometimes came to incompatible conclusions about fundamental aspects of the earth – its structure, its composition, and its history.

In several of the most famous and bitter conflicts, it was the geological protagonists rather than the geophysical ones who were later vindicated. Yet, by the mid-twentieth century, the geophysical tradition – expanded to include the oceans and atmospheres as well as the solid earth – was clearly ascendant, if not entirely dominant. The unifying theory of plate tectonics was substantiated largely on the basis of geophysical evidence, and geologists embraced many of the techniques, instruments, and assumptions of geophysics (or geochemistry). Why did earth scientists turn so firmly toward geophysics? Although advances in geophysical knowledge of course contributed to the rise of that discipline, and geophysical research in the second half of the twentieth century has proved very fruitful, the ascendance of geophysics was not primarily the result of prior intellectual success. Rather, it was the result of an abstract epistemological belief in the primacy of physics and chemistry, coupled with strong institutional backing for geophysics premised on its concrete applicability to perceived national-security needs.

TRADITIONS AND CONFLICT IN THE STUDY OF THE EARTH

Historically, geologists had not been directly concerned with the interior of the earth, for their methods restricted them to the study of materials at the surface. Yet the earth's internal structure and processes were implicitly at stake in any geological work, for the interpretation of tilted stratigraphic sequences and deformed rocks in mountain belts required geologists to invoke the processes of the earth's interior. With the rise of industrialization in Europe and North America, understanding the structures beneath became an explicit demand as geologists increasingly engaged in the pursuit of valuable earth materials at depth. The study of volcanic rocks likewise caused geologists to speculate about the earth's interior: Were molten rocks derived from a molten interior? Or did internal processes, such as pressure release, lead to local melting of otherwise solid rocks?

Volcanoes and hot springs convinced many eighteenth- and early-nineteenth-century observers that the interior of the earth must be partly or wholly liquid, a view implicit in versions of contraction theory that explained surface topography by the collapse of a shrinking crust into a molten substrate. In the 1840s, however, William Hopkins (1793–1866) suggested that the precession and nutation of the earth's axis was inconsistent with a fluid earth: The solid crust must be close to 1,000 miles thick to account for the earth's rigid behavior. Lord Kelvin (1824–1907), Hopkins's student, later expanded this reasoning to argue that the existence of the ocean tides proved conclusively that the earth was entirely solid; were it not, it would deform along with the surface waters, and there would be no tides. From this line of reasoning came Kelvin's famous pronouncement that the earth as a whole was more rigid than a globe of solid glass, and probably more rigid than a globe of steel.[1]

These traditions soon clashed in the late-nineteenth-century debate over the age of the earth. This is the most famous historical clash between geology and geophysics over epistemic and evidentiary standards, but it was neither the first nor the last. In the 1850s, John Forbes argued with William Hopkins (and later John Tyndall) over the mechanisms of glacier motion. Forbes argued on the basis of field observation that glaciers flowed like a stream, with some parts moving faster than others and the whole mass deforming internally. Although they looked solid, glaciers were really fluid. Hopkins argued instead on theoretical grounds that glaciers slid downhill as solid objects lubricated by a melted layer at the base. While Hopkins's arguments were theoretically plausible, geologists argued that that was not what happened in nature. Fieldwork later vindicated the geological position.[2]

A similarly revealing controversy engaged geologists and geophysicists over the structure of mountain ranges and the earth beneath them. Detailed field mapping in the Swiss Alps and elsewhere suggested lateral displacement of huge slabs of rocks over vast distances within mountain ranges.[3] In pondering what conditions might permit such displacements, or *nappes,* Swiss geologist Albert Heim (1849–1937) suggested that a plastic "zone of flow" underlies the earth's solid crust. This concept generated difficulties for geophysicists

---

[1] Joe D. Burchfield, *Lord Kelvin and the Age of the Earth* (Chicago: University of Chicago Press, 1974); Crosbie Smith and M. Norton Wise, *Energy and Empire: A Biographical Study of Lord Kelvin* (Cambridge: Cambridge University Press, 1989), pp. 552–78 and 600–2; Stephen G. Brush, *Transmuted Past: The Age of the Earth and the Evolution of the Elements from Lyell to Patterson* (Cambridge: Cambridge University Press, 1996); Naomi Oreskes, *The Rejection of Continental Drift* (New York: Oxford University Press, 1999).

[2] Bruce Hevly, "The Heroic Science of Glacier Motion," *Osiris,* 11 (1996), 66–8.

[3] Mott T. Greene, *Geology in the Nineteenth Century* (Ithaca, N.Y.: Cornell University Press, 1982), pp. 192–220; Rudolf Trümpy, "The Glarus Nappes: A Controversy of a Century Ago," in *Controversies in Modern Geology,* ed. D. W. Muller, J. A. McKenzie, and H. Weissert (London: Academic Press, 1991), pp. 397–8.

committed to a solid earth. An example is found in the work of Osmond Fisher (1817–1914), who at midcentury was attempting to mathematize the concept of terrestrial contraction and thereby demonstrate its sufficiency as an explanation for the earth's surface features. Instead, he proved the reverse: Mathematical analysis showed that thermal contraction was incapable of causing observed global differences in elevation.

Led by this unexpected result to reexamine his assumptions, Fisher concluded that geophysical theory was underdetermined, because even "known" constraints could often "be satisfied in more ways than one."[4] An example was the constraint of rigidity. The tides that for Kelvin were proof of a solid earth were for Fisher only proof of a *mostly* solid earth. If the crust were solid by virtue of its low temperature and the core by virtue of its high pressure, there could be a crossover zone where temperatures were high enough to cause melting but pressures were low enough to sustain a liquid (albeit perhaps highly viscous). Kelvin's objections notwithstanding, Fisher argued, the earth might yet contain an internal fluid layer somewhere between the crust and the core, accommodating the geological evidence of surface dislocations.

Fisher's work spoke to the heart of the matter: a recurring tension between phenomenological evidence and theoretical explanation, and between alternative theoretical accounts of weakly understood phenomena. Nowhere was this tension more evident then in the debate over isostasy, an idea that emerged from nineteenth-century geodetic surveys. Discrepancies between distances measured on the basis of triangulation and those computed on the basis of astronomical observation led John Pratt (1809–1871), a Cambridge-trained mathematician, to compute the expected deflection of a plumb bob based on the observable mass of the Himalayas. Pratt discovered that the measured discrepancies were less than they should have been – as if part of the mountain range were missing. George Biddell Airy (1801–1892), the astronomer royal of the United Kingdom, suggested an explanation: The surface mass of the Himalayas was gravitationally "compensated" for by low-density roots beneath, akin to icebergs floating at sea. At some unknown depth, the weight of the overlying rocks would be the same everywhere, a condition American geologist Clarence Dutton named *isostasy* – equal standing. Pratt, however, suggested an alternative interpretation: Isostasy is achieved by subterranean density variations that compensate for surface topography. In Airy's model, the crust had constant density and variable thickness; in Pratt's model, constant thickness and variable density.[5]

---

[4] Osmond Fisher, *Physics of the Earth's Crust* (London: Macmillan, 1881), p. 270; David S. Kushner, "The Emergence of Geophysics in Nineteeth Century Britain," PhD diss., Princeton University, 1990; Smith and Wise, *Energy and Empire*, pp. 573–8; Oreskes, *Rejection of Drift*, pp. 25–9.

[5] Sir George Biddell Airy, "On the Computation of the Effect of Attraction of Mountain Masses as Disturbing the Apparent Astronomical Latitude of Stations in Geodetic Surveys," *Philosophical Transactions of the Royal Society of London*, 145 (1855), 101–4; J. H. Pratt, "On the Constitution of the

Fisher's fluid zone accommodated Airy's floating continents and Heim's surface displacements, and many geologists and geodesists, particularly in Europe, embraced it. But the Pratt model provided an alternative account for those committed to a fully solid earth. By the early twentieth century, many scientists perceived both the importance and the extent of the intellectual conflict between these two theoretical camps. In the year that Dutton coined the term isostasy, the American astronomer and geodesist Robert Woodward (1849–1924) – later director of the Carnegie Institution of Washington – summarized the state of scientific knowledge regarding the earth's interior. In an address to the American Association for the Advancement of Science, entitled "The Mathematical Theories of the Earth," Woodward concluded that the nature of the earth's interior was a "vexed question ... still lingering on the battle fields of scientific opinion." Revisiting a metaphor that Kelvin had used self-referentially – Dryden's vain king who "thrice slew the slain" – he concluded that the battle over the interior constitution of the earth would yet be "fought o'er again."[6] Woodward's comments were more prescient than he imagined: Debate continued not merely for several years but for several decades.

GEOLOGY, GEOPHYSICS, AND CONTINENTAL DRIFT

The argument over the earth's interior was recapitulated in the early twentieth century in the context of continental drift. The case for moving continents was primarily phenomenological: Fossil assemblages and stratigraphic and structural relations suggested that the earth's continents had once been united, and paleoclimatic indicators in rocks revealed changes that could not be accounted for by secular variation. Such evidence suggested that the continents had moved, both separately and together.

Like their Alpine predecessors, the advocates of drift faced the problem of moving enormous slabs of rock – in this case, whole continents. When Irish geologist John Joly (1857–1933) first began to argue in favor of continental-scale dislocations – several years before Alfred Wegener (1880–1930) proposed continental drift – he explicitly invoked resistance to Heim's work as a cautionary parable. At first, geologists did not believe in the existence of nappes, Joly noted, because they could see no explanation for them. But ultimately they were convinced by the strength of the empirical evidence.[7] So it should be for continental mobility: The continents, Wegener and Joly argued, could

---

Solid Crust of the Earth," *Philosophical Transactions of the Royal Society of London*, 161 (1871), 335–57; Oreskes, *Rejection of Drift*, pp. 23–5.

[6] Robert S. Woodward, "The Mathematical Theories of the Earth," *American Journal of Science*, 38 (1889, 3d ser.), 343–4, 352.

[7] John Joly, *Radioactivity and Geology: An Account of Radioactive Energy on Terrestrial History* (London: Archibald Constable, 1909), pp. 143–4.

move because the substrate beneath them was fluid. In *The Origin of Continents and Oceans* Wegener invoked isostasy, which since the 1870s had been confirmed by far more data. Plainly, continents could float in hydrostatic equilibrium if and only if the substrate in which they were imbedded behaved as a fluid. If the substrate were fluid, then the continents could, at least in principle, move through it.

By this time, however, geophysicists had gone beyond their earlier theoretical arguments for a solid earth. The growth of instrumental seismology had given them access to the earth's interior – albeit indirectly – and from the rate of propagation of earthquake waves, seismologists such as Harold Jeffreys in the United Kingdom and James Macelwane in the United States had calculated the viscosity of the earth's interior. Their results supported Kelvin's rigid earth. Once again the notion of a fluid zone seemed to be refuted by geophysical arguments. Wegener countered that many materials behave in a rigid manner in response to short, sharp blows, but in a plastic manner when the applied pressures are small, steady, and slow. Glass, for example, or wax. The response of the earth to short-duration seismic events was not necessarily indicative of its behavior over geological time. Wegener concluded, like Darwin and Lyell before him, that the key to understanding earth history was the element of time, "insufficiently appreciated in previous literature, but... of the greatest importance in geophysics."[8]

Wegener's arguments were not accepted. Most earth scientists – particularly geophysicists – vociferously rejected the idea of moving continents before the development of the theory of plate tectonics in the 1960s. Existing histories have tended to credit the delayed acceptance of continental mobility either to a lack of "proof" or to a lack of causal explanation. Historians and philosophers have emphasized the role of geophysical data not available to Wegener – particularly paleomagnetic and refined seismic data – in establishing plate tectonics. Scientists have emphasized the question of kinematic and dynamic accounts. In both cases, the available data and theoretical support are deemed insufficient to have constituted proof of moving continents.[9]

---

[8] Alfred L. Wegener, *The Origin of Continents and Oceans*, 3d ed. trans. G. A. Skerl (London: Methuen, 1924), pp. 130–1, and 4th ed. trans. John Biram (1929; New York: Dover, 1966), pp. 54–9.

[9] Allan Cox, ed., *Plate Tectonics and Geomagnetic Reverals* (San Francisco: W. H. Freeman, 1973); Ursula B. Marvin, *Continental Drift: The Evolution of a Concept* (Washington, D.C.: Smithsonian Institution Press, 1973); Seiya Uyeda, *The New View of the Earth: Moving Continents and Moving Oceans*, trans. Masako Ohnuki (San Francisco: W. H. Freeman, 1978); Henry Frankel, "Alfred Wegener and the Specialists," *Centaurus*, 20 (1976), 305–24; Frankel, "Why Continental Drift Was Accepted by the Geological Community with the Confirmation of Harry Hess' Concept of Sea-floor Spreading," in *Two Hundred Years of Geology in America*, ed. C. J. Schneer (Hanover, N.H.: University of New England Press, 1979), pp. 337–53; Frankel, "The Continental Drift Debate," in *Resolution of Scientific Controversies: Theoretical Perspectives on Closure*, ed. A. Caplan and H. T. Engelhardt, Jr. (Cambridge: Cambridge University Press, 1985), pp. 312–73; Robert Muir Wood, *The Dark Side of the Earth* (London: Allen and Unwin, 1985); Homer E. LeGrand, *Drifting Continents and Shifting Theories* (Cambridge: Cambridge University Press, 1988); Rachel Laudan and Larry Laudan,

These interpretive positions tend toward the presentist and the apologetic, and both obscure important historiographic issues that become evident when the broader context and history of the earth sciences are considered. Two important points have been raised by recent work. First, the causal mechanism accepted by geophysicists in the 1960s to explain plate tectonics – convection currents in the earth's plastic substrate below the rigid crust – was proposed and widely discussed in the 1920s and 1930s by advocates of drift. The mechanism that has been deemed crucial for the acceptance of plate tectonics was available in the debate over continental drift. Second, and crucial for the historiographic issues being raised here, the phenomenological evidence of continental drift, like the phenomenological evidence of a plastic substrate on which the idea depended, was retrospectively accepted in the light of plate tectonics. Although plate tectonics was established on the basis of geophysical data, these data in the end led to the same result as the previously rejected geological arguments. Sociologically, the geophysical data proved to be more powerful in moving men, but epistemically they proved to be equivalent with respect to moving continents.[10]

In these recurring debates, a familiar pattern emerges. Geologists argued from qualitative and phenomenological evidence, geophysicists from quantitative and theoretical evidence. Both sides affirmed the superiority of their methods and denied the claims of the other: Geophysicists argued for the greater rigor of mathematical analysis and dismissed empirical counterarguments; geologists defended the accuracy of their observations and frequently dismissed theoretical claims that challenged their conclusions.

Today, most earth scientists agree that the geophysical methods were "better" – more quantitative, more theoretically grounded, and therefore, in some sense, more "scientific" – and many historians have explicitly or implicitly accepted this verdict. However, this assessment merits reconsideration. For in each of these major debates – the age of the earth, the nature of the crustal substrate, the mobility of continents – the geological arguments were later vindicated, the geophysical ones shown to be in error. By the standards of contemporary knowledge, geologists were right when they insisted that the earth was older than Lord Kelvin's calculations allowed, they were right when they held that the substrate behaved in a fluid manner, and those who insisted on the reality of continental movement in the face of geophysical opposition were right, too.

Yet, as Robert Woodward predicted, the thrice-slain combatants did rise up again, not merely to fight but to rule the kingdom. If the rise of geophysics was not based on prior intellectual success, then what was it based on? Understanding the rise of geophysical and geochemical approaches to the earth

"Dominance and the Disunity of Method: Solving the Problems of Innovation and Consensus," *Philosophy of Science*, 56 (1989), 221–37.
[10] Oreskes, *Rejection of Drift*, p. 307.

sciences requires a broader investigation of both the epistemic commitments and institutional affiliations of earth scientists in the twentieth century.

THE DEPERSONALIZATION OF GEOLOGY

Well before Alfred Wegener drew upon field geological data to argue his theory of continental drift, a significant group of geologists, particularly in the United States, were moving away from reliance on field methods and toward the methods of physics and chemistry. They did so in the hope of making their science more *potent*. They manifested their concerns and desires in strongly articulated worries about the intuitive practices of geology, in efforts to recreate geology as a laboratory science, and in a broad pattern of attempting to make geological field practices more like those of the laboratory. By midcentury, this pattern was evident in other areas of earth science as well, indeed across many scientific disciplines.

By the 1870s, reasoning from physics and chemistry was evident in the work of many of North America's most important geologists. Clarence Dutton, Clarence King, T. C. Chamberlin, and G. K. Gilbert all emphasized the application of physics and chemistry to the understanding of earth processes and structures. Chamberlin (1843–1928) was one of the earliest scientists to consider the role of atmospheric chemistry in climate change, and his cosmological theory was arguably more influential among astronomers than among geologists.[11]

Charles Van Hise (1857–1918), a pioneer of chemical and physical analysis of rocks and later president of the University of Wisconsin, argued that geology should be no more or less than "the science of the physics and chemistry of the Earth."[12] Van Hise agreed with G. K. Gilbert's argument that understanding emerges by viewing a problem simultaneously from multiple perspectives. But whereas Gilbert had spoken primarily in terms of individual understanding – using the analogy of the field geologist locating himself by triangulation – Van Hise extended the metaphor to the disciplinary community. Because the historical perspective was already established in geology, the physical and chemical perspectives needed to be brought on par. Scientists needed to triangulate conceptually among geology, physics, and chemistry to gain a clear picture of the earth. The complexity of earth processes, which was sometimes used as an argument against quantification, was for Van Hise the strongest argument for it: Only by quantification could one evaluate the relative importance of various contributory forces.

[11] Stephen G. Brush, "A Geologist among Astronomers: The Rise and Fall of the Chamberlin-Moulton Cosmogony," *Journal for the History of Astronomy*, 9 (1978), 1–41 and 77–104; Stephen J. Pyne, *Grove Karl Gilbert: A Great Engine of Research* (Austin: University of Texas Press, 1980).

[12] Charles R. Van Hise, "The Problems of Geology," *Journal of Geology*, 12 (1904), 590–3; John W. Servos, *Physical Chemistry from Ostwald to Pauling: The Making of a Science in America* (Princeton, N.J.: Princeton University Press, 1990), pp. 227–9.

Van Hise's vision for geology was in part fulfilled by the establishment of the Geophysical Laboratory of the Carnegie Institution of Washington (CIW). Founded in 1907, the laboratory became a major center for research into the melting, crystallization, and optical properties of minerals, the magnetic properties of rocks, the variation of gravity, the origins of lunar craters, and many other aspects of the physics and chemistry of Earth (and Moon).[13] But for many scientists, the role of physics and chemistry lay not only in providing factual or conceptual constraints but also in suggesting "more perfect methods." In parallel with the desire to apply the principles of physics and chemistry, some geologists expressed an ancillary desire to make geology more nomological – that is, less descriptive and more law based. Van Hise expressed vexation at the vagueness of geological theorizing, wanting definite "rules of the game." Uniformitarianism was one such rule; it gave geologists grounds on which to interpret the geological record in terms of presently observable processes. But it did little to illuminate the forces behind those processes. For this, Van Hise and others wanted "the reduction of [geology] to order under the principles of physics and chemistry."[14]

The desire for a nomological geology and the confusions and contradictions it generated is clearly seen in the work of Walter Bucher (1889–1965). His well-known 1933 book (reprinted in 1957), *The Deformation of the Earth's Crust*, consisted of the articulation of forty-six laws of crustal deformation. But these were not laws in any sense that physicists or philosophers would have understood. Rather, as Bucher admitted, they were "essential geological facts" assembled into "carefully worded generalizations."[15] Then why call them laws? Bucher gave two reasons. The first was to make them impersonal – that is, law-*like*. The second was to stimulate thinking. Bucher's use of the term was a planned provocation – to stimulate his colleagues into a stance of consciously entertaining and testing specific, well-articulated theoretical ideas, to foster an atmosphere in which the idea of having laws seemed as natural in geology as it did in physics. Bucher's treatise was not a success – his laws and opinions are today more forgotten than refuted – but his impulses are revealing, for his work speaks to the tension many geologists felt: on the one hand, committed to a field-based enterprise grounded in experience and

---

[13] John W. Servos, "To Explore the Borderland: The Foundation of the Geophysical Laboratory of the Carnegie Institution of Washington," *Historical Studies in the Physical and Biological Sciences*, 14 (1984), 147–86, and *Physical Chemistry*; Nathan Reingold, "National Science Policy in a Private Foundation," in *Science, American Style* (New Brunswick, N.J.: Rutgers University Press, 1991), pp. 190–223; Hatton S. Yoder, "Scientific Highlights of the Geophysical Laboratory, 1905–1989," *The Carnegie Institution of Washington Annual Report of the Director of the Geophysical Lab* (1989), 143–203; and Gregory A. Good, ed., *The Earth, the Heavens* and the *Carnegie Institution of Washington* (Washington, D.C.: The American Geophysical Union, 1994), also printed as *History of Geophysics*, 5 (1994), 1–252.

[14] Van Hise, "Problems of Geology," p. 615.

[15] Walter H. Bucher, *The Deformation of The Earth's Crust* (Princeton, N.J.: Princeton University Press, 1933), pp. v–vii.

observation of the natural world and, on the other hand, feeling that science should be made of firmer stuff.

Whereas Walter Bucher strove to derive universal principles from his fieldwork, others strove to move geological work out of the field and into the laboratory. By the mid-1930s, the CIW's Geophysical Laboratory had become one of the world's leading locales for laboratory investigations of geological processes, and work done there inspired scientists at other American institutions. At Harvard, for example, Reginald Daly joined forces with Percy Bridgman to raise funds for a high-pressure laboratory to determine the physical properties of rocks under conditions prevailing deep within the earth. At Princeton, Richard Field joined forces with the U.S. Navy to measure gravity at sea. The application of physics and chemistry to the earth was also advanced at the CIW's Department of Terrestrial Magnetism, where scientists pursued geomagnetism, isotopic dating, and explosion seismology. By midcentury, the origins of igneous and metamorphic rocks had been explained, the age of the earth accurately determined, and the behavior of rocks under pressure elucidated, largely through the application of instrumental and laboratory methods.[16]

Van Hise and his colleagues had advocated laboratory methods as a complement to, not a replacement for, field geology, but their successors increasingly viewed the problem in terms of competing alternatives – in terms of new methods replacing the old. When sedimentologist Francis J. Pettijohn (b. 1904) joined the geology department at the University of Chicago in 1929, he was surprised to find faculty in white lab coats. Pettijohn consciously shifted his focus from field studies of ancient sedimentary rocks to laboratory analysis of modern sediments in response to colleagues who conveyed the message that fieldwork was "something we are trying to get away from."[17] In retrospect, Pettijohn interpreted his experience as an encounter with the ideology of quantification, but the larger historical context suggests that events at Chicago were part of a broader move in the earth sciences from the field to the laboratory, reflecting an idealization of the epistemic values of exactitude and control that laboratory work embodies.[18] There *are* aspects of fieldwork that can be quantified, but most Chicago geologists were not striving to make their work in the field more quantitative; they were striving to remove it from the field altogether.

Those who took the extreme view and wished to remove geology entirely from the field were not successful – fieldwork continued to play a role in the earth sciences throughout the twentieth century, and still does

---

[16] Good, *The Earth, the Heavens*.
[17] F. J. Pettijohn, *Memoirs of an Unrepentant Field Geologist* (Chicago: University of Chicago Press, 1984), p. 207.
[18] On exactitude and control as epistemic and moral values, see Kathryn Olesko, *Physics as a Calling: Discipline and Practice in the Konigsberg Seminar for Physics* (Ithaca, N.Y.: Cornell University Press, 1991), and M. Norton Wise, ed., *The Values of Precision* (Princeton, N.J.: Princeton University Press, 1995).

today. But, like Bucher's attempt at nomothesis, their efforts speak to feelings of need. Geology had long been a personal science, in which individuals experienced the natural world with their own eyes, hands, and feet. One walked the ground, collected samples, examined and observed. One developed geological intuitions through the unconscious analysis of experience. Geological evidence was hardly ever mathematical; it was almost always circumstantial. Arthur Holmes – brilliant in both the laboratory and the field – argued in 1929 that the "circumstantial evidence of geology is not likely to lead us far astray, so long as we read it right."[19] But others were not so sanguine, for how did one know if evidence was being read right? The traditional answer was to "read" it for yourself. Accordingly, professors advised their students to be greedy for experience: H. H. Read famously professed that the best geologist was the one who had seen the most rocks; Charles Lyell was oft quoted as having advocated "travel, travel, travel."

The logic of seeing for believing was clear enough, but the practicalities were another matter. How could a science be built in which everyone had to see everything? Martin Rudwick has described how, in nineteenth-century Britain, when Henry De la Beche and Roderick Murchison clashed on the interpretation of field evidence, members of the Geological Society of London repaired to Devon to examine the disputed strata. Such field excursions were common then and remained significant throughout the twentieth century. But as geology grew as a science, and particularly as it grew in North America where scientists and sites were widely scattered, such direct approaches became impractical. Even if one did go to the field to see for oneself, single outcrops were rarely revealing. Geological interpretations were built on the amalgamation of observations – widely scattered and observed over many field seasons – and they were not infrequently undermined when new evidence became available.[20]

In 1937, American geodesist William Bowie (1872–1940) advocated an epistemological reversal of the roles of nature and laboratory, a reversal that by now has become so commonplace among earth scientists that they scarcely question it: that the earth be viewed as a "natural laboratory." When considering the earth, he suggested, "one is observing the working of the greatest laboratory on earth, with nature as the operator."[21] Laboratories were once viewed as places where men tried to recapitulate the operations of nature; now the operations of nature were being cast as a recapitulation of the work of men.

---

[19] Arthur Holmes, "A Review of the Continental Drift Hypothesis," *Mining Magazine*, 40 (1929), 205–9, 286–8, and 340–7, at p. 347.
[20] Martin J. S. Rudwick, *The Great Devonian Controversy* (Chicago: University of Chicago Press, 1985); Julie Newell, "American Geologists and their Geology, 1780–1865," PhD diss., University of Wisconsin, 1993.
[21] William Bowie, "Scientist to Weigh the Floating Earth Crust," *New York Times*, 20 September 1925, p. xx.

THE EMERGENCE OF MODERN EARTH SCIENCE

Field scientists promoted the values of authenticity, accuracy, and completeness; laboratory scientists promoted the values of exactitude, precision, and control. Each group affirmed the strengths of its methodological approach and implicitly or explicitly denied the strengths of the alternatives. But, in the twentieth century as compared with the nineteenth, the balance had tipped. Fieldwork was no longer the backbone of earth science. Nor were these patterns limited to studies of the solid earth. By the early twentieth century, new instruments and techniques borrowed from physics and chemistry began to transform other closely related fields, particularly meteorology and oceanography. What came to be known by the collective phrase "earth sciences" by the 1960s and 1970s reflected not only the intellectual unity of the object of study but also an increasingly unified methodology.

Meteorology in the mid-nineteenth century had a well-developed empirical tradition: Forecasters used large numbers of previous weather patterns to study the development of storms as the basis for weather prediction. The development of rapid telegraphic communication of meteorological data led to a more synoptic science – a significant advance – but did little to increase its theoretical content. Forecasters generally remained skeptical, if not disdainful, of theorists. The empirical tradition was thus attacked in the late nineteenth century by researchers who hoped to reduce meteorological systems to problems of physics and hydrodynamics. Chief among these was Norwegian physicist Wilhelm Bjerknes (1862–1951), whose polar front concept gave physical interpretation to the behavior of storm systems and the interactions of air masses. Bjerknes's explicit goal was to transform meteorology through the application of physical principles and standardized measurements; he drew on his training in mathematical physics as a source of authority for this desideratum. Bjerknes was successful: Mathematical and physical methods became the dominant practice in meteorology, as Carl-Gustav Rossby, Jerome Namias, and other researchers increasingly focused on such physical problems as the global circulation of the atmosphere.[22]

Bjerknes believed that the future behavior of weather systems could be deterministically calculated, much as one could compute the positions of the planets by knowing their orbits and initial conditions. This problem was taken up in the 1920s by British mathematician Lewis Fry Richardson (1881–1953), who attempted to make numerical forecasts using partial differential equations and, after World War II, digital computers. Although Richardson made great advances in empirical forecasting, and increasingly accurate twenty-four-hour forecasts eventually resulted from this work, his

---

[22] Robert Marc Friedman, *Appropriating the Weather: Vilhelm Bjerknes and the Construction of a Modern Meteorology* (Ithaca, N.Y.: Cornell University Press, 1989); see also James R. Fleming, *Meteorology in America* (Baltimore: Johns Hopkins University Press, 1990).

deterministic hopes were dashed by the work of Edward Lorenz (b. 1917). Lorenz, a mathematician and meteorologist, found that small effects, such as local thunderstorms or minor temperature fluctuations, could introduce very large perturbations in meteorological models. This realization – known today as the "butterfly effect" – was a key element in the development of chaos theory.[23]

While Bjerknes relied primarily on his training in physics for the development of dynamic meteorology, both chemistry and physics played a fundamental role in developing studies of the outer atmosphere and solar system. T. C. Chamberlin's early-twentieth-century cosmogony was the last significant geological contribution in a field increasingly dominated by astrophysics and geochemistry. Data drawn from these disciplines – the distribution of atomic abundances, the nature of interstellar clouds, the rotational velocities of sunlike stars – were central to the influential cosmogonal theories developed by astrophysicist Gerard P. Kuiper and geochemist Harold C. Urey in the 1950s. Magnetic fields and the physics of small particle accretions similarly informed the cosmogonies of Hannes Alfvén (1908–1995) and Victor Safronov (1917–1999) in the following decade.

Advances in physics and planetary science also contributed to the recognition of meteorite impacts as a fundamental geologic force, an overt departure from the dominant geological reasoning of the nineteenth and early twentieth centuries. Although classically trained geologists, particularly Eugene M. Shoemaker (1928–1997) and Robert Dietz (1914–1995), provided key contributions in establishing this concept, most geologists resisted the idea of impacts as somehow violating uniformitarianism. Explanation "without aid of comets" had been a mantra of nineteenth-century geologists seeking to avoid accounts that seemed suggestive of supernatural or divine intervention. With the success of Lyell's uniformitarianism and Darwin's application of it to the problem of evolution, catastrophism – once a respected interpretive position – was discredited, particularly among Anglo-American geologists. In the twentieth century, extraterrestrial phenomena remained vaguely suspect. Although field relations and descriptive mineralogy were used by Shoemaker and Dietz in their interpretations, methods derived from astronomy, geochemistry, and high-pressure laboratory physics proved most effective in advancing general acceptance of meteorite impacts.[24]

Oceanography (like geomagnetism) reflects less a history of conflicting traditions than an extension of physical methods and instruments into realms not otherwise accessible. While there were state-sponsored surveys in the nineteenth century – notably the British *Challenger* expedition – these were

---

[23] Frederick Nebeker, *Calculating the Weather: Meteorology in the Twentieth Century* (San Diego, Calif.: Academic Press, 1995), p. 36; quotation from Friedman, *Appropriating the Weather*, p. 46.

[24] Ronald E. Doel, *Solar System Astronomy in America: Communities, Patronage, and Interdisciplinary Research, 1920–1960* (Cambridge: Cambridge University Press, 1996), pp. 151–87.

episodic and infrequent; physical oceanography in its early years was confined largely to coastal research. The first half of the twentieth century, however, witnessed the growth of systematic studies of global ocean circulation. This work was inspired by parallel studies in meteorology, as with Fridtjof Nansen's borrowing of the geostrophic approximation to assess physical circulation, and as students of Bjerknes, such as Harald Sverdrup (1888–1957), moved into oceanography. After World War II, oceanographic research adapted many instrumental approaches developed in meteorology, particularly the use of digital computers to solve problems in fluid dynamics. Furthermore, many of the concerns of physical oceanographers – ocean surface waves, tidal fluctuations, the energy spectrum of internal waves, and the fluctuating mesoscale circulation – reflected the research traditions of classical physics. Geomagnetism, a province of specialized instrumentation closely linked to electromagnetic theory, was similarly less transformed than incorporated into the modern earth sciences community. While surveys of magnetic intensity were expanded during the twentieth century – the world-circling voyages of the *Carnegie* were joined by aerial magnetometer surveys by the late 1930s and by planetary magnetosphere studies two decades later – laboratory-based studies of paleomagnetism and mathematical-deductive theory became the core research traditions in this field.[25]

Seismology was similar to oceanography. By necessity a theoretical and instrumental field, seismology expanded dramatically at the turn of the twentieth century. Theoreticians, such as R. C. Oldham, Harold Jeffreys, James Macelwane, Beno Gutenberg, and Inge Lehmann, advanced the analysis of seismic wave propagation, while Jesuit scientists took on seismological recording and interpretation as a specialized practice, setting up stations in China, Madagascar, Lebanon, Australia, and the United States.[26] Seismology, like oceanography, was a field science, but its fieldwork was instrumental rather than directly observational, its descriptions were quantitative rather than qualitative, and the properties being measured were the physical properties of the earth.

Geological traditions were not wholly absent from these fields of research by the mid-twentieth century, just as the practice of field geology did not vanish at research institutions. In oceanography in the 1950s and 1960s, bathymetric mapping of the seafloor was done largely by individuals with geological training, who relied on their geological knowledge and intuitions to interpret and interpolate between soundings. In the 1970s, the undersea submersible *Alvin* allowed researchers to conduct fieldwork not entirely unlike that on

---

[25] Myrl C. Hendershot, "The Role of Instruments in the Development of Physical Oceanography," in *Oceanography: The Past,* ed. Mary Sears and Daniel Merriman (New York: Springer-Verlag, 1980), pp. 195–203, and Robert P. Multhauf and Gregory Good, *A Brief History of Geomagnetism and a Catalog of the Collections of the National Museum of American History* (*Smithsonian Studies in History and Technology,* 48) (Washington, D.C.: Smithsonian Institution Press, 1987).
[26] Carl-Henry Geschwind, "Embracing Science and Research: Early Twentieth-Century Jesuits and Seismology in the United States," *Isis,* 89 (1998), 27–49.

land. In seismology, knowledge of rocks was relevant to the interpretation of seismic stratigraphy. Geology also played a significant role in the *Apollo* lunar landings: Astronauts were given geological training to aid their sample collection, and a professional geologist, Harrison "Jack" Schmitt, flew on the final *Apollo* mission. Traditional field practices have likewise played a role in the photographic mapping of other planets and satellites in the solar system, where geologists have reconstructed planetary histories by studying cratering patterns and other morphological characteristics. Yet these examples are perhaps the exceptions that prove the rule; by the late 1950s, instrumental styles and physico-chemical approaches to Earth (and other planets) had become defining. Geology, where it contributed, generally did so in a supportive role.[27]

Such a brief survey does not exhaust the range of chemical and physical approaches applied to studies of the earth since the nineteenth century. What it does illustrate, however, is the extent to which these traditions came to dominate the practice of the earth sciences by the end of the twentieth century. Van Hise's goal was largely achieved.

EPISTEMIC AND INSTITUTIONAL REINFORCEMENT

The changes described here were manifestations of the epistemic commitments of scientists who studied the earth in the late nineteenth and early twentieth centuries. Yet scientists cannot achieve their abstract goals without concrete support; the emergence of the modern earth sciences depended as well on patronage. The ascendance of geophysics reflected, and was fundamentally shaped by, the demands of the second industrial revolution, and particularly by the needs of military patrons during World War II and the Cold War. Greatly increased funds for geophysical and geochemical studies of the earth influenced the development of research institutions and university graduate programs, supported new instrumental practices, and increased professional opportunities for individuals trained in these techniques.

Shifting patterns of patronage for the geological sciences were already apparent by the turn of the century. The principal support for geology in the mid-nineteenth century lay in geological surveys pursued in aid of mineral exploration and land surveying. But this began to wane by the 1890s, particularly in the United States, where the closure of the frontier and the sharp economic depression of 1893 led to curbs in federal expenditures for science (which also affected geophysics at the U.S. Coast and Geodetic Survey and the U.S. Naval Observatory). Generous funding for descriptive geological

[27] Don E. Wilhelms, *To A Rocky Moon: A Geologist's History of Lunar Exploration* (Tucson: University of Arizona Press, 1993); Naomi Oreskes, "La lente plongée vers le fond des océans," *Science et Vie* (March 1998), 84–90.

surveys was never fully reestablished, and in the early twentieth century, private funds began to flow in new directions.

Petroleum corporations invested heavily in geophysical research as they realized the value of gravity and seismic refraction studies for locating oil and natural gas deposits. Geophysics became a major presence in academic earth science departments, as instrumentation developed in aid of prospecting was also used to advance theoretical understanding of the earth. Maurice Ewing (1906–1974), who in the 1950s built the Lamont-Doherty Geological Observatory at Columbia University almost entirely on navy contracts, began his career in the 1920s working in the Texas oil fields where he was introduced to seismic refraction studies. Ewing made seismic wave refraction the topic of his PhD dissertation; having once advanced understanding in this highly applied field, he turned the technique toward basic geological questions, such as the structure of the ocean basins.

Even more influential than direct industrial needs were the managers of the Rockefeller and Carnegie Foundations, themselves offshoots of the industrial age. Like the leaders of the Carnegie Institution of Washington, program managers at the Rockefeller Foundation emphasized the intellectual superiority of controlled laboratory work. In the late 1920s, managers Warren Weaver and Max Mason, both trained in the physical sciences, consciously excluded geology from major gifts for scientific research to American universities, declaring it insufficiently "fundamental." But they did fund geophysics in the United States and Europe.[28]

As important as industry funding for geophysics was, however, it cannot be viewed as decisive in tipping the balance of geoscience research toward geophysics. For industry also supported traditional geology. Petroleum companies funded research in stratigraphy, sedimentology, and paleontology; mining companies paid for studies in petrology, mineralogy, and crystallography. Industrial funding of both geological and geophysical research remained strong into the mid-twentieth century; what tipped the balance in favor of geophysics was national security. By midcentury, industrial support was overtaken by military funding, and new areas of geophysical research – for example, paleomagnetics – were stimulated above all by their relevance to national security concerns.

World War II marked a turning point in the relationship between earth scientists and military patrons. While geophysicists and oceanographers in the 1920s and early 1930s had approached the U.S. military for access to ships and submarines to study the ocean basin, the military, for the most part, had expressed only modest reciprocal interest. This changed in the mid-1930s, as a number of key discoveries demonstrated the strategic value of geophysics and

---

[28] Robert E. Kohler, *Partners in Science: Foundations and Natural Scientists, 1900–1945* (Chicago: University of Chicago Press, 1990), pp. 157–8, 202, 256–7, and C. H. Smyth to H. Alexander Smith, 22 December 1925, Smith papers, Mudd Library, Princeton University.

oceanography. One example is the invention of the bathythermograph and discovery of sound channeling. Working at the behest of the U.S. Navy, geophysicist Althelstan Spilhaus (1911–1998) developed an improved bathythermograph – a device invented by his thesis advisor, meteorologist Carl-Gustav Rossby – to study the effect of water temperature on acoustic transmissions. Spilhaus found that sound waves are refracted through the thermocline (a zone of rapid temperature decrease a few hundred meters below the surface), creating an acoustic shadow zone in which submarines may be effectively hidden. This led to the discovery by Maurice Ewing of sound channeling – the phenomenon whereby sound waves travel virtually unattenuated near the base of the thermocline – making it possible to send signals over vast distances. The SOFAR (Sound Fixing and Ranging) system, used during World War II to track downed pilots and later as a basis for submarine navigation systems, was based on this discovery, which encouraged the navy to support Ewing in particular and geophysics in general in the postwar years.[29]

The bathythermograph is one example; there are many others. Physical oceanographers pioneered new techniques to forecast ocean swell. Meteorologists developed improved methods to forecast weather conditions for critical military operations, such as the Allied invasion of Normandy. As the Cold War intensified in the late 1940s, and as the United States developed the nuclear triad in the 1950s, military–earth science relationships became deeply entwined. Geophysics and oceanography were viewed as essential for protecting submarines; solid earth geophysics was relevant to land-based missile guidance; meteorology was pertinent to the performance of airborne weaponry. Mapping the topography of the ocean floor became a high priority for antisubmarine warfare operations. Major new weapons systems, particularly the guided missile, inspired gravity and geomagnetic studies to aid missile navigation and targeting, meteor astronomy to probe the characteristics of the upper atmosphere, and ionospheric physics to aid in-flight communication and tracking. Geophysicist Joseph Kaplan (1902–1991), summarizing this relationship, declared, "Weapons and important military tools which have recently been perfected – airplanes, submarines, radar – can be used to advantage only when the conditions under which they must operate are recognized."[30] The conditions under which they operated were widely understood, by scientists and military officers alike, to be geophysical, meteorological, and oceanographic.

---

[29] Gary E. Weir, *Forged in War: The Naval Industrial Complex and American Submarine Construction, 1940–1961* (Washington, D.C.: U.S. Government Printing Office, 1993); Naomi Oreskes, "Weighing the Earth from a Submarine: The Gravity-Measuring Cruise of the U.S.S. S-2," in *The Earth, the Heavens*, pp. 53–68; Oreskes, "Laissez-tomber: Military Patronage and Women's Work in Mid 20th-Century Oceanography," *Historical Studies in Physical and Biological Sciences*, 30 (2000), 373–92.

[30] J. Kaplan to C. G. Rossby, 3 July 1944, Office of the Director Files, Scripps Institution of Oceanography Archives; see also Ronald E. Doel, "Earth Sciences and Geophysics," in *Science in the Twentieth Century*, ed. John Krige and Dominque Pestre (Paris: Harwood, 1997), pp. 361–88.

Perhaps no earth science field was more affected by strategic concerns than seismology, which saw sudden, enormous growth in the late 1950s in response to underground nuclear weapons testing. The Limited Nuclear Test Ban Treaty negotiated between the United States and the Soviet Union in 1963 caused a dramatic increase in the number of seismic stations and the training of researchers in this field, as seismology was used to differentiate underground nuclear tests from natural earthquakes.[31] Geochemistry also saw substantial growth during the Cold War, as the U.S. Atomic Energy Commission supported research on sources of uranium for nuclear weapons, the dispersion of radionuclides in the oceans and atmosphere, and nuclear waste disposal and fuel reprocessing.

There is much historical work to be done to understand the full range of military-geophysical collaboration in the twentieth century and its military, scientific, and political effects, as well as to study these relationships beyond the United States. What is clear from work done to date is that the significance of this relationship was not simply one of increased practical application, but of vastly increased funding for geophysical and geochemical work, which spawned a greatly expanded institutional base and largely determined the priorities of the discipline. New curricula in physical oceanography were funded by the Office of Naval Research by the late 1940s, while greatly increased levels of military contracts allowed existing geophysical departments to swell. Enterprising geophysicists, such as Ewing and Kaplan, used navy contracts to create university-affiliated geophysical institutes, further expanding the market for scientists trained in geophysical techniques. The International Geophysical Year (1957–8) was organized by Lloyd Berkner (1905–1967) and other geophysicists, who fully appreciated the links between geophysics and national security concerns. Involving tens of thousands of scientists from sixty-six nations at a cost of over $1 billion, the IGY illuminates the extent to which external factors bolstered, solidified, and directed geophysical research in the mid- to late twentieth century. The ascendancy of geophysics and geochemistry in the second half of the twentieth century was firmly linked to forces operating far outside the discipline and even outside the scientific community.[32]

---

[31] Kai-Henrik Barth, "Science and Politics in Early Nuclear Test Ban Negotiations," *Physics Today*, 51 (March 1998), 34–9.

[32] Allan A. Needell, "From Military Research to Big Science: Lloyd Berkner and Science-Statesmanship in the Postwar Era," in *Big Science: The Growth of Large-Scale Research*, ed. Peter Galison and Bruce Hevly (Stanford, Calif.: Stanford University Press, 1989), pp. 290–311; Doel, "Earth Sciences"; Barton Hacker, "Military Patronage and the Geophysical Sciences in the United States: An Introduction," *Historical Studies in the Physical and Biological Sciences*, 30 (2000), 309–14; James Rodger Fleming, "Storms, Strikes, and Surveillance: The U.S. Army Signal Office, 1861–1891," *Historical Studies in the Physical and Biological Sciences*, 30 (2000), 315–32; Martin Leavitt, "The Development and Politicization of the American Helium Industry, 1917–1941," *Historical Studies in the Physical and Biological Sciences*, 30 (2000), 333–48; Ronald Rainger, "Science at the Crossroads: The Navy, Bikini Atoll, and American Oceanography in the 1940s," *Historical Studies in the Physical and Biological Sciences*, 30 (2000), 349–72; Deborah Warner, "From Tallahassee to Timbuktu: Cold War Efforts

While much has improved since Mott Greene called the history of geology "terra incognita," much remains to be done to understand the complex history of the earth sciences.[33] Recent work has focused on the United States and on the twentieth century; there is much to know about other times and other countries and many large historical questions still to answer. Why, for example, did geophysics and geochemistry develop primarily as branches of earth science, rather than as branches of physics and chemistry? How did institutional affiliations and patronage shape the contours of geophysical knowledge? Why was it not until the late nineteenth century that the institutional and intellectual barriers between the geological sciences and the oceanic and atmospheric sciences began to be breached? And why have the geophysical sciences been particularly impervious to the participation of women and other minority groups?

What is clear from existing work is that the ascendancy of geophysics in the twentieth century was not exclusively or even primarily the result of prior intellectual success. Rather, it was the result of an abstract epistemological commitment to "rigor" that can be traced back to the nineteenth century, combined with the concrete applicability of geophysics to national security concerns that became paramount in the mid-twentieth century. Scientists would not have gone down the path of geophysics and geochemistry had they not expected it to be fruitful, of course. But other paths that might also have proved fruitful were never pursued or were actively enfeebled by the concentration of financial, logistical, and human resources into physics-based approaches to the earth.[34]

The role of governmental support in the growth of geophysics highlights the issue of patronage of the earth sciences in general. Existing work on nineteenth-century geology, particularly in the United Kingdom, has tended to emphasize the "gentlemanly" aspects of the field tradition.[35] But while geology's elite practitioners may have been men of independent means, geology

---

to Measure Intercontinental Distances," *Historical Studies in the Physical and Biological Sciences*, 30 (2000), 393–416; Naomi Oreskes and James R. Fleming, "Why Geophysics?" *Studies in the History and Philosophy of Modern Physics*, 31 (2000), 253–7; Naomi Orskes and Ronald Rainger, "Science and Security before the Atomic Bomb: The Loyalty Case of Harald U. Sverdrup," *Studies in the History and Philosophy of Modern Physics*, 31 (2000), 309–70; and John Cloud, "Crossing the Olentangy River: The Figure of the Earth and the Military-Industrial Academic Complex, 1947–1972," *Studies in the History and Philosophy of Modern Physics*, 31 (2000), 371–404.

[33] Mott T. Greene, "History of Geology," in *Historical Writing on American Science*, ed. Sally Gregory Kohlstedt and Margaret W. Rossiter, *Osiris*, 2d ser., 1 (1985), 97–116, at p. 97.

[34] Ray Siever, "Doing Earth Science Research during the Cold War," in *The Cold War and the University*, ed. Noam Chomsky et al. (New York: New Press, 1997), pp. 147–70. On the enfeeblement of intellectual traditions by Cold War priorities, see Michael A. Bernstein, "American Economics and the National Security State, 1941–1953," *Radical History Review*, 63 (1995), 8–26.

[35] Rudwick, *Devonian Controversy*; James A. Secord, *Controversy in Victorian Geology: The Cambrian-Silurian Dispute* (Princeton, N.J.: Princeton University Press, 1986); David Oldroyd, *The Highlands Controversy: Constructing Geological Knowledge through Fieldwork in Nineteenth-Century Britain* (Chicago: University of Chicago Press, 1990).

as a whole was widely promoted for its commercial and military value.[36] Geological and geodetic surveys were set up by governments across the globe to make better maps in aid of exploration and conquest, and to delineate commercially valuable earth materials like limestone, coal, and, later, oil and gas. A disproportionate emphasis on gentlemanly origins may obscure a larger point: The earth sciences have long been supported by governments for commercial and strategic reasons. And as the reasons have changed, the loci of support have also changed. Geology in the nineteenth century was critical for industrialization, no less than geophysics in the twentieth century was critical for detecting submarines and verifying the limited test ban. At the beginning of the twenty-first century, the earth sciences once again have a role to play in the issues of the day, as the atmospheric and oceanographic sciences come to the fore in the wake of postindustrial environmental concerns. With each of these changes in subject matter has come a set of changed methodological and epistemic expectations. The earth sciences are a particularly good place to see the ways in which broader social demands have influenced not just the subjects but also the methods and values of science.

[36] Rachel Laudan, "William Smith: Stratigraphy without Paleontology," *Centaurus,* 20 (1976), 210–26; Paul Lucier, "Commerical Interests and Scientific Disinterestedness: Consulting Geologists in Antebellum American," *Isis,* 86 (1995), 245–67.

Part VI

PROBLEMS AND PROMISES AT THE END OF THE TWENTIETH CENTURY

29

SCIENCE, TECHNOLOGY, AND WAR

Alex Roland

Hephaestus, arms maker to the gods, was the only deity with a physical disability. Lame and deformed, he caricatured what his own handiwork could do to the human body. Not until the later twentieth century, however, did his heirs and successors attain the power to inflict such damage on the whole human race. Nuclear weapons lent salience to the long history of military technology. The Cold War contest between the United States and the Soviet Union attracted the most attention and concern, but in the second half of the twentieth century, science and technology transformed conventional warfare as well. Even small states with comparatively modest arsenals found themselves stressed by the growing ties and tensions between science and war.

The relationship between science, technology, and war can be said to have a set of defining characteristics: (1) *State funding* or patronage of arms makers has flowed through (2) *institutions* ranging from state arsenals to private contracts. This patronage purchased (3) *qualitative improvements* in military arms and equipment, as well as (4) *large-scale, dependable, standardized production*. To guarantee an adequate supply of scientists and engineers, the state also underwrote (5) *education and training*. As knowledge replaced skill in the production of superior arms and equipment, a cloak of (6) *secrecy* fell over military technology. The scale of activity, especially in peacetime, could give rise to (7) *political coalitions;* in the United States these took the form of the military-industrial complex. The scale also imposed upon states significant (8) *opportunity costs* in science and engineering that were often addressed by pursuit of (9) *dual-use technologies*. For some scientists and engineers, participation in this work posed serious (10) *moral questions*.

These characteristics emerged in three historic periods. State funding, institutions, and qualitative improvements appeared in the era of historic warfare before 1500 A.D. Mass production, formal education, secrecy, and political coalitions arose in the era between the introduction of siege artillery in the fifteenth century and nuclear weapons in 1945. Opportunity costs, dual-use technologies, and moral questions gained prominence in the Cold War (1947–91).

A historical survey such as this necessarily achieves chronological scope at the expense of historical specificity. Science is taken here to mean systematic study of the physical world. When science is applied or directed to systematic manipulation of the material world, it verges on technology. But technology need not be science based; indeed, it has not been throughout most of history. Engineering in its modern sense is a product of the eighteenth century, but as far back as the ancient world, engineers were those who worked engines of war, such as catapults and ballistae. The account that follows ranges widely through recorded history, but it concentrates on the United States in the twentieth century because more historical research has been done on that setting than on any other.

PATRONAGE

War is one of the chief reasons that states have chosen to support science and technology. The first craftsmen to specialize in the production of weapons or fortifications no doubt attracted the first patronage. State sponsorship of military research and development appeared in the Mediterranean world of classical Greece, but it languished under the Romans and throughout much of the Middle Ages.[1] It reemerged in the Renaissance, with the wealthy city states of northern Italy patronizing such men as Leonardo da Vinci (1452–1519), an avowed expert in military technologies ranging from naval architecture to ordnance.

By the fifteenth century, gunpowder was changing the relationship in the West between technology, the state, and war. Historian William McNeill argues that Europe invented a unique form of free enterprise in the early modern period, one that imposed market forces on the production of new weaponry.[2] Rulers paid high prices for the new gunpowder technology, especially siege guns. With these they reduced the castles of their neighbors and converted feudal obligations of service into obligations to pay taxes. With the revenue collected, sovereigns bought more and better weapons and subdued more competitors, until they had achieved a monopoly of armed force within their borders. Thus, weapons yielded political power; political power coerced revenue; and revenue bought more weapons.

---

[1] Werner Soedel and Vernard Foley, "Ancient Catapults," *Scientific American*, 240 (March 1979), 150–60; J. G. Landels, *Engineering in the Ancient World* (Berkeley: University of California Press, 1978), pp. 99–132; Brian Craven, *Dionysius I: Warlord of Sicily* (New Haven, Conn.: Yale University Press, 1990), pp. 90–7; Lionel Casson, *Ships and Seamanship in the Ancient World* (1971; repr., Princeton, N.J.: Princeton University Press, 1986), pp. 97–135.

[2] William H. McNeill, *The Pursuit of Power: Technology, Armed Force and Society since 1000 A.D.* (Chicago: University of Chicago Press, 1982). Charles Tilly makes a similar argument in *Coercion, Capital, and European States, AD 990–1992* (Cambridge, Mass.: Blackwell, 1992), except that he concentrates more on the role of capital and less on the role of technology.

Emergent European monarchs used this formula not only to build nation-states at home. They exported it around the world. Beginning in the late fifteenth century, side-gunned sailing vessels allowed Europeans to dominate the world's oceans and littorals.[3] The favorable trading relationship flowing from this domination funded successive generations of military technology. In the late nineteenth century, new technologies, such as steamships, railroads, and the telegraph, spread European control into the hinterlands of Africa and Asia.[4]

As the power of this relationship impressed itself upon Western governments, state support for both science and technology grew, accelerating forces already at work in civil society. States adopted patent policies to protect and encourage invention. New institutions, such as scientific academies, promoted still more research. The Scientific Revolution lent new force and prestige to the advancement of science. The Industrial Revolution multiplied the productivity of technology, just in time to arm and equip the mass armies unleashed by the democratic revolutions of the late eighteenth and early nineteenth centuries.

The French Revolution, in fact, was a hothouse of state-supported scientific and technological development for purposes of war.[5] Antoine Lavoisier (1743–1794) improved upon gunpowder manufacture before himself succumbing to the revolution. Gaspard Monge (1746–1818) taught a science-based curriculum in military schools, wrote treatises on military manufacture, and served as minister of the navy and confidante to Napoleon (1769–1821). Lazare Carnot (1753–1823), father of the engineer and thermodynamicist Sadi Carnot (1796–1832), oversaw the mobilization of French science and industry for purposes of war, including innovations in mass production and interchangeable parts. Napoleon himself rose to be emperor of the French from his schooling in one of the technical branches: artillery. Not since Marshall Sebastien Le Prestre de Vauban (1633–1707) in the age of Louis XIV (1639–1715) had technical specialists risen to such heights.

Military support for science and technology continued during the Pax Britannica (1815–1914), but it never regained the levels experienced in France during the wars of the revolution and Napoleon. In fact, World War I represented something of a setback, at least in the United States. American industry mobilized slowly, and the military services paid little attention to the scientists and engineers it put in uniform during the war. Large sums of money were invested in research and development, and significant advances were made in such fields as radio, sonar, and munitions. But disharmony

---

[3] Carlo M. Cipolla, *Guns, Sails and Empire: Technological Innovation and the Early Phases of European Expansion, 1400–1700* (New York: Pantheon, 1965); Geoffrey Parker, *The Military Revolution: Military Innovation and the Rise of the West, 1500–1800* (Cambridge: Cambridge University Press, 1988).

[4] Daniel R. Headrick, *The Tools of Empire: Technology and European Imperialism in the Nineteenth Century* (New York: Oxford University Press, 1981).

[5] Ken Alder, *Engineering the Revolution: Arms and Enlightenment in France, 1763–1815* (Princeton, N.J.: Princeton University Press, 1997).

between the government and the military on the one hand, and science and industry on the other, arrested the enterprise well short of its potential.[6]

European combatants in World War I experienced similar relations with science and technology. Advances in weaponry came increasingly from private firms, in contrast to government arsenals, but those firms exercised only limited control over government policies.[7] Germany is often portrayed as realizing most fully the military potential of science and technology, but its greatest advances were in the chemical industry – munitions and gas warfare. In other fields, technological innovation had little impact. The machine gun, a product of prewar development by private manufacturers, shared with artillery a domination of the European battlefield that neither the tank nor the airplane could overcome. At sea, the submarine, another product of private, prewar development, challenged the supremacy of surface vessels, but finally succumbed to the age-old system of convoy and to hastily developed innovations in underwater acoustics and depth charges.

World War II proved to be a very different experience.[8] In the United States, MIT engineer Vannevar Bush (1890–1974) convinced President Franklin Roosevelt (1882–1945) to create a mechanism for mobilizing science in war, leaving the scientists in their laboratories and funding them through contracts. The Office of Scientific Research and Development (OSRD) spent only $270 million on R&D during the war. The military services in the same period spent $1,710 million, not counting military salaries or the R&D financed from production and procurement funds. Also not counted is the $2 billion spent on the Manhattan Project.[9]

The results were revolutionary. Chemists developed new fuels, paints, hydraulic fluids, and explosives. Physicists improved acoustics, ballistics, rockets, fire control, communications, and sensing devices. Psychologists studied human engineering, propaganda, personnel screening, training, and combat fatigue. Physicians and biological scientists researched everything from malaria control to mass production of penicillin.[10] British development of the multicavity magnetron under the leadership of Henry Tizard (1885–1959) made possible microwave radar; the United States exploited this breakthrough during the war, producing 120 different kinds of radar, including the proximity fuse. U.S. work on radar was exceeded in scope and impact only by the massive Manhattan Project, which in six years converted the scientific theory of controlled nuclear fission into the weapons that fell on Hiroshima and Nagasaki.

[6] Carol S. Gruber, *Mars and Minerva: World War I and the Uses of Higher Learning in America* (Baton Rouge: Louisiana State University Press, 1975).
[7] David Stevenson, *Armaments and the Coming of War: Europe, 1904–1914* (Oxford: Oxford University Press, 2000).
[8] Daniel Kevles, *The Physicists: The History of a Scientific Community* (New York: Vintage, 1979).
[9] U.S. Bureau of the Census, *Historical Statistics of the United States, Colonial Times to 1957* (Washington, D.C.: GPO, 1961), p. 613.
[10] James Phinney Baxter, 3rd, *Scientists against Time* (Boston: Little Brown, 1946).

World War II went beyond mere production in quantity; it systematically improved the quality of weapons. Indeed, it introduced whole new categories of weapons. Jet engines, liquid-fuel rockets, the proximity fuse, and the atomic bomb moved from concept to application in the course of the war. World War II produced, for the first time in history, a substantial rearming of combatants; the victors emerged from the war with a different arsenal than they had at the outset.

Other nations also mobilized science and technology for war. Britain's great contributions, the multicavity magnetron and gaseous diffusion of uranium, were handed over to the United States because of Britain's vulnerability to air attack. Germany improved upon its record in World War I by directly supporting development programs, such as Wernher von Braun's ballistic missiles, although it failed to sustain a crash program to develop an atomic bomb. Japan did have an atom bomb program, but it lacked sufficient resources, both human and material, to bring it to fruition. The Soviet Union emphasized quantitative advances over qualitative, but it supported competitive design bureaus in an attempt to foster innovation.

This experience transformed the relationship between science, the state, and war. It convinced the major industrial states that the next war would be won not by industrial production in factories, but by scientific and technological research in laboratories. The world wars had been wars of industrial production; future wars would be won by qualitative improvements in arms and equipment. Furthermore, the threat of nuclear weapons delivered by airplanes or missiles in a matter of hours or minutes meant that states no longer had the luxury of mobilizing after war was declared. They must instead remain in a permanent state of readiness. The arsenal for the next war had to be invented, developed, and deployed today. Therefore, science and technology were themselves permanently mobilized.

In the United States, this conversion of military thinking was overdetermined. World War II had demonstrated both what American scientists and engineers could do and what the Germans might have done had the war dragged on. The demobilization following World War II convinced the military services that Americans would not tolerate a large, standing armed force; the services would have to match the sheer numbers of the Warsaw pact military establishment with technology. In the same vein, automated and sophisticated weaponry promised to improve the survivability of Americans in combat, minimizing the casualties of which democracies seemed so intolerant. Finally, the services concluded from Vannevar Bush's initiative to create a postwar National Research Establishment that if they did not set the agenda for military science and technology, the scientists and engineers might.[11]

---

[11] Some of these motives are explored in Michael Sherry, *Preparing for the Next War: American Plans for Postwar Defense, 1941–1945* (New Haven, Conn.: Yale University Press, 1977).

U.S. expenditures on research and development increased by an order of magnitude during World War II, from $99.1 million in 1940 to $1,564.5 million in 1945 (both in 1940 dollars); the military component rose from $35.3 million to $504.5 million.[12] After a brief slowdown at the end of the war, the funding of military research and development began to grow again in 1949. Seldom did it fall, since then, below 50% of all federal funding for research and development.[13] In constant (1987) dollars, spending increased more than sevenfold, from $3.8 billion in 1949 to $29.3 billion in 1995. In the period from 1953 to 1984, the federal government supported well over half the total funding for research and development in the United States, almost two-thirds of the funding for basic research. On average, something in excess of one-fourth of all research and development in the United States in the period of the Cold War was supported by the military. Were the space program and nuclear research added to these figures, the percentages would be even higher.

Not all countries, of course, experienced the Cold War similarly. The Soviet Union made comparable investments in military research and development, even though it had to devote a larger portion of its smaller gross domestic product to the enterprise. Even at that, its military technology tended to be more derivative of Western developments and less dependent on basic research and scientific advances. No other countries attempted to keep pace with the superpowers. Between 1984 and 1993, the United States committed 72 percent of its total government expenditures for R&D to military objectives, while members of the European Union devoted only 28 percent to the same purposes.[14]

INSTITUTIONS

Institutions mediate the military support of science and technology. The earliest arms makers and fortification builders were probably freelancers, who provided goods and services for pay. During its revolution, France provided more support for science and technology in its schools and bureaus than it did in other institutions. Early schools for artillerists and military engineers led to the creation of the Ecole Polytechnique in 1794. Other military academies, such as West Point in the United States, imitated the French model. The mathematics and physics of war, from ballistics to strength of materials, became part of the shared knowledge of well-educated officers.

---

[12] David C. Mowery and Nathan Rosenberg, *Technology and the Pursuit of Economic Growth* (Cambridge: Cambridge University Press, 1991), p. 124.

[13] *Historical Statistics of the United States*, p. 613; *Budget of the United States Government, Fiscal Year 1997: Historical Tables* (Washington, D.C.: GPO, 1996), p. 149.

[14] Ruth Leger Sivard, *World Military and Social Expenditures 1996*, 16th ed. (Washington, D.C.: World Priorities, 1996), p. 8.

Arsenals and armories continued to provide support for science and engineering, by promoting improvements in gunpowder in Europe and perfecting the "American System of Manufacture" in the United States. This system, the precursor to modern mass production, combined division of labor, the assembly line, reliance on jigs and patterns, machine tools, and interchangeable parts in the manufacture of small arms, before such techniques were economically competitive. Because the military placed a high value on interchangeable parts, which allowed for rapid repair of equipment in the field, it was willing to support a technology that the market would not. Only later did this technology spill into the commercial realm, culminating in the Fordism of the early twentieth century.[15]

At the end of the nineteenth century, the military still looked to its own arsenals and private industry for the technological development that was accelerating change in warfare. In World War I, governments of the major combatants turned increasingly to universities. In some cases, as with Fritz Haber's (1868–1934) chemical research at the Kaiser Wilhelm Institute in World War I, the scientist was supported for work in his own laboratory, even after being inducted into military service. In the United States, however, scientists were recruited to work in government laboratories and offices. Poor results led the United States to create the OSRD in World War II.

After the war, the military services created their own institutions to harness the potential of science and technology: They set up advisory committees. They expanded their World War II research laboratories, such as the Applied Physics Laboratory at the Johns Hopkins University. They supported specialized laboratories at universities around the country, such as Lincoln Laboratory at MIT. They continued to fund individual researchers in universities and industry to conduct both basic and applied research. They sponsored think tanks such as RAND Corporation and the Institute for Defense Analyses to conduct specialized military research. And they reorganized their own infrastructure to create administrative and management positions to oversee research and development.

The institutionalization of military research and development reflected another trend that generally shaped science and technology in the modern era. Galileo flourished in an age of the lone researcher, often living off independent wealth or personal patronage, controlling a small library and modest experimental means. In contrast, U.S. physicist and Secretary of Defense Harold Brown (b. 1927) rose to prominence at the Lawrence Livermore Laboratory. The growing complexity of science and technology, especially big science and high technology, forced research into large institutions. There,

---

[15] Otto Mayr and Robert C. Post, eds., *Yankee Enterprise: The Rise of the American System of Manufacture* (Washington, D.C.: Smithsonian Institution Press, 1981); David A. Hounshell, *From the American System to Mass Production, 1800–1932: The Development of Mass Manufacturing Technology in the United States* (Baltimore: Johns Hopkins University Press, 1984).

teams of investigators combined disciplinary talents in projects built upon intricate funding assumptions and high promise of results.

QUALITATIVE IMPROVEMENTS

Throughout most of human history, states fought one another in weapons symmetry. Both sides deployed similar arms and equipment, and for the most part, military professionals resisted technological change. Only slowly did political and military leaders come to appreciate the value of qualitatively superior military technology. The introduction of gunpowder in the West initiated growing enthusiasm for new technology, but the English did not replace the longbow with firearms until 1689. Military conservatism worked against the adoption of new technology right up until World War II. Scientists, engineers, inventors, and craftsmen who proposed innovations in military technology most often directed their proposals to the civilian government. Abraham Lincoln (1809–1865) was the most avid supporter of weapons innovation in the American Civil War, and Vannevar Bush's OSRD received its authorization and its power directly from President Roosevelt.

World War II convinced the last of the skeptics. Indeed, it provoked a complete reversal of behavior. Traditional military conservatism gave way to an apparently reckless, competitive enthusiasm for technological innovation. The services still argued over which technology to develop – missiles or airplanes, nuclear or fossil fuels, propellers or jet propulsion, solid or liquid fuels, inertial or stellar guidance – but virtually all agreed that qualitative improvement was the new desideratum. The growth in funding for military research and development and the proliferation of institutions devoted to this end reflected a thorough conversion of the American military that was mirrored around the world. The services actually struggled with one another for the exclusive right to develop and apply new technologies, such as ballistic missiles and satellites. In the process, they fueled the growth of the military-industrial complex in the United States and comparable political coalitions elsewhere.

Improvements in military technology, therefore, evolved at different rates over history. Throughout the Middle Ages, the evolution was slow and episodic. From 1500 to 1945, the pace of change accelerated in many Western countries, restrained somewhat by resistance from the military and by the imposition of secrecy. Since World War II, the pace has quickened still further, producing the "electronic battlefield" of computers, sensors, smart weapons, instant communication, global navigation and positioning by satellite, and countless other multipliers of the speed, scope, and destructiveness of combat. The strategic nuclear weapons that continue to hang over the world like a sword of Damocles are the most powerful symbol of the qualitative improvements in military technology that characterize modern warfare. All are the gift of science and technology.

These qualitative improvements exact a high price. In 1997 a single airplane, the U.S. B-2 stealth bomber, cost in excess of $2 billion, more than the total military budget of most countries of the world. Under the right circumstances, that airplane can be destroyed by a single Stinger missile, also the product of American research and development. The Stinger cost about $100,000 in 1997, more on the black market. By controlling the sale and distribution of such weapons, the United States and other developed nations that support large arms industries could shape war around the world. The products of science and technology have come to define large-scale, organized violence between states.

LARGE-SCALE, DEPENDABLE, STANDARDIZED PRODUCTION

The earliest weapons were comparatively simple, durable, and interchangeable with those of friend or foe. Complex weapons systems, such as the chariot or the warship, were not unknown in ancient times, but they were the exception. By and large, craft skills sufficed to produce most premodern weapons, and the quantity required was constrained by the facts that armies were limited in size and weapons could be recycled. Warfare was labor intensive in the ancient world, but not resource intensive.

The great sailing fleets of the seventeenth and eighteenth centuries taxed the timber resources of England and other competitors. But it was really the introduction of gunpowder that began to impose demands upon the productive capacity of nations. Armies could no longer live off the land. They had to produce and transport their own ordnance and ammunition. From the few thousand rounds, over forty-eight days, required for the final siege of Constantinople in 1453, artillery came to demand two million rounds, in two days of preparation fire, before the Germans launched their 1916 assault on Verdun. Meanwhile, the democratic revolutions had increased army sizes from the tens of thousands that fought at the decisive battle of Blenheim in 1704 to the 700,000 that Napoléon had under arms in 1808. The arms and ammunition required by these soldiers accelerated the Industrial Revolution, sparked the American system of manufacture, and contributed to the development of mass production. Along the way, the scientific management of human activity pioneered by Frederick Taylor (1865–1915), first developed at Watertown Arsenal, spread through American industry and overseas as well.

The world wars were wars of industrial production. They mobilized what Thomas P. Hughes calls large-scale technological systems to generate the tanks, ships, guns, and planes that are the backbone of industrialized warfare.[16] In 1940, President Roosevelt set the impossible goal of

---

[16] Thomas P. Hughes, *Networks of Power: Electrification in Western Society, 1880–1930* (Baltimore: Johns Hopkins University Press, 1983).

manufacturing 50,000 warplanes a year, more than the total number produced in World War I, and then watched United States industry surpass that target. The battle of the Atlantic turned for a while on the issue of whether or not the United States could manufacture ships faster than German submarines could sink them. The Allies lost a total of 23 million tons of merchant shipping in World War II, 14 million of it to submarines. But during the war, the United States produced 57 million tons, over one-third of the total tonnage of shipping existing at the beginning of the war. Scientific and technical improvements in antisubmarine warfare contributed to Allied victory in the battle of the Atlantic, but the sheer productive capacity of American industry was as important as any other factor.

Since World War II, production of arms and equipment for the military has been caught in a tension between quantity, quality, and cost. High-tech products, such as aerospace vehicles, require expensive research and development that can seldom be spread over large production runs. Demanding military specifications for fail-safe performance in the stressful environment of combat drive production costs up. And yet combat is a profligate consumer of munitions and equipment.

Cost also weighs in the balance between people and machines. The United States and other democracies gravitate toward machine warfare in part because of aversion to their own casualties. Trading equipment for lives, these states field remotely controlled vehicles, "stand-off" weapons, protective armor, and dense fields of fire to keep their own troops out of harm's way. This kind of war is profligate in its use of innovation and resources. It proved effective in the Gulf War against Iraq in 1991 but failed to overcome the surreptitious, indirect tactics of the guerrillas in Vietnam.

EDUCATION AND TRAINING

Throughout most of history, soldiers and sailors learned their craft by apprenticeship; they learned to fight by fighting. Although some privileged members of the warrior class grew up in martial environments with military tutors, most learned in the barracks or on campaign. Military engineers were probably the single significant exception before modern times, although little is known about their professional development.

Again, gunpowder precipitated change. Officers preparing for careers in artillery and engineering required more scientific and technical knowledge than their peers, leading to the establishment of schools. From the French artillery schools of the 1720s to the Ecole Polytechnique of the revolution, formal training in science-based engineering came to be seen as essential not only to the education of officers but to civic development as well.

Governments did not, however, look to the graduates of these schools for innovations in science and technology. With some notable exceptions, such as William Congreve (1670–1729; rockets), Thomas Rodman (1815–1871; ordnance), and Frank Whittle (1907–1996; jet propulsion), serving military officers, no matter how well trained, have not pioneered new military technologies. These continue to come from the civilian sector. Until the middle of the twentieth century, the major sources were private inventors and industry. Since World War II, universities and government-sponsored laboratories have added their contributions. The education and training of the scientists and engineers who staff these institutions has, therefore, become a matter of national military policy.

Two issues are paramount. First, to ensure that there will be enough scientists and engineers for military work, governments have underwritten their education and training. In France, for example, graduates of the Ecole Polytechnique, particularly Paul Vieille, played pivotal roles in developing the first smokeless high-explosive powder in the 1880s.[17] In the United States, the national security crisis surrounding the launch of the Soviet satellite *Sputnik* in 1957 precipitated passage of the National Defense Education Act the following year. This law provided federal funding for training in science and engineering. Government research grants to universities also ensure that students will be supported. And research contracts and grants for military research and development ensure that the graduates are drawn to defense work in sufficient numbers to do the nation's business.

That last source of funding raises the second major issue, the militarization of the academic community. In fiscal year 1995, the Department of Defense listed two universities (MIT and Johns Hopkins) and two of MIT's spin-offs (MITRE Corporation and Draper Laboratories) as being among the top fifty defense contractors, as measured in dollar volume of prime contracts.[18] Military funding on such a scale can distort the curriculum and the research agenda of institutions of higher education.[19]

SECRECY

Until the early modern period in European history, ideas about weaponry circulated freely.[20] The only known exception, Greek fire, reportedly came to the Byzantines in the late seventh century from a Syrian engineer and was kept by them as a state secret for centuries.[21] By the time of the Renaissance,

---

[17] I am indebted to Seymor Mauskopf for this example.
[18] *The World Almanac, 1997* (Mahwah, N.J.: World Almanac Books, 1996), p. 180.
[19] Stuart W. Leslie, *The Cold War and American Science: The Military-Industrial-Academic Complex at MIT and Stanford* (New York: Columbia University Press, 1993); David F. Noble, *Forces of Production: A Social History of Industrial Automation* (New York: Knopf, 1984); Paul Forman and Jose M. Sanchez-Ron, eds., *National Military Establishments and the Advancement of Science and Technology* (Dordrecht: Kluwer, 1996), esp. 9–14 and 261–326.
[20] Pamela Long and Alex Roland, "Military Secrecy in Antiquity and Early Medieval Europe: A Critical Reassessment," *History of Technology*, 11 (1994), 259–90.
[21] Alex Roland, "Secrecy, Technology, and War: Greek Fire and the Defense of Byzantium," *Technology and Culture*, 33 (October 1992), 655–79.

however, authors were regularly publishing books in which they professed to know secret inventions, many of them military, which they were not at liberty to disclose. This widespread phenomenon had complex historical roots, including the spread of knowledge in printed books written in vernacular languages, the birth of modern Western capitalism, and the emergence of the nation-state. Military knowledge had market value in such an environment.

Two reasons were often advanced for keeping military technology secret. First, the devices wrought such horror and destruction that it ill served humankind to unleash them upon the world. This concern, for example, moved Leonardo to withhold his design for a submarine. As the modern era unfolded and gunpowder made war still more destructive, one might have expected this scruple to gain more purchase. But it did not. The lure of financial gain overcame the qualms of many inventors. Some professed to believe that their creations would eliminate war altogether. Only in the Cold War did moral scruple reemerge as a major constraint on military researchers.

The second reason advanced for military secrecy gained weight during the modern period. Instead of the weapons symmetry that had characterized most warfare throughout history, research and development now promised to lend significant military advantage to the side with superior science and technology. Individual inventors, therefore, sought patent protection or guarantees of privilege when sharing their ideas with governments. And governments themselves increasingly imposed secrecy upon their own arsenals and laboratories. By the end of the twentieth century, international espionage focused as much on the technology of the enemy as it did on his force structure and strategy.

The secrecy surrounding military weaponry reached something of an apex during World War II, when United States General Leslie Groves (1896–1970) imposed upon the Manhattan Project standards of secrecy that denied information even to members of the project. Participating scientists, such as Los Alamos Director J. Robert Oppenheimer (1904–1967), protested that this "compartmentalization" of information conflicted with normal scientific practice and slowed their work.[22] Richard Feynman (1918–1988), for example, delighted in circumventing Groves's draconian restrictions.[23] But Groves's policies prevailed, a clear harbinger of the secrecy that would surround weapons development in the Cold War.

Military research and development during the Cold War further strained the relationship between the scientific community and the state. As military funding of research and development grew, scientists experienced more onerous constraints on their freedom to publish. Furthermore, the definition of national security broadened during the Cold War, coming to encompass such

---

[22] Richard Rhodes, *The Making of the Atomic Bomb* (New York: Simon and Schuster, 1986), pp. 449, 454, 539–40, 552, et passim.
[23] Richard Feynman, *Surely You're Joking, Mr. Feynman: Adventures of a Curious Character* (New York: W. W. Norton, 1985), pp. 115–20, 137–55.

fields as optics, computers, microelectronics, composite materials, superconductivity, and biotechnology.[24] Accepting government funding for research in these areas could limit a scientist's freedom to publish his or her results. But in some areas, the military services were the only significant source of funding. In microelectronics and nuclear physics, for example, it was difficult to identify research projects that did not have military implications.

POLITICAL COALITIONS

Behind the imposing arsenals of nuclear and conventional weapons maintained by the superpowers in the Cold War lay an infrastructure that stretched from university and government research laboratories to industrial giants, such as McDonnell Douglas Aircraft and Electric Boat Corporation, and from Soviet research institutes and design bureaus to the scientific production associations and the Military-Industrial Commission. Their operations could not be mobilized quickly in time of war; they had to be permanently engaged. Furthermore, both states promoted competition within their separate infrastructures, pitting company against company and bureau against bureau. National security could not be dependent upon a single source of any vital technology.

States, therefore, sustained peacetime establishments for both production and research and development. The scale of this infrastructure depended upon the state's perception of the danger it faced. This profoundly important question of public policy – how much defense is enough – might have been limited to the arena of politics, to be debated on practical and philosophical grounds. But the answer had important implications for large segments of industry and the research community. Inevitably, therefore, these establishments found themselves drawn into the politics of defense spending and strategy.

In his Farewell Address in 1961, President Dwight Eisenhower (1890–1969) labeled the result the "military-industrial complex." In private, he called it the "delta of power." Both terms scored the alliance among the defense industry, the military services, and Cold Warriors in Congress. Eisenhower added that "public policy itself could become the captive of a scientific/technological elite."[25] His science advisors insisted that Eisenhower later disavowed this warning, one that was clearly at odds with his expressed appreciation for the

---

[24] Board on Army Science and Technology, Commission on Engineering and Technical Systems, National Research Council, *Star 21: Strategic Technologies for the Army of the Twenty-First Century* (Washington, D.C.: National Academy Press, 1992), pp. 277–80; Herbert N. Foerstel, *Secret Science: Federal Control of American Science and Technology* (Westport, Conn.: Praeger, 1993); "Secrecy in University-Based Research: Who Controls? Who Tells?" special issue of *Science, Technology, and Human Values*, 10 (Spring 1982).

[25] Dwight Eisenhower, Farewell Address, *Public Papers of the Presidents of the United States: Dwight D. Eisenhower, 1960–1961* (Washington, D.C.: GPO, 1961), pp. 1038–9.

scientists who counseled him.[26] Nevertheless, Eisenhower's public concern called attention to a politicization of science that flowed directly from its enhanced role in national security. The growing dependence of scientists and their institutions on military funding eroded their ability to view defense policy dispassionately and impartially. Many scientists, especially those who specialized in military technologies, held important posts in government or on government advisory committees. Independent voices remained, such as the National Science Foundation and the American Association for the Advancement of Science, but often it was the scientists most closely aligned with government policies who offered the most influential counsel.

One result, in the United States, was that government policy was shaped in mutually contradictory and perhaps mistaken directions. Many observers have noted the maldistribution of research funds between defense-related fields and all others. And a few observers believe that the undue influence accorded to physical scientists, largely on the basis of their role in the Manhattan Project and other military development programs, led to a science policy in the United States when a technology policy was really needed.[27]

OPPORTUNITY COSTS

The concentration of national resources on military research and development imposed on industrial societies opportunity costs that are beyond precise calculation. What price did society pay for investing its scientific and engineering talent in war instead of devoting it to more peaceful and productive activities? Scientists studying nuclear radiation and fallout might better have been investigating the causes of environmental degradation. Study of strategic bombing might have been better directed to techniques of urban renewal. Improved desalinization of seawater might do more to help people live in peace than a whole fleet of ballistic missile submarines.

Similarly, a larger percentage of national treasure might have gone into basic research instead of to the applied or directed research required by the military. Many scientists believed that basic research generated the seed corn from which technology sprouted, but critics argued that technology contributed as much to science as it got in return. In the 1960s, the U.S. Department of Defense studied the conceptual origins of twenty weapons systems, concluding that basic research contributed little. The National Science Foundation responded with its own study, TRACES, arguing that basic research lay behind many of the country's most important technologies.[28] Some

---

[26] James R. Killian, Jr., *Sputnik, Scientists, and Eisenhower* (Cambridge, Mass.: MIT Press, 1977), pp. 237–9.
[27] Deborah Shapley and Rustum Roy, *Lost at the Frontier: U.S. Science and Technology Policy Adrift* (Philadelphia: ISI Press, 1985).
[28] Raymond S. Isenson, *Project Hindsight Final Report* (Washington, D.C.: Office of the Director of Defense Research and Engineering, 1969); Illinois Institute of Technology Research Institute,

military agencies, such as the Office of Naval Research, supported some basic research. Throughout most of the Cold War, however, "mission agencies," such as the Department of Defense, were enjoined to leave basic research to the National Science Foundation. The Mansfield amendment of 1970 made this injunction explicit. Although it was repealed the following year, the services thereafter took pains to clothe their basic research in the mantle of applied research and development. Basic research, therefore, struggled along with comparatively modest funding, much of it coming from universities and private foundations.

The controversy revolved around the issue of autonomy raised by Vannevar Bush at the end of World War II. He insisted that the OSRD had succeeded in the war because the scientists were empowered to set their own agenda. He wanted government funding after the war to continue that formula. President Harry Truman (1884–1972), however, insisted on accountability. Most government funding of research and development would be dependent upon demonstrated contribution to the public welfare. Throughout the Cold War, national security was the public good that attracted the most funding. Had that investment not been made in research and development for national security, it is not clear that the government would have chosen to spend comparable amounts on the scientists' agenda. While there may have been an opportunity cost involved in focusing on military research and development, it is impossible to discern what opportunities were foregone.

DUAL-USE TECHNOLOGIES

As the Cold War proceeded, three factors pushed military research and development toward dual-use technologies, that is, technologies with both military and civilian applications. First, nonweapons technologies, such as computers, became more and more important to military preparedness. Second, the rising cost of research and development placed a premium on technologies whose development could be supported in part by market forces and whose purchase price could experience economies of scale through mass production for a commercial market. Third, the stalemate produced by the huge nuclear arsenals of the superpowers convinced most political leaders that international conflict between industrialized states would have to be resolved in the future by means other than war. Economic competition became increasingly important as the threat of armed conflict among the great powers declined.

Dual-use technologies are as old as the Roman road network, which was built both to facilitate the movement of legions to the frontier and to promote commerce; the interstate highway system served the same purposes in the United States. In the modern world, the military has often pioneered the

Technology in Retrospect and Critical Events in Science (TRACES), 2 vols. (Washington, D.C.: National Science Foundation, 1968).

development of technologies that were not yet commercially viable. Interchangeable parts are one example, computers another. In both cases, governments saw a military advantage that was worth a premium consumers were not yet willing to pay. As the technology evolved, its cost often dropped to the point where commercial applications became practical. Examples of this "spin-off" from military research and development range from radar to jet aircraft to global positioning satellites.

Of course, the reverse has always been true as well. Research and development conducted for commercial purposes often proves militarily useful. The transistor, for example, was developed by Bell Laboratories to meet the anticipated demand for telephone switching in the period after World War II. It is hard to think of a technology that has had a greater impact on warfare in the second half of the twentieth century. Radio and the airplane were similarly invented for commercial purposes but developed more quickly because of early adoption by the military. Most innovations in military technology come from outside the military – from industry, universities, and private inventors – and many are intended primarily for commercial consumption and only secondarily for military use. Only a few specialized industries such as shipbuilding have found that they can prosper over time by relying exclusively on a military customer. The consolidation of the aircraft manufacturing industry in the United States in the last years of the twentieth century is a case in point.

Some science and technology, while not consciously dual-use, nonetheless falls in a gray area between civilian and military. Civilian space activities, for example, originated in the Cold War competition between the United States and the Soviet Union. The earliest spacecraft were launched on military rockets, and the space race that drove Americans to the moon in the 1960s was a continuation of the Cold War by other means.[29] Commercial nuclear power likewise spun off from military developments, without ever being able to escape its roots. In the United States, the same agency – first the Atomic Energy Commission and then the Department of Energy – exercised authority over military and civilian nuclear research and development. Furthermore, the by-products of commercial nuclear power plants remain a source of military concern around the world because of their potential conversion to weaponry.

Military research and development also produced spin-offs that contributed to civilian science and technology. Exploration in the nineteenth century, such as the expeditions of Army Captain Merriwether Lewis (1774–1809) and William Clark (1770–1838) in the Louisiana Purchase and of Navy Lieutenant Charles Wilkes (1798–1877) in the Antarctic and the Pacific Oceans, helped open those areas to other Americans. The United States

---

[29] Walter A. McDougall, *The Heavens and the Earth: A Political History of the Space Age* (New York: Basic Books, 1985).

weather service originated in the army. More recently, military research in such areas as ballistic missile targeting, submarine navigation and tracking, detection of underground nuclear explosions, and command and control have contributed to geophysics, seismology, oceanography, and commercial navigation of ships, planes, and land vehicles. Spin-off from military research may not address the same research areas that a strictly civilian agenda would have dictated, but nonetheless it has made a substantial contribution.

Dual-use technologies illustrate the complex ways in which science and the military have come to find themselves in symbiosis. Science and technology contribute to military purposes even as the military provides funding, institutions, and rationale for scientific and technological development in general. Science and technology developed for the military find commercial applications, while civilian science and technology are bent to military purposes.

MORALITY

History offers little evidence that military work troubled scientists and engineers before the modern period. By the end of the Middle Ages, however, the introduction of gunpowder seems to have given pause to at least some researchers. Leonardo was just one of many scientists and engineers who elected to keep their ideas secret, in the interests of humanity. Robert Boyle (1627–1691) attempted to dissuade the heirs of Cornelius Drebbel (1572–1634) from peddling that inventor's plans for a submarine.[30]

Others have professed to believe that morality dictated the introduction of new weapons, for war would disappear if it finally became horrible enough. Under this banner, Robert Fulton (1765–1815) marketed submarines, torpedoes, and naval steamships in France, England, and the United States. Far from disappearing, however, war became consistently more deadly and destructive throughout the modern era, culminating in World War II. Ironically, the weapons introduced at the end of that conflict appear to us to have produced the stalemate of the Cold War. Finally, a weapon had been introduced so horrible that nations eschewed its use. The nuclear peace that followed is attributed by some to the existence of these weapons.

Ironically, those weapons have also produced the greatest moral crusade against war yet conducted by the scientific community. In 1946, veterans of the World War II Manhattan Project began publishing the *Bulletin of the Atomic Scientists,* with its ominous doomsday clock on the cover, warning of imminent nuclear cataclysm. Manhattan Project veterans also founded the Federation of Atomic Scientists in 1945, now the Federation of American

---

[30] Alex Roland, *Underwater Warfare in the Age of Sail* (Bloomington: Indiana University Press, 1978), pp. 41, 50.

Scientists; this group monitors science, technology, and public policy. In 1957, twenty-two scientists from ten countries met in Pugwash, Nova Scotia, to discuss the threat posed by nuclear weapons. Since then, hundreds of Pugwash conferences, symposia, and workshops have brought together scholars and public figures to address the world's problems, most of them tied to military science and technology. In 1995, the Pugwash conferences and their president, Joseph Rotblat, shared the Nobel Peace Prize. The community that released the nuclear genie has led the crusade to contain it.

30

SCIENCE, IDEOLOGY, AND THE STATE
Physics in the Twentieth Century

Paul Josephson

In the past century, the state has assumed a central role in fostering the development of science. Through direct action, such as subsidies and stipends, and indirect action, such as tax incentives, the modern nation-state supports research in universities, national laboratories, institutes, and industrial firms. Political leaders recognize that science serves a variety of needs: Public health and defense are the most visible, with research on radar, jet engines, and nuclear weapons among the most widely studied. Scientists, too, understand that state support is crucial to their enterprise, for research has grown increasingly complex and expensive, involving large teams of specialists and costly apparatuses. In some countries, philanthropic organizations have underwritten expenses. In communist countries, where the state took control of private capital in the name of the worker, the government was virtually the only source of funding.

The reasons for state support of research seem universal, bridging even great differences in the ideological superstructures that frame economic and political desiderata. Some reasons are tangible, such as national security, but some are intangible, including the desire to prove the superiority of a given system and its scientists through such visible artifacts as hydropower stations, particle accelerators, and nuclear reactors. Whether we consider tangible or intangible issues, capitalist or socialist economies, authoritarian or pluralist polities, the role of the state and its ideology is crucial in understanding the genesis of modern science, its funding, institutional basis, and epistemological foundations.

Although often analyzed through the prism of notorious cases of interference in normal science under authoritarian regimes, ideological considerations have been no less important in shaping scientific research in democracies. In the modern United States, spokespersons explicitly drew the connection between democracy and science in flood control, space, and nuclear energy programs. David Lillienthal, an early leader of the Tennessee Valley Authority in the 1930s, believed that the TVA dams, hydropower stations, and reservoirs were a symbol of the vitality of democracy. For Presidents John Kennedy (for whom the race to put a man on the moon would demonstrate the superiority of the American way of life) and Ronald Reagan (for whom the Star Wars "peace shield" would protect this way of life from evil communism) and others, success in science and technology went hand in hand with the democratic ends of individual rights, freedom, and peace.[1]

Some ideologues developed the view that science in their respective countries must differ significantly from that in other countries in terms of epistemology, object of study, and organization of research. In the most extreme cases, the Soviet Union and Nazi Germany, belief in the uniqueness of national science and fear of its ideological contamination from various outside forces (for example, tainted ideology or persons) justified sharp restrictions on scientific contacts with persons and ideas from abroad. The ideologization of science by these forces contributed to the determination of what was "good" science and what was "bad" science, leading in the case of physics to short-term, but highly disruptive, prohibitions against the reception of relativity theory and quantum mechanics, resulting in the loss of autonomy of scientists, impingement on academic freedom, and the nearly complete exclusion of valid public concerns in the resolution of scientific controversies.

SOVIET MARXISM AND THE NEW PHYSICS

Physicists in Moscow, Petrograd, Kiev, Kharkiv, and elsewhere greeted the Russian revolution in 1917 with the hope that the new regime would support their research programs in a way that the tsarist regime had not. Throughout the 1920s, while the Communist Party strove to control the direction of academic life, it provided sufficient funding for scientists to establish a series of new institutes and embark on new research programs. Such facilities as the Leningrad Physical Technical Institute quickly gained international recognition under the direction of the dean of Soviet physics, Abram Ioffe (1880–1960).

For the Soviet leadership, science and technology would ensure the modernization of the country. Vladimir Lenin (1870–1924) believed in science and technology, notably electrification and the modern factory, as a panacea for the USSR's economic backwardness. Joseph Stalin (1879–1953), who rose to unquestioned power by 1929, went a step further, proclaiming the existence of socialist science and technology, which were indispensable for building "socialism in one country." Given "hostile capitalist encirclement," the nation required the development of indigenous science and technology, independent of the influences of bourgeois society. The party leadership saw

[1] Thomas Hughes, *American Genesis* (New York: Viking, 1989), pp. 360–76.

science primarily in material terms, that is, for its potential contribution to the economy. This science would emphasize planning to avoid duplication of research effort; applied science at the expense of fundamental research to ensure results of value to proletarian society; and strict ideological control of scientific activity to guarantee the promotion of science commensurate with the values of the working class. With the exception of visits of small numbers of Western scientists, from the mid-1930s onward, the conformity of science to ideology led to the international isolation of Soviet science.

For physicists, the most shocking aspect of Stalinist science policy concerned the effort to control the ideological, and hence epistemological, content of the new physics. The new physics – relativity theory and quantum mechanics – gave rise to epistemological paradoxes that confronted scientists everywhere: A series of phenomena failed to fit neatly into the Newtonian system. They included the physics of the very small (subatomic particles, such as electrons and the alpha, beta, and gamma particles emitted by radioactive atoms) and of the very fast (for example, visible light, ultraviolet light, and x rays). Max Planck (1858–1947) and Albert Einstein (1879–1955) contributed to a revolution in physics that would give rise to quantum mechanics, relativity theory, nuclear physics, and astrophysics. Experiments confirmed the interrelation of continuous and discrete phenomena, such as light that manifests wave and particle properties, as well as the existence of matter-energy. Quantum mechanics required the synthesis of statistical and dynamic laws to describe the behavior of subatomic phenomena, and it pointed to the inherent difficulty of accounting for the interaction of the subject and the object in subatomic processes, including measurement itself. This was the "uncertainty" principle. When we observe a macroscopic object, the perturbation in its behavior introduced by our observation is negligible. In the microworld, measurement influences behavior. We can know either location or momentum with complete precision, but not both at once. Finally, the new physics involved new understanding of atomic structure, leading to nuclear physics. In 1932, James Chadwick reported discovery of a neutrally charged particle in the atomic nucleus, the neutron. Taken with the discovery of other particles, this result enabled physicists to understand why the nucleus is stable in most circumstances, but decays or undergoes fission in others, and the neutron explains the existence of isotopes.

The new physics, and especially the uncertainty principle, disturbed a number of philosophers by suggesting the limits of, or inherent subjectivity in, human knowledge. Those who rejected the new physics wondered out loud whether mathematical formalisms adequately described the real, physical world. Was theoretical physics no more than an intellectual exercise that had little to do with reality? Many older physicists simply could not abandon Newtonian, mechanical explanations. For Soviet Marxists, it was disturbing that relativity required a rejection of so many classical notions, even the indestructible and unchanging atom, and of mass itself. They believed that the

Soviet philosophy of science, dialectical materialism, provided the benchmark by which to gauge all epistemological questions. Dialectical materialism is based on three principles: "All that exists is real; this real world consists of matter-energy; and this matter-energy develops in accordance with universal regularities or laws." Since Marx and Engels provided only general outlines of their view of the relationship between the materialist philosophy of nature and modern science, frequently in notebooks and unpublished essays, it remained for Soviet writers to clarify the details.[2] One of the details was how Engels's three dialectical laws of nature (the interpenetration of opposites, the negation of the negation, and the transformation of qualitative changes into quantitative ones and vice versa) would be applied to an understanding of the new physics. An example of the first law might be a magnet with a north and south pole, or the wave particle duality of light. However, not all of the philosophical disputes of the 1930s were so neatly reduced to a consideration of the applicability of these laws to modern science, although many of the participants in the debates believed that they were.

The representatives of the two major trends of Marxist philosophy, the "Deborinites" and "Mechanists," organized a number of research institutes in the mid-1920s that aimed to become more conversant with recent advances in the sciences, train young communist workers in the ways of modern science, and attract natural scientists to their fold. The Deborinites believed that the epistemological questions that had arisen in response to the major developments in physics in the first third of the century demonstrated the compatibility of modern physics with dialectical materialism.[3] The Mechanists took exception to many of the Deborinites' positions. They believed that all processes in the external world could be explained in terms of the laws of classical mechanics. They referred to the works of leading Marxist scholars, Engels and Lenin in particular, on mathematics, physics, chemistry, and biology in order to demonstrate that mechanical processes in both the organic and inorganic worlds reduce to matter in motion, subject to the concept that all qualitative differences are differences of quantity. They also confused physical with philosophical relativism.

The epistemological debates that concerned physicists and philosophers might never have intersected were it not for the "Great Break" in the Soviet Union. The Great Break of the late 1920s and early 1930s was a self-proclaimed revolutionary abandonment of all vestiges of bourgeois society, a period of forced collectivization of agriculture and rapid industrialization, of cultural revolution, and of the first rumblings of the purges. Cultural revolution was intended to lead to the replacement of so-called bourgeois specialists with

---

[2] Loren Graham, *Science, Philosophy, and Human Behavior in the Soviet Union* (New York: Columbia University Press, 1987), pp. 24–67.
[3] David Joravsky, *Soviet Marxism and Natural Science, 1917–1931* (New York: Columbia University Press, 1961), pp. 279–97.

scientists of proletarian social origin and worldview. Cultural revolution included attempts by communist cadres to take over the administration of scientific research institutes, economic enterprises, and educational institutions, as they penetrated the ranks of those organizations.[4]

The Great Break had a significant impact on physics. To ensure that science, no less than art and literature, had proletarian credentials, the Communist Party established formal study circles of materialist outlook and working-class members. The study circles contributed to the joint effort of physicists and philosophers to find common ground. But they were mainly the Party's vehicle for the proletarianization of science. The proletarianization of physics was connected with the notion that the working class needed to create proletarian institutions to replace bourgeois ones, including science, its methodology, and its epistemology. As philosophical disputes unfolded, this meant that the new physics, as a breeding ground for idealism, had to be eradicated, in part through the spread of dialectical materialism into research settings in the minds and bodies of workers whom the Party had advanced to radicalize the institutes.[5] The seemingly abstruse epistemological debates among the Mechanists and dialecticians now came to have great significance for scientists: The upshot was that Stalinist ideologues acquired the power to tell scientists which approaches were ideologically acceptable.

One of the major figures of this ideological struggle was Arkady Timiriazev (1880–1955), a professor of physics at Moscow State University, a party member, and an anti-Semite. Timiriazev was distinguished for his unceasing devotion to Newtonian classical mechanics. Like Phillip Lenard in Germany, Timiriazev used the hypothesis of an ether that filled the universe to explain the transmission of electromagnetic energy mechanically through space. Timiriazev was troubled by the increasing role of statistical laws, the "mathematization" of matter, and the apparent rejection of causality. Timiriazev resorted to political intrigues, gossip, and innuendo to achieve his goal of the official condemnation of relativity, reserving his most hostile commentary for such Jewish theoreticians as Leonid Mandelshtam (1879–1944), Iakov Frenkel (1894–1952), and future Nobel laureate Lev Landau (1908–1968).[6]

On the other side of the ideological disputes in the Moscow State University physics department stood Boris Gessen, a middling physicist, if devout Marxist, and such first-rate scientists as Igor Tamm (1895–1971) and Leonid

---

[4] Archive of the Academy of Sciences (hereafter A AN), f. 364, op. 4, ed. khr. 28, l. 127; and Archive of Moscow State University (hereafter A MGU), f. 46, op. 1., ed. khr. 29, k. 1, ll. 109, and ed. khr. 42, k. 26. On cultural revolution, see Sheila Fitzpatrick, ed., *Cultural Revolution in Russia, 1928–31* (Bloomington: Indiana University Press, 1978).

[5] A AN, f. 351, op. 1, ed. khr. 82, ll. 1–6, 23–6, and ed. khr. 161, ll. 1–2. On the Marxist study circles, see Paul Josephson, *Physics and Politics in Revolutionary Russia* (Berkeley: University of California Press, 1991), pp. 203–8.

[6] A MGU, f. 201, op. 1, ed. khr., 366, k. 19, and f. 225, op. 1, ed. khr. 23, and A AN, f. 351, op. 2, ed. khr. 39, l. 6. See *Mekhanisticheskoe estestvoznania i dialekticheskii materializm* (Vologda, 1925) for a major Mechanist tract.

Mandelshtam. Gessen headed up the physics section of the Communist Academy of Sciences and was the dean of the university physics department until his arrest in the fall of 1936. He disappeared in the purges in 1937, soon after having been denounced in a public forum for his idealist failings.[7] Gessen defended the new physics as commensurate with Marxist theory of the dialectic. For example, Gessen concluded that the dialectical law of the interpenetration of opposites was reflected in the wave-particle nature of light; the existence of matter-energy; the complementarity of statistical and dynamic laws; and the vital, new understanding of the dialectical nature of relationships between subject and object. Gessen pointed out that the new physics modified the concepts of absolute space and time, which had taken on metaphysical eminence in classical physics.[8]

In spite of Gessen's efforts to show how the new physics and Soviet ideology meshed neatly, the ideological desiderata of the Great Break ensured his defeat. For the Great Break concerned more than esoteric points of dialectical materialism. Control of the scientific establishment was at stake. Militant Stalinist communists feared the power, knowledge, and authority of physicists, especially the leaders of the physics community whose careers dated to the tsarist era. Stalinist ideologues used Timiriazev's doubts about the new physics to claim that Gessen and other physicists were enemies of Soviet power.[9] Further, in the mid-1930s the Great Terror spread throughout Soviet society. Most likely ten million died; ten to fifteen million were interned at one point or another in Stalin's labor-camp system. The terror hit scientists hard. Dozens of leading scholars were arrested; hundreds of midlevel physicists, too, lost their careers or lives. The terror struck the entire discipline, in Moscow, Kharkiv, Dnepropetrovsk, and above all in Leningrad with its leading theoreticians. The rational science of central planning had given way to xenophobic, irrational proletarian science.

While also engaging in the rhetoric of proletarian science on occasion, Ioffe himself now courageously stepped forward.[10] In the leading theoretical

---

[7] "Lichnoe delo B. M. Gessena," A AN, f. 364, op. 3a, ed. khr. 17, ll. 1, 3, 4–6, 8–10; f. 154, op. 4, ed. khr. 30; f. 351, op. 1, ed. khr. 63, ll. 34–5, ed. khr. 74, and op. 2, ed. khr. 26, l. 69–70; f. 355, op. 2, ed. khr. 71; f. 364, op. 4, ed. khr. 24, ll. 130–2; op. 4, ed. khr. 24, l. 130–4; f. 354, op. 4, no. 1, l. 19; and A MGU, f. 225, op. 1, ed. khr. 40, ll. 1–3. Boris Gessen is best known for a paper he delivered to the Second International Congress of the History of Science in London in 1931, entitled "On the Social and Economic Roots of Newton's *Principia*," which stimulated the development of "externalism" in the history of science. See Loren Graham, "The Socio-Political Roots of Boris Hessen: Soviet Marxism and the History of Science," *Social Studies of Science*, 15 (1985), 705–22.

[8] Gessen, *Osnovnye idei teorii otnositel'nosti* (Moscow: n.p., 1928), pp. 64–5. Gessen wrote two earlier articles in which he questioned Timiriazev's uninformed attack of relativity theory, "Ob otnoshenii A. Timiriazeva k sovremennoi nauke," *Pod znamenem marksizma*, no. 2–3 (1927), 188–99; and "Mekhanicheskii materializm i sovremennaia fizika," *Pod znamenem marksizma*, no. 7–8 (1928), 5–47.

[9] Josephson, *Physics and Politics*, pp. 228–32, 252–61.

[10] *Zhurnal tekhnicheskoi fiziki*, 8 (1937), 884. On the purges, see Robert Conquest, *The Great Terror* (New York: Collier Books, 1968). On the impact on Soviet physicists, see Gennady Gorelik and Viktor Frenkel, *Matvei Petrovich Bronstein and Soviet Theoretical Physics in the Thirties*, trans.

journal of the Party in 1937, he took issue with both individuals and tactics. He questioned the traditional Bolshevik rule of disputation that "he who is not with us is against us." He demonstrated the fallacy of continued adherence to a Newtonian view of electromagnetic phenomena. He concluded by lumping together the antirelativists in the USSR with their Nazi "allies," Stark and Lenard, calling them reactionary and anti-Semitic.[11] Ioffe and other mainstream physicists referred to the many significant Soviet achievements in electrification, communications, and metallurgy as proof that the Mechanists' hypotheses were anachronistic and based on shoddy work.

World War II provided only brief pause from ideological attack. During the Cold War, Party officials grew more vigilant over what they saw as deviations from ideological norms. This period of vigilance is known as the *Zhdanovshchina,* after Andrei Zhdanov, the politburo member responsible for cultural affairs. Jingoistic efforts to demonstrate Russian priority in all fields resulted. Writers, artists, musicians, and scientists had to avoid being accused of "kowtowing" before the West, that is, showing any taint of what was termed bourgeois culture in their work. Some were accused of "cosmopolitanism," a code word suggesting ties to an international Jewish conspiracy. In physics, those who had never trusted, let alone understood, the new physics used the Zhdanovshchina to continue to attack relativity theory, quantum mechanics, and their supporters at home and abroad.[12]

Timiriazev's allies among Moscow State University physicists renewed their insistence that idealism pervaded physics. Hearing their calls, the central committee Party apparatus instructed dozens of institutes to hold open meetings to expose traitors to the Soviet way of thinking among physicists.[13] These meetings paralleled the effort between November 1948 and May 1949 to establish an agenda for a national conference to condemn the new physics. This meeting would be like the one in biology held in the summer of 1948 that had given Trofim Lysenko the authority to force genetics underground. The university physicists directed their actions largely against Leningrad physicists, mainly academy physicists, and in what must have haunted those who knew anything about Nazism, against Jewish theoreticians. (According to an apocryphal story, physicists working on the atomic bomb project got wind of the idea, called Lavrenti Beria, head of the secret police, and informed him that a bomb could not be constructed without taking note of relativity

Valentina Levina (Basel: Birkhäuser Verlag, 1994); V. V. Kosarev, "Fiztekh, gulag i obratno," in V. M. Tuchkevich, ed., *Chteniia pamiati A. F. Ioffe* (St. Petersburg: Nauka, 1990) and Josephson, *Physics and Politics,* 308–17.

[11] Abram Ioffe, "O Polozhenii na filosofskom fronte sovetskoi fiziki," *Pod znamenem marksizma,* 11–12 (1937), 133–43.

[12] For more on Lysenko and Lysenkoism, see David Joravsky, *The Lysenko Affair* (Cambridge, Mass.: Harvard University Press, 1970); Zhores Medvedev, *The Rise and Fall of T. D. Lysenko* (New York: Columbia University Press, 1969); and Valery N. Soyfer, *Lysenko and the Tragedy of Soviet Science,* trans. Leo Gruliow and Rebecca Gruliow (New Brunswick, N.J.: Rutgers University Press, 1994).

[13] For a stenographic account of the meeting at Ioffe's institute, see Archive of the Leningrad Physical Technical Institute, f. 3, op. 1, ed. khr. 195.

theory and the equivalence of matter and energy.) The conference luckily was not convened.[14] But mainstream physicists understood where power resided and the importance of ideological considerations in their work.

The fact of the matter is that physicists were able to maintain a degree of autonomy because of their achievements in basic and applied science. They participated in the genesis of quantum mechanics. The industrialization effort ensured that budgetary allocations for their institutes would grow significantly. The importance of physics for communications, electrification, metallurgy, and other industrial programs provided a shield to physicists. The financial support given to theoretical departments was tacit acknowledgement by the authorities of the validity of theoretical endeavors, so long as they were accompanied by applied pursuits. With the death of Stalin in 1953, the physicists reasserted their primacy in the resolution of scientific conflicts according to international norms.

ARYAN PHYSICS AND NAZI IDEOLOGY

Under National Socialist rule, German physicists, too, fell prey to ideological forces that influenced the content and direction of their work, and often their career paths. If in the Soviet Union the criticism was Marxian and class based, in Nazi Germany it was racially based. Physicists tended to be conservatives who, in spite of the large numbers of Jews among them, embraced the strong anti-Semitic, antidemocratic, imperialistic, and nationalistic currents that dated in German science to the Wilhelmian empire. Most scientists never trusted the Weimar leadership and welcomed the National Socialists to power.

The Nazi state took a special, if intermittent, interest in science and technology. The Nazis recognized the historical greatness of German engineering feats in the chemistry industry and the new superhighway system, the Autobahn. They used biology to promote a racially pure Third Reich. Members of the Nazi Party and their representatives in the scientific establishment argued that there was a special Aryan science. All science (and all morality and truth) was judged by its accordance with the interest and preservation of the *Volk*, which consisted of a metaphysical belief in an essential German people of organic purity and their historic mission to control world civilization. Tied to *völkisch* beliefs was a rejection of democratic government, since the state alone could claim to reflect the will of the people. Aryan science was applied and technical, its supporters claimed, not overly mathematical, theoretical, and formalistic.

An attack on the physics profession followed the promulgation of the Nazi race laws. If the episode of Aryan physics was short in duration, it was an extreme case of science being shaped by ideology, accompanied by firings

[14] A AN, f. 596.

and arrests. If Aryan physics was empty scientifically, politically it was a threat to scientific institutions and careers. The intellectual migration that resulted devastated German physics, with perhaps one-quarter of German physicists forced from their jobs, mostly by the laws excluding Jews from the civil service (and state institutes and universities) beginning in 1933, without protest or expression of outrage by their colleagues.[15] Physicists motivated by professional jealousies were able to take advantage of the situation to advance their careers. Theoretical physics lost its privileged position. Temporarily, an anachronistic and mechanistic view of physics predominated; it required the rejection of the new physics and repudiation of such Jewish representatives as Einstein. In this environment, even Werner Heisenberg (1901–1976), plainly a German patriot, was attacked for his support of modern physics, while "Aryan" physics was touted by its supporters as the only true German physics.

The effort to create an "Aryan science" freed from Jewish influence was especially prominent from 1933 to 1939. The leading physicists who remained to serve their nation – including Nobel Prize winners Max Planck (1918), Werner Heisenberg (1932), Phillip Lenard (1905), and Johannes Stark (1919) – were left to sort out what role modern theories, created in part by Jews such as Einstein, should play under a totalitarian regime founded on principles of racial purity. Lenard (1862–1947) and Stark (1874–1957) rejected these modern theories and published a large number of anti-Semitic speeches to tout Aryan physics. They were experimentalists who had made their careers early in the century by elaborating the Newtonian worldview. Academic physics appointments heretofore worldwide had been dominated by experiment; as theory became useful, it opened a niche in academia for bright outsiders. Hence Jews, who were allowed to take German civil service positions only beginning in the late 1860s, tended to move into these career paths as universities opened positions in theoretical physics. Stark and Lenard resented what they believed was the diminution of the importance of experiment, and they quarreled with more recent notions of light quanta, relativity, and quantum mechanics. Stark responded to these feelings and a series of professional disappointments by writing *Die gegenwärtige Krisis in der deutschen Physik* (The present crisis in German physics), which attacked relativity and quantum theory.[16]

Lenard was also a good candidate to embrace Nazism. His plodding, conservative approach was more conducive to work in experimental physics, not the rapidly unfolding areas of theory. Lenard carried deep-seated hostility toward Jews. He blamed Germany's defeat in World War I in part on Jews, and he hated the Weimar republic. He resented Einstein and the acclaim

---

[15] Alan Beyerchen, *Scientists Under Hitler: Politics and the Physics Community in the Third Reich* (New Haven, Conn.: Yale University Press, 1977), pp. 43–7. See Ruth Sime, *Lise Meitner: A Life in Physics* (Berkeley: University of California Press, 1996), pp. 134–209.
[16] David Cassidy, *Uncertainty: The Life and Science of Werner Heisenberg* (New York: W. H. Freeman, 1992), pp. 342–3, and Beyerchen, *Scientists Under Hitler*, pp. 103–15.

accorded relativity. Like Timiriazev, he reserved most of his hatred for the "abolition of the ether" demanded by relativity theory. Like older physicists in many countries, he believed it was outrageous to replace the comfortably mechanical concept of a fluid bearing light waves with nothing more substantial than a set of equations. Lenard affixed the label of "Jewish" to concepts with which he did not agree in the fight to save the ether. He also turned toward consideration of the role of racial heritage in science. True science was experimental, national, and racially pure, he determined. The culmination of his contemplations was the four-volume *Deutsche Physik* (1936–7), in which *völkisch* concepts were front and center. All these writings were geared toward freeing physics from "Jewish Marxist domination" and fighting against the "Jewification of German science."[17]

Inasmuch as it was a political, more than a scientific, movement, "Aryan physics" did not describe a standard approach to physical laws of nature. Still, there were central features of Aryan physics. It was anthropologically/racially based, with all leading concepts, its adherents unflinchingly claimed, originating among Aryan-German contributors. Experiment and observation were considered the only true bases of knowledge. Since experimental, Aryan physics was highly useful for technology and industry, the better to promote economic self-sufficiency. The *völkisch* nature of this physics stemmed from the belief that the Nordic race had created not only mechanics but all experimental science. Nordic researchers had a penchant for observation, repetition, modesty, "joy in struggling with the object – joy in the hunt." The Jew, in contrast, had a predilection for theory and abstraction, spearheaded the effort to abolish the ether conception, and constituted a threat to Aryan science.[18]

Aryan physics had little appeal among most physicists owing to its rejection of relativity and quantum mechanics, but it captured Nazi attention. Stark and Lenard had been among the few leading scientists who supported Adolf Hitler (1889–1945) during his brief imprisonment in 1924, following the Munich Beer Hall Putsch in November 1923. Stark praised Hitler in print before the latter's rise to power, referring to his anti-Semitic ideas and his autobiographical tract *Mein Kampf* as evidence that Hitler was not a demagogue but a "great thinker." In March 1933, Lenard wrote directly to Hitler, offering his services in personnel decisions affecting physics. Stark urged his fellow German Nobel laureates to join in a public declaration of support for Hitler in preparation for an August 1934 plebiscite; they declined his offer. Hitler could hardly forget the support two Nobel Prize winners had given him after he assumed supreme control of the government in 1933.[19]

---

[17] Philipp Lenard, *Über Relativitätsprinzip, Äther, Gravitation* (Leipzig: Verlag S. Hirzel, 1920). Another work characteristic of racist physics was L. W. Helwig's four-volume *Die Deutsche Physik* (1935). See Cassidy, *Uncertainty*, pp. 342–4; Beyerchen, *Scientists*, pp. 79–95; and Albert Speer, *Inside the Third Reich*, trans. Richard and Clara Winston (New York: Collier Books, 1970), p. 288.

[18] Beyerchen, *Scientists*, pp. 123–40.

[19] Johannes Stark, *Adolf Hitlers Ziele und Persönlichkeit* (Munich: Deutscher Volksverlag, 1932); Mark Walker, "National Socialism and German Physics," *Journal of Contemporary History*, 24 (1989), 64,

During the early years of the Third Reich, Lenard, Stark, and their associates blocked the attempt to appoint Heisenberg as successor to his Munich university teacher Arnold Sommerfeld, a well-merited appointment already approved by the university and the Bavarian Ministry of Culture. The high point of the Aryan physics movement was 1936 when Stark, Lenard, and their supporters attacked mainstream German physicists, including Heisenberg, in the *Völkische Beobachter*, the semiofficial paper of the party, and in speeches. The SS (*Schutzstaffel*) newspaper *Das Schwarze Korps* called Heisenberg and other German physicists "white Jews," that is, individuals of Aryan heritage under the influence of Judaism, in this case, Einstein's physics. The attacks led a significant number of students to decline to study with Heisenberg and other theoreticians. The attacks on Heisenberg stirred the German physics community to retaliate, prompting the government to reappraise modern theoretical physics. It took Heisenberg several years of careful treading and fortuitous personal contacts to overcome these attacks. Still, he had a number of disconcerting experiences.[20]

Surely it was difficult for patriotic scientists to determine the right course in response to Nazism. Many honest Germans believed that by using quiet diplomacy, rather than visible public protest, they might moderate Hitler's behavior. Others welcomed National Socialism's "call to national cultural renewal, unity, and glory." Still others assumed that the Nazis were a short-lived regime, not representative of the true spirit of Germany, and strove to remain apolitical.[21] As in the USSR, where the Party's ideological scrutiny required careful rendering of physical concepts or reference to individuals considered anathema to the state, so in Nazi Germany the state was always capable of using its power in an arbitrary and capricious fashion against scientists. Fortunately, the Nazi Party came to believe that the dispute between representatives of the new physics and Aryan physics was a professional, intramural dispute, not a political one, with both groups loyal to the regime, and this proved to be what saved the new physics in Nazi Germany.[22]

SCIENCE AND PLURALIST IDEOLOGY: THE AMERICAN CASE

Anti-Semitism existed in United States physics, too, but there was no official policy. Rather, an equating of science with inevitable progress and a belief that

---

and *National Socialism and the Quest for Nuclear Power* (Cambridge: Cambridge University Press, 1989), pp. 61–2.

[20] Walker, *Nuclear Power*, pp. 61–2, and "German Physics," 63–4, 66; Cassidy, *Uncertainty*, pp. 384–93; Herbert Mehrtens, "Irresponsible Purity: The Political and Moral Structure of the Mathematical Sciences in the National Socialist State," in Monika Renneberg and Mark Walker, eds., *Science, Technology, and National Socialism* (Cambridge: Cambridge University Press, 1994), pp. 324–38.

[21] On the challenges facing Max Planck, see John Heilbron, *The Dilemmas of an Upright Man* (Berkeley: University of California Press, 1986), pp. 149–203.

[22] Walker, "German Physics," 69–75, 79–85.

science, like the economy, ought to operate according to laissez-faire principles dominated the ideology of science. Until the Great Depression, from the point of view of business, the scientist applied knowledge so that a given corporation might turn a higher profit. For the federal government, apparent constitutional prohibitions against the financing of research limited involvement in the scientific enterprise to low-level military, health, and regulatory responsibilities until World War II. The physicist was expected to set forth "facts," not enunciate political opinion, let alone engage directly in political activities. For the scientist, the freedom to pursue the truth, unfettered by social, political, or financial concerns, would signal arrival in the promised laboratory. So long as industry, government, and foundations underwrote their increasingly expensive research, they would contribute to progress (for example, cyclotrons, they claimed, promised medical applications).[23]

Many American scientists abandoned their optimism in the 1930s. Industry had applied and marketed their discoveries, but the public saw them as responsible for the loss of jobs and their replacement by modern machines (unemployment). If critics of science were in a minority, many scientists still took criticism to heart. Adherents of technocracy wondered if science held a key to the future of democracy. Some scientists agitated for social responsibility, especially when abuses by Nazi scientists came to light. Although they learned vividly the dangers of authoritarian rule from German émigré scientists, this did not prevent some American scientists from seeing the USSR, a state avowedly devoted to science and technology, as a panacea and the example to follow.[24]

Physicists earned postwar influence with policy makers and accolades from the public for building atomic weapons that brought an end to the war. The military director of the Manhattan (atomic bomb) Project, General Leslie Groves, strove to prevent physicists from discussing the moral aspects of their atomic weapons research by adopting a policy of compartmentalization of small teams of experts in different locations. Beginning with the Franck Report (1944), which questioned the need to use atomic bombs on Japan, and continuing through such organizations as the Federation of American Scientists, activist scientists rejected government control of research, questioned its infringements on academic freedom in the name of national security, and protested growing military expenditures and failure to come to grips with the arms race. During the McCarthy era (the early 1950s), the government questioned many of these scientists' loyalty on specious, anticommunist, ideological grounds, a phenomenon that reached its moral nadir in the revocation of the security clearance of J. Robert Oppenheimer (1902–1967), the scientific leader of the Manhattan Project.[25]

---

[23] Daniel Kevles, *The Physicists* (New York: Knopf, 1978).
[24] Peter Kuznick, *Beyond the Laboratory: Scientists as Political Activists in 1930s America* (Chicago: University of Chicago Press, 1987).
[25] Martin Sherwin, *A World Destroyed* (New York: Knopf, 1975). On the scientists' movement and awakening of moral concern about nuclear weapons, see Alice Kimball Smith, *A Peril and a Hope*

The bomb projects in Germany (during World War II) and the USSR (1943 onward) also had a profound effect on the ways in which physicists related to the regimes. But the threat of legal or professional retaliation prevented them from voicing moral opposition openly. Werner Heisenberg argued that German physicists had concentrated on building a reactor, not a bomb. But it seems clear that he had avoided moral issues while serving the Nazi regime; in fact, administrative, rather than moral, factors interfered with the Nazi bomb. Andrei Sakharov (1921–1989) and a few other scientists opposed Soviet efforts to deploy a huge nuclear arsenal. But, denied channels available to scientists in other countries to oppose nuclear weapons, both Soviet and German scientists resolved to enjoy the largesse bestowed on them for research and did not speak out on the morality of nuclear weapons in the USSR and Nazi Germany, respectively.[26]

In the United States, too, most physicists found it difficult to wean themselves from the seemingly endless source of government funds that Manhattan (and national security) came to signify. Postwar prosperity and Cold War largesse meant reliance on the government's military research priorities. Physicists assumed a growing role as policy makers in the General Advisory Committee of the Atomic Energy Commission, the National Security Council, the president's science advisory committee, the Congressional Office of Technology Assessment, and other ad hoc committees. For most physicists, this relationship indicated that, like the American political system itself, science operated according to democratic principles, and that scientists could, without difficulty, provide a factual basis on which decision makers could determine rational policies. The political scientist Donald K. Price argued that scientists comprised a fourth estate, whose activities ensured the fostering of democratic institutions in society as a whole. In the same vein, political philosopher Michael Polanyi alluded to a republic of science.[27] For most Americans, physicists were saviors, not purveyors of doom, and atomic weapons would secure the nation against communist threats.[28]

In the 1960s, the authority of physicists began to erode for a series of reasons. One was growing awareness of the danger of nuclear fallout from atmospheric weapons tests. In addition, the publication of Rachel Carson's *Silent Spring* (1962) documented the fact that savior pesticides and herbicides

---

(Chicago: University of Chicago Press, 1965). In 1953 the American Chemical Society rejected the membership application of Irène Joliot-Curie (1897–1956) because of her connections with the French Communist Party. See Margaret Rossiter, "'But She's an Avowed Communist!' L'Affaire Curie at the American Chemical Society, 1953–1955," *Bulletin for the History of Chemistry*, 20 (1997), 33–41. My thanks to Mary Jo Nye for bringing this article to my attention.

[26] Samuel Goudsmit, *ALSOS* (New York: Henry Shuman, 1947); Thomas Powers, *Heisenberg's War: The Secret History of the German Bomb* (New York: Knopf, 1993), pp. 430–52; Walker, *National Socialism and Nuclear Power*, pp. 229–33; Andrei Sakharov, *Memoirs*, trans. Richard Lourie (New York: Knopf, 1990), pp. 97–7, 197–209, 215–18; David Holloway, *Stalin and the Bomb* (New Haven, Conn.: Yale University Press, 1994).

[27] Don K. Price, *The Scientific Estate* (Cambridge, Mass.: Harvard University Press, 1965). See also Michael Polanyi, "The Republic of Science," *Minerva*, 1 (1962), 54–73, and Harvey Brooks, *The Government of Science* (Cambridge, Mass.: MIT Press, 1968).

[28] Paul Boyer, *By the Bomb's Early Light* (New York: Pantheon, 1985).

were, in many cases, "elixirs of death." Second, reports on the Vietnam War showed that physicists were responsible for having created antipersonnel weapons and the electronic battlefield. Third, whereas the United States could find engineering solutions to put a man on the moon, it could not solve a wide range of more important problems, such as poverty. Americans had desired to put a man on the moon by the end of the 1960s not only for the purposes of science but also for the ideological purpose of demonstrating the superiority of the American system over the Soviet. There is also evidence that similar forces of technocracy had acted to push the American and Soviet space programs far beyond what political leaders initially believed was feasible.[29]

These factors contributed to the creation of public interest science – groups of citizens and scientist advocates who promote science in the human, not governmental, that is, largely military, interest. Still, physicists commanded substantial resources to conduct research in the big sciences of space, high energy physics, and nuclear power, failing only in the 1990s to secure funds for the Superconducting Super Collider and fusion reactors, although retaining billions of dollars for space. The failures were due largely to the end of the Cold War, rather than to change in the ideology that equated science with progress.

THE IDEOLOGICAL SIGNIFICANCE OF BIG SCIENCE AND TECHNOLOGY

Ideology is an important consideration, not only in theoretical physics but also in the technologies of physics research and the artifacts of engineering that result from the application of physical knowledge. Some big engineering projects have limited input from physics. But in many electrification, metallurgical, construction, and hydrological projects, research in solid state physics, material sciences, or geophysics finds its way quickly translated into practice. Nation-states support science and engineering directly in huge construction projects, not only at universities and institutes that carry out basic research. Finally, these projects have been the locus of significant concentrations of financial and manpower resources, and for all these reasons merit attention.

Some analysts argue that technology is value-neutral, serving the rational ends of achieving a desired outcome in "the one best way." The one-best-way distinction is crucial, for it implies that given any engineering problem, the solution will be based on universal engineering calculations that employ the scientific method: Rockets and jets the world over resemble each other because other designs would not fly. All hydroelectric stations, subways, bridges, and skyscrapers share essential materials, structural elements, and

---

[29] Walter A. McDougall, *The Heavens and the Earth* (New York: Basic Books, 1985).

components, or they would not stand. This in part explains why engineering, like science, can flourish in authoritarian regimes, even in the absence of concerned groups of citizens that may make it more socially responsible.[30]

Technologies are also symbols of national achievement. They demonstrate the prowess of a nation's scientists and engineers. They are central to national security strategies. They serve foreign policy purposes through technology transfer. Especially in the twentieth century, nations have embraced large-scale technologies as symbols of the legitimacy of the polity and economy of a particular nation. Technologies, thus, have what has been called "display value": social, cultural, and ideological significance, not merely imposing physical presence.

What distinguishes, then, large-scale technological systems in authoritarian regimes? First, the authoritarian state is the prime mover in technological development. The state harnesses the efforts of engineers and scientists to its programs for economic self-sufficiency and military might, shaping what areas merit study. In exchange for funding, experts are held accountable for producing results, often as specified in national planning documents. Failure to meet targets may trigger personal reprisals. Second, a highly centralized and bureaucratized system of funding and monitoring ensures accountability of the scientist and engineer to state goals. Third, technologies in totalitarian regimes are characterized by gigantomania, for example, Nazi armaments minister Albert Speer's plans for wide-gauge (four-meter) railroad tracks with two-story-high cars, or Stalin's seven Moscow skyscrapers that resemble gothic wedding cakes, or speed and distance records in aviation achieved at great risk by Soviet aviators in the 1930s.[31] Gigantomania often results in waste of labor and capital resources. In totalitarian regimes, projects seem to take on a life of their own, so important are they for cultural and political ends, as opposed to the ends of engineering rationality. Of course, it might be said that projects and bureaucracies everywhere seem to take on a life of their own, becoming institutions in search of a mission. But centralization of power in bureaucracies in totalitarian regimes enables one or a few institutes to gain the unassailable power to define scientific and engineering orthodoxy. Owing to this momentum, it is more difficult to derail economically unfeasible and environmentally dangerous projects than it would be in pluralist regimes.

To take one example, the Soviet Union embraced large-scale technologies as a means of converting a peasant society into a well-oiled machine of workers dedicated to the construction of communism. They believed that large-scale technologies would marshal scarce resources efficiently and provide the appropriate forum for the political and cultural education of a burgeoning working class. Soviet leaders had the utmost confidence in the

---

[30] Loren Graham, *What Have We Learned about Science and Technology from the Russian Experience?* (Stanford, Calif.: Stanford University Press, 1998).
[31] Kendall Bailes, *Technology and Society under Lenin and Stalin: Origins of the Soviet Technical Intelligentsia, 1917–1941* (Princeton, N.J.: Princeton University Press, 1978), chap. 14.

ability of technology to transform nature and bring freedom to Soviet citizens. Constructivist visions of the communist future found expression in Lenin's electrification, Stalin's canals and hydropower stations, Khrushchev's atomic energy program, and Brezhnev's Siberian river diversion project. There were glorious chapters in the history of large-scale technology in the USSR, including the pioneering conquests of the atom and the cosmos.[32]

Similarly, in Nazi Germany, Hitler desired immense monuments to his rule and the glory of the Third Reich. He called upon Albert Speer (1905–1981) to be chief architect behind the projects of the next millennium. Speer designed the Nuremburg fields and stadiums for military exercises as symbols of Nazi power. If completed, the fields would have filled more than six square miles. All the structures would have been much larger than the ancient stadiums of Athens. Hitler also ordered Speer to rebuild an Aryan Berlin, a project which, if completed, would have demonstrated the insignificance of the individual next to Nazi avenues and buildings. Speer designed the future Reich headquarters to hold nearly 200,000 persons standing.

THE NATIONAL LABORATORY AS LOCUS OF IDEOLOGY AND KNOWLEDGE

Display value and military designs came together in the major institutional innovation of the physical sciences in the twentieth century, the national laboratory, a crucial site for observing the interaction among the state, ideology, and science. The impetus for the national laboratory came from the government's interest in security issues or from business seeking competitive advantage in international markets, with government subsidies for expensive research. But scientists recognized the laboratory, with its relative ease of access to funding, expensive equipment, and talented researchers, as a means of pursing a variety of research ends. The national laboratory enables them simultaneously to undertake basic and applied tasks and frees them from obligations to teach, in order to have time to experiment. There is a melding of utopian constructivist vision, financial wherewithal, institutional support, and the logical development of a field, leading to research and engineering projects that are exceedingly difficult to curtail. Institutional momentum characterizes national laboratories no less than big science and technology generally.

The most famous national laboratories have been connected with the development of weapons of mass destruction. In the United States, they include Los Alamos National Laboratory, where the first atomic bomb was designed;

---

[32] Paul Josephson, " 'Projects of the Century' in Soviet History: Large Scale Technologies from Lenin to Gorbachev," *Technology and Culture,* 36, no. 3 (July 1995), 519–59, and "Rockets, Reactors and Soviet Culture," in Loren Graham, ed., *Science and the Soviet Social Order* (Cambridge, Mass.: Harvard University Press, 1990), pp. 168–91.

Oak Ridge National Laboratory, now a major facility in a variety of scientific fields, but originally built to separate uranium isotopes; and Lawrence Livermore National Laboratory, connected with the hydrogen bomb and the strategic defense initiative.[33] In the USSR, they are Arzamas-16, where Soviet physicists designed nuclear weapons; Chernogolovka, a physics research center; and Cheliabinsk, a nuclear fuel facility. The modern national laboratory has scores of laboratories, tens of thousands of employees, with sometimes hundreds of doctoral candidates writing dissertations based on a single experiment, and broad research vistas. National laboratories are large not only in size, multinational profile, and budget but also in the extent of their contacts with other areas of human activity (political, industrial, university), which requires them to engage in continuous political and social justification, an activity with significant ideological overtones.[34]

Among the first national laboratories were the Kaiser Wilhelm Institutes for Chemistry and for Physical Chemistry (later the Max Planck Institutes), which were intended to maintain scientific excellence in Germany. They officially opened in October 1912. The institutes survived the economic turmoil of the first years of the Weimar republic and the political interference of the Nazi years, including the dismissal of all Jewish employees.[35] They were involved in military efforts during World War II, but Nazi programs for atomic bombs and rockets were carried out in an ad hoc arrangement of institutions and scientists, under a variety of jurisdictions, and with vacillating attention of the leaders. In spite of being a poor weapon, the V-2 rocket, the first large guided rocket, foreshadowed the Manhattan Project and other big science of the postwar years as a paradigm of state mobilization in forcing the invention of new military technologies and the rise of a military-industrial complex.[36]

In the United States and the Soviet Union, teams of scientists were able to work largely without bureaucratic squabbles and wavering support on the part of political leaders toward the achievement of their goals in newly founded national laboratories.[37] In both cases, expenditures for military R&D grew significantly in the Cold War, tying researchers in universities and industrial laboratories to military projects through contracts, grants, or outright

---

[33] For an ethnographic study of Livermore Laboratory and the interplay of cultures and truths of weapons designers and antinuclear activists, see Hugh Gusterson, *Nuclear Rites* (Berkeley: University of California Press, 1996).

[34] Peter Galison and Bruce Hevly, eds., *Big Science* (Stanford, Calif.: Stanford University Press, 1992).

[35] Kristie Macrakis, *Surviving the Swastika* (New York: Oxford University Press, 1994).

[36] Speer, *Inside the Third Reich*, pp. 363–70, 409–10; Michael J. Neufeld, "The Guided Missile and the Third Reich: Peenemünde and the Forging of a Technological Revolution," in *Science, Technology, and National Socialism*, ed. Monika Renneberg and Mark Walker (Cambridge: Cambridge University Press, 1994), pp. 51–66; and Neufeld, "Weimar Culture and Futuristic Technology: The Rocketry and Spaceflight Fad in Germany, 1923–33," *Technology and Culture*, 31 (1990), 725–52.

[37] Among the outstanding works on the atomic bomb projects, see David Holloway, *Stalin and the Bomb* (New Haven, Conn.: Yale University Press, 1994); Robert Jungk, *Brighter Than a Thousand Suns*, trans. James Cleugh (New York: Harcourt Brace, 1958); Richard Rhodes, *The Making of the Atomic Bomb* (New York: Simon and Schuster, 1986); Walker, *National Socialism;* and Sherwin, *A World Destroyed*.

budgetary allocations. As demonstrated by Paul Forman, the enormously increased resources for basic research in the United States from 1945 to 1960 were intended primarily to increase the security of the United States, not to increase physicists' knowledge.[38] Similarly, in the 1980s under the banner of the Strategic Defense Initiative (so-called Star Wars), the United States government provided billions of dollars for solid state, laser, computer, and other physics research, which tied tens of thousands of researchers to foreign policy, pushing them toward technique and away from basic science.

Some analysts claim that these national laboratories needlessly diverted funds away from important fields of fundamental research and away from important areas of human activity, such as medicine, environment, and education. The ability of laboratory officials and scientists in national laboratories to open new areas of research divorced from the original purpose has contributed to their growth and longevity. For example, Oak Ridge National Laboratory, originally established for isotope separation and then adapted to reactor development, more recently found it lucrative to move into nuclear medicine and genomic research. The continued funding of these laboratories and agencies at high levels indicates their importance for national military and ideological ends.

Most scholars agree that some universalist ethic infuses science. But as the cases of twentieth-century Soviet, German, and American physics indicate, national science or nationalism in science also exists, as investigation of the relationship among science, ideology, and the state shows. For the Soviet and German citizen, the state insisted upon specific views of science and its place in the ideological superstructure. Proletarian science in the Soviet Union and Aryan science in Nazi Germany shared an essential belief that national science tied to state-determined goals was the only true science, and that national science was superior to the science practiced by members of the international scientific community as measured by methodology, philosophical implications, and research emphasis. The treatment of such dissidents as the physicists Andrei Sakharov and Iuri Orlov indicates the extent to which the power of scientists was limited.[39] Yet science in those totalitarian regimes often moved ahead because of state support, the quality of scientific organizations, and the desire of most scientists to steer clear of politics for research, even if it lagged behind other nations according to traditional measures of scientific excellence, such as publication in refereed journals, peer review, grant

---

[38] Paul Forman, "Behind Quantum Electronics: National Security as Basis for Physical Research in the United States, 1940–1960," *Historical Studies in the Physical and Biological Sciences*, 18 (1987), 149–229, and "Into Quantum Electronics: The Maser as 'Gadget' of Cold-War America," in *National Military Establishments and the Advancement of Science and Technology*, ed. Paul Forman and J. M. Sanchez-Ron (Dordrecht: Kluwer, 1996), pp. 261–326.

[39] Sakharov, *Memoirs,* and Yuri Orlov, *Dangerous Thoughts,* trans. Thomas P. Whitney (New York: William Morrow, 1991).

applications, scientific citations, or membership in national and international scientific organizations.

Ideology is also important in shaping the physical sciences in pluralist regimes, such as the United States, where the normal practice is for scientific disputes to be aired openly, although egos may be bruised in the process. Competition between schools of research inspires the confidence of scientists everywhere that they are establishing "facts" independent of political or personal issues.[40] Granted, in pluralist systems, scientists may go outside of their normal professional channels to air disputes in the political arena. Controversies over the morality of the bomb, the definition of wetlands, fetal tissue research, and the workings of Star Wars antimissile technologies demonstrate that political, ethical, ideological, and economic forces shape scientific debates. But in totalitarian regimes, there are taboo subjects proscribed primarily, if not solely, by ideological, not ethical, considerations. Researchers who venture into those areas risk job security and personal freedom. Individual scientists, ideologues, and administrators gain the power to define what is "good science" in a way that limits academic freedom.

Ideology also helps determine the position that specialists maintain between the public and government. In all systems, the public holds scientists in esteem for their contributions to the understanding of nature and the improvement of the quality of life. Occasionally, this esteem is tempered by technological advance resulting in the displacement of workers; immoral or unethical activities, such as uninformed and hurtful research on human or animal subjects; and the unintended, unanticipated, or wrongly ignored consequences of science, such as radioactive fallout and mutagenic results of drugs like thalidomide or food additives. Scientists themselves generally see a kind of universalism operating in science, that is, the pursuit of the "truth," which enables them to communicate with one another toward the ends of peace and rationalism when governments interfere. Yet ideology and state politics present barriers to internationalism and rationalism.

---

[40] Polanyi, "Republic of Science."

31

COMPUTER SCIENCE AND THE COMPUTER REVOLUTION

William Aspray

Until the mid-1950s, the word "computer" commonly referred to a woman employed in operating a calculating machine in a business office or a scientific calculating laboratory. With the invention in 1945 of the stored-program computer, several months after the Second World War ended, and with the publicity surrounding the introduction in 1952 of the first commercial computer (the Universal Automatic Computer, or UNIVAC), the word computer became associated with a machine, rather than a human.

This machine had three attributes that rendered prior calculating technologies obsolete in less than two decades. The electronic switching of its components eventually made the computer billions of times faster than its mechanical ancestors. The digital storage of information enhanced precision to practically unrestricted levels. The stored program capability, that is, the ability to store instructions as well as data inside the machine and to have the machine process those instructions during the course of a computation without human intervention, had two advantages: First, it enabled almost any computer to be used as a universal machine, in other words, to carry out virtually any computation possible by a machine. Second, stored programming was critical to the automation of the computational process, so that the overall speed of computation could reflect the electronic speed of the components.

COMPUTING BEFORE 1945

The first calculators were built in the seventeenth century by natural philosophers – the three most famous were designed by Gottfried Wilhelm Leibniz (1646–1716) and Wilhelm Shickard (1592–1635) for scientific uses, and by Blaise Pascal (1623–1662) for accounting purposes. These were desk calculators, mechanical devices that could be placed on a desk and used to do addition and subtraction, and sometimes multiplication and division.[1] Until the second half of the nineteenth century, such calculating devices were cabinet curiosities. They were custom-made in small numbers, they did not work well, and they were not gainfully employed for either scientific work or business. In the last quarter of the nineteenth century, a number of technical improvements were introduced: Reliable mechanisms were perfected for handling carries in addition, easy entry of numbers, and printing of results. Calculators began to be mass-produced and introduced into businesses.[2] By the 1920s thousands of desk calculators were being employed around the world in many businesses and in a few scientific enterprises. Electromechanical relays, developed for the telephone industry, were adapted for use in the higher-end desk calculators of the 1920s and 1930s to increase their speed.

Desk calculators were by no means the only early calculating devices. Another line of development was the punched-card tabulating systems developed by the American inventor Herman Hollerith (1860–1929) for processing the returns from the 1890 U.S. census. These tabulating systems were continually improved over the next fifty years and used in high-end businesses, as well as by government agencies. They began to be used occasionally in the 1930s for statistical and astronomical research, most notably at Columbia University to calculate tables of the motion of the moon.

A third line of machine development was the analog device, which measured, rather than counted, the results.[3] (The slide rule is an example of an analog device in that you gain the results of a multiplication by measuring the position of the slide against the fixed part of the rule.) One important class of analog machines was the tide predictor, which was used throughout the British empire in the late nineteenth century to calculate the height of tides at a given time and place. Lord Kelvin (1824–1907) devised one of the most successful tide predictors. The other name generally associated with analog calculating devices is Vannevar Bush (1890–1974), who built a number of devices at MIT in the 1920s and 1930s, primarily for use in designing networks and equipment for the electric power industry.[4] Analog devices were the calculating device of choice for engineering, which became a much more mathematical discipline in the 1890s, with the development of alternating current systems. There were two reasons for this: First, analog devices were better suited than desk calculators or punched-card tabulators for the continuous-variable problems that engineers commonly had to solve;

---

[1] Peggy A. Kidwell and Paul E. Ceruzzi, *Landmarks in Digital Computing* (Washington, D.C.: Smithsonian Institution Press, 1994).
[2] James W. Cortada, *Before the Computer* (Princeton, N.J.: Princeton University Press, 1993).
[3] Allan G. Bromley, "Analog Computing Devices" in *Computing Before Computers*, ed. William Aspray (Ames: Iowa State University Press, 1990), pp. 156–99.
[4] Karl L. Wildes and Nilo A. Lindgren, *A Century of Electrical Engineering and Computer Science at MIT, 1882–1982* (Cambridge, Mass.: MIT Press, 1985).

and second, engineers felt more comfortable building analog equipment and using it to measure results, rather than learning how to reduce their problems to mathematical equations and using mathematical algorithms to solve them on a desk calculator.

The last major kind of calculating device prior to the computer was the scientific calculator.[5] A small number of one-of-a-kind devices of this sort were built in the 1930s and early 1940s. The most important were built by Konrad Zuse (1910–1995) in Germany for aeronautical engineering (destroyed during the war), by Howard Aiken (1900–1973) at Harvard with generous assistance in engineering and financing from IBM and used in the Allied war effort, and by George Stibitz (1904–1995) at Bell Telephone Laboratories for internal use. These were intended to be high-performance machines for automating large numbers of arithmetical calculations, which they accomplished with either electromechanical or much-faster electronic switching elements. They were calculators, in contrast to computers, because they were unable to store their instructions or modify the course of a calculation on the basis of intermediate results without human intervention.

While prior to 1945 there had been a few examinations of the history of calculation and calculating machines, this subject took on new meaning once computers became available and their value to science and commerce became apparent.[6] There have basically been two approaches: In the first, computing practitioners, and later professional historians, studied calculating machines that were regarded as antecedents to the computer. These early historical studies identified (and sometimes forced) genealogical lines of calculating machines leading to the modern computer – and, unfortunately, sometimes neglected developments that did not lead directly to the computer. In these studies, Charles Babbage (1792–1871) was prominently mentioned for his role as the inventor of a machine – never completed – known as the Analytical Engine, a mechanical device functionally similar to the stored-program computer. In this line of historical analysis, the technical features – whether inchoate or fully fashioned – receive more importance than the dissemination and use of the technology. While this kind of scholarship is unfashionable today among historians, much can be learned from the best of it about the design of early calculating technologies.[7]

A historical reexamination of the pre-1945 era was undertaken during the 1980s and continues today. This second approach had its origins in the important reinterpretation of business history by Alfred Chandler. In documenting the rise of big business in the late nineteenth and early twentieth

---

[5] Paul E. Ceruzzi, *Reckoners: The Prehistory of the Digital Computer, from Relays to the Stored Program Concept, 1935–1945* (Westport, Conn.: Greenwood Press, 1983).
[6] E. M. Horsburg, ed., *Handbook of the Napier Tercentenary Celebration or Modern Instruments and Methods of Calculation* (original ed., Edinburgh: G. Bell and Sons and the Royal Society of Edinburgh, 1914; reprint, Los Angeles: Tomash Publishers, 1982; now distributed by MIT Press).
[7] Michael R. Williams, *A History of Computing Technology*, 2d ed. (Los Alamitos, Calif.: IEEE Computer Society Press, 1997).

centuries, Chandler described how ownership and management were separated and office machines introduced as part of the professional tool set of managers. In the *Control Revolution,* James Beniger was the first to examine how calculating machinery was involved in this Chandlerian revolution in business. Since then, JoAnn Yates and Martin Campbell-Kelly have given careful historical analysis of important case studies in which nonmachine procedures as well as machinery were used to meet the information needs of large enterprises, such as insurance agencies, bank clearing houses, and telegraph offices; while James Cortada has examined the history of calculating machines in the context of other business machines and business machine manufacturers.[8] In the light of this research, it is easy to understand how the desk calculator came to be mass-produced at about the same time as the typewriter, the cash register, and the dictating machine. This research also suggests why the business machine manufacturers, such as International Business Machines (IBM), Remington Rand, Burroughs, and National Cash Register (NCR), were among the successful entrants into the computing manufacturing industry in the 1950s.

DESIGNING COMPUTING SYSTEMS FOR THE COLD WAR

The story of the creation of the computer has been told many times.[9] A project was undertaken at the University of Pennsylvania during the Second World War to build an electronic calculating device, the ENIAC (Electronic Numerical Integrator and Computer), to calculate firing tables needed for directing the operation of new guns and shells. ENIAC was not completed until a few months after the war ended, and it originally lacked full-fledged stored programming capability. Nevertheless, it was extremely important to the future of the field. Its successful completion in 1945 convinced the government and the scientific community of the feasibility and value of computers. The follow-on EDVAC (Electronic Discrete Variable Automatic Computer) project, to remedy some of the shortcomings of the ENIAC design, led in 1945 to the description of the stored-program concept that was embodied in all subsequent computers.

Dozens of projects to build computers emerged during the postwar decade. Among the first to come to completion were those at Manchester University and the British National Physical Laboratory, under the intellectual direction

[8] James R. Beniger, *The Control Revolution* (Cambridge, Mass.: Harvard University Press, 1986); JoAnn Yates, *Control Through Communication* (Baltimore: Johns Hopkins University Press, 1989); Martin Campbell-Kelly, "Large-Scale Data Processing in the Prudential, 1850–1930," *Accounting, Business, and Financial History,* 2 (1992), 117–39.
[9] Nancy Stern, *From ENIAC to UNIVAC: An Appraisal of the Eckert-Mauchly Computers* (Bedford, Mass.: Digital Press, 1981); William Aspray, *John von Neumann and the Origins of Modern Computing* (Cambridge, Mass.: MIT Press, 1990); Martin Campbell-Kelly and William Aspray, *Computer: A History of the Information Machine* (New York: Basic Books, 1996).

of M. H. A. Newman (1897–1984) and Alan Turing (1912–1954), who had used electronic machines to break codes during the Second World War; and at Cambridge University, under Maurice Wilkes (b. 1913), who was inspired by lectures he had heard in Philadelphia about EDVAC at a summer-school course in 1946 that included many of the early computer designers. In the United States, many of these early computers were built with government, mainly Cold War, funding for military applications. The most important was the Whirlwind computer built at MIT, which was the starting point for the computer-driven SAGE (Semiautomatic Ground Environment) missile-defense system.[10] Whirlwind introduced numerous technical innovations, and it demonstrated the possibility of building in the reliability and speed required to do real-time calculation. This was the first computer that could process and respond to data as quickly as it was received from the outside "real" world. In the case of SAGE, data came from radar and observer stations for the location, direction, and speed of aircraft, so as to be able to track and intercept unfriendly aircraft as they flew into U.S. air space. (Real-time computing was later critical to such applications as controlling manufacturing processes and operating airline reservation systems.) This period of one-of-a-kind computers built by users ended in the mid-1950s, when an industry arose to manufacture standardized and custom computer equipment.

Widespread experimentation in the design of the overall system and its various components occurred during this first postwar decade. The most urgent problem – solved eventually by magnetic core devices developed for the Whirlwind – was a memory device that could store large amounts of information reliably for long periods of time, that was economical to build and maintain, and that could retrieve or store data rapidly enough not to slow the overall operation of the computer.

Although the basic design of computers had stabilized by the beginning of the commercial computer era, extraordinarily rapid-paced innovation of two types has continued until the present day. First, component innovations made by companies in the computer and semiconductor industries over the next forty years enabled increases in speed, reliability, and storage capacity and decreases in energy consumption, size, and cost – improvements of more than a million-fold on each of these measures. Second, changes in the mode of operation of computers originated primarily through research conducted in academic laboratories supported by government funding, although the refinement and dissemination of this research in commercial products came primarily through the industrial sector. High-level programming languages, real-time computing, time-sharing, networking, and graphical user interfaces are important examples of innovations originating in the academic sector that were disseminated by the industrial sector.

[10] Kent C. Redmond and Thomas M. Smith, *Project Whirlwind* (Boston: Digital Press, 1980).

While the original emphasis was on building computer hardware, over time, software became increasingly important. It is not coincidental to the steadily growing importance of software that the hardware company IBM was the dominant company in the computer industry of the 1960s and 1970s, whereas the software company Microsoft became dominant in the 1990s. Software written by users to carry out a specific application can best be studied by historians as part of a study of users and applications. But another aspect of software history – about making computers usable – deserves mention here. Although it is not generally seen in this way, the supply-side history of software can be regarded as a process to automate the use of computers – compilers that enable the computer, rather than the human user, to determine how to execute programs; debugging tools to automate the search for syntactical errors; operating systems to manage the storage of information inside the machine; and programming languages to make the machine speak in a human-like language.

Another major theme in the history of software is the search for methods to manage the complexity of large, real-world software development. In the first half of the 1960s, the memory size and processing speed of computer hardware increased tenfold, making possible much more powerful and complex programming projects. The problem, labeled the "software crisis" at a North Atlantic Treaty Organization conference in 1967, was that techniques for writing software had not improved nearly as rapidly as hardware development. As a result, especially in large applications, software development schedules slipped repeatedly, costs skyrocketed, errors were hard to locate and correct, and revisions of the software were hard to implement. In order to write the operating system for the 360 family of computers, which involved more than a million lines of program code, IBM decided simply to throw more programmers at the task. As Fred Brooks (b. 1931), head of this operation, related in *The Mythical Man-Month,* additional programmers added during the course of the project increased communication and management problems, which actually slowed the writing of the software.[11] Beginning in the 1970s, there was an effort to develop a field of software engineering to manage such complexity.[12] Many different tools and working practices were suggested, such as structured design, formal methods, and development models. None has been a panacea.

The history of computers is closely related to the history of electronics, especially of semiconductors. These two industries have driven each other since the 1960s. The computer industry has been the largest consumer of semiconductors, and it has greatly influenced the research and development agendas of the chip manufacturers. Innovations in semiconductors have led to

---

[11] Frederick P. Brooks, Jr., *The Mythical Man-Month* (Reading, Mass.: Addison Wesley, 1975).
[12] William Aspray, Reinhard Keil-Slawik, and David L. Parnas, *The History of Software Engineering* (Dagstuhl Seminar Report 153) (Schloss Dagstuhl, Germany: Internationales Begegnungs- und Forschungszentrum für Informatik, 1996).

extraordinary price-performance advances in computers. Transistors brought the price, size, and reliability of computers to a point at which they were affordable not only by government and large business organizations, but also by medium and eventually small businesses. The integrated circuit made possible the minicomputer and the supercomputer, thus expanding the realm of computer usage. The continued decrease in scale of circuitry on a chip led in the early 1970s to the microprocessor, the computer on a chip, and so to the advent of the personal computer and the embedding of computers into all kinds of industrial and domestic products.

It is widely acknowledged that the federal government, especially the military and the energy laboratories, were critical to the development of computers. Enormous sums were spent on computerized missile-defense programs, computer networks for tying together military researchers and military organizations, computer simulations for atomic weapon design, computer-aided design of military aircraft, computerized logistical systems for the coordination of troops and supplies, and so forth. Various military and other government organizations supported computer research and development, both directly with grants and contracts to university researchers and indirectly by providing a secure market for the products of the computer manufacturers. Nevertheless, the role of the military in shaping computing technology continues to be debated today. Some scholars, such as Paul Edwards, regard the influence as profound.[13] Others believe that government support simply gave academic researchers the freedom to pursue research programs that were already of interest to them.

Recent developments in computing open an entirely new set of questions for the historian. The personal computer and the Internet had their origins in technological developments of the late 1960s, but they did not come into their own until the 1980s. The computer phenomenon of the 1980s and 1990s was markedly different from that of the early decades. There continued to be mainframe, mini-, and supercomputers – as well as operating systems, programming languages, and applications software for them. But what captures the historian's interest in this recent period is not the continuation of this old style of computing, but the new styles associated with the personal computer and the Internet: the rapid pace of innovation, the tumultuous but extraordinary economic opportunities and challenges, and the personalization of computing.

BUSINESS STRATEGIES AND COMPUTER MARKETS

If the most studied aspect of the history of computing is the technology itself, the second most studied aspect is the business history of the computer

---

[13] Paul N. Edwards, *The Closed World* (Cambridge, Mass.: MIT Press, 1996).

manufacturers, software manufacturers, and secondary computer industries, such as service providers and manufacturers of peripherals.[14] The supply side of computing – both technological and business aspects – has been investigated much more extensively than the demand side.

In the second half of the 1950s, a computer industry began to crystallize, selling large, stand-alone computers (mainframes). A company that entered this business generally had one of four backgrounds: It was a start-up firm organized by people with engineering backgrounds (Control Data, Digital Equipment); or it was an established firm from one of three industries: business machine manufacturers (IBM, Remington Rand, NCR, Burroughs), electrical equipment manufacturers (General Electric [GE], Honeywell, Radio Corporation of America [RCA], Philco, Sylvania), or defense contractors (Ramo Wooldridge, Texas Instruments). The business machine companies were by far the most successful, owing generally to the fact that they had a ready market and established relations with customers who became the purchasers of their computers.

Most of the industry's history until the mid-1980s can be seen as actions taken by IBM, which at times held as much as 80 percent of the market, and responses from other companies to them. IBM had a profitable tabulating equipment business that was its main source of revenue until the 1960s, and so it moved more slowly into the mainframe business than some other companies, notably Remington Rand (later Sperry Rand, now Unisys). Even so, by the end of the 1950s, IBM had become the market leader in computers.

The most important event in the computer industry prior to the personal computer was IBM's development in the early 1960s of the 360 family of computers. Prior to that development, each computer had its own software and peripherals, which were generally incompatible with those for other computers, even if the computers were manufactured by the same company. For the manufacturers, it meant that development, support, repair, and maintenance costs for the plethora of architectures, programming languages, printers, and other peripherals were very expensive. For the users, it meant major trouble and expense in running their software and maintaining their data when their computers became old or their needs outgrew them. System 360 changed everything about the business, promising a fully compatible family of computer products, with standardized software and peripherals – and the company mostly delivered on this promise, although it spent an unprecedented amount of money ($500 million) in the development process.

A major restructuring of the industry took place in the wake of the 360 announcement. No company could survive in other than a market niche if it did not sell its own entire line of compatible computers, peripherals, and

[14] Franklin M. Fisher, James W. McKie, and Richard B. Mancke, *IBM and the U.S. Data Processing Industry* (New York: Praeger, 1983); Emerson W. Pugh, *IBM: Shaping an Industry and Its Technology* (Cambridge, Mass.: MIT Press, 1995).

software; but neither could it survive unless it differentiated itself from IBM. One method of differentiation was to develop a line of computers, perhaps designed to appeal to some particular application areas, that were compatible within the family but incompatible with the IBM products. This is what GE attempted, trying to take advantage of IBM's weakness in time-sharing. RCA employed the different strategy of differentiating on price. It used reverse engineering to build products compatible with the 360 family, but offered its products with comparable performance at lower prices than IBM products. Both strategies were fraught with peril, and a number of companies, including RCA, GE, and other powerful competitors, succumbed. Highly profitable secondary computer industries sprouted to provide IBM-compatible software and peripherals. And in the following decade of the 1970s, a major niche industry for minicomputers, affordable to the individual researcher or to a small business, came of age.

Another wholesale transformation came about with the development of the personal computer (PC). The first of these machines appeared in the late 1970s. They were made possible by the invention of the microprocessor (the central processor on a silicon chip), developed at Intel in 1970 and independently elsewhere a little later. IBM was not among the first companies to manufacture personal computers – this was done principally by small, undercapitalized start-ups – but IBM served the microcomputer industry well by legitimating the personal computer as a product that the business world could trust and buy when it came out with its own PC in 1980.

Although rapid innovation in personal computer hardware continued, with the incorporation of larger memories and ever-more-powerful microprocessors each year, the hardware business stabilized into a mature appliance business, much like television sets, by the late 1980s. The majority of the action, and the profits, were in the personal computer software business. This industry, at first comprising almost entirely start-ups, established barriers to entry based on the large number of lines of code in its application programs and on extensive marketing operations. As a result, there was a shakeout of the industry into a few major players (Microsoft, Novell, Lotus, etc.) holding 80 percent of the business. This industry grew so much and so quickly that Microsoft surpassed the profitability of IBM, which did not fare so well because its traditional mainframe business was being undermined by personal computers. The success of the personal computer is personified in Bill Gates, the principal founder of Microsoft, who became one of the world's wealthiest individuals, as Microsoft dominated the market by skillfully exploiting its monopolistic advantages – every bit as forcefully as IBM did in its heyday.

On the international scene, the story has also been largely one of reactions to IBM and the development and protection of indigenous national computer industries. In the 1960s and 1970s, the French government regarded IBM as a national enemy. The British government strong-armed a series of mergers of its computer manufacturers, hoping to build a company with sufficient scale to fight off IBM.[15] The Brazilian government established an indigenous national company industry with strict import rules, but it has not been able to keep up with the pace of innovation. The Japanese have succeeded perhaps best of all against the American companies, mainly by using legal and cultural means to protect a strong home market.[16]

IBM was fond of saying that it was not in the business of selling computers but, rather, of selling solutions. IBM had learned to operate in this manner in the 1920s and 1930s, when it was installing punched-card tabulating systems into businesses. The punched-card systems required that the customers reorganize their way of doing business, and IBM became adept at learning the operations of their customers' businesses and how to "rerationalize" them with IBM equipment at the center of those operations. This knowledge of a customer's business differentiated IBM from many of its competitors, who focused simply on building machines.

It is clear both that computers had to be integrated into a work environment to be effective, and that the needs of users shaped the computers that the computer manufacturers built. Perhaps the best historical study along these lines has been that of Donald MacKenzie, who has shown how scientists at the Los Alamos and Livermore laboratories had a voice in the design of commercially manufactured supercomputers, which resulted in their being designed to maximize use in atomic weapons simulation.[17] In another interesting study, Jan van den Ende has historically examined how computer technologies were incorporated into typical work environments in the Netherlands between 1900 and 1965, showing the wide variation in computing technologies employed, functions automated, adaptation to work culture, and organization of labor, as one moved from scientific calculations to data processing to production control.[18] Until more historians pay attention to the demand side of computing, and its interaction with the supply side, we will have incomplete and one-sided knowledge of the computerization of society.

COMPUTING AS A SCIENCE AND A PROFESSION

In 1967, when an office for computer science was established in the National Science Foundation, a foundation official quipped, "What next, car science?" Indeed, there have been many questions raised inside and outside the computing profession about the nature of computing as an intellectual discipline. Is it merely a discipline that studies and builds machines? When faced with

---

[15] Martin Campbell-Kelly, *ICL: A Business and Technical History* (Oxford: Oxford University Press, 1989).
[16] Marie Anchordoguy, *Computers Inc.* (Cambridge, Mass.: Harvard University Press, 1989).
[17] Donald MacKenzie, "The Influence of the Los Alamos and Livermore National Laboratories on the Development of Supercomputing," *Annals of the History of Computing*, 13 (1991), 179–201.
[18] Jan van den Ende, *The Turn of the Tide* (Delft, Netherlands: Delft University Press, 1994).

this same question in the nineteenth century, the mechanical engineers answered that their field was more than the study of mechanical devices, that it was rooted in the physical laws of thermodynamics. A similar answer has been given by some computer scientists, who claim that their field is rooted in the physical laws of information (which, as Claude Shannon [b. 1916], Norbert Wiener [1894–1964], and others have shown, are closely related to the laws of thermodynamics).

Computer science can be seen as an amalgam of three or four intellectual traditions, with courses representing each of these traditions often taught in a single university department. There are purely mathematical studies, with their roots in mathematical logic, concerning such theoretical topics as the nature of computability and complexity.[19] There is a distinct engineering tradition of building computer hardware. There is also an experimental science tradition, closely associated with what Herbert Simon (b. 1916) has called the "science of the artificial." Artificial intelligence researchers, for example, build artifacts in the laboratory to test theories of learning, speech recognition, vision, and so forth.[20]

A fourth possible type of computer science study, software engineering, is harder to classify; in some ways, it is akin to a business and management school discipline, seeking out methods for organizing large teams of people to work together effectively to produce results answering to certain specifications involving reliability, maintainability, and cost. In another way, software engineering is a kind of engineering discipline. To emphasize this point of view, some practitioners have consciously adopted terminology from the organization of factory work: the production line, the software manufactory, the clean room, and others. But as Michael Mahoney has observed, these allusions to historical manufacturing processes, such as Fordism, are rife with historical inaccuracies.[21] Software engineers themselves continue to debate what their field is about.

The battle to define computing as a science, or merely as a service to the scientific and engineering (and business) communities, has been fought out in professional societies, funding agencies, and, most of all, universities. The potential of the computer was recognized early, and most of the major U.S. professional computing societies had their origins in the 1950s when there were only a few working computers in the United States. These included the Association for Computing Machinery (ACM), which was an ill-chosen name for this society of mostly academics, many with mathematical orientations; the Society of Industrial and Applied Mathematics (SIAM), which was interested mainly in numerical analysis, scientific computation,

---

[19] Michael S. Mahoney, "Computer Science: The Search for a Mathematical Theory," in *Science in the Twentieth Century*, ed. John Krige and Dominique Pestre (Amsterdam: Harwood, 1997), chap. 31.
[20] Pamela McCorduck, *Machines Who Think* (San Francisco: W. H. Freeman, 1979).
[21] Michael S. Mahoney, "The Roots of Software Engineering," *CWI Quarterly*, 3, no. 4 (1990), 325–34.

and industrial applications of computers; the computing committee of the American Institute of Electrical Engineers (AIEE), which was comprised mainly of electrical engineers interested in analog computers for electric power applications; and the computing group of the Institute of Radio Engineers (IRE), which was mainly a group of electronics engineers who were interested in engineering aspects of computer design. The AIEE and IRE merged in 1963 to form the Institute of Electrical and Electronics Engineers (IEEE), and the Computer Society established by the IEEE is today the largest professional computing organization in the world, with approximately 100,000 members. In the 1950s, each of these groups held meetings and established conferences, which were the main distribution channels for technical information about computers. Over time, as their memberships grew and the field expanded, special-interest groups and specialized journals appeared. ACM and the IEEE Computer Society today have dozens of journals, most with an international authorship and readership.

The National Bureau of Standards and the Office of Naval Research were early supporters of computing research in the United States, and in recent years, the National Air and Space Administration, the National Institutes of Health, and other organizations have provided significant funding. The main two supporters, however, were the National Science Foundation (NSF) and the Defense Advanced Research Projects Agency (DARPA). DARPA provided large, team-oriented, multiyear grants to a select few universities to develop breakthrough technologies that might have value to the Department of Defense.[22] The budgets were sufficiently large to carry out the strategy successfully, and DARPA-sponsored research produced major advances in time-sharing, networking, artificial intelligence, and computer graphics. The interest of NSF in the health of the entire scientific community led it to a somewhat different mix of programs.[23] It ran a facilities program from the late 1950s to the early 1970s that brought the first computers to hundreds of U.S. colleges and universities. NSF's research grant program awarded smaller grants to many researchers distributed across the entire academic sector; it did not focus on a few areas but spread its efforts across a spectrum of topics suggested by proposers and vetted as good science through peer review. In the 1980s, NSF established a program to introduce experimental research to the many computer science departments that could not afford it. This program

[22] Arthur L. Norberg and Judy E. O'Neill, *Transforming Computer Technology* (Baltimore: Johns Hopkins University Press, 1996).
[23] William Aspray, Bernard O. Williams, and Andrew Goldstein, "The Social and Intellectual Shaping of a New Mathematical Discipline: The Role of the National Science Foundation in the Rise of Theoretical Computer Science and Engineering," in *Vita Mathematica: Historical Research and Integration with Teaching*, ed. Ronald Calinger (M.A.A. Notes Series) (Washington, D.C.: Mathematical Association of America, 1996); William Aspray and Bernard Williams, "Arming American Scientists: The Role of the National Science Foundation in the Provision of Scientific Computing Facilities," *Annals of the History of Computing*, 16, no. 4 (Winter 1994); William Aspray and Bernard Williams, "Computing in Science and Engineering Education: The Programs of the National Science Foundation," *IEEE Electro / 93 Proceedings* (1993), 234–40.

deserves major credit for retaining faculty and students in the university in the face of tremendous opportunities in industry for personal wealth and better research facilities.

Many of the first computers in the United States were built at universities. It was not until the mid-1950s that a computer industry was established and universities started buying, rather than building, their own computers. In the late 1940s and 1950s, several universities (notably Harvard, Carnegie Mellon, and Michigan) established interdisciplinary graduate programs in computing. The first PhD program formally in computer science was established at Purdue in 1962. Others soon followed, and by the mid-1970s, there were programs at 100 universities. In most cases, computer science evolved from existing programs in either mathematics or electrical engineering. Departments emerging from mathematics, such as Stanford's, generally had a strong program in numerical analysis or in logic-oriented, theoretical computer subjects, such as complexity theory. Many mathematicians were skeptical, however, about the value of computing and generally had other priorities for spending the large sums of money required to support a computer and computing program. Electrical engineering departments were generally more supportive because they had an interest in computers as both designers and users, and because they were accustomed to the large costs and organizational aspects of laboratories with capital-intensive equipment. The electrical engineering groups were naturally interested in the design of computers (circuit design, computer architecture, etc.), but they were also interested in a number of theoretical subjects (systems theory, control theory, information theory, fuzzy logic). On some campuses, computing emerged in multiple places. At one time, Michigan had five computing programs; and on the Berkeley campus, there was a major battle for resources until the administration forced the computing group affiliated with the mathematics department to become part of the computing group in electrical engineering.

How have the four approaches to computer science research – mathematical studies, hardware engineering, artificial intelligence, and software engineering – fared over time? Until the 1980s, the number of mathematically oriented theoreticians in a computer science department correlated strongly with the reputation of the department. This is presumably an artifact of the high premium placed on a hard scientific core while the field was fighting for scientific recognition on campus, in the funding agencies, and in the National Academy of Science. Since then, however, the engineering aspects of computing have become increasingly important. One indicator is the number of departments that have renamed themselves as "computer engineering" or "computer science and engineering" departments. The reception of that portion of computer science that studies the "sciences of the artificial" has varied over time. Researchers received generous funding in the 1950s and 1960s for machine translation of language and speech recognition; but in the 1970s, this research and the field of artificial intelligence generally came under assault for its inability to deliver on the promises of the early years, and much of the federal support was abruptly terminated. In the 1980s and 1990s, expert systems and neural nets have proved to be of practical value, and there has been renewed interest in this field. Practical applications have attracted most of the funding – to the dismay of researchers who are more interested in learning about the fundamental nature of intelligence. Software engineering is perhaps the most controversial because the methodologies developed have generally not been scientifically rigorous, and there is much skepticism about their efficacy in solving the "software crisis." Only one top-ranked computer science department in the United States, at Carnegie Mellon, has invested heavily in software engineering faculty, while some of the leading departments at other universities have avoided appointments in this area, skeptical of its value.

OTHER ASPECTS OF THE COMPUTER REVOLUTION

Computing, like electronics and biotechnology, is one of the technical fields that have been most closely associated with Silicon Valley (near San Francisco) and Route 128 (near Boston). These areas have thrived as places for high-tech development because of their proximity to major research universities, the presence of a highly skilled workforce, the proximity of specialty suppliers and services, the development of new kinds of funding (venture capital) that understands how the high-tech firms operate, and the opportunities for the technical workforce to remain fluid by changing jobs without relocation. Certain technical niches have been concentrated in particular regions, such as minicomputers along Route 128, and some even believe that there are characteristic styles of computing associated with these different regions.[24] Geographers and historians, such as Anna Lee Saxenian and Bill Leslie, have made important studies of the geographic regions of high-tech development.[25]

The computer has had a significant impact on mathematics.[26] For example, numerical analysis was brought back to life after many decades of stagnation, as the problems and methods of numerical analysts were all reconsidered. The classical interest in truncation errors was abandoned, methods for solving linear systems and inverting matrices were reevaluated on how efficiently

---

[24] See, for example, Richard Sprague, "A Western View," *Communications of the ACM*, 15 (July 1972), 686–92.
[25] Anna Saxenian, *Regional Advantage: Culture and Competition in Silicon Valley and Route 128* (Cambridge, Mass.: Harvard University Press, 1994); Stuart W. Leslie, *The Cold War and American Science* (New York: Columbia University Press, 1993).
[26] William Aspray, "The Transformation of Numerical Analysis by the Computer," in *History of Modern Mathematics*, vol. 2, ed. John McCleary and David Rowe (Boston: Academic Press, 1990); "The Mathematical Reception of the Computer," in *Studies in the History of Mathematics*, ed. E. R. Philips (Washington, D.C.: Mathematical Association of America, 1987), pp. 166–94.

they could be implemented on computers, round-off errors took on a new importance, and new investigations were made into nonlinear phenomena and partial differential equations. Computers began to be used by mathematicians to test critical cases of general problems so as to gain intuition that would help them in their search for a general solution. Computers were also used to resolve numerous cases in large proofs, such as in the 1976 proof for the famous Four Color Problem, which asserted that four colors suffice to color a map so that no two contiguous countries are of the same color.

The impact of the computer on other scientific disciplines was equally profound.[27] Computers were used to control laboratory equipment, collect and analyze massive amounts of test data, and portray results visually. Chemists used computers to organize and keep track of the numerous chemical molecules and compounds. Nuclear physicists, aerospace engineers, and others use computers for modeling in cases where testing is too expensive, too dangerous, or otherwise not possible. Computers have enabled physicists to introduce a new set of nondeterministic approaches, such as Monte Carlo methods for studying subatomic particles.

The strong images that are evoked when we think of hackers, nerds, young entrepreneurs working in their garages, and ordinary people's fascination with the Internet suggest the strength of the computer as a cultural force. Enormously popular and revealing studies have been written by journalists, such as the one by Steven Levy on the hacker community or that of Tracy Kidder on the development environment for a minicomputer.[28] Sociologist Rob Kling, anthropologist Sherry Turkle, business historian Shoshana Zuboff, classicist Jay Bolter, and many other specialists in the social sciences and the humanities have discussed the meaning of computers to children, to institutions, and to the American public.[29] Students of Donna Haraway are introducing the methods of postmodern cultural studies to computing history in their studies of cyborgs and cyberspace. These studies have not yet been closely connected with those of the mainly technological, economic, and business historians who have traditionally studied computing, but it seems inevitable that this eventually will occur.

In the desk calculator era, the scientist (typically, male) who wanted to make an extensive scientific calculation would select the numerical approach to be taken and write out the algorithm that was to be followed, but the actual calculations would be carried out by a woman or a team of women

---

[27] Peter Galison, *How Experiments End* (Chicago: University of Chicago Press, 1987).

[28] Steven Levy, *Hackers* (New York: Dell, 1984); Tracy Kidder, *Soul of a New Machine* (Boston: Little, Brown, 1981).

[29] Suzanne Iacano and Rob Kling, "Changing Office Technologies and Transformations of Clerical Jobs: A Historical Perspective," in *Technology and the Transformation of White-Collar Work*, ed. E. Kraut (Hillsdale, N.J.: Lawrence Erlbaum Associates, 1987), pp. 53–75; Sherry Turkle, *The Second Self* (New York: Simon and Schuster, 1984); Shoshana Zuboff, *In the Age of the Smart Machine* (New York: Basic Books, 1984); David Jay Bolter, *Turing's Man* (Chapel Hill: University of North Carolina Press, 1984).

sitting at desk calculators. When the first postwar calculating devices were developed, the division of labor was continued with only slight modification. The scientists had essentially the same role as before. The women (the "computers," as mentioned earlier), instead of physically working the desk calculating machine, would "code" the problems for the computer, that is, write up the instructions line by line, set the beginning values of the variables for the calculation, and do other preparatory work. This work had a higher skill level than simply working a desk calculator, but the reward system still strongly favored the male engineers and scientific users. The introduction of stored programming, programming tools, and high-level programming languages somewhat automated the programming process; coder positions were largely eliminated, and programming became a much more male-dominated profession than it had previously been – practiced either by the scientists themselves or by (typically, male) programmers. The 1980s saw a slow but steady increase in the number of women who became computing researchers, but in the mid-1990s, (for unexplained reasons) this trend was reversed. Today, less than 20 percent of the people in the educational pipeline who become computer researchers are women. Most studies of women in computer history have been biographies of individual women pioneers in computing, not more general studies of gender and labor issues.[30]

With the maturing of the personal computer in the 1980s and the Internet in the 1990s, the computer is no longer a tool only of government and big business. The rapidly decreasing price means that computing technology has become widely disseminated across Western societies and is beginning to make inroads in Africa, China, and India. Performance for a given price has increased rapidly, meaning that individuals and small businesses are now empowered by computing tools on their desks. There is an active research community in the academic and industrial sectors, and there seems to be no letup in invention and innovation in the foreseeable future. These advances in computing technology and science are being rapidly appropriated by the scientific community to enhance its research – notably today in the area of visualization of scientific phenomena.

The new world of computers, however, poses certain challenges. Ample evidence has shown that monopolies create increased cost for consumers and inferior products; and the concentration of power in the hands of a few companies poses serious concerns. The legal system is having a hard time adapting its case law to issues of privacy on the Internet and to intellectual property rights related to software. Ethical questions have been raised for two decades about privacy issues and the limits that should be placed on the

---

[30] See, for example, Charlene Billings, *Grace Hopper: Navy Admiral and Computer Pioneer* (Hillsdale, N.J.: Enslow Publishers, 1989); David Alan Grier, "Gertrude Blanch of the Mathematical Tables Project," *Annals of the History of Computing*, 19, no. 4 (October–December 1997), 18–27; Paul E. Ceruzzi, "When Computers Were Human," *Annals of the History of Computing*, 13, no. 3 (1991), 237–44.

applications of computers, for example, in the area of artificial intelligence. Today, ethical and political questions are frequently raised about universal service: If the computer is such a positive tool for humanity, then what should and can be done to make it affordable and available to all people?

There are also economic considerations, which have policy implications about the merits of investment in computer technology. One issue is the so-called productivity paradox.[31] Most people believe that computers have led to tremendous productivity gains, but in the repeated efforts to measure it, economists have found only modest gains in blue-collar work and essentially no gain in white-collar work. If this is so, why should companies expend capital on computer technology? Social return is another question. When a nation invests in computing research, what is the overall payoff to the nation in wealth created and jobs produced? Economists have had a hard time documenting this social return on computing to be any higher than return on capital invested in other ways. If this is the case, why should a nation spend its scarce resources to invest in computing research programs for its universities and industries?

These economic issues are of particular interest in the United States today because of the changing rationale for federal support of computing research. From 1945 until the 1980s, federal expenditures on computing were usually justified in terms of the Cold War. With the disintegration of the Iron Curtain, this rationale has lost its argumentative force and has been replaced by appeals to national economic competitiveness. During the Reagan presidency, the principal competitor was seen to be Japan, whose industry was rapidly cutting into U.S. market share in semiconductors and whose "Fifth Generation" computing project had the goal of making similar gains in the computing field. U.S. industry won back much of the semiconductor business during the 1990s, and the Fifth Generation plan was largely a failure for the Japanese computer industry. The paranoia about foreign competition in these industries has lessened, but federal expenditure on computing is still justified primarily in terms of economic development, not national defense. It is probably true that the computer has created real gains in productivity and social return, but that the economics has not yet evolved to a point where it can measure these gains.

---

[31] Daniel E. Sichel, *The Computer Revolution: An Economic Perspective* (Washington, D.C.: Brookings Institution Press, 1997); Thomas K. Landauer, *The Trouble with Computers* (Cambridge, Mass.: MIT Press, 1995); Paul A. David, "The Dynamo and the Computer: An Historical Perspective on the Modern Productivity Paradox," *American Economic Review,* 80 (May 1990), 355–61.

32

THE PHYSICAL SCIENCES AND THE PHYSICIAN'S EYE
Dissolving Disciplinary Boundaries

Bettyann Holtzmann Kevles

When medical technology met computers in the last third of the twentieth century, the conjoining triggered changes almost as radical as the ones that followed the discovery of x rays in 1895. As in that earlier revolution, the greatest change was in the realm of vision. Whereas x rays and fluoroscopy allowed physicians to peer into the living body to see foreign objects, or tumors and lungs disfigured by tuberculosis (TB), the new digitized images locate dysfunction deep inside organs, like the brain, that are opaque to x rays. The initial medical impact of these new devices, like the x ray before it, was in diagnosis.

Wilhelm Conrad Röntgen's (1845–1923) announcement of the discovery of x rays in 1896 was probably the first scientific media event. Within months, x-ray apparatus was hauled into department stores, and slot machine versions were installed in the palaces of kings and tsars, and in railroad stations for the titillation of the masses. Although the phenomenon had been discovered by a physicist who had no interest in either personal profit or any practical application, it was obvious to physicians and surgeons, as well as to those who sold them instruments, how the discovery could help make diagnoses.

The advantages seemed so great that, for the most part, purveyors of x- ray machines were either oblivious to the dangers of radiation or able to find alternative explanations for burns and ulcerating sores that kept appearing. Even so, with the exception of military medicine, exemplified in the United States by the use of x rays during the Spanish-American war, the machines were not employed routinely in American hospitals for at least a decade after their discovery.[1] This was, in part, due to conservatism within the medical community and, in part, to the fragility and unreliability of early x-ray tubes. These gas-filled tubes, used until William David Coolidge (1873–1975) developed the vacuum x-ray tube at General Electric in 1913, were unstable,

---

[1] Joel Howell, *Technology in the Hospital* (Baltimore: Johns Hopkins University Press, 1995).

unreliable, and inconsistent in the amount of radiation they produced.[2] The widespread acceptance of the vacuum tube, together with improved image resolution from filters and grids, made x rays a staple of the hospital-centered medicine that followed World War I.[3] By the 1920s, for the first time, healthy people were routinely having their chests screened with x rays for TB, their injured arms and legs x-rayed after automobile or skiing accidents, and their teeth x-rayed at regular dental check-ups. The x ray became, and remains, the most frequently used diagnostic instrument.

As medicine organized by specialty after World War I, radiologists were the only doctors who shared a technique, rather than an interest in a part of the body (like cardiologists) or a kind of disease (like rheumatologists). Machine dependent, radiologists tended to work closely with engineers at companies like General Electric in New York or Picker in Ohio. They were interested in reducing the size of machines and the time needed to take and develop images. They were also interested in eliminating exposure of both patients and physicians to ionizing radiation, since by this time, it was impossible to ignore studies suggesting that all exposure to radiation ought to be minimized.[4]

In the first decades after World War II, x-ray technology evolved to include specially dedicated instruments for imaging breast tissue without exposing the patient to life-threatening quantities of radiation, as well as intensifiers that eliminated the need for radiologists to use red glasses. Since then, the quality of x-ray images constantly improved, even as the amount of radiation to which patient and technician are exposed was reduced. Yet these modern radiographs were still created by the passage of x radiation through a patient onto a screen or film, much the same as Roentgen's original pictures.[5]

The daughter technologies that emerged from the linkage of x rays with computers, including the technology of computerized tomography (CT), which uses x rays, differ in kind from traditional x rays. They were welcomed in the medical community, which had grown accustomed to seeing into the living body. But the new technologies followed different routes from one another, and from the original x ray, as they journeyed from the laboratory to the clinic. Unlike the x ray, a chance discovery, the new technologies were the product of years of effort on the part of determined individuals who had to convince the world that these machines were both possible and worth the investment. Funded in part by government grants in the United States and Great Britain, the new instruments usually arose from small-scale projects in a handful of medical centers. Few of the scientists involved in developing

---

[2] Ruth Brecher and Edward Brecher, *The Rays: A History of Radiology in the United States and Canada.* (Baltimore: Williams and Wilkins, 1969).
[3] Paul Starr, *The Social Transformation of American Medicine* (New York: Basic Books, 1982).
[4] Bettyann Holtzmann Kevles, *Naked to the Bone: Medical Imaging in the Twentieth Century* (New Brunswick, N.J.: Rutgers University Press, 1997); Gilbert F. Whitemore, "The National Committee on Radiation Protection, 1928–1960: From Professional Guidelines to Government Regulation," PhD diss., Harvard University, 1986.
[5] Brecher, *The Rays.*

these devices in the postwar decades anticipated the scale of the revolution they were bringing about.

Before World War II, the only part of the electromagnetic spectrum used in medicine was ionizing radiation – x rays and nuclear radiations from radium. Wartime research brought physicists, chemists, and engineers into a variety of weapons programs. The resulting research in nuclear physics and materials science, as well as radar (radio detecting and ranging) and sonar (sound navigation ranging), provided young scientists with knowledge and skills that had both immediate and long-term implications for medicine. During the first postwar decades, veterans of these projects turned their attention to medical instrumentation. Between 1945 and 1985, they created machines that harnessed the x rays in a new way, found medical uses for short-lived radioisotopes created in nuclear reactors, and turned radiation-free experiments in nuclear magnetic resonance and sonography into imaging technologies.

In medicine, the rapid appearance after 1972 of new diagnostic imaging machines gave the impression of a technological engine racing inevitably toward a common goal. The truth, however, is somewhat different. New technologies did emerge in the first postwar decades, but lack of communication among the scientists in different fields probably slowed the eventual development of clinically useful machines. There was nothing inevitable about the avalanche of inventions. It is hard to imagine how the imaging revolution that happened so dramatically would have occurred were it not for the peculiar concatenation of events that transpired in England in the 1950s and 1960s.[6]

The histories that follow – the stories of x-ray-based CT and magnetic-field-based MRI (magnetic resonance imaging) – illustrate how the ideas for certain inventions may be "in the air" at a certain time, how their development depends on the tenacity of their inventors, and how their success emerges from serendipity and the fickleness of the marketplace. The interactions of physicians with physicists, engineers, and even astronomers exemplify the breakdown of disciplinary boundaries that characterizes contemporary research.[7]

ORIGINS OF CT IN ACADEMIC AND MEDICAL DISCIPLINES

Computerized tomography existed in the imaginations of a handful of dreamers before it became a reality in 1972. As early as 1921, doctors were increasingly

---

[6] Stuart S. Blume, *Insight and Industry: On the Dynamics of Technological Change in Medicine* (Cambridge, Mass.: MIT Press, 1992).

[7] For the physics of CT, see Steve Webb in *From the Watching of Shadows: The Origins of Radiological Tomography* (Bristol, England: Adam Hilger, 1990). A broader examination of a larger cluster of imaging technologies, including CT and MRI, is Stuart S. Blume's *Insight and Industry*.

frustrated by their inability to see beneath the rib cage into the lungs and heart. The problem inspired inventors in four countries over a period of ten years, including André E. M. Bocage in France and Bernard Ziedes des Plantes in the Netherlands, to patent machines based on the fact that x rays of pregnant women did not show a fetus if the fetus was moving. All the patents provided that either the x-ray source, the patient, or both move, thus blurring out bones on top of the organ doctors wanted to see. They called these images tomographs – a word coined from the Greek *tomo* for section or slice.[8] These tomographs were useful diagnostically, spurring inventors in the postwar years to get better tomographs by reconstructing the image of an internal slice from one-dimensional data using the new power of computers. By the mid-1950s, four different men in separate scientific and medical fields had begun searching for a way to reconstruct an image of a slice or cross section of an object using x rays and computer reconstructions of data.

The first researcher, Ronald Bracewell (b. 1921), a solar astronomer at Stanford University in California, was mapping sunspots – areas of intense microwave emissions – with radio telescopes in 1955. Because he could not focus a radio antenna on a localized spot on the sun, he approached the problem by getting a strip of data from a line in one dimension, from which he reconstructed a two-dimensional map using the mathematical algorithms of Fourier transforms. He published these results in an Australian physics journal in 1956. Then in 1967, while using radio waves to map the moon's brightness, he reconstructed a similar image using formulas computationally more economical of the then-limited computer capacity than Fourier transforms. These formulas were eventually picked up by scientists using computers for a variety of other tasks.[9]

At almost the same time, William Oldendorf (1925–1993), a neurologist at the University of California in Los Angeles, was struck by the inadequacy of pneumoencephalography, the only way at the time to get an image inside the skull. The procedure sent air into the brain through an injection in the spinal column; the result was a crude image accompanied by enormous suffering. Looking for a better way to see the brain's interior, Oldendorf, who was also an engineer, was inspired by a problem posed to a colleague with whom he met regularly. The local orange growers' cooperative wanted an apparatus that could sort frostbitten oranges from good ones. The bad oranges looked fine from the outside, but dehydrated segments hid beneath healthy skin. The dehydrated segments reminded Oldendorf of bad segments, such as tumors, hidden beneath the skulls of his patients.

Oldendorf's colleague had mused about using some kind of x ray to sort the oranges but had given up. Oldendorf realized that a solution to the problem of bad oranges could apply to images of the human brain. Mulling

---

[8] Kevles, *Naked to the Bone*, pp. 108–10.
[9] Ibid., pp. 147–8.

it over, he figured that he could measure the radio density of a point within an object that is not homogeneous by sending a beam of x or gamma rays through it so that the rays would strike a detector on the opposite side. If he then rotated the beam and the detector about the same axis within a plane, the beam would pass through the object from many angles, pinpointing a singular spot. By moving the object along the path of a line, eventually he would be able to measure the density along that single line and reconstruct the density relationships of points inside the object.

Oldendorf retired to his home laboratory to test this theory. There, in 1959, he built a model. Using forty-one identical iron nails, one aluminum nail, and a plastic block (with holes for the nails), he placed this phantom "head" on a toy electric-train flatcar and track (borrowed from his young sons). He used a gamma-ray source within a lead shield, rather than x rays (because gamma rays were easier to control), and scanned all the points in the plane by using rotation – isolating a point – and translation – moving the point along a line. (Translate/rotate would become the catchwords of CT scanning.) Mounted on a 16 rpm phonograph turntable, the "machine" moved the point of intersection of the axis and the beam through the model at 80 millimeters per hour. It sent a collimated beam of high-energy particles through a plane in the model head. The beam located both the iron and aluminum nails inserted in the center of the block. The particles emerged, struck a photon detector, and were counted, and a recognizable pattern was displayed as a two-dimensional image.

He explained the way his machine, and all CT scans, would work: "An observer standing stationary in a forest might have a difficult time viewing a distant person because that person might be blocked by trees in between. But if the observer begins to move through the forest, while at the same time looking in the direction of the distant person, then the trees in the foreground would seem to move past, while the distant individual would seem to stay still." Using this analogy, Oldendorf said that the distant person represents the nails in the center of his model, and the trees the line of nails obscuring it. The observer's line of sight is like the gamma-ray beam that is continuously pointed through the surrounding ring of nails at the interior nail. As the gamma source circles, the nails in front and behind the central nail momentarily absorb gamma rays, deleting them from the beam, creating the equivalent of the blurring motion of trees in the forest. The interior nail itself, located at the center of motion of the gamma-ray source, absorbs gamma rays continuously.[10]

Despite its apparent simplicity, Oldendorf's model incorporated the fundamental concepts of all later computerized tomographic scanners – except for the modern digital computer. At one point, Oldendorf conferred with Robert Beck, an imaging scientist at the University of Chicago, who recalls

[10] Ibid. (quoting from unpublished memoir, Mrs. Stella Oldendorf, 1995), p. 331.

how in those precomputer days, he reckoned Oldendorf would need 28,000 simultaneous equations to get the information to reconstruct an image. So he told him, "Forget it!"[11]

Oldendorf did not have the computational tools to interpret the quantity of data he would acquire. But he had demonstrated that to measure the radio density – the ability to absorb radiation – of a point, he had to uncouple the effects of radiation in one point from all the other points on the same plane. He had worked out back projections to reconstruct a two-dimensional display of images and had tried it out successfully with his jerry-built model. Believing it was simply a matter of scaling up the dimensions and sensitivity of this crude apparatus before he could similarly scan a head, and realizing that with patients he would need x rays instead of gamma rays, he applied for a patent for such a machine in 1960.[12] The following year he published the results of his experiment in *Bio-Medical Electronics,* the transactions of the Institute of Electrical and Electronic Engineers, of which he was a member. When he received the patent in 1963, he approached a host of x-ray manufacturers but was roundly dismissed. One corporation responded that "even if it could be made to work as you suggest, we cannot imagine a significant market for such an expensive apparatus which would do nothing but make radiographic cross-sections of the head."[13]

Also in the late fifties, Alan Cormack, (1924–1998), while still living in his native South Africa, got a call from Capetown's Groote Schuur Hospital for a physicist to monitor radiation therapy. Cormack, then a university lecturer, took the extra job. There, from his desk in the radiology department, he was appalled at how haphazardly radiotherapy was designed. It was based on the absorption of radiation by homogenous matter approximating human tissue as if there were no differences in the absorption of bone, muscle, or lung tissue. It struck him that what was needed was a set of maps of absorption coefficients for the different tissues in different sections of the body.[14]

He was thinking about therapy, of course, not imaging, and he was concerned with unnecessary overexposure to radiation. This led him to ponder the notion of body maps. And from this thought he began to search for a way to map the body using x rays. He was not yet consciously thinking about extracting images, and he had not yet heard the word "tomograph." Images, the logical by-product of his quest, did not interest him at this point as much as solving the mathematical problem of measuring the absorption of x rays along lines through inhomogeneous tissues of the body.

This "line-integral" problem haunted Cormack, and during the next year, 1957, he tried an experiment using a gamma-ray source on a circular model (or phantom) to test the theory he was developing. What happened next

---

[11] Ibid. (quoted from conversation with author, Spring 1993), p. 151.
[12] William H. Oldendorf, *The Quest for an Image of the Brain* (New York: Raven Press, 1980), p. 85.
[13] Kevles, *Naked to the Bone,* pp. 148–53.
[14] Alan Cormack, interview with the author, Medford, Mass., 16 March 1993.

was a technician's error from which he was quick to profit. He had asked the university machinists to make a symmetrical phantom out of a uniform disc of aluminum surrounded by a wooden ring. When he got an image, he found an anomaly in the data near the center. He questioned the machinists and discovered that the phantom was not uniform; the machinists had put a peg of slightly different density at the center of the disc. This "error" on the machinist's part revealed that he, Cormack, was on the right track. His scanner had actually detected the small density difference. This unexpected bit of detection kept him working on the line-integral problem and formed the kernel of the paper he published in 1963. By this time he had moved to the United States, to Boston and Tufts University, and he relegated the line-integral problem to a kind of mental attic space where he continued to mull over it in spare moments. He had come to think of it proprietorially as *his* problem. But sensing that someone must have solved it already, he wrote to mathematicians on three continents to find out who. Years later he learned that he had tried the wrong mathematicians. "His" problem had, indeed, been solved, and more than once, first by a Dutch physicist in 1905, and later in 1917 by the Austrian mathematician Johann Radon.

In Boston Cormack continued to work on the problem, and in 1963 he experimented with a model that included a phantom designed with irregular symmetry; using a computer he reconstructed images of asymmetrical phantoms. He published the results and, like Oldendorf in California, who received his patent in 1963 from the U.S. Patent Office, tried to drum up outside support. The single inquiry he received after publication was from the University of Neuchatel where a representative of the Swiss Avalanche Research Center wondered if Cormack's approach could predict the depth of snow.[15]

ORIGINS OF CT IN PRIVATE INDUSTRY

Meanwhile in Britain, an engineer at EMI (Electrical and Musical Industries Limited), was following a different course. The precise details are hard to track because they occurred in the laboratory of a private corporation, which sought patent protection immediately and for which secrecy was key. There is no doubt, however, that the man behind the patents was Godfrey Newbold Hounsfield (b. 1919), a research engineer who had been developing a computer for EMI in the early 1950s, and who had an interest in pattern recognition.

Founded in 1898 as The Gramophone Company (a deliberate inversion of "phonogram," Edison's term for a recording), EMI was formed when the Gramophone Company merged with two other recording companies in

---

[15] Personal communication with Alan Cormack. The letter was from Claude Jaccard.

1931.[16] The new company's research and development program had assisted in the development of television in the 1930s. In the 1950s, EMI had encouraged research in transistors and early computers, but had never lost sight of its signature product – musical recordings. EMI sold classic and popular music in whatever form the public wanted: hi-fi, stereo, and tapes. In the late sixties, the success of their Beatles recordings accounted for over half of the company's considerable assets, whereas electronics accounted for less than a quarter of sales. Medical instruments were virtually nonexistent.

Hounsfield, assigned in the fifties to streamline the company's commercial British computer, successfully redesigned it to work on transistors. But EMI needed cash for its rapidly diversifying record business and sold the computer facility. At this point, Hounsfield was told to find a new project, and he opted for pattern recognition, a problem related to EMI's primary mission of television, recording and playing back information. He was attracted by the puzzle and unfazed by the threat of complexity. He believed that "most of these problems are just... using common sense, and then proving it by maths afterwards."[17]

Hounsfield's wartime work included two very different intellectual feats: One comprised storing image information on linear TV scans; the other positioned radar where the topographical features of the landscape were displayed on a cathode ray tube screen. Hounsfield later recalled thinking over these problems on "a long country ramble" when the seeds of the CT scanner began to grow in his mind.[18] If many measurements were made through an object at various angles, he realized, the information provided could be used to reconstruct the image. Although that would take thousands of mathematical equations, he was confident that the computers he had so recently worked with could handle them. Hounsfield worked for a business and knew that anything he suggested had to have a practical application. It seemed obvious that the practical application was radiology and that the object to be scanned would be a patient. The computers he used could also do more than solve equations; they could store pictures, bits that could be presented as sets of pixels. Like Cormack, he realized that ordinary x rays are inefficient because their random scatter contributes no information to the film, and because the superimposition of other images – bones and soft tissue, for instance – makes some images difficult to read. A CT scan, as Hounsfield envisaged it, would provide more information than an ordinary x-ray image. Each scan would look like a cross-sectional cut, and a series made closely together could be built up into what looks like a three-dimensional image. The major question

[16] Charles Suskind, "The Invention of Computed Tomography," in *History of Technology*, vol. 6, ed. A. Rupert Hall and Norman Smith (London: Mansell Publishing, 1981); also, Sir Godfrey Hounsfield, telephone conversation with author in London, 20 May 1994.

[17] Charles Suskind, "The Invention of Computed Tomography," in *History of Technology*, vol. 6, p. 47. Personal conversation with author, August 1997.

[18] Sir Godfrey Hounsfield, telephone conversation with author, 20 May 1994.

was whether he could get a fine enough image – and one free of noise – without zapping the patient with an overdose of radiation.[19]

He wanted to try, and his ideas were the basis for EMI's 1968 British patent: "A Method of and Apparatus for Examination of a Body by Radiation such as X or Gamma Radiations."[20] At this point, Hounsfield's solution used much of the same reasoning as those of the Australian, South African, and American inventors. He asked EMI to let him build a model. The mathematics that had obsessed Cormack did not worry Hounsfield; he knew there were algorithms, including Fourier transforms and Bracewell's published work, for reconstructing data from projections. As it turned out, he rejected them all in favor of a simple "iterative" algebraic technique because he found it personally satisfying. This was fine for making the first images but was exceedingly slow and did not take advantage of the computer's potential. At this stage, however, in 1971, Hounsfield was interested simply in proving that he could do it at all. Later he would streamline the process, but in 1971 he enjoyed having hands-on control.[21]

EMI, however, hesitated and refused to provide more money without some evidence of a market. The next step was a visit to Britain's Department of Health and Social Security (DHSS), where Hounsfield explained that his proposed device would be useful for detecting tiny tumors. When the suggestion of mass screening for tiny growths did not seem to excite the imagination of the health officer, Hounsfield proposed that the machine could see inside the brain! This approach struck the right chord. The idea of "seeing" into the brain offered the promise of economic savings, since most brain problems necessitated expensive exploratory surgery. But more than the pocketbook was involved. Since the discovery of x rays, the prospect of seeing the living brain in action had stirred the popular imagination.

Soon Hounsfield was teamed with a distinguished neuroradiologist who, recognizing the potential of the invention, put Hounsfield in touch with a neurosurgeon who had been exploring other efforts of imaging the brain. Throughout 1968 and 1969, Hounsfield worked on models using gamma rays – as Oldendorf had – instead of x rays. The gamma-ray machine took nine days to scan the subject and two and a half hours to process the data on a computer. He replaced the gamma rays with x rays, cutting the scan time to nine hours. As he continued to refine the system, he moved from artificial models to a pig's head (he once inadvertently left it, neatly wrapped, on the London underground) and, eventually, human organs.

With funds raised from the government and from EMI's Beatles profits, the project headed for realization. Inasmuch as secrecy was a priority, it is now difficult to discover how much was known, by whom, and at what time.

---

[19] Sir Godfrey Hounsfield, telephone conversation, 24 October 1997.
[20] Ibid.
[21] Personal conversation with Hounsfield, London, 20 May 1994.

However, it is clear that on 1 October 1971, Hounsfield scanned the first patient, a forty-one-year-old woman who had symptoms suggesting a brain tumor somewhere in the frontal lobe.

With a rubber cap on her head the woman lay to one side of a plastic water-filled box. All of the earliest machines had placed the patient's head *in* water, which was necessary because its density is closer than that of air to the density of bone. By excluding air, Hounsfield reduced the range of information that would have to be processed and, thus, the number of calculations that the computer – with very limited power compared with later models – would have to make. A collimated beam of x rays was sent through her head; these were picked up by a scintillation detector on the other side. As both the source of the rays and the detector were moved linearly, information was gathered from 160 points, or "scan passes," and stored in the computer. Rotating the unit through one degree around the patient's head, Hounsfield collected another 160 points of data. Altogether, this first machine gathered information from 180 data points on magnetic tape, a total of 28,000 readings. Through it all, the patient had to remain still, her head against the water-filled box the entire time, a total of fifteen hours.[22] But the amount of radiation in a single beam was very small, and the total radiation, even after 15 hours, was about what a patient would receive from a routine radiograph of the gastrointestinal track.

The data tape was sent across London to a computer where it was processed. The results were in turn, processed by another computer that produced a cross-sectional image, which was photographed from the monitor's screen. The photograph was finally carried back to the surgeon, who easily saw a tumor in the patient's left frontal lobe, and excised it.

The achievement of the first CT scanner triggered improvements so rapidly that within three years, the machine had passed through three generations. The newer instruments imaged the entire body, used multiple detectors that reduced the time needed to obtain a single image from twenty minutes to twenty seconds – the time a patient could hold his or her breath – and then to a single second. CT continued to become faster and more flexible as computer capacity grew capable of processing more data more rapidly. From the start, CT enabled physicians to image bone, cartilage, and muscle in increasingly narrow slices or cross sections. Whether the brain, the liver, or the knee, it enabled physicians to see the body as if a surgeon had made an incision. CT altered the nature of trauma treatment.

Cormack and Hounsfield shared the Nobel Prize in Physiology or Medicine in 1979 for the development of CT. The award was controversial. Hounsfield seemed an obvious choice, though from the perspective of scientists, what he had done was good engineering, not science. He had not derived any original algorithm or invented the collimated beams or receptors that furnished the raw data. Yet he had designed a machine that worked, even

---

[22] Telephone conversation with Hounsfield, 25 October 1997.

though theory and experience suggested it could not succeed. The theory for the mathematics had been articulated by Cormack, which would have pleased Roentgen, who had once said, "The physicist in preparing for his work needs three things: mathematics, mathematics and mathematics."[23]

Oldendorf, disappointed, remarked that he had paid the price for being twenty years ahead of his time, and he turned to other research. Bracewell, who had never thought himself a contender, continued his work in astronomy but joined the editorial board of the new *Journal of Computed Axial Tomography* in 1977. Whatever the politics of the prize, the fact that the 1979 Nobel Prize in Physiology or Medicine went to an engineer and a nuclear physicist shows how thoroughly diagnostic medicine and pure and applied physics had already merged.

CT is not the ultimate imaging mode, however: It does not image soft tissues or structures, such as tumors, inside bone. It also subjects patients to the same quantity of ionizing radiation as they would receive from a complete gastrointestinal examination using barium. During its early years, the only competition came from nuclear medicine, a complex technology in which trace amounts of radioactive substances are injected into the bloodstream and, as they spread throughout the body, are mapped by external detectors. At this time, exposure to some form of ionizing radiation seemed to be the unavoidable toll for extracting images noninvasively from inside the living body.

FROM NUCLEAR MAGNETIC RESONANCE TO MAGNETIC RESONANCE IMAGING

The radiation toll disappeared in the 1980s with the entry of magnetic resonance imaging into the clinic. MRI removed radiation from the equation and imaged soft tissues invisible to x rays. MRI benefited from Cormack's obsession with finding the mathematics to feed his computer. Although the nature of MRI signals differs altogether from the way CT obtains data, the target – the human body – is the same. The problem of reconstructing images from a mass of data from within the body is also similar. In the decade separating the development of the two technologies, computer power had grown enormously and mathematicians had developed formidable new algorithms to process the data. These were the first of the major gifts CT gave to MRI. The second was the conviction among the groups racing to construct a clinically useful machine that the same hospitals that had paid $700,000 or more for a CT scanner would pay at least that much for a machine with a giant magnet that would most likely require expensive special-site preparation.

Although MRI, like CT, makes computer-reconstructed images of the interior of the living body, their scientific roots are entirely different. MRI

---

[23] George Sarton, "The Discovery of X Rays," *Isis*, 26 (1936), 362.

grew from the realization that the interior of the atom's nucleus could be manipulated, an idea first suggested by the physicist Wolfgang Pauli. In 1924 he proposed that the nuclei of certain atoms have angular momentum, or "spin," and under certain conditions they display magnetic properties. Evidence of this nuclear magnetism was detected in frozen hydrogen by two Soviet scientists in 1937, the same year that the American physicist I. I. Rabi (1898–1988) actually measured the magnetic moment (or spin) of the nucleus, for which he coined the phrase "nuclear magnetic resonance," or NMR. Two years later, Rabi measured the moments of the proton and deuteron. The next year, 1940, Felix Bloch (1905–1983) used a variation of this method to measure the neutron moment, and in 1945, he explored the question of whether the nuclear transitions could be detected by electromagnetic methods, which he called "nuclear induction," using radio-frequency fields. During this time, Bloch did not seem interested in the commercial possibilities of his research, a situation that would change as he saw the advantages of cooperating with Varian Associates, an organization with close ties to Stanford University.[24]

In the beginning, before 1980, there was only NMR. It was the province of physicists like Rabi, who won the 1944 Nobel Prize in Physics for his experimental measurement of nuclear magnetic resonance. Eight years later, the Nobel Prize, in physics again, went to Felix Bloch and Edward Purcell (b. 1912) for independent, but almost simultaneous, publication in *The Physical Review* in 1946 of descriptions of their methods for measuring NMR in bulk matter. Both Bloch and Purcell began with the knowledge that nuclei with odd numbers of protons, neutrons, or both will align themselves like little compasses when exposed to a strong magnetic field. Then, when an alternating magnetic induction is turned on at the radio frequency of the particular atom – its resonance frequency – the protons in the nuclei resonate.

In most laboratory NMR, and later in medical imaging, the nucleus imaged is hydrogen, because as the major constituent of water, it is the most prevalent element in the human body.[25] Eventually, when NMR was adapted to sophisticated imaging systems, it became possible to derive additional images from sophisticated manipulations of the radio signals. This was done by pulsing them (turning the frequency on and off rapidly) and by taking advantage of other qualities of the spinning nuclei.

NMR was almost immediately adopted by chemists who saw it as an excellent tool for chemical analysis of any substance. In the first decades after World War II, NMR was the interdisciplinary offspring of physics and chemistry and was used to explore test-tube-size samples of homogeneous inorganic substances. It had nothing to do with medicine, and twenty-five

---

[24] Timothy Lenoir and Christopher Lecuyer, "Instrument Makers and Discipline Builders: The Case of Nuclear Magnetic Resonance," *Perspectives On Science*, 3, no. 2 (Fall 1995), 284–9.
[25] Felix W. Wehrli, "The Origins and Future of Magnetic Resonance Imaging," *Physics Today* (June 1992), 34–42.

years would elapse before its medical applications were realized. But during these years, the instrument makers in the NMR field built larger and larger magnets so that when MRI became possible, the enormous magnets needed for clinical-size instruments could and would be made by these companies.

MRI grew on the shoulders of NMR through the epiphany of one man, Raymond Damadian (b. 1936), and the zeal and determination of another, Paul Lautebur (b. 1929). As with other scientific discoveries, there was nothing inevitable about the leap from NMR spectroscopy to medical imaging. But once it had been demonstrated that NMR could yield images of previously hidden regions within the body, the race for priority began.

In 1947 there was no evidence that it was safe for a human being to be subjected to powerful magnetic fields. Purcell had inserted his head into a magnet with an NMR field (with a force of 2 tesla) in 1948 and reported feeling a buzzing emitted by the metal fillings in his teeth and also tasting metal. But this limited exposure was inconsequential in comparison with exposing a patient for an hour inside such a field. It was a mystery what, if anything, magnetic fields do to organic tissue. Medical researchers assumed that magnetic fields are harmless since life evolved and has flourished in one. NMR fields have to be as homogenous as possible, and so chemists often spin their samples in order to expose them to a uniform field.

During the 1950s and 1960s, chemists expanded the use of NMR to organic substances until it became possible to examine larger tissue samples. Building on these results, Damadian, a physician at Downstate Medical Center in Brooklyn, New York, came to suspect that malignant tissue would differ from healthy tissue in a way that could be discerned by NMR. He brought several rat subjects to a laboratory outside Pittsburgh, NMR Specialities, to test his theory. His objective was finding a way to detect cancer at an early stage, and NMR seemed to hold that promise.[26]

At the same time, a chemist at the State University of New York at Stony Brook, Lautebur, had taken a summer job working for NMR Specialities, which was in a precarious state financially. He was not impressed with the idea of examining specimens during what he assumed would be surgery, to see whether there was a sign of cancer. He was interested in NMR data from very small tissue samples and recalled that when an NMR machine is not finely adjusted, all sorts of weird shapes and lumps and line splitting appear, artifacts that have to be eliminated. It occurred to him that those lumps and bumps carried information, not only about the magnetic field but about the sample as well.

It was this insight that provoked his question: "Was there some way one could tell exactly where an NMR signal came from?" His answer: "By using magnetic field gradients." If a magnetic field varies from one point in

---

[26] Raymond Damadian, interview with author, April 1993; Sonny Kleinfield, *A Machine Called Indomitable* (New York: Times Books, 1985).

the object to another, he reasoned, the resonant frequency that is directly proportional to the strength of the magnetic field will vary the same way. So, for example, if he made a magnetic field increase a little from his left ear to his right ear, the left ear would have one resonant frequency and the right ear would have a different one. With that in mind, he plotted out the resonance frequencies and deduced that he would see a little ripple on one side for one ear and a little ripple on the other side for the other ear. That would give one dimension of information by reducing all the complexity in his head between his two ears to a single trace. But a single trace is not the same as a full image. He could get a full image by applying magnetic field gradients in different directions. How could he work back in three dimensions to a simple scan? Then the answer came to him. "I ran out to a drug store that evening and got a notebook – the best I could find – and wrote down these notes which then I had witnessed, September 2, 1971."[27]

This was the beginning of what is now known as one-dimensional imaging. Lautebur could translate those single points of data from different places along the magnetic gradient into spatial information – a qualitative step beyond the kind of linear image Damadian was getting at this same time, which had no spatial dimension. A few days later Lauterbur figured out a better way, indeed, the kind of method Hounsfield later published for CT. Lauterbur used an algebraic reconstruction technique for projective reconstruction and thought he had invented a whole new applied mathematics field. Of course he had not. But he had discovered a way to create images from NMR. He published this discovery in *Nature,* where it immediately attracted enormous attention within the specialized precincts of medicine and biochemistry.[28]

But just as Lauterbur had been unaware of Bracewell's and Cormack's mathematics, so the physicist Peter Mansfield (b. 1933) at the University of Nottingham had been unaware of what Lauterbur was up to. Perhaps it was a function of the increased specialization of science, as each discipline had its own journals, convened its own meetings, and used its own increasingly private vocabulary. Thus, Mansfield, a physicist, could not have been expected to read about chemistry, even in an interdisciplinary journal like *Nature.*

Mansfield's objective was remote from the physician Damadian's, which was to map areas of malignant tissue with NMR, and different from that of the chemist Lauterbur, which was to focus on imaging liquids. In 1973, Mansfield published "NMR Diffraction in Solids," which, in principle, would allow the imaging of solids to the level of atomic structure. Using a different vocabulary from that of Lauterbur, Mansfield later recalled: "We came up with an elaborate mathematical explanation. What we did was we made a

---

[27] Paul Lautebur, interview with author, Urbana, Ill., 2 December 1992.
[28] Paul Lautebur, "Image Formation by Induced Local Interactions: Examples Employing Nuclear Magnetic Resonance," *Nature,* 242 (1973), 190–1.

lattice. A model lattice of the material with a much coarser gradient. The principles behind that, of course, are exactly that this would have applications in biology."[29]

When he presented his results at a physics conference in Poland in 1974, Mansfield thought he was telling the world for the first time about imaging. Someone in the audience asked if he was aware of Paul Lauterbur's work. "What work? I didn't know anything about it. It was a bit of a bombshell to me." Mansfield had come up with the idea of a gradient, which he had described in terms of the physics of solids, rather than in the biological framework of largely liquid materials: "With the benefit of evolution and time, we introduced ideas in a mathematical framework for imaging so-called k-space. And when you talk to people today about how imaging works, they always talk about k-space trajectories." K-space trajectories are a way in which physicists talk about imaging in what Mansfield calls "reciprocal space" instead of real space.

In 1974 Mansfield turned his research away from solids and began imaging liquids – that is, the human body. He was not the only scientist in Britain working with NMR. To the north in Aberdeen, medical physicist John Mallard was soon to get excellent images of a freshly stunned mouse. And in London, EMI was also well along in NMR imaging. By this time, laboratories in Britain and the United States were racing each other for NMR images with sharper and higher resolution. Most of the funds for this research came from private sources. Small grants came from the National Institutes of Health, but only for basic science and not instrumentation, although by this time it was not always easy to separate the two.

MRI AND THE MARKETPLACE

By the early 1980s, machines were in clinics in the United States and England, and the technology, which had been called NMR, had been renamed MRI. Some attribute the change to an effort to avoid using the word "nuclear," a term the public associated with bombs; others believe it represented the attempt by radiologists to distinguish their speciality from the realm of nuclear medicine. Magnetic resonance technology continued to evolve; its images became more and more precise, and in 1991 fast MRI – fMRI – using Mansfield's ideas as well as formulas developed at Bell Laboratories, began to image metabolic function in addition to anatomy. fMRI can track activity in the brain, while MRS, magnetic resonance spectrography, can image elements besides hydrogen in particular regions of the body.

---

[29] This and subsequent quotations by Sir Peter Mansfield, are from a conversation with the author in Nottingham, England, 18 May 1994.

To Lauterbur, CT ought to have been obsolete as soon as MR was in the marketplace. He points out that it can do the same thing as x rays, but without subjecting the patient to ionizing radiation. It can image areas otherwise obscured by bone. Most of those involved in developing MRI were convinced from the start that it was better than CT, and they begrudged the arrival of CT on the market first.

Paul Lauterbur confessed to a favorite thought experiment:

> If suppose, for whatever reasons, somebody worked out how to do magnetic resonance imaging and it had suddenly blossomed before anyone figured out that you could do similar things with x rays, and then someone else came to the NIH and to GE and said "you know, we can do the same sort of things with x rays. It's going to give people a tremendous radiation dose and the bone is going to obscure a lot of the soft tissue detail and you can't really see as clearly the differences in soft tissues, and you can only figure out how to get transverse planes in the head instead of all these 3-Ds and all the different slices that show up so beautifully in the MRI, but gee, we'd like to develop it anyway." Would the federal government have smiled on that? Would the grants have been forthcoming? Would there be companies willing to sink their money in it? Not likely. It would have been stillborn.[30]

Perhaps. But would MRI ever have reached the marketplace if CT had not prepared the minds and pocketbooks of hospital administrators? After spending $400,000 each for a couple of CT scanners in 1973, which they replaced within three years by $500,000 machines, the idea of hospitals spending $100,000 for a magnet alone did not seem outrageous. MRI proved that CT's algorithms were transferable and infinitely malleable with the vastly expanded capacity of new generations of computers. These same algorithms were adapted to nuclear imaging machines – SPECT (single positron emission computed tomography) and PET (positron emission tomography) – machines that extracted images from radioactive isotopes injected into the bloodstream, and by the late 1980s to ultrasound machines, which had become a vital part of diagnostic cardiology, urology, and obstetrics. Without CT's dramatic entry into the medical market, its successor machines would most likely have appeared in time, but it is hard to imagine how they would have fared in an atmosphere of managed care and cost cutting.

It took almost two decades between the time the first patent for a tomograph was issued in France in 1921 and the time it was manufactured in the United States in 1938. Computerized tomography appeared in the visions of Oldendorf and Cormack in 1960, but CT was not realized as a clinically useful machine until 1971. The delay between inspiration and production in each case was a matter of economic factors, including prospective demand,

---

[30] Paul Lautebur, interview with author.

the ability of the medical community to invest, the structure of the medical insurance system in the United States, and the ability of hospital budgets to finance the experiment. There was no sizable investment anywhere outside of the United States in the production of these expensive machines. It was also dependent, in the case of CT, on the parallel evolution of computers.

But even if all these elements had been in place, had it not been for EMI's windfall of Beatles earnings and its naïveté in the medical market, the go-ahead to Hounsfield might never have occurred. Timing is everything, of course, and the delay of a decade might have made such an investment unthinkable in a medical world already under the gun in the United States after 1976 to tighten its belt. The tremendous leaps in diagnostic imaging would probably have happened, but more as incremental jumps than in the revolutionary way that they occurred. When MEG, magneto encephalography – an expensive machine that images magnetic waves from the brain and reveals the delicate operations of the auditory system – was developed in the early 1990s, it moved slowly; it is only now beginning to appear in laboratories involved in mapping the human brain. The years of large investments in technological innovation may have passed; what investment dollars remained in the late 1990s seemed to go into the development of cheaper versions of existing imaging technology.

THE FUTURE OF MEDICAL IMAGING

Computers enable medical imagers to process enormous quantities of data rapidly and to display an almost endless series of permutations and possibilities without obscuring artifacts. The expansion of computer memory in the last quarter of the twentieth century transformed the possible applications of medical imaging machines in surgery, therapy, and diagnosis.

Even here, where x rays – the familiar black and white images of teeth and bone – still comprise 80 percent of all diagnostic images, they, too, are rapidly disappearing, at least in their traditional form on film. By the end of the 1990s they, too, had been transformed by high-powered computers. To new ways of digitizing x rays, explored initially as a solution to the problem of storing vast files of x-ray films, have been added the benefits of a reduction in the amount of radiation to which patients are exposed. Equally important is that technicians are now able to magnify trouble spots on these reconstructed images.

A final example of the impact of the physical sciences on medicine at the end of the twentieth century, one that illustrates the permeability of disciplinary boundaries, is a new way of scrutinizing mammograms – x-ray films. In 1994, astronomers at the Space Telescope Science Institute examined images returned from space of a remote galaxy. They had been taken through the Hubble Space Telescope's original malformed mirror, prior to the 1993 mission that corrected the hardware. The astronomers were working on software to clarify the earlier images and discovered that a radiologist at nearby Georgetown University Medical school was attacking a similar problem. Both tasks focused on white spots on digital images. In the case of the remote galaxy, the Hubble staff had to discriminate the white flecks of cosmic ray hits from faint stars. The mammographers wanted to eliminate everything from the soft tissue images except the white dots, which they recognized as microcalcifications, precursors of cancer. The same techniques that enabled the astronomers to eliminate the spots allowed the radiologists to keep them. The flaw in Hubble's original mirror turned out to be a window of opportunity for early computer detection of cancer.[31]

Diagnostic imaging, however, while a major enterprise, is only part of the story. Visual breakthroughs have also dominated the surgical revolution. The brain is now operated upon with stereotactic surgery, a system in which a detailed three-dimensional image is created by combining CT and MR images that allow surgeons to pin-point precisely the tissue they plan to excise. The development of miniature television cameras has also enabled surgeons to see exactly what they are operating on in other parts of the body with scarcely any incision. Surgeons can insert these miniature cameras and a light to be able to see the area in question enlarged on a television monitor, and can then insert and use miniature surgical instruments, sometimes remotely controlled, while watching the monitor. The new surgical techniques include endoscopy, whereby the surgeon enters the esophagus and stomach via the throat, often using optical fibers as a tool for both guiding and operating on the area to be treated; and laparoscopy, whereby the body is entered via the abdominal cavity through a keyhole opening near the naval. Not only do surgeons now see the interior of the organs they will be operating on, but they see it magnified so that they can monitor their progress inside the deep recesses of the patient's body from entry to final suture.

Imaging, first diagnostic and than applied, has revolutionized surgical practice. To this last category the physical sciences have added the benefits of lasers and fiber optics. Medicine has become a special kind of applied science. The art and craft of the physician are still important, but the tools are new and demand a degree of sophistication unknown to most physicians in the past.

The separation of disciplines that characterized the way the physical sciences worked during the middle decades of the twentieth century became obsolete in its closing decade. New interdisciplinary programs came into existence, and medicine was the beneficiary. Whereas in the years preceding

---

[31] Personal communication, Robert Hanisch, Space Telescope Science Institute, Baltimore, August 1997.

1970, mathematicians, physicists, astronomers, and physicians published in separate journals, later interdisciplinary journals reflected a new scientific field of imaging science. The visual culture that grew to dominate much of twentieth-century medicine is part of the expansion, through technology, of the visual spectrum that promised to continue altering medicine, as imaging brought ever smaller and remote aspects of the living body to light.

33

GLOBAL ENVIRONMENTAL CHANGE AND THE HISTORY OF SCIENCE

James Rodger Fleming

"Global environmental change," three words heard with increasing frequency in both science and policy circles, is shorthand for the inevitability of change in the geosphere-biosphere. It also expresses the realization that human activities have now reached the level of a planetary force. Since 1945, we have grown increasingly apprehensive about a number of global environmental issues, including population, energy consumption, pollution, and the health of the biosphere. At the beginning of a new millennium, instead of standing firmly on the technoscientific foundations of our "enlightened" predecessors, we find ourselves apprehensive about global environmental change, teetering on the uncertainties of a new century and unsure about the future quality and even habitability of the global environment.[1]

Much of the concern is rightfully focused on changes in the atmosphere caused by human activities. Only a century after the discovery of the stratosphere, only five decades after the invention of chlorofluorocarbons (CFCs), and only two decades after atmospheric chemists warned of the destructive nature of chlorine and other compounds, we fear that ozone in the stratosphere is being damaged by human activity. Only a century after the first models of the carbon cycle were developed, only three decades after regular carbon dioxide (CO_2) measurements began at Mauna Loa Observatory, and only two decades after climate modelers first doubled the CO_2 in a computerized atmosphere, we fear that the earth may experience a sudden and possibly catastrophic warming caused by industrial pollution.

These and other environmental issues were brought to our attention by the work of scientists and engineers, but the problems (and the responsibility for finding solutions to them) belong to us all. Recently, humanists, policy-oriented social scientists, public officials, and diplomats have turned their attention to the complex human dimensions of global change. There

---

[1] For a complete account of the issues discussed here, see James Rodger Fleming, *Historical Perspectives on Climate Change* (New York: Oxford University Press, 1998).

has been a rising tide of literature – scholarly works, new journals, textbooks, government documents, treaties, popular accounts – some quite innovative, others derivative and somewhat repetitious. This has resulted in growing public awareness of environmental issues, new understanding of global-change science and policy, widespread concerns over environmental risks, and recently formulated plans to intervene in the global environment through various forms of social and behavioral engineering, and possibly geoengineering. Global change is now at the center of an international agenda to understand, predict, protect, and possibly control the global environment.

One fundamental aspect of global environmental change – the historical dimension – has not been adequately addressed. The literature in the rapidly growing field of environmental history tends to have a *local* focus. There are a number of excellent studies of the problems of particular places or regions. The field has also developed a semicanonical narrative about the rise of the environmental movement from Henry David Thoreau to the Earth Summit in 1992.[2] On the other hand, most of the literature on *global* environmental change is ahistorical and is rather narrowly focused on scientific and policy responses to current issues. There are, of course, notable exceptions.[3] In the field of climate studies, for example, some scientists work in collaboration with historians, archaeologists, and anthropologists to reconstruct the temperature and rainfall records of the past. Others use the available scientific data to explore the effect of climatic variations on past societies. Some scientists have given serious consideration to the history of their fields.[4]

[2] See for example Richard White, "American Environmental History: The Development of a New Historical Field," *Pacific Historical Review* (August 1985), 297–335; and [John Opie], "History and the Environment," in *Environmental Protection: Solving Environmental Problems from Social Science and Humanities Perspectives*, ed. Nancy Coppola et al. (Dubuque, Iowa: Kendall/Hunt, 1997), pp. 1–70.
[3] Some of the more interesting exceptions include John A. Dutton, "The Challenges of Global Change," in *Science, Technology, and the Environment: Multidisciplinary Perspectives*, ed. James Rodger Fleming and Henry A. Gemery (Akron, Ohio: University of Akron Press, 1994), pp. 53–111, which has a scientific focus; Harold K. Jacobson and Martin F. Price, *A Framework for Research on the Human Dimensions of Global Environmental Change* (Geneva: Unesco, 1991), which discusses the contributions of social scientists, but excludes history and the humanities; Mats Rolén and Bo Heurling, eds., *Environmental Change: A Challenge for Social Science and the Humanities* (Stockholm: Norstedts, 1994), which is much broader, but is restricted to Swedish examples; and Leo Marx, "The Humanities and the Defense of the Environment," Working Paper No. 15 (Cambridge, Mass.: MIT Program in Science, Technology, and Society, n.d., ca. 1990), 32 pp., which offers a fruitful approach for humanists.
[4] For a scientific reconstruction of the historical climate record, see Raymond S. Bradley and Philip D. Jones, eds., *Climate Since* A.D. *1500* (London: Routledge, 1992); and Philip D. Jones, Raymond S. Bradley, and Jean Jouzel, eds., *Climatic Variations and Forcing Mechanisms of the Last 2000 Years* (Berlin: Springer, 1996). On historians' interpretations of climatic changes see, for example, Emmanuel Le Roy Ladurie, *Times of Feast, Times of Famine: A History of Climate since the Year 1000*, trans. Barbara Bray (Garden City, N.Y.: Doubleday, 1971); Robert I. Rotberg and Theodore K. Rabb, eds., *Climate and History: Studies in Interdisciplinary History* (Princeton, N.J.: Princeton University Press, 1981); T. M. L. Wigley, M. J. Ingram, and G. Farmer, eds., *Climate and History: Studies in Past Climates and Their Impact on Man* (Cambridge: Cambridge University Press, 1981); and H. H. Lamb, *Climate, History and the Modern World*, 2d ed. (London: Routledge, 1995). Scientists exploring history include John Imbrie and Katherine Palmer Imbrie, *Ice Ages: Solving the Mystery* (Short Hills, N.J.: Enslow Publishers, 1979); and the entire issue of *Ambio*, 26, no. 1 (Feb. 1997).

Given the gaps in the existing literature, historians of science can make distinctive contributions to our understanding of global environmental change. This is particularly true because, on time scales of decades to centuries, *ideas about the global environment are changing along with the global environment itself.* This chapter examines theories of climatic change from the Enlightenment to the late twentieth century. Because of the complexity of the issues and the deliberately broad temporal coverage, only an outline of major trends and developments will be presented here. In the interest of further simplification, the focus will be primarily on climate and one climatic factor – temperature.[5] The chapter begins by examining the transformation of the literary tradition of the Enlightenment into the scientific discourse of the late nineteenth century. A brief overview of a number of competing theories of climatic change, circa 1900, sets the stage for a more detailed analysis of the rise of anthropogenic climate concerns, especially those centering around the role of carbon dioxide. Such "macrohistory" provides a much-needed context for more detailed studies of climatology and climatic change. This is not, as Mott Greene would say, the "nth" thesis on Darwinism or "n + 1" on Newton.[6] Rather, it is a first step in a much larger project to examine in detail the history of global change research.

ENLIGHTENMENT

The concern about climatic change, both from natural causes and human activity, is not at all new. While the debate about "global warming" has been prominent in recent years, there have been many other climatic concerns throughout history. For example, Abbé Jean-Baptiste Du Bos (1670–1742), member (later perpetual secretary) of the French Academy, discussed climate change and linked it to cultural changes in his 1719 *Réflexions critiques sur la poësie et sur la peinture.*[7] In this work, ostensibly an essay on aesthetics, Du Bos argued that artistic genius flourished only in countries with suitable climates (always between 25 and 52 degrees north latitude), that the rise and decline of the creative spirit in particular nations can be explained by changes in climate, and that the climate of Europe and the Mediterranean

---

[5] Justification for focusing on climate change is provided by Robert G. Fleagle, *Global Environmental Change: Interactions of Science, Policy, and Politics in the United States* (Westport, Conn.: Praeger, 1994). Fleagle cites the leadership of meteorological institutions and the prominence of weather- and climate-related issues in the development of the field.
[6] Mott T. Greene, "History of Geology," in *Historical Writing on American Science,* ed. Sally Gregory Kohlstedt and Margaret Rossiter, *Osiris,* 2d ser., 1 (1985), 97–116, examines both the opportunities and the pitfalls of working in new historical fields.
[7] Abbé Jean-Baptiste Du Bos, *Réflexions critiques sur la poësie et sur la peinture,* 2 vols. (Paris, 1719). English translation by Thomas Nugent, *Critical Reflections on Poetry, Painting and Music,* 3 vols. (London, 1748). According to Nugent, "there have been very few books published of late years that have met with a better reception, or attained to a greater reputation in the learned world, than the following Critical Reflections."

area had become warmer than in ancient times. In explaining changes on the Italian peninsula in particular, Du Bos suggested: "[T]here has been such a prodigious change in the air of Rome and the adjacent country, since the time of the Caesars, that it is not at all astonishing there should be a difference between the present and ancient inhabitants."[8] Du Bos attributed both cultural differences among nations and differences within the same nation in different eras to environmental changes:

> I conclude ... that as the difference of the character of nations is attributed to the different qualities of the air of their respective countries; in like manner the changes which happen in the manners and genius of the inhabitants of a particular country, must be imputed to the alterations of the qualities of the air of that same country. Wherefore as the difference observable between the French and Italians, is assigned to the difference there is between the air of France and Italy; so the sensible difference between the manners and genius of the French of two different ages, must be attributed to the alteration of the qualities of the French air.[9]

Du Bos's basic argument may be encapsulated as follows: As the grapes of one particular region or year produce a characteristic vintage, so the inhabitants of a particular nation in a given epoch represent a cultural vintage distilled from the overall quality of the air and soil. Only the most favored nations and epochs have produced superior cultural distillations, while most have produced table wines or vinegars.[10] He cited four examples of "illustrious ages" that gave rise to extraordinarily creative cultures: Greece under Philip of Macedon, Rome under Julius and Augustus Caesar, sixteenth-century Italy at the time of Popes Julius II and Leo X, and his own – seventeenth-century France under Louis XIV. Du Bos's idea that climate affected culture was derived in part from the writings of ancient philosophers, geographers, and historians, but also from more proximate sources, such as the work of Jean Bodin, John Barclay, Fontenelle, and Sir John Chardin. Du Bos, in turn, influenced other famous authors, including Edward Gibbon, Johann Gottfried Herder, and Montesquieu (whom Du Bos sponsored for a position in the French Academy).[11]

David Hume (1711–1776) followed Du Bos explicitly on the issue of climate change. In his essay "Of the Populousness of Ancient Nations" (ca. 1750), Hume noted that the advance of cultivation in the nations of Europe had

---

[8] Armin Hajman Koller, *The Abbé Du Bos – His Advocacy of the Theory of Climate: A Precursor of Johann Gottfried Herder* (Champaign, Ill.: Garrard Press, 1937), pp. 26, 98.

[9] Du Bos, *Critical Reflections*, vol. 2 (1748), p. 224.

[10] These ideas were developed further by John Arbuthnot, *An Essay Concerning the Effects of Air on Human Bodies* (London, 1733).

[11] Koller, *Abbé Du Bos*, pp. 67–8, 109–10. For more on climate in the eighteenth century, see Clarence J. Glacken, *Traces on the Rhodian Shore: Nature and Culture in Western Thought from Ancient Times to the End of the Eighteenth Century* (Berkeley: University of California Press, 1967), p. 434. See also Marian J. Tooley, "Bodin and the Mediaeval Theory of Climate," *Speculum*, 28 (1953), 64–83; and E. Fournol, *Bodin prédécesseur de Montesquieu* (Paris, 1896).

caused a gradual change in climate in the previous two millennia. Moreover, he thought similar, but much more rapid, changes were occurring in the Americas:

> Allowing, therefore, this remark [of Du Bos] to be just, that Europe is become warmer than formerly; how can we account for it? Plainly, by no other method, than by supposing that the land is at present much better cultivated, and that the woods are cleared, which formerly threw a shade upon the earth, and kept the rays of the sun from penetrating to it. Our northern colonies in America become more temperate, in proportion as the woods are felled.[12]

The ideas of Du Bos and his followers dominated climatic discourse in the second half of the eighteenth century, generating a powerful vision of the climates of Europe and America as shaping the course of empire and the arts, the concerted efforts of innumerable individuals, in turn, shaping the climate itself. By the end of the eighteenth century, Enlightenment thinkers had come to the following conclusions regarding climate change, culture, and cultivation:

1. Cultures are determined or at least strongly shaped by climate.
2. The climate of Europe had moderated since ancient times.
3. The change was caused by the gradual clearing of the forests and by cultivation.
4. The American climate was undergoing rapid and dramatic changes caused by settlement.
5. The amelioration of the American climate would make it more fit for European-type civilization and less suitable for the "primitive" native cultures.

In response to these precepts, Thomas Jefferson (1743–1826) advocated a practical policy: "Measurements of the American climate should begin immediately, before the climate has changed too drastically. These measurements should be repeated... once or twice in a century, to show the effect of clearing and culture towards the changes of climate."[13]

LITERARY AND SCIENTIFIC TRANSFORMATION: THE AMERICAN CASE

Early settlers in North America found the atmosphere more changeable, the climate harsher, and the storms more violent than in the Old World.[14]

---

[12] David Hume, "Of the Populousness of Ancient Nations," in David Hume, *Essays: Moral, Political, and Literary*, ed. T. H. Green and T. H. Grose, 2 vols. (London, 1875), 1: 432–9.

[13] Thomas Jefferson to Lewis E. Beck, 16 July 1824, in Albert Ellery Bergh, ed., *The Writings of Thomas Jefferson*, vol. 15 (Washington, D.C.: Thomas Jefferson Memorial Association of the United States, 1907), pp. 71–2.

[14] For details, see James Rodger Fleming, *Meteorology in America, 1800–1870* (Baltimore: Johns Hopkins University Press, 1990), pp. 2–3; and Karen Ordahl Kupperman, "The Puzzle of the American Climate in the Early Colonial Period," *American Historical Review*, 87 (1982), 1270.

Explaining why this was so in a region situated farther south than most European nations was a major problem in natural philosophy. Colonials thought rainfall and temperature patterns were changing as the forests were cleared. There was no general agreement, however, about the direction or magnitude of the change.[15] While many hoped the American climate was becoming warmer due to the efforts of settlers, the more philosophically minded thought that many years of observations would be necessary to settle the issue. Jefferson's *Notes on the State of Virginia*, which includes a patriotic defense of the natural phenomena of the New World, presented an apology for the harsh climate and suggested that it was being improved by settlement:

> A change in our climate . . . is taking place very sensibly. Both heats and colds are become much more moderate within the memory even of the middle-aged. Snows are less frequent and less deep. . . . The elderly inform me, the earth used to be covered with snow about three months in every year. The rivers, which then seldom failed to freeze over in the course of the winter, scarcely ever do so now.[16]

Inspired in part by Benjamin Franklin's suggestion that extensive measurements of the climate would be necessary to resolve the issue, Jefferson advised his correspondents to keep weather diaries and send them to the American Philosophical Society.[17] This was the beginning of more systematic data collection. Within two decades, other groups, including college professors in New England, the General Land Office, and the U.S. Army Medical Department, had begun to monitor the climate at diverse locations across the country.[18]

The Enlightenment view of climatic change was rebutted in two distinct ways: literary and scientific. The literary response was spearheaded by Noah Webster (1758–1843); the scientific response came from climatologists, who subjected the growing body of thermometric data to statistical analysis. In his essay of 1799, "On the Supposed Change in the Temperature of Winter," Webster criticized Europeans and Americans who were writing on climatic change for their loose citation of sources, both ancient and contemporary, and the improper inferences they drew from these citations. The force of Webster's critique, however, was blunted by his own indecision on the question of climatic change and cultivation. After a careful rereading of the sources, Webster convinced himself that the climate, if it had not changed outright,

---

[15] William Cronon, *Changes in the Land: Indians, Colonists, and the Ecology of New England* (New York: Hill and Wang, 1983), pp. 122–6, discusses the ecology of these changes.

[16] Thomas Jefferson, *Notes on the State of Virginia* (Paris, 1785; reprint, Gloucester, Mass.: Peter Smith, 1976), p. 79.

[17] Many of the diaries have been preserved. See, e.g., Stephen J. Catlett, ed., *A New Guide to the Collections in the Library of the American Philosophical Society* (Philadelphia: American Philosophical Society, 1987), entry 718.

[18] Fleming, *Meteorology in America*, pp. 9–19 and passim.

was indeed more variable and had, in fact, rearranged itself in response to cultivation.[19]

The republication in 1843 of Webster's essay on climatic change motivated Dr. Samuel Forry (1811–1844) to conduct an analysis of weather data gathered since 1814 at more than sixty locations by the Army Medical Department. Forry concluded that: (a) climates are stable and no accurate thermometric observations warrant the conclusion of climatic change; (b) climates are susceptible of melioration by the changes wrought by the labors of man; but (c) these effects are much less influential than physical geography: oceans, lakes, mountains, dimensions of continents over latitude, and so forth.[20] A similar argument was presented by Lorin Blodget (1823–1901), author of *Climatology of the United States* (1857), who used temperature data from both the Army Medical Department and the Smithsonian Institution to argue that climates must be assumed permanent until proven changeable. For Blodget, vegetation was an effect, not a cause, of climate. Rather than changing the climate, cleared and cultivated lands, unless maintained constantly, will inevitably revert to a state of nature dictated by the climate. The only reliable way to judge climatic change was in the thermometric record, and, according to Blodget: "No series of temperature observations worthy of confidence extends further back in the United States than 78 years. We find from the Philadelphia observations that from 1771 to 1814 the mean annual heat has hardly risen $2°.7$, an increase that may be fairly credited to the extension of the town. This increase may also be due to accident, &c."[21]

A decade later, Charles A. Schott (1826–1901), an assistant in the U.S. Coast Survey well versed in the newly emerging field of statistical data analysis, prepared two innovative monographs on the rainfall and temperature of the United States, using records gathered by the Smithsonian Institution, the Army Medical Department, the Lake Survey, the Coast Survey, the states of New York and Pennsylvania, and private journals extending back into the eighteenth century.[22] Schott prepared a harmonic analysis of the temperature data to examine secular changes in the climate, concluding that

---

[19] Noah Webster, "On the Supposed Change in the Temperature of Winter," *Memoirs of the Connecticut Academy of Arts and Sciences*, 1, pt. 1 (1810), 216–60; two distinct essays read in 1799 and 1806 before the academy, reprinted in Webster, *A Collection of Papers on Political, Literary, and Moral Subjects* (New York, 1843), pp. 119–62.

[20] Samuel Forty, "Research in Elucidation of the Distribution of Heat over the Globe, and especially of the Climatic Features peculiar to the Region of the United States," *American Journal of Science and Arts*, 47 (1844), pp. 18–50, 221–41, especially p. 239.

[21] Lorin Blodget, *Climatology of the United States* (Philadelphia: J. B. Lippincott, 1857), chap. 17, pp. 481–92, quotations at pp. 481, 484.

[22] Charles A. Schott, "Tables and Results of the Precipitation, in Rain and Snow, in the United States and at Some Stations in Adjacent Parts of North America and in Central and South America," *Smithsonian Contributions to Knowledge*, 18, Article II (1872); Schott, "Tables, Distribution, and Variations of the Atmospheric Temperature in the United States and Some Adjacent Parts of North America," *Smithsonian Contributions to Knowledge*, 21, Article V (1876); Schott's manuscript is in RG 27, National Archives.

there is nothing in these curves to countenance the idea of any permanent change in the climate having taken place, or being about to take place; in the last 90 years of thermometric records, the mean temperatures showing no indication whatever of a sustained rise or fall. The same conclusion was reached in the discussion of the secular change in the Rain-Fall, which appears also to have remained permanent in amount as well as in annual distribution.[23]

Cleveland Abbe (1838–1916), chief scientist in the U.S. Army Signal Office, the national weather service of the time, agreed with Schott and Loomis that the old debates about climatic change had finally been settled. In an article titled "Is Our Climate Changing?" Abbe defined the climate as "the average about which the temporary conditions permanently oscillate; it assumes and implies permanence."[24] Alluding to the recent discovery of the ice ages, Abbe conceded that "great changes have taken place during geological ages perhaps 50,000 years distant; but no important climatic change has yet been demonstrated since human history began." He continued:

> It will be seen that rational climatology gives no basis for the much-talked-of influence upon the climate of a country produced by the growth or destruction of forests, the building of railroads or telegraphs, and the cultivation of crops over a wide extent of prairie. Any opinion as to the meteorological effects of man's activity must be based either upon the records of observations or on *à priori* theoretical reasoning.... The true problem for the climatologist to settle during the present century is not whether the climate has lately changed, but what our present climate is, what its well-defined features are, and how they can be most clearly expressed in numbers.[25]

Thus, the shift was complete, circa 1890, from literary to empirical studies of climate. It is important to remember, however, that this transformation was not the end of climatic determinism, as evidenced, for example, by the work of Ellsworth Huntington.[26]

SCIENTIFIC THEORIES OF CLIMATIC CHANGE

The debate over climatic change caused by human activities ended just about the time when scientists discovered that the earth had experienced ice ages and interglacials – tremendous advances and retreats of the glaciers over

---

[23] Schott, "Tables, Distribution, and Variations of the Atmospheric Temperature," p. 311. Of similar opinion were Elias Loomis and H. A. Newton, "On the Mean Temperature, and On the Fluctuations of Temperature, at New Haven, Conn., Lat. 41° 18′ N., Long. 72° 55′ W. of Greenwich," *Transactions of the Connecticut Academy of Arts and Sciences*, 1, pt. 1 (1866), 194–246.
[24] Cleveland Abbe, "Is Our Climate Changing?" *Forum*, 6 (1889), 678–88, quotation at 679.
[25] Ibid., pp. 687–8.
[26] On Huntington see Fleming, *Historical Perspectives on Climate Change*, pp. 95–106.

geological time periods. These discoveries, especially the need to explain multiple glaciations, produced a plethora of complex, but highly speculative, theories of climatic change involving astronomical, physical, and geological factors. Joseph Adhémar, James Croll, Svante Arrhenius, T. C. Chamberlin, and many others attempted explanations based on the behavior of the oceans, the earth's orbital elements, and the global carbon budget.[27] In the mid- to late nineteenth century, infrared radiation was being measured at increasingly long wavelengths, first by Macedonio Melloni, with his "thermal telescope" in the very near infrared, and then by Samuel P. Langley, with his bolometer, at wavelengths of about five microns.[28] John Tyndall (1820–1893) was doing pioneer work on the absorption and emission properties of atmospheric constituents, particularly aqueous vapor and carbonic acid (H_2O and CO_2). Tyndall thought that changes in the amount of radiatively active gases in the atmosphere could have produced "all the mutations of climate which the researches of geologists reveal."[29]

In 1896, the noted Swedish chemist Svante Arrhenius (1859–1927) published a long memoir "On the Influence of Carbonic Acid in the Air Upon the Temperature of the Ground." His theory explained the glacial periods and other great climatic changes by the ability of carbon dioxide to absorb infrared radiation emitted from the earth's surface. He argued that variations in the trace components of the atmosphere could have a very great influence on the overall heat budget. His calculations, which were based on a very limited understanding of infrared radiation, indicated that a halving of the percentage of carbon dioxide in the air would lower the temperature of the earth's surface by 4 degrees; on the other hand, a doubling of the percentage of carbon dioxide in the air would raise the temperature of the earth's surface by 4 degrees. Arrhenius argued that a reduction of atmospheric carbon dioxide levels of 55 to 62 percent would be sufficient to cause glaciation at 40 to 50 degrees north latitude.

As Elisabeth Crawford has shown, Arrhenius's paper did not stem from his concern over increasing levels of CO_2 in the atmosphere from the burning

---

[27] Joseph Alphonse Adhémar, *Révolutions de la mer, déluges périodiques* (Paris, 1842); James Campbell Irons, *Autobiographical Sketch of James Croll, with Memoir of His Life and Work* (London, 1896). On Arrhenius and Chamberlin, see Fleming, *Historical Perspectives on Climate Change*, pp. 74–93; and *The Legacy of Svante Arthenius: Understanding the Greenhouse Effect*, ed. H. Rodhe and Robert J. Charlson (Stockholm: Royal Swedish Academy of Sciences, 1988), especially pp. 9–32.

[28] On Melloni (1798–1854), see E. S. Barr, "The Infrared Pioneers II: Macedonio Melloni," *Infrared Physics*, 2 (1962), 67–73. For Langley (1834–1906): Samuel P. Langley, "Observations on Invisible Heat Spectra and the Recognition of Hitherto Unmeasured Wavelengths, Made at the Allegheny Observatory," *Philosophical Magazine*, 21 (1886), 394–409. See also J. T. Kiehl, "A History of the Development of Atmospheric Radiation, 1800–1930," typescript, National Center for Atmospheric Research, Boulder, Colo., 1986.

[29] John Tyndall, "On the Absorption and Radiation of Heat by Gases and Vapours, and on the Physical Connexion of Radiation, Absorption, and Conduction," *Philosophical Magazine*, 4th ser., 22 (1861), 169–94, 273–85; and Fleming, *Historical Perspectives on Climate Change*, pp. 112–29.

of fossil fuels.³⁰ He regarded "volcanic exhalations" as the chief source of carbonic acid in the atmosphere. Industry played a minor role. By his estimate, the world's current production of coal, if transformed into carbonic acid, would correspond to only one one-thousandth of the CO_2 in the atmosphere. There was really no buildup, however, since anthropogenic carbon emissions were just compensated by the formation of limestone and other minerals. Despite Arrhenius's recently growing reputation as a "father" of the greenhouse effect, we should understand that his work was motivated by a desire to explain the ice ages. Rather than being unique or especially prophetic about the effects of a CO_2 doubling, his results were only superficially similar to the results of today's climate models. One might judge that he got the right answer for the wrong reasons.³¹ As one of his biographers pointed out, "theoretical explanations of poorly known natural systems display a high mortality rate when confronted with accumulated evidence."³² Such was the fate of Arrhenius's geophysical work, which served primarily as a catalyst for the more empirically based investigations of others.

About a decade later, in 1903, Arrhenius, who came from a cold climate, noted that burning fossil fuels might help prevent a rapid return to the conditions of an ice age or inaugurate a new carboniferous age of enormous plant growth:

> We often hear lamentations that the coal stored up in the earth is wasted by the present generation without any thought of the future.... [However,] [b]y the influence of the increasing percentage of carbonic acid in the atmosphere, we may hope to enjoy ages with more equable and better climates, especially as regards the colder regions of the earth, ages when the earth will bring forth much more abundant crops than at present, for the benefit of a rapidly propagating mankind.³³

Thus, rising CO_2 levels and rising population were considered good, and expanded fossil fuel consumption, rather than clearing and cultivation, was proposed as the agent of beneficial changes in the climate. In short, until recent decades, most scientists did not believe that increased CO_2 levels would result in global warming, because it was thought that a small amount of the gas

---

[30] Elisabeth Crawford, *Arrhenius: From Ionic Theory to the Greenhouse Effect* (Canton, Mass.: Science History Publications, 1996). See also Spencer Weart, "The Discovery of the Risk of Global Warming," *Physics Today* (Jan. 1997), 34–40.

[31] See, for example, the special issue of *Ambio*, 26, no. 1 (Feb. 1997); see also J. E. Kutzbach, "Steps in the Evolution of Climatology: From Descriptive to Analytic," in *Historical Essays on Meteorology, 1919–1995*, ed. James Rodger Fleming (Boston: American Meteorological Society, 1996), p. 357.

[32] Gustaf O. S. Arrhenius, "Svante Arrhenius' Contribution to Earth Science and Cosmology," in *Svante Arrhenius: till 100-årsminnet av hans födelse* (Uppsala: Almqvist and Wiksells, 1959), pp. 76–7.

[33] Svante Arrhenius, *Worlds in the Making: The Evolution of the Universe*, trans. H. Borns (New York: Harper and Brothers, 1908), p. 63.

would absorb all the available long wave radiation; in this view, any additional increases in CO_2 would augment plant growth but would not change the radiative heat balance of the planet. This is quite different from both the Enlightenment pastoral view that clearing and cultivation lead to beneficial changes in the climate, and the current view that industrial emissions and massive deforestation cause a harmful, pollution-induced "super greenhouse effect." In fact, until recent decades, increased CO_2 was not considered to be an important agent of climate change.[34]

By 1900, most of the chief theories of climate change had been proposed, if not yet fully explored: changes in solar output, changes in the earth's orbital geometry, changes in terrestrial geography (including the form and height of continents and the circulation of the oceans), and changes in atmospheric transparency and composition, in part due to human activities.[35] Of course, there were many others. William Jackson Humphreys (1862–1949), author of *Physics of the Air* and a strong proponent of the theory that volcanic dust was the leading cause of ice ages, thought that none of the current theories was adequate: "Change after change of climate in an almost endless succession, and even additional ice ages, presumably are still to be experienced, though ... when they shall begin, how intense they may be, or how long they shall last no one can form the slightest idea."[36] Most scientists of the time supported only one or another of the major mechanisms of climatic change; some grudgingly admitted that other mechanisms might play a secondary role.

In the 1930s, the Serbian astronomer Milutin Milanković (1879–1958) outlined a comprehensive "astronomical theory of the ice ages" caused by periodic changes in the earth's orbital elements, a topic that was debated until the 1980s.[37] Evidence for glaciation in low latitudes was explained by Wladimir Köppen and Alfred Wegener as the result of continents drifting northward under climate zones controlled mainly by latitude.[38] Although this theory was not accepted by geologists, it now is seen as a first step in paleoclimatic reconstruction. Atmospheric heat budgets were first constructed early in the twentieth century by William Henry Dines and George Clark Simpson,

---

[34] E.g., W. J. Humphreys, *Physics of the Air*, 2d ed. (Philadelphia: J. B. Lippincott, 1920), and Richard Joel Russell, "Climatic Change through the Ages," in U.S. Dept. of Agriculture, *Climate and Man: Yearbook of Agriculture 1941* (Washington, D.C.: U.S. House of Representatives, 1941), pp. 67–97.

[35] Many of these theories are surveyed in C. E. P. Brooks, *Climate Through the Ages: A Study of the Climatic Factors and Their Variations*, 2d ed., rev. (New York: McGraw Hill, 1949).

[36] W. J. Humphreys, "Volcanic Dust and Other Factors in the Production of Climatic Changes and Their Possible Relation to Ice Ages," *Journal of the Franklin Institute*, 176 (1913), 132.

[37] On Milanković, see John Imbrie and Katherine Palmer Imbrie, *Ice Ages* (Short Hills, N.J.: Enslow Publishers, 1979), and A. Berger, "Milankovitch Theory and Climate," *Reviews of Geophysics*, 26 (1988), 624–57. Excerpts from his autobiography with comments by his son appear in *Milutin Milanković, 1879–1958* (Katlenburg-Lindau, F.R.G.: European Geophysical Society, 1995).

[38] Wladimir Köppen and Alfred Wegener, *Die Klimate der geologischen Vorzeit* (Berlin: Gebruder Borntraeger, 1924). See also Martin Schwarzbach, *Alfred Wegener: The Father of Continental Drift*, trans. Carla Love (Madison, Wis.: Science Tech, 1986), pp. 86–101.

among others.[39] Measurements of infrared radiation at longer wavelengths – including the 8–12 micron atmospheric "window" – and at finer band resolutions were completed in the 1930s.[40] And in 1938, G. S. Callendar read a paper to the Royal Meteorological Society that argued that CO_2 from fossil fuel consumption had caused a modest but measurable increase in the earth's temperature of about one quarter of a degree in the previous fifty years.[41] All of these issues, especially whether the earth would experience a new ice age or would warm because of greenhouse gas emissions, continued to be debated after 1940.

GLOBAL WARMING: EARLY SCIENTIFIC WORK AND PUBLIC CONCERN

The role of anthropogenic carbon dioxide in climatic change was reevaluated by Guy Stewart Callendar (1897–1964), a British steam engineer and amateur meteorologist. In 1949 Callendar acknowledged the "checquered history" of CO_2: "[I]t was abandoned for many years when the preponding influence of water vapour radiation in the lower atmosphere was first discovered, but was revived again a few years ago when more accurate measurements of the water vapour spectrum became available."[42] Noting that humans had long been able to intervene in and accelerate natural processes, Callendar pointed out that humanity was now intervening heavily into the slow-moving carbon cycle by "throwing some 9,000 tons of carbon dioxide into the air each minute."[43]

In a remarkable series of papers, published between 1938 and 1961, Callendar reexamined the role of anthropogenic carbon dioxide in the "global warming" being experienced at the time. He pointed out that fuel combustion had generated some 150,000 million tons of carbon dioxide in the previous half century and that three-quarters of this had remained in the atmosphere.

---

[39] See, for example, W. H. Dines, "The Heat Balance of the Atmosphere," *Quarterly Journal of the Royal Meteorological Society*, 43 (1917), 151–8; and G. C. Simpson, "Some Studies in Terrestrial Radiation," *Memoirs of the Royal Meteorological Society*, 2 (1928), 69–95. A review paper is Garry E. Hunt, Robert Kandel, and Ann T. Mecherikunnel, "A History of Presatellite Investigations of the Earth's Radiation Budget," *Reviews of Geophysics*, 24 (1986), 351–6.

[40] For example, Louis Russell Weber, "The Infrared Absorption Spectrum of Water Vapor Beyond 10μ," PhD diss., University of Michigan, 1932; and Paul Edmund Martin, "Infrared Absorption Spectrum of Carbon Dioxide," *Physical Review*, 41 (1932), 291–303. On infrared radiation in the atmosphere circa 1950, see L. Goldberg, "The Absorption Spectrum of the Atmosphere," in *The Earth as a Planet*, ed. G. P. Kuiper (Chicago: University of Chicago Press, 1954), 434. ff.

[41] G. S. Callendar, "The Artificial Production of Carbon Dioxide and Its Influence on Temperature," *Quarterly Journal of the Royal Meteorological Society*, 64 (1938), 223–40.

[42] G. S. Callendar, "Can Carbon Dioxide Influence Climate?" *Weather*, 4 (1949), 310–14; quotation at 310.

[43] Callendar, "The Composition of the Atmosphere through the Ages," *Meteorological Magazine*, 74 (1939), 38.

His 1939 paper contains an early statement of the now-familiar claim that humanity is conducting a "grand experiment" and has become an "agent of global change." Callendar considered it a "commonplace" that humanity had sped up natural processes and had interfered with the carbon cycle:

> As man is now changing the composition of the atmosphere at a rate which must be very exceptional on the geological time scale, it is natural to seek for the probable effects of such a change. From the best laboratory observations it appears that the principal result of increasing atmospheric carbon dioxide... would be a gradual increase in the mean temperature of the colder regions of the earth.[44]

According to Callendar: "The five years 1934–38 are easily the warmest such period at several stations whose records commenced up to 180 years ago."

In a 1958 paper on the amount of carbon dioxide in the atmosphere, Callendar noted the "close agreement" between the cumulative amount of fossil fuel consumption and the rise in measured ambient CO_2 concentrations. He considered this agreement possibly coincidental, but potentially significant, pending the outcome of further investigations. His figures indicate a rate of increase of CO_2 of about 25 percent per century, not far from modern estimates. Callendar also pointed out that the rate of CO_2 increase had been accelerating recently, perhaps due to the expansion of industry.[45] His value for the amount of CO_2 in the atmosphere was about 325 parts per million (ppm); this is in basic agreement with the modern "Keeling curve," which started at 315 ppm in 1957.

By 1961, Callendar had concluded that the trend toward higher temperatures was significant, especially north of 45 degrees latitude; that increased use of fossil fuels had caused a rise of the concentration of CO_2 in the atmosphere; and that increased sky radiation from the extra CO_2 was linked to the rising temperature trend.[46] Callendar's work, contrary to the assertions of some, was not "largely ignored because of World War II," nor was he quite the obscure figure others make him out to be.[47] In 1944, Gordon Manley noted Callendar's valuable contributions to the study of climatic change and provided support for what Roger Revelle later called the "Callendar effect,"

---

[44] Ibid.
[45] Callendar, "On the Amount of Carbon Dioxide in the Atmosphere," *Tellus*, 10 (1958), 243–8.
[46] Callendar, "Temperature Fluctuations and Trends over the Earth," *Quarterly Journal of the Royal Meteorological Society*, 87 (1961), 1–11.
[47] M. D. Handel and J. S. Risbey, "An Annotated Bibliography on the Greenhouse Effect and Climate Change," *Climatic Change*, 21, no. 2 (1 June 1992), 97–255, say that Callendar's work was "quickly ignored as World War II intervened and Northern Hemisphere surface temperatures began to decline in the 1940s." Spencer Weart, "From the Nuclear Frying Pan into the Global Fire," *Bulletin of the Atomic Scientists* (June 1992), 19–27, mentions Callendar's 1938 article to the Royal Meteorological Society but indicates that his work was obscure and no one really cared. Spencer Weart, "Global Warming, Cold War, and the Evolution of Research Plans," *Historical Studies in the Physical Sciences*, 27 (1997), 319–56, again emphasizes Callendar's obscurity and amateur status.

linking the "global warming" of the first half of the twentieth century to industrial emissions of CO_2.

In the 1950s, several developments combined to increase public awareness of geophysical issues. Many people were certain that atmospheric nuclear testing was changing the earth's weather. Weather bureau officials dismissed such speculation, arguing that the impact of the tests on the atmosphere was primarily local and temporary. Radioactive fallout posed far more insidious dangers to human health and environmental quality. Radioactive materials in the environment, however, provided new tools that enabled ecologists and geophysicists to trace the flow of materials through the biosphere, atmosphere, and oceans. The International Geophysical Year (1957–8) provided an organizational and financial boost to academic geophysics, including meteorology. The successful launch of the Soviet IGY satellite *Sputnik*, however, combined with the failure of the U.S. Vanguard program, precipitated a crisis in public confidence, a "race" to close a perceived missile gap, and an increase in Cold War tensions. Some even wanted to use weather control as a weapon of war.

"Global warming" was firmly on the public agenda in the late 1940s and early 1950s, as Northern Hemisphere temperatures reached an early-twentieth-century peak. Concerns were expressed both in the scientific and popular press about changing climates, rising sea levels, loss of habitat, and shifting agricultural zones. In 1950, the *Saturday Evening Post* asked, "Is the World Getting Warmer?" Topics of climatic speculation cited in the article included a warmer planet; rising sea levels; shifts of agriculture; the retreat of the Greenland ice cap and other glaciers; changes in ocean fisheries, perhaps the result of changes in the Gulf Stream; and the possible migration of millions of people displaced by climate change. The article quoted Hans Ahlmann, a climatologist at Stockholm University, who was of the opinion that "if older people say that they have lived through many more hard winters in their youth, they are stating a real fact." Thomas Jefferson would have concurred. In fact, it seems there is little that is actually new or unique in popular climate discourse. Ahlmann also was concerned about the unprecedented rate of change. He pointed out that the climate was now changing so fast that "each new contribution to the subject is out of date almost as soon as it is published." Perhaps he also meant to say that climatology was experiencing unprecedented rates of change. *Today's Revolution in Weather*, a 1953 compilation of news items on weather extremes and global warming, reiterated popular concerns over the social consequences of global warming. The compiler, economic forecaster William J. Baxter, predicted a climate-induced real estate boom in the north and advised, "Go north-west young man."

On a more serious note, in 1956 the infrared physicist Gilbert Plass (b. 1920) noted that humanity was conducting an uncontrolled experiment by releasing CO_2 into the atmosphere: "If at the end of this century, measurements show that the carbon dioxide content of the atmosphere has risen appreciably and at the same time the temperature has continued to rise throughout the world, it will be firmly established that carbon dioxide is an important factor in causing climatic change." A year later this sentiment was popularized by Roger Revelle (1909–1991).[48] Global warming, however, had not yet become an enduring policy issue.

GLOBAL COOLING, GLOBAL WARMING

In the 1970s, fear of sudden "global cooling" and the possibility of a return to an ice age climate brought atmospheric scientists and the Central Intelligence Agency together in an attempt to determine the geopolitical consequences of a failure of the Soviet grain harvest.[49] The leading culprits in this cooling were thought to be particulates from industrial sources, increased cirrus clouds due to jet airplane contrails, and the configuration of the earth's orbital elements according to the astronomical theory of the ice ages. The popular press was filled with articles on the advance of the glaciers.[50]

The RAND Corporation, fearful that the United States might be harmed either inadvertently or maliciously by changes in the climate, had developed a program on "climate dynamics for environmental security" by 1970. Indeed, in the decades following World War II, many meteorologists and their military patrons were convinced that weather and climate control by cloud seeding was entirely feasible and that they were in a race with the Russians for control of the environment. The sentiment of the time is captured in a statement of Professor Henry G. Houghton, Chairman of the Department of Meteorology at MIT: "I shudder to think of the consequences of a prior Russian discovery of a feasible method of weather control. . . . An unfavorable modification of our climate in the guise of a peaceful effort to improve Russia's climate could seriously weaken our economy and our ability to resist."[51]

In the second half of the twentieth century, electronic computers and satellites have offered new privileged perspectives on climate issues. Shortly after the development of numerical weather prediction, a computer model

---

[48] Gilbert N. Plass, "Effect of Carbon Dioxide Variations on Climate," *American Journal of Physics*, 24 (1956), 387; Roger Revelle and Hans E. Suess, "Carbon Dioxide Exchange between Atmosphere and Ocean and the Question of an Increase in Atmospheric CO_2 during the Past Decades," *Tellus*, 9 (1957), 19. Revelle's phrase "geophysical experiment" is in U.S. House of Representatives, Committee on Appropriations, *National Science Foundation – International Geophysical Year* (Washington, D.C., 1956), p. 473.

[49] See, for example, Lowell Ponte, *The Cooling* (Englewood Cliffs, N.J.: Prentice Hall, 1976); and U.S. Central Intelligence Agency, "A Study of Climatological Research as It Pertains to Intelligence Problems," and "Potential Implications of Trends in World Population, Food Production, and Climate," reprinted in the Impact Team, *The Weather Conspiracy: The Coming of the New Ice Age* (New York: Ballentine, 1974).

[50] For example, Francis Bello, "Climate: The Heat May Be Off," *Fortune* (Aug. 1954), 108–11, 160, 162, 164; and Betty Friedan, "The Coming Ice Age," *Harper's Magazine*, 217 (Sept. 1958), 39–45.

[51] Henry G. Houghton, "Present Position and Future Possibilities of Weather Control," in *Final Report of the United States Advisory Committee on Weather Control*, vol. 2, p. 288, as quoted in *Newsweek* (13 January 1958), 54.

known as Nile Blue was developed by the Advanced Research Projects Administration in the U.S. Department of Defense (DARPA). It was hoped that this model could be used to test the sensitivity of the climate to major perturbations, including Soviet tinkering and effects that could result from a major environmental war. In 1967, Syukuro Manabe and Richard T. Wetherald published the computerized equivalent of Arrhenius's much earlier exercise in cosmic physics.[52] And one of the earliest public reports on artificial earth satellites confided that "eye-in-the-sky" satellites, designed by the Army Signal Corps, would be used to monitor the earth's weather from space, support the efforts of the climate modelers, monitor changes in global weather patterns and heat budgets (perhaps natural, perhaps the result of Russian tinkering), and trace the effects of atmospheric nuclear tests.

Although the cooling mechanisms – industrial particulates, contrails, volcanic eruptions, and the astronomical theory – are still factors in ongoing debate, the dominant concern since the late 1980s has been "global warming." In 1988, NASA scientist James Hansen announced to Congress and the world: "Global warming has begun."[53] Hansen went on to report that, at least to his satisfaction, he had seen the "signal" in the climate noise and that we were in for a hell of a warming, perhaps in the form of a runaway greenhouse effect. This revelation has been accompanied by a shift in our relationship to the earth's atmosphere. The "sheltering sky" has lost its meaning; it has turned menacing even when the winds are calm. How can we enjoy a day at the beach if we know that because of stratospheric ozone depletion a sunburn could lead to skin cancer? Were the killer hurricanes Gilbert, Hugo, and Andrew the result of human intervention in the climate? Probably not. Still, there is no way around it for the realist or the skeptic: Humanity is now a geophysical agent, more menacing than the old nemeses – volcanoes, hurricanes, or tsunamis. While most geophysical agents are local or at least short-lived, humanity's industrial emissions are long-lived, chronic, and upward trending; and they threaten (by most accounts) the very habitability of the planet.

Global environmental change is a hybrid of nature, politics, and discourse. Apprehensions about the climate did not begin in 1988 or even in 1896. Humankind's relationship to nature and the environment is both culture-bound and historically contingent; this includes our own current apprehensions and fears.[54]

The Abbé Du Bos's book on poetry and painting argued for a link between changes in the climate and the rise and fall of creative genius. His theory

---

[52] Syukuro Manabe and Richard T. Wetherald, "Thermal Equilibrium of the Atmosphere with a Given Distribution of Relative Humidity," *Journal of the Atmospheric Sciences*, 24 (1967), 241–59; see also Manabe, "Early Development in the Study of Greenhouse Warming: The Emergence of Climate Models," *Ambio*, 26, no. 1 (1997), 47–51.
[53] *New York Times*, 24 June 1988, p. 1.
[54] See, for example, Bruno Latour, *We Have Never Been Modern* (Cambridge, Mass.: Harvard University Press, 1993).

of climatic influence, although based on shaky philosophical and literary foundations, lived on in the work of Montesquieu, Gibbon, and others, and it found a receptive audience among American colonists and early patriots who hoped that the climate of the New World was being improved by settlement and cultivation. In the nineteenth and twentieth centuries, this culture-bound discussion of climatic change was superseded by more objective (but still culture-bound) attempts to examine the atmosphere and its changes. The modern, scientific description of weather and climate has been gradually established since about the mid-nineteenth century. Like most sciences, it has focused on understanding, prediction, and control – attempting to reduce atmospheric phenomena to their equations of motion, chemical constituents, or other "manageable" components. The atmosphere, however, is not so easily characterized.

Recently, pessimistic forecasts of economic and other dislocations related to global change have forced social scientists and policy makers to return to the human dimensions of the atmosphere. Major environmental treaties, including the Montreal Protocol and the Framework Convention on Climate Change, have been stimulated by growing climate apprehensions. As of 2001 neither the United States nor the United Kingdom had ratified the Kyoto Protocol of 1997, and climate negotiations were at a standstill. There is a flood of new literature on managing planet Earth.[55] Isn't it time for historical, literary, and other humanistic explorations and reevaluations of environmental change as well?

As both our technical prowess and our capacity to pollute increase, it is crucial that we understand how civilizations perceive – and have perceived, relate to – and have related to, the natural environment. The history of science has a particularly valuable contribution to make by elucidating the cultural roots of environmental issues. This history is one of understanding (and misunderstanding), foreboding, and intervention. We need to learn how people in the past have apprehended global change. The result should be a better understanding of the science and policy of global change, a better understanding of the role of global change education in the modern world, and a view of the human dimensions of global changes rendered more complete by a study of the past.

---

[55] For example, "Managing Planet Earth," special theme issue of *Scientific American,* September 1989, and National Academy of Sciences, *Policy Implications of Greenhouse Warming* (Washington, D.C.: National Academy Press, 1991).

INDEX

Abbe, Cleveland, 641
Abbott, Edwin A., 101–2, 466
Abel, Niels Henrik, 123, 476
abiogenesis, 45
Abstract Impressionism, 214, 215
abstraction, and art, 193–4, 214–15
Académie des Sciences (Paris), 45, 85, 176–81, 227
accelerators, and particle physics, 371, 372
acidity, hydrogen theory of, 259
action, and quantum mechanics, 332–4
action-at-a-distance theories, 312, 323
Adet, Pierre-Auguste, 179
Adhémer, Joseph, 642
Advanced Research Projects Agency (ARPA), 424, 649
affinity: and chemical classification, 178, 247–8; and theory of chemical structure, 266
aggregate theory, 431, 433
Agnesi, Maria, 54
Agnes Scott College, 62
Ahlmann, Hans, 647
Aiken, Howard, 600
air engine, 291
Airy, George Biddell, 159, 281, 313, 541, 542
Albrecht, Andreas, 535
alchemy, 176
alcohol, and elemental composition, 258
Alexandroff, Paul S., 130
Alfvén, Hannes, 550
Algarotti, Francesco, 70
algebra, 117, 125, 474. *See also* mathematics
alizarin, 271
allotropy, 256
alpha-helix concept, 441, 442
Alpher, Ralph, 530
Alpher-Herman-Follin theory, 530–1
Altarelli, Guido, 392–3

altazimuth circle, 155
alternating current (AC) systems, 320
Alvin (undersea submersible), 551
amateur tradition, in mathematics, 125
American Association for the Advancement of Science, 574
American Association of University Women, 67
American Astronomical Society, 66–7
American Chemical Society, 66, 67, 402, 403, 408–9, 591n25
American Institute of Electrical Engineers (AIEE), 609
American Mathematical Society, 66
American Physical Society, 66, 68, 151
American Research and Development Corporation, 150–1
American system of manufacture, 567, 569
Ampère, André-Marie, 279, 312, 477, 478
analog device, 599–600
analogy: and analytical approach to physics, 92; and chemical classification, 250; and metaphor in nuclear physics, 207
Analytical Engine, 600
Anderson, Carl D., 367, 380
Andromeda galaxy, 525
Ångström, A. J., 282
aniline dyes, 270
animal electricity, 233–6
anthologies, of scientific poetry and literature, 102–3
anthropic principle, 50, 51
antiatomism, 239, 241, 254
antifoundationalism, 48
antiphlogistonist theory, 177, 189
anti-Semitism, in U. S. physics, 589
Apollo lunar landing, 552
Appleton, Edward, 416

apprenticeship, and scientific education in England, 145
Arago, François, 274
Archive for the History of Quantum Physics (AHQP), 331
argument from design, 50
Aristotelianism, 22–3
Aristotle, 21, 23, 39n11, 219, 220, 451, 463
arithmetization, of analysis, 477
Armstrong, Henry, 267
Army Signal Corps, 649
Arnold, Matthew, 85, 93
aromatic compounds, 248
Arrhenius, Svante, 17, 240, 353, 642–4
art, representation and abstraction in, 193–4, 214–15. *See also* illustration; visual imagery
artificial intelligence, 608, 610–11, 614
Artin, Emil, 117
Aryan science, 15, 586–9, 596
Arzamas-16 laboratory, 595
Association for Computing Machinery (ACM), 608
Association for Women in Science and Engineering (WISE), 68
Astbury, William T., 439, 441
asteroids, 155
Aston, Francis William, 362, 417
astronomy: and big science, 170–3; cosmologies and cosmogonies of space and time, 522–37; developments in since eighteenth century, 12–14; different goals of in late eighteenth and nineteenth centuries, 160–5; and electromagnetic spectrum, 165–7; instrumentation and history of, 6; and literature, 97, 98, 100–103, 104; and Martian canal controversy, 39n8; and plasma physics, 417; and popularization of science, 79, 80–1; of position, 154–9; and space programs, 167–70. *See also* astrophysics; cosmology; extraterrestrial life; planets; stars; Sun; universe
astrophysics: and astronomy, 516–18; and particle physics, 393; and reflecting telescopes, 162–4; and stellar models, 518–21; and stellar spectroscopy, 514–16. *See also* astronomy; stars; Sun; universe
asymmetric synthesis argument, 261–2
asymptotic freedom, and grand unified theory, 393
Atkinson, Robert, 520
Atomic Energy Commission, 130, 423, 424, 555, 576, 591
atomic-molecular hypothesis, 240
atomic notation, 185, 186

atomic weights, 7, 241–3, 253–4
atomism, 7, 241, 255–6
atoms: Bohr's theory of, 197, 200, 336–9; chemical versus physical concepts of, 238–9; and gases, 239–41; representation of, 213, *214f*; and spectroscopy, 12–13
Austin, Louis W., 416–17
Australia, and women in physical sciences, 68
authoritarian regimes, 579, 593
autobiography, and relationship between literature and physical sciences, 108
Avery, Oswald T., 443
Avogadro, Amedeo, 242, 252, 265

Baade, Walter, 532
Babbage, Charles, 16, 125, 461, 600
Bacher, Robert, 368
background radiation, 533–4, 535
Bacon, Francis, 96
Badash, Lawrence, 353
Bahcall, John, 171
Baily, Francis, 509
Baire, René, 479
Baker, H. F., 125
Balfour, Arthur, 84
Ball, Robert, 80
Balmer, Johann K., 283
Balzac, Honoré de, 101
Banach, Stefan, 486
Bancroft, Wilder D., 408, 431
Banks, Sir Joseph, 74
Banville, John, 102
Bardeen, John, 419
Barrow, John D., 51n42
Barth, Karl, 47, 49
Bartow, Edward and Virginia, 66
Bassi, Laura, 54, 60
bathythermograph, 554
Baudelaire, Charles, 101
Baxter, William J., 647
Bayesianism, 31–2
Beagle (ship), 81
Bearman, David, 442
Beaumont, Léonce Elie de, 538
Beck, Robert, 619–20
Beckett, Samuel, 102
Becquerel, Antoine Henri, 352
Becquerel, Edmond, 511
Béguyer de Chancourtois, Alexandre Emile, 251
Behe, Michael, 50n37
Bell, John S., 29, 349
Bell Laboratories, 11, 419, 421–2, 424, 576
Beniger, James, 601
Bennett, J. A., 159

benzene, 180, 248, 266–7, 269
Bergman, Torbern, 176–7
Bergmann, Max, 431
Beria, Lavrenti, 585
Berkeley, George, 29, 452–4
Berkner, Lloyd, 555
Berlin Academy of Sciences, 54
Bernal, John D., 441
Bernard, Claude, 219
Bernoulli, Daniel, 471, 473, 490
Bernstein, Michael A., 556n34
Berthelot, Marcellin, 185, 239–40
Berthollet, Claude-Louis, 177, 242, 279
Berzelius, J. J., 7, 181, 227, 242, 243–4, 245, 256, 257, 430
Bessel, Friedrich Wilhelm, 156, 158–9, 507
beta decay, 378–9, 383
beta-ray controversy, 363, 367
Bethe, Hans, 368, 381, 414, 419, 519, 520
Bewick, Thomas, 75
Bhabha, Homi, 423
Bianchi, Luigi, 124
Bieberbach, Ludwig, 128
big bang theory, 520, 528–9, 530–2, 533–4, 536
Big Science: ideological significance of, 592–4; and postwar physics, 366
biochemistry, 229
biography, and women in physical sciences, 56–7. *See also* autobiography
biology: and macromolecules, 437–40; and physical sciences in seventeenth and eighteenth centuries, 221–4. *See also* life sciences; molecular biology
Biot, Jean-Baptiste, 274, 279
Birkhoff, George D., 127
Bjerknes, Wilhelm, 549
Bjerrum, Niels, 336
Bjorken, James, 390
Black, Joseph, 231
blackbody spectrum, 332–3, 335, 503
Blackett, Patrick Maynard Stuart, 362
Blake, William, 96, 97–8
Bliss, Gilbert Ames, 127
Bloch, Felix, 414, 626
Blodget, Lorin, 640
Blodgett, Katharine, 63
blood circulation, analyses of, 222–4, 226, 229–30
Boas, Marie, 107
Bocage, André E. M., 618
Bôcher, Maxime, 483
Bochner, Salomon, 484, 487
Boerhaave, Hermann, 224–5
Bohm, David, 349, 419, 421

"Bohm diffusion," 419
Bohr, Niels, 9, 197, 203, 204, 209, 283, 337–9, 340–1, 348, 357, 359, 369, 378, 397
Bohr Institute (Copenhagen), 368
Bohr-Kramers-Slater (BKS) theory, 341, 342–3
Bolter, Jay, 612
Boltzano, Bernhard, 474
Boltzmann, Ludwig, 12, 57, 239, 304, 333, 493–4, 495, 498–9, 501–2, 503, 523
Bolyai, János, 464
Bolza, Oskar, 127
Bondi, Hermann, 531
Bonnet, P. O., 123
Bono, James J., 107
books, and popularization of science, 75, 77, 80, 86. *See also* literature; textbooks
bootstrapping, and physics, 31
Borel, Emile, 130, 479
Borelli, Giovanni Alfonso, 221–2, 223
Borges, Jorge Luis, 102
Born, Max, 341, 343, 347
Bose, J. Chandra, 286
Bose, Satyendra Nath, 344–5
Bothe, Walther, 342–3, 361
Bottazzini, Umberto, 474n20–1
Bourbaki, Nicolas, 117
Boussinesq, Joseph, 498
Bowie, William, 548
Boyle, Robert, 221, 230, 577
Bracewell, Ronald, 618, 625
Bragg, Lawrence, 441
Bragg, William Henry, 335
Branly, E., 286
Braque, Georges, 214
Brecht, Bertolt, 102
Brewster, David, 281, 511
Brezhnev, Leonid, 594
Bridgman, Percy, 34, 35, 547
Brillouin, Léon, 414
Brioschi, Francesco, 124
British Association for the Advancement of Science (BAAS), 80–1, 82, 290–1, 292, 295, 316, 408, 462
British Museum, 77
British National Physical Laboratory, 601–2
Brock, W. H., 175n7
Brodie, Benjamin, 75
Broglie, Louis de, 201, 345, 406
Broglie, Maurice de, 342
Bromberg, Joan L., 423
bromine, 180
Bronowski, Jacob, 93–4
Brooke, John, 38, 259n10
Brookhaven facility, 371

Brooks, Fred, 603
Brooks, Harriet, 59, 63
Brouwer, L. E. J., 128, 479
Brown, Alexander Crum, 250, 267
Brown, Harold, 567
Brownian motion, 499
Brown University, 129
Brücke, Ernst, 235
Brush, Stephen, 491
Bryn Mawr College, 61
bubble chamber, 209–12, 371
Bucher, Walter, 546–7
Buchner, Eduard, 229
Büchner, Ludwig, 43–4, 45
Buchwald, Jed Z., 8, 276, 277
Buckle, Henry Thomas, 494–6
Buckley, Oliver E., 424
Bud, R., 145
Buffon, Georges de, 538
Bulletin of Atomic Scientists, 577
Bultmann, Rudolf, 47
Bunsen, Robert, 258, 281, 282, 512
Burbidge, E. Margaret, 64, 67, 520
Burbidge, Geoffrey, 520
Burnell, Jocelyn Bell, 64
Burroughs Corp., 601
Bush, Vannevar, 129, 564, 565, 568, 575, 599
business history, and computers, 604–7. *See also* industry
Butlerov, Aleksandr M., 248, 250, 264, 267
Bykov, G. V., 264n26

calculators, 598–601
calculus, 114, 124–5, 126, 453, 469. *See also* mathematics
California Institute of Technology, 129
Callendar, C. S., 16, 645
Calvinism, and theories of energy, 299, 302, 309. *See also* religion
Calvino, Italo, 102
Cambridge Analytical Society, 125
Cambridge Mathematics Journal, 290, 462
Cambridge Philosophical Society, 76
Cambridge University, 450, 460–2
Cameron, Alastair, 520
Campbell, Lewis, 317
Campbell, Robert, 496
Campbell-Kelly, Martin, 601
Cannizzaro, Stanislas, 183, 184–5, 240, 242
Cannon, Annie J., 13, 58, 64, 517
Cantor, Geoffrey, 45n25
Cantor, Georg, 5, 273, 477
Cantorian set theory, 114
carbon: and chemical classification, 245; and theory of respiration, 232

carbon dioxide, and global warming, 634, 642–6, 647–8
Carnap, Rudolf, 28, 29–30, 31
Carnegie (ship), 551
Carnegie Foundation, 553
Carnegie Institution of Washington (CIW) Geophysical Laboratory, 546, 547
Carnot, Lazare, 118, 563
Carnot, Sadi, 41, 291, 297, 298, 299, 309, 500, 563
Carothers, Wallace H., 11, 433
Carr, Emma Perry, 61
Carrington, Richard, 510
Carson, Rachel, 591–2
Caspersson, Torbjoerr O., 443
Catalá, Rafael, 108
catastrophe theory and catastrophism, 131, 550
category theory, 131
cathode rays, 240, 325
Cauchy, Augustin-Louis, 12, 116, 275, 456–7, 472, 474–5
causality, Hume's critique of, 29
Cavendish, Margaret, 54
Cavendish Laboratory, 63, 360, 362, 363–4, 365, 417
Cayley, Arthur, 125, 462, 464, 467
Cedering, Siv, 108
Center for Materials Science and Engineering (MIT), 424
Central Intelligence Agency (CIA), 648
Centre Européen pour la Recherche Nucléaire (CERN), 371–2, 392
Centre National de la Recherche Scientifique (CNRS), 68
Cepheid variable stars, 524, 525
Cézanne, Paul, 193
Chadwick, James, 359, 365, 378, 379, 581
"chain reaction," and nuclear fission, 369
Challenger expedition, 550
Chalmers, Thomas, 38, 302
Chamberlain, T. C., 545, 550, 642
Chambers, Robert, 45, 77
Chambers, William, 77
Chandler, Alfred, 600–1
chaos theory, 550
Chapple, J. A. V., 99, 105
Chaptal, J. A. C., 279
charmed quark, 390
Chase, Martha, 443
Chasles, M., 123
Chatelet, Emilie du, 54
Cheliabinsk Laboratory, 595
chemical physics, 10, 394–412
Chemical Reviews, 409–10
Chemical Revolution, 54
Chemical Society of London, 66

chemical types: and classification, 245–6; theories of, 259–62
chemistry: atomism and chemical classification, 237–54; and digestion, 224–30; and early-nineteenth-century imagination, 4; and literature, 98, 99–100; and macromolecules, 429–45; and nuclear magnetic resonance, 626, 627; and physics, 10; popularization of, 75; reformation and standardization of nomenclature, 6, 174–90; religion and materialism, 44; and respiration, 230–3; and spectroscopy of celestial bodies, 511–12; and technical education in England, 146–7; and themes in history of science, 7–9; and theory of chemical structure, 255–71; and theory of chemical type, 245–7, 250, 260–2. *See also* chemical physics; geochemistry; polymer chemistry; quantum chemistry
Chevreul, Michel-Eugène, 245
Chew, Geoffrey, 385–6
Chisholm, Grace, 65
chlorine, 180, 258–9
chlorofluorocarbons (CFCs), 634
Choquet, Gustave, 486
Choquet-Bruhat, Yvonne, 66
Christianity, Paine's critique of, 37–8. *See also* religion
Christoffel, E. B., 482
Civil War, American, 568
Clairaut, Alexis-Claude, 454
Clancy, Tom, 99
Clapeyron, Emile, 291, 294
Clark, Latimer, 315
Clark, William, 81, 576
class, and education system in Great Britain, 145
classification: and astronomy, 13, 517–18; and chemical nomenclature, 178, 243–54
Clausius, Rudolf, 42, 239, 298, 490–1, 494, 522–3
Clebsch, A., 122
Clerke, Ellen and Agnes, 66
climate and climate change: and computer revolution, 16; culture and discussions of, 649, 650; development of scientific theories on, 641–5; and Enlightenment, 636–8; and global warming, 16–17, 634–6, 645–50; and literary and scientific history of U.S., 638–41. *See also* meteorology
cloud chamber, 358, 362, 367, 371
CNO cycle, 520
Cockcroft, John, 365–6
cognitive historiography, 10
Cold War: and computers, 601–4, 614; and geophysics, 554; and mathematics, 129–32; and national laboratories, 595–6; and nuclear weapons, 350, 370, 423; and relationship between science, technology, and war, 15, 561, 562, 566, 572–3, 575–7; and research on thermonuclear fusion and semiconductor electronics, 11; and solid-state physics, 421; and space programs, 168; state ideology and new physics, 585, 591; and women in physical sciences, 68
Cole, Henry, 78
Coleman, Sidney, 388
Coleridge, Samuel Taylor, 100, 101
Collins, Harry, 363
colloids, and macromolecules, 431, 432
Colloque de la Liaison Chimique, 409
colonialism: and public culture, 81; and state sponsorship of military research and development, 563
colors, of quarks, 389, 391, 393
Columbia University, 553
Combe, George, 83
Commission on Nomenclature of Organic Chemistry (1930), 186–8
communication: and mathematics, 117–20; patterns of and construction of scientific languages, 6; polymer chemistry and protein research, 11–12. *See also* language; radio; telegraph and telegraphy
comparability, and scientific method, 34
complementarity: and quantum mechanics, 348; and radiation, 9; and visual imagery in physics, 204, 213
completeness, and realism in physics, 27
Compton, Arthur, 342
computerized tomography (CT), 16, 616, 617–25, 630–1
computers and computer science: attributes of modern, 598; before 1945, 598–601; and business history, 604–7; and climate models, 648–9; and Cold War, 601–4; and earth sciences, 549, 551; impact of on scientific disciplines, 612–14; and mathematics, 130, 611–12; and medicine, 615, 631; military research and development, 575; and quantum chemistry, 399; as science and profession, 607–11; and scientific revolution, 16
Comte, Auguste, 1, 13, 174, 507–8
Condorcet, M. J. A., 118
Conference on Molecular Quantum Mechanics (Colorado), 409
Configurations (journal), 106
confinement, of quarks, 391
confirmation, and hypothetico-deductive method, 30–2
Congrene, William, 571
Congressional Office of Technology Assessment, 591

Conrad, Joseph, 102
constructionism, and history of science, 9, 290
contextualism, and history of science, 9, 290
continental drift, 14, 542–5
continuity: and Euler's concept of function, 476; as principle of geometry, 456
contraction theory, in geology, 540, 541
conventionalism, 24–5
convergence, and mathematical functions, 476
Conversi, Marcello, 383
Cook, James, 81
Cooke, W. F., 312
Coolidge, William David, 615
Cooper, James Fenimore, 100
Coover, Robert, 102
Cori, Gerty and Carl, 65
Cormack, Alan, 620–1, 624–5
Cornell, E. S., 278n14
corporations, and scientific and engineering education in U.S., 149. *See also* business history; industry
correspondence principle, 199, 340–1
Corry, Leo, 117
Cortada, James, 601
Cortazár, Julio, 102
Cosmic Background Explorer (COBE) satellite, 535–6
cosmic rays, 371, 380, 385
cosmogony, 40, 303
cosmology: and big bang theory, 530–1, 533–4; developments in eighteenth century, 12–14; developments in twentieth century, 534–7; galaxies and nebula until 1925, 523–5; and grand unified theory, 393; and literature, 100–3; nineteenth-century heritage of, 522–3; nonrelativistic, 529; and radio astronomy, 532–3; and relativity, 525–6, 537; and steady-state challenge, 531–2; and universe, 526–9. *See also* astronomy; astrophysics
cosmopolitanism, and Soviet Union, 585
Cosslett, Tess, 105
Coulomb, Charles, 311
Coulomb force, 208, 212
Coulson, Charles A., 11, 395, 396, 398–400, 401, 409
Couper, Archibald Scott, 248, 263
Courant, Richard, 116, 129, 484
covalent bonds and covalency, 254, 397
Cranefield, Paul, 236
Crawford, Elisabeth, 642–3
Crell, Lorenz, 76
Crelle, August Leopold, 458
Cremer, Erika, 59
Cremona, Luigi, 124

Critchfield, C. L., 520
Croll, James, 642
Crookes, William, 84, 87
Crosland, Maurice, 175
Crowe, Michael, 38
Crowther, J. G., 365
crystallography, 248, 249, 260, 415, 432–3
Crystal Palace, 77
Cubism, 214, 215
Cultural revolution, in Soviet Union, 582–3
culture: avant-garde and representation in physics, 215; and climate change, 637, 638, 649, 650; and computers, 612; cultural studies and history of science, 9; of physics and reductionism, 403; of professional mathematicians in postmodern era, 131–2; and relations between literary scholars and scientists, 92–5. *See also* art; literature; music; popular culture; public culture
Curie, Eve, 56
Curie, Irène. *See* Joliot-Curie, Irène
Curie, Marie, 4, 54, 56–7, 58, 62, 63, 65, 68, 71, 142, 352, 353, 354, 355, 358–9, 360
Curie, Pierre, 57, 65, 142, 352, 353, 354, 355
Curtis, Heber, 525
Cuvier, Georges, 45, 81, 260, 538
cyclotrons, 590

Dagognet, François, 176, 179
daguerreotypes, 163
Dale, Peter Allan, 105
d'Alembert, Jean LeRond, 454, 470, 482
Dalton, John, 7, 181, 238, 241, 250, 254, 255–6
Daly, Reginald, 547
Damadian, Raymond, 627, 628
Dana, James Dwight, 538
Darboux, Gaston, 124
Darrigol, Olivier, 378
Darwin, Charles, 42, 44, 81, 303, 497, 519
Darwin, Erasmus, 85
Darwinism, 80, 82. *See also* evolution
Dashkova, Ekaterina Romanovna, 54
Dauben, Joseph, 114
Daudel, Raymond, 406
Davis, J. L., 57
Davisson, Clinton, 348
Davy, Humphry, 73, 85, 100, 180, 244, 256, 279
Debierne, André, 353
Deborinites, and Mechanists, 582
Debye, Peter J. W., 339, 437
Dedekind, Richard, 477
deduction, and scientific method, 21–2
Defense Advanced Research Projects Agency (DARPA), 609

Déjerine-Klumpe, Augusta, 66
De la Beche, Henry, 548
Delbrück, Max, 443
DeLillo, Don, 102
Delta function, 479–81
democracy, 579–80, 590
Democritus, 238, 497
Deny, Jacques, 486
Department of Defense, 424, 574, 575, 649
Department of Health and Social Security (DHSS), 623
depersonalization, of geology, 545–8
Descartes, René, 22, 219, 221, 273, 452
descriptive geometry, 455, 457
desubstantiation, and visual representations, 214–15
determinism, of nature, 497
DeVorkin, David H., 168, 516n15
Dewar, James, 250
dialectical materialism, 582, 584
Dick, Thomas, 80
Dicke, Robert, 50, 533
Dickinson, Emily, 101
Dickson, Leonard, 127
Dietz, Robert, 550
Dieudonné, Jean, 485n50
differential calculus, 469
differential equations, 481–4
digestion, and chemistry, 224–30
Dines, William Henry, 644
Dirac, Paul A. M., 10, 50, 343, 377, 402–3, 412, 479, 481, 529
direct current (DC) systems, 320
Dirichlet, Johann Peter Gustav Lejeune, 471–3, 477, 478
disciplines and discipline-building: and chemical nomenclature, 174–90; education and industrial performance, 133–53; and interdisciplinary programs, 632–3; and mathematics, 113–32; medicine and physical sciences, 631–2; and nuclear physics, 355–60, 368–70; and observational astronomy, 154–73; and quantum chemistry, 394, 400, 407–10, 411; and themes in history of science, 2, 5–6, 7; and visual imagery in physics, 191–215
discontinuity: and late-nineteenth-century developments in geometry and analysis, 12; and quantum mechanics, 334–6, 347
disintegration theory, and radioactivity, 354, 362, 367
dislocation theory, in solid-state physics, 415
"display value," 593, 594–6
division of labor, in astronomy, 13, 173, 509, 517

DNA, and molecular biology, 11–12, 438, 441, 443–5
Dobbs, Betty Jo Teeter, 96
Döbereiner, Johann Wolfgang, 251
Doel, Ronald, 39n8
Doppler shift, 527
Dorpat refractor, 157–8
double bonds, and chemical structure, 265–6
double-helix model, of DNA, 444
Draper, J. W., 511
Drebbel, Cornelius, 577
duality, as principle of geometry, 456
dual-use technologies, and military, 575–7
DuBois-Reymond, Emil, 235, 236
Du Bos, Abbé Jean-Baptiste, 636–7, 638, 649–50
Duhem, Pierre, 24, 30, 33, 47n28, 240
Duhem-Quine problem, 30, 32
Dulong, Pierre Louis, 242
Dumas, Jean-Baptiste, 232, 240, 243, 245, 258, 259, 260
Dupin, Charles, 123, 455
Dupont Chemicals, 11
Dürrenmatt, Friedrich, 102
Dutton, Clarence, 541, 542, 545
dyes, and textile industry, 270–1
Dyson, Freeman, 51n41, 207, 377, 384

Earth: debate on age of, 14, 82, 303, 540; debate on end of, 39–43; traditions and conflict in study of, 539–42
earth sciences: emergence of modern, 549–52; physics and natural history traditions in, 14. See also geology; geophysics
Earth Summit (1992), 635
Eastern Europe, and women in physical sciences, 68
eclipses, 508–9
Eco, Umberto, 102
Ecole Normale Supérieure, 124
Ecole Polytechnique, 116, 123, 139, 450, 455, 457, 458, 566, 571
Ecoles des Arts et Métiers, 139–40
economics: and climate change, 650; and computer research, 614. See also opportunity costs
Eddington, Arthur Stanley, 50, 417, 518–19, 521, 526
Edge, David, 167n38
Edgeworth, Francis, 490
Edison, Thomas, 320
Edsall, John T., 440n21, 442
education, scientific: geometry and English liberal tradition, 460–2; and national differences in research and industrial

education (*cont.*)
  development, 5, 133–53; and popularization of science in Great Britain, 72–3; and relationship between science, technology, and war, 570–1; and women in physical sciences, 55, 60, 62, 70–1. *See also* apprenticeship; technical education; training; universities
EDVAC (Electronic Discrete Variable Automatic Computer), 601, 602
Edwards, Paul, 604
Egorov, Dmitri, 129
Ehrenfest, Paul, 336, 339
Einstein, Albert: and Bohr's theory of atom, 339; and electrodynamics, 9; and molecular weights, 437; and Pythagoreans, 23; and quantum discontinuity, 334–6; and relativity theory, 9, 13, 35, 326–7, 525–6, 581; and scientific method, 22; and tensor calculus, 125; and visual imagery in physics, 194–5, 196, 199
Einstein-Podolsky-Rosen experiment, 29
Eisenhower, Dwight, 573–4
electrical engineering, 8. *See also* engineering
electrical measurement, 316, 317–19
Electrical and Musical Industries (EMI), 16
Electric Boat Corporation, 573
electricity: history of theory and practice, 311–27; and popularization of science, 83. *See also* animal electricity; electrical measurement; electromagnetism
electrochemical dualism, theory of, 244, 245, 257–9
electrodynamics, 9
electromagnetic spectrum, and astronomy, 165–7
electromagnetism: and animal electricity, 235; Einstein and visual imagery, 194–5; and energy, 304–8; and history of electricity, 311–14; and Maxwellians, 321–4; and nuclear magnetic resonance, 626; reformulation of Maxwell's theory of, 8–9; and theory of heat, 292–3; and theory of light, 284–6
electronics industry, 420–1, 603–4
electrons, 324–7, 378
electrovalency, and inorganic chemistry, 254
electroweak theory, 211–12
Eliot, George, 86, 100
Ellis, Charles, 363
Ellis, Robert Leslie, 462
Elsasser, Walter, 345–6
Elster, Julius, 353
Emerson, Ralph Waldo, 101
EMI (Electrical and Musical Industries Limited), 621, 629, 631
emission theory, of light, 272–7
empiricism, and scientific method, 21, 22, 29, 31
enantiomorphism, 248–9

Encyklopädie der mathematischen Wissenschaften (Klein, 1895), 128
endoscopy, 632
Energeticist school (Germany), 309
energy: and electromagnetism, 304–8; and models of sun, 513; and natural philosophy, 8, 296; and physics in nineteenth century, 296–304, 308–10; religion and views on conservation of, 41–2, 44; statistical interpretation of conservation of, 9
energy flux formula, 322
Engels, Friedrich, 122, 582
engineering: and analog device, 599–600; and big science, 593; and computer revolution, 16; corporations and education in U.S., 149; definition of in France, 139, 142. *See also* technical education
England. *See* Great Britian
ENIAC (Electronic Numerical Integrator and Computer), 601
Enlightenment: climatology and climate change, 636–8, 639; and geometry, 454; and religion, 36; and scientific activity in France, 118
entropy, 334, 494, 502, 523
environmental history, 635
environmentalism: atmospheric and oceanographic sciences and, 557; and global warming, 635; and public-interest science, 17
epistemology: and geology in early twentieth century, 14, 552–7; and mathematics, 113; and new physics, 581; and positivism, 28–9
Epstein, Paul, 339
Erasistratus, 219
ergodic theory, 499–500
"Erlanger Program," 467
error, and statistical methods, 491–4
eschatology, 43
essentialism, and gender analysis of physical sciences, 69
ether, theories of, 26, 324–7
ethics, and computers, 613–14. *See also* morality
Euclid, 115, 449, 451
Euclidean geometry, 24–5
eugenics, 88
Euler, Leonhard, 116, 118, 273, 469–70, 473
European Space Agency, 170–2
European Union, and women in physical sciences, 68
evidence, and scientific method, 29–32
evolution: astronomy and concepts of biological and thermodynamics, 13; geological evidence for, 42; religion and physical sciences, 44, 45, 49, 497; and theories of energy, 303. *See also* Darwinism

Ewing, Maurice, 553, 554
exceptionalism, and women in physical sciences, 54–71
existentialism, 47
expeditions, and popularization of science, 81. *See also Challenger* expedition; *Beagle*
experimental traditions, and scientific method, 32–5
extraterrestrial life, 37–9

Fairbairn, William, 291
Faraday, Michael, 45n25, 73, 74, 89, 259, 290, 295, 312–14
Faraday Society, 408, 434
fast magnetic resonance imaging (fMRI), 629
Fechner, G. T., 314
Federal Bureau of Investigation (FBI), 417
Federation of American Scientists, 577–8, 590
Federation of Atomic Scientists, 577
feminist philosophy, 71
fermentation, 229
Fermi, Enrico, 367, 378, 379
Fermi interactions, 383
Feuerbach, Ludwig, 43, 86
Feynman, Richard, 91–2, 108–9, 207, 384, 572
Fiedler, Wilhelm, 125
Field, Cyrus, 8, 315–16, 319
Field, Richard, 547
field equations, 10
field geology, 547–8
field theory, 312, 323. *See also* quantum field theory
Fifth Generation computing project, 614
Fischer, Emil, 439
Fischer, Ernst, 485
Fisher, Osmond, 541
FitzGerald, George F., 285, 309, 321–2, 323
Fitzgerald, Penelope, 102
FitzGerald-Lorentz contraction, 325
Fizeau, A., 283
Fizeau, Hippolyte, 318
flavors, of quarks, 388, 389, 390–1
Fleagle, Robert G., 636n5
Flinders, Matthew, 81
fluid dynamics, and oceanography, 551
Follin, James, 530
Fontenelle, Bernard de, 70
Forbes, James, 280, 511–12
Forbes, John, 540
Fordism, 567
Forman, Paul, 331, 332, 420, 596
formulas, and chemical classification, 250–1
fossil fuels, and global warming, 646
Foster, Michael, 220
Foucault, Léon, 161, 281

Foucault, Michel, 178
foundationalism, 11
Four Color Problem, 612
Fourcroy, Antoine-François de, 177
Fourier, Joseph B. J., 82, 119, 290, 473, 479–80, 500
Fourier series, 129–30
Fourier transforms, 618
Fowler, Ralph H., 404, 406
Fowler, William, 520
Fox, Robert, 133
Framework Convention on Climate Change, 650
France: and chemical nomenclature, 185; and geometry, 450, 454–8; industrial development and scientific education and research, 138–43; mathematics in postrevolutionary, 114–15, 116, 118–19; and national tradition in mathematics, 123, 124; and nuclear physics, 360–1; and organic chemistry, 267; and quantum chemistry, 406, 409; and state support for military science and technology, 566, 571. *See also* French Revolution
Franck, James, 338
Franck Report (1944), 590
Franco-Prussian War (1870), 73
Frankel, Eugene, 275
Frankland, Edward, 247–8, 250, 261, 262, 267
Franklin, Benjamin, 83, 639
Franklin, Rosalind, 12, 58, 62, 444
Fraunhofer, Joseph, 157, 158, 280, 508, 511, 512
Frechet, Maurice, 485
Freeman, Joan, 63, 68
free will, 44, 496–8
French Revolution, 74, 118–19, 450, 454–7, 563
Frenkel, Iakov, 583
Fresnel, Augustin, 8, 274, 275, 276, 277, 287
Freundlich, Herbert, 431
Fricke, Robert, 122
Friedel, Charles, 186–7
Friedman, Herbert, 168–9
Friedmann, Alexander A., 13, 527, 528
Friedmann equations, 527
Friedrichs, Kurt Otto, 484, 486
Frisch, Otto, 369
Frölich, Herbert, 414
Fruton, Joseph S., 2
Fulton, Robert, 577
functions, mathematical: delta form of, 479–81; and differential equations, 481–4; Dirichlet's concept of, 471–3; and distributions, 484–7; Euler's concept of, 469–70; generality of, 477–9; and physics, 470–1; and rigor, 474–7. *See also* mathematics

galaxies, and cosmology, 523–5
Galen, 219, 226, 230
Galileo Galilei, 40, 191–3, 194, 209, 508, 567
Galison, Peter, 371, 375–6
gallium, 181, 252
Galois, Evariste, 123
Galton, Francis, 87, 88, 492
Galvani, Luigi, 233–4
galvanometers, 235
Gamow, George, 364, 367, 530–1, 536
Garnett, William, 301, 317
Garvan, Francis P., 67
gases, kinetic theory of, 239–41, 265, 490–3
gauge theory, 388, 391–3
Gauss, Carl Friedrich, 55, 115, 312, 458, 464, 474
Gaussian distribution, 491
Gay-Lussac, Joseph-Louis, 241–2, 245, 256, 258–9
Geiger, Hans, 337, 342–3, 356, 361
Geitel, Hans, 353
Gell-Mann, Murray, 380n10, 384, 388
General Electric Company, 415, 606, 615
geochemistry, 550, 555, 556. *See also* geology
Geoffroy, Etienne, 238
Geological Society of London, 82, 548
geology: physics and chemistry of earth and, 538–57; physics and natural history traditions in, 14; and religion, 42. *See also* geochemistry; geophysics
geomagnetism, 551
geometry: and British tradition in mathematics, 125; and conventionalism, 24–5; definition and theory of distributions and non-Euclidean, 485; discontinuity and revolution in late-nineteenth-century developments in, 12; eighteenth-century background of, 450–4; and English liberal education, 460–2; Euclidean and non-Euclidean compared, 462–4; and French Revolution, 454–7; and German universities, 458–60; non-Euclidean and revolution in mathematics, 115, 177; origins of Euclidean, 449–50; transitions in late nineteenth century, 464–7. *See also* mathematics
geophysics: and continental drift, 542–5; and debate on Darwinism, 82; geology compared to, 538–9; and petroleum industry, 553. *See also* geology
Georgetown University Medical School, 632
Gergonne, Joseph Diez, 123, 456
Gerhardt, Charles, 181–2, 183, 246–7, 260–1
Germain, Sophie, 55
germanium, 181, 252
Germany: and astronomy instruments, 156, 158; and chemistry, 267; geometry and universities, 458–60; industrial development and scientific education and research, 134–8; and mathematics education, 116–17, 119–23, 127–9; and national laboratories, 595; and national tradition in mathematics, 124, 126; national values and ideologies and scientific research, 15; Nazism and Aryan physics, 586–9; and nuclear physics, 360, 361; and nuclear weapons, 591; and quantum chemistry, 405–6; women and universities in 1920s and 1930s, 59, 71; World War I and military research and development, 564; World War II and military research and development, 565. *See also* Nazism; World War I; World War II
Gessen, Boris. *See* Hessen, Boris
Giacconi, Riccardo, 169
Gibbs, Josiah W., 126, 304, 499–500
Giesel, Friedrich Oskar, 353
gigantomania, 593
Gilbert, G. K., 545
Gilkey, Langdon, 49
Gillispie, Charles, 491
Gingerich, Owen, 107
Gladstone, J. H., 84
glassmaking, and astronomy instruments, 156–7
Gleditsch, Ellen, 64n24
global warming, 16–17, 634–6, 645–50
Gmelin, Leopold, 227–8, 250
Goeppert-Mayer, Maria, 58, 65, 66, 372
Goethe, Johann Wolfgang von, 100
Gold, Thomas, 531
Goldstone, Jeffrey, 377, 387
Gooding, David, 312–13
Gordon, Lewis, 297
Gosling, Raymond G., 444
Göttingen University, 368
Goudsmit, Samuel, 342
government: and computer research, 604; and women in physical sciences, 64. *See also* politics; state; *specific countries*
Grabiner, Judith, 114, 474n21
Graebe, Carl, 237n1, 268, 271
Graham, Loren R., 15
grandes écoles, 139, 140, 457
grand unified theories (GUT), 393, 534
Grassmann, H. G., 126
Grattan-Guinness, Ivor, 114
gravitational constant, 50, 51
"Great Break," in Soviet Union, 582–3, 584
Great Britain: and astronomy instruments, 156; and chemistry, 267; and geology, 556; and geometry, 450, 460–2; industrial development and scientific education and research, 143–7; and national tradition in mathematics, 125–6;

and nuclear physics, 361–2; and
popularization of science in nineteenth
century, 72–90; and quantum chemistry,
406–7. *See also* Scotland
Great Depression, and government support for
science in U.S., 590
Great Exhibition (1851), 72
Great Terror (Soviet Union), 584
Greece, classical, and state sponsorship of
military research and development, 562
Greek fire, 571
Green, George, 126, 275
Greene, Mott, 556, 636
Green's theorem, 482–3
Gregory, Duncan, 462
Gregory, Frederick, 3–4
Gregory, William, 84
Gray, Stephen, 96
Grimaux, Edouard, 186
Groves, Gen. Leslie, 572, 590
Grubb, Howard, 161
Guagnini, Anna, 133
Guinand, Pierre-Louis, 156–7
Gulf War, 570
Gunn, James, 534
gunpowder, 562, 568, 569, 570, 572, 577
Gusterson, Hugh, 595n33
Gutenberg, Beno, 551
Guyton de Morveau, Louis-Bernard, 176–8

Haber, Fritz, 567
Hachett, J., 123
Hacking, Ian, 375, 377
Hadamard, Jacques, 485
hadrons, 385, 386, 388, 390, 391
Haeckel, Ernst, 44
Hahn, Otto, 56, 57, 354, 369, 487
Hale, George Ellery, 164, 172, 516
Hales, Stephen, 224, 229
Halley's Comet, 170
Hamilton, William Rowan, 125, 126
Hammarsten, Einar, 443
Hampson, William, 354
Handel, M. D., 646n47
Hankel, Hermann, 472, 477
Hansen, James, 649
Haraway, Donna, 612
Hardy, G. H., 125
Hardy, Thomas, 88, 101
Harnack, Axel, 484
Harries, Carl D., 431
Harrison, Anna Jane, 61
Hartle, James, 535
Hartmann, Hermann, 406

Hartree, Douglas, 406
Harvard Observatory, 163, 517
Harvard University, 547
Harvey, William, 219, 221, 222
Hassenfratz, Jean-Henri, 179
HASTRO (listserv), 107
Hawking, Stephen, 534, 535
Hawthorne, Nathaniel, 100
Hayashi, Chushiro, 530
Hayles, N. Katherine, 107
Haynes, Roslynn, 104
Hearnshaw, J. B., 510n7
heat: mechanical value of, 290–6; theory of flow,
500. *See also* energy
Heaviside, Oliver, 8, 126, 309, 321, 322–3, 325,
480–1
Heaviside function, 480–1
"heavy water," 366
Heilbron, John, 331
Heim, Albert, 540
Heims, Steve, 130
Heine, Eduard, 476, 477
Heisenberg, Werner: and Aryan physics, 587,
589; and atomic structure, 341, 346, 347–8;
and indeterminism in quantum mechanics, 9;
nuclear weapons and morality, 591; and
quantum chemistry, 10, 395, 404; and
quantum field theory, 380; and solid-state
theory, 414; and visual imagery in physics,
200–1, 202, 203–4, 206–7, 209; and wave
mechanics, 364
Heitler, Walter, 10, 395, 401, 403, 414
heliometers, 158
helium, and nuclear force, 206
Hellman, Hans, 406
Helmholtz, Hermann von, 42, 43, 73, 85, 233,
235, 285, 296, 300, 314, 465–6, 513
Helwig, L. W., 588n17
hemoglobin, 442
Hendry, John, 331
Henry Draper Memorial, 517
Hentschel, Klaus, 282n23
Herman, Robert, 530
Hermite, Charles, 122, 123, 478
Herschel, Caroline, 55, 66
Herschel, Frederick William, 277–8, 279
Herschel, John, 77, 89, 125, 158–9, 281, 461, 491,
492, 511, 512–13
Herschel, William, 37, 66, 80, 96, 160, 287–8,
506, 508, 514, 521, 522
Hershey, Alfred D., 443
Hertz, Gustav, 338, 416
Hertz, Heinrich, 8, 285–6, 288, 321, 323
Hertzsprung, Ejnar, 518

Hertzsprung-Russell (H-R) diagram, 518
Herzog, Reginald, 433
Hess, Kurt, 431
Hess, Victor, 357
Hesse, Mary, 491
Hessen, Boris, 15, 583, 584
heteroatoms, 263
heterogenesis, 45
hidden-variable theory, 349
Higgs mechanism, 392
high energy physics, 372–3
Hilbert, David, 116–17, 127–8, 450, 467, 484, 486
Hinrich, Gustavus Detlev, 251–2
Hinton, Charles H., 466
historiography: of chemistry, 237; of cosmology, 537; of earth sciences, 543–4, 556; of mathematics, 113–14, 132; of U.S. industry, research, and education relations, 147–8
history, of science: and astronomy in nineteenth and twentieth centuries, 505–8; contextualist and constructionist methods in analysis of, 9, 290; and development of physics during twentieth century, 375–7; and global environmental change, 635–6; and interrelations of literature and science, 98, 106–8; and nuclear physics, 373–4; polymer chemistry and molecular biology, 429–30; and postmodernism, 1; and quantum chemistry, 400, 411–12; and quantum mechanics, 331–2; and scientific disciplines, 2; themes and interpretive frameworks of, 3–17; and visual imagery in physics, 212–15. *See also* historiography
Hitler, Adolf, 588, 589, 594
Hobbler, Icie Macy, 67n36
Hoddeson, Lillian, 376
Hodgkin, Dorothy Crowfoot, 58, 62, 65
Hoffmann, E. T. A., 100
Hofmann, August W., 182, 185, 187, 244, 246, 250, 253, 262, 270
Holland, Henry, 86
Holleman, A. F., 187
Hollerith, Herman, 599
Holmes, Arthur, 523, 548
Holmes, Frederic Lawrence, 231n25, 239
Holton, Gerald, 107
home economics, 64
Hooft, Gerard 't, 376, 388
Hopkins, William, 299, 538, 540
Houghton, Henry G., 648
Hounsfield, Godfrey Newbold, 621–5, 631
Houtermans, Fritz, 520
Hoyle, Fred, 520, 531

H-theorem, 502
Hubble, Edwin, 13, 525, 526–7, 528
Hubble law, 527, 529
Hubble Space Telescope, 6, 14, 165, 170–3, 536, 632
Hückel, Erich, 395, 405
Hückel, Walter, 405
Hufbauer, Karl, 163n25, 508–9
Huggins, Margaret, 65
Huggins, Williams, 65, 282, 515
Hughes, Thomas P., 320, 569
human freedom, and statistical methods, 44, 496–8
humanism, and critiques of modern science, 17. *See also* neohumanism
humanities: sciences and interdisciplinary scholarship, 94–5; and "two cultures" concept, 4
Humason, Milton, 165, 526
Humboldt, Alexander von, 81
Hume, David, 29, 637–8
Humphreys, William Jackson, 644
hurricanes, 649
Hurwitz, Adolf, 116
Huxley, Aldous, 93
Huxley, T. H., 73, 79, 82, 85, 86, 89, 93
Huygens, Christiaan, 273
hydrogen: and chemical classification, 245, 248; and nuclear magnetic resonance, 626; and stars' interiors, 520; and theory of acidity, 259; and theory of respiration, 232
hypothetico-deductive method, 30–2

iatromechanism, 220, 221, 222
IBM (International Business Machines), 601, 603, 605–7
ice ages, 17, 643, 644, 648
idealism, and new physics in Soviet Union, 583, 585
ideology, and influence of state on physics, 579–97
I. G. Farben laboratories, 415
Ihde, Aaron J., 237
Illiopoulos, John, 390
illustration, and popularization of science, 75, 80, 82, 89
imagery, and construction of scientific languages, 6. *See also* visual imagery
Immerwahr, Clara, 65
immortality, of soul, 44
indeterminism, 9
induction, and scientific method, 21, 22, 23, 31
industrialization, and geology, 539, 557
Industrial Revolution, 73, 97, 563, 569

industry: and chemical structure theory, 270–1; and computerized tomography, 621–5; and computers, 604–7; and earth sciences, 14, 553, 557; and national differences in scientific education and research, 5, 133–53; and scientific innovation, 8; technocracy and physics in U.S., 590; thermonuclear fusion and semiconductor electronics, 11; and weapons, 569–70. *See also* business history; corporations; petroleum industry; textile industry
inference, to best explanation, 26
inflationary theory, and cosmology, 535–6
infrastructures, and military research and development, 573
Ingold, Christopher, 241
innovation: industry and scientific, 8; and military technology, 576
Institute for Defense Analyses, 567
Institute of Electrical and Electronics Engineers (IEEE), 609
Institute of Radio Engineers (IRE), 609
Institute for Theoretical Physics (Copenhagen), 364
Institut für Plasmaphysik (IPP), 425
Institut du Radium, 361
Institut für Radiumforschung (Vienna), 357
institutions and institutionalization: and geology, 552–7; and research on radioactivity, 355–60; and state support for military science and technology, 566–8; and women in physical sciences, 60–1, 71. *See also* laboratories; observatories; scientific academies; universities
instrumentalism, 24, 26
instruments and instrumentation: and discipline building, 5–6; and optics, 287–8; and particle physics, 371; and positional astronomy, 154–9; and radioactivity, 362–8. *See also* telescopes
integrated circuits, 604
interdisciplinary scholarship: and connection between sciences and humanities, 94–5, 103–6; and medicine, 632–3
interferometers, 167, 325
internalism, and history of mathematics, 114
International Association of Chemical Societies, 187
International Atomic Energy Agency (IAEA), 424
International Conference of Chemistry (1889), 186–7
International Conference on Peaceful Uses of Atomic Energy, 423
International Geophysical Year (1957–8), 555, 647

internationalism: ideology and state politics, 597; and language of chemistry, 189; and research on radioactivity, 357, 359
International Journal of Quantum Chemistry, 394, 408
International Union of Pure and Applied Chemistry (IUPAC), 186, 189, 190
Internet, 604, 613
inverse-square law, 313
iodine, 180
Ioffe, Abram, 580, 584–5
ionization, and atoms, 240
ionospheric physics, 416, 554
Ireland, popular science and education, 73
irreducibility, of biochemical complexity, 50
irreversibility, doctrine of, 302–3, 333, 334
island universe theory, 524, 525
isomerism, 248–50, 256, 257, 259, 266
isomorphism, 248, 252
isostasy, 541–2, 543
isotopic spin, 380, 385, 386
Italy: and national tradition in mathematics, 124–5; and state sponsorship of military research and development, 562

Jacob, Margaret, 96
Jacobi, Carl G. J., 116, 122, 458–9
Jahnke, Hans Niels, 470n6
James, Frank, 282n21
James, William, 84
Jammer, Max, 331
Jansky, Karl, 166
Japan: and computer industry, 614; and space program, 170; and state sponsorship of military research and development, 565
Jeans, James, 519
Jefferson, Thomas, 16–17, 638, 647
Jeffreys, Harold, 543, 551
Jenkin, Fleeming, 301, 308, 318
Jensen, Hans Daniel, 372
Jensen, K. A., 175n7
Jews and Judaism: and emigration of scientists from Nazi Germany, 368, 586, 587, 588, 589; state ideology of Soviet Union and new physics, 585
John Paul II, Pope, 49
Johns Hopkins University, 15, 567, 571
Johnson, Samuel, 96–8
Joliot, Frédéric, 65
Joliot-Curie, Irène, 58, 65, 361, 591n25
Joly, John, 542–3
Jordan, Camille, 115
Jordan, Pascual, 343, 377–8, 529
Jorgensen, C. K., 175n7
Jorissen, W. P., 187

Josephson, Paul, 15
Joule, James, 79, 292–5, 299
Journal of Chemical Physics, 407–8
journals, and popularization of science, 76, 86–7, 88, 89
Joyce, James, 102

Kaiser Wilhelm Institutes for Chemistry and Physical Chemistry, 415, 432, 441, 567, 595
KAM (Kolmogorov-Arnold-Moser) theory, 130
Kandinsky, Wassily, 215
Kanigel, Robert, 108
Kant, Immanuel, 37, 46–7, 196, 459
Kapteyn, Jacobus Cornelius, 517, 524
Karlsruhe Congress (1860), 181–6, 189, 190, 240, 242, 252, 253, 265
Karrer, Paul, 431
Kauffman, G. B., 175n7
Kay, William, 359
Keck Foundation, 172
Keill, James, 223
Kekulé, August, 7–8, 181, 183–4, 240, 247, 250, 262, 263, 264, 265, 266–7, 432
Kelvin, Lord. *See* Thomson, William
Kemble, Edwin, 336
Kendrew, John C., 439, 441–2
Kennedy, John F., 580
Kennelly-Heaviside layer, 416
Kenney, Richard, 108
keratin, 441
Khinchin, A. Ya., 129–30
Khrushchev, Nikita, 594
Kidder, Tracy, 612
Kidwell, Peggy, 513
Kim, Mi Gyung, 243
"kinematic relativity," 529
King, Clarence, 545
Kirch, Christoph, 66
Kirch, Maria Winkelmann, 54, 66
Kirchhoff, Gustav R., 281, 282, 333, 480, 512
Kirsch, Gerard, 362
Kistiakowsky, Vera, 68
Klein, Felix, 116, 122, 126, 127–8, 467
Klein, Martin, 331
Klein, Oskar, 383
Klemperer, Otto, 363
Kling, Rob, 612
Knoespel, Kenneth, 96
knowledge: and historiographical views of mathematics, 113–14; industrial development and capitalization of, 134; Kuhnian model of growth in scientific, 375; laboratories as locus of, 594–6
Knowles, James, 87

Koestler, Arthur, 102
Kohlrausch, Rudolph, 318
Kohn, Hedwig, 61
Kolbe, Hermann, 247, 261, 263, 268–9
Kolmogorov, A. N., 130
König, Wolfgang, 136
Königsberg Observatory, 156
Köppen, Wladimir, 644
Körner, Wilhelm, 267–8
Kossel, Walther, 254, 338
Kovalevsky, Sonya, 54, 56, 57
Kragh, Helge, 2
Kramers, Hendrik, 209, 340, 384
Krönig, Auguste, 490
Kronig, Ralph, 342, 415
K-space trajectories, 629
Kuhn, Thomas S., 3, 8, 9, 48, 107, 114, 275, 289–90, 331, 332, 376, 536
Kuhn, Werner, 436
Kuiper, Gerard P., 550
Kummer, E. E., 116
Kyoto Protocol (1997), 650

labor. *See* division of labor
laboratories: and astronomy, 13; and corporate research in U.S., 148; and geology, 547–8; as locus of ideology and knowledge, 594–6; and radioactivity research, 364; and women in physical sciences, 55, 62–3
Lacaille, Abbé Nicholas-Louis de, 155
Lacan, Jacques, 189
Lacroix, Sylvestre François, 116, 457–8, 472, 473
Ladd-Franklin, Christine, 61
Ladies' Diary, The (magazine), 70
Lagrange, J. L., 119, 472, 473, 474, 482, 484
Lakatos, Imre, 376, 476n28
Lamarck, J.-B., 45
Lamb, Willis, 383–4
Lambert, Johann Heinrich, 118, 463–4
Lamb shift, 384
Lamont-Doherty Geological Observatory, 553
Lanczos, Cornelius, 526
Landé, Alfred, 341
Landers 1 and *2* (satellites), 170
Land Grant Act (1862), 149
Langley, Samuel P., 283, 642
Langmuir, Irving, 416, 442
language: and identity of scientific discipline, 174; imagery and construction of scientific, 6; theory of and chemical nomenclature, 184, 189. *See also* classification; communication; nomenclature
Lankford, John, 163n28, 516n15
laparoscopy, 632

Laplace, Pierre-Simon, 37, 41, 80, 115, 231, 232, 273–4, 290, 514, 521, 522
Laplacian cosmos, 40–1
Lardner, Dionysius, 77
Lark-Horovitz, Karl, 419
Larmor, Joseph, 9, 289, 324
Lassell, William, 160–1
Laszlo, Pierre, 189
Latin America, and women in physical sciences, 69
Latour, Bruno, 375
Laugwitz, Detlef, 487
Laurent, Auguste, 182, 185, 242, 246–7, 250, 259–61
Lautebur, Paul, 627–8, 629, 630
Lavoisier, Antoine-Laurent, 74, 76, 177–9, 180, 184, 231–3, 244
Lavoisier, Marie Anne, 54
Lawrence, Ernest, 366, 419
Lawrence Livermore Laboratory, 567, 595
Lawson, Robert, 359
leadership, and mathematics education in Germany, 121
Leavis, F. R., 93
Leavitt, Henrietta, 13, 58, 524
Le Bel, Joseph-Achille, 249, 268
Lebesgue, Henri, 130, 479
Lecoq de Boisbaudran, Paul-Emile, 252
lectures, and popularization of science, 74–5
Lee, Tsung Dao, 383, 386
legal system, and computers, 613
Legendre, Adrien-Marie, 123, 454
Lehmann, Inge, 551
Leibniz, Gottfried Wilhelm, 453
Lemaître, Georges, 13, 526, 527, 528
Lenard, Phillip, 15, 337, 583, 585, 587–8, 589
Lenin, Vladimir, 580
Leningrad Physical Technical Institute, 580
Lennard-Jones, John Edward, 397, 409
Leonardo da Vinci, 562, 572, 577
lepton, 392, 393
Leray, Jean, 484
Lermontova, Julia, 59
Leslie, John, 278
Leslie, Stuart W. (Bill), 611
Levere, Trevor, 105
LeVerrier, U. J. J., 159
Levi-Civita, Tullio, 124
Levy, Steven, 612
Lewis, D. C., 486
Lewis, Gilbert Newton, 254, 397
Lewis, Meriwether, 81, 576
"lexicons," and Kuhnian revolutions, 377
Libby, Leona Woods Marshall, 65

Lick Observatory (California), 163
Lie, Sophus, 122
Liebermann, Carl, 271
Liebig, Justus von, 7, 137, 232, 245, 247, 256, 258, 259, 260
Leibniz, Gottfried Wilhelm, 598
life: changes in concepts of in seventeenth century, 219–20; origin of, 44–5; question of extraterrestrial, 37–9; and space programs, 170
life sciences, relations with physical sciences and concepts of vital processes, 219–36. See also biology
light: electromagnetic theory of, 284–6, 317–18; history of nineteenth-century theories of, 8; wave theory of, 272–7
Lightman, Alan, 102
Lillienthal, David, 579–80
Limited Nuclear Test Ban Treaty, 555
Lincoln, Abraham, 568
Linde, Andrei, 535
linear combination of atomic orbitals (LCAO), 397
Linnaeus, Carl, 177
Liouville, Joseph, 123
literary criticism, 94
literature: and physical sciences, 91–109; and popularization of science, 85; and "two cultures" concept, 4. See also autobiography; biography; books; poetry; science fiction
Littlewood, J. E., 125
Livingston, Stanley, 368
Lobachevsky, Nicholai Ivanovich, 464
Locke, David, 107
Locke, John, 29
Lockyer, J. Norman, 87, 282, 515, 521
Lodge, Oliver, 44n22, 285, 309, 321, 322–3
Lombroso, Cesare, 88
London, Fritz, 10, 395, 401, 415
London Mathematical Society, 125
Lonsdale, Kathleen, 58, 65
Lorentz, Edward, 550
Lorentz, H. A., 8–9, 201–2, 324, 335
Los Alamos Laboratory, 11, 422–4, 594–5
Loschmidt, Josef, 500, 501, 502
Louis XIV, 563
Lovell, Bernard, 166–7
Löwdin, Per-Olov, 394
Lower, Richard, 222–3
Ludwig, Carl, 230, 235
Lundgreen, Peter, 133
Lusin, N. N., 129–30
Lyell, Charles, 42, 299, 303, 538, 548
Lyman, T., 284
Lysenko, Trofim, 585

MacCullagh, James, 275
Macelwane, James, 543, 551
Mach, Ernst, 26, 29, 193, 241, 501
Mackay, Charles, 84
MacKenzie, Donald, 607
MacKinnon, Edward, 331
Maclaurin, Colin, 453, 454
MacMillan, William, 529
Macquart, L. C. H., 226
Macquet, Pierre-Joseph, 176–7
"macrohistory," 636
"macromolecular chemistry," 431
macromolecules, and organic chemistry, 11, 429–45
madder dye, 271
Mädler, Johann, 523
magazines, and popularization of science, 89
Magellanic Clouds, 524
Magendie, François, 234
magnetic resonance imaging (MRI), 625–6, 627, 629–31
magnetism, and popularization of science, 83. *See also* electromagnetism
magneto encephalography (MEG), 631
magneto-ionic theory, 416
Mahoney, Michael, 608
Maiani, Luciano, 390
Makinson, Rachel, 64
Makower, Walter, 356
Mallard, John, 629
Mallarmé, Stéphane, 100
Malpighi, Marcello, 221
Malthus, T. R., 500
Malus, Etienne, 123, 274, 287
Manabe, Syukuro, 649
Manchester Literary and Philosophical Society, 293
Manchester University, 601–2
Mandelshtam, Leonid, 583–4
Manhattan Project, 369–70, 418–19, 564, 572, 574, 577, 590
Manley, Gordon, 646
Mansfield, Peter, 628–9
Mansfield amendment of 1970, 575
Marcet, Jane, 4, 55, 75
Marconi, Guglielmo, 286, 323
Marić, Mileva, 63
Marie Curie Fellowship Association, 68n43
Mark, Herman F., 11, 433–4, 436, 440–2
Markley, Robert, 96
Mark-Meyer theory, 434
marriage, and women in physical sciences, 65–6
Mars, 39, 169–70
Marsden, Ernest, 337, 356

Marshall, Robert, 383
Martí, José, 101
Marx, Karl, 582
Maschke, Heinrich, 127
Mason, Max, 553
Massachusetts Institute of Technology (MIT), 15, 62, 149, 150–1, 419, 567, 571
mass production, 569
"mass renormalization," 384
materialism: and positivism of Vienna Circle, 28; religion and physics, 43–5; and theories of energy, 302
Mathematical Reviews, 131
mathematics: and changes in production and communication, 117–20; and computers, 611–12; and concept of function in analysis, 468–87; developments in since eighteenth century, 12–14; and discipline building, 5; historiography of, 113–14, 132; and literature, 98; and national traditions, 123–7; physical sciences and models of, 22–5; pure and applied in Cold War era, 129–32; and representation of atom, 213; and research schools in Germany, 120–3, 127–9; texts and contexts of, 114–17. *See also* algebra; calculus; geometry
Mauna Kea Observatory, 172
Maury, Antonia, 13, 65, 517–18
Maury, Carlotta, 66
mauve, and textile dyes, 270
Max-Planck-Institut für Festkörperforschung, 427
Maxwell, James Clerk: and British tradition in mathematics, 125; and electrical measurement, 317–19; and electromagnetic theory, 284–5, 288, 304–9; and kinetic theory of gases, 239, 490; and link between religion and science, 4, 44, 305; and North British group, 301; and poetry, 100; and realism, 25; and statistical methods, 12, 489, 491, 492–3, 495–6
Maxwellians, 321–4
Maxwell's equations, 322, 323
Mayer, Joseph, 65, 410
Mayer, Julius Robert, 42, 296, 513
Mayer, Maria Goeppert. *See* Goeppert-Mayer, Maria
Mayow, John, 230
McCarthy era, and physics in U.S., 590
McDonnell Douglas Aircraft, 573
McKie, Douglas, 244
McNeill, William, 562
McRae, Robert James, 279–80
Meadows, A. J., 104
measurement: formal theory of, 24; and validity in observations, 33–5. *See also* electrical measurement

Index

mechanical philosophy, 221
mechanics, conventionalism and principles of, 25
mechanics' institutes, 75, 145
Mechanists, and Deborinites, 582, 583
medicine: and computerized tomography, 617–21; and computer revolution, 16, 615; and future of medical imaging, 631–3; and magnetic resonance imaging, 625–6, 629–31; and nuclear magnetic resonance, 626–9; and x rays, 615–17
Megaw, W. John, 69
Mehra, Jagdish, 331
Mehrtens, Herbert, 114n1
Meitner, Lise, 54, 56, 57, 65, 356, 359, 363, 369
Melloni, Macedonio, 279, 280, 288, 642
Melville, Herman, 100
Mendel, Lafayette B., 63
Mendeleyev, Dmitry, 184–5, 252
mentors, and women in physical sciences, 62–3
Meredith, George, 101
meridian circles, 155
Merz, John Theodore, 289, 492
Mesmer, Anton, 83
mesmerism, 83
mesons, 380, 383, 385
"mesoscopic history," 376
metallurgy, 415
metaphors, and visual imagery in physics, 199–200, 207
meteorites and meteorite impacts, 550
meteoritic hypothesis, 513, 515
meteorology, 549–50, 554. See also climate and climate change
metrical geometry, 467
Metzger, Hélène, 59
Meyer, Kurt H., 11, 434
Meyer, Lothar, 252, 253
Meyer, Stefan, 353, 357
Meyer, Victor, 268
Michelson, Albert A., 9, 325
microbrownian motions, 436, 437
microphysics, 537
microprocessors, 604, 606
Microsoft, 603, 606
microwave radiation, 419–20
Mikusinski, Jan, 487
Milanković, Milutin, 644
military: and applied mathematics, 129, 130; and climatology, 648; and computers, 604; and earth sciences, 14, 553–5, 557; and institutionalization of research and development, 566–8; laboratories and gender stereotypes, 63; and national laboratories, 594–6; and plasma physics, 418, 422. See also Army Signal Corps; Navy, U.S.; war
Military Industrial Commission, 573
military-industrial complex, in U.S., 568
Milky Way, 523–5
Millikan, Robert, 367, 529
Mills, Robert, 385, 386, 387
Milne, Edward, 529
mineralogy, 415
Mitchell, Maria, 55, 61
Mitscherlich, Eilhard, 228, 242, 250
Mittag-Leffler, Gösta, 57
Möbius, August, 460
modernity and modernism: and critiques of modern science, 16; and nuclear physics, 369; religion and physics, 48
molecular biology, 11–12, 429–30, 437–45
molecular-orbital theory, 11, 395, 397–8, 399, 402
molecules: and relationship between quantum chemistry and chemical physics, 10; and spectroscopy, 12–13, 400, 408; types of and chemical classification, 245–6
Mondrian, Piet, 215
Monge, Gaspard, 118–19, 123, 455, 457, 460, 563
Montreal Protocol, 650
Moon, and space program, 169
Mooney, Rose, 61
Moore, Eliakim Hastings, 127
Moore, Robert Lee, 127
morality, and military research and development, 577–8. See also ethics
"moral statistics," 490
Morley, E. W., 325
morphine, 180
Morris, Errol, 108
Morse, S. F. B., 312
Moseley, Henry Gwynn, 253, 287, 338
Mott, Nevill, 11, 415
mountain ranges, and debate between geologists and geophysicists, 540–1
Mount Holyoke College, 61
Mount Palomar Observatory, 532
Mount Wilson Observatory, 164, 165
Mulkay, Michael, 167n38
Müller, Johannes, 228, 235
Müller, Walther, 363
Mulliken, Robert S., 336, 395, 397–8, 409
Murchison, Roderick, 548
Museum of Natural History (Paris), 77
museums, and popularization of science, 77, 78
music: mathematical functions and theory of, 471; physics of, 85
myoglobin, 442

Nabokov, Vladimir, 102
Nambu, Yoichiro, 387
Namias, Jerome, 549
Nansen, Fridtjof, 551
Napier, James Robert, 301
Napoleon Bonaparte, 139, 563
narrative, and structure of scientific texts, 94
Nassar, Salwa, 62n18
National Academy of Sciences, 66, 610
National Aeronautics and Space Administration (NASA), 168, 170–2, 609
National Bureau of Standards, 416–17, 609
National Cash Register (NCR), 601
National Defense Education Act (1958), 571
national differences: and application of quantum mechanics to chemistry, 11; impact of values and ideologies on scientific research, 15; and relationship between science education, research, and industry, 5; and traditions in mathematics, 122, 123–7; and women in physical sciences, 69. *See also* nationalism
National Institutes of Health, 609, 629
nationalism: and big science, 593; and chemical language, 181; and national laboratories, 596. *See also* national differences; state
National Science Foundation, 130, 574, 575, 607
National Security Council, 591
natural history: and chemical classification, 243; and popularization of science, 76, 79–80
natural philosophy: and chemical classification, 243; and electromagnetism, 312; and end of world, 39–40; and energy, 8, 296, 301; Newtonian science and literature, 95–6; and Newton's mathematics, 451; and positivism, 28; and solar science, 506; and universities in U.S., 149
natural selection. *See* evolution
natural theology: and anthropic principle, 50, 51; and geometry education in England, 461, 462; and plurality of worlds, 38; and popularization of science, 75, 76
nature: Engels's dialectical laws of, 582; Scientific Revolution and view of, 52; theories of energy and views of, 302
Nature (journal), 628
Naval Research Laboratory, 168–9
Navy, U. S., and earth sciences, 547, 554, 555. *See also* military; submarines
Nazism: and Aryan physics, 586–9; and big science, 593, 594; and mathematics, 128
nebulae, and cosmology, 523–5
nebular hypothesis, 512–14, 522
Neddermeyer, Seth, 380

Needell, Allan, 332
Néel, Louis, 426
Ne'eman, Yuval, 388
Nelson, Richard, 148, 152
neohumanism, 5, 119
neoorthodoxy, Barthian, 49–50
Nernst, Walther, 335–6, 358, 529
Neu, John, 107
Neumann, Franz, 314, 458–9
Neurath, Otto, 28
neutrino, 366
neutrons, 365, 366, 379
"New Astronomy," 506
Newlands, Alexander Reina, 251
Newman, M. H. A., 602
"new math" movement, 130–1
Newnham College, 62
newspapers, and popularization of science, 88–9
Newton, Isaac, 40, 93, 95, 115, 450–2, 466, 510, 525
Newtonian science, and literature, 95–8
New York University, 129
Nichol, J. P., 80
Nicholson, Edward, 270
Nicholson, John, 338
Nietzsche, Friedrich, 523
nitrogen, and chemical nomenclature, 179
Nobili, Leopoldo, 279
Noble, David, 148
Noddack, Ida, 59
Noether, Emmy, 59, 117
nomenclature, chemical, 174–90
nomological geology, 546
nonlinear dynamics, 427
"normal science," and chemical nomenclature, 186
North Atlantic Treaty Organization (NATO), 603
North British group, and views of energy and nature, 8, 301, 302–3, 308, 310
Novalis (Friedrich Leopold, Baron von Hardenberg), 100
novels, and physical sciences, 100–1
Novikov, Igor, 534
Noyes, A. A., 150, 151
"nuclear culture," 355
nuclear force, and visual imagery in physics, 205–8
nuclear magnetic resonance (NMR), 626–9
nuclear physics: and big bang cosmology, 530; radioactivity and history of, 10, 350–74
nuclear power, 576. *See also* Atomic Energy Commission

nuclear weapons: and development of nuclear physics, 350, 370, 373; fallout and environmental quality, 647; and geophysics, 554–5; and history of quantum physics, 10; and plasma physics, 418–19, 422–4, 426; and relationship between science, technology, and war, 561, 562, 564–5, 568, 575, 577–8; and state ideology of U.S., 590–2; and women as physicists, 65
nucleonics, 370
nucleosynthesis, 520
nucleus theory, and chemical types, 259–60
Nugent, Thomas, 636n7
Nye, Mary Jo, 185, 237, 396

Oak Ridge National Laboratory, 595, 596
OBAFGKMRNS sequence, 517
observatories, and astronomy, 13, 80. *See also* instruments and instrumentation; telescopes
oceanography, 550–2, 554, 557
Odling, William, 246, 251, 262–3
Oersted, H. C., 312
Office of Naval Research (ONR), 424, 555, 575, 609
Office of Scientific Research and Development (OSRD), 420, 564, 567, 568, 575
ohm, 318
Ohm, Georg Simon, 500
Oken, Lorenz, 79
Olby, Robert, 430n2
Oldendorf, William, 618–20, 625
Oldham, R. C., 551
olefins, 265
Olesko, Kathryn, 156
operationalism, and scientific method, 35
Oppenheimer, J. Robert, 93, 380, 572, 590
opportunity costs, and military research and development, 574–5
optics, 23, 287–8. *See also* astronomy; instruments and instrumentation; light
organic chemistry, 254, 430–5, 445
Orlov, Iurii, 596
Oseen, C. W., 484
Ostriker, Jeremiah, 535
Ostwald, Wilhelm, 241, 309, 431, 432, 501
oxygen: atomic weight of, 253; and chemical nomenclature, 179, 180
ozone layer, 634, 649

Pagels, Heinz, 51
Paget, Rose, 63
Paine, Thomas, 37
Pallas, P. S., 81
Pancini, Ettore, 383

pantheism, 51n42
Paracelsus, 219
Paris Exposition (1900), 72, 89
Paris Observatory, 159
Parkes, Samuel, 75
Parkinson, James, 81
Parr, R. G., 412
particle physics: cognitive historiography and developments in, 10; and cosmology, 535; differentiation of from nuclear physics, 370–3; and standard model, 382–8
Partington, James R., 237
Pascal, Blaise, 598
Pasteur, Louis, 4, 45, 248
patriarchy, and gender analysis of physical sciences, 69
patronage: and astronomy, 172; and geology, 552–3; and relationship between science, technology, and war, 562–6
Patterson, A. M., 188
Paul VI, Pope, 49
Pauli, Wolfgang, 202, 341–2, 367, 378–9, 396, 415, 626
Pauling, Linus, 11, 395, 398, 400, 403, 409–10, 415, 439, 440, 441
Pax Britannica (1815–1914), 563
Paxton, Joseph, 77
Payne, Cecilia, 13, 58, 519–20
Peacock, George, 125, 461
Pearson, Karl, 503
Pecquet, Jean, 221, 222
Peebles, James, 533, 534
Peierls, Rudolph, 415
Peirce, Charles Sanders, 503
Pender, John, 319
Penrose, Roger, 534
Penzias, Arno, 533
pepsin, 228–9, 441
periodic system, and chemical classification, 251–4
Perkin, W. H., Jr., 267, 270
Perkowitz, Sidney, 95
Pérouse, Galaup de la, 81
perpetual motion, 41
Perrin, Jean, 241n10, 406
personal computer (PC), 606, 613
Perutz, Max F., 439, 441–2
Pestalozzi, Johann Heinrich, 459
Petit, Alexis Thérèse, 242
Petrini, H., 482
petroleum industry, 553
Petruccioli, Sandro, 331
Pettersson, Hans, 362
Pettijohn, Francis J., 547

Philippines, and women in physical sciences, 69
philosophy: of cosmology, 537; feminist and critique of science, 71; of language and chemical nomenclature, 176, 178; and non-Euclidean geometry, 466; and visual imagery in physics, 195–6. *See also* epistemology; natural philosophy
photography, and astronomy, 13, 163, 509, 517
phrenology, 83
physicalism, and positivism, 28
physical sciences: contributions of women to, 54–71; historical relations between life sciences and, 219–36; and literature, 91–109; popularization of in nineteenth-century Great Britain, 72–90; and public cultures, 3–4; and Pythagoreans, 24; and religion in nineteenth and twentieth centuries, 36–53; and scientific method, 21–35; and themes in history of science, 7–9. *See also* chemistry; geology; physics
physicotheology, 251
physics: atomic and molecular sciences in twentieth century, 9–12; and concept of mathematical function, 470–1; and development of quantum field theory, 375–93; and electrical theory and practice, 311–27; and inductivism, 23; and mathematics, 128; and medicine, 615–33; and oceanography, 551; optics and radiation in nineteenth century, 272–88; quantum theory and atomic structure in early twentieth century, 331–49; state and ideology in twentieth century, 579–97; and statistical methods, 12, 488–504; and theories of force, energy, and thermodynamics in eighteenth century, 289–310; and visual imagery, 6, 191–215. *See also* astrophysics; chemical physics; nuclear physics; particle physics; plasma- and solid-state physics; quantum physics; solar physics
Physikalisch-Technische Reichsanstalt (PTR), 138, 333, 361
Piazzi, Giovanni, 6, 154–5
Picard, Emile, 124
Picasso, Pablo, 194, 214
Piccioni, Oreste, 383
Pickering, Andrew, 376
Pickering, Edward, 13, 163, 338, 517, 518
Pickett, Lucy, 61
Pius XII, Pope, 49
Planck, Max, 4, 9, 57, 332, 333–4, 335, 503, 581, 587
Planck's constant, 199, 203, 204
planets, and debate on plurality of worlds, 37–9, 80. *See also* Earth; Mars; Venus

plasma- and solid-state physics: common themes in reviews of, 11, 413; conceptual and contextual roots of, 413–17; consolidation of, 425–7; formative years of, 420–4; and models of scientific growth, 427–8; and World War II, 417–20
Plass, Gilbert, 647–8
plate tectonics, theory of, 543–4
Plato, 93
Playfair, Lyon, 78
Plücker, Julius, 460
pluralism: and natural theology, 38; and realism in physics, 27; and relationship of industry to scientific education and research, 153; state ideology and physics in U.S., 589–92, 597
pneumoencephalography, 618
Pockels, Agnes, 58–9
Poe, Edgar Allan, 100
poetry: and physical sciences, 91–2, 93, 98, 100–1; and popularization of science, 85–6. *See also* literature
Poincaré, Henri, 24, 124, 326, 478, 499
point-contact transistor, 421–2
point-set topology, 131
Poiseuille, Jean Léonard Marie, 229
Poisson, S. D., 123, 290
Polanyi, Michael, 591
polarization, and emission theory of light, 276
politics: coalitions and military research and development, 573–4; and popularization of science, 88; state ideologies and physics, 597. *See also* Cold War; state
polonium, 181
polymer chemistry, 11–12, 429–30, 439–40, 444–5
polymorphism, 256
Poncelet, Jean Victor, 123, 456
Pope, Alexander, 95
Popper, Karl, 30
popular culture: and dissemination of science, 98; and relation between religion and science, 48n32. *See also* culture; public culture
popularization, of science: and Great Britain in nineteenth century, 72–90; and role of women, 4
positional aromatic isomerism, 268–9
positional astronomy, 154–9
positivism, 1, 28–9, 193, 356
positron, 367, 379
positron emission tomography (PET), 630
postmodernism: and cultural impact of computers, 612; and discussions about religion and science, 48, 52; and history of science, 1; and radical relativism, 53n46
Pouchet, Félix, 45

Powell, Wilson, 383
Poynting, J. H., 321
pragmatism, 11
Pratt, John, 541–2
prediction, and complementarity in atomic physics, 204
predictive models, of global warming, 648
Preece, W. H., 322
Presbyterianism, 8. *See also* religion
Preuss, Heinz-Werner, 406
Prévost, Pierre, 281
Price, Donald K., 591
Priestley, Joseph, 74, 231
Princeton University, 425, 547
Pringsheim, Hans, 431
"probabilistic revolution," 488
probability theory, 130, 488, 489, 499
Proctor, Richard, 80
productivity paradox, 614
projective geometry, 466–7
Project Sherwood, 423
proletarian science, 583, 596
proteins: and molecular biology, 439, 440–2; and polymer chemistry, 11–12
protons, 359–60, 378
Prout, William, 243
Ptolemy, 451
public culture: literature and physical sciences, 91–109; and physical sciences after 1800, 3–4; and popularization of science in Great Britain, 72–90. *See also* culture; popular culture
public interest science, 17, 592
Pugwash conferences, 578
Pulkovo Observatory, 158
Pummerer, Rudolf, 431
Puppi, Gianpietro, 383
Purcell, Edward, 626, 627
Purdue University, 610
Pushkin, Alexander, 101
Pycior, Helena, 57
Pynchon, Thomas, 100, 102
Pythagoreanism, 23–4, 27, 52–3

qualitative improvements, and relationship between science, technology, and war, 568–9
quantum chemistry, 10–11, 394–412
quantum chromodynamics (QCD), 391
quantum discontinuity, 334–6
quantum field theory, 377–82. *See also* field theory
quantum gas, 344–6
quantum mechanics: and action, 332–4; crisis in early twentieth century, 341–4; and Delta function, 481; and history of science, 331–2; and indeterminism, 9–10; and national values and ideologies, 15; and new physics in Soviet Union, 581, 586; and quantum chemistry, 395–6, 400–1, 409; religion and physics, 47, 50, 51; and solid-state theory, 413–14, 417; and statistical methods, 488, 500; and transformation properties of equations, 347; and visualization in physics, 200, 204, 205
quantum physics, and visual imagery, 212–13
Quetelet, Adolphe, 489, 490, 492
Quine, Willard von Orman, 30

Rabi, Isidor, 383–4, 626
radiation: and computerized tomography, 624, 625; and continuous spectrum, 277–80; history of nineteenth-century theories of, 8; quantum gas and wave mechanics, 344–6; and relativity, quantum, and nuclear theories, 9; and x-ray images, 616, 620. *See also* radioactivity
Radiation Laboratory (RadLab), 419
radicals: and chemical classification, 246, 247; and electrochemical dualism, 257–9
radio, 323–4, 416–17. *See also* radio astronomy
radioactivity: atomic theories and decay of, 240; and history of nuclear physics, 10, 350–2, 373–4; institutionalization and research on, 355–60; and instruments, 362–8; and internationalism, 360–2; and "political economy" of radium, 352–5. *See also* radiation
radio astronomy, 166–7, 522–3
radiology, 616, 620, 622
radium, 352–5
Radon, Johann, 621
Ramsauer, Carl, 346
Ramsay, William, 252, 353–4, 361
Ramsden, Jesse, 155
RAND Corporation, 16, 567, 648
Rankine, William J. Macquorn, 298, 301, 308–9, 523
rational formulas, and chemical classification, 244, 257
rationalism: ideology and state politics, 597; and Pythagoreanism, 23–4; and scientific method, 22
Ray, John, 50n38
Rayleigh, Lord (John William Strutt), 252
RCA, 606
Read, H. H., 548
Reagan, Ronald, 580, 614
real analysis, 129–30
realism: and physico-mathematical sciences, 25–7; and visual imagery in physics, 6
reasoning, styles of scientific, 377

Réaumur, René-Antoine, 225, 226
Rechenberg, Helmut, 331
recurrence paradox, and statistical methods, 501, 502
reductionism, 7, 219, 356, 374, 400–4
reflecting telescope, 160–2, 164–5
refractors, and telescopes, 157–8
Regnault, Victor, 34–5, 308
relativism: and postmodernism, 1
relativity theory: and Aryan physics, 587–8; and astronomy, 13, 525–6, 537; Einstein's theory of, 195, 196, 199; electrons and ether, 324–7; and new physics, 581–2; and quantum theory, 384–5; religion and physics, 47; and scientific method, 35; and twentieth-century theories of matter and radiation, 9
religion: and physical sciences in nineteenth and twentieth centuries, 3–4, 36–53; and positivism of Vienna Circle, 28. *See also* Calvinism; Christianity; Presbyterianism
Remington Rand Corp., 601, 605
Renaissance, and state sponsorship of military research and development, 562, 571–2
renormalization theory, 376–7
representation: in art and science of early twentieth century, 193–4; and visual imagery in physics, 208–9
research: impact of national values and ideologies on, 15; and national differences in scientific education and industrial development, 5, 133–53. *See also* laboratories; observatories; *specific disciplines*
residues, theory of molecular, 246
resonance theory, 10–11, 396, 398–9
respiration, and chemistry, 230–3
Revelle, Roger, 646, 648
reversibility paradox, 501, 523
revolutions, scientific: and astronomy in early twentieth century, 13; and computer science, 16; examples of Kuhnian, 376–7; German universities and intellectual, 458; and mathematics, 12, 115; and theoretical cosmology, 525. *See also* Scientific Revolution
Ricardo, David, 500
Ricci-Curbastro, Gregorio, 124
Richards, Ellen, 61, 65, 66
Richards, Joan, 114
Richards, Robert, 65
Richardson, Lewis Fry, 549–50
Richardson, R. G., 129
Richelot, Friedrich, 122
Richter, Jeremias Benjamin, 243
Riemann, Bernhard Georg, 121–2, 124, 464–5, 466, 482

Riesz, Frigyes, 485
Righi, A., 286
rigor, and mathematical analysis, 12, 470, 474–7
Rimbaud, Arthur, 101
Ringer, Fritz, 133, 136–7, 144
Risbey, J. S., 646n47
Ritter, Johann Wilhelm, 244, 278
RNA (ribonucleic acid), 444
Roberts, Dorothea Klumpe, 66
Roberts, G. K., 145
Robertson, Howard, 526
Robinson, Abraham, 487
Rocke, Alan, 184, 238–9
Rockefeller Foundation, 14, 364
Roddenberry, Gene, 99
Rodman, Thomas, 571
Romanticism, in poetry and physical sciences, 93
Röntgen, Wilhelm Conrad, 286, 352, 615
Roosevelt, Franklin, 564, 568, 569–70
Rorty, Richard, 48
Roscoe, Henry, 267
Rosenberg, Nathan, 148, 152
Rosenbluth, Marshall, 383
Ross, Andrew, 95
Rossby, Carl-Gustav, 549, 554
Rosse, William Parsons (Lord Oxmantown), 80, 160–1
Rotblat, Joseph, 578
Rothko, Mark, 215
Rowland, Henry A., 282
Royal Astronomical Society, 80
Royal College of Chemistry, 145
Royal Greenwich Observatory, 64, 159
Royal Society of London, 54, 66, 72, 74–5, 294, 295, 462
rubber, and macromolecular chemistry, 432–3, 436–7, 440
Rubbia, Carlo, 392
Rudwick, Martin, 548
Russell, Bertrand, 467
Russell, Henry Norris, 518, 520
Russia. *See* Soviet Union
Russian revolution (1917), 580
Rutherford, Ernest, 197, 353, 354, 356–7, 360, 361, 378, 417
Rutherfurd, Lewis M., 515
Ryle, Martin, 532–3

Saccheri, Girolamo, 463
Sachse, Ulrich, 249
Sackur, Otto, 344
Safronov, Victor, 550

Index

SAGE (Semiautomatic Ground Environment), 602
Saha, Meghnad, 519
Saint-Hilaire, Etienne Geoffroy, 45
Sakata, Shoichi, 383
Sakharov, Andrei, 11, 423, 591, 596
Salam, Abdus, 211, 387
Salmon, George, 125, 462
Sandage, Allan, 532
Sato, Mikio, 487
Saxenian, Anna Lee, 611
scandium, 181, 252
Schaffer, Simon, 375
Scheele, Karl Wilhelm, 245, 278
Scheiner, Christoph, 508
Schiaparelli, Giovanni, 39
Schmidt, Erhard, 485
Schmieden, Curt, 487
Schmitt, Harrison, 552
Scholnick, Robert, 104
Schott, Charles A., 640–1
Schottky, Walter, 416
Schramm, David, 534
Schrödinger, Erwin, 201–3, 346, 364, 437–8
Schröter, J. H., 156
Schubert, G. H., 45
Schumann, V., 284
Schuster, Arthur, 283
Schwabe, Heinrich, 509
Schwann, Theodor, 228–9
Schwartz, Laurent, 484–5, 486–7
Schwartz, Richard B., 96
Schwarzchild, Karl, 339
Schweikart, Ferdinand, 464
Schwinger, Julian, 207, 384
Sciama, Dennis, 533, 534
science fiction, 87, 99, 102, 104
scientific academies, and popularization of science, 81
scientific calculator, 600
scientific materialism, 45
scientific method, theories of, 3, 21–35, 87
scientific naturalism, 488
Scientific Revolution, 52, 563. See also revolutions
"scientist," origins of term, 73, 78
Scotland: Calvinism and views of nature, 302; and energy physics, 309; shipbuilding and engineering education, 291; universities and scientific education, 72
Scott, Charlotte Angas, 66
Scott, David, 80
Secchi, Angelo, 515
secrecy, and military research and development, 571–3

Sedgwick, Adam, 299
Seebeck, Thomas J., 278
Segre, Corrado, 124
seismology, 14, 543, 551, 555
Seitz, Frederick, 415
semiconductor electronics, 11, 418, 419, 421–2, 602, 603–4
Sequin, Armand, 231–2
Serber, Robert, 380
Servos, John, 151, 411
Shannon, Claude, 608
Shapley, Harlow, 524, 525
Shelley, Mary, 86, 100
Shelley, P. B., 100, 101
Shelter Island Conference, 382–3, 384, 399, 409
Sherman, Albert, 402
Sherrill, Mary, 61
Shickard, Wilhelm, 598
Shoemaker, Eugene M., 550
Sibum, Otto, 295
Sidgwick, Henry, 84
Sidgwick, N. V., 406–7
Siegel, Daniel M., 281n20
Signer, Rudolf, 443, 444
Silberstein, Ludwik, 524
Silliman, Benjamin, 76
Sime, Ruth, 57
Simon, Herbert, 608
Simpson, George Clark, 644
Singer, Rudolf, 11–12
single positron emission computed tomography (SPECT), 630
Sitter, Willem de, 525–6
Slater, John Clarke, 11, 342, 395, 414
Slavings, Ricky L., 516n15
Slipher, Vesto, 524
S-matrix theory, 10, 381–2, 384, 385–6
Smith, Clarence, 190
Smith, Crosbie, 43
Smith, Jonathan, 105
Snow, C. P., 85, 93, 94
Sobel, Dava, 108
Sobolev, Sergei, 484, 486–7
social constructivism, and history of science, 290
social sciences, and statistical methods, 488, 489–90, 494–8. See also sociology
Société Centrale de Produits Chimiques, 353
societies, and popularization of science, 75–6, 77
Society of Industrial and Applied Mathematics (SIAM), 608–9
Society for Literature and Science (SLS), 94, 106
Society for the Promotion of Engineering Education, 149, 150
Society for Psychical Research, 84

sociology, of astronomy, 521. *See also* social sciences
Socolow, Elizabeth, 108
Soddy, Frederick, 253–4, 337, 353–4, 361
SOFAR (Sound Fixing and Ranging) system, 554
software, computer, 603, 606, 608, 611
Sokal, Alan, 95
solar physics, 508–16
solid-state physics. *See* plasma- and solid-state physics
Solvay, Ernest, 336
Solvay Congress, 357–8, 369, 399n13
Somerville, Mary, 4, 55
Somerville College, 62
Sommerfeld, Arnold, 339, 340–1, 357
Sophie Newcomb College, 61
sound channeling, 554
Soviet Union: and big science, 593–4; and cosmology, 532; ideology of state and new physics, 580–6; and mathematics in Cold War era, 129; and national laboratories, 595; and nuclear weapons, 591; and relationship between science, technology, and war, 561, 562, 565, 566, 573; and space program, 168, 169, 170; and theory of chemical structure, 264; and theory of resonance, 399. *See also* Cold War
space, geometrical concept of, 449, 465–7
space programs: and astronomy, 167–70; and environmentalism, 647; and military technology, 576; and state ideology of U.S., 592
Space Telescope Science Institute, 631–2
space-time singularity, 534
Spallanzani, Lazzaro, 225–6
Spanish-American War, 615
specialization, of education in Great Britain, 72
spectroscopy: and advances in astronomy, 505, 509; and quantum mechanics, 336; and solar physics, 510–12, 514–16; and spectrum analysis, 280–4; and study of molecules and atoms, 12–13, 240, 400; and x rays, 345
spectrum: analysis of and development of spectroscopy, 280–4; radiation and idea of continuous, 277–80
Speer, Albert, 593, 594
Spilhaus, Althelstan, 554
spiritualism, 84–5, 466
Spitzer, Lyman, Jr., 171, 423
Spivak, Michael, 130
Sponer-Franck, Hertha, 61
spontaneous symmetry breaking (SSB), 387, 388, 389
Sprat, Thomas, 96

Spurzheim, J. G., 83
Sputnik (satellite), 168, 571, 647
Stalin, Joseph, 580, 586, 593, 594
standardization, and chemical classification, 251–4
standard model, in particle physics, 382–8, 391–3
Stanford Linear Accelerator, 389–90
Stanford University, 610
Stark, Johannes, 15, 335, 337, 585, 587, 588, 589
stars: astrophysics and stellar models, 518–21; cataloging and classification of, 517–18; and developmental sequences, 506; and plasma physics, 417; and stellar spectroscopy, 514–16. *See also* Sun
Stas, Jean Servais, 243
state: and patronage for military research and development, 562–6; physics and ideology of in twentieth century, 579–97. *See also* authoritarian regimes; democracy; totalitarianism; *specific countries*
statistical mechanics, 499
statistical methods: in astronomy and astrophysics, 516–17; and physics, 12, 488–504
Staudinger, Hermann, 430–3, 434–5, 436, 438
steady state universe, 531–3
Steenbeck, Max, 416
Stefan, Josef, 493
Steigman, Gary, 534
Steinberger, Jack, 383
Steiner, Jakob, 459–60
Steinhardt, Paul, 535
stellarator, 425
Stendahl (Marie-Henri Beyle), 101
Stent, Gunther S., 443
Stephen, James Fitzjames, 496
stereochemistry, 248–50, 268
stereoisomerism, 268–9
stereotypes, of women in physical sciences, 58, 70–1
Stevenson, Edward C., 380–1
Stevin, Simon, 40
Stewart, Balfour, 87, 281, 301
Stibitz, George, 600
Stirling, Robert, 291
stoichiometry, 239, 243
Stokes, George Gabriel, 126, 281, 283–4, 295, 476
Stokes, Navier, 484
Stone, Marshall, 486
Stoner, Edmund, 341
Stoney, G. J., 282
Strassmann, Fritz, 369
Strategic Defense Initiative, 596
Strauss, David, 86

Street, Curry, 380
string theory, 31
Strömgren, Bengt, 520
structural realism, 27
structure theory, in chemistry, 7–8, 248, 249, 255–71
Struve, F. G. W., 158
Stückelberg, Ernest, 380
subjectivity: and Bayesianism, 32; and postmodernism, 1
submarines, 564, 570, 572, 577
submarine undersea cables, 314–16, 319–20, 323
suicide, and statistical methods in social sciences, 489–90
Sun: and concepts of biological and thermodynamic evolution, 13; and solar physics, 508–16, 618; and theories of energy, 303. *See also* stars
sunspots, 508, 509–10, 618
Superconducting Super Collider (SSC), 373, 392, 592
surgery, 632
Suslin, M. Ya., 129–30
Sutton, M. A., 282n21
Svedberg, The(odor), 433, 442
Sverdrup, Harald, 551
Swainson, William, 77
Swan, William, 281
Swift, Jonathan, 95
Swinburne, Algernon, 101
Swiss Avalanche Research Center, 621
Switzerland, and scientific education for women, 62–3
Sylvester, James Joseph, 125, 462
symmetry, as style of scientific reasoning, 377, 386. *See also* spontaneous symmetry breaking (SSB)

tabulating systems, and development of computer, 599, 607
Tait, Peter Guthrie, 87, 125, 301, 304–5, 309, 497
Tamm, Igor, 11, 423
taxonomy: and chemical classification, 252; and stellar classification, 517–18
Tayler, Roger, 533
Taylor, Frederick, 569
technical education: and France, 141; and Germany, 137, 138, 139; and Great Britain, 143–4. *See also* engineering
Technische Hochschulen, 135–8, 267
Technische Mittelschulen, 136, 137
technocracy, 590
technology: and advances in astronomy, 505; relationship between science, war, and, 561–78; and Technische Hochschulen in Germany, 137
Teilhard de Chardin, Pierre, 51n42
telegraph and telegraphy, 8, 286, 314–16, 319–20, 322, 323–4, 549
telescopes, 6, 13, 80, 157, 160–2, 164–5. *See also* observatories
Teller, Edward, 11, 415
Tennessee Valley Authority, 579–80
Tennyson, Alfred, 82, 85–6, 101
test function method, 483
Tetrode, Hugo, 344
textbooks: and functional analysis, 486; and quantum chemistry, 410; and popularization of science, 73, 87; and radioactivity, 357; and thermodynamics, 304
textile industry, and chemistry, 270–1
theology. *See* natural theology; religion
theory: and astrophysics, 506; and "experiment" in nuclear physics, 358, 373; and phenomenological evidence in geology, 541; and scientific method, 33–4. *See also* grand unified theories
Theory of Everything, 53
thermodynamics, laws of: and atomism, 241; and cosmology, 522, 523; and macromolecular chemistry, 436; and popularization of science, 82, 88; and quantum mechanics, 333; and religion, 40, 42, 43; and solar physics, 512–14; and spectroscopy, 281; and statistical methods, 494, 495, 501; and theories of energy, 299, 303, 304; and wave theory of light, 280
thermonuclear fusion, 11, 418, 422–4, 426–7
Thompson, Benjamin (Count von Rumford), 74
Thompson, Francis, 101
Thomson, George Paget, 11, 422
Thomson, James, 4, 8, 44, 95, 291, 297, 298–9
Thomson, Joseph John, 84, 240, 285, 325, 337, 356
Thomson, Thomas, 243
Thomson, William (Lord Kelvin): and British tradition in mathematics, 125; and electrical theory, 313; and electromagnetic theory of light, 284; and evolution, 42, 43; gases and second law of thermodynamics, 497; and history of Earth, 82, 88, 538, 540, 541; and link between religion and science, 4, 8; and materialist worldview, 81; and mechanical value of heat, 290–2, 295, 308; and physics of energy, 296–7, 299, 304, 309; and realism, 25; and reversibility paradox, 500; and telegraph, 315–16; and theory of solar energy, 513; and tide predictors, 599
Thoneman, Peter, 423

Thoreau, Henry David, 635
tide predictor, 599
Tiedemann, Friedrich, 227–8
Tilly, Charles, 562n2
timescales, and solar physics, 513
Timiriazev, Arkady, 583, 584
Tinsley, Beatrice, 68
Tiomno, Jaime, 383
Tipler, Frank J., 51n42
Tizard, Henry, 564
Tokyo Institute of Physical and Chemical Research, 415
Tolman, Richard, 526
Tomonaga, Sin-itoro, 207
Toshiko, Yuasa, 68
totalitarianism, 15, 593, 597
TRACES, 574
training, and relationship between science, technology, and war, 570–1. *See also* apprenticeship
transistors, 604
Traweek, Sharon, 71, 95
Treiman, Sam, 385
Trenn, Thaddeus, 353
Truman, Harry, 575
Turing, Alan, 602
Turkey, and scientific education of women, 69
Turkle, Sherry, 612
Turner, Martha A., 104
Twain, Mark, 101
"two cultures" concept, 4, 93, 96, 98, 109
two-meson hypothesis, 383
Tyndall, John, 43, 80–1, 300, 302, 497, 642
type theory, in chemistry, 245–7, 250, 260–2

Uhlenbeck, George, 342
Uhuru (satellite), 169
Ulam, Stanislaw, 129
ultraviolet light, 278
uncertainty principle, 581
underrecognition, of women in physical sciences, 66–7
undulatory theory, of light, 273
uniformitarianism, 299, 303, 546, 550
Union of the Chemical Societies, 186
Union internationale de chimie pure et appliquée (UICPA), 186, 187, 188
Unisys, 605
United States: climate change and history of, 638–41; industrial development and scientific education and research, 147–52; and institutionalization of astrophysics, 164; and military-industrial complex, 568; and national traditions in mathematics, 126–7; pluralist ideology and physics, 589–92; and quantum chemistry, 404–5; and relationship between science, technology, and war, 561, 562, 565–6, 567, 570, 571, 574; and solid-state physics, 421; and space program, 168, 169, 170–2; state ideology and physics, 579; and women in physical sciences, 69, 71. *See also* Cold War
unity, and realism in physics, 27
universalism, and national laboratories, 596
universe: age of, 14, 523, 536; and cosmology, 526–9; Einstein's model of, 525–6
universities: and computer science, 16, 608–11; geometry and German, 458–60; geometry and mathematical analysis, 12; industrial development and scientific education in U.S., 147–50; and mathematics education in Germany, 120–3; and military contracts, 15; and popularization of science in Great Britain, 72, 73; and quantum chemistry, 409; and scientific education in England, 144, 146; and Technische Hochschulen in Germany, 135; and women in physical sciences, 56, 59, 60, 61–3. *See also* education
University of California at Berkeley, 610
University of Chicago, 127, 547
University of London, 73
University of Michigan, 610
University of Neuchatel, 621
University of Pennsylvania, 601
University of Texas, 127
University of Zurich, 62–3
Unsöld, Albrecht, 417
uranium, 352
Urey, Harold C., 366, 407, 550
Ussher, James, 81

Vail, Alfred, 312
valence and valence-bond theory, 11, 262–5, 395, 397–8, 399, 402
Valéry, Paul, 100
validity, and experimentation, 33
van den Broek, Antonius, 338
van den Ende, Jan, 607
van de Graaff, Robert J., 366
Van Helden, Albert, 164
Van Hise, Charles, 14, 545–6
van der Waals, Johannes Diderick, 240
van der Waerden, B. L., 117, 486
van't Hoff, Jacobus H., 185, 240, 250–1, 268
Varian Associates, 626
variation, and statistical methods, 491–4
Vassar College, 60, 61
Vauban, Sebastien Le Prestre, 563
Veblen, Oswald, 127

Index   677

vector analysis, 126
Venera 7 (satellite), 169
Venus, and space program, 169
Verguin, F. E., 270
Verne, Jules, 87, 101
Very Large Array (VLA), 167, 172
vibrating string, and mathematical function, 470–1, 472, 482
vibration theory, of light, 273
Vielle, Paul, 571
Vienna Circle, 28
Vierordt, Carl, 230
Vietnam War, 570, 592
Viking 1 and *Viking 2* (satellites), 169–70
viscosity law, 435–6
visual imagery, in physics, 6, 191–215
visualization and visualizability, and Kantian terminology, 196–7
vitalism, 7, 219, 244–5
Vogel, Hermann Carl, 515
Vogt, Karl, 45
volcanoes, 539–40
Volkmann, Alfred, 230
Volk and *völkisch* concepts, and Nazi ideology, 586, 588
Volta, Alessandro, 234, 311
Volterra, Vito, 485
von Baeyer, Adolf, 249, 432
von Braun, Wernher, 565
von Kármán, Theodor, 129
von Laue, Max, 287
von Liebig, Justus, 73
von Mises, Hilda Geiringer, 59
von Neumann, John, 130, 347, 481, 486
von Reichenbach, Karl, Baron, 84
von Schweidler, Egon, 353, 357
von Seidel, Philipp, 476
von Steinheil, K. A., 161
von Utzschneider, Joseph, 156
von Weizsäcker, Carl Friedrich, 520, 528–9
Vorilong, William, 37n1
V-2 rocket, 595

Walden, Paul, 237n1
Walker, W., 150, 151
Walton, Ernest, 365
Wang, Jessica, 15
war, relationship between science, technology, and, 14–15, 561–78. *See also* Civil War; Cold War; Franco-Prussian War; French Revolution; military; nuclear weapons; Spanish-American War; Vietnam War; World War I; World War II
Ward, Mary, 86

water: and chemical nomenclature, 181–2; and theory of respiration, 231. *See also* "heavy water"
Waterston, John James, 513
Watson, James D., 444
wave equation, 470–1
wave function, and quantum chemistry, 395–7
wavelengths, 166, 167, 286
wave mechanics, 201–3, 344–6, 364
Weart, Spencer, 646n47
Weaver, Warren, 129, 553
Weber, Ernst Heinrich, 229–30
Weber, Wilhelm, 229, 305, 308, 312–14, 318
Webster, Noah, 639–40
Wegener, Alfred, 542–3
Weierstrass, Karl, 57, 116, 124, 476, 478–9
Weierstrassian analysis, 116, 121–2
Weinberg, Steven, 3, 211, 387, 534
Weizmann, Chaim and Anna, 66
Welker, Heinrich, 424
Wellesley College, 61
Wells, H. G., 4, 88, 101, 355, 466
Wentzel, Gregor, 380
Werner, Abraham Gottlob, 538
Werner, Alfred, 249
Wertheim, Margaret, 53n46
Wertzien, Karl, 181
Westfahl, Gary, 104
Westfall, Richard, 220
Wetherald, Richard T., 649
Weyl, Hermann, 347, 349, 483, 484
Wheaton, Bruce R., 287n30, 331
Wheatstone, Charles, 312
Wheeler, John, 51, 382
Wheland, George, 398
Whewell, William, 4, 5, 38, 73, 80, 274, 276, 305–7, 461
Whirlwind computer, 602
Whitehouse, Wildman, 315–16
Whiting, Sarah, 61
Whitman, Walt, 101
Whitney, Mary, 61
Whittle, Frank, 571
Wick, Frances, 61
Wiechert, Emil, 324
Wiener, Norbert, 130, 484, 487, 608
Wigner, Eugene, 403, 415
Wilberforce, Samuel, 78
Wilczek, Frank, 392
Wilkes, Charles, 576
Wilkins, Maurice, 11–12, 444, 602
Williams, M. E. W., 158
Williamson, Alexander W., 182, 187, 246, 247, 261–2

Wilson, Charles Thomas Rees, 358
Wilson, E. B., 410
Wilson, Robert, 533
Wislicenus, Johannes, 251, 253, 268
Wöhler, Friedrich, 244, 256, 258, 259
Woker, Gertrud, 59
Wolf, J. Rudolf, 509–10
Wollaston, William Hyde, 239, 241, 250, 256, 274, 280, 511
women: and computers, 612–13; and division of labor in astronomy, 13, 173, 509, 517; exclusion of from scientific education and organizations, 4; and physical sciences, 51–3, 54–71; and research on radioactivity, 354
women's liberation movements, 67–70, 71
"women's work," 62, 63–5
Woodward, Robert, 542, 544–5
Woolf, Virginia, 102
Wordsworth, William, 101
World War I: and development of science, 90; and relations between French and German mathematicians, 128; and scientific research in France, 143; and state sponsorship of military research and development, 563–4, 567, 569; and women in physical sciences, 65, 71
World War II: and astronomy, 166, 168; and computer revolution, 16; and earth sciences, 553–5; and nuclear physics, 350, 359, 369–70, 373; plasma- and solid-state physics and, 417–20; and quantum chemistry, 406; and relationship between science, technology, and war, 14–15, 567, 568, 569–70, 572, 577; and research on thermonuclear fusion and semiconductor electronics, 11; state ideology of Soviet Union and new physics, 585; and state sponsorship of military research and development, 564–6; and synthetic rubber research, 437; and women in physical sciences, 65; and x-ray research, 617

Wranglers, and mathematics in Great Britain, 125–6
Wrinch, Dorothy M., 442
Wu, C. S., 58, 65, 66
Wu, Yuan, 65
Wünsch, C. E., 278
Wurtz, Charles Adolphe, 181, 182, 246, 262, 263, 267
Wynn-Williams, Eryl, 363

x rays: and astronomy, 168–9; and electromagnetic theory of light, 286–7; and medicine, 615–17, 631; and spectroscopy, 345

Yahil, Amos, 535
Yang, Chen Ning, 383, 385, 386, 387
Yates, JoAnn, 601
Yeats, W. B., 100
Yerkes Observatory, 164, 165
Young, Thomas, 229, 273, 278, 280
Young, Will, 65
Yudkin, Michael, 93
Yukawa, Hideki, 207, 380

Zacharias, Jerrold, 420
Zeeman, Pieter, 324, 339, 340, 341
Zeeman effect, 341
Zel'dovich, Yakov, 534
Zentralblatt, 131
Zermelo, Ernst, 501
Zhanovshchina, 585
Ziedes des Plantes, Bernard, 618
Zilsel, Edgar, 28
Zola, Emile, 101
Zöllner, Johann Carl Friedrich, 162
Zsigmondy, Richard A., 431
Zuboff, Shoshana, 612
Zuse, Konrad, 600
Zweig, George, 388
Zwicky, Fritz, 527